Biology and Management of Inland Striped Bass and Hybrid Striped Bass

Biology and Management of Inland Striped Bass and Hybrid Striped Bass

Edited by

James S. Bulak

South Carolina Department of Natural Resources
1921 Van Boklen Road, Eastover, South Carolina 29044, USA

Charles C. Coutant

Oak Ridge National Laboratory (retired)
120 Miramar Circle, Oak Ridge, Tennessee 37830, USA

James A. Rice

Department of Biology, North Carolina State University
Campus Box 7617, Raleigh, North Carolina 27695, USA

American Fisheries Society Symposium 80

Proceedings of the Symposium
"Striper 2009: Inland Striped Bass and Hybrid Striped Bass Management"
Held in Nashville, Tennessee, USA
31 August–2 September 2009

American Fisheries Society
Bethesda, Maryland, USA
2013

Suggested citation formats follow.

Entire book

Bulak, J. S., C. C. Coutant, and J. A. Rice. 2013. Biology and management of inland striped bass and hybrid striped bass. American Fisheries Society, Symposium 80, Bethesda, Maryland.

Chapter in book

Gustaveson, W., and G. Blommer. 2013. History of striped bass management in the Colorado River. Pages 15–23 *in* J. S. Bulak, C. C. Coutant, and J. A. Rice, editors. Biology and management of inland striped bass and hybrid striped bass. American Fisheries Society, Symposium 80, Bethesda, Maryland.

Printed in the United States of America on acid-free paper.

Library of Congress Control Number 2013932981

ISBN 978-1-934874-36-3
ISSN 0892-2284

American Fisheries Society Web site: www.fisheries.org

American Fisheries Society
5410 Grosvenor Lane, Suite 110
Bethesda, Maryland 20814

Contents

Growth and Condition

Population and Harvest Evaluation

Stocking Evaluations

Natural Reproduction

Harvest Regulations

Conflicts

Economics

Challenges and Opportunities

Preface

At the 139th Annual Meeting of the American Fisheries Society (AFS) in Nashville, Tennessee, the Southern Division's Striped Bass Technical Committee held a 2-day symposium, "Striper 2009: Inland Striped Bass and Hybrid Striped Bass Management." It was our pleasure to assist with organizing and chairing the symposium and serving as editors for this publication. That well-attended symposium and this book of peer-reviewed chapters that is its formal record attest to the unique aspects of the biology and management of striped bass *Morone saxatilis* and striped bass hybrids that are restricted to inland waters.

The striped bass is a celebrated and well-known sport fish, native to the East and Gulf coasts of North America. In much of its range, it is anadromous. In southern portions of its range, generally south of Cape Hatteras, North Carolina, it primarily occupies rivers and estuaries. As Van Horn (Chapter 1) summarizes, dam closures in South Carolina in the 1940s and in North Carolina–Virginia in the 1950s demonstrated that striped bass could succeed in a purely freshwater environment.

This fortuitous discovery coincided with the rapid expansion of dam construction and reservoir filling throughout much of the United States. The potential to establish fisheries for a large, open-water sport fish in reservoirs that were overpopulated with pelagic forage fish, primarily gizzard shad *Dorosoma cepedianum* and threadfin shad *D. petenense*, was an inviting prospect for many fisheries managers. Widespread stocking followed the development of hatchery production techniques, to the point where many freshwater reservoirs now support fisheries for striped bass or their hybrids.

Nonetheless, striped bass populations established in the 1960s and 1970s were not uniformly successful. Warm, eutrophic reservoirs often had troubling summer die-offs of adults. In other reservoirs, growth rates were slow and trophy-sized fish found in some systems failed to materialize. Concurrently, marine-estuarine striped bass populations on the East and West (introduced in the late 1800s) coasts were experiencing population declines. At the 112th AFS Annual Meeting at Hilton Head, South Carolina in 1982, the symposium "Striped Bass: Environmental Risks in Fresh and Salt Water" explored a variety of hypotheses and relevant evidence (*Transactions of the American Fisheries Society* 114:1–151). A major theme of that symposium, as stated in the published preface, "was the search for commonality of environmental stressors and biological susceptibility among the geographically diverse stocks."

Much has been sorted out since that time, both in freshwater and saltwater. In 2005, the AFS Southern Division's Striped Bass Technical Committee, under the leadership of chairman Dan Wilson, judged that a major synthesis of biology and management of freshwater striped bass and their hybrids was overdue. The committee developed an initial list of topics and a symposium for the 2009 Nashville Annual Meeting, in the midst of Tennessee's reservoirs brimming with striped bass and hybrids, was established as a goal. The editors scoped the symposium in mid-2006 and the Striped Bass Technical Committee and editors met at the Oklahoma Biological Research Station on Lake Texoma in August 2006 (hosted by Jeff Boxrucker) to further refine the list of topics that needed to be addressed.

It was decided that a reference book with chapters useful for guiding current and future management would be most valuable. It would not be a techniques manual, per se, but provide examples of state-of-the-art work in relevant topics. Speakers and chapters on major themes were invited, grouped as History, Habitat, Growth and Condition, Population and Harvest Evaluation, Stocking Evaluations, Natural Reproduction, Harvest Regulations, Conflicts, Economics, and Challenges and Opportunities. All papers were peer reviewed, usually by one reviewer who attended the symposium and one who did not but who had relevant expertise. We extend our thanks to these many reviewers. Each editor reviewed and edited each paper (excluding their own), including reviewer comments and revisions, although chapters were divided among the editors for correspondence.

Our definition of "inland" deserves comment. We have focused on populations in freshwater lakes, reservoirs, and rivers, as well as populations (such as the Gulf of Mexico stocks) that are almost exclusively riverine even though they have free access to the ocean. We are particularly pleased to have a thorough summary of the history of the Gulf stocks, whose decline has led to a complicated history of stocking, stock identification, and management (Chapter 3, Long et al.). We do not attempt to cover marine stocks of the East or West coasts, although the decades-long environmental bioassay in inland waters has suggested causes for some population difficulties in these areas.

Much of the contents reflect the extensive work in the Southeast, where the notion of introducing striped bass to inland waters originated. However, Gustaveson and Blommer (Chapter 2) and Stewart and Burrell (Chapter 25) relate the story of introduction, spread, and management of striped bass in the Colorado River basin.

The role of summer temperature and dissolved oxygen as controlling, and often limiting, factors in the success of striped bass in inland waters has earned prominence in management. The biological side of the story is given in a series of chapters: the history of developing and testing the "temperature–oxygen squeeze hypothesis" from the mid-1970s through 2005 (Chapter 4, Coutant), the contributing role of forage availability (Chapter 5, Thompson and Rice), the unique role of metalimnetic hypoxia in some systems (Chapter 6, Rice et al.), and a case study comparing behavior of striped bass and hybrid striped bass in a common environment (Chapter 7, Kilpatrick and Ney). Ruane et al. (Chapter 8) and Taylor et al. (Chapter 9) demonstrate the feasibility of using water quality modeling to estimate the temporal and spatial abundance of temperature and dissolved oxygen conditions that create habitat limitations. Mobley et al. (Chapter 10) show that oxygen can be added to hypoxic reservoirs to maintain suitable thermal–oxygen habitat in summer.

Growth and condition of striped bass and their hybrids often reflect the suitability of habitats and effectiveness of management. Three chapters cover this topic: a survey of striped bass growth and condition in relation to physical habitat characteristics in southeastern reservoirs (Chapter 11, Wilson et al.), comparisons of growth among hybrid striped bass populations across North America (Chapter 12, Schultz et al.), and a case study comparing the growth and ecology of juvenile striped bass and hybrid striped bass in a Virginia reservoir (Chapter 13, Rash and Ney).

Assessing population characteristics and harvest are essential for successful fishery management. Six chapters evaluate methods for accomplishing such estimation, including tagging methods (Chapter 14, Hightower and Pollock), gill-net catches (Chapter 15, McRae et al.), hydroacoustics (Chapter 16, Hightower et al.), creel surveys (Chapter 17, Schlechte et al.), use of anglers to supplement data collection (Chapter 19, Wilson), and application of population models for assessment and management (Chapter 20, Allen and Guinn). In a case study of hybrid striped bass in an Indiana reservoir, Hoffman et al. (Chapter 18) provide estimates of a wide variety of population metrics.

With stocking so crucial to maintaining non-self-reproducing populations in many inland waters, three chapters deal with striped bass stocking evaluations. Sutton et al. (Chapter 21) review the operative biotic and abiotic factors affecting stocking success, Yow et al. (Chapter 22) review fingerling transportation practices affecting survival, and Darden et al. (Chapter 23) describe the use of microsatellite markers for evaluating stocking success. Hoffman et al. (Chapter 18) also provide an evaluation of hybrid striped bass year class strength as a function of fingerling stocking rates.

Successful natural reproduction can occur in inland waters where spawning habitat is available. Lamprecht et al. (Chapter 24) discuss natural reproduction and associated complexities in three reservoir systems in the Southeast. In the Colorado River basin, successful natural reproduction was not expected and its surprising occurrence complicated management, which was initially conceived as being thoroughly controllable by stocking. Stewart and Burrell (Chapter 25) discuss the dispersion of striped bass in a part of the Colorado River basin and effects on fish management, expanding on the history of this fishery provided by Gustaveson and Blommer in Chapter 2.

Harvest regulations have been a traditional mainstay for managing fish populations and fisheries, and striped bass and their hybrids are no exception. Collier et al. (Chapter 26) review the regulations governing striped bass and hybrid striped bass fisheries while Schultz et al. (Chapter 27) describe size-based mortality caps of hybrid striped bass. Catch-and-release mortality as a complicating factor of management is discussed by Bettinger and Wilde (Chapter 28).

Conflicts have arisen because striped bass and their hybrids do not occur in waters devoid of other valued fish species, and hybrid striped bass may pose a genetic threat to reproducing striped bass stocks. Miranda and Raborn (Chapter 29) discuss the potential for direct and indirect interactions between striped bass and other sport fish, while the more sociological aspects of managing user conflicts are presented by McMullin (Chapter 30). Harrell (Chapter 31) explores biological and ethical concerns regarding release of striped bass hybrids into tributary waters with reproducing striped bass stocks.

The economic value of striped bass (and hybrid striped bass) fisheries is often the bottom line for those justifying expenditures to stock and manage an introduced species with assumed high value to anglers. Whitehead (Chapter 32) provides an analysis of the economics of recreational fisheries for these fish. In some systems, concerns about contaminants can have negative impacts on striped bass or hybrid striped bass fisheries; Jakus (Chapter 33) addresses this issue and outlines an approach for conducting benefit–cost analyses of fish consumption advisories.

Finally, we are indebted to Phil Bettoli (Chapter 34) for providing his views of the challenges and opportunities for studying and managing striped bass and their hybrids in the 21st century.

Hopefully, this book will not only provide useful information on the selected topics, but will also emphasize that striped bass and hybrid striped bass are important inland sport fish that are much better understood now than when they were first introduced. However, many questions remain to be answered. Our work on this effort underscores that hybrid striped bass are much less understood than striped bass and demand further inquiry. As for striped bass, managers now have the tools to be able to address the suitability and management of this species in specific water bodies across the nation. We expect that further refinement of the use of this species will occur in the decades ahead.

This book would not be possible without the stimulus and support of the Striped Bass Technical Committee of the Southern Division of the American Fisheries Society (SDAFS). The committee maintains a Web site at http://sdafs.org/sbtc/ that summarizes annual activities and synthesizes information relevant to striped bass and hybrid striped bass management. The Web site includes abstracts from a 2005 SDAFS symposium on striped bass and a 2008 SDAFS workshop on striped bass sampling. The committee has had changing membership over the course of planning, organizing, and conducting the symposium, as well as the preparation of this book. However, success of this effort is due in large part to continued enthusiasm by the committee chairs during this period: Dan Wilson (2004–2008), Jim Bulak (2008–2012), and Tim Churchill (2012–

present). We especially acknowledge Scott Van Horn who chaired the committee from 2000 to 2004 and changed the committee's emphasis from aquaculture to management. We thank Debby Lehman, AFS book production coordinator, for copyediting, layout, and cover design, and Aaron Lerner, AFS publications director, for invaluable assistance throughout the editorial process. Cover art by Duane Raver was graciously shared by Jim Bulak.

James S. (Jim) Bulak
Charles C. (Chuck) Coutant
James A. (Jim) Rice

January 2013

Symposium Contributors

We acknowledge and greatly appreciate financial support from the following organizations for publication of this book.

Platinum ($1,000+)

Striped Bass Technical Committee, Southern Division, American Fisheries Society

Southeast Region, U.S. Fish and Wildlife Service
Panama City Field Office, U.S. Fish and Wildlife Service
Welaks National Fish Hatchery, U.S. Fish and Wildlife Service

Arkansas Game and Fish Commission

South Carolina Department of Natural Resources

Texas Parks and Wildlife

Harry Hampton Memorial Wildlife Fund

Gold ($750–$999)

Duke Energy

Silver ($500–$749)

Fisheries Management Section, American Fisheries Society
Midlands Striper Club, South Carolina
North Central Division, American Fisheries Society
Oklahoma Chapter of the American Fisheries Society
Santee-Cooper Country
Striped Bass Conservation Coalition
Striper Kings, Greenville, South Carolina
Tennessee Chapter of the American Fisheries Society
Texas Chapter of the American Fisheries Society

Bronze ($100–$499)

Arkansas Chapter of the American Fisheries Society
Appalachian Aquatics
Jordan Lake Striper Club, North Carolina
South Carolina Chapter of the American Fisheries Society
Virginia Chapter of the American Fisheries Society

Symbols and Abbreviations

The following symbols and abbreviations may be found in this book without definition. Also undefined are standard mathematical and statistical symbols given in most dictionaries.

A	ampere
AC	alternating current
Bq	becquerel
C	coulomb
°C	degrees Celsius
cal	calorie
cd	candela
cm	centimeter
Co.	Company
Corp.	Corporation
cov	covariance
DC	direct current; District of Columbia
D	dextro (as a prefix)
d	day
d	dextrorotatory
df	degrees of freedom
dL	deciliter
E	east
E	expected value
e	base of natural logarithm (2.71828…)
e.g.	(exempli gratia) for example
eq	equivalent
et al.	(et alii) and others
etc.	et cetera
eV	electron volt
F	filial generation; Farad
°F	degrees Fahrenheit
fc	footcandle (0.0929 lx)
ft	foot (30.5 cm)
ft^3/s	cubic feet per second (0.0283 m^3/s)
g	gram
G	giga (10^9, as a prefix)
gal	gallon (3.79 L)
Gy	gray
h	hour
ha	hectare (2.47 acres)
hp	horsepower (746 W)
Hz	hertz
in	inch (2.54 cm)
Inc.	Incorporated
i.e.	(id est) that is
IU	international unit
J	joule
K	Kelvin (degrees above absolute zero)
k	kilo (10^3, as a prefix)
kg	kilogram
km	kilometer
l	levorotatory
L	levo (as a prefix)
L	liter (0.264 gal, 1.06 qt)
lb	pound (0.454 kg, 454g)
lm	lumen
log	logarithm
Ltd.	Limited
M	mega (10^6, as a prefix); molar (as a suffix or by itself)
m	meter (as a suffix or by itself); milli (10^{-3}, as a prefix)
mi	mile (1.61 km)
min	minute
mol	mole
N	normal (for chemistry); north (for geography); newton
N	sample size
NS	not significant
n	ploidy; nanno (10^{-9}, as a prefix)
o	ortho (as a chemical prefix)
oz	ounce (28.4 g)
P	probability
p	para (as a chemical prefix)
p	pico (10^{-12}, as a prefix)
Pa	pascal
pH	negative log of hydrogen ion activity
ppm	parts per million
qt	quart (0.946 L)
R	multiple correlation or regression coefficient
r	simple correlation or regression coefficient
rad	radian
S	siemens (for electrical conductance); south (for geography)

SD	standard deviation
SE	standard error
s	second
T	tesla
tris	tris(hydroxymethyl)-aminomethane (a buffer)
UK	United Kingdom
U.S.	United States (adjective)
USA	United States of America (noun)
V	volt
V, Var	variance (population)
var	variance (sample)
W	watt (for power); west (for geography)
Wb	weber
yd	yard (0.914 m, 91.4 cm)
α	probability of type I error (false rejection of null hypothesis)
β	probability of type II error (false acceptance of null hypothesis)
Ω	ohm
μ	micro (10^{-6}, as a prefix)
′	minute (angular)
″	second (angular)
°	degree (temperature as a prefix, angular as a suffix)
%	per cent (per hundred)
‰	per mille (per thousand)

American Fisheries Society Symposium 80:1–13, 2013

A Brief History of Inland Striped Bass Management

Scott L. Van Horn*
921 Danbury Drive, Durham, North Carolina 27704, USA

Abstract.—The term "inland striped bass" *Morone saxatilis* is used to describe populations established within freshwater reservoirs and their headwaters and indigenous populations living wholly within the freshwater portion of coastal rivers. In 1941, the gates closed on a project impounding the Santee and Cooper rivers, isolating a population of striped bass in a freshwater reservoir for the first time. This population thrived and biologists theorized that in reservoirs, striped bass could make use of additional forage fish by increasing predation pressure on large gizzard shad *Dorosoma cepedianum*, interact little as pelagic predators with existing reservoir game fish, and create a new sport fishery. Interest in duplicating the Santee-Cooper experience in other reservoirs created a demand for striped bass that was ultimately met by developing a successful striped bass culture protocol. Expanding striped bass culture capacity allowed the rapid expansion of inland striped bass around the country. As the new reservoir fisheries developed, stocking rates, growth rates, and mortality rates were related to harvest and catch statistics to inform evolving management strategies. Research in both reservoirs and coastal rivers also focused on understanding striped bass and hybrid striped bass habitat requirements, impacts on associated fish populations, and genetic considerations. U.S. Fish and Wildlife Service figures from 2006 indicated that 25 million anglers fished for striped bass, white bass, and their hybrids in freshwater (excluding the Great Lakes) for 420 million days and estimated the total trip and equipment expenses at US$24.6 billion. Some controversy has followed inland striped bass management. A segment of reservoir striped bass anglers is increasingly strident in its demands for ever-increasing striped bass stocking rates while another segment of the angler population remains suspicious that the exotic introduction threatens more traditional reservoir fisheries. In coastal river populations, genetic tools have demonstrated that well-intentioned augmentation stocking has threatened striped bass populations indigenous to the Gulf of Mexico.

Introduction

The history of management of inland striped bass *Morone saxatilis* (and its hybrids with white bass *M. chrysops*) is a complex story with a serendipitous beginning in South Carolina. Ecologically, inland striped bass populations can be divided between those in impoundments and endemic populations in coastal rivers that rarely, if ever, enter marine waters. Inland striped bass management is a story of collaboration between science and management to meet perceived ecological and fisheries goals. It is a story with significant local and regional economic considerations, especially within the tourism and sport fishing industries. Inland striped bass management in reservoirs can also be viewed as a social experiment. Estimates of directed effort spent fishing for reservoir striped bass document the sport's

* Corresponding author: vanhornsc@nc.rr.com

popularity with anglers, but fisheries management agencies have continually been challenged by striped bass angling advocates and detractors alike. Angler attitudes have evolved as striped bass fisheries and their supporting habitats have changed. Finally, much of inland striped bass management can be viewed from the perspective of exotic introduction ecology.

The Beginnings

While striped bass had been introduced successfully into California estuaries as early as 1879 (Raney et al. 1952), the first attempted freshwater introduction, in New Jersey in the mid-1930s, was unsuccessful (Surber 1958). Then, in the middle of the Great Depression, the Santee–Cooper Project was conceived as a way to bring jobs and economic prosperity to inland South Carolina. The Santee and Cooper rivers were impounded and connected to create hydroelectric power and a water-based navigation route to the coast. The gates closed on the dams in 1941, isolating a population of striped bass in the freshwater reservoirs. The presence of numerous small striped bass in the system by 1949 implied successful striped bass reproduction. By the early 1950s, a striped bass sport fishery had developed within the system (Harrell et al. 1990).

The Wateree and Congaree rivers, forming the headwaters of the newly impounded system, provided sufficient quantities and duration of flow to support successful spawning and larval development of the landlocked striped bass population. Conditions within the two conjoined reservoirs of the project, Lakes Marion and Moultrie, supported survival and growth. Scruggs and Fuller (1955) verified that the population was landlocked and self-sustaining.

A second freshwater reservoir fishery for striped bass developed in the late 1950s at John H. Kerr Reservoir (Kerr Lake) on the border of Virginia and North Carolina. Kerr Lake was created in 1952 as an impoundment of the Roanoke River and simultaneously backed up the lower reaches of the Dan River, a major Roanoke River tributary. Successful striped bass spawning was documented in both upstream rivers (Whitehurst and Carwile 1982). It is likely that some adult striped bass were trapped behind the newly constructed dam. Striped bass fry were also stocked in Kerr Lake in 1953–1956 (Surber 1958).

The creation of self-sustaining reservoir striped bass populations at Santee–Cooper and Kerr reservoirs occurred during an active period of reservoir construction in the United States that peaked in the 1960s. The focus of the state agencies responsible for managing fisheries resources naturally turned to address the new reservoir resources. The Southern Division of the American Fisheries Society (SDAFS) created its Reservoir Committee in 1957. The U.S. Fish and Wildlife Service made additional fish hatchery space available and created the National Reservoir Research Program.

Striped bass were quickly promoted as a candidate to exploit an "empty niche" opportunity identified as the combination of available open-water habitat and an underutilized prey base of juvenile and adult gizzard shad *Dorosoma cepedianum*. Early work inventorying fish populations in reservoirs often found abundant populations of large, old gizzard shad. Biologists recognized that long-lived gizzard shad that grew too large to contribute to the prey base for typical reservoir game fish such as largemouth bass *Micropterus salmoides* could ultimately reduce reservoir game fish production. Biologists also identified the absence of any open-water predators in their large, new reservoirs. Striped bass stocking was seen as a way to create a new sport fishery that would also benefit other reservoir game fish by increasing the availability of smaller gizzard shad. The rush to duplicate the Santee–Cooper and Kerr Reservoir experiences began (Table 1).

Overcoming Striped Bass Culture Constraints

The most immediate problem for reservoir managers eager to create new landlocked striped bass populations was finding fish to stock. In the 1950s, Arkansas, South Carolina,

TABLE 1. First recorded striped bass introduction or successful scientific collection of introduced striped bass by time period and state. Stocking events were included whether or not they were successful. Asterisks denote coastal stocking. Compiled by the author from multiple sources.

<1900	1900–1949	1950–1959	1960–1969	1970–1979	1980–1989	1990–1999	2000–2009
California*	Hawaii*	Arkansas	Alabama	Iowa	Kansas	North Dakota	Michigan
	Oregon*	Arizona	Colorado	Illinois	Minnesota	Wisconsin	
	Washington*	Maryland	Florida	New Mexico	New York		
	New Jersey	North Carolina	Georgia	Ohio	West Virginia		
		Oklahoma	Kentucky	Pennsylvania			
		South Carolina	Louisiana	Utah			
		Tennessee	Missouri				
		Virginia	Mississippi				
			Nebraska				
			Nevada				
			Texas				

North Carolina, Kentucky, Maryland, Georgia, and Oklahoma all tried stocking wild fish, sometimes purchased from saltwater commercial fishermen, always without success (Bailey 1975). Oklahoma may have had the most ambitious adult fish stocking program, releasing 4,000–5,000 striped bass adults in 1955 (Surber 1958). There were no documented successful attempts to introduce striped bass in inland waters using adult fish.

Timely access to striped bass gametes from ripe, mature adults had limited hatchery production through the early 1960s to a small hatchery in Weldon, North Carolina. The Weldon site was successful because it was located on the principal spawning grounds of the Albemarle Sound striped bass population in the Roanoke River. Hatchery managers simply purchased ripe eggs and milt from anglers during the spawning season fishery and were able to produce larvae. South Carolina constructed its striped bass hatchery at Monck's Corner in 1961 and developed the technique for inducing nonripe fish to spawn using human growth hormone injections (Stevens et al. 1965; Stevens 1966, 1967). Bishop (1967) further developed this technique in Tennessee. Biologists at Monck's Corner were concurrently working with hybrids of striped bass trying to find a fish that might overcome problems of fragile larvae and limited access to ripe gametes.

Even as advances in larval production were made, establishing striped bass and hybrid striped bass populations in reservoirs with larval stocking proved highly unreliable (Bailey 1975). Stocking larval striped bass dominated much of the 1960s. Newly hatched striped bass larvae swim in the water column where they will live off their still attached yolk sac until their mouthparts develop in around 5 d. Typically, larvae were stocked into reservoirs at this life stage to avoid having to feed them. Arkansas, Louisiana, Kentucky, North Carolina, South Carolina, Tennessee, and Texas all had larval stocking programs.

The successful proliferation of striped bass populations in inland waters would wait for the development of pond-cultured striped bass juveniles, often called fingerlings, typically stocked into reservoirs at about 50 mm total length. Much of the impetus for the advances in striped bass culture came from a small subcommittee of the SDAFS Reservoir Committee interested in advancing striped bass as a reservoir management tool. When the first striped bass fingerling production occurred in 1968 at a federal hatchery in Edenton, North Carolina, it represented the culmination of the work and ideas of many biologists around the southeastern United States. By 1973, nine state hatcheries were producing more than 10 million striped bass fingerlings (Stevens 1974).

To further develop and communicate the technologies needed to introduce striped bass and their hybrids into reservoirs, the striped bass subcommittee of the SDAFS Reservoir Committee was elevated to full technical committee status as the SDAFS Striped Bass Committee in 1970. This committee produced the *Guidelines for Striped Bass Culture* in 1976 (Bonn et al. 1976), making knowledge of the new striped bass culture technology widely available. The committee also played a key role in developing stocking procedures and stocking rates for reservoirs (Harrell et al. 1990). By 1981, with successful striped bass and hybrid striped bass culture finally routine and widespread, striped bass and their hybrids occupied at least 289 reservoirs and nearly 4 million acres (1,615,000 ha) of water in 34 states (Axon and Whitehurst 1985). Where striped bass or their hybrids have been stocked in inland waters of the United States can be found at the U.S. Geological Survey Nonindigenous Aquatic Species Web site (http://nas.er.usgs.gov/queries/). The U.S. Geological Survey maps affirm that most reservoir striped bass are still found in the Southeast. There are a number of important populations in the West, notably along the Colorado River (Gustaveson et al. 1984; Gustaveson and Blummer 2013, this volume; Stewart and Burrell 2013, this volume). Much less widely distributed as a reservoir sport fish, striped bass hybrids are used intensively primarily in Texas,

Alabama, Florida, West Virginia, the Midwest, and the Carolinas.

Management History

Reservoirs

The first goal of every reservoir-stocking program was simply to establish a population and then sustain it when it was clear that natural reproduction was impossible. The SDAFS Striped Bass Committee documented the quantity of fingerlings produced annually in the Southeast and stocking rates by location into the 1980s. Early stocking rates varied from 0.5 to 178 fingerlings per acre (1.24–441 per hectare; Bailey 1975). Eventually, annual fingerling stocking rates in reservoirs would vary between 3 and 20 fingerlings per acre (7.5–49.5 per hectare). Agencies looking for stocking efficiencies and, perhaps responding to pressure from angling groups to stock more striped bass in reservoirs, began to relate striped bass stocking strategies (e.g., varying numbers of fish released, single versus multiple-release points, open water releases, smaller versus larger fingerlings) to subsequent striped bass year-class strength or surrogates for abundance (Moore et al. 1991; Van Horn et al. 2001). Even in those few locations where striped bass successfully reproduced in reservoirs, stocking became a consideration to support or improve striped bass fisheries (Bulak et al. 1995). Notable exceptions to this generalization exist in several western reservoirs where striped bass are spawning within the reservoirs, eggs are surviving on the lake bottoms, and year-class strength from natural reproduction is consistently excellent (Gustaveson et al. 1984; Gustaveson and Blommer 2013, this volume).

One of the ecological objectives of stocking striped bass was to manage gizzard shad and forage production. Early food habits work conducted in the Santee–Cooper system (Stevens 1958) and subsequently often repeated (Combs 1979; Filipek and Tommey 1987; Moore et al. 1987; Matthews et al. 1988) provided evidence that reservoir striped bass were indeed foraging on clupeids. Jenkins and Morais (1979) used predator gape size and prey body size to develop an index of prey availability to predators in reservoir environments. They were among the first to caution that not all reservoirs might be capable of supporting striped bass populations due to prey inadequacy.

As the number of reservoir populations of striped bass increased, so did the amount of data on growth and body condition; variation was soon apparent. Striped bass growth and body condition (as a surrogate for prey sufficiency) eventually began to inform decisions about striped bass stocking rates. A variety of traditional fisheries gears (Boxrucker et al. 1995) were applied in new habitats, and hydroacoustic equipment became available to help describe and quantify striped bass forage availability. Bioenergetics approaches were developed that would help quantify striped bass predator–prey relationships (Hartman and Brandt 1995; Cyterski et al. 2003; Thompson et al. 2005; Thompson and Rice 2013, this volume).

Competition with other reservoir piscivores (especially game fish) was an early concern and closely related to developments in forage availability. When Stevens (1958) and others found few game fish in striped bass stomachs, some felt the prediction implied in the empty niche theory had been supported. Low levels of striped bass predation on other game fish were widely interpreted as absence of competition. A forage-availability study by Jenkins and Morais (1979) suggested that competition for forage among multiple reservoir predators might be possible but nevertheless concluded that competition with striped bass was not evidenced in sport fish harvest data from 11 southeastern reservoirs. Gilliland and Clady (1984) and Moore (1988) found limited or no diet overlap between striped bass and their hybrids and largemouth bass. However, some anglers remained convinced that the addition of striped bass to reservoirs was detrimental to other reservoir game fish (especially largemouth bass, crappies *Pomoxis* spp., walleye *Sander vitreus*, and white bass). This suspicion reached its most virulent form at Norris Reservoir (Churchill et al. 2002).

In the most sophisticated study to date, Raborn et al. (2002) applied a bioenergetics approach at Norris Reservoir to conclude that direct predation on or competition for available prey by striped bass has small (<15%) impacts on game fish production. Controversies continue, however, despite considerable evidence regarding all modes of potential interactions (Miranda and Raborn 2013, this volume).

Noticeable numbers of dead striped bass and hybrid striped bass in Cherokee Reservoir in the 1970s and other locations precipitated a line of inquiry relating summer temperature and dissolved oxygen profiles to striped bass mortalities. Coutant (1985) hypothesized that in many reservoirs and estuarine systems, temperature and dissolved oxygen levels in summer were not meeting the physiological requirements of larger striped bass. The hypothesis successfully explained summer body condition loss and mass striped bass mortalities in many reservoirs, rivers, and estuaries. A number of subsequent studies essentially corroborated the concept at other locations (Douglas and Jahn 1987; Matthews et al. 1989; Schaffler et al. 2002, Wilkerson and Fisher 1997; Zale et al. 1990; Bettoli 2005; Coutant 2013, this volume). More recent work by Thompson et al. (2005) and Thompson and Rice (2013) suggested that poor summer habitat impacts on striped bass growth and condition could be mitigated in part by high forage availability.

Early attempts to evaluate temperature and dissolved oxygen distribution in reservoirs relied on plotting point measurements of water column temperature and dissolved oxygen profiles. As water quality computer models became more available and sophisticated, biologists were able to extend the utility of their field data by applying models that predicted spatial and temporal patterns of temperature and dissolved oxygen concentration as functions of limnological and meteorological factors (Ruane et al. 2013, this volume). Some of the earliest whole-reservoir models were used by utility companies to evaluate power plant designs and predict impacts of operations on reservoir water quality, for evaluation by regulatory agencies. These advances have allowed investigators to identify and predict oxygen stratification patterns that can isolate striped bass and create summer kills (Rice et al. 2013, this volume). Reservoir temperature and dissolved oxygen regimes can sometimes be managed by manipulating the electric power production process. Examples include varying power plant cooling water intake depth and quantity or by managing the amount and depth of water releases through reservoir dams for hydroelectric production (Duke Power Company 1986). Oxygen diffusers that were initially installed to meet water quality standards in reservoir release water can have beneficial consequences for striped bass populations both above and below affected dams (Mobley 1997; Mobley et al. 2013, this volume; Ruane et al. 2013, this volume). Improvement in localized temperature and oxygen habitat around oxygen diffusers can have unintended consequences, however, as high concentrations of striped bass can result in high vulnerability to anglers (Mobley et al. 2013).

Determining the effects of varying stocking strategies on striped bass populations, linking forage availability to striped bass growth and condition, quantifying competition, estimating the impacts of summer kill events, or developing mortality rates required developing some estimate of striped bass abundance. Traditionally, estimates of relative abundance have been made with gill nets, although gill-net catch rates have often been prohibitively variable (McRae et al. 2013, this volume). Shoreline seining and electrofishing for age-0 striped bass have also been used in reservoirs, but also with problematic sampling variability. The South Carolina Department of Natural Resources has estimated striped bass fishing mortality rate in the Santee–Cooper system using tag returns (J. S. Bulak, South Carolina Department of Natural Resources, personal communication). Several studies have estimated striped bass fishing and natural mortality rates using telemetry techniques (Hightower et al. 2001; Young and Isely 2004; Thompson et al. 2007).

Fisheries managers and striped bass anglers have been particularly concerned about hooking mortality of released fish as it might relate to total striped bass mortality and size and creel regulations. Hooking mortality studies were conducted in saltwater and freshwater and in varying seasons evaluating the effects of variables such as bait type (e.g., natural bait, artificial lures), temperature, and hooking locations (e.g., jaw, esophagus). Wilde et al. (2000) and Bettinger and Wilde (2013, this volume) summarized much of the data, suggesting that hooking mortality is highest at warmer water temperatures and for fish caught on natural baits.

Striped bass regulations have evolved to address what managers have learned about striped bass total and fishing mortality rates, hooking mortality rates, body condition and growth, and some anglers' perception of striped bass as a trophy animal. The demand for more and larger fish while some populations showed changes towards smaller sizes has generally moved regulations in the direction of higher size limits and lower creel limits. High summertime hooking mortality rates are increasing the number of states suspending size limits during the warm weather months. The public's perception that more restrictive size limits will give anglers more larger striped bass is often inconsistent with the poor growth and high mortality rates (including poor catch and release survival) where striped bass are heavily fished and temperature–oxygen habitat is poor to marginal. Protective size and creel limits play no role in some of the western reservoirs where striped bass densities are so high that competition for available forage severely limits striped bass growth. In lightly fished systems with excellent habitat, more restrictive size limits are sometimes employed to conserve existing populations of larger, older striped bass.

Hybrid striped bass have been stocked intermittently in inland waters since Stevens produced the first experimental cross between female striped bass and male white bass (palmetto bass) in 1965 (Bayless 1968; Ware 1975). The reciprocal hybrid (sunshine bass) has also been produced and stocked. Responses to an informal 2005 query indicated that hybrids were being stocked in inland waters in Arkansas, Georgia, Illinois, Indiana, Kentucky, Mississippi, Missouri, South Carolina, Tennessee, and Virginia (C. C. Coutant, unpublished). Hybrids are similar in appearance to striped bass but tend to grow faster in their first 2 years of life, and will weigh in excess of 4.5 kg at maturity in 2–3 years (Logan 1968; Williams 1970). They are thought to be more tolerant of high temperatures and thus are desired for habitats that are thermally marginal for striped bass. They are often managed with the same size and creel limits used for striped bass when the two types of fish are found together to relieve confusion over fish identification. Size and creel limits are otherwise often less restrictive for striped bass hybrids, reflecting differences in trophy fish potential and differing management objectives. Collier et al. (2013, this volume) provide an excellent overview of the use of regulations to manage striped bass and their hybrids in inland waters.

Finally, as reservoir striped bass angling became more popular from its inception into the 1970s and 1980s, striped bass clubs began to form to promote the sport and advocate for their fisheries. Most clubs are associated with particular reservoirs, and they are most common in the southeastern states. The late 1980s and 1990s were marked by conflicts between club-associated angler advocates for better reservoir striped bass fishing and state agencies responsible for managing their fisheries. Some of the more widely recognized angler–agency conflicts occurred at Lake Norris (Churchill et al. 2002), Smith Mountain Lake (M. Duval, Minnesota Department of Natural Resources, personal communication), Lake Norman (Van Horn 2003), and Lake Greenwood (G. Hayes, South Carolina Department of Natural Resources, personal communication). Miranda and Raborn (2013, this volume) provided a biological perspective on these conflicts.

There was a discernible pattern to many of these conflicts. Most reservoir striped bass populations were established in the late 1960s and 1970s. Anglers new to the idea of fishing for striped bass in these new habitats were few in number and inefficient fishers. As the new striped bass populations matured, they contained the older, larger fish expected in lightly exploited fisheries. Successful fisheries encouraged additional fishing pressure. Expertise was rapidly acquired and shared (often within fishing clubs), and fishing mortality rates (including hooking mortality) began to climb. Few striped bass anglers practiced catch and release in the 1970s and 1980s. As a result, large striped bass became less common. These problems were exacerbated in reservoirs with moderate to poor striped bass habitat and close proximity to large populations of anglers. In some locations, water quality improvements in upstream rivers motivated by the federal Clean Water Act reduced reservoir nutrient inputs and decreased fish production (Ney 1996). More and bigger striped bass was an unachievable goal with lower production, higher harvest rates, habitat restrictions for larger fish, and hooking mortality that worked against protective size limits. Angler demands to raise stocking rates risked reducing growth rates and pushing the goal of larger fish even farther away. Anglers that were convinced striped bass were related to the decline of their favorite fisheries were alarmed at the additional demands for increased striped bass stocking rates and restrictive size and creel limits.

Reservoir fisheries managers turned to human dimensions specialists and natural resource policy experts for assistance in handling conflicting fishery uses of reservoirs. In 2002, the SDAFS Striped Bass Committee held a workshop in North Carolina that was attended by a natural resource policy expert from Duke University's Nicholas School of the Environment and a conflict resolution specialist, Dr. Steve McMullin, from Virginia Tech. The exercise demonstrated that reservoir striped bass managers were moving towards a more sophisticated consideration of the end users of their product, the anglers.

Coastal Rivers

The advancements in hatchery production of striped bass also had implications for coastal river populations. Techniques for successful larval production were first worked out at the Weldon Hatchery in the 1880s (Worth 1884). The motivation was to find a hatchery product that could be used to mitigate anticipated overexploitation of coastal river stocks by commercial and sport fish interests. Weldon striped bass larvae were used primarily to stock the Roanoke River until the establishment of striped bass populations in Lakes Marion and Moultrie on the Santee–Cooper system signaled the beginnings of reservoir striped bass management. The development of striped bass fingerling production raised the scope and effectiveness of coastal river striped bass introductions, especially in the southeast and along the Gulf of Mexico. The motivation for supplemental stocking of striped bass into southern coastal rivers was the desire by fisheries managers to protect those populations from commercial and sport fishing overexploitation, yet it has had unintended negative consequences for the native stocks of striped bass in tributaries of the Gulf of Mexico (Long et al. 2013, this volume).

While many of the same advances in management concepts and technologies that benefit reservoir striped bass management have some application in managing coastal river stocks, advances in the field of genetics would prove especially important. Unfortunately, as the genetic tools became available, biologists recognized that many of the coastal riverine striped bass stocks were endemic, completing their life cycle without leaving their riverine environment. Wirgin et al. (1991) found an indigenous Gulf of Mexico strain of striped bass still present in the Apalachicola-Chattahoochee-Flint River system in Georgia, Alabama and Florida. Bulak et al. (2004) found evidence that some coastal riverine populations of striped bass were genetically distinct (endemic). Thus, the riverine populations of striped bass in southern states appear not to be anadromous, have little natural genetic interchange among river basins, and are truly "inland striped bass." Both studies urge appropriate broodfish selection and gene conservation strategies to preserve the genetic integrity of the remaining endemic striped bass populations.

Widespread use of annually stocked striped bass hybrids in reservoirs raises the possibility that downstream escapement may create competition and genetic introgression with coastal river striped bass stocks (Avise and VanDenAvyle 1984; Patrick and Moser 2001). North Carolina has discontinued the use of hybrid striped bass in the Cape Fear River and declined their use within Neuse River impoundments as a precautionary measure (J. Kornegay, North Carolina Wildlife Resources Commission, personal communication). Harrell (2013, this volume) considers the degree of risk associated with stocking hybrid striped bass into drainages where reproducing striped bass are found.

Conclusion

Striped bass culture advances from the 1960s and 1970s have given fisheries managers the tools to use hatchery-reared striped bass and hybrid striped bass in nonnative environments and to augment coastal striped bass populations. Fisheries biologists have gained new tools and refined their knowledge and management skills over the past four decades. While mistakes were made, important successes have been documented. Figures from the U.S. Fish and Wildlife Service's recent National Survey of Fishing, Hunting, and Wildlife-Associated Recreation (U.S. Department of Interior, Fish and Wildlife Service, and U.S. Department of Commerce, U.S. Census Bureau 2006) indicate that 25 million anglers fished for striped bass, white bass, and their hybrids in freshwater (excluding the Great Lakes) for 420 million days and estimate the total trip and equipment expenses at $24.6 billion. However, reservoir and coastal river habitats are changing, angler expectations are evolving, and our understanding of the real consequences of using striped bass hatchery products in our environment continues to improve. From that perspective, inland striped bass management is still a young endeavor posing many new challenges. Bettoli (2013, this volume) summarized these challenges, which are amply described in each paper in this volume.

References

Avise, J. C., and M. J. VanDenAvyle. 1984. Genetic analysis of reproduction of hybrid white bass × striped bass in the Savannah River. Transactions of the American Fisheries Society 11:563–570.

Axon, J. R., and D. K. Whitehurst. 1985. Striped bass management in lakes with emphasis on management problems. Transactions of the American Fisheries Society 114:8–11.

Bayless, J. D. 1968. Striped bass hatching and hybridization experiments. Proceedings of the Annual Conference Southeastern Association of Game and Fish Commissioners 21(1967):233–244.

Bailey, W. M. 1975. An evaluation of striped bass introductions in the southeastern United States. Proceedings of the Annual Conference Southeastern Association of Game and Fish Commissioners 28(1974):54–68.

Bettinger, J. M., and G. R. Wilde. 2013. Catch-and-release mortality of inland striped bass and hybrid striped bass. Pages 473–499 *in* J. S. Bulak, C. C. Coutant, and J. A. Rice, editors. Biology and management of inland striped bass and hybrid striped bass. American Fisheries Society, Symposium 80, Bethesda, Maryland.

Bettoli, P. W. 2005. The fundamental thermal niche of adult landlocked striped bass. Transactions of the American Fisheries Society 114:305–314.

Bettoli, P. W. 2013. Challenges and opportunities in studying and managing striped bass and its hybrids in the 21st century. Pages 575–588 *in* J. S. Bulak, C. C. Coutant, and J. A. Rice, editors. Biology and management of inland striped bass and hybrid striped bass. American Fisheries Society, Symposium 80, Bethesda, Maryland.

Bishop, R. D. 1967. Evaluation of the striped bass (*Roccus saxatilis*) and white bass (*Roccus chrysops*) hybrids after two years. Proceedings of the Annual Conference Southeast ern Association of Game and Fish Commissioners 21:245–253.

Bonn, E. W., W. M. Bailey, J. D. Bayless, K. E. Erickson, and R. E. Stevens. 1976. Guidelines for striped bass culture. American Fisheries Society, Southern Division, Striped Bass Committee, Bethesda, Maryland.

Boxrucker, J., P Michaletz, M. J. VanDenAvyle, and B. Vondracek. 1995. Overview of gear evaluation study for sampling gizzard shad and threadfin shad populations in reservoirs. North American Journal of Fisheries Management 15:885–890.

Bulak, J. S., D. S. Wethy, M. G. White III. 1995. Evaluation of management options for a reproducing striped bass population in Santee–Cooper system, South Carolina. North American Journal of Fisheries Management 15:84–94.

Bulak, J. S., C. S. Thomason, K. Han, and B. Ely. 2004. Genetic variation and management of striped bass populations in the coastal rivers of South Carolina. North American Journal of Fisheries Management 24:1322–1329.

Churchill, T. N., P. W. Bettoli, D. C. Peterson, W. C. Reeves, and B. Hodge. 2002. Angler conflicts in fisheries management: a case study of the striped bass controversy at Norris Reservoir, Tennessee. Fisheries 27:10–19.

Collier, W. R., P. W Bettoli, G. D. Scholten, and T. N. Churchill. 2013. Regulation of striped bass and hybrid striped bass fisheries in the United States. Pages 449–460 *in* J. S. Bulak, C. C. Coutant, and J. A. Rice, editors. Biology and management of inland striped bass and hybrid striped bass. American Fisheries Society, Symposium 80, Bethesda, Maryland.

Combs, D. L. 1979. Food habits of adult striped bass from Keystone reservoir and its tailwaters. Proceedings of the Annual Conference Southeastern Association of Fish and Wildlife Agencies 32(1978):571–575.

Coutant, C. C. 1985. Striped bass, temperature, and dissolved oxygen: a speculative hypothesis for environmental risk. Transactions of the American Fisheries Society 114:31–61.

Coutant, C. C. 2013. When is habitat limiting for striped bass? Three decades of testing the temperature–oxygen squeeze hypothesis. Pages 65–91 *in* J. S. Bulak, C. C. Coutant, and J. A. Rice, editors. Biology and management of inland striped bass and hybrid striped bass. American Fisheries Society, Symposium 80, Bethesda, Maryland.

Cyterski, M., J. J. Ney, and M. Duval. 2003. Estimation of surplus biomass of clupeids in Smith Mountain Lake, Virginia. Transactions of the American Fisheries Society 132:361–370.

Douglas, D. R., and L. A. Jahn. 1987. Radiotracking hybrid striped bass in Spring Lake, Illinois, to determine temperature and oxygen preferences. North American Journal of Fisheries Management 7:531–534.

Duke Power Company. 1986. Lake Norman annual report. Charlotte, North Carolina.

Filipek, S. P., and W. H. Tommey. 1987. The food habits of adult striped bass from Lake Hamilton, Arkansas, before and during an extreme drawdown. Proceedings of the Annual Conference Southeastern Association of Fish and Wildlife Agencies 38(1984):327–334.

Gilliland, E., and M. Clady. 1984. Diet overlap of striped bass × white bass hybrids and largemouth bass in Sooner Lake, Oklahoma. Proceedings of the Annual Conference Southeastern Association of Fish and Wildlife Agencies 35(1981):317–330.

Gustaveson, A. W., T. D. Pettengill, J. E. Johnson, and J. R. Wahl. 1984. Evidence of in reservoir spawning of striped bass in Lake Powell, Utah–Arizona. North American Journal of Fisheries Management 4:540–546.

Gustaveson, W., and G. Blommer. 2013. History of striped bass management in the Colorado River. Pages 15–23 *in* J. S. Bulak, C. C. Coutant, and J. A. Rice, editors. Biology and management of inland striped bass and hybrid striped bass. American Fisheries Society, Symposium 80, Bethesda, Maryland.

Harrell, R. M., J. H. Kirby, and R. V. Minton, editors. 1990. Culture and propagation of striped bass and its hybrids. American Fisheries Society, Southern Division, Striped Bass Committee, Maryland.

Harrell, R. M. 2013. Releasing hybrid *Morone* in natural waters with congeneric species: implications and ethics. Pages 531–549 *in* J. S. Bulak, C. C. Coutant, and J. A. Rice, editors. Biology and management of inland striped bass and hybrid striped bass. American Fisheries Society, Symposium 80, Bethesda, Maryland.

Hartman, K. J., and S. B. Brandt. 1995. Comparative energetics and the development of bioenergetics models for sympatric estuarine piscivores. Canadian Journal of Fisheries and Aquatic Sciences 52:1647–1666.

Hightower, J. E., J. R. Jackson, and K. H. Pollack. 2001. Use of telemetry methods to evaluate natural and fishing mortality of striped bass in Lake Gaston, North Carolina. Transactions of the American Fisheries society 130:557–567.

Jenkins, R. M., and D. I. Morais. 1979. Predator-prey relations in the predator-stocking-evaluation reservoirs. Proceedings of the Annual Conference Southeastern Association of Fish and Wildlife Agencies 32(1978):141–157.

Logan, H. J. 1968. Comparison of growth and survival rates of striped bass and striped bass × white bass hybrids under controlled environments. Proceedings of the Annual Conference Southeastern Game and Fish Commissioners 21(1967):260–263.

Long, E. A., C. L. Mesing, K. J. Herrington, R. R. Weller, and I. I. Wirgin. 2013. Restoration of Gulf striped bass: lessons and management implications. Pages 25–63 *in* J. S. Bulak, C. C. Coutant, and J. A. Rice, editors. Biology and management of inland striped bass and hybrid striped bass. American Fisheries Society, Symposium 80, Bethesda, Maryland.

Matthews, W. J., L. G. Hill, D. R. Edds, J. J. Hoover, and T. G. Heger. 1988. Trophic ecology of striped bass, *Morone saxatilis*, in a freshwater reservoir (Lake Texoma, USA). Journal of Fish Biology 33:273–288.

Matthews, W. J., L. G. Hill, D. R. Edds, and F. P. Gelwick. 1989. Influence of water quality and season on habitat use by striped bass in a large southwestern reservoir. Transactions of the American Fisheries Society 118:243–250.

McRae, B. J., J. S. Bulak, B. E. Taylor, and C. T. Waters. 2013. Evaluation of the use of gill nets for monitoring reservoir striped bass fisheries. Pages 263–278 *in* J. S. Bulak, C. C. Coutant, and J. A. Rice, editors. Biology and management of inland striped bass and hybrid striped bass. American Fisheries Society, Symposium 80, Bethesda, Maryland.

Miranda, L. E., and S. W. Raborn. 2013. Interactions between striped bass and other game fish in reservoirs. Pages 501–519 *in* J. S. Bulak, C. C. Coutant, and J. A. Rice, editors. Biology and management of inland striped bass and hybrid striped bass. American Fisheries Society, Symposium 80, Bethesda, Maryland.

Mobley, M. 1997. TVA reservoir aeration diffuser system. American Society of Civil Engineers, ASCE Waterpower 97, Washington, D.C.

Mobley, M., E. Shallenberger, M. W. Beutel, P. Gantzer, and B. Sak. 2013. Oxygen diffusers to create and maintain summer fish habitat. Pages 197–208 *in* J. S. Bulak, C. C. Coutant, and J. A. Rice, editors. Biology and management of inland striped bass and hybrid striped bass. American Fisheries Society, Symposium 80, Bethesda, Maryland.

Moore, C. M. 1988. Food habits, population dynamics, and bioenergetics of four predatory fish species in Smith Mountain Lake, Virginia. Doctoral dissertation. Virginia Polytechnical Institute and State University, Blacksburg.

Moore, C. M., R. J. Neves, and J. J. Ney. 1991. Survival and abundance of stocked striped bass in Smith Mountain Lake, Virginia. North American Journal of Fisheries Management 11:393–399.

Moore, C. M., R. J. Neves, J. J. Ney, and D. K. Whitehurst. 1987. Utilization of alewives and gizzard shad by striped bass in Smith Mountain Lake, Virginia. Proceedings of the Annual Conference Southeastern Association of Fish and Wildlife Agencies 39(1985):108–115.

Ney, J. J. 1996. Oligotrophication and its discontents: effects of reduced nutrient loading on reservoir fisheries. Pages 285–295 *in* L. E. Miranda and D. R. DeVries, editors. Multidimensional approaches to reservoir fisheries management. American Fisheries Society, Symposium 16, Bethesda, Maryland.

Patrick, W. S., and M. L. Moser. 2001. Potential competition between hybrid striped bass (*Morone saxatilis* × *M. Americana*) and striped bass (M. saxatilis) in the Cape Fear estuary, North Carolina. Estuaries and Coasts 24:425–429.

Raborn, S. W., L. E. Miranda, and M. T. Driscoll. 2002. Effects of a simulated removal of striped bass from a southeastern reservoir. North American Journal of Fisheries Management 22:406–417.

Raney, E. C., E. F. Tresselt, E. H. Hollis, V. D. Vladykov, and D. H. Wallace, editors. 1952. The striped bass (*Roccus saxatilis*). Bulletin of the Bingham Oceanographic Collection Yale University 14.

Rice, J. R., J. S. Thompson, J. A. Sykes, and C. T. Waters. 2013. The role of metalimnetic hypoxia in striped bass summer kills: consequences and management implications. Pages 121–145 *in* J. S. Bulak, C. C. Coutant, and J. A. Rice, editors. Biology and management of inland striped bass and hybrid striped bass. American Fisheries Society, Symposium 80, Bethesda, Maryland.

Ruane, R. J., G. E. Hauser, and A. F. Sawyer. 2013. Water quality modeling as a management tool for striped bass and hybrid striped bass. Pages 161–179 *in* J. S. Bulak, C. C. Coutant, and J. A. Rice, editors. Biology and management of inland striped bass and hybrid striped bass. American Fisheries Society, Symposium 80, Bethesda, Maryland.

Schaffler, J. J., J. J. Isely, and W. E. Hayes. 2002. Habitat use by striped bass in relation to seasonal changes in water quality in a southeastern reservoir. Transactions of the American Fisheries Society 131:817–827.

Scruggs, G. D., Jr., and J. C. Fuller, Jr. 1955. Indications of a freshwater population of striped bass, *Rocus saxatilis* (Walbaum), in Santee–Cooper Reservoirs. Proceedings of the Annual Conference Southeastern Association of Game and Fish Commissioners 8(1954):64–69.

Stevens, R. E. 1958. The striped bass of the Santee–Cooper Reservoir. Proceedings of the Annual Conference Southeastern Association of Game and Fish Commissioners 11(1957):253–264.

Stevens, R. E. 1966. Hormone-induced spawning of striped bass for reservoir stocking. Progressive Fish-Culturist 28:19–28.

Stevens, R. E. 1967. A final report on the use of hormones to ovulate striped bass, *Roccus saxatilis* (Walbaum). Proceedings of the Annual Conference Southeastern Association of Game and Fish Commissioners 18(1964):525–538.

Stevens, R. E. 1974. Current and future considerations concerning striped bass culture and managment. Proceedings of the Annual Conference Southeast Association of Fish and Wildlife Agencies 25:69–73.

Stevens, R. E., O. D. May, Jr., and H. J. Logan. 1965. An interim report on the use of hormones to ovulate striped bass (*Roccus saxatilis*). Proceedings of the Annual Conference Southeastern Association of Game and Fish Commissioners 17(1963):226–237.

Stewart, W. T., and M. Burrell. 2013. Striped bass dispersion and effects on fisheries management in Lakes Mohave and Pleasant, Colorado River basin. Pages 431–448 *in* J. S. Bulak, C. C. Coutant, and J. A. Rice, editors. Biology and management of inland striped bass and hybrid striped bass. American Fisheries Society, Symposium 80, Bethesda, Maryland.

Surber, E. W. 1958. Results of striped bass (*Roccus saxatilis*) introductions into freshwater impoundments. Proceedings of the Annual Conference Southeastern Association of Game and Fish Commissioners 11(1957):273–276.

Thompson, J. S., J. A. Rice, D. S. Waters, J. C. Taylor, J. E. Hightower, L. A. Davias, and P. S. Rand. 2005. Energetics of reservoir striped bass populations: phase 1 final report. North Carolina Wildlife Resources Commission, Federal Aid in Sport Fish Restoration Project F-68–4, Raleigh.

Thompson, J. S., and J. A. Rice. 2013. The relative influence of thermal experience and forage availability on growth of age 1–5 striped bass in two southeastern reservoirs. Pages 93–120 *in* J. S. Bulak, C. C. Coutant, and J. A. Rice, editors. Biology and management of inland striped bass and hybrid striped bass. American Fisheries Society, Symposium 80, Bethesda, Maryland.

Thompson, J. S., D. S. Waters, J. A. Rice, and J. E. Hightower. 2007. Seasonal fishing and natural mortality of striped bass in a southeastern reservoir. North American Journal of Fisheries Management 27:681–694.

U.S. Department of Interior, Fish and Wildlife

Service, and U.S. Department of Commerce, U.S. Census Bureau. 2006. 2006 national survey of fishing, hunting, and wildlife-associated recreation. U.S. Fish and Wildlife Service, Washington, D.C.

Van Horn, S. L. 2003. Lake Norman striped bass management: a conflict between fishery managers and anglers. Proceedings of the Annual Conference Southeastern Association of Fish and Wildlife Agencies 55(2001):316–323.

Van Horn, S. L., B. K. Baker, M. Rash. 2001. Growth and condition response of Lake Norman striped bass to increased stocking rates and more restrictive harvest regulations. Proceedings of the Annual Conference Southeastern Association of Fish and Wildlife Agencies 53(1999):193–199.

Ware, F. J. 1975. Progress with *Morone* hybrids in fresh water. Proceedings of the Annual Conference Southeastern Association of Game and Fish Commissioners 28(1974):48–54.

Whitehurst, D. K., and C. N. Carwile. 1982 Kerr Reservoir striped bass study. Virginia Commission of Game and Inland Fisheries, Project Completion Report, Federal Aid in Fish Restoration, Project F-34-R, Richmond.

Wilde, G. R., M. L. Muoneke, P. W. Bettoli, K. L. Nelson, and B. T. Highsmith. 2000. Bait and temperature effects on striped bass hooking mortality in freshwater. North American Journal of Fisheries Management 20:810–815.

Wilkerson, M. L., and W. L. Fisher. 1997. Striped bass distribution, movements, and site fidelity in Robert S. Kerr Reservoir, Oklahoma. North American Journal of Fisheries Management 17:677–686.

Williams, H. M. 1970. Preliminary studies of certain aspects of the life history of the hybrid (striped bass × white bass) in two South Carolina reservoirs. Proceedings of the Annual Conference Southeastern Game and Fish Commissioners 24:424–430.

Wirgin, I. I., C. Grunwald, S. J. Garte, and C. Mesing. 1991. Use of DNA fingerprinting in the identification and management of a striped bass population in the southeastern United States. Transactions of the American Fisheries Society 120:273–282.

Worth, S. G. 1884. Report upon the propagation of striped bass at Weldon, N. C., in the spring of 1884. Bulletin of the United States Fish Commission 4(15):225–230.

Young, S. P., and J. J. Isely. 2004. Temporal and spatial estimates of adult striped bass mortality from telemetry and transmitter return data. North American Journal of Fisheries Management 24:1112–1119.

Zale, A. V., J. D. Wiechman, R. I. Lochmiller, and J. Burroughs. 1990. Limnological conditions associated with summer mortality of striped bass in Keystone Reservoir, Oklahoma. Transactions of the American Fisheries Society 119:72–76.

American Fisheries Society Symposium 80:15–23, 2013

History of Striped Bass Management in the Colorado River

Wayne Gustaveson* and Georg Blommer
Utah Division of Wildlife Resources
Post Office Box 1446, Page, Arizona 86040, USA

Abstract.—The discovery of landlocked populations of striped bass *Morone saxatilis* in Santee–Cooper Reservoir, South Carolina and Kerr Reservoir, Virginia prompted a rush to stock striped bass in other inland waters of the United States, including impoundments in the Colorado River. Fisheries managers responsible for Colorado River waters studied existing literature and predicted that it would be unlikely for successful natural reproduction of striped bass in these systems. Striped bass population development proved unique in the Colorado River system, which is marked by nutrient-poor, well-oxygenated waters with limited forage. Natural reproduction did occur in these reservoirs despite the lack of current previously thought to be essential for successful reproduction, resulting in high survival. Developing populations were sometimes overabundant to the point of forage elimination from pelagic zones. Lack of prey limited growth and temporarily reduced reproduction. Eventually forage returned, increasing striped bass growth and maturity, which led to more reproduction (a "boom and bust" cycle). Planned low-impact, low-abundance adult trophy fisheries produced by managed stocking were replaced by high-abundance juvenile fisheries with high catch rates produced by natural reproduction. In most years, juvenile striped bass living in warm surface waters proved to have the competitive advantage over adults for limited forage.

Colorado River Historical Overview

The preimpoundment Colorado River was characterized by harsh environmental conditions ranging from muddy spring floods exceeding 8,495 m^3/s to flows less than 8 m^3/s in drought plagued summer months. Hoover Dam was completed in 1935, creating Lake Mead, the first impoundment on the river and one of the largest reservoirs by volume in the United States (Figure 1). At full pool, Lake Mead is 63,900 ha with a maximum depth of 177 m and maximum length of approximately 184 km. Two smaller but significant reservoirs followed. Lake Havasu (1938) and Lake Mohave (1952) were constructed downstream (Stewart and Burrell 2013, this volume). The last impoundment was completed in 1963 when Glen Canyon Dam was closed, forming Lake Powell (Utah and Arizona), which is 300 km long with a surface area of 65,000 ha when full.

The Colorado River watershed is mostly devoid of agricultural and commercial development, which has been the main cause of nutrient enrichment in many of the nation's impoundments and river systems. Colorado River impoundments are characterized by low phosphorus and nitrogen levels. The new lakes filled deep canyons carved out by the Colorado River with limited littoral zones. Seven native fish species that thrived in the river prior to impoundment were poorly suited for life in the new lake environment. Introduced nonnative

* Corresponding author: wayne@wayneswords.com

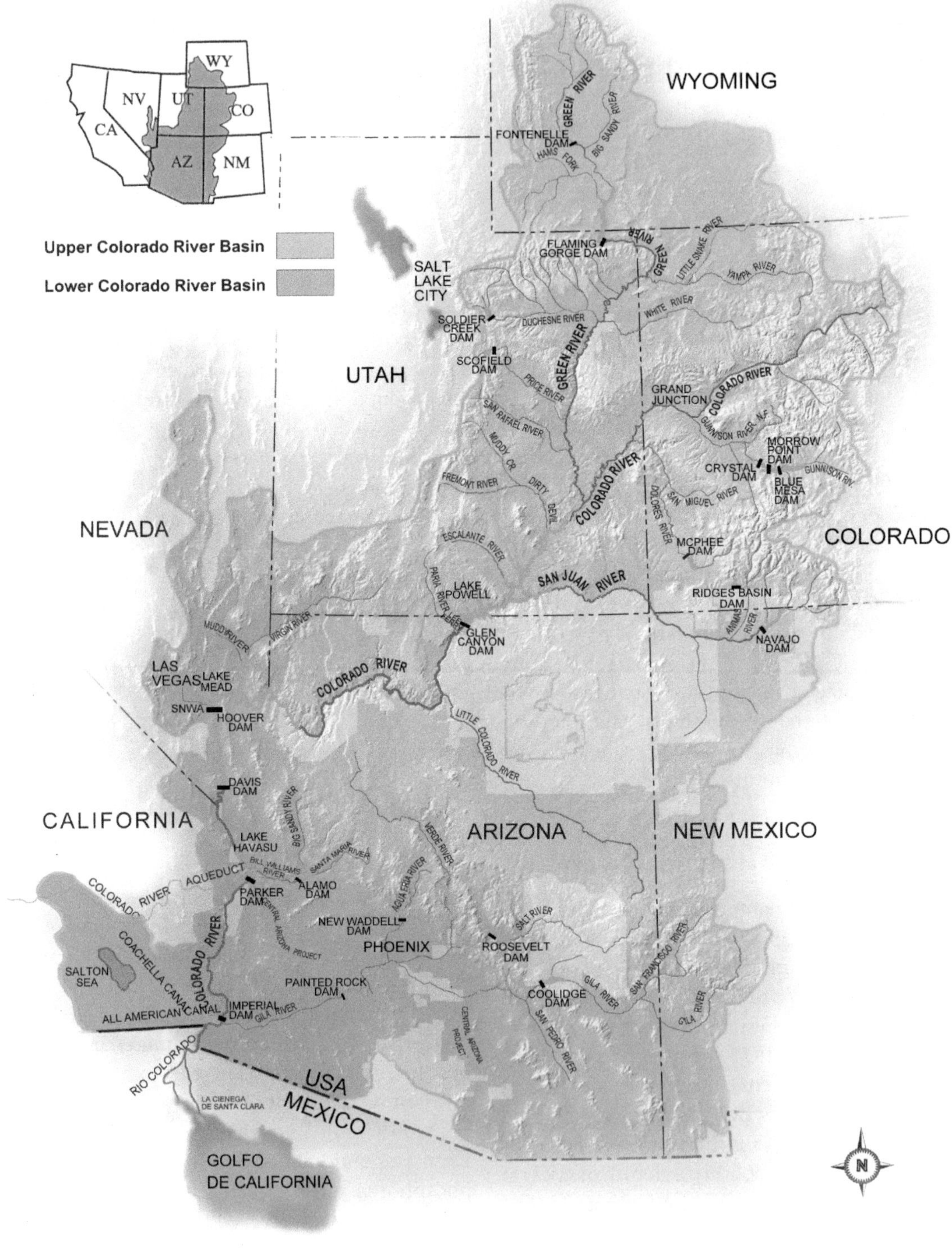

FIGURE 1. Map of Colorado River drainage with large reservoirs included.

sport and forage fish species dominate the fish communities that developed in these reservoirs.

Striped Bass History

Many new impoundments were constructed in the United States between 1930 and 1970. Striped bass *Morone saxatilis* trapped in Santee–Cooper Reservoir (South Carolina) impoundments and Kerr Reservoir (Virginia–North Carolina) survived and reproduced (Van Horn 2013, this volume). Land-locked striped bass provided fisheries managers a top-level predator that produced charismatic trophy fish for anglers while controlling overabundant forage fish populations. A nationwide rush to stock striped bass in reservoirs ensued as fish managers hoped to reap the biological and recreational rewards. Biologists in the Southeast quickly determined that most striped bass populations in reservoirs failed to reproduce because the semibuoyant eggs would settle to the bottom without supporting water currents and die due to siltation or lack of dissolved oxygen. Barkuloo (1970) proclaimed that "striped bass eggs and larvae must be suspended by current to prevent settling and death by suffocation." This finding guided managers in the Colorado River basin who contemplated introduction of striped bass.

Biologists at the California Department of Fish and Game stocked the first striped bass into the Colorado River in 1959 (St. Amant 1959). Young-of-year striped bass were seined from the San Joaquin River Delta and transported overland to the Colorado River near Blythe, California. All stocked striped bass were descendants of fish that were transported from the East Coast in 1879 and stocked in the delta near San Francisco Bay. Collaborative stocking efforts from 1962 to 1969 by state wildlife agencies in Arizona, California, and Nevada created thriving striped bass populations in Lake Mead, Lake Havasu, and the connecting river reaches (Edwards 1974). Popular striped bass sport fisheries developed, which included trophy-sized adults.

Striped bass were stocked by the Utah Division of Wildlife Resources in Lake Powell in 1974. It was initially thought that "reproduction would not be sufficient to maintain a sport fishery" (R. Stone, Utah Division of Wildlife Resources, personal communication). Hatchery rearing ponds were built near Lake Powell to meet the anticipated annual stocking requirement. The initial 1974 stocking in Lake Powell came from Virginia, California, and North Carolina genetic stock (May and Hepworth 1975). Over time, most of the fish stocked in Lake Powell came from North Carolina and the California Delta stock (Hepworth et al. 1976; Gustaveson et al. 1979).

Natural Reproduction

Based on the literature (Bailey 1975), biologists stocking Colorado River reservoirs believed that most reservoir striped bass populations were not self-sustaining. Most thought that fingerling introductions would be required to support a continuing sport fishery. However, in the late 1970s, reports from Arizona and Nevada suggested that evidence of striped bass natural reproduction had been found. Striped bass reproduction in Lake Mead was first noted in 1973 and in every year since (Allan and Roden 1978). Possible spawning sites included areas where river current could suspend eggs and larvae, as the literature suggested, but spawning sites were also suspected in reservoir locations with minimal current. The Lake Mead analysis by Allan and Roden (1978) referenced an important but overlooked hatchery experiment conducted by Bayless (1968) confirming that "suspension of striped bass eggs by water current is not necessary for a successful hatch provided eggs are not subjected to suffocation by silt or water quality."

Evidence of striped bass natural reproduction in Lake Powell was first discovered in 1979 and annually thereafter. Initially, it was thought that reproduction was confined to the Colorado River above the lake where river current could suspend eggs and larvae. The spawning site was in or below Cataract Canyon, a 19-km gorge containing 23 sets of rapids. Striped bass apparently used less than 20 km of river above the reservoir because ripe striped bass adults were collected below but not above Cataract Canyon during spawning season (Persons and Bulkley 1982).

During spring 1979, striped bass spawning was discovered in the lower reservoir near the dam. In spring of most years, prespawning striped bass aggregated near Glen Canyon Dam. These fish seemed attracted to the current created as water was drawn through the dam penstocks. With time and warming, the aggregation left the 165-m-deep dam forebay and moved to nearby coves where spawning occurred. Most fish left the prespawning staging area simultaneously during early May as water surface temperature reached 16–19°C, which is the generally accepted peak spawning temperature range (Setzler et al. 1980). Spawning sites were located near the dam where floating masses of dead (unfertilized) eggs were clearly visible. J. D. Bayless (South Carolina Wildlife and Marine Resources Department, personal communication) found that unfertilized eggs would float but that immediately after fertilization, eggs would sink 0.3 m in 27 s. Settling rate slowed with time, but eggs still descended 0.3 m every 60 s some 24 h after fertilization (Gustaveson et al. 1984).

An oxygen-temperature profile taken in spawning coves near the dam showed oxygen levels of 8.4 mg/L near the surface and 13.2 mg/L on the substrate in 9 m of water. Siltation was insignificant in these coves on the rock and sand substrate during the brief 48-h incubation period (Hardy 1978). Thus, eggs settling on the bottom of Lake Powell had adequate oxygen for normal development.

Collection of larval striped bass 18–22 mm total length (TL) with midwater trawl and meter tow net samples confirmed successful reproduction of striped bass near the dam. The collected larvae were determined not to be derived from the river upstream of the reservoir. Prolarvae are capable of swimming at 4 d of age and could be expected to travel great distances if assisted by strong mainstream reservoir currents. However, studies of density currents using total dissolved solids as an indicator showed a weak density current in the reservoir that could not have assisted in moving larvae from the inflow any closer to the dam than 190 km (D. Merritt, U.S. Bureau of Reclamation, personal communication). Instead, striped bass larvae were captured 295 km downstream from the inflow. If spawning occurred only at inflow areas and larval fish were distributed by reservoir currents, then larval fish should have been found throughout the reservoir. In contrast, there was a preponderance of young striped bass at two distinct locations, the inflow and again at the dam. It was evident from these analyses that these young striped bass found near the dam were the result of successful in-reservoir spawning (Gustaveson et al. 1984).

Over the years, it has been confirmed that striped bass spawning occurs annually in Lake Powell, Lake Mead, Lake Mohave (Liles 1985), and Lake Havasu. Eggs and larval striped bass also pass through the dams to provide additional recruits to downstream lakes and canals as water is delivered from the Colorado River to locations in California, Nevada and Arizona (Stewart and Burrell 2013).

Population Dynamics

Initial growth of Lake Powell and Lake Mohave striped bass was comparable to other fast growing cohorts in the United States (Liles 1985; Gustaveson 1999). Striped bass spawning was first discovered in Lake Powell in 1979, and stocking was terminated that same year. Threadfin shad *Dorosoma petenense* the only pelagic forage fish in Lake Powell, were well established before striped bass were introduced, having been stocked in 1968 and 1969 to provide forage for reservoir centrarchids and the anticipated introduction of striped bass (Gustaveson 1999). By 1982, threadfin shad stocks were reduced, likely by predation from the rapidly growing striped bass population. From 1982 to 1990, pelagic shad remained in low abundance and were often eliminated entirely from open water, presumably by striped bass predation (Figure 2). Shad persisted in warm

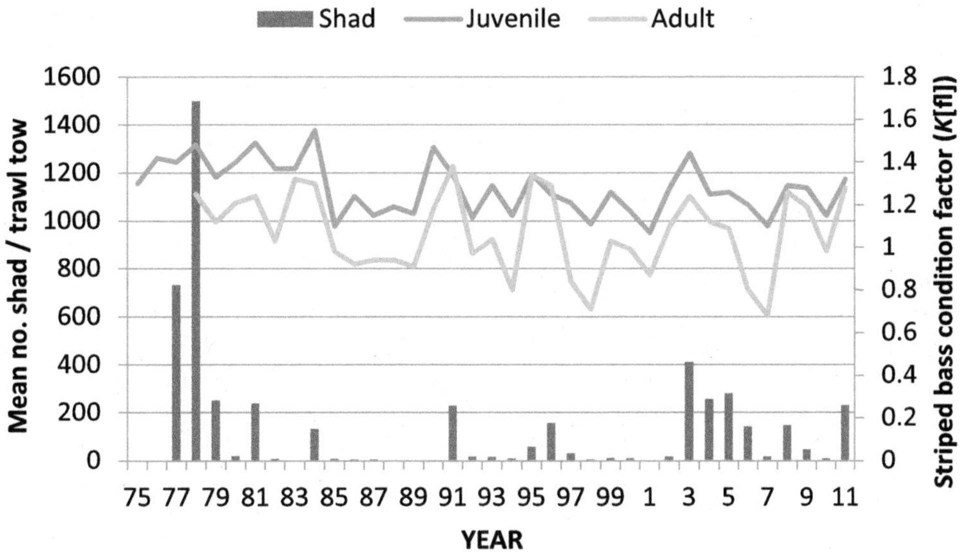

FIGURE 2. Pelagic shad abundance (vertical bars) compared to adult and juvenile striped bass condition, *K*(fl) (*K* factor based on fork length), 1976–2011, Lake Powell.

thermal refuges in turbid water in the backs of many canyons where striped bass foraging success was limited. Growth of striped bass decreased during periods of low forage (Gustaveson 1999). Adult striped bass suffered malnutrition while young striped bass maintained body condition on a zooplankton diet (Figure 2). As striped bass numbers peaked and forage fish abundance declined, back-calculated total length of each age-class declined 100–125 mm compared to lengths observed in the first 5 years of striped bass expansion. Striped bass were shorter and did not live as long, and some never gained sexual maturity. Most of the population was composed of immature fish 300–400 mm TL, which were as young as age 2 or as old as age 6 (Gustaveson 1999). When Lake Powell shad abundance increased in 2002, striped bass growth and condition immediately improved.

Striped bass condition (*K* factor based on fork length [*K*(fl)]) in Lake Powell (Gustaveson 1999) correlated with shad abundance (Figure 2). The initial expanding population (1975–1981) was consistently healthy, with *K*(fl) near 1.3 despite cyclical swings in shad abundance. Prior to striped bass establishment, shad exhibited a population peak every third year with two moderate production years in between. Shad population peaks in 1981 and 1984 resulted in high cohort survival, increased growth, and improved physical condition of striped bass. Condition of striped bass declined in 1982 and 1985–1989 to levels where adult fish were thin and juvenile fish smaller than expected.

Juvenile striped bass were able to maintain higher condition than adults partly because they occupied the same niche as shad in the warm epilimnion. When adult and juvenile striped bass competed for limited shad resources that were located in the pelagic epilimnion, smaller fish thrived while condition of adults declined. Adult striped bass preferred cooler temperatures and were forced into deeper, cooler water of the metalimnion and away from shad (Coutant 1985, 2013, this volume; Thompson and Rice 2013, this volume). Juvenile striped bass resided in and tolerated the warm surface layers. The ontogenetic separation of striped bass size-classes favored juvenile fish at the expense of adults.

Colorado River striped bass fisheries are thus characterized by a boom and bust cycle.

The initial stocked or naturally reproduced striped bass fed on abundant threadfin shad populations, grew rapidly with high condition factors, matured, and spawned. Shad populations then declined, and striped bass were less well fed. Spawning decreased when schooling adult striped bass with poor body condition probably were no longer able to produce viable eggs (similar to that observed by Coutant 1987 in Tennessee). Adult striped bass numbers declined in years of forage shortage, but a relatively small successful year-class was produced annually as a few females exceeding 10 kg spawned with young precocious males. These fast-growing females grew to larger size and were able to consume pelagic common carp *Cyprinus carpio* and sport fish of a size unavailable to the majority of schooling striped bass due to mouth gape limitations. Threadfin shad abundance eventually rebounded, allowing rapid growth, followed by spawning of the remaining striped bass.

Similar population swings were seen in each impoundment from Lake Powell to Lake Havasu. In most years, juvenile striped bass dominated the fish assemblage, with larger numbers of adults occurring on occasion. Lake Mohave has the best habitat conditions for a sustained population of larger adult fish in the upper lake, which is fed by the constantly cold oxygenated tailwater released from Hoover Dam (Stewart and Burrell 2013). Shad forage has been absent from Lake Mohave for the past decade, resulting in a striped bass population composed of fish less than 350 mm existing on plankton and trophy adults that forage on stocked rainbow trout and other fish species not available to smaller stripers. Fish size and condition fluctuates in Lakes Mead, Havasu, and Powell, depending on forage abundance during each particular year.

The unintentional introduction of gizzard shad *D. cepedianum* into Lake Powell in 2000 added an additional pelagic forage fish that may affect the boom and bust cycle. Gizzard shad were present in a load of largemouth bass from a federal fish hatchery in Texas, which were stocked into Morgan Lake, New Mexico. A thriving gizzard shad population was established and remains in Morgan Lake today (Knowles 2002). Some gizzard shad escaped into the San Juan River during flood events and moved downstream to Lake Powell. Gizzard shad apparently from Lake Powell were found in Lake Mead in 2008. This larger pelagic forage fish likely will pass through the dams and river system until it invades all Colorado River impoundments downstream from Lake Powell, which will in turn influence the predator–prey relationships for striped bass in each of these reservoirs. From the experience in Lake Powell, it appears that additional forage provided by gizzard shad will be advantageous to striped bass in forage-poor years and waters. Gizzard shad adults eat benthos and grow beyond the mouth gape of all but the largest stripers in the system. In 2009, for instance, gizzard shad added significantly to the forage base of Lake Powell. The ratio of gizzard shad to threadfin shad captured in midwater trawl samples was almost equal. How the boom-bust cycle will be affected in the Colorado River basin is unclear. Both threadfin and gizzard shad occupy many Southeast reservoirs where striped bass and striped bass hybrids currently thrive without such a boom and bust population cycle.

Management Implications

Striped bass were introduced into the Colorado River basin 50 years ago. Results differed from expectations of those who planned the introductions, both for population management via stocking and for development of a fishery for trophy-sized striped bass. Natural reproduction outpaced the planned stocking program. A few large trophy fish (20–27 kg) were taken shortly after the initial introductions but are now uncommon. Nonetheless, some striped bass exceeding 13 kg are still taken each year from the various impoundments.

More importantly, striped bass weighing from 1 to 2 kg are very abundant, depending on the lake and year, and provide a popular fishery. The trophy fish goal has been replaced by quality angling where many fish are caught. In Lake Powell, there is no striped bass creel limit, allowing those so inclined to harvest more than 100 fish per day (Figure 3). In Lake Mohave, there is no limit on striped bass less than 500 mm. Liberal creel limits exist in all Colorado River lakes. In years of low forage, creel survey catch rates exceeding two fish per hour are common. Management plans for such prolific populations that can overwhelm the forage base require maximizing harvest. Innovative plans to balance predator and prey numbers include liberal catch regulations combined with information dissemination to encourage more anglers to participate in harvesting striped bass.

There have been other inland locations where natural reproduction, either known in advance or unexpected, has dictated management plans for striped bass. The most well known are Lake Texoma (Texas and Oklahoma), John S. Kerr (Virginia and North Carolina), and the Santee–Cooper reservoirs (South Carolina) (Lamprecht et al. 2013, this volume). In each case, a fishery for young fish has predominated over a fishery for large, trophy-sized fish.

A Web page, www.wayneswords.com, is devoted to providing Lake Powell fishing and recreational information to the angling public. The goal is to provide detailed and timely information to Lake Powell fishers, allowing them to catch more fish, especially striped bass. As 3 million visitors come to Lake Powell each year, those that would not normally fish while enjoying a recreational trip are encouraged to try fishing. The Web site has been successful as evidenced by many fishers who have received fishing information and found it helpful. They are now submitting personal, successful fishing techniques and locations to the Web site, keeping the information current and broad-based. It is common for results of a successful fishing trip to be displayed on the Internet in a matter of hours. Arriving anglers with Internet or "smart phone" access can immediately use the

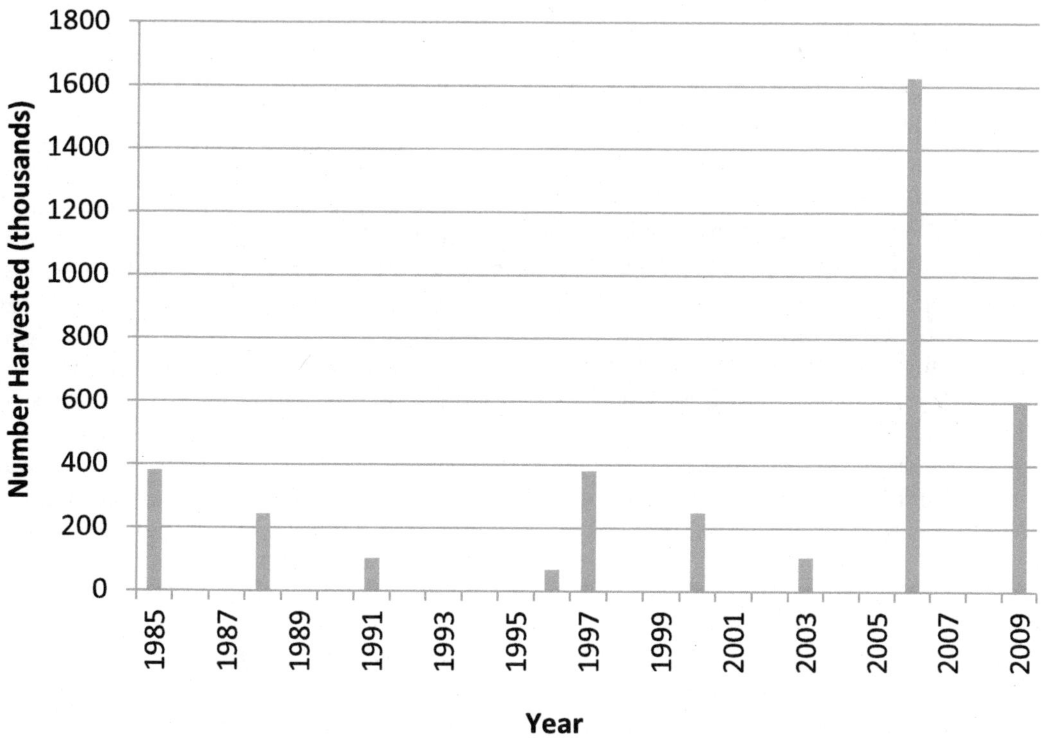

FIGURE 3. Angler harvest of striped bass in Lake Powell, Utah–Arizona, estimated by creel survey.

information about a hot fishing technique or location. In the age of catch and release elsewhere, the opportunity to "catch and keep" in Colorado River reservoirs is a bonus.

Anglers of all skill levels come from across the United States to harvest striped bass during times of high abundance. Colorado River reservoirs often have a mix of shad and striped bass schools in surface feeding events (striper boils) that are exciting to the fishing public and glamorous enough to entice nonfishers to join in the sport. The end result is increased harvest and license sales. At Lake Powell, harvest is limited only by the will of the angler to care for the number of fish caught. Since striped bass are often the largest fish ever caught by many anglers, the fishery is extremely popular. In 2006, 1.67 million striped bass were harvested (Blommer and Gustaveson 2007). Comparatively, in 2009, a high forage abundance year, harvest was estimated to be 700,000 striped bass at Lake Powell.

In summary, striped bass reproduce within the main stem reservoirs and the connecting Colorado River and its tributaries. Low nutrient loading provides adequate oxygen on the substrate for settling eggs to hatch. This makes striped bass populations within the Colorado River system nearly unique. Management of this prolific species must address adequate harvest to prevent stunting or death by starvation. Fisheries are cyclical with great population fluctuations. Predator populations periodically recover and rebound creating periods of low or high fish harvest, depending on prey population abundance. Fisheries vary but are generally classified as successful and highly sought after by those desiring a unique experience of fishing for a highly favored game fish in the incredibly beautiful red rock country of the desert Southwest.

References

Allan C. A., and D. L. Roden. 1978. Fish of Lake Mead and Lake Mohave. Nevada Department of Wildlife, Biological Bulletin No. 7, Reno.

Bailey, W. M. 1975. An evaluation of striped bass introductions in the southeastern US. Proceedings of the 28th Annual Conference Southeastern Association of Game and Fish Commissioners 28(1974):54–68.

Barkuloo, J. M. 1970. Taxonomic status and reproduction of striped bass (*Morone saxatilis*) in Florida. U.S. Bureau of Sport Fisheries and Wildlife, Technical Paper 44, Washington, D.C.

Bayless, J. D. 1968. Striped bass hatching and hybridization experiments. Proceedings of the 21st Annual Conference Southeastern Association of Game and Fish Commissioners 21(1967):233–244.

Blommer G. L., and A. W. Gustaveson. 2007. Lake Powell fisheries investigations. Utah Division of Wildlife Resources, Completion report, May 2004–April 2007, Salt Lake City.

Coutant, C. C. 1985. Striped bass, temperature, and dissolved oxygen: a speculative hypothesis for environmental risk. Transactions American Fisheries Society 114:31–61.

Coutant, C. C. 1987. Poor reproductive success of striped bass from a reservoir with reduced summer habitat. Transactions of the American Fisheries Society 116:154–160.

Coutant, C.C. 2013. When is habitat limiting for striped bass? Three decades of testing the temperature–oxygen squeeze hypothesis. Pages 65–91 *in* J. S. Bulak, C. C. Coutant, and J. A. Rice, editors. Biology and management of inland striped bass and hybrid striped bass. American Fisheries Society, Symposium 80, Bethesda, Maryland.

Edwards, G. B. 1974. Biology of the striped bass Morone saxatilis in the lower Colorado River (Arizona–California–Nevada). Master's thesis. Arizona State University, Phoenix.

Gustaveson, A. W. 1999. Cyclical population dynamics of self-sustaining striped bass in Lake Powell, Utah-Arizona, 1974–1998. Utah Division of Wildlife Resources, Salt Lake City.

Gustaveson, A. W., T. D. Pettengill, M. J. Ottenbacher, and R. S. Stone. 1979. Lake Powell post impoundment investigations. Utah Division of Wildlife Resources, Annual Performance Report January 1978–December 1978, Publication Number 79–8, DJ Project F-28-R-7, Salt Lake City.

Gustaveson, A. W., T. D. Pettengill, J. E. Johnson, and J. R. Wahl. 1984. Evidence of in-reservoir spawning of striped bass in Lake Powell, Utah–Arizona. North American Journal of Fisheries Management 4:540–546.

Hardy, J. D., Jr., 1978. Development of fishes of the Mid Atlantic Bight: an atlas of the egg, larval and juvenile stages. Volume III. U.S. Fish and Wildlife Service, Biological Services Program, FWS/OBS-78/12, 394 p, Washington, D.C.

Hepworth, D. K., A. W. Gustaveson, R. Biggins. 1976. Post impoundment investigations. Utah Division of Wildlife Resources, Annual Performance Report January 1975, Publication Number 76–21, DJ Project F-28-R-4, Salt Lake City.

Knowles, S. 2002. Fish and Wildlife blunders in Lake Powell. Salt Lake Tribune (August 27).

Lamprecht, S. D., M. D. Mauck, and V. J. DiCenzo. 2013. Natural reproduction: how it has affected striped bass management of John H. Kerr and Santee–Cooper reservoirs and Lake Texoma. Pages 413–430 *in* J. S. Bulak, C. C. Coutant, and J. A. Rice, editors. Biology and management of inland striped bass and hybrid striped bass. American Fisheries Society, Symposium 80, Bethesda, Maryland.

Liles, T. A. 1985. Status of striped bass in Lake Mohave. Arizona Game and Fish Department, Federal Aid Completion Report, Kingman.

May, B., and D. K. Hepworth. 1975. Lake Powell post-impoundment investigations. Annual Progress report. Utah Division of Wildlife Resources, Publication No.76-3, Salt Lake City.

Persons, W. R., and R. V. Bulkley. 1982. Feeding activity and spawning time of striped bass in the Colorado River inlet, Lake Powell, Utah. North American Journal of Fish Management 2:403–408.

Setzler, E. M., W. R. Boynton, K.V. Wood, H. H. Zion, L. Lubbers, N. K. Mountford, P. Frere, L. Tucker, and J. A. Mihursky. 1980. Synopsis of biological data on striped bass, *Morone saxatilis* (Walbaum). NOAA (National Oceanic and Atmospheric Administration) Technical Report NMFS (National Marine Fisheries Report) 433.

St. Amant, J. A. 1959. Striped bass introduced into the Colorado River. California Department of Fish and Game 45:353.

Stewart, W. T., and M. Burrell. 2013. Striped bass dispersion and effects on fisheries management in Lakes Mohave and Pleasant, Colorado River basin. Pages 431–448 *in* J. S. Bulak, C. C. Coutant, and J. A. Rice, editors. Biology and management of inland striped bass and hybrid striped bass. American Fisheries Society, Symposium 80, Bethesda, Maryland.

Thompson J. S., and J. A. Rice. 2013. The relative influence of thermal experience and forage availability on growth of age 1–5 striped bass in two southeastern reservoirs Pages 93–120 *in* J. S. Bulak, C. C. Coutant, and J. A. Rice, editors. Biology and management of inland striped bass and hybrid striped bass. American Fisheries Society, Symposium 80, Bethesda, Maryland.

Van Horn, S. L. 2013. A brief history of inland striped bass management. Pages 1–13 *in* J. S. Bulak, C. C. Coutant, and J. A. Rice, editors. Biology and management of inland striped bass and hybrid striped bass. American Fisheries Society, Symposium 80, Bethesda, Maryland.

American Fisheries Society Symposium 80:25–63, 2013

Restoration of Gulf Striped Bass: Lessons and Management Implications

Eric A. Long* and Charles L. Mesing
Florida Fish and Wildlife Conservation Commission
Post Office Box 59, Midway, Florida 32343, USA

Karen J. Herrington
U.S. Fish and Wildlife Service
1601 Balboa Avenue, Panama City, Florida 32405, USA

Robert R. Weller
Georgia Department of Natural Resources, Wildlife Resources Division
2024 Newton Road, Albany, Georgia 31701, USA

Isaac I. Wirgin
New York University School of Medicine, Institute of Environmental Medicine
57 Old Forge Road, Tuxedo, New York 10987, USA

Abstract.—Since the 1970s, the only known naturally reproducing population of native Gulf of Mexico (Gulf) striped bass *Morone saxatilis* occurs in the Apalachicola–Chattahoochee–Flint River system (ACF) in Florida, Georgia, and Alabama. To augment its depleted population, low numbers of fry and fingerlings of Atlantic coast ancestry were released into the ACF between 1965 and 1976. Restoration of Gulf striped bass was initiated in 1980 when putative Gulf fingerlings spawned from Apalachicola River (Gulf) broodfish were stocked back into the ACF. Since the initial stocking, approximately 10 million phase-I (25–50 mm) and 900,000 phase-II (150–250 mm) fingerlings have been released into Lake Seminole and the Apalachicola River, with hundreds of thousands more released into upstream reservoirs. Low levels of successful natural reproduction in the ACF were documented in 9 of the 10 years that natural reproduction was evaluated. Marked stocked fish have typically comprised 75–100% of fall age-0 samples. After stocking was initiated, striped bass harvest estimates increased as much as 10-fold during peak-season creel surveys conducted in the tailrace of Jim Woodruff Lock and Dam. A comparison of Atlantic-origin and Gulf striped bass co-stocked into an adjacent river-reservoir system over a 5-year period indicated no consistent differences in relative survival or growth through age 4. Gulf striped bass occupied coolwater thermal refuges during summer. Enhancement of thermal refuge habitats was successful, but results were short-lived. Small populations of Gulf striped bass, dependent on stocking of hatchery fish, now exist in several Gulf of Mexico tributary systems where adequate habitat is present. Genetic analysis of both mitochondrial and nuclear genomes revealed that a high percentage of fish from the ACF exhibit mitochondrial DNA (mtDNA) haplo-

* Corresponding author: eric.long@myfwc.com

types and nuclear DNA (nDNA) alleles that are absent in Atlantic populations. However, significant introgression of Atlantic nDNA alleles was documented in the extant population. Knowledge of the life history of Gulf striped bass was improved as a result of this multi-state collaboration as well as a large stocking program, new Gulf broodfish repositories, extensive genetic cataloged database, and expanded trophy fisheries. ACF Gulf striped bass restoration goals and objectives were defined, adjusted, and revised throughout the collaborative process to meet the concerns and management needs of all participating agencies.

Introduction

Striped bass *Morone saxatilis* were reported from the river drainages and estuaries of the Gulf of Mexico (Gulf) during the late 1800s (Goode and Bean 1879; Bean 1883; Goode 1884). The historical range of striped bass in the Gulf region is considered to be from the Suwannee River, Florida, westward to the rivers of the Lake Pontchartrain basin, Louisiana (Pearson 1938; Merriman 1941; Barkuloo 1961, 1967; GSMFC 2006; Figure 1). Although striped bass is an anadromous species, populations at the northern and southern extremes of its range are generally potamodromous. Within the Gulf region, there are no records of specimens taken from coastal waters, and striped bass are riverine or estuarine in nature (Barkuloo 1967; McIlwain 1980). It is likely that striped bass populations within tributaries of the Gulf of Mexico were riverine and endemic to individual rivers (Barkuloo 1967; McIlwain 1980) because there is little documentation of occurrence in open Gulf waters and very limited evidence of interchange between river systems.

Striped bass in Gulf rivers were first differentiated from Atlantic coast striped bass by lateral line scale counts (LLSCs). Striped bass in the Apalachicola (Florida) and Alabama (Alabama) rivers exhibited mean LLSCs (66.7) significantly higher than those found in Atlantic coast populations (54.4–62.2; Raney and Woolcott 1955), and minimal overlap of LLSCs was observed between fish from the two coasts (Goode and Bean 1879; Bean 1883; Brown 1965; Barkuloo 1970). Barkuloo (1970) concluded that the native striped bass population from the Apalachicola–Chattahoochee–Flint (Florida–Georgia–Alabama) River system (ACF) was a distinct race from Atlantic coast striped bass and could be identified using LLSCs. The Gulf States Marine Fisheries Commission (GSMFC 2006) adopted “race” for referring to the unique form of striped bass found in the ACF or descended from the ACF population, and the term “race” is used herein.

Historical accounts indicate that Gulf race striped bass were never abundant. The largest populations likely occurred in the ACF and Mobile-Alabama-Tombigbee (Alabama) River system. In 1957, following closure of Jim Woodruff Lock and Dam (JWLD), forming Lake Seminole (Florida–Georgia) at the headwaters of the Apalachicola River, many young and adult striped bass were taken by anglers fishing in the tailrace (Barkuloo 1961). Young striped bass were so abundant that many were destroyed as a nuisance by anglers. Barkuloo (1961) concluded that the presence of young fish demonstrated natural reproduction and that striped bass angler success indicated that an important sport fishery might be developed in the tailrace area. An apparent reduction in Gulf striped bass occurred after the closure of JWLD in 1957. Following the closure of JWLD, Georgia Department of Natural Resources (GADNR) fisheries personnel (GADNR 1969; Gennings et al. 1970) described a declining striped bass fishery in Lake Seminole, which was attributed to the cessation of spawning runs and to gill netting in the reservoir. By the mid-1960s, native striped bass were considered extirpated from most river drainages in the Gulf region (McIlwain 1980; GSMFC 2006). Barkuloo

(1961, 1967) reported that striped bass occurred in all major rivers along Florida's Gulf coast, but populations in rivers outside the ACF were too small to maintain a sport fishery. By the 1970s, the only known remaining naturally reproducing population along the Gulf was in the ACF (Wooley and Crateau 1983). The decline of striped bass in Gulf drainages was attributed to the construction of dams that blocked spawning migrations and access to thermal refuge habitat, water quality degradation, and other anthropogenic impacts.

With the development of techniques to artificially propagate striped bass, fisheries managers began stocking striped bass of Atlantic origin into many Gulf rivers and reservoirs in the mid-1960s in efforts to re-establish or augment populations, and stocking Atlantic race fish continues in some Gulf of Mexico tributaries to the present. Striped bass (Atlantic race) stocking programs for the ACF were initiated by GADNR (GADNR 1969) for Lake Seminole in 1965 and Lake Blackshear (Flint River, Georgia) in 1967. The Florida Game and Fresh Water Fish Commission (now the Florida Fish and Wildlife Conservation Commission [FFWC]) also released fingerlings of Atlantic origin into the Apalachicola River in 1976 (Crateau et al. 1980). All fish stocked into the ACF were progeny of broodfish collected from the Santee–Cooper system, South Carolina. In total, about 1.8 million Atlantic fry (<15 mm) and 125,000 fingerlings or adults (25–750 mm total length [TL]) were stocked into the ACF from 1965 to 1976 (Pasch et al. 1973; Nicholson et al. 1986; Keefer 1986). Introduction of hybrid striped bass (palmetto bass [male white bass *M. chrysops* × female striped bass] and sunshine bass [female white bass × male striped bass]) into ACF reservoirs was also initiated during the 1970s and continued throughout the course of this restoration project.

In the 1980s, fisheries managers working on the ACF became interested in the conservation and restoration of the native striped bass population. The collection of several very large striped bass (27–32 kg) and the hypothesis by Wooley and Crateau (1983) that Gulf fish were longer-lived and better adapted to surviving the warm summer climate than introduced Atlantic striped bass brought attention to Gulf striped bass in the ACF. This interest was further motivated by Wooley and Crateau (1983), who estimated that the population of adult striped bass (>381 mm TL) in the Apalachicola River was between 1,500 and 2,000 individuals, with approximately 43% native Gulf fish (identified by LLSCs), 51% introduced Atlantic fish, and 6% intergrades. The U.S. Fish and Wildlife Service (FWS) began stocking putative Gulf striped bass into the ACF in 1980. Broodfish were collected from the ACF system, primarily in the JWLD tailrace, and identified as Gulf striped bass based on LLSC; only fish with a LLSC of 65 or greater were accepted for hatchery propagation. However, the ability to discriminate Gulf, Atlantic, and intergrade striped bass using LLSCs was challenged after 1983 and 1984 year-class phase-II (150–250 mm TL) fish exhibited LLSCs ranging from 60 to 71, even though they were progeny of high LLSC (>65) broodfish (Florida Fish and Wildlife Conservation Commission, unpublished data). Concerns of inbreeding depression resulted in a relaxation of the LLSC criteria from 65 to 63 scales so that additional broodfish could be used to make crosses in the hatchery. The scale count criterion was later discontinued and Gulf striped bass are currently defined as any striped bass, or progeny, originating from the ACF.

One of the major questions at the inception of this native restoration program was the genetic integrity of striped bass from the ACF when compared with those from Atlantic coast systems. Specifically, were genetic characters unique to striped bass from Gulf coast drainages, and could these be used as heritable markers to unequivocally distinguish Gulf striped bass in the ACF from introduced Atlantic fish? The identification of unique genetic characters in ACF fish would provide the justification for efforts to restore a genetically unique fish along the Gulf coast. By the early 1980s, techniques were being developed to inspect mtDNA for sequence variation using restriction fragment-length polymorphism analysis (RFLP). Thus, genetic description and cataloging became a major component of the Gulf coast striped bass restoration effort.

Biologists and fisheries managers from Florida, Georgia, Alabama, and the FWS first met together in 1982 to discuss the management of striped bass and hybrid striped bass in the ACF at the first annual ACF *Morone* workshop. In 1987, representatives of conservation agencies in Alabama, Florida, and Georgia and the regional director of the FWS entered into a cooperative agreement for Gulf striped bass restoration in the ACF. The emphasis of the agreement was to restore a self-sustaining population of (Gulf) striped bass to the maximum extent possible. Under the agreement, the ACF Striped Bass Technical Committee, comprised of representatives from each agency, was formed to develop and implement a striped bass restoration plan for the ACF. The agencies also agreed to provide hatchery space, technical support, and other personnel. Later, representatives from Mississippi, Louisiana, the Gulf States Marine Fisheries Commission, New York University School of Medicine, Gulf Coast Research Laboratory (a facility of the University of Southern Mississippi), and the U.S. Army Corps of Engineers became participants of the workshop series, and Gulf restoration efforts expanded to include all states within the species' historical range. The ACF *Morone* workshop has now met for 29 consecutive years (as of 2011) and is one of the longest running cooperative fishery restoration efforts in the country.

The first striped bass management plan for the ACF was finalized in 1996 (ACF Striped Bass Technical Committee 1996) and established a general goal of enhancing the native Gulf striped bass population within the river system. Project objectives included determining hatchery production requirements and optimum stocking rates and sizes; monitoring broodfish collections and stocking success; monitoring the sport fishery; identifying, enhancing, or protecting thermal refuge habitat; establishing broodfish repositories outside of the ACF; and participating in public outreach. The plan was revised in 2004 (ACF Striped Bass Technical Committee 2004) to include a goal of restoring and maintaining a population of native Gulf striped bass "leading to a self-sustaining population that will (1) provide a broodfish source for the ACF and other Gulf state restoration programs; (2) support recreational fishing opportunities at optimum yield levels consistent with carrying capacity of available, restored, and enhanced habitat; and (3) maximize natural reproduction and recruitment of Gulf striped bass."

Concurrent with development of Gulf striped bass restoration strategies in the ACF, the Gulf States Marine Fisheries Commission's (GSMFC) Anadromous Fish Subcommittee authored the "Striped Bass Fishery Management Plan" (Nicholson et al. 1986) for the Gulf of Mexico. This management plan set goals of achieving and maintaining optimum sustainable yield for striped bass throughout their historical range, determining the validity of the Gulf striped bass as a race, and restoring or maintaining Gulf striped bass populations in suitable rivers at levels that would meet state restoration programs Gulf-wide. The GSMFC plan was revised in 2006 (GSMFC 2006) and established restoration and management goals across the Gulf region. Gulf-wide, river-specific restoration goals for striped bass include striped bass fisheries—lower Mississippi River (Louisiana–Mississippi); striped bass put-grow-and-take fisheries—Wolf and Jourdan rivers, Biloxi and Tchouticabouffa rivers, and Old Fort Bayou (Mississippi); Gulf race put-grow-and-take fisheries—Tangipahoa and Tchefuncte rivers (Louisiana), Perdido River (Alabama), Blackwater and Yellow rivers (Florida), and Ochlockonee River (Florida–Georgia); and self-sustaining Gulf race populations—Pearl River (Louisiana–Mississippi), Pascagoula River (Mississippi), Tallapoosa River (Alabama, between R.L. Harris Dam and Lake Martin), Escambia–Conecuh rivers (Florida–Alabama), Choctawhatchee River (Florida–Alabama), and the ACF.

The goals of this paper are to present data addressing the major components of the 1996 and 2004 ACF striped bass restoration and evaluation plans, to assess how restoration efforts have met the objectives, and to discuss the future status of the Gulf race of striped bass.

Methods

The ACF is located in northwestern Florida, western Georgia, and eastern Alabama (Figure 1). The system drains approximately 50,700 km^2 (Light et al. 1998). There are four federal dams and nine private hydroelectric dams on the Chattahoochee River. The Flint River has two private hydroelectric dams near Albany, Georgia, and there are approximately 168 km of unimpeded river between Jim Woodruff Lock and Dam and the Flint River Hydroelectric Project (Albany Dam). The JWLD was constructed at the confluence of the Chattahoochee and Flint rivers, forming Lake Seminole on the Florida–Georgia border. Lake Seminole is a 15,200-ha run-of-the-river reservoir with a mean and maximum depth of 3.0 and 9.1 m, respectively, with a stable water level that typically varies in elevation less than 1 m/year. The nonnative, invasive plant hydrilla *Hydrilla verticillata* was first observed in Lake Seminole in 1967 and now dominates the submersed plant community, covering as much as 64% of the surface area (USACOE 1998; D. Morgan, U.S. Army Corps of Engineers, personal communication). The JWLD outlet becomes the Apalachicola

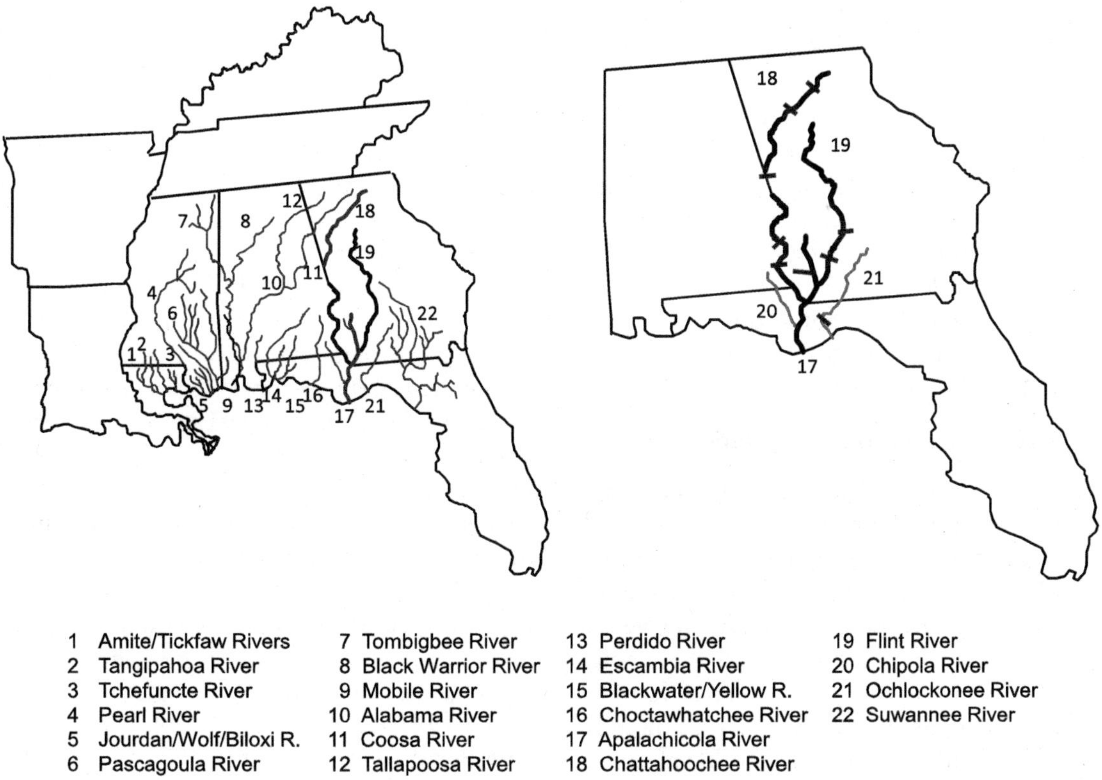

FIGURE 1. Left: Map of the Gulf of Mexico coast from Florida to Louisiana with the major tributaries that made up the historical range of Gulf striped bass. Right: Florida, Georgia, and Alabama with the location of the Apalachicola-Chattahoochee-Flint River system (17–20) and the Ochlockonee River (21), and the major dams located on these river systems. Six smaller dams (not indicated) also occur in the middle portion of the Chattahoochee River.

River, which flows approximately 171 km south to Apalachicola Bay and the Gulf of Mexico. The Apalachicola River is the largest in Florida in terms of discharge, with a mean annual discharge of 620 m^3/s measured at the U.S. Geological Survey gauge at Chattahoochee, Florida (Light et al. 2006). Its largest tributary is the Chipola River, which enters approximately 45 km upstream of the Apalachicola River mouth. The Chipola River, which was dammed at the mouth of Dead Lake from 1968 until 1988 (Hill et al. 1996), flows through lime rock topography and is highly influenced by springs and groundwater. Ambient water temperature in the upper Chipola River remains cooler than in the Apalachicola River during summer months.

From 1982 until 1996, Gulf striped bass restoration efforts were determined by state agency management and research needs, and efforts were reviewed by the ACF Striped Bass Technical Committee and at the annual ACF *Morone* workshop. The first ACF Striped Bass Restoration and Evaluation Plan (hereafter referred to as the ACFP) was completed in September 1996, and the revised ACFP was completed in 2004. The following items describe methods employed to address objectives and tasks of the two plans, but also incorporate activities collected prior to issuance of the 1996 ACFP.

Broodfish collections and indices

Objectives of both ACFPs (1996, 2004) were to annually produce phase-I and phase-II fingerlings to meet participating agency and restoration goals. Tasks included the collection of broodfish for hatcheries and conducting genetic analysis of the broodfish. To meet these objectives, striped bass broodfish were collected in several locations in the ACF basin. These included the tailrace of JWLD (i.e., Apalachicola River), the tailrace of George Andrews Lock and Dam (GALD; Chattahoochee River), and sections of the Flint River. Other tasks included a goal of meeting or exceeding an annual average of 0.7–1.0 adult fish (minimum size of 457 mm TL) per hour of electrofishing during broodfish collections in the ACF (1996) or to evaluate relative abundance of adult fish (2004). To meet these tasks, collected fish were measured (mm TL) and weighed. Mean catch per unit effort (CPUE) was calculated as mean number captured per hour (fish/h) of electrofishing time for small (>1.4 kg), medium (4.5–9.1 kg), and large (>9.1 kg) broodfish. We evaluated the potential association between time and mean catch rates overall and by size using Spearman correlation analysis. Only broodfish data collected since 1990 were used for our analysis as data collected prior to 1990 often lacked quantitative sampling effort. Fin clips were taken for genetic analysis from all fish that were transported to hatcheries. Sagittal otoliths were removed from broodfish that died or were sacrificed at the hatchery for age determination (Heidinger and Clodfelter 1987; Secor et al. 1995).

Stocking evaluation and young-of-year indices

Objectives of both ACFPs (1996, 2004) included determining the optimum stocking density, stocking locations, and sizes of stocked fish for restoration efforts in the ACF. Initial target numbers (1996) for striped bass annual stocking into the ACF were (all numbers are phase I unless specified otherwise) 500,000—Lake Seminole, 100,000—upper Apalachicola River, 170,000—Lake Blackshear, 100,000 phase II—lower Apalachicola River, 60,000—Bartlett's Ferry Lake (Georgia–Alabama), and 100,000—Walter F. George Lake (Georgia–Alabama). Stocking priority and numbers were reviewed annually at the ACF *Morone* workshop and adjusted as needed to fulfill study objectives for individual reservoirs, conservation agencies, or goals of the 1996 ACFP. An additional task of the 1996 ACFP was to establish broodfish repositories outside of the ACF by annually stocking phase-I fingerlings into Smith Lake (Alabama; 40,000), Lake Talquin (Florida; 250,000), Blackwater River (Florida; 100,000) and a domestic brood source housed in the National Fish Hatchery system (10,000). Gulf race phase-I (25–50 mm TL) fingerlings were stocked into Lake Seminole in 1980 and 1982 and then annually beginning in 1986, except in 1997 and 2006 when natural reproduction was evaluated. Phase-I fingerlings were stocked periodically into the upper Apalachicola River (UAR; U.S. Army Corps of Engineers navigation mile 77.5–106.4), lower Apalachicola River (LAR; navigation mile 0.0–20.0), and Apalachicola Bay (ARB) from 1982 to 1996. Age-0 phase-II (125–300 mm TL) fingerlings were also stocked into the LAR, ARB, Apalachicola River, and Intracoastal Waterway (ICW) near White City, Florida, during 16 years between 1980 and 2008.

Additional components of the 1996 and 2004 ACFPs were to develop an index to monitor survival, growth, condition, and age structure of striped bass in the ACF. Tasks included evaluating post-stocking survival and designing and implementing a relative abundance index of age-0 fish. To meet these tasks, approximately 50–100 fish from each delivery of phase-I fish to a stocking site (1992–2000) were held in receiving water in a 76-L aerated aquarium for 48–72 h to estimate short-term survival of stocked fish. Approximately 10–20 phase-II fish per delivery were also placed into live cars at stocking locations in the LAR and ICW.

Beginning in 2001, all phase-I fingerlings were marked with oxytetracycline (OTC; Secor et al. 1991) to determine the relative abundance of wild and hatchery fish. Phase-II fish were marked either with internal anchor tags, dart tags, T-bar tags, coded wire tags, fin clips, or OTC.

Relative abundance of stocked fish was monitored using electrofishing. Beginning in 1985, four fixed shoreline stations in Lake Seminole, UAR, and LAR were sampled for 10 min (Smith-Root Type 6A, GPP 5.0, or GPP 7.5 set to DC output) at night (Boynton et al. 1981) to estimate the relative abundance of age-0 fish. Stations were located on sand habitat in 0.6–1.2 m of water (Setzler et al. 1980; Boynton et al. 1981; Van Den Avyle et al. 1983). In 1985, sampling was conducted monthly from June through December. From 1986 to the present, except 1997, sampling was not initiated until the fall. Post 1986, stations were typically sampled twice, approximately 30 d apart, between late September and mid-November. The UAR and LAR were not sampled every year.

Catch-per-unit-effort (CPUE) values for age-0 fish were calculated as number of fish per minute (fish/min), and mean CPUE values were calculated for Lake Seminole, the UAR and LAR individually, and for all samples combined. We tested for differences in mean catch rates using Welch's t-test for unequal variances, and we evaluated the potential association between mean catch rates and time using Spearman correlation analysis. In addition, CPUE values of OTC-marked stocked age-0 fish and wild fish collected in Lake Seminole samples (2001–2004) were compared using Welch's t-test, as were CPUE values of wild age-0 fish collected in the absence of stocking (1985, 1997, and 2006) and values for wild fish collected during years when marked phase-I fish were stocked (2001–2004). In order to compare age-0 relative abundance in Lake Seminole from year to year, an age-0 relative abundance value (RAV) was calculated as the mean CPUE divided by the number stocked (Van Den Avyle and Higginbotham 1980) and then rescaled by multiplying by 100,000, similar to Brooks et al. (2002). Total length and weight were measured from collected fish. Based on comparison to previously aged fish (Florida Fish and Wildlife Conservation Commission, unpublished data), we assumed that all fish less than 250 mm TL were age 0, used otoliths to confirm the age of fish between 250 and 350 mm TL, and assumed that all fish greater than 350 mm TL were age 1or older. Catch per unit effort was calculated for all age-0 fish and separately for hatchery and wild age-0 fish. Wild age-0 fish were subtracted from the total catch prior to calculating the RAV for the 2001–2004 year-classes.

Relationships between mean annual CPUE of age-0 striped bass in Lake Seminole, the number of phase-I fish stocked annually and annual density of hydrilla (hectares) were calculated with multiple linear regression. Catch-per-unit-effort data were transformed as $\log_e(x + 0.0001)$ to normalize the distribution. The linear regression assumptions of normality and constant variance were evaluated by examining the normal quantile-quantile plot of residuals and the Studentized residuals plot. Both plots indicated that the normality and constant variation assumptions were met.

Mean relative weight (W_r) was calculated for age-0 fish (>150 mm TL; Brown and Murphy 1991a) using the standard weight (W_s) equation developed by Brown and Murphy (1991a). A correlation coefficient (r) was computed to evaluate the relationship, if any, between mean W_r and the number of fish stocked.

Creel census surveys

An objective of the 1996 ACFP was to restore striped bass populations in the ACF to a level that would meet management objectives of partner agencies. Agency objectives included development of recreational and trophy fisheries. This objective was omitted from the 2004 ACFP, but an objective to continue collecting fishery-dependent data was included. To meet these objectives, creel census surveys were conducted at several locations in the ACF. In the tailrace of JWLD, angler catch and harvest were monitored by continuing springtime roving creel surveys having nonuniform probability (Malvestuto 1983; Young 1987); they were initiated in 1980 and were conducted through 2009. Anglers were surveyed during a 3-h period (1000–1300, 1300–1600, or 1600–1900 hours eastern time) five times every 2 weeks—three times on weekdays and twice on weekend days. Surveys were conducted for 14 weeks between late February and early June. The area surveyed covered a river reach from the drive-up area at the powerhouse downstream for 11.3 km. Field data were statistically expanded to provide estimates of striped bass total catch (beginning in 1992), harvest, effort (hours), and angler success (striped bass catch or harvest per hour).

Randomly stratified roving creel surveys with nonuniform probability were initiated on the LAR and ICW in 1990. Anglers were surveyed during a 4-h period (morning or afternoon) six times every 2 weeks—three times on weekdays and three times on weekend days. Surveys were conducted for 10 or 12 weeks during the fall (October–December) and 16 weeks during the spring (March—June). Fall surveys were discontinued in 1999 and spring surveys in 2004. Field data were statistically expanded similarly to the JWLD tailrace survey. The area surveyed included the Apalachicola River from navigation mile 0.0 (U.S. Highway 98 Bridge) upstream 16.4 km, the major distributaries in the river delta, and the ICW from the LAR to Port St. Joe (approximately 40 km), Florida.

A nonuniform probability access creel survey (Malvestuto 1983) was conducted in the tailrace of George Andrews Lock and Dam on the Chattahoochee River and in the tailrace of Albany Dam on the Flint River from 1995 to 1998. Surveys were conducted for 26 weeks from November through April, which corresponded to the open season for striped bass angling and harvest at both locations. Survey periods were 2 weeks in length and were stratified by weekday and weekend. Four weekend days and six weekdays were sampled per period. The survey was conducted for 6 h, either in the morning or the afternoon. Access sites on the east and west side of the river were used for both surveys.

Thermal refuge habitat

Objectives of the 1996 and 2004 ACFPs were to pursue projects that would protect, enhance, or restore important striped bass habitats in the ACF. Because success of striped bass restoration in the ACF was assumed to depend on availability of summer thermal refuges, based on research on striped bass habitat usage conducted elsewhere (Coutant and Carroll 1980; Cheek et al. 1985; Coutant 1985), tasks of these plan objectives included identifying, protecting, or restoring/enhancing coolwater habitats in the ACF. To help meet these objectives, on four occasions, striped bass were tagged with temperature-sensitive radio or ultrasonic transmitters, and movements within sections of the ACF were monitored to locate thermal refuges and determine usage (Van Den Avyle and Evans 1990; Baker and Jennings 2001; Long 2001). Transmitters used on striped bass released into Lake Blackshear were also equipped with a 12-h mortality indicator. Fish used for telemetry studies were collected from the Apalachicola or Flint rivers or the ICW or were striped bass that had been spawned and housed at FWS national fish hatcheries. Transmitters were surgically implanted similar to Hart and Summerfelt (1975) or were attached externally (Baker and Jennings 2001). Fish were usually located by boat; however, fixed wing airplanes and/or helicopters also were used.

To sustain population enhancement efforts, several springs, spring runs, and coolwater creek tributaries in the Apalachicola and Flint rivers were renovated to increase flow or to enlarge thermal refuge volume by increasing pooling of cool water. In the Apalachicola River, five refuge creek and spring run tributaries identified by telemetry were de-snagged and excavated using either a dragline or clamshell dredge to provide 1 m of depth at typical low summer flows. The creeks were sampled by electrofishing during summer months, pre- and postrenovation, to determine utilization by striped bass. Between 1997 and 2007, eight springs in the Flint River between Lake Seminole and the Albany Dam were renovated using a pneumatic suction dredge. Annually, from 1997 through 2008, when ambient water temperature exceeded 27°C (typically June through September), monthly direct scuba counts of striped bass were conducted to evaluate the use of these springs and two nonrenovated springs as thermal refuges. During these surveys, two divers would position themselves as unobtrusively as possible within the opening of the spring, or on the edge of the refuge in the water-mixing zone, and count the number of striped bass observed during a 15-min time interval. Observed fish were categorized as either larger or smaller than 9.0 kg.

Gulf and Atlantic striped bass performance evaluation

Prior to issuance of the 1996 ACF plan, discussions at the *Morone* workshop prompted initiation of a study to compare growth and survival of Gulf and Atlantic race striped bass under similar conditions. A performance evaluation of co-stocked Gulf and Atlantic fish was conducted on Lake Talquin, a 6,561-ha run-of-the-river reservoir on the Ochlockonee River, Florida (Figure 1). Equal numbers of phase-I extant Gulf and Atlantic phase-I fingerlings were stocked into Lake Talquin from 1988 to 1993, and relative survival and growth were monitored through 1996. All stocking was done during daylight hours. Gulf fish originated from the ACF, and Atlantic fish were shipped as fry from the South Carolina Department of Natural Resources hatchery at Monks Corners (Santee–Cooper River system) to Welaka National Fish Hatchery for grow-out. Mitochondrial DNA from broodfish was analyzed to provide genetic tags to unequivocally identify the race of all sampled fish. All Atlantic fish exhibited the *Xba* I-1 haplotype, and all Gulf fish subsequent to the 1988 year-class exhibited the *Xba* I-2 haplotype unique to Gulf race fish (Wirgin et al. 1989). Gulf fish from the 1988 year-class that had the *Xba* I-1 haplotype D-1 also were characterized using nuclear DNA (nDNA) genotypes.

Striped bass from multiple year-classes were collected from Lake Talquin using experimental gill nets (38–254 mm stretch mesh, 2.4–4.3 m deep, and 91.4 m long) set overnight during November and December. Late fall was selected for sampling to help ensure that striped bass had vacated thermal refuges and would be redistributed within the reservoir. Mitochondrial DNA genotypes and race were determined according to Wirgin (1987). Total length, weight, and LLSCs were recorded for each fish. Otoliths were removed from all fish for age-class estimation (Heidinger and Clodfelter 1987; Secor et al. 1995). Relative survival of Gulf and Atlantic fish was compared using Pearson chi-square analysis to test for differences ($P < 0.05$) in the sample ratio of Gulf and Atlantic striped bass collected within each co-stocked year-class from ages 1 to 5. Because equal numbers of Gulf and Atlantic fish were stocked annually, we expected equal numbers of each race in our annual samples. As a result of small sample sizes, all lengths and weights at given ages for each race, and for all year-classes, were pooled to test for differences in mean length (millimeters total length) at age for each race using the Student's t-test. Differences in relative condition were tested by pooling data from the 1988 and 1989 year-classes for each race and comparing using analysis of covariance (ANCOVA) to determine differences in natural log transformed length–weight relations (least squared regressions). The comparison was limited to the 1988 and 1989 year-classes because they were the only ones from which sufficient numbers of fish older than age 4 were collected.

Genetic characterization

Since 1983, live fish or tissue samples (frozen liver or blood, fin clips preserved in 95% ethanol, or dried scales) from ACF and other Gulf coast striped bass were retained for genetic analysis. DNA isolated from these tissues was used to differentiate Gulf from Atlantic fish, characterize broodfish (1996 and 2004 ACFP) and wild fish, identify stocked fish, identify genetic tags (1996 ACFP), and compare the extant population with archived samples collected before the introduction of Atlantic striped bass (2004 ACFP). Several techniques, including mtDNA RFLP analysis (Wirgin 1987; Wirgin et al. 1989), nDNA fingerprinting (Wirgin et al. 1991), single nucleotide polymorphism (SNP; Wirgin and Maceda 1991), and microsatellite analysis (Roy et al. 2000; Wirgin et al. 2005) were developed and applied to the ACF population to assist in this restoration effort.

Results

Broodfish collections and indices

Since 1990, relative abundance has varied temporally (Figure 2). In the Apalachicola River, CPUE for all broodfish trended upward from 1990 (2.73 fish/h; SE = 0.48) until 1998 (11.90 fish/h; SE = 3.75), reached its high in 2003, and reached its low in 2007, the only year that CPUE did not meet the ACFP target of 0.7–1.0 fish/h. For the full period (1990–2009), there was no association between mean CPUE over time, but from the subset of 1990 to 2003, there was a significant positive association with all broodfish collected in the Apalachicola River and time ($r = 0.68$; $P = 0.002$). Mean CPUE for only medium-sized (4.5–9.1 kg) broodfish had similar trends to total broodfish mean CPUE, but no significant correlations were detected with medium or large broodfish collected over time in the Apalachicola River in either time set. In the Chattahoochee and Flint rivers, abundance of all broodfish tended to increase from 1990 to 2009 (Figure 2). Mean CPUE values for fish exceeding 1.4 kg increased significantly ($r = 0.81$; $P < 0.0001$) and ranged from 0.07 (1992) to 7.25 (2009; Figure 2). Mean CPUE of medium-sized broodfish in the Flint and Chattahoochee rivers also tended to increase over time ($r = 0.68$; $P = 0.02$). There was no association between mean CPUE of large broodfish and time, which ranged from 0.0 (1992) to 0.92 (2009), hitting peaks in 1997 and 1998 and again in 2009 (Figure 2).

Broodfish age and sex were indexed by fish sacrificed at hatcheries. Between 1993 and 2009, 94 female and 125 male striped bass broodfish from the ACF were sacrificed at the hatcheries and aged (Table 1). Females ranged from age 2 to 13 and males ranged from age 2 to 7. Sixty-five percent of the females were ages 4–6, and 67% of the males were ages 3 and 4. The youngest females were typically collected toward the end of the spawning season. The largest female collected was 21.6 kg, and the largest male was 16.2 kg

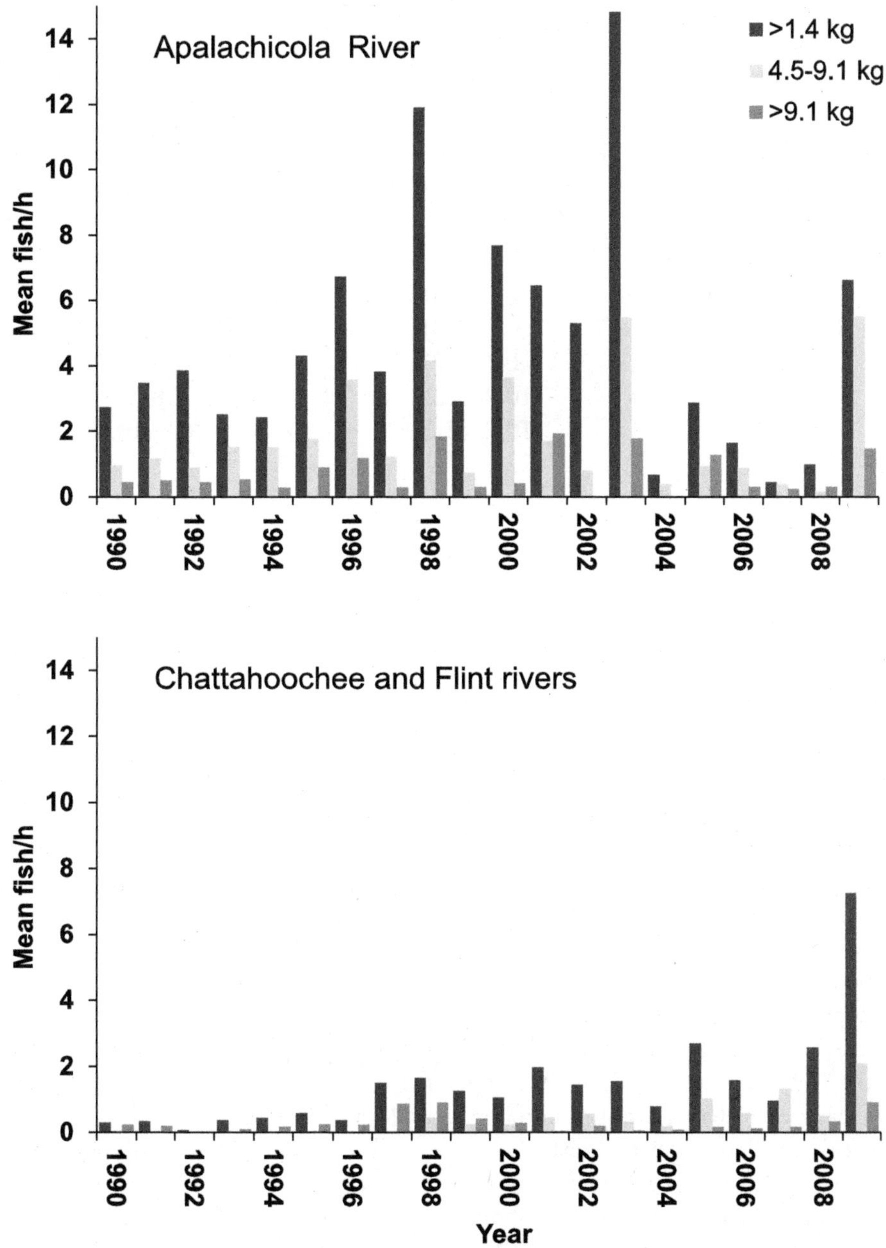

FIGURE 2. Mean catch per unit effort (fish/h) of striped bass caught for broodfish in the Aplalachicola (top) and the Chattahoochee-Flint rivers (bottom), 1990–2009, grouped by size (all >1.4 kg, 4.5–9.1 kg, and >9.1 kg).

and was not aged. Large males were typically released because hatchery managers preferred not to handle males greater than 9.1 kg, so males older than age 7 may have been collected and released.

Stocking evaluation

During the 28 years (1980–2008) since the first Gulf striped bass were stocked, more than 11,600,000 fingerlings were stocked into Lake Seminole, the Apalachicola River, and Apala-

TABLE 1. Numbers, weight (kg) range, and mean weight per age-class of Apalachicola-Chattahoochee-Flint striped bass broodfish sacrificed at hatcheries, 1993–2008. Numbers in parentheses represent one standard error.

Age	Female *N*	Female Weight range (kg)	Female Mean weight (kg)	Male *N*	Male Weight range (kg)	Male Mean weight (kg)
2	1	1.8		14	1.2–2.6	1.9 (0.5)
3	10	2.1–6.9	3.7 (1.3)	41	1.7–4.7	3.1 (0.8)
4	21	3.6–7.6	5.7 (1.0)	43	1.2–7.0	4.4 (1.1)
5	28	4.6–14.0	8.4 (2.2)	20	4.2–11.5	6.6 (2.0)
6	16	7.6–14.6	10.2 (2.2)	2	7.8–9.3	8.6 (1.1)
7	8	7.2–14.0	10.2 (2.2)	5	6.8–10.6	8.0 (1.5)
8	3	8.6–18.1	12.1 (2.2)			
9	1	21.6				
10	1	15.9				
11	4	11.0–15.9	14.1 (2.3)			
12	0					
13	1	21.0				

chicola Bay. Stocking numbers set in the 1996 ACFP were achieved for Lake Seminole for 3 of the 7 years from 1998 to 2004 that fish were stocked and were within 85% of the target number during the four other years (Table 2). Phase-II stocking numbers were achieved for the Apalachicola River during 5 of the 8 years from 1997 and 2004 and within 60% of the target for the other 3 years (Table 2). The goal of stocking 100,000 phase-I fingerlings into the LAR was dropped at the 1997 *Morone* workshop. Stocking numbers typically exceeded the target numbers for Lake Blackshear and Bartlett's Ferry Lake, but Lake Walter F. George was placed low on the priority list at ensuing *Morone* workshops and did not receive any fish until 2001and a full complement until 2004. Stocking phase-I fingerlings into Lakes Talquin and Smith and into the Blackwater River successfully resulted in establishment of broodfish repositories in these systems. In addition, a domestic broodfish source was developed at Warm Springs and Mammoth Springs national fish hatcheries (NFH).

Phase-I fingerlings were typically stocked between late April and mid-May at approximately 30 d posthatch. The total numbers of phase-I fish annually stocked into Lake Seminole ranged from 0 to 760,000; the average number stocked was 363,000 (24/ha). Short-term survival of phase-I fish stocked into the ACF from 1992 to 2000 ranged from 68% to 99% and averaged 86%.

There was no association between mean catch rates of age-0 fish and time. The mean CPUE for age-0 fish from Lake Seminole ranged considerably, from 0.08 (SE = 0.05) fish/min in 1993 to 4.64 (SE = 1.72) fish/min in 2000 (Figure 3).

There was a marginally significant difference (t = 2.52; df = 2; P = 0.06) between CPUE of wild fish in years of stocking and no stocking. Catch rates of wild fish were higher in years when no fish were stocked (mean of 0.5 fish/min) than years when fish were stocked (mean of 0.1 fish/min). Mean CPUE values of wild fish during years when no stocking occurred were 0.19 (SE = 0.05), 0.66 (SE = 0.48), and 0.52 (SE = 0.33) fish/min in 1985, 1997, and 2006, respectively. From 2001 to 2004, catch rates of OTC-marked stocked fish (range of 0.03–1.60 fish/min) were significantly higher (t = 2.93; df = 3; P = 0.03) than those of wild fish

TABLE 2. Number (rounded to the nearest 1,000) of phase-I and phase-II fingerling Gulf striped bass stocked into Lake Seminole and the Apalachicola River, 1980–2008. Number stocked per hectare (no./ha) for Lake Seminole is rounded to the nearest whole number. Stocking numbers are averaged for only the years in which stocking occurred.

	Phase I					Phase II		
	Lake Seminole		Appalachicola River			Lake Seminole		Apalachicola River
Year	number (x 1,000)	no./ha	upper (no. × 1,000)	middle (no. × 1,000)	lower (no. × 1,000)	number × 1,000)	no./ha	number × 1,000)
1980	100	7						13
1981								
1982			38					11
1983	80	5	53					
1984			20					
1985								
1986	273	18			180			13
1987	545	36						
1988	321	21	1					3
1989	516	34			135			
1990	374	25						17,000
1991	133	9			127			
1992	374	25	32		526			21
1993	200	13	55	50	430			30
1994	392	26	90					30
1995	345	23	60					54
1996	730	48	72					92
1997							129	
1998	533	35					148	
1999	428	28					150	
2000	520	34					149	
2001	476	31					110	
2002	199	13					74	
2003	494	33				46	3	71
2004	623	41				24	2	63
2005	145	10				32	2	59
2006							20	1
2007	169	11				42	3	
2008	372	24				55	4	14
Total	8,342	550	421	50	1,398	219	15	1,251
Average	363		24	47	280	37	2.5	62.5

(range 0.00–0.20 fish/min). Oxytetracycline-marked fish comprised 78–100% of the age-0 fish collected from Lake Seminole.

The abundance of age-0 striped bass in Lake Seminole (1986 to 2006) was positively correlated with the number of fish stocked and

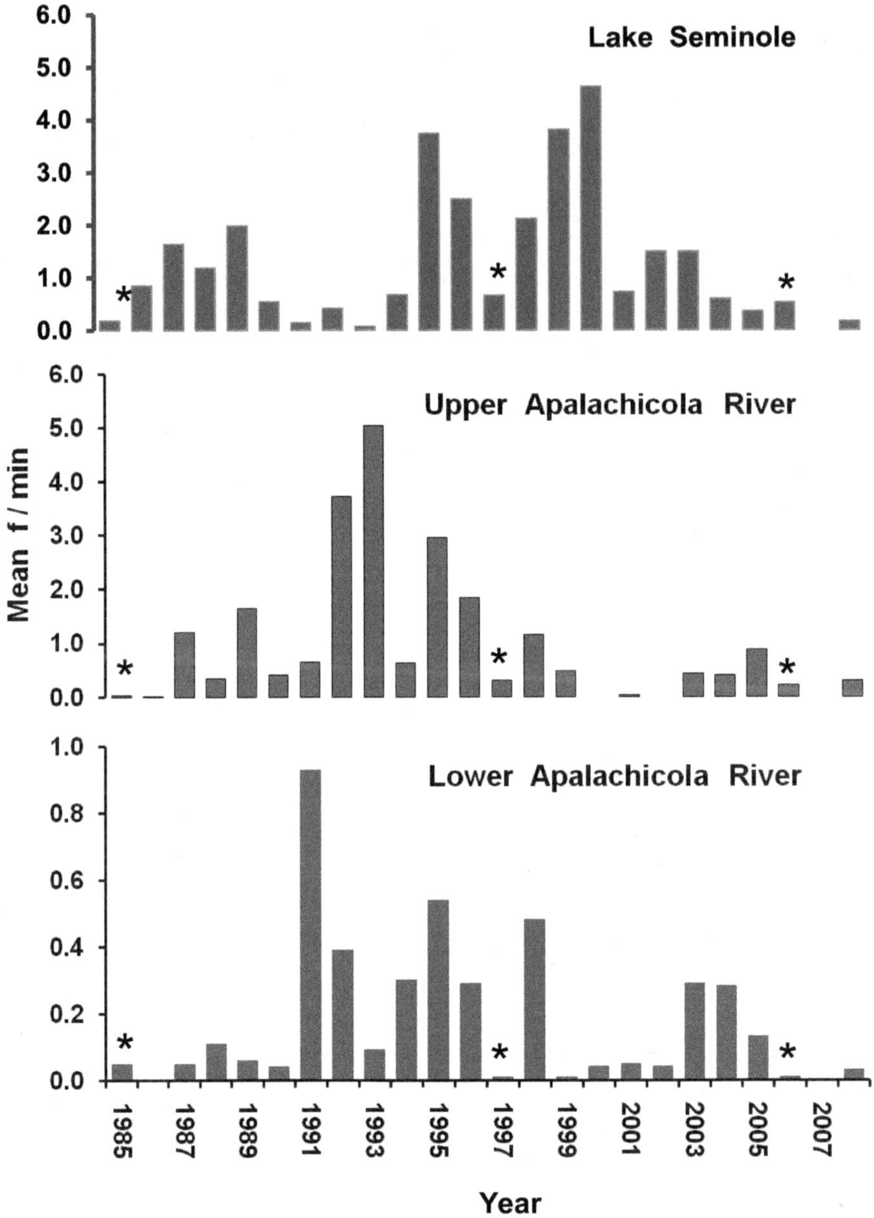

FIGURE 3. Mean catch per unit effort (fish/min [f/min]) of age-0 striped bass electrofished from fixed sites in Lake Seminole (top), the upper Apalachicola River (middle), and lower Apalachicola River (bottom), from 1985 through 2008. * denotes years when no phase-I stocking occurred. No electrofishing samples were collected in 2007.

negatively correlated with the surface area of hydrilla coverage (USACE 1998; D. Morgan, U.S. Army Corps of Engineers, personal communication). Multiple regression analysis described the relationship as

$$\log_e(\text{CPUE} + 0.0001) = 1.2534 + -0.0003689(\text{hydrilla}) + 0.00000403(\text{stocked})$$

(Adjusted $R^2 = 0.635$, $P < 0.001$).

Together, the surface area of hydrilla and the

number of fish stocked explained about 63% of the variation in abundance of age-0 striped bass in Lake Seminole. Including hydrilla coverage in the model added significance over and above the variance explained by the number of fish stocked alone, which was 45%.

Age-0 relative abundance values (RAVs) for Lake Seminole exhibited wide variation among years, ranging from 0.04 in 1993 to 1.08 in 1995 (Figure 4). Relative abundance value was also significantly and negatively correlated (r = -0.59; df = 14; $P < 0.05$) with surface area of hydrilla. During the years 1986–1989, phase-I stocked year-classes that recruited well into the tailrace fishery had RAVs above 0.30 and we defined this value as indicative of a good year-class. Relative abundance value exceeded 0.30 in 11 of the 21 years

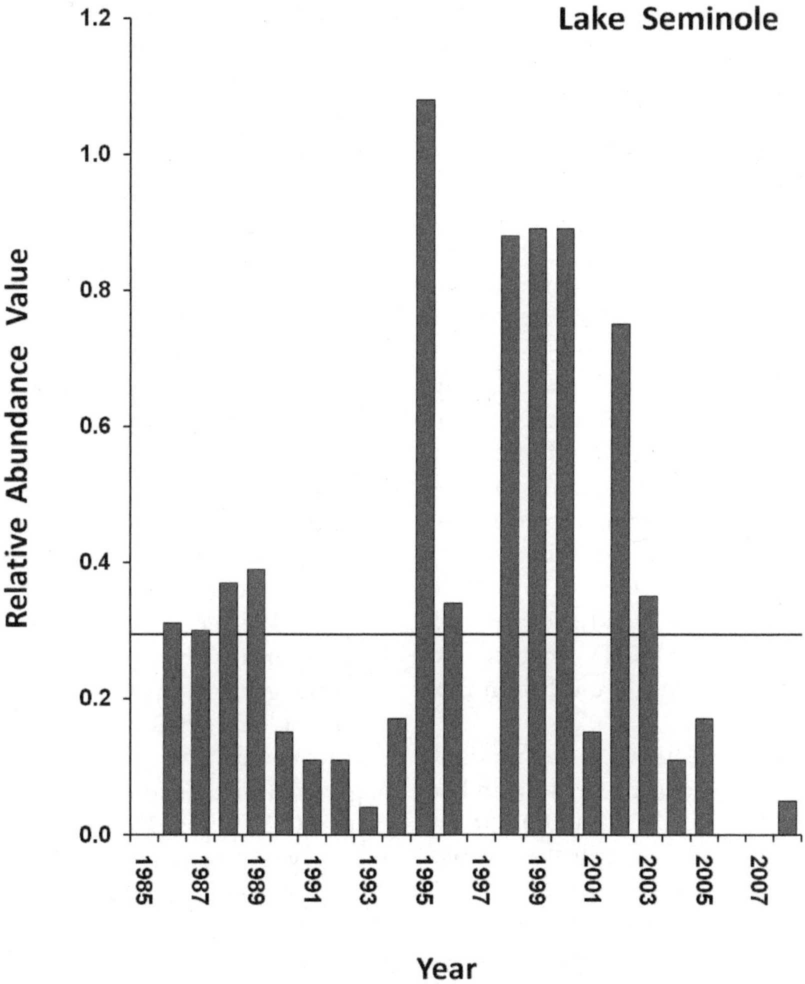

FIGURE 4. Relative abundance values (RAVs) of age-0 striped bass electrofished from Lake Seminole from 1986 through 2008. RAV was calculated as mean catch per unit effort (CPUE; fish/min) divided by the number stocked and multiplied by 100,000. The horizontal line represents the threshold level (0.3) considered to represent good survival through the fall. RAVs were not calculated for 1985, 1997, and 2006 because no phase-I striped bass were stocked. From 2000 through 2008, when all stocked fish were marked with oxytetracycline, wild striped bass were subtracted from total sample prior to calculating CPUE. No electrofishing samples were collected in 2007.

(1986–2008) that striped bass fingerlings were stocked into Lake Seminole.

Age-0 striped bass CPUE in the UAR and LAR trended higher in the 1990s, when phase-I fingerlings were being stocked, than during the 1980s or 2000s, when no stocking occurred (Figure 3). From 1985 to 2008, mean CPUE in the UAR ranged from 0.01fish/min (SE = 0.06; 1986) to 5.03 fish/min (SE = 1.91; 1993). Catch per unit effort in the UAR was highest during years that phase-I fingerlings were stocked into the upper river (1992, 1993, and 1995). However, mean values were above average (1.08 fish/min) during years that fingerlings were not stocked and floodgates at JWLD were open during the late spring or summer. In the lower Apalachicola River, mean catch rates ranged from 0.0 to 0.93 fish/min (SE = 0.51; 1991). Catch rates were higher during years that phase-I fingerlings were stocked, but stocking did not always result in higher CPUE than in nonstocking years. Mean CPUE in the LAR was almost always lower than in the UAR. Although CPUE in river samples tended to be higher during years when stocking occurred, stocking phase-I fingerlings did not contribute appreciably to the adult population and was discontinued in the LAR in 1993 and the UAR in 1996. From 2001 to 2008, stocked (Lake Seminole and upstream reservoirs) OTC-marked fish comprised 76–100% of the age-0 fish in river samples.

Mean W_r of age-0 fish in Lake Seminole was significantly correlated with numbers stocked (r = 0.59; df = 22; $P < 0.01$) but was not correlated with RAV or hydrilla coverage. Mean relative weight (W_r) derived for individual year-classes ranged from 74 to 97 (Figure 5), with only one year-class (1993) within the optimal 100 ± 5 range. From 1985 to 2008, mean total length of age-0 striped bass collected from Lake Seminole and the Apalachicola River during fall electrofishing ranged from 126.9 to 231.3 mm TL (Figure 6). Mean total length was not correlated with number stocked, RAV, or hydrilla.

Approximately 1,251,000 phase-II fingerlings were stocked into the Apalachicola River at various locations during 20 years between 1980 and 2008 (Table 2), with annual numbers stocked ranging from 3,000 (1988) to 149,000 (2000). Beginning in 2003, 20,000–60,000 (1–4/ha) phase-II fish were stocked annually into Lake Seminole. Short-term survival (48 h) of phase-II fish held in live cars at the stocking sites from 1995 to 2001 ranged from 61% to 94% and averaged 83%. Phase-II tag return data were collected from 1986 to 1988, 1993 to 1996, and 2000. A total of 26,477 phase-II fish received external tags with phone numbers, and 747 (2.8%) tag returns were reported. Angler returns were low for most year-classes (1–4%) but did reach 10% during 1 year. The vast majority of tagged fish were caught near the stocking location within a few months of release, and then tag returns dwindled to zero per month. Only seven externally tagged fish were captured two or more years poststocking. Phase-II stocking was considered unsuccessful and was discontinued in Lake Seminole and the Apalachicola River.

Creel census surveys

Upper Apalachicola River.—The creel survey results demonstrated that many anglers targeted striped bass during the 14-week survey period. Between 1982 and 2009, the number of striped bass anglers interviewed ranged from 24 to 416 and averaged 166 (Table 3), comprising 2–30% of the anglers interviewed each year. During the surveys conducted from 1980 to 2009, directed angling effort estimates for striped bass ranged from 1,098 (SE = 318) to 9,485 (SE = 2,936) hours, averaging 3,696 h.

Annual harvest estimates for striped bass during early stocking years from 1980 to 1992 increased 10-fold, from 152 (SE = 130) to 1,505 (SE = 474; Table 3). Prior to 1992, released striped bass were not documented in the survey data, and it was assumed that under prevailing harvest regulations, all caught fish were harvested. Catch and harvest estimates in 1992 (1,570 and 1,505, respectively) and 1993 (307 and 285, respectively; Table 3) supported

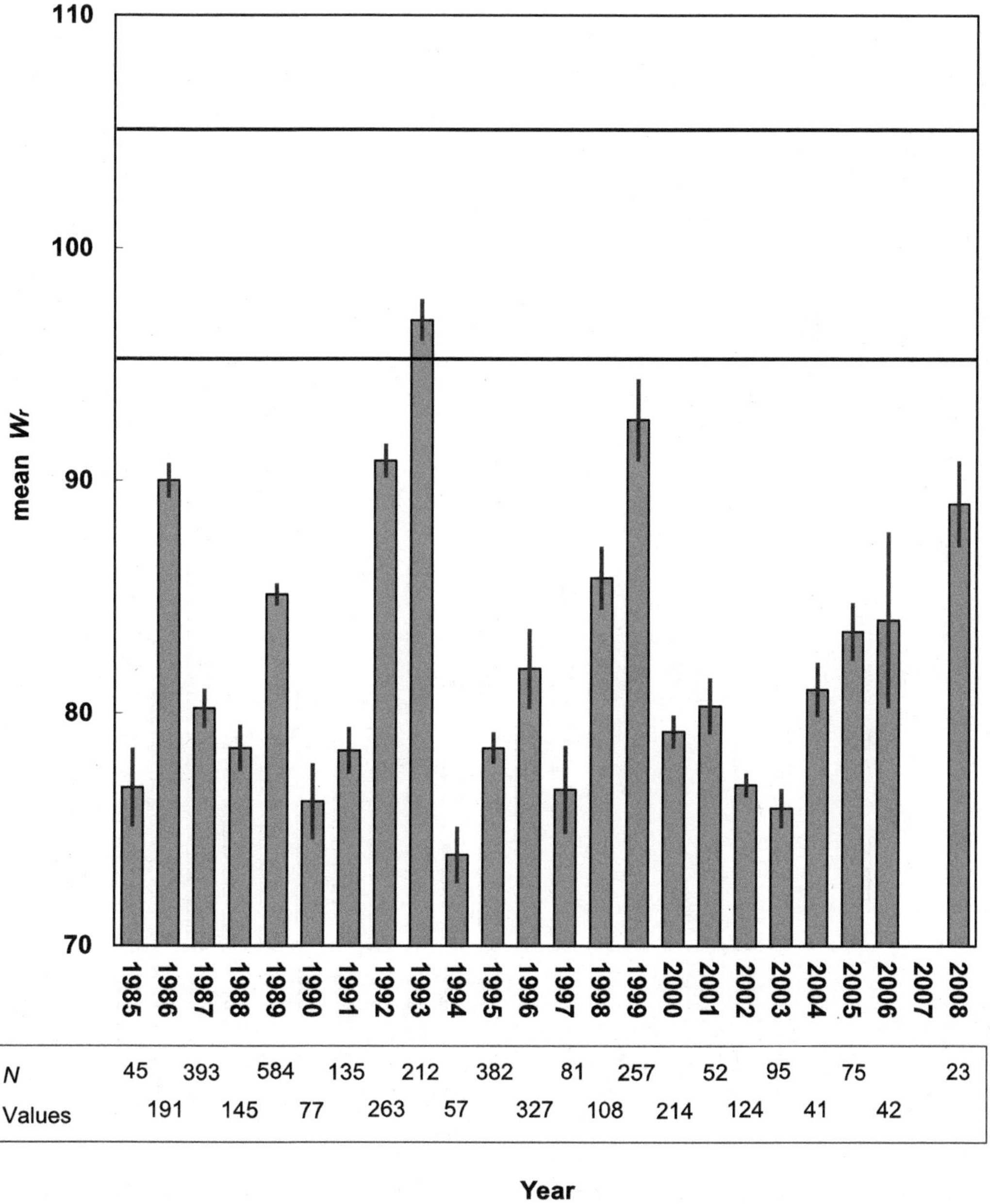

FIGURE 5. Mean relative weights (W_r) of age-0 striped bass greater than 150 mm total length electrofished from Lake Seminole and the Apalachicola River during September–December, 1985 to 2008. Vertical lines represent one standard error. Horizontal lines represent the optimal range of W_r (95–105; Brown and Murphy 1991).

this assumption. Following a change in harvest regulation (1994 to 2009) that reduced the bag and implemented a minimum size, estimates of striped bass total catch ranged from 311 (SE = 156) to 1,457 (SE = 346) and averaged 688 fish (Table 3). During this period, harvest estimates ranged from 134 (SE = 49) to 796 (SE = 214) and averaged 379 fish (Table 3).

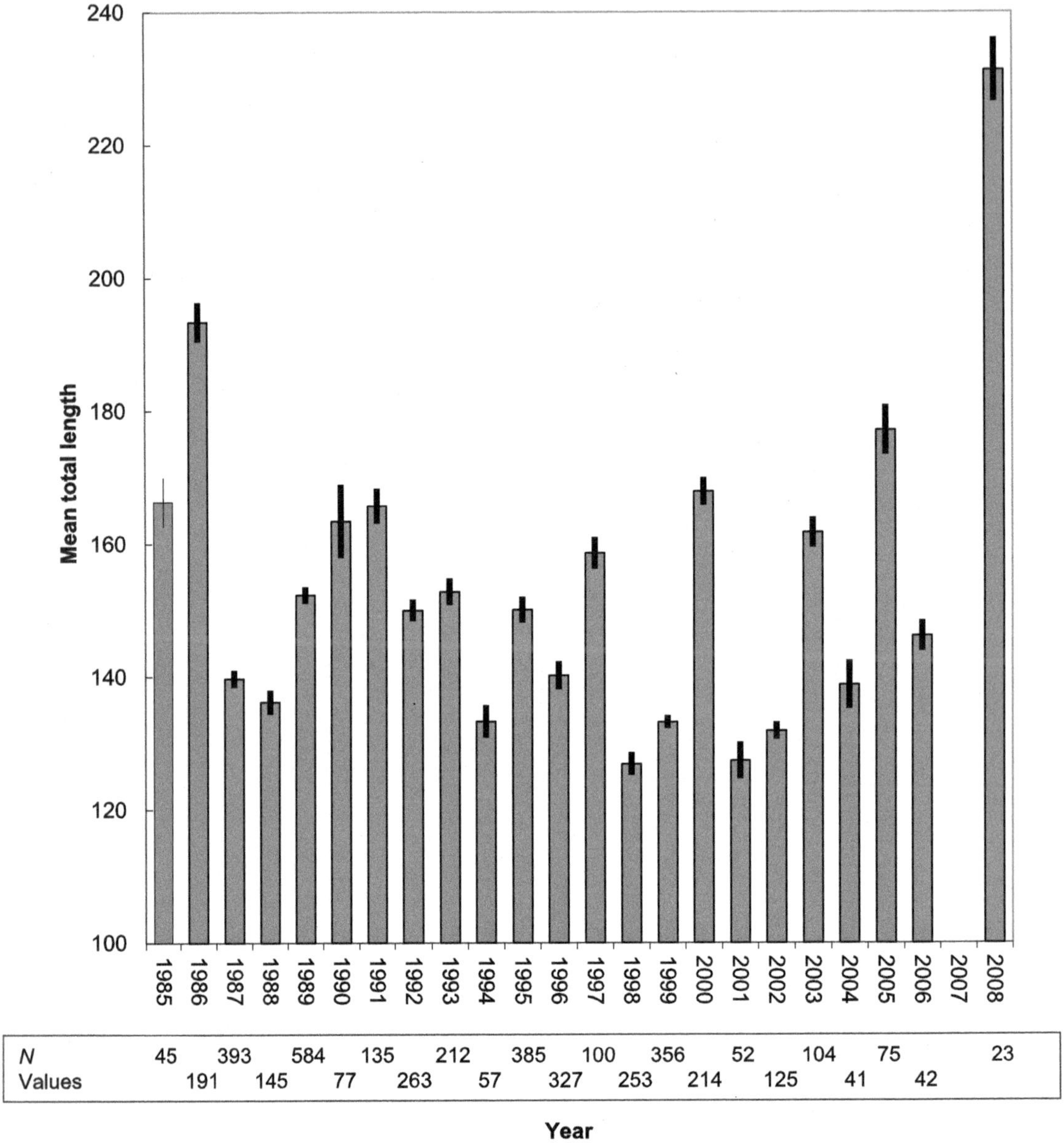

FIGURE 6. Mean total lengths of age-0 striped bass electrofished from Lake Seminole and the Apalachicola River during September–December, 1985 to 2008. Vertical lines represent one standard error.

Lower Apalachicola River/Intracoastal Waterway.—Few anglers targeted or harvested striped bass in the LAR/ICW area of the ACF. During spring (1990–2004) and fall (1990–1999) creel surveys, the number of anglers interviewed ranged from 250 to 550 and 225 to 502, respectively (Table 4). The most striped bass anglers encountered during any survey period was 24, which resulted in the highest estimate of effort (653 h; SE = 347; Table 4). Estimates of striped bass total catch ranged from 2 (SE = 2) to 1,458 (SE = 990) in the spring and 11 (SE = 12) to 1,521 (SE = 749) in the fall (Table 4). No harvest was

TABLE 3. Upper Apalachicola River 14-week spring creel census survey estimates, 1980–2009. Numbers in parentheses represent one standard error. Data for number of anglers and striped bass anglers were not available for 1980 and 1981.

Year	Number of anglers	Striped bass anglers	Harvest (SE)	Total catch (SE)	% released	Angling hours (SE)	Harvest success (SE)	Total success (SE)
1980			152 (130)			1,480 (514)	0.04 (0.03)	
1981			182 (60)			629 (162)	0.29 (0.05)	
1982	1,459	56	703 (467)			1,594 (720)	0.60 (0.13)	
1983	1,012	24	366 (124)			750 (274)	0.81 (0.51)	
1984	1,192	46	149 (68)			1,098 (318)	0.60 (0.20)	
1985	1,250	31	374 (118)			919 (156)	0.36 (0.02)	
1986	1,368	236	877 (337)			6,145 (682)	0.10 (0.03)	
1987	1,157	159	827 (129)			3,272 (507)	0.17 (0.02)	
1988	1,590	227	674 (177)			3,300 (1,275)	0.19 (0.05)	
1989	1,003	223	622 (121)			5,519 (1,032)	0.11 (0.02)	
1990	1,367	416	1,140 (205)			9,108 (1,049)	0.11 (0.03)	
1991	988	298	911 (161)			9,485 (2,936)	0.07 (0.02)	
1992	1,043	247	1,505 (474)	1,570 (516)	4	9,255 (3,378)	0.15 (0.04)	0.15 (0.04)
1993	1,077	179	285 (65)	307 (58)	7	3,163 (499)	0.06 (0.02)	0.07 (0.01)
1994	969	285	210 (69)	886 (104)	76	5,876 (703)	0.03 (0.01)	0.13 (0.02)
1995	1,009	297	134 (49)	311 (156)	57	5,951 (1,024)	0.02 (0.01)	0.03 (0.01)
1996	932	119	316 (97)	929 (226)	66	2,609 (474)	0.16 (0.03)	0.44 (0.09)
1997	518	271	196 (56)	823 (153)	76	4,180 (522)	0.05 (0.01)	0.20 (0.03)
1998	813	104	297 (49)	364 (74)	18	1,593 (380)	0.15 (0.07)	0.19 (0.09)
1999	896	158	500 (112)	700 (133)	29	4,250 (793)	0.12 (0.07)	0.15 (0.08)
2000	872	241	742 (189)	914 (225)	19	6,583 (2,200)	0.08 (0.02)	0.10 (0.02)
2001	1,078	34	722 (211)	1,199 (260)	40	968 (293)	0.22 (0.09)	0.34 (0.11)
2002	739	213	796 (214)	1,457 (346)	45	3,505 (457)	0.12 (0.03)	0.23 (0.05)
2003	1,155	206	257 (95)	598 (146)	57	4,275 (596)	0.06 (0.02)	0.17 (0.04)
2004	749	136	322 (81)	353 (82)	9	4,109 (651)	0.10 (0.02)	0.10 (0.02)
2005	1,085	51	303 (100)	313 (100)	5	1,716 (487)	0.25 (0.06)	0.25 (0.06)
2006	894	59	272 (95)	395 (166)	31	1,845 (489)	0.19 (0.03)	0.28 (0.06)
2007	810	129	236 (60)	389 (89)	39	3,139 (370)	0.07 (0.01)	0.10 (0.02)
2008	754	93	163 (61)	196 (64)	17	2,431 (325)	0.08 (0.02)	0.10 (0.03)
2009	670	102	150 (78)	219 (90)	32	2,135 (385)	0.09 (0.04)	0.12 (0.04)

recorded during five spring surveys and three fall surveys. Typically, striped bass were caught incidentally by anglers fishing for other species, and most anglers reported releasing their catch because the fish were shorter than the 457-mm-TL minimum size limit. During the 15 years that the spring surveys were conducted, only an estimated 1,484 (15%) of the 9,831 striped bass total catch estimate was harvested.

Flint and Chattahoochee rivers.—Creel surveys directed at the striped bass fishery in the Flint and Chattahoochee rivers were conducted from 1995 through 1998. The total number of angler trips during the 26-week creel surveys conducted at the Albany Dam tailrace on the Flint River ranged from 1,741 to 5,553 and averaged 3,855 (Georgia Department of Natural Resources, unpublished data). The estimated number of hours of directed fishing ef-

TABLE 4. Lower Apalachicola River spring and fall creel survey results. Numbers in parentheses represent one standard error.

Year	Number of anglers	Striped bass anglers	Harvest (SE)	Total catch (SE)	% released	Angling hours (SE)	Harvest success (SE)	Total success (SE)
				Spring creel surveys				
1990	393	3	25 (23)	84 (39)	70	109 (39)	0.02 (0.01)	0.65 (0.13)
1991	389	3	6 (4)	6 (4)	0	16 (12)	0.29 (0.04)	0.29 (0.04)
1992	291	0	0	2 (2)	100	0	0	0
1993	250	0	0	74 (81)	100	0	0	0
1994	298	0	0	151 (88)	100	0	0	0
1995	321	24	6 (6)	6 (6)	0	653 (347)	0.12 (0.04)	0.01 (0.04)
1996	861	1	365 (216)	1,311 (319)	72	133 (103)	2.00 (0.26)	2.00 (0.26)
1997	890	1	74 (57)	141 (99)	48	13 (10)	1.00 (0.13)	1.00 (0.13)
1998	855	0	8 (6)	379 (232)	98	0	0	0
1999	446	4	0	623 (104)	100	157 (73)	0	0.89 (0.24)
2000	397	0	0	1,458 (990)	100	0	0	0
2001	376	5	25 (19)	713 (268)	96	594 (576)	0	0
2002	452	1	2 (2)	430 (166)	99	253 (255)	0	0.11 (0.17)
2003	481	4	59 (65)	374 (246)	84	313 (343)	0	0.63 (0.12)
2004	421	0	0	0	0	0	0	0
				Fall creel surveys				
1990	301	19	381 (345)	595 (535)	36	332 (206)	0.83 (0.18)	1.28 (0.27)
1991	274	10	41 (45)	47 (53)	13	107 (65)	0.27 (0.16)	0.36 (0.21)
1992	296	8	11 (12)	44 (28)	75	67 (42)	0.13 (0.05)	0.13 (0.05)
1993	228	6	20 (22)	43 (48)	53	222 (161)	0.10 (0.02)	0.20 (0.05)
1994	225	6	0	11 (12)	100	659 (547)	0	0.01 (0.02)
1995	417	2	0	479 (209)	100	61 (48)	0	1.50 (0.24)
1996	502	10	256 (183)	1,521 (749)	83	256 (183)	0.56 (0.91)	1.56 (0.80)
1997	473	1	166 (95)	441 (211)	62	33 (23)	0	0.33 (0.04)
1998	421	3	39 (24)	755 (460)	95	94 (64)	0.57 (0.19)	1.16 (0.92)
1999	446	2	0	323 (199)	100	34 (27)	0	0

fort for striped bass averaged 801 h and ranged from 442 to 1,056 h. Expanded estimates of harvest for striped bass ranged from 15 to 37 fish and averaged 28 striped bass harvested. During the same time period, the total number of angler trips for the creel census conducted at the GALD on the Chattahoochee River ranged from 4,758 to 7,437 and averaged 6,134 angler trips. The number of hours of directed fishing effort for striped bass ranged from 0 to 466 h and averaged 192 h. Expanded estimates of harvest for striped bass ranged from 0 to 70 and averaged 30 fish. The timing of these surveys corresponded with the open season (1 November–31 May) for the harvest of striped bass. Harvest regulations for striped bass in these waters were 15 fish in combination, including striped bass, hybrid striped bass, and white bass, only two of which could be over 559 mm TL.

Thermal refuge habitat

Apalachicola River.—A total of 66 striped bass ranging in size from 0.7 to 9.6 kg were surgically implanted with temperature sensitive radio or ultrasonic transmitters and released into the UAR (N = 13; 2.5–9.6 kg; ages 2 and older), LAR and ICW (0.7–3.8 kg; age 1 N = 1, age 2 N = 42, age 3 N = 1), and Chipola River (N = 9; 1.8–5.4 kg). Ages of tagged fish released into the LAR and ICW were estimated based on comparison with aged fish of similar length and weight.

Ten radio-tagged striped bass released into the UAR located nine coolwater thermal refuges, which included two within-bank springs (22–25.5°C) and seven spring runs or canopied coolwater creeks (20–24°C) and began occupying these refuges when ambient temperature ranged between 23°C and 25.0°C. Medium (7.0–7.5 kg) and large (>9.0 kg) fish began to occupy thermal refuges between 28 April and 11 May. The smallest fish (2.7–3.8 kg) began using the refuges between 24 and 29 May. After taking residence in the refuge, only one fish (3.0 kg) was located outside of a refuge (twice), although several fish did move between two adjacent refuges located approximately 775 m apart. Only one fish (7.75 kg) survived over the summer of 1990. This fish remained in a refuge until 17 October (158 d after the first refuge relocation) when ambient temperature was 23.9°C. Three fish released into the UAR provided no data.

Nine tagged fish released into the LAR and ICW moved at least 160 km upstream to the UAR. Four were relocated in coolwater refuges in the upper river, four were relocated in the tailrace where several refuges occur, and one was harvested by an angler. One radio-tagged fish moved to the upper Chipola River where it was observed in a refuge by divers but was not relocated by telemetry. This fish was identified as a study fish by the trailing radio antenna and external tag. There was no apparent relationship between size (1–2.7 kg) and movement upstream. Of the 23 fish that remained in the LAR and ICW and were located at least once, four died (located 2–38 times) and six were caught by anglers (located 3–28 times). Several fish were relocated early in the study, disappeared, and were relocated in late summer or early fall. It was not clear whether these fish left the lower system and returned or occupied areas where the transmitter signal was occluded (e.g., deep holes or higher salinity). Of the 44 fish tagged (12 radio and 32 ultrasonic) and released into the LAR and ICW, 14 were never relocated or remained at the release site for several days and then disappeared, including five of the largest (2.0–3.8 kg) fish tagged. Fish tagged and released in this study were not relocated in thermal refuges in the lower part of the Apalachicola River system. However, several preferred areas were identified during telemetry that included bridge pilings and tidal mixing zones. Fish utilized these preferred areas, even though ambient temperature reached as high as 33°C during the summer. No fish tagged with ultrasonic transmitters (which could be detected in saline waters) were located outside of the main river, ICW, or distributary streams in the more saline areas of Apalachicola Bay (LAR) or St. Joe Bay (ICW).

Tagged fish released into the Chipola River were located in seven springs (21.6–22°C) and six coolwater creeks (22.8–25°C) that provided thermal refuge. While these fish preferred the discrete coolwater refuges, they frequently moved throughout the upper 85 km of the Chipola River, occupying water temperatures that ranged from 23°C to 26°C throughout the summer.

Lake Blackshear.—A total of 33 striped bass ranging in size from 2.7 to 15.0 kg were internally (22) or externally (11) fitted with temperature-sensing radio transmitters and released into Lake Blackshear (Flint River) during the fall of 1998 (N = 16), spring of 1999 (N = 11), and spring of 2000 (N = 6; Baker and Jennings 2001) to determine habitat use, movement patterns, and survival of adult striped bass in Lake Blackshear. Eight fish (25%) were detected at least once within 2 months of release but not during the remainder of the study and were considered to be missing. Mortality switches on the transmitters indicated that all of the 25 fish that remained within Lake Blackshear died during the summer following their release. Only five fish were ever detected within thermal refuges in the reservoir, and these fish survived longer than those that were never located within a thermal refuge.

In addition to the telemetry studies, known thermal refuges in Lake Seminole and the Flint River were qualitatively monitored during the summertime for striped bass usage by scuba divers. During monthly surveys of 10 Flint River springs (1997–2008), the average number of large fish (>9.0 kg) observed per dive ranged from 0.09 to 2.15 and the average number of small and medium fish (<9.0 kg) ranged from 1.4 to 19.8. The highest counts for both large and small striped bass occurred from 1999 to 2000, followed by a gradual decline through 2005 and then increases during the most recent years (Georgia Department of Natural Resources, unpublished data).

Thermal refuge renovation

Five spring runs and canopied coolwater creeks in the upper Apalachicola River identified as thermal refuges by telemetry were renovated by removing snags and excavating sediments from the mouths of the creeks during 2001. The amount of substrate material removed ranged from 190 to 1,900 m^3. The excavations partially filled in within 1 to 2 years, and as low summertime discharges persisted in the Apalachicola River because of drought conditions, availability of these refuges for striped bass returned to prerenovation levels (Florida Fish and Wildlife Conservation Commission, unpublished data).

Gulf and Atlantic striped bass performance evaluation

We did not find consistent differences in survival or growth between Gulf and Atlantic races of striped bass when co-stocked into Lake Talquin. From 1988 to 1992, a total of 277,000 Gulf and Atlantic race phase-I striped bass were co-stocked into the reservoir. During the study, size and health of fish at stocking varied within and between year-classes, resulting in highly variable survival and growth to age 1. Atlantic fingerlings tended to be larger at stocking, whereas sizes of Gulf fingerlings were more variable (900–4,400/kg) than Atlantic fingerlings (1,850–4,400/kg) among years. Incomplete air bladder inflation was observed for both races but appeared to be more prevalent among Gulf fish. For unknown reasons, one Gulf mtDNA haplotype (B2) that was stocked in 3 of the 4 years was absent from samples collected at age 1. Although numerous fish were collected from other year-classes, obtained sample sizes through age 4 limited the comparison of relative survival to the 1988, 1989, 1991, and 1992 year-classes.

Relative survival of Gulf and Atlantic race striped bass was similar from age 1 to age 5. Results of Pearson chi-square analysis revealed no difference ($P > 0.05$) in the number of Gulf and Atlantic fish collected for three of four year-classes evaluated (Table 5). However, Atlantic striped bass from the 1992 year-class demonstrated significantly higher (Table 5; $\chi^2 = 25.034$, df = 3, $P = 0.0001$) relative survival at age 3 (Atlantic $N = 34$; Gulf $N = 16$) and age 4 (Atlantic $N = 47$; Gulf $N = 6$).

Although we observed some significant differences in mean total length and weights between races, these differences were not consistent from ages 1 to 6. Early differences may have been linked to the size at stocking. Atlantic phase-I fingerlings were generally larger than Gulf fingerlings at the time of stocking, and we observed significantly greater mean total length ($P = 0.004$) and total weights ($P = 0.03$) for all Atlantic fish sampled at age 1 (Table 6). However, by age 4, Gulf striped bass had significantly ($P < 0.006$) greater mean total length than Atlantic fish, but no significant differences ($P > 0.05$) were observed at ages 5 or 6. ANCOVA revealed no differences ($P > 0.05$) in the elevation of the regression lines between races ($P = 0.155$) for the 1988 and 1989 year-classes. Also, no significant difference in the length–weight natural log regressions of the two races was found, suggesting similar growth and body conditions during the evaluation. These composite results suggested that Gulf

TABLE 5. Number of Gulf and Atlantic striped bass phase-I fingerlings stocked into Lake Talquin for the 1988, 1989, 1991, and 1992 year-classes and number of Gulf and Atlantic fish per year-class recaptured at ages 1 through 5 during November–December gill netting. The ratios of number of Gulf and Atlantic fish collected were evaluated using Pearson chi-square analysis ($P < 0.05$). Significant difference is indicated for the year class by *.

		Total number collected				
Race	Number stocked	Age 1	Age 2	Age 3	Age 4	Age 5
	Year-class 1988 (χ^2 value = 5.76		df = 4.0	P = 0.22)		
Atlantic	100,000	18	21	39	31	22
Gulf	100,000	24	34	37	24	16
	Year-class 1989 (χ^2 value = 0.20		df = 2.0	P = 0.91)		
Atlantic	50,000	28	25	18	14	5
Gulf	50,000	19	17	10	3	4
	Year-class 1991 (χ^2 value = 1.33		df = 2.0	P = 0.51)		
Atlantic	42,000	23	17	15	4	5
Gulf	42,000	34	17	14	6	3
	*Year-class 1992 (χ^2 value = 25.03		df = 3.0	P < 0.0001)		
Atlantic	85,000	29	47	36	47	–
Gulf	85,000	29	42	14	6	–

and Atlantic striped bass co-stocked into Lake Talquin performed similarly in terms of growth and condition from 1988 to 1996.

Genetic characterization

A pivotal question in justifying the restoration program was whether the ACF still harbored a population of native striped bass that was genetically distinct from Atlantic coast populations. Advances in genetic analyses in the decades since stocking began have identified unique features of the Gulf genotype, but also significant introgression by the Atlantic race into the current population. In the mid-1980s, allozymes had failed to reveal sufficient levels of genetic variation in striped bass to address population structure questions and the use of mtDNA was in its infancy. Using RFLP analysis, Wirgin (1987) demonstrated that approximately 55% of Gulf striped bass from the ACF exhibited a mtDNA haplotype with the restriction enzyme *Xba* I that was absent in Atlantic coast samples. Over a 24-year period (1983–2007), 40% to 79% of Gulf broodfish collected annually from the ACF exhibited the unique *Xba* I haplotype (Wirgin et al. 1997; I. I. Wirgin, unpublished data). Gulf striped bass mtDNA also showed a high frequency of a length variant (length A) that was present, but in low frequencies, in Atlantic populations. Additional studies also demonstrated that the Gulf-unique mtDNA *Xba* I haplotype was absent in Atlantic populations that extended from New Brunswick to Georgia and included the Santee–Cooper source of Atlantic-race introductions into the ACF (Wirgin et al. 1990, 1993). As a result, descendants of broodstock with this haplotype were preferred for reintroduction back into the ACF. Further studies that analyzed mtDNA in striped bass from other Gulf tributary rivers in Mississippi, Louisiana, and Texas

TABLE 6. Total numbers and mean lengths and weights of Gulf and Atlantic striped bass, by age-class, collected by gill netting from Lake Talquin during November–December 1988–1996. Standard deviations are provided. Differences were analyzed using Student's *t*-test and significant differences ($P < 0.05$) are denoted by *.

Race	Total number collected	(%)	Mean total length (mm)	SD	Mean weight (kg)	SD
			Age 1 (N = 327)			
Atlantic	150	46	*454	37	*1.18	0.29
Gulf	177	54	441	46	1.12	0.37
			Age 2 (N = 263)			
Atlantic	125	48	596	33	2.59	0.49
Gulf	138	52	590	35	2.53	0.51
			Age 3 (N = 216)			
Atlantic	134	65	688	34	4.06	0.69
Gulf	82	35	693	36	4.07	0.62
			Age 4 (N = 136)			
Atlantic	96	71	722	33	4.69	0.78
Gulf	40	29	*742	38	4.86	0.96
			Age 5 (N = 56)			
Atlantic	32	57	786	33	5.75	0.83
Gulf	24	43	781	30	5.37	0.87
			Age 6 (N = 17)			
Atlantic	9	53	788	34	5.96	0.96
Gulf	8	47	797	38	5.86	0.65

where the Atlantic race had been stocked failed to detect the Gulf-specific *Xba* I haplotype (Wirgin et al. 1997). Because the ACF was the only system along the Gulf to harbor striped bass with the *Xba* I mtDNA haplotype, it was presumed to have the only remaining stocks of true Gulf striped bass. Because of these results, efforts to restore Gulf populations with supplemental stocking were dependent on the ACF as the only source of Gulf fish having the unique genetic characteristics.

Although a high percentage of ACF fish had unique mtDNA haplotypes, the mtDNA genome comprises less than 0.1% of total genetic material in cells and its maternal heritability differs from that of biparentally transmitted nDNA. Therefore, it is possible that fish exhibited predominantly Gulf mtDNA haplotypes in the background of predominantly Atlantic nuclear genomes. Two different nDNA approaches, each interrogating different forms of nDNA variation, addressed this possibility, and both provided the same answer. First, two of four heterologous probes used with nDNA fingerprinting (multi-loci minisatellite analysis) revealed that more than 90% of ACF fish exhibited individual minisatellite DNA fragments that were absent from Atlantic striped bass (Wirgin et al. 1991). Second, striped bass-specific SNP markers, identified from a striped bass genomic DNA library, and using RFLP with Southern blot analyses, identified several polymorphisms that were unique to fish from the ACF population (Wirgin and Maceda 1991). Over the ensuing years, continued analysis of broodstock from the ACF, and additional Atlantic fish, confirmed that several SNP alleles were unique to the ACF population (Wirgin, unpublished data).

In summary, three different DNA approaches, focusing on two different genomes, demonstrated that the ACF population harbored a high percentage of fish that were genetically distinguishable from all Atlantic fish. Also, both minisatellite and microsatellite analyses revealed high levels of allelic diversity in the ACF, comparable to, if not higher than, diversity observed in all Atlantic populations investigated (Wirgin et al. 2005).

Although these studies confirmed that the ACF was genetically differentiated from Atlantic populations, they did not evaluate the genetic impact of the introduction of Atlantic fish into the ACF. To address this issue, museum-archived samples from the ACF whose collection predated Atlantic introductions were screened. MtDNA analysis of the *Xba* I haplotype showed that the allele frequencies did not differ significantly between the prestocking fish and extant ACF populations, indicating that significant introgression of the Atlantic mtDNA haplotype into the ACF population had not occurred (Wirgin et al. 1997). However, subsequent work with nuclear markers demonstrated significant introgression of nDNA into the Gulf genome.

The introgression of Atlantic striped bass nDNA alleles into the Gulf genome was addressed by comparing the frequencies of informative microsatellite DNA polymorphisms in archived ACF samples from prior to any stocking to the frequencies in the extant ACF population. Five microsatellite loci were identified from a striped bass genomic DNA library (Roy et al. 2000) that all exhibited fixed (two loci) or highly significant allelic frequency differences (three loci) between the extant ACF and Atlantic collections from the Hudson River, Chesapeake Bay, Santee–Cooper system, and St. Mary's River, Florida–Georgia (Wirgin et al. 2005). These diagnostic loci were then used as markers to evaluate the extent of nDNA introgression into the extant ACF population.

DNA was isolated from dried striped bass scales archived from the ACF ($N = 48$) prior to stocking and also the extirpated St. Johns River, Florida population ($N = 40$), which historically was the southernmost population along the Atlantic coast (Wirgin et al. 2005). When this DNA was analyzed for microsatellite variation at three (two fixed and one almost fixed) of the diagnostic microsatellite loci discussed in the preceding paragraph, fixed allelic differences were found at two of three loci between the archived ACF and St. Johns River samples. At these two loci, 45% and 67% of fish from the archived ACF sample exhibited genotypes that were absent from the St. Johns River sample. At the third locus, 80% of prestocking ACF fish exhibited genotypes that were observed in only 7% of St. John River specimens. Comparison of allelic frequencies at these three microsatellite loci in the prestocking versus extant ACF collection yielded highly significant differences ($P < 0.001$). Greater allelic diversity at all three loci was observed in the extant compared to the archived prestocking ACF samples, and all alleles were those commonly found in Atlantic populations. Results of a mixture model (ADMIX 2) suggested that at these three loci, the proportion of Atlantic alleles in extant ACF fish was approximately 51%, indicating that significant introgression of Atlantic nDNA alleles had occurred in the ACF population (Wirgin et al. 2005).

Discussion

Restoration of Gulf striped bass in the ACF system has been ongoing for 32 years (as of 2011), as conservation agencies from Alabama,

Georgia, Florida, and the federal government have worked together to define, protect, and enhance the population of this unique race of striped bass. Knowledge of the life history of Gulf striped bass was improved as a result of this partnership. A large cooperative stocking program was implemented, and cooperative monitoring and evaluation programs were conducted. Alternative broodfish repositories were developed outside of the ACF in Lake Talquin, the Blackwater River, and Smith Lake and Lake Martin (Alabama). Genetic cataloging of broodfish and captive fish was accomplished. Thermal refuge habitats important to adult striped bass were identified, protected, and enhanced or rehabilitated. Recreational fisheries were developed or augmented in several reservoirs in the ACF (Walter F. George Lake, West Point Lake, Bartlett's Ferry Lake, and Lake Blackshear), and enhanced trophy fishing opportunities resulted in the catches of new state record fish. Restoration efforts in the ACF have been a catalyst for expansion of Gulf striped bass re-establishment throughout the Gulf region, including the Mobile-Alabama-Tombigbee system (Alabama), the Pascagoula and Pearl rivers (Mississippi), and the Tangipahoa and Tchefuncte rivers (Louisiana). Propagated Gulf striped bass fingerlings are now available for stocking programs across the Gulf region.

Stocking Gulf striped bass phase-I fingerlings into the ACF successfully resulted in an initial increase in the adult population evidenced by significant increases in broodfish catch rates in the tailrace of JWLD as well as in the Chattahoochee and Flint rivers and a 10-fold increase in estimated harvest in the JWLD tailrace fishery during the 1990s. From 1997 to 2000, estimated harvest and success rates for the UAR creel survey increased to the point that striped bass were the most sought species in the tailwater fishery. In addition, recreational fisheries were developed in the Chattahoochee and Flint rivers. Continued stocking in Lake Seminole and the Apalachicola River did not result in further enhancement to broodfish relative abundance or in catch and harvest estimates in the JWLD tailrace or in sustaining the elevated numbers of fish observed during the 1990s. Instead, broodfish CPUE and catch estimates oscillated off of their lows and highs through the remainder of this study. Age-0 CPUE in Lake Seminole suggested several weak year-classes during the mid-1990s, which likely affected broodfish relative abundance during the late 1990s and early 2000s. In addition, severe drought conditions in the ACF basin beginning in 1999 would have resulted in fewer fish being discharged from Lake Seminole and reduced thermal refuge capacity (e.g., reduced groundwater discharge and disconnection of tributaries from the channel) in the upper Apalachicola River. In contrast, in the Chattahoochee and Flint rivers, where thermal refuge habitat is more abundant, broodfish CPUE values continued to increase through the 2000s despite weak year-classes in Lake Seminole and the potential impacts (e.g., reduced groundwater discharge and water withdrawals for irrigation) to thermal refuges that resulted from drought conditions.

Available evidence suggests that natural reproduction remained at a relatively low level in the ACF during the study period despite the stocking program and enhancement in the broodfish population. Although CPUE of wild age-0 fish in Lake Seminole was higher in 1997 and 2006 than in 1985 (nonstocking years), catch rates during nonstocking years were similar to those observed during stocking years that relative abundance values indicated were weak year-classes (RAV < 0.30). Higher catch rates of wild age-0 fish during nonstocking years than during stocking years (2001–2004) also suggested that density dependent mechanisms could be affecting survival of wild fish during some years when stocking occurs. Catch per unit effort was above average (1.45 fish/min) in 2002 and 2003 and wild fish comprised about 13% of the total, while catch rates were well below average in 2001 and 2004 and wild fish comprised 0% and 25% of the total. Natural reproduction was also documented in the Chattahoochee River upstream of West

Point Lake, Alabama–Georgia (Hess and Jennings 2001, 2002) following the introduction of Gulf striped bass into West Point Lake in the 1990s, but only during one year. It is likely that stocking will be required to sustain higher numbers of adult fish in the population than wild fish abundances observed prior to restoration efforts.

Most wild, age-0 fish sampled during fall electrofishing were collected in Lake Seminole, suggesting that successful natural reproduction was primarily occurring in the Flint or Chattahoochee rivers, supporting Keefer's (1986) assessment that successful striped bass reproduction occurs in the Flint River. The predominance of OTC-marked, age-0 hatchery fish (stocked only above JWLD) collected during fall electrofishing in the Apalachicola River demonstrated that a large proportion of age-0 fish in the river were discharged to the Apalachicola River through turbines, flood gates, or the lock at JWLD. From 2001 through 2008, unmarked age-0 fish were collected from the Apalachicola River only in the presence of OTC-marked hatchery fish. Since most age-0 fish collected in the river are taken from the two uppermost sites below JWLD, it is likely that unmarked age-0 fish represent migrants from Lake Seminole rather than wild fish spawned in the Apalachicola River.

Other investigators have reported a negative relationship between phase-I stocking density and survival. Van Den Avyle and Higginbotham (1980) found that stocking density and survival were inversely correlated, though not significantly, for striped bass fingerlings stocked into Watts Bar Reservoir, Tennessee. These authors hypothesized that increased mortality at higher stocking rates may have been influenced more by the smaller size at stocking when stocking numbers were higher than by density dependent variables. Moore et al. (1991) found a significant inverse correlation between the survival and numbers of phase-I fingerlings stocked into Smith Mountain Lake, Virginia. Our analysis demonstrated a positive relationship between numbers stocked and relative abundance of age-0 striped bass in fall electrofishing samples. However, our results may have differed had we sampled during the spring following stocking when fish were age 1 (e.g., Moore et al. 1991).

Studies that evaluated the merits of stocking phase-I and phase-II striped bass or hybrid striped bass fingerlings into coastal rivers (Yeager 1988; Wallin and Van Den Avyle 1995; Wallin et al. 1995) have suggested that stocking phase-II fingerlings into coastal rivers is more efficient than stocking phase-I fingerlings (Yeager 1988). Stocking fingerlings into brackish areas resulted in higher initial survival than stocking into freshwater (Wallin and Van Den Avyle 1995). However, Wallin et al. (1995) reported that survival to age 2 was higher for fingerlings stocked into freshwater despite early survival advantages of fingerlings stocked into brackish water. Our results suggested that although initial survival (48–72 h; 83%) of fingerlings stocked into the UAR, LAR, ICW, and ARB was similar to that reported by others, survival to adulthood for both phase-I and phase-II fish stocked into riverine or brackish habitat was poor. Although we did not evaluate the reasons for poor survival, we believe that factors included predation, and in the case of phase-II fingerlings, extended (8–10 months) culture on artificial feed reduced their fitness for foraging on natural food and may have affected their ability to digest and assimilate natural food.

The invasion and expansion of hydrilla in Lake Seminole appeared to negatively affect the reproductive potential and carrying capacity of age-0 striped bass in the reservoir. Relative abundance of age-0 striped bass in Lake Seminole was inversely correlated with surface area of hydrilla. Dense hydrilla may impact primary productivity by filtering out nutrients, thus reducing phytoplankton (Carter et al. 1988; Jones 1990), or adversely impacting zooplankton communities (Maceina and Shireman 1985; Carter et al. 1988; Jones 1990; Tsai et al. 1991; Michaletz and Bonneau 2005). The effects of hydrilla on plankton communities would be most pronounced during summer months (Carter et al. 1988; Jones 1990), and the impact would be greatest on juvenile fish by reducing preferred sand habitat, available prey, and feeding efficiency (Maceina and Shireman 1985). Loss of sand habitat may concentrate age-0 striped bass into smaller areas of preferred habitat, resulting in poor growth, reduced condition, or starvation. We observed slower growth and low W_r values of age-0 striped bass in Lake Seminole after hydrilla expansion. In addition, the alteration or reduction in zooplankton populations may impact important forage fish species (Maceina and Shireman 1985; Michaletz and Bonneau 2005) such as threadfin shad *Dorosoma petenense*, gizzard shad *D. cepedianum*, skipjack herring *Alosa chrysochloris*, and Alabama shad *A. alabamae* that are important prey for older juvenile and adult striped bass. A reduction in shad populations could also result in increased predation on age-0 striped bass by other piscivores such as largemouth bass *Micropterus salmoides* (Michaelson et al. 2001), particularly given the close proximity of dense hydrilla to available sand habitat.

From a management perspective, the inverse relationship between relative abundance of age-0 fish and hydrilla density in Lake Seminole could be used to adjust annual stocking numbers based on the expected density of hydrilla. Although the effects of expanded hydrilla appear to be detrimental to young-of-year survival, the relationship suggests that losses could be offset by increasing stocking numbers, as occurred in the 1990s. For example, years having RAVs above 0.30 and high hydrilla estimates in the fall were typically associated with higher stocking rates (>33 fish/ha). Conversely, when fall hydrilla estimates were low or declining, stocking rates as low as 15 fish/ha resulted in RAVs similar to years when twice as many fish were stocked. In future years, stocking rates could be adjusted at the spring *Morone* workshop to adapt to expected hydrilla conditions in the reservoir.

Creel surveys conducted on the upper Apalachicola River, Chattahoochee River, and Flint River demonstrated directed effort towards striped bass and that stocking helped develop or enhance recreational fisheries below dams in the system. From 1997 to 2000, estimated harvest and success rates for the UAR creel survey increased to the point that striped bass were the most sought species in the tailwater fishery. Stocking phase-I and phase-II fingerlings into the lower Apalachicola River, Bay, and Intracoastal Waterway resulted in very little directed fishing effort towards striped bass.

The striped bass population in the Apalachicola River is largely dependent on hatchery fish stocked into Lake Seminole and the subsequent emigration of fish through JWLD at some time during their life history. Downstream emigration of *Morone* species has been well documented in the ACF. Young (1987) described the establishment of a palmetto bass fishery in the tailrace of JWLD through emigration of this hybrid stocked into Lake Seminole, but he did not address the life stage at which palmetto bass exited the reservoir. Mesing et al. (1999) investigated the movement of palmetto bass and sunshine bass stocked into Lake Seminole and documented emigration of both hybrid crosses from the lake through JWLD into the tailrace and recruitment into the fishery as early as age 1. Our results demonstrated that age-0 striped bass emigrated to the tailrace during periods of both low and high discharge, but we did not address emigration of subadult or adult fish through the dam. Other studies (Van Den Avyle and Higginbotham 1980; Henley 1998; Hightower et al. 2001; Thompson et al. 2007) found that adult striped bass emigrated downstream through turbines or locks from other reservoirs in the Southeast. Trash racks likely prevent passage of most adult fish downstream through the turbines at JWLD; however, adult fish emigrate through JWLD via the lock or floodgates during high water events. Van Den Avyle and Evans (1990) reported that the greatest number of contacts with radio-tagged adult striped bass in the lower portions of Lake Seminole occurred between November and January, a time when floodgates at JWLD are most likely to be open. We attempted to correlate broodfish indices and creel catch/harvest estimates in the tailrace with monthly discharge and lock usage at JWLD. Significant correlations occurred only when April discharge was included in the analysis, which indicated to us that we were only seeing the immediate effects of water discharge on angling or broodfish electrofishing efficiency during the peak concentration of striped bass in the tailrace and not the overall effects of discharge or lockages on emigration from Lake Seminole. Stewart and Burrell (2013, this volume) discuss the effects of emigration through dams, though the rate of emigration has not been quantified.

Striped bass along the Gulf coast occur on the southern fringe of their range in North America, and their habitats and existence in many Gulf of Mexico rivers has always been tenuous. Striped bass do not occur in peninsular Florida, except in the St. Johns River, their southernmost range on the Atlantic coast. Ware (1971) attempted to establish striped bass in five central Florida lakes and Lake Talquin (northwest Florida) by stocking phase-I Atlantic (Moncks Corners origin) fingerlings during 1968 and 1969. He reported excellent growth, with the most rapid growth occurring during cooler fall and winter months, and survival to age 2. However, at age 1, summer die-offs beginning in June were reported for the central Florida lakes. Die-offs were associated with low coefficients of condition (K ranged from 1.41 to 1.78) that was attributed to parasitism by the marine nematode *Goezia* sp. (Ware 1971; Gaines and Rogers 1972). The vector of infection was assumed to be diet supplementation using a marine herring during fingerling culture. However, die-offs were not reported for Lake Talquin, which was stocked with the same hatchery products. Die-offs in the central Florida lakes were more likely the result of thermal stress that may have also exacerbated the effects of parasitism. The survival of fish stocked into Lake Talquin was more likely due to the presence of coolwater habitat. Summer die-offs are also atypical for striped bass in the St. Johns River, an Atlantic population that utilizes coolwater habitats (springs and streams) during the summer, which has also been parasitized with Goezia sp. (Gaines et al. 1973; Florida Fish and Wildlife Conservation Commission, unpublished data). Stocking striped bass into central Florida lakes was discontinued after 1969 in lieu of more successful stocking with sunshine bass (Ware 1975), which tolerate warmer temperatures.

The hypothesis of a thermal niche for striped bass and size/age dependent thermal partitioning among striped bass was proposed by Coutant and Carroll (1980). After extensive literature review, Coutant (1985) further developed the hypothesis and concluded that cool, oxygenated water was a requirement for adult (age 2–3 or >5 kg) striped bass during hot summer months. Other researchers (Dudley et al. 1977; Cheek et al. 1985; Matthews 1985; Matthews et al. 1989; Zale et al. 1990; Schaffler and Isely 2002; Young and Isely 2002) provided supporting evidence that adult striped bass, particularly large individuals, select coolwater refuges when thermal habitat is constrained during summer in southeastern and southwestern rivers and reservoirs and suggested that cool, oxygenated water was important for large striped bass. Wooley and Crateau (1983), Moss (1985), Lamprecht and Shelton (1988), Van Den Avyle and Evans (1990), and Weeks and Van Den Avyle (1998) reported that striped bass in Gulf coast tributaries similarly select coolwater refuges during summer months, and their period of residence was longer than reported for other southeastern reservoirs and river systems (Lamprecht and Shelton 1988; Van Den Avyle and Evans 1990; Weeks and Van Den Avyle 1998). Our telemetry results demonstrated that adult Gulf striped bass in the Apalachicola River began occupying coolwater refuges as early as April and remained in the refuges through mid-October, consistent with findings by Van Den Avyle and Evans (1990) that striped bass in Lake Seminole and the Flint River began using coolwater refuges in May and occupied them as late as November.

Many state fisheries agencies that stock striped bass into freshwater lakes have reported the mortality of adult striped bass due to thermal stress from high temperatures and low dissolved oxygen concentrations as the most frequent management problem for this species (Axon and Whitehurst 1985). Nine agencies identified these issues as limiting factors for striped bass survival (Axon and Whitehurst 1985). Matthews (1985) conducted a survey of states that stock striped bass into reservoirs and found that summer die-offs of adult striped bass were reported for 27 of 80 reservoirs. Large adult fish (>5.0 kg) were most often included in the mortality reports, although mortalities of fish as young as age 1 were also included. In many southeastern and southwestern reservoirs, thermal refuges consist of a hypolimnetic stratum of cool, oxygenated water that often becomes thinner, smaller, and more confined as summer temperatures increase and the season is prolonged (Matthews et al. 1985, 1989; Zale et al. 1990; Van Horn et al. 1998; Jackson and Hightower 2001; Thompson et al. 2010). In each study, the stratum of preferred temperature and oxygen disappeared by mid to late August, and striped bass occupied suboptimal conditions for a short period during the remainder of the summer. Only Zale et al. (1990) reported significant mortality, but the duration of suboptimal conditions in Keystone Reservoir was longer than reported for other study lakes. In the ACF, and other Gulf tributaries that have been dammed, reservoirs are typically shallow and become isothermal early in the summer. In these systems, striped bass are reliant on springs and coolwater creeks for thermal refuge during the summer. Although these refuges may provide consistent habitat throughout the summer and fall, their size and availability may be a limiting factor for the adult population size (Wooley and Crateau 1983; Coutant 1987b; Lukens 1988; Van Den Avyle and Evans 1990; Weeks and Van Den Avyle 1998). Our telemetry results in the Apalachicola River indicated a high percentage of mortality among all sizes of fish during the summer that was not observed by Van Den Avyle and Evans (1990) in Lake Seminole and the Flint River.

Van Den Avyle and Evans (1990) reported that 96% of their summertime (May–October) locations of adult striped bass (3.2–30.0 kg) were in coolwater refuges in Lake Seminole and the Flint River, and the only exceptions were during a brief period of exceptionally high discharge during August. Although thermal refuges may ameliorate the effects of high summer temperatures, extended residence in refuges may be detrimental to Gulf striped bass because these summer habitats are typically void of forage species (Weeks and Van Den Avyle 1998). Striped bass may lose as much as 22% of their body weight while occupying refuges (Wooley and Crateau 1983; McDaniel et al. 1993; Florida Fish and Wildlife Conservation Commission, unpublished data). Ware (1971) suggested that a coefficient of condition (K) less than 1.70 would result in mortality of subadult striped bass. Loss of weight and condition during the summer may contribute to reduced spawning success the following spring (Coutant 1987a). Striped bass in the ACF and other Gulf tributaries have a short period of weight recovery between vacating the refuges (mid-October to early November) and the spawning season (mid to late March). There is also only a short period between the spawning season and occupation of refuges.

While our telemetry studies did not define the thermal tolerances of Gulf striped bass, they were consistent with findings by earlier investigators (Cheek et al. 1985; Coutant 1985; Moss 1985; Lukens 1988; Van Den Avyle and Evans 1990; GSMFC 2006) that coolwater refuges are important, if not essential, habitats for over-summer survival of adult striped bass. Our observations indicated that some small striped bass (1–3 kg) were able to over-summer in the brackish portion of the lower Apalachicola River and Intracoastal Waterway, without using known thermal refuges, and occupied habitats similar to those reported by Haeseker et al. (1996). However, this telemetry study also documented that small fish (age 2) that remained in the freshwater main stem of the river moved upstream and occupied coolwater refuges during the summer. While we cannot quantify the importance of thermal refuges to the ACF population, if reproduction by large individuals contributes disproportionately to recruitment, then thermal refuge availability may be a limiting factor for the striped bass population in the ACF, as well as for populations in other Gulf of Mexico tributary rivers.

Accessibility to tributaries that are known thermal refuges in the Apalachicola River has been impacted by entrenchment of the river channel and decreased water discharge through JWLD. Current water management and upstream consumptive use in the ACF basin has resulted in lower discharge into the Apalachicola River, particularly during summer months and periods of drought when coolwater refuge is most important. Riverbed entrenchment typically occurs below dams (American Rivers 2002; Poff and Hart 2002; Pizzuto 2002) and, in the case of the Apalachicola River, has been accelerated by navigation maintenance practices such as dredging, bendway easing, rock removal, and construction of training dikes (Light et al. 2006). Entrenchment of the riverbed has resulted in the need for higher water discharge through JWLD during summer months to maintain connectivity between thermal refuge tributaries and the main channel. When discharge from JWLD is less than 312 m^3/s, thermal refuge creeks and spring runs in the upper river become disconnected (Light et al. 1998), resulting in the loss of important habitat in the Apalachicola River during summer months. When discharge is less than 255 m^3/s, only three thermal refuge creeks or spring runs are connected to the main channel; below a discharge of 142 m^3/s (the prescribed minimum flow for JWLD), only one refuge stream is connected to the main channel (Light et al. 1998). Loss of thermal refuge habitat in the upper Apalachicola River emphasizes the need to maintain suitable water levels in the river during summer months and to investigate upstream passage of striped bass via the lock at JWLD, or by other means, for access to thermal refuges above the dam. It also underscores the need to protect thermal refuge habitat not only in the Apalachicola River, but in Lake Seminole and the Flint River as well, as a step toward conserving the last remnant of a naturally reproducing population of Gulf race striped bass along the Gulf of Mexico.

Thermal refuges provided by springs and other groundwater sources in the Flint River may have been impacted also by consumptive water use in the ACF. Demands for water have been increasing in this area to accommodate population growth in municipal areas and farmland irrigation in rural areas. Center-pivot irrigation has become widespread in southwestern Georgia and northwestern Florida, and the negative effects on spring discharges have been well documented (Albertson and Torak 2002; Golladay et al. 2007). The numbers of Gulf striped bass using the 10 monitored thermal refuges in Lake Seminole and the Flint River has varied over time. The severe drought that occurred from 1999 through 2002, in conjunction with increased groundwater usage in the basin, resulted in decreases in flow in some of the monitored springs and may be directly related to a decline in striped bass habitation in some thermal refuges. Georgia Department of Natural Resources fisheries personnel observed reversal of flow within a thermal refuge spring on the Flint River shortly after a center-pivot system approximately 1.5 km away began pumping groundwater for irrigation (J. Kilpatrick, Georgia Department of Natural Resources, personal communication). More recently, the numbers of fish observed using springs in the Flint River has been gradually increasing.

Renovation of thermal refuges appears to have provided only short-term gains as many renovated areas degraded relatively quickly, especially during drought conditions. There was no measurable increase in use by striped bass in eight refuges in the Flint River and Lake Blackshear following renovation. However, the removal of material constricting the flow of cool water should have increased the quantity of thermal refuge area available to striped bass. During the telemetry study in Lake Blackshear in the summer of 1999, two of three thermal refuges in the reservoir that were renovated by GADNR did not flow (Baker and Jennings 2001). Since the 2001 study, these springs have still not flowed, probably because of hydrostatic pressure, continued drought, and groundwater withdrawals (Weeks and Van Den Avyle 1998; Georgia Department of Natural Resources, unpublished data).

A co-stocking experiment conducted in Lake Talquin, Florida suggested that there were no major differences in performance of Gulf and Atlantic race striped bass. Atlantic race striped bass showed significantly better survival at ages 3 and 4 for a single year-class; however, no other differences in survival were noted across the four tested cohorts. Differences in mean length and mean weight at age between Atlantic and Gulf fish also were not consistent throughout the study. Atlantic fish were larger at age 1, which was attributed to differences in size at stocking. By age 4, Gulf fish were larger, but the difference was not significant at ages 5 and 6. It is likely that these differences in growth were not biologically significant relative to habitat limitations.

Indiscriminate stocking of nonnative striped bass conspecifics where natural reproduction persists has resulted in genetic consequences that are probably irreversible. Although the ACF striped bass population remains genetically unique from Atlantic populations, a pure Gulf race no longer exists. The long-term effects of introgression of Atlantic-origin alleles into the Gulf striped bass genome may never be fully known.

Genetic analysis has been an integral component of the ACF restoration program since its inception. The long-term commitment and support to the genetic component of this project was essential for successfully completing genetic components of the 1996 and 2004 ACF-Ps. The detection of numerous mtDNA and nDNA markers should prove informative in future performance evaluation tests. Additional diagnostic markers will provide more flexibility in the design of future studies. Darden et al. (2013, this volume) provide an up-to-date description of microsatellite markers and resultant genetic analysis that are available for striped bass. All DNAs and/or tissues analyzed in this program have been stored in a repository and provide a valuable resource. Molecular research techniques evolve over time, and questions that prove refractory to analysis can often be addressed in the future.

This effort showed that the analysis of DNA from archived collections in museums or other sources can provide a wealth of information regarding the genetic composition of extirpated populations or comparisons of extirpated with extant populations. Such information will permit retrospective analysis of long-term trends in genotypic composition of populations. For example, we can envision the use of such materials to compare the effective population size of the original versus extant ACF population. Minimizing the likelihood of the development of inbreeding depression in small populations such as in the ACF is a primary objective of conservation genetics programs. This can be achieved by preventing severe population declines and resulting bottlenecks. In cases such as the ACF where augmentation is used to supplement natural reproduction, hatchery programs can be optimized to prevent decreases in effective population size. Current restoration efforts seek to maximize genetic diversity in the ACF population and broodfish repositories by accepting progeny from all crosses originating from the ACF, regardless of genotypic or meristic characteristics, and stocking progeny from as many different crossings as feasible. Based on more than 25 years of genetic analyses, the ACF population appears to be more genetically diverse than Atlantic populations, perhaps resulting from its initial genetic isolation. More recently, introductions of nonnative Atlantic fish have also increased genetic diversity.

Lessons and management implications

Collaborative group discussions, planning, and cost sharing among multi-state and federal agencies provided a forum to address a declining native Gulf striped bass population in the ACF and ultimately other river systems along the Gulf of Mexico. Without the cooperative efforts of the conservation agencies involved with fish management on the ACF, many tasks of the ACFPs could not have been attempted or successfully completed. Collaborative efforts were also beneficial in establishing and maintaining small populations, through stocking, in several other river and reservoir systems along the Gulf of Mexico.

Monitoring programs used in the ACF restoration program demonstrated that we were able to enhance the Gulf striped bass population in the ACF through a large stocking program. However, benefits obtained through stocking large numbers of fish have not been consistently sustained through time and have not resulted in the ability of the population to self-sustain at a desired level that is substantially larger than previously observed from natural reproduction alone.

Striped bass along the Gulf coast occur on the southern extent of their range. Limited historical references suggest that populations were small, and given the ecological constraints (e.g., high temperature regime) and the alterations to riverine environments (e.g., segmentation, dredging, and water withdrawals), populations will likely remain small despite stocking large numbers of fish. The most challenging aspects of our restoration program were the recognition by all parties involved that historic populations were inherently small, the acceptance of the role of altered habitat on restricting Gulf striped bass populations, and that feasible opportunities for restoration of degraded habitat were limited.

Large perturbations to the environment, such as the construction of dams that block access to spawning grounds, disrupt spawning processes, and limit access to thermal refuges, were harmful and difficult to remediate. However, less obvious impacts, such as the diversion or deforestation of a canopied creek, or well and irrigation withdrawals from aquifers, may be equally detrimental to coolwater habitat that is beneficial to the continued existence of small adult populations. We were able to meet objectives of the ACF plans that pursued restoration or enhancement of important known striped bass habitat areas by physically renovating several thermal refuges in the Apalachicola and Flint rivers. However, in dynamic river systems, such as the ACF, renovation efforts have been small and short-lived. Protection and enhancement of remaining functional thermal refuges will require a broader water-management strategy encompassing dam releases and groundwater supplies as well as periodic maintenance of habitat enhancement projects.

The observed low levels of striped bass natural reproduction in the ACF resulted in a reliance on stocking large numbers of hatchery fish into the system in lieu of addressing the more challenging issues of segmented rivers and lost or degraded habitats. Stocking has resulted in the introgression of genetic material from the Atlantic race into the Gulf genome. Genetic analyses of archived and extant striped bass from the ACF demonstrated that there is no longer a pure Gulf race striped bass population. However, because the ACF is the only natural population to still harbor a vestige of the Gulf genome, efforts to protect and restore the unique ACF population, as well as those where ACF progeny have been introduced, are warranted. Florida fish managers have committed to stocking only Gulf striped bass into Gulf of Mexico tributaries within their historic range. Likewise, fish managers in Alabama have committed their stocking program in all river systems to Gulf striped bass (N. Nichols, Alabama Wildlife and Freshwater Fisheries, personal communication). Fish managers in Mississippi and Louisiana have also expressed a desire to stock only Gulf striped bass but with the caveat that they cannot sacrifice their established striped bass fisheries when Gulf fingerlings are not available for stocking. To this end, fish managers in Mississippi are developing a broodfish repository in Ross Barnett Reservoir.

The ACF Striped Bass Restoration and Evaluation Plan will be revised in 2012. As a result of the work described herein, sufficient information now exists to redefine appropriate management, research, and restoration goals in the revised plan.

Acknowledgments

This project was funded in part by grants from the U.S. Fish and Wildlife Service (Federal Grant #1448-0004-95-9155), the Gulf States Marine Fisheries Commission (Subcontract #749-97-001), Federal Aid in Sport Fish Restoration (F-39), the Florida Fish and Wildlife Conservation Commission, and the Georgia Department of Natural Resources. Special thanks are owed to D. Frugé and A. Brown and his staff at Welaka NFH; V. Mudrak, G. Looney, C. Echevarria, and G. Moyer and their staff at Warm Springs NFH; J. Breelan and his staff at Carbon Hill NFH; R. Campbell and his staff at Private John Allen NFH; N. Nichols and his staff at the ALDCNR Marion SFH; D. Yeager, B. Arnold, R. Simmons, and the rest of their staff at the FFWC Blackwater FRDC; C. Starling, H. Revell, and their staff at the FWCC Richloam SFH; M. Newman , T. Marshall, and their staff at the GADNR SCSFH; J. Swearingen, C. Robbins, and their staff at the GADNR CSFH; F. Ware, B. Rousseau, R. Land, M. Spelman, M. Hill, G. Wood, L. Wachter, T. Pittman, D. Heinke, P. Smith, J. McLaughlin, and D. Barfoot of FFWC; G. Carmody, F. Parauka, L. Jenkins, J. Hemming, and their staff at the FWS Panama City field office; and M. Thomas, D. Partridge, H. Wilson, T. Feltman, S. Steiner, T., Ingram, J. Kilpatrick, J. Tannehill, and D. Barber of GADNR. This project also is deeply indebted to L. Maceda of the New York University Institute of Environmental Medicine for her tireless work in genetic analysis. We also thank P. Behnke, E. Irwin, an anonymous reviewer, and the editors for their critiques, which improved this manuscript.

The findings and conclusions in this article are those of the author(s) and do not necessarily represent the views of the U.S. Fish and Wildlife Service.

References

ACF (Apalachicola-Chattahoochee-Flint) Striped Bass Technical Committee. 1996. ACF striped bass restoration and evaluation plan. ACF Striped Bass Technical Committee, Panama City, Florida.

ACF (Apalachicola-Chattahoochee-Flint) Striped Bass Technical Committee. 2004. ACF striped bass restoration and evaluation five-year plan. ACF Striped Bass Technical Committee, Panama City, Florida.

Albertson, P. N., and L. J. Torak, 2002. Simulated effects of ground-water pumpage on stream-aquifer flow in the vicinity of federally protected species of freshwater mussels in the lower Apalachicola-Chattahoochee-Flint River basin (subarea 4), southeastern Alabama, northwestern Florida, and southwestern Georgia. U.S. Geological Survey, Water Resources Investigations Report 02-4016, Atlanta.

American Rivers. 2002. The ecology of dam removal: a summary of benefits and impacts. American Rivers, Washington, D.C.

Axon, J. R., and D. K. Whitehurst. 1985. Striped bass management in lakes with emphasis on management problems. Transactions of the American Fisheries Society 114:8–11.

Baker, T. L., and C. A. Jennings. 2001. Survival, habitat use, and movement patterns of adult striped bass in Lake Blackshear, Georgia. In D. Frugé, editor, Restoration of striped bass in three Gulf of Mexico river systems. U.S. Fish and Wildlife Service, Fisheries Stewardship Initiative Project, Final Report, Atlanta.

Barkuloo, J. M. 1961. Distribution and abundance of striped bass (*Roccus saxatilis*, Walbaum) on the Florida Gulf Coast. Proceedings of the Annual Conference Southeastern Association of Game and Fish Commissioners 15(1961):223–226.

Barkuloo, J. M. 1967. Florida striped bass. Florida Game and Fresh Water Fish Commission, Fishery Bulletin No 4, Tallahassee.

Barkuloo, J. M. 1970. Taxonomic status and reproduction of striped bass (*Morone saxatilis*) in Florida. U.S. Department of the Interior, U.S. Bureau of Sport Fish and Wildlife, Technical Paper 44, Washington, D.C.

Bean, T. H. 1883. On the occurrence of the striped

bass in the lower Mississippi Valley. Proceedings of United States National Museum 7:242–244.

Boynton, W. R., T. T. Polgar, and H. H. Zion. 1981. Importance of juvenile striped bass food habits in the Potomac estuary. Transactions of the American Fisheries Society 110:56–63.

Brooks, R. C., R. C. Heidinger, R. J. H. Hoxmeier, and D. H. Wahl. 2002. Relative survival of three sizes of walleyes stocked into Illinois lakes. North American Journal of Fisheries Management 22:995–1006.

Brown, B. E. 1965. Meristic counts of striped bass from Alabama. Transactions of the American Fisheries Society 94:278–279.

Brown, M. L., and B. R. Murphy. 1991a. Standard weights (W_s) for striped bass, white bass, and hybrid striped bass. North American Journal of Fisheries Management 11:451–467.

Carter, V., J. W. Barko, G. L. Godshalk, and N. B. Rybicki. 1988. Effects of submersed macrophytes on water quality in the tidal Potomac River, Maryland. Journal of Freshwater Ecology 4:493–501.

Cheek, T. E., M. J. Van Den Avyle, and C. C. Coutant. 1985. Influences of water quality on distribution of striped bass in a Tennessee River impoundment. Transactions of the American Fisheries Society 114:67–76.

Coutant, C. C. 1985. Striped bass, temperature, and dissolved oxygen: a speculative hypothesis for environmental risk. Transactions of the American Fisheries Society 114:31–61.

Coutant, C. C. 1987a. Poor reproductive success of striped bass from a reservoir with reduced summer habitat. Transactions of the American Fisheries Society 116:154–160.

Coutant, C. C. 1987b. Thermal preference: when does an asset become a liability? Environmental Biology of Fishes 18(3):161–172.

Coutant, C. C., and D. S. Carroll. 1980. Temperatures occupied by ten ultrasonic-tagged striped bass in freshwater lakes. Transactions of the American Fisheries Society 109:195–202.

Crateau, E. J., P. Moon, and C. M. Wooley. 1980. Apalachicola River striped bass project. U.S. Fish and Wildlife Service, Fisheries Resources Project, Annual Progress Report, Panama City, Florida.

Darden, T. L., F. Sessions, and M. R. Denson. 2013. Use of genetic microsatellite markers to identify factors affecting stocking success in striped bass. Pages 395–411 *in* J. S. Bulak, C. C. Coutant, and J. A. Rice, editors. Biology and management of inland striped bass and hybrid striped bass. American Fisheries Society, Symposium 80, Bethesda, Maryland.

Dudley, R. G., A. W. Mullis, and J. W. Terrell. 1977. Movements of adult striped bass (*Morone saxatilis*) in the Savannah River, Georgia. Transactions of the American Fisheries Society 106:314–322.

GADNR (Georgia Department of Natural Resources). 1969. Evaluation of striped bass introductions in lakes Blackshear and Seminole. Georgia Department of Natural Resources, Game and Fish Division, Statewide Fisheries Investigations, Annual Progress Report, Atlanta.

Gaines, J. L., Jr., and W. A. Rogers. 1972. Fish mortalities associated with *Goezia* sp. (Nematoda: Ascaroidea) in central Florida. Proceedings of the Annual Conference Southeastern Association of Game and Fish Commissioners 25(1971):496–497.

Gaines, J. L., Jr., F. J. Ware, and W. A. Rogers. 1973. A summary of findings on the nematode, *Goezia* sp. in the southeastern United States. Proceedings of the Annual Conference Southeastern Association of Game and Fish Commissioners 26(1972):334–335.

Gennings, R. M., R. D. Gasaway, R. W. Taylor, J. L. Major, and L. E. McSwain. 1970. Study number XI: evaluation of striped bass introduction in lakes Blackshear and Seminole. Georgia Department of Natural Resources, Game and Fish Division, Statewide Fisheries Investigations, Annual Progress Report, Atlanta.

Golladay, S. W., D. W. Hicks, and T. K. Muenz. 2007. Stream flow changes associated with water use and climatic variation in the lower Flint River basin, southwest Georgia. Pages 479–482 *in* T. C. Rasmussen, G. D. Carroll, and A. P. Georgakakos, editors. Proceedings of the 2007 Georgia Water Conference. University of Georgia, Institute of Ecology, Athens.

Goode, G. B. 1884. The fisheries and fishery industries of the United States: natural history of useful aquatic animals. Volume 1, section

III. U.S. Commission of Fish and Fisheries, Washington, D.C.

Goode, G. B., and T. H. Bean. 1879. Catalogue of a collection of fishes sent from Pensacola, Florida, and vicinity, by Mr. Silas Stearns, with descriptions of six new species. Proceedings of United States National Museum 2:121–156.

GSMFC (Gulf States Marine Fisheries Commission). 2006. The striped bass fishery of the Gulf of Mexico, United States: a regional management plan. GSMFC, Ocean Springs, Mississippi.

Haeseker, S. L., J. T. Carmichael, and J. E. Hightower. 1996. Summer distribution and condition of striped bass within Albemarle Sound, North Carolina. Transactions of the American Fisheries Society 125:690–704.

Hart, L. G., and R. C. Summerfelt. 1975. Surgical procedures for implanting ultrasonic transmitters into flathead catfish (*Pylodictis olivaris*). Transactions of the American Fisheries Society 104:56–59.

Heidinger, R.C., and K. Clodfelter. 1987. Validity of the otoliths for determining age and growth of walleye, striped bass, and smallmouth bass in power plant cooling ponds. Pages 241–251 *in* R. C. Summerfelt and G. E. Hall, editors. Age and growth of fish. The Iowa State University Press, Ames.

Henley, D. T. 1998. Inter-river migration of striped bass in western Kentucky. Proceedings of the Annual Conference Southeastern Association of Fish and Wildlife Agencies 50(1996):62–78.

Hess, B. J., and C. A. Jennings. 2001. Reproductive success of stocked striped bass (*Morone saxatilis*) in the Chattahoochee River above West Point Lake: implications for the *Morone* stocking program. University of Georgia, Fisheries Stewardship Project, Final Report, Atlanta.

Hess, B. J., and C. A. Jennings. 2002. Striped bass in trout waters of the upper Chattahoochee River, Georgia: can these two fisheries coexist? Proceedings of the Annual Conference Southeastern Association of Fish and Wildlife Agencies 54(2000):107–117.

Hightower, J. E., J. R. Jackson, and K. H. Pollack. 2001. Use of telemetry methods to estimate natural and fishing mortality of striped bass in Lake Gaston, North Carolina. Transactions of the American Fisheries Society 130:557–567.

Hill, M. J., E. A. Long, and S. Hardin. 1996. Effects of dam removal on Dead Lake, Chipola River, Florida. Proceedings of the Annual Conference Southeastern Association of Fish and Wildlife Agencies 48(1994):512–523.

Jackson, J. R., and J. E. Hightower. 2001. Reservoir striped bass movements and site fidelity in relation to seasonal patterns in habitat quality. North American Journal of Fisheries Management 21:34–45.

Jones, R. C. 1990. The effect of submersed aquatic vegetation on phytoplankton and water quality in the tidal freshwater Potomac River. Journal of Freshwater Ecology 5:279–288.

Keefer, L.C. 1986. Early life history of striped bass in the Flint River. Georgia Department of Natural Resources, Game and Fish Division, Southwest Region Fisheries Investigations, Final Report, F-28–13, Atlanta.

Lamprecht, S. D., and W. L. Shelton. 1988. Spatial and temporal movements of striped bass in the upper Alabama River. Proceedings of the Annual Conference Southeastern Association of Fish and Wildlife Agencies 40(1986):266–274.

Light, H. M., M. R. Darst, and J. W. Grubbs. 1998. Aquatic habitats in relation to river flow in the Apalachicola River floodplain, Florida. U.S. Geological Survey, Professional Paper 1594, Washington, D.C.

Light, H. M., K. R. Vincent, M. R. Darst, and F. D. Price. 2006. Water level decline in the Apalachicola River, Florida, due to long-term changes in stage-discharge relations from 1954 to 2004, and effects on floodplain habitats. U.S. Geological Survey, Scientific Investigations Report 2006-5173, Washington, D.C.

Long, E. A. 2001. Striped bass restoration in the Apalachicola River. In D. Frugé, editor, Restoration of striped bass in three Gulf of Mexico river systems. U.S. Fish and Wildlife Service, Fisheries Stewardship Project, Final Report, Atlanta.

Lukens, R. L. 1988. Habitat criteria for striped bass stocked into rivers of the northern Gulf of Mexico. Gulf States Marine Fisheries Commission, W-B No. 1, Oceans Springs, Mississippi.

Maceina, M. J., and J. V. Shireman. 1985. Influence of dense hydrilla infestation on black crappie growth. Proceedings of the Annual Conference Southeastern Association of Fish and Wildlife Agencies 36(1982):394–402.

Malvestuto, S. P. 1983. Sampling the recreational fishery. Pages 397–419 *in* L. A. Nielsen and D. L. Johnson, editors. Fisheries techniques. American Fisheries Society, Bethesda, Maryland.

Matthews, W. J. 1985. Summer mortality of striped bass in reservoirs of the United States. Transactions of the American Fisheries Society 114:62–66

Matthews, W. J., L. G. Hill, and S. M. Schellhaass. 1985. Depth distribution of striped bass and other fish in Lake Texoma (Oklahoma–Texas) during summer stratification. Transactions of the American Fisheries Society 114:84–91.

Matthews, W. J., L. G. Hill, D. R. Edds, and F. P. Gelwick. 1989. Influence of water quality and season on habitat use by striped bass in a large southwestern reservoir. Transactions of the American Fisheries Society 118:243–250.

McDaniel, C. K., L. E. Snyder, and L. L. Connor. 1993. Impacts of thermal stress on the condition of striped bass. Proceedings of the Annual Conference Southeastern Association of Fish and Wildlife Agencies 45(1991):361–369.

McIlwain, T. D. 1980. Striped bass in coastal waters, South Atlantic and Gulf. Pages 37–43 *in* H. Clepper, editor. Marine recreational fisheries. Proceedings of the Fifth Annual Marine Recreational Fisheries Symposium. Sport Fishing Institute, Washington, D.C.

Merriman, D. 1941. Studies on the striped bass (*Roccus saxatilis*) of the Atlantic coast. U.S. Fish and Wildlife Service, Fishery Bulletin 50, Washington, D.C.

Mesing, C. L., E. A. Long, I.I. Wirgin, and L. Maceda. 1999. Age, growth, and movement of 2 *Morone* hybrids in the Apalachicola River system, Florida. Proceedings of the Annual Conference Southeastern Association of Fish and Wildlife Agencies 51(1997):123–134.

Michaelson, D. P., J. J. North, and T. M. Sutton. 2001. Largemouth bass predation on stocked striped bass in Smith Mountain Lake, Virginia. North American Journal of Fisheries Management 21:326–332.

Michaletz, P. H., and J. L. Bonneau. 2005. Age-0 gizzard shad abundance is reduced in the presence of macrophytes: implications for interactions with bluegill. Transactions of the American Fisheries Society 134:149–159.

Moore, C. M., R. J. Neves, and J. J. Ney. 1991. Survival and abundance of stocked striped bass in Smith Mountain Lake, Virginia. North American Journal of Fisheries Management 11:393–399.

Moss, J. L. 1985. Summer selection of thermal refuges by striped bass in Alabama reservoirs and tailwaters. Transactions of the American Fisheries Society 114:77–83.

Nicholson, L. C., I. B. Byrd, E. Createau, J. A. Huff, V. Minton, M. Powell, G. E. Saul, F. Ware, and A. Williams. 1986. Striped bass fishery management plan (Gulf of Mexico). Gulf States Marine Fisheries Commission, Publication No. 16, Ocean Springs, Mississippi.

Pasch, R. W., R. M. Gennings, L. E. McSwain, R. W. Taylor, J. H. Smith, and C. C. Mann. 1973. Study XI: evaluation of striped bass introduction in lakes Blackshear and Seminole. Georgia Department of Natural Resources, Game and Fish Division, Statewide Fisheries Investigations, Final Report, F-21–5, Atlanta.

Pearson, J.C. 1938. The life history of the striped bass, or rockfish *Roccus saxatilis* (Walbaum). U.S. Bureau of Fisheries Bulletin 28:825–851.

Pizzuto, J. 2002. Effects of dam removal on river form and process. Bioscience 52:683–691.

Poff, N. L., and D. D. Hart. 2002. How dams vary and why it matters for the emerging science of dam removal. Bioscience 52(8):659–668.

Raney, E. C., and W. S. Woolcott. 1955. Races of the striped bass, *Roccus saxatilis* (Walbaum), in southeastern United States. Journal of Wildlife Biology 19:444–450.

Roy, N., L. Maceda, and I. Wirgin. 2000. Isolation of microsatellites in striped bass *Morone saxatilis* (Teleostei) and their preliminary use in population identification. Molecular Ecology 9:827–829.

Schaffler, J. J., and J. J. Isely. 2002. Habitat use by striped bass in relation to seasonal changes in water quality in a southern reservoir. Transactions of the American Fisheries Society 131:817–827.

Secor, D. H., M. G. White, and J. M. Dean. 1991. Immersion marking of larval and juvenile hatchery-produced striped bass with oxytetracycline. Transactions of the American Fisheries Society 120:261–266.

Secor, D. H., T. M. Price, and H. T. Hornick. 1995. Validation of otoliths-based ageing and a comparison of otoliths and scale-based ageing in mark-recaptured Chesapeake Bay striped bass, *Morone saxatilis*. Fishery Bulletin 93:186–190.

Setzler, E. M., W. R. Boynton, K. V. Wood, H. H. Zion, L. Lubbards, N. K. Mountford, P. Frere, L. Tucker, and J. A. Mihursky. 1980. Synopsis of biological data on striped bass, *Morone saxatilis* (Walbaum). FAO (Food and Agriculture Organization of the United Nations) Fish Synopsis 121.

Stewart, W. T., and M. Burrell. 2013. Striped bass dispersion and effects on fisheries management in Lakes Mohave and Pleasant, Colorado River basin. Pages 431–448 *in* J. S. Bulak, C. C. Coutant, and J. A. Rice, editors. Biology and management of inland striped bass and hybrid striped bass. American Fisheries Society, Symposium 80, Bethesda, Maryland.

Thompson, J. S., D. S. Waters, J. A. Rice, and J. E. Hightower. 2007. Seasonal natural and fishing mortality of striped bass in a southeastern reservoir. North American Journal of Fisheries Management 27:681–694.

Thompson, J. S., D. S. Waters, and J. A. Rice. 2010. Striped bass habitat selection rules in reservoirs without suitable summer habitat offer insight into consequences for growth. Transaction of the American Fisheries Society 139:1450–1464.

Tsai, C., M. Wiley, and A. Chai. 1991. Rise and fall of the Potomac River striped bass stock: a hypothesis of the role of sewage. Transactions of the American Fisheries Society 120:1–22.

USACOE (U.S. Army Corps of Engineers). 1998. Volume II appendices. Lake Seminole, FL-GA-AL hydrilla action plan: final supplement to the master plan and final supplement to the environmental impact statement. U.S. Army Engineer District, Inland Environment Section, Mobile, Alabama.

Van Den Avyle, M. J., and J. Evans. 1990. Temperature selection by striped bass in a Gulf of Mexico coastal river system. North American Journal of Fisheries Management 10:58–66.

Van Den Avyle, M. J., and B. J. Higginbotham. 1980. Growth, survival and distribution of striped bass stocked into Watts Bar Reservoir, Tennessee. Proceedings of the Annual Conference Southeastern Association of Fish and Wildlife Agencies 33(1979):361–370.

Van Den Avyle, M. J., B. J. Higginbotham, B. T. James, and F. J. Bulow. 1983. Habitat preferences and food habits of young-of-the-year striped bass, white bass, and yellow bass in Watts Bar Reservoir, Tennessee. North American Journal of Fisheries Management 3:163–170.

Van Horn, S.L., J.R. Finke, and. Degan. 1998. Summer habitat selection of striped bass in Lake Norman. Proceedings of the Annual Conference Southeastern Association of Fish and Wildlife Agencies 50(1996):91–97.

Ware, F. J. 1971. Some early life history of Florida's inland striped bass, *Morone saxatilis*. Proceedings of the Annual Conference Southeastern Association of Game and Fish Commissioners 24(1970):439–447.

Ware, F. J. 1975. Progress with *Morone* hybrids in fresh water. Proceedings of the Annual Conference Southeastern Association of Game and Fish Commissioners 28(1974):48–54.

Wallin, J. E. and M. J. Van Den Avyle. 1995. Interactive effects of stocking site salinity and handling stress on survival of striped bass fingerlings. Transactions of the American Fisheries Society 124:736–745.

Wallin, J., M. J. Van Den Avyle, and T. Sinclair. 1995. Juvenile striped bass nursery habitat and abundance index in the lower Savannah River. University of Georgia, Georgia Cooperative Fish and Wildlife Research Unit, Federal Aid in Sport Fish Restoration F-52, Final Report, Athens.

Weeks, G. N., and M. J. Van Den Avyle. 1998. Selection of thermal refuges by striped bass in a Gulf of Mexico coastal river. Proceedings of the Annual Conference Southeastern Association of Fish and Wildlife Agencies 50(1996):51–61.

Wirgin, I. I. 1987. Molecular evolution in the fish genus *Morone*. Doctoral dissertation. The City University of New York, New York.

Wirgin, I. I., D. Currie, N. Roy, L. Maceda, and J. R. Waldman. 2005. Introgression of nuclear DNA (nDNA) alleles of stocked Atlantic coast striped bass with the last remaining native Gulf of Mexico population. North American Journal of Fisheries Management 25:464–474.

Wirgin, I. I., C. Grunwald, S. J. Garte, and C. Mesing. 1991. Use of DNA fingerprinting in the identification and management of a striped bass population in the southeastern United States. Transactions of the American Fisheries Society 120:273–282.

Wirgin, I. I., and L. Maceda. 1991. Development and use of striped bass-specific RFLP probes. Journal of Fish Biology 39:159–167.

Wirgin, I., L. Maceda, J. Stable, and C. Mesing. 1997. An evaluation of Atlantic coast striped bass mitochondrial DNA in a Gulf of Mexico population using formalin-preserved museum collections. Molecular Biology 6:907–916.

Wirgin, I. I., L. Maceda, J. R. Waldman, and R. Crittenden. 1993. Use of mitochondrial DNA polymorphisms to estimate the relative contributions of the Hudson River and Chesapeake Bay striped bass stocks to the mixed fishery on the Atlantic coast. Transactions of the American Fisheries Society 122:669–684.

Wirgin, I. I., R. Proenca, and J. Grossfield. 1989. Mitochondrial DNA diversity among populations of striped bass in the southeastern United States. Canadian Journal of Zoology 67:891–907.

Wirgin, I. I., P. Silverstein, and J. Grossfield. 1990. Restriction endonuclease analysis of striped mitochondrial DNA: the Atlantic coastal migratory stock. Pages 475–491 *in* N. C. Parker, A. E Giorgi, R. C. Heidinger, D. B. Jester, Jr., and G. A. Winans, editors. Fish-marking techniques. American Fisheries Society, Symposium 7, Bethesda, Maryland.

Wooley, C. M., and E. J. Crateau. 1983. Biology, population estimates, and movement of native and introduced striped bass, Apalachicola River, Florida. North American Journal of Fisheries Management 203:383–394.

Yeager, D. M. 1988. Evaluation of phase I and phase II hybrid striped bass in the Escambia River, Florida. Proceedings of the Annual Conference Southeastern Association of Fish and Wildlife Agencies 41(1987):41–47.

Young, N. 1987. Establishment of a hybrid *Morone* fishery in the Apalachicola River, Florida. Proceedings of the Annual Conference Southeastern Association of Fish and Wildlife Agencies 38(1984):450–454.

Young, S. P., and J. J. Isely. 2002. Striped bass annual site fidelity and habitat utilization in J. Strom Thurmond Reservoir, South Carolina–Georgia. Transactions of the American Fisheries Society 131:828–837.

Zale, A. V., J. D. Wiechman, R.L. Lochmiller, and J. Burroughs. 1990. Limnological conditions associated with summer mortality of striped bass in Keystone Reservoir, Oklahoma. Transactions of the American Fisheries society 119:72–76.

American Fisheries Society Symposium 80:65–91, 2013

When Is Habitat Limiting for Striped Bass? Three Decades of Testing the Temperature–Oxygen Squeeze Hypothesis

CHARLES C. COUTANT*
Oak Ridge National Laboratory (retired)
120 Miramar Circle, Oak Ridge, Tennessee 37830, USA

Abstract.—Striped bass *Morone saxatilis*, originally a coastal and estuarine species, has been introduced in reservoirs in the southeastern and western United States. Although such stocking established many successful fisheries, there were troublesome die-offs of adult striped bass (3–9 kg, generally >5 kg) in some waters, usually in late summer. In contrast, juveniles and small adults thrived. In response to these patterns, my students and I conducted several years of telemetry studies of adult striped bass, primarily in Cherokee and Watts Bar reservoirs, Tennessee, and laboratory studies of juvenile temperature selection. In 1985, I published the "temperature–oxygen squeeze" hypothesis to explain mortalities of large fish on the basis of limited availability of cool (<25°C), oxygenated (>2 mg/L) water in summer while juveniles successfully occupied a warmer thermal niche (>25°C). We now have more than 20 years of research and management since 1985, primarily across the Southeast, in which the published hypothesis has, explicitly or not, been tested, generally confirmed, and applied to management. This retrospective paper reviews the studies our team conducted to develop and test the hypothesis and about 20 years of relevant studies by others that have added important nuances, addressed lingering issues, and turned a controversial idea into generally accepted understanding and management practice. Nonetheless, issues remain for understanding the effects of poor summer habitat on striped bass, such as why some studies show striped bass occupying warmer temperatures without mortalities and the role of prey availability in survival of fish obliged to occupy warm water. Other papers in this volume augment and extend the saga of progressively developing knowledge that this paper recalls of striped bass habitat requirements, thermal niche segregation by size (or age), and management constraints and opportunities.

Introduction

Fisheries students are generally taught the scientific method of posing testable hypotheses as a framework for research and subsequent management of aquatic resources. However, we find few examples in the fisheries literature where a hypothesis is developed, published, and tested and a retrospective analysis conducted to see if the hypothesis was correct (or how it was modified with increased information). We have this opportunity to evaluate a hypothesis I published in 1985 to explain summer die-offs of striped bass *Morone saxatilis* generally larger than 5 kg in southeastern reservoirs (Coutant 1985).

The striped bass is an eastern North American coastal species originally occupying a range from the St. Lawrence River in the north to

* Corresponding author: ccoutant3@comcast.net

northern Florida in the south. Populations also occurred along the coast of the Gulf of Mexico from northern Florida to Texas. In the late 1800s, they were introduced into Pacific coast waters, where they expanded into the San Francisco Bay system. With an accidental isolation of a spawning run of adult striped bass in the Santee–Cooper Dam Project in South Carolina (damming the adjacent Santee and Cooper rivers) in the late 1950s, biologists realized that this species could fully carry out its life cycle in freshwater (Scruggs and Fuller 1955). Hydropower development was occurring rapidly throughout the Southeast at that time, and biologically productive reservoirs were increasingly populated by gizzard shad *Dorosoma cepedianum* and other forage fish. The notion of stocking large-bodied striped bass in these reservoirs for both trophy fisheries and shad control took hold among fisheries managers. Hatchery techniques were developed for artificial spawning in tanks and raising young striped bass in ponds. Stocking occurred, and soon anglers were landing 10–15 kg adult striped bass. This history is recorded in papers presented at a 1982 striped bass symposium published as a special 1985 issue of *Transactions of the American Fisheries Society* (14[1]) and reprised in Van Horn (2013, this volume).

Despite the widespread success of striped bass stockings in the early 1970s, there were some troubling signs. Striped bass, mostly 3–9 kg but almost always over about 2.5 kg, would be found at the surface in summer in obvious stress, and many died (Coutant 1978). These summer die-offs became a regular annual occurrence in some reservoirs, such as Cherokee Reservoir on the Holston River in eastern Tennessee and others in the Southeast (Matthews 1985). Many dying fish were "skinny" with low condition factors indicative of poor feeding, and they often exhibited red sores on their skins. Fishery managers were at a loss to explain the mortalities, and an Interagency Cherokee Task Force invoked the usual candidates—pollution by metals, pesticides, and power plants (Coutant 1978). High temperature was generally ruled out because young striped bass, assumed to be representative of the species, thrived in warm hatchery ponds.

At that time, I was leading a team of researchers at Oak Ridge National Laboratory conducting temperature-effects studies of fish and other aquatic life in order to better manage thermal discharges from electric power stations. Striped bass were a species of national concern for power plant cooling-water impacts. Controversies over effects of power plant cooling systems on striped bass, particularly in the Hudson River, New York, prompted us to focus some of our team's research on the temperature requirements and temperature-related behavior of this species, which also occupied local reservoirs. Nearby Cherokee Reservoir had the Tennessee Valley Authority's John Sevier steam electric station in its upper reach, suggesting a possible influence of its thermal discharge on mortalities there. A significant component of our research involved the species' temperature selection and avoidance in the field using radio and ultrasonic temperature-sensing telemetry.

Based on the results of our field studies in Tennessee with striped bass greater than about 2.5 kg, I hypothesized that adult striped bass have important water quality habitat limitations—they appeared to require (and select, if available) cool water less than about 25°C to survive through the summer and were often squeezed into restricted habitats by avoidance of warm surface water and poorly oxygenated deeper water (Coutant 1978, 1980, 1985; Figure 1). When the cool, oxygenated habitat of a reservoir hypolimnion is progressively eliminated through the summer, the striped bass are forced into available thermal refuges with suitable (or nearly so) temperature and oxygen conditions or into physiologically marginal or unacceptable habitat. These habitat limitations may cause striped bass to starve over time by isolation from prey, become vulnerable to disease, or experience summer mortality. The hypothesis was formulated based on the behavior of a range of sizes of adult striped bass equipped with temperature-sensing transmitters in ver-

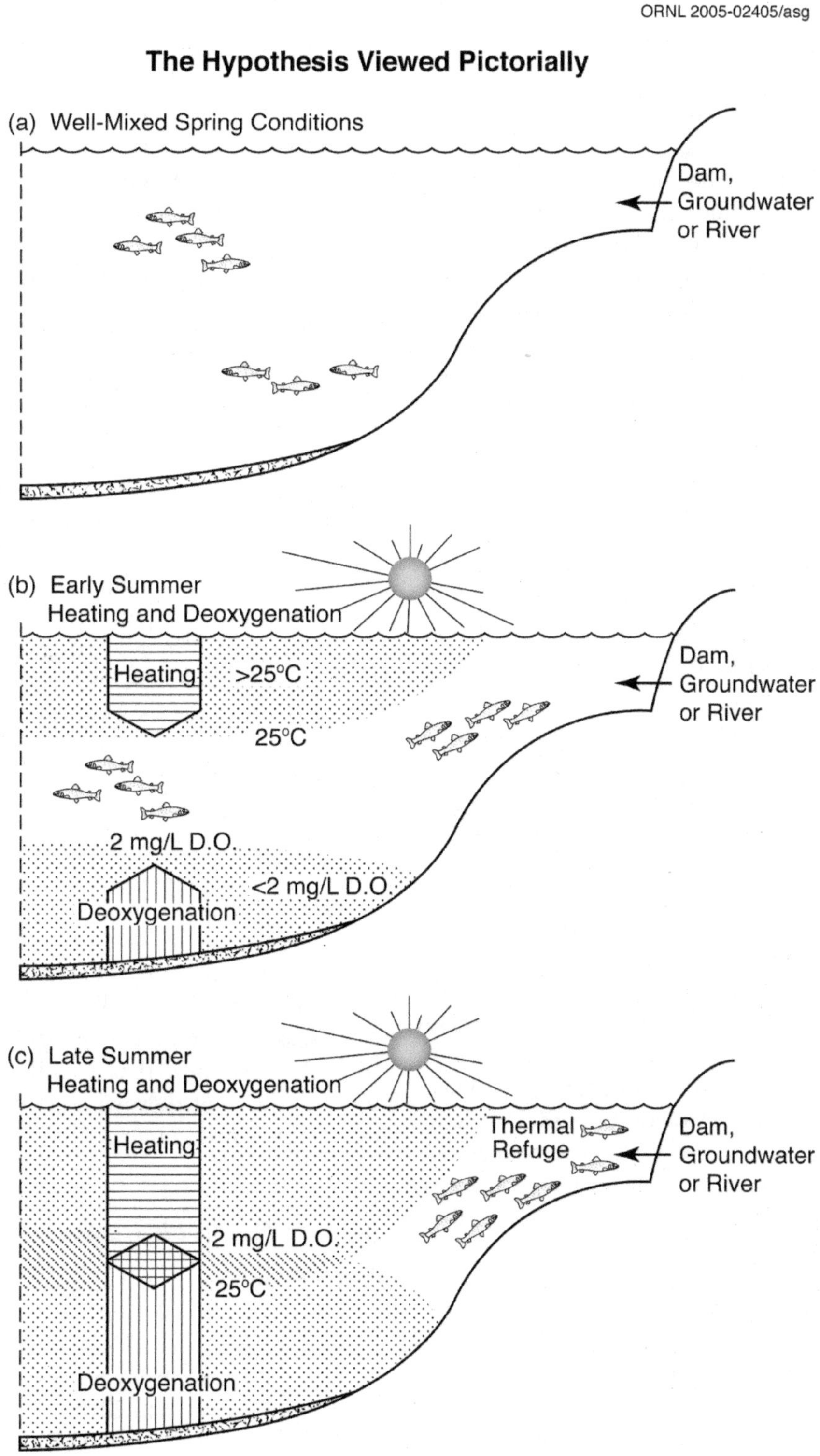

FIGURE 1. The temperature–oxygen squeeze hypothesis for large striped bass viewed pictorially. (Modified from Coutant 1985)

tical thermal gradients in two rock quarries in 1974 and 1975 (Coutant and Carroll 1980), in Cherokee Reservoir in 1976 to 1978 (Waddle et al. 1980; Schaich and Coutant 1980), and in Watts Bar Reservoir, Tennessee in 1979 and 1980 (Cheek et al. 1985). The field studies were augmented with laboratory and other field studies, primarily of juveniles, which showed quite different thermal responses by young fish (Cox and Coutant 1981; Coutant et al. 1984). The aggregate of our studies on juveniles and adults in the laboratory and field indicated that as fish size increased, their tolerance for warm water decreased.

In this paper, I recount the history of our early studies and the development of the temperature–oxygen squeeze hypothesis of habitat limitation and review numerous subsequent studies that have sought similar information in other waters and (explicitly or not) tested the hypothesis. Much of the hypothesis has been confirmed and, with increasing information, expanded in useful ways (Thompson et al. 2010; Thompson and Rice 2013, this volume; Rice et al. 2013, this volume). When this work is seen from the general perspective of ecological resources and habitat niches (e.g., Magnuson et al. 1979), the hypothesis could be better stated as the thermal niche–oxygen squeeze hypothesis, for it is now clear that medium-sized (~3–8 kg) and large (>~8 kg) adult striped bass have a coolwater fundamental thermal niche both behaviorally and physiologically, whereas their juveniles have a warmwater fundamental niche (Coutant 1986; Bettoli 2005). These weight ranges are imprecise, as the change in thermal habitat preference is apparently a continuum and its expression is confounded in the field by other factors. I have highlighted key conceptual additions to the hypothesis as the research results of 20-some years unfolded after our studies ended. In addition to recognizing the biological habitat limitations of striped bass, it has become clear that the critical features of the physical–chemical habitat and its seasonal change can be understood and modeled (Chang et al. 1992; Ruane et al. 2013, this volume; Taylor et al. 2013, this volume), and in some cases, remedial measures can be adopted to manage relevant habitat limitations (Mobley et al. 2013, this volume). This retrospective is largely terminated with literature from the mid-2000s because others in this volume report their own work since then.

Development of the Hypothesis

We initiated basic studies of temperature requirements of striped bass in the laboratory and field in the 1970s. To determine temperature preference under field conditions, 430- to 626-mm-long fish (0.91–3.52 kg) with temperature-sensing transmitters were followed in two enclosed rock quarries in Oak Ridge, Tennessee, with adequate dissolved oxygen at all depths (Coutant and Carroll 1980). These fish occupied the warmest temperatures available more than 1.5 m deep through spring and 20–24°C (centering near 22°C) in summer. In contrast, 80–270-mm juvenile fish (approximately 0.25–0.72 kg) generally selected 24–27°C in summer in laboratory gradient tanks and in one of the quarry lakes (Coutant et al. 1984). Laboratory studies of juvenile growth rates found optimal growth temperatures near 24°C, with some growth occurring to 33.5°C (Cox and Coutant 1981). The close proximity we found for preferred and optimum growth temperatures in juvenile striped bass indicated a physiological basis, as observed for other species (Jobling 1981). We concluded that juveniles and adults likely had different thermal requirements, with larger fish needing cooler water for physiological reasons.

Cherokee Reservoir studies

Our first confirmation that adult striped bass in a large reservoir preferred cool water and the first suggestions of a temperature–oxygen "squeeze" came from telemetry studies of habitat selection in 12,222-ha Cherokee Reservoir, Tennessee (Coutant 1978; Schaich and Coutant 1980; Waddle et al. 1980). Cherokee Reservoir is an 86-km-long upland storage reservoir on the Holston River, a tributary of the Tennessee River; it is eutrophic and stratified thermally and with respect to dissolved oxygen during the period May to October. Our initial results from Cherokee Reservoir indicated that larger fish (more than about 5 kg) in summer were crowded into cool springs and small, cool tributaries (less than 20°C) when reservoir temperatures were warm (above 27°C) and deeper, cooler layers were often below 3 mg/L of dissolved oxygen (Coutant 1978). Anglers concentrated there, and fish for our tagging often were captured on hook and line from these areas. Malnutrition and disease were common, and there were summer die-offs of often-emaciated fish in the 3–9-kg range in the main body of the reservoir (Coutant 1978). No records of size structure of the reservoir striped bass population were available, but anglers noted that fish greater than 9 kg were rarely caught. In retrospect, there were also hints of such behavior from another telemetry study of striped bass, conducted in Percy Priest Reservoir, a Tennessee reservoir located just outside Nashville on the Stones River, a tributary of the Cumberland River (Stooksbury 1977), although not recognized by the author at the time. Stooksbury observed seasonal changes in location of transmitter-equipped (location only) striped bass consistent with seasonal changes in reservoir temperature and dissolved oxygen, which we suspected were the driving factors for striped bass habitat selection.

Additional studies over 2 years in Cherokee Reservoir (Schaich and Coutant 1980; Waddle et al. 1980) showed that 2–5-kg striped bass selected the warmest temperatures available at 1–3 m depth in March–May and 15.5–25°C at depths of greater than 8 m in June and early July. By late July, the fish avoided the hypolimnion with oxygen less than 3 mg/L and surface temperatures generally above 22°C and moved to cool (15–22°C), oxygenated (above 5 mg/L) springs or creek channels. Some fish made excursions into the main reservoir at temperatures up to about 25°C. Fish moved out of refuges in autumn when reservoir temperatures approximated 22–24°C. There was a strong suggestion of progressively cooler temperature preferences with increasing size among the adult fish we tagged and monitored (Figure 2). Striped bass of 5 kg preferred cooler temperatures (16–22C) than 1-kg fish (20–24C).

We wondered if the apparently stressful nature of Cherokee Reservoir's warm temperatures and poor dissolved oxygen would influence the reproductive success of striped bass. The importance of cool, oxygenated water was demonstrated by retrospectively comparing hatchery production records for spawning stock collected from three Tennessee reservoirs—Cherokee, Watts Bar, and Norris—with differing water quality (Coutant 1987a). Spawning and early development in hatchery tanks were carried out under the same controlled conditions for fish from each source. Records showed that spawners taken from temperature–oxygen-limited Cherokee Reservoir consistently yielded poor hatchery survival of eggs and larvae while spawners from the other two reservoirs, which were less habitat-limited, produced eggs and larvae with good survival. These results were consistent with stress having a physiological (likely bioenergetic) basis that affected egg maturation. There were clear implications for the sources of hatchery spawners and also for potential negative effects on natural reproduction in systems where that is important but suitable summer habitat is limited.

I used our work and relevant evidence from striped bass stocks elsewhere to synthesize the thermal niche hypothesis with a summer temperature–dissolved oxygen squeeze for striped bass (Coutant 1980, 1985, 1986, 1987b). I described the evidence for progressively cooler thermal niches as striped bass grow and illustrated the effects of summer temperature and oxygen dynamics on striped bass habitat selection in reservoirs (Figure 1). I noted that the strong temperature preference of adult striped bass could be an asset for survival (occupying temperatures near a physiological optimum) or a liability (forcing fish into restricted thermal refuges with limited resources) depending on

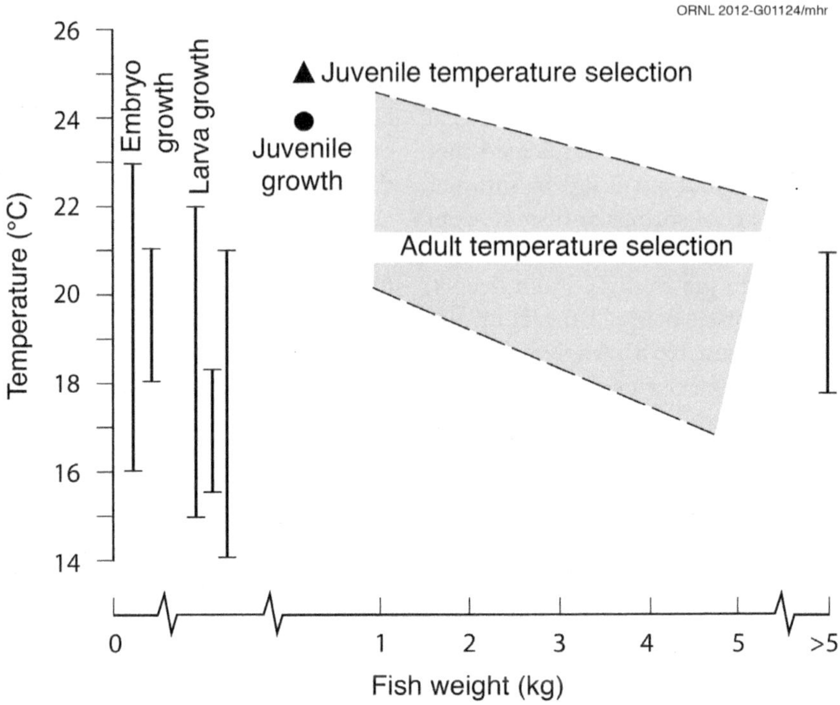

FIGURE 2. The changing thermal niche with size of striped bass. Bars at left: thermal ranges for embryo-larval stages according to different authors; all are in the spring spawning temperature range (Setzler et al. 1980). Juvenile temperature selection is from Coutant et al. (1984). Optimal juvenile growth is from Cox and Coutant (1981). The shaded ranges of temperature selection for adults up to 5 kg in the field are from Schaich and Coutant (1980) and Waddle et al. (1980). The thermal preference bar for fish larger than 5 kg is from Cheek et al. 1985. (Modified from Coutant 1987b)

the available habitats and water quality (Coutant 1987b).

As our research progressed from its start in the late 1970s, it generated both positive and negative comments from resource managers. Many were outspokenly antagonistic toward our work; see papers in *Transactions of the American Fisheries Society* 14(1). The controversy prompted a survey of other reservoirs in the Southeast that documented the common (but not universal) occurrence of summer die-offs of inland striped bass (Matthews 1985).

Hybrid striped bass (white bass *M. chrysops* × striped bass) were also of interest as we conducted our studies with pure striped bass. They were present in Cherokee Reservoir, but we did not study them in detail. Casual observations indicated that they did not exhibit the signs of summer stress shown by striped bass. We observed few dead hybrids at times when striped bass were dying. These observations suggested that pairing the thermally sensitive striped bass with the warmwater white bass apparently yields a hybrid that is more tolerant of high temperatures yet attains much the same size and angler favor as striped bass. Other investigators have provided some support for this notion with respect to temperature preference responses, although better physiological tolerance by adults has yet to be confirmed (Kilpatrick and Ney 2013, this volume).

Our tests of the hypothesis

We tested our ideas of seasonal thermal–oxygen habitat limitation with temperature–telemetry studies in another reservoir, Watts Bar, Tennessee, on the Tennessee River (Cheek et al. 1985).

Watts Bar Reservoir is a 15,440-ha mainstream impoundment in the Tennessee River that includes the confluence of the Clinch River. We found that 24°C and oxygen concentrations of about 4 mg/L appeared to determine the summer spatial distribution of striped bass 2–13.5 kg (mean = 6.3 kg) fitted with temperature-sensing transmitters. During winter and early spring when temperature and dissolved oxygen were vertically and horizontally quite uniform, striped bass were mobile and widely distributed. As the reservoir warmed in summer, fish were less mobile and were progressively limited to areas in tributary arms where temperature was less than 24°C (mean = 20°C, SD 2.1°C) and dissolved oxygen exceeded 4 mg/L. Notable among these areas were deepwater discharges in the tailwaters of upstream impoundments (Melton Hill Dam on the Clinch River and Fort Loudon Dam on the Tennessee River), a cool tributary of the Tennessee River (Tellico River) and groundwater inflows. Striped bass were restricted to these areas until fall when the whole reservoir cooled. There appeared to be abundant prey for striped bass in the numerous large, cool refuges.

Studies by Others

Our research and the controversial nature of our results stimulated studies of striped bass habitat selection by researchers at many locations across the southeastern United States (Figure 3). These were not always rigorous tests of the hypothesis, but researchers sought a general description of behavior in different environments. Many of their reported findings, as discussed individually below, were consistent with various components of the hypothesis. These studies showed that adult striped bass preferentially occupied water cooler than 24–25°C with sufficient oxygen if it was available and supported a dependence of temperature preferences and tolerance on fish size.

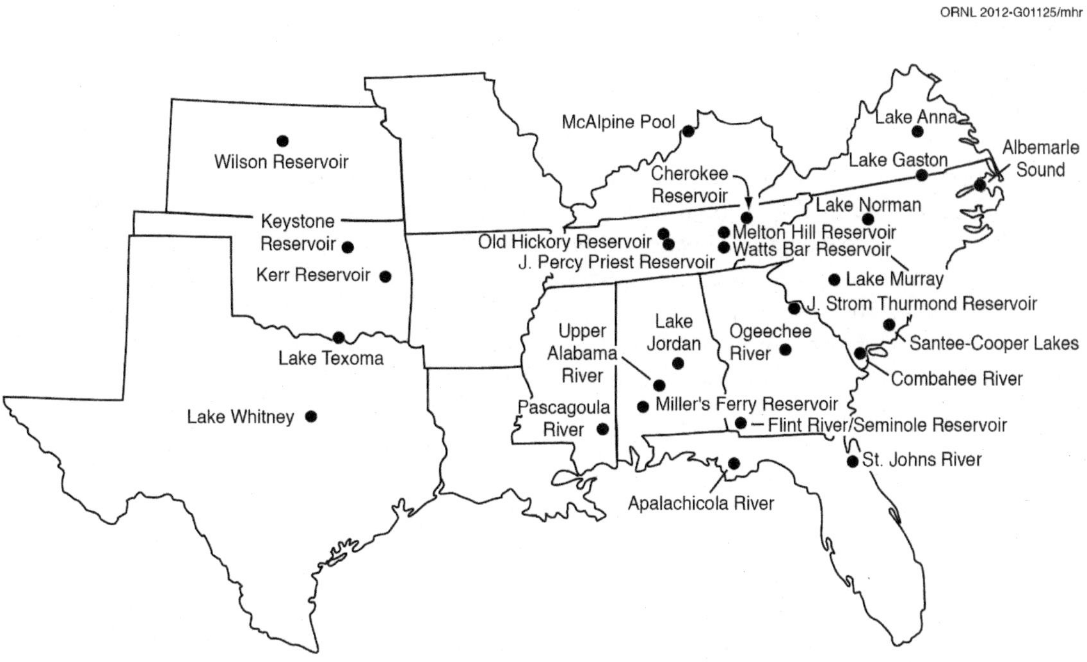

FIGURE 3. Locations primarily in the southeastern United States where field studies cited in this paper have been conducted on the summer distribution of striped bass in relation to water temperature and dissolved oxygen.

The consequences of occupying warm temperatures when cooler water was not available have, however, been more variable than the mortalities expected based on the original hypothesis. Although the largest striped bass (more than about 10 kg) incurred heavy mortalities, moderately sized adult striped bass could survive temperatures warmer than 25°C (generally up to 29–30°C) for a limited duration, if oxygen concentrations and food supplies were adequate. Some mortality seems to be largely the result of progressive bioenergetic deficit over a period of time under high temperatures and poor feeding, forced upon them by avoidance of low dissolved oxygen, with disease occurring secondarily. Reservoirs with chronic warm epilimnetic temperatures, low hypolimnetic oxygen concentrations, and no thermal refuges tend to be dominated by young ages and small sizes, with striped bass larger than 9–10 kg being rare or not found. The relevant studies are summarized below, organized by key components of the hypothesis that they address or additional concepts that emerged from them.

Systems with true refuges

A number of studies have documented seasonal selection of habitat by striped bass that is largely based on water temperature. Nearly all studies supported the general thermal habitat preferences identified in the original hypothesis, if fish had access to oxygenated water in the preferred temperature range. Thermal refuges (a limited area of the reservoir with suitable temperature and dissolved oxygen habitat) in summer were often observed but varied considerably among locations, ranging from discrete cool locations, such as tributaries, to middepth metalimnia. Studies where there were true thermal refuges showed striped bass behavior most consistent with the original hypothesis that was developed from locations with refuges.

In Alabama, tributaries and dam tailwaters were found to be thermal–oxygen refuges for adult striped bass when ambient reservoir temperatures approached 27°C (Moss 1985). In this study in the Alabama River, Lake Jordan, and Millers Ferry Reservoir, six striped bass (2.7–5.9 kg) tagged with transmitters sought refuges in cool (near 26°C) tributaries, although they would move for short periods (typically a few hours, occasionally longer) through reservoir temperatures of near 30°C (Moss 1985). Notably, it was the smallest fish that spent the most time in warm water, and condition factors for fish in refuges were highest for the small sizes, suggesting to Moss that the larger fish may have been subject to greater stress. Further studies by Lamprecht and Shelton (1988) on 5–9-year-old fish (721–914 mm or about 5–9 kg based on weights of Moss's fish) in the upper Alabama River system confirmed that adult striped bass preferentially occupied the Thurlow Dam tailwater (Tallapoosa River; maximum temperature 24°C) rather than the adjacent lower Coosa River (maximum temperature 33°C). The fish moved to the tailwater in spring when the Tallapoosa River warmed to 16–17.5°C and departed when the river cooled to 19.5°C (1981) or 18.5°C (1982).

Striped bass in the St. John's River, Florida, congregated in summer in relatively cool thermal refuges in springs but were dispersed throughout the system in winter (McDaniel et al. 1993). During summer, surface water temperatures in the river proper often exceeded 29°C while temperatures in thermal refuge areas ranged from 22.2°C to 28.8°C. Fish of a wide range of sizes, from 0.2 to 5.4 kg, used the refuges. Summer fish were less robust (lower weight per unit length) than winter fish at all lengths, especially the larger fish, but no summer die-offs were reported.

Adult striped bass in the Combahee River, South Carolina, appeared to use the upper region of the river as a large thermal refuge in summer (Bjorgo et al. 2000). Fish equipped with transmitters wintered in the tidally influenced river at 9–18°C, then moved upriver in late April and May where temperatures were near 18°C. They remained in the upper river in summer when it was as much as 5°C cooler than the lower river (>27°C). Fish moved downstream again when lower river temperatures ranged from 19°C to 27°C. No discrete thermal refuges were identified in the upper river other than the cool river itself (Bjorgo et al. 2000). Six 3.8–7.6-kg adult striped bass tracked in the Ogeechee River, Georgia, in summer inhabited a known coolwater spring (temperatures not reported), although most of the 13 tracked fish remained in the lower river estuary that is typically above 25°C then (Hess et al. 1999).

A discrete seasonal cycle of spatial distribution was found for adult striped bass in J. Strom Thurmond Reservoir, South Carolina–Georgia (Young and Isely 2002). Temperature-sensitive radio transmitters showed that as the reservoir became thermally stratified in May, fish vacated warm tributaries. From June to October, all tagged striped bass were found within the historical river channel. By August, most were found in the upper section of the reservoir where optimal habitat was provided by cool, artificially oxygenated hypolimnetic discharges from Richard B. Russell Dam. In October, fish redistributed throughout the reservoir. Average monthly temperatures and dissolved oxygen concentrations in areas used by striped bass were 19.0–20.4°C and 4.9–6.4 mg/L during May–October. Striped bass avoided temperatures above 25.1°C and dissolved oxygen concentrations less than 2.3 mg/L. There were mortalities in mid- to late summer in the middle section of the reservoir where striped bass were trapped in unsuitable summer habitat of the middle and lower sections of the reservoir, although some mortalities were attributed to high angling pressure in the Richard B. Russell Dam tailwater (Young and Isely 2004).

In Lake Murray, South Carolina, 61 adult striped bass 597–914 mm in total length (likely about 2.5–9 kg) equipped with temperature-sensing transmitters moved in summer to either the upstream tailwater, a particular cool embayment in the lower reservoir, or to a thermal refuge in the Saluda River (Schaffler et al. 2002). In August, fish that remained in Lake Murray occupied temperatures of 16–18°C, when oxygen was 2.5–4.3 mg/L (mean 2.7); in September, fish were found most often in or just above the thermocline at temperatures of 20–24°C with oxygen concentrations of 2.5–5 mg/L. Several striped bass occupied temperatures of 24–26°C where oxygen was greater than 5 mg/L late in September. After the reservoir began to cool, the striped bass began to disperse.

In Kerr Reservoir, Oklahoma, adult striped bass in two size-groups (1.2–2.4 kg and 2.9–15 kg) exhibited a distinct summer "thermal refuge season" (Wilkerson and Fisher 1997). The transmitter-equipped fish in both sizes moved to the Illinois River tributary in summer where temperatures were cool then because of upstream hypolimnetic discharge (often below 20°C). Fish moved to the river refuge when reservoir temperatures exceeded 22°C and left when the reservoir had cooled to 14–22°C.

Lakes Marion and Moultrie, South Carolina (Santee–Cooper system) apparently have two populations of striped bass, one that relies on a tributary thermal refuge and one that remains in the reservoirs in summer. Bettinger (2008) showed that a tailwater upstream of the Santee–Cooper lakes (Saluda River downstream of Lake Murray, South Carolina) serves as thermal refuge for larger fish during the summer, though some percentage of fish choose to oversummer in the downstream lakes, Marion and Moultrie. Braschler et al. (1989) showed that when 2.5–8-kg striped bass that remained in the lakes had only a narrow range of high temperatures available, they would still select areas with cooler water, even though the temperature differences were small. These lakes were not stratified in summer, with surface temperatures approaching 30°C and a small (<3°C) surface-to-bottom gradient. Oxygen concentrations were nearly always above about 4 mg/L. Transmitter-equipped fish generally occupied deep water that averaged 2°C cooler than surface temperatures in the lower reservoirs, for the 2 months that fish were tracked. On 9 of 61 occasions, when water column temperatures averaged 29.0°C, fish were located in restricted areas of springs, sink holes, and creek beds. However, fish in these locations were only 0.3°C cooler than nearby bottom temperatures and 1°C cooler than average water column temperatures. Because the population appeared to be thriving, the authors concluded that striped bass adults were able to tolerate water temperatures in excess of 28°C if adequate dissolved oxygen was present.

Systems without true refuges

In other studies, there were no clear thermal refuges available to striped bass. These studies supported the original hypothesis of thermal preferences and avoidance of low dissolved oxygen in summer, but fish successfully inhabited waters warmer than 25°C when avoiding low dissolved oxygen and no oxygenated water at temperatures in the preferred range were available (Figure 4).

A series of studies on Lake Texoma, Oklahoma–Texas, helped define the seasonal habitat selection of striped bass in another large southeastern reservoir with no cool tributaries for refuge. The general summer pattern for transmitter-equipped adults (mean weight 6.9 kg; Summers 1982), echo-located "subadults and adults" (Matthews et al. 1985) and gill-netted striped bass 1.36 kg to above 2.27 kg (Matthews et al. 1989) was for the larger fish to congregate in middepth near the dam and to be progressively squeezed vertically into the most suitable available habitat between hypoxia in the hypolimnion and high temperatures in the epilimnion. In July and August, most of the larger fish occupied zones just above where oxygen was depleted and where temperatures were near 28°C. Surface temperatures at that time were 28–30°C. Despite exposure to temperatures higher than 25°C, there were no apparent mortalities of striped bass.

Other reservoir studies confirmed a general pattern of summer movement toward deep waters near the dam as headwaters warmed and no cool tributaries were available, even when water near the dam was warmer than 25°C. For example, Hampton et al. (1988) tracked ultrasonic-tagged adults less than 4.5 kg in well-mixed Wilson Reservoir, Kansas, which had a thriving population of stocked striped bass. Fish in summer generally occupied middepths near the dam, with little movement. Oxygen was always high at all depths. Temperatures were

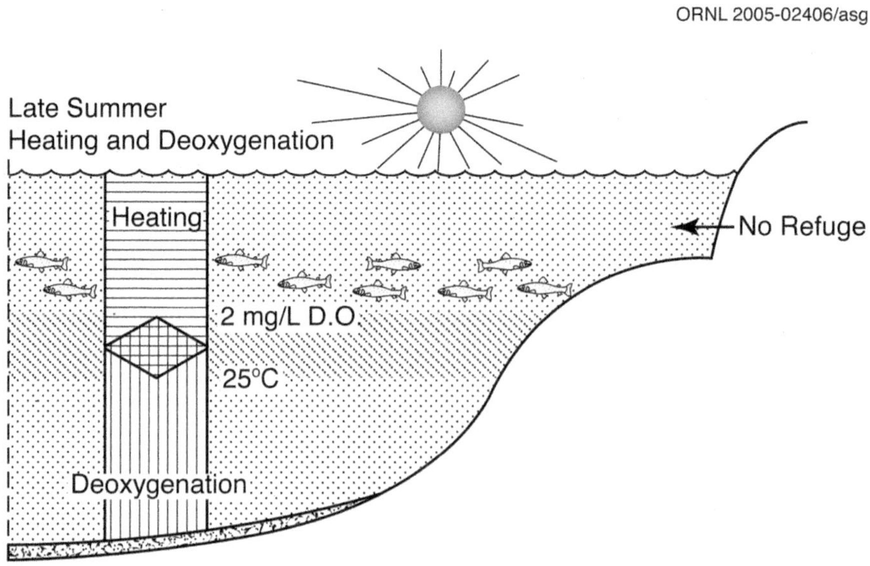

FIGURE 4. Summer temperature–oxygen squeeze for large striped bass when no thermal refuges are available in tributaries or cool springs and fish are restricted to the metalimnion in marginally suitable water quality habitat.

above 25°C from mid-July to mid-August and reached 27.1°C, yet no mortalities were noted. If tagged fish were representative of the sizes in the reservoir, the population (stocked since 1965) was composed of mostly small fish. In another study, adult striped bass (3.2–8.6 kg) were tracked in Lake Whitney, Texas, by Farquhar and Gutreuter (1989). Summer mortalities had occurred but not in the year of this study. Summer distribution was limited to near the dam. Fish were generally in the coolest water available in the vertical profile in summer (27–29°C) that contained dissolved oxygen greater than 4 mg/L. The average temperature of water occupied by fish with implanted temperature-sensing tags in July 1983 was 27.0°C. In early September, the cool water was depleted by water withdrawals and the reservoir became nearly isothermal at 28–30°C, which lasted until cooling began in late September.

Grimes (1987, 1993) assessed seasonal movements and summer tolerances in relation to a thermal discharge from a nuclear power station on Lake Anna, Virginia, which introduced horizontal thermal gradients. Thirty adult striped bass (about 2–5 kg) with temperature-sensing transmitters were found in the warm waters of three diked heat treatment coves in winter, and most moved to the deepest parts of the main reservoir in summer (the coves were above 28°C for 3 weeks in summer, and there were some mortalities there). In the main reservoir, they were found for 2 months in a 1-m-thick stratum of the metalimnion with temperatures of 27–29°C and dissolved oxygen between 2 and 7 mg/L, with no apparent mortalities, remaining there except for periods of cool summer rain when they moved temporarily to cooler tributaries. Coefficients of condition declined in summer to below 1, however, compared to being uniformly above 1 in other months. Summer ovary weights, condition factors, and body lengths and weights were significantly lower in heated Lake Anna than in nearby Smith Mountain Lake, and the composition (but not quantity) of egg vitellolipids and vitelloproteins differed.

In early studies of Lake Norman, North Carolina, Lewis et al. (1979) and Lewis (1985) reported adult striped bass emaciated in late summer and fall and dead and stressed individuals with *Epistylis* infections. Gill netting through the reservoir found striped bass (192–710 mm; probably about 0.5–4 kg) in warmer surface waters from March through May at temperatures of 17.0–24.5°C (Lewis 1985). In June through September, the fish were collected from the metalimnion (temperature 22–26°C; dissolved oxygen [DO] 2–7 mg/L) followed by dispersal at overturn in October. Selected temperatures in July (20–24°C; 4–5 mg/L DO) and August (24–26°C; 3–6 mg/L DO) indicated progressive movement to higher temperatures. There was different habitat selection according to fish size, as summarized below. More recent studies of Lake Norman are presented in Rice et al. (2013).

Adult striped bass in Lake Gaston, Virginia–North Carolina did not show a distinct seasonal movement pattern, in contrast to most other studies (Jackson and Hightower 2001). They occupied habitats throughout the reservoir in summer, likely experiencing conditions of 27–28°C and oxygen near 2 mg/L for less than a month. No mortalities were seen, but striped bass were in poorer condition than many populations in similar reservoirs.

Habitat selection according to size

Several studies further supported differences in habitat selection by size noted in our studies, beyond the differences between first-year and older fish emphasized in the original hypothesis. Size- or age-specific habitat limitations can became a factor in evaluating water bodies for their suitability for striped bass as a stocked fishery, in particular explaining why some waters produced trophy sizes while others produced mostly smaller striped bass.

In Lake Texoma, Oklahoma–Texas, there was different habitat use by gill-netted striped bass in three size categories over a 6-year period that was attributed to differing thermal preferences (Matthews et al. 1989). Fish larger than 2.27 kg were never found in up-lake areas in

June–September when surface temperature exceeded 22°C. Medium-sized fish 1.36–2.27 kg were significantly less abundant there at times when temperatures exceeded 22°C than in other months. Small (but not age 0) fish less than 1.36 kg remained abundant in the warm up-lake areas all year.

Telemetry showed that four different size-groups of striped bass selected distinct thermal habitats in Old Hickory Reservoir, Tennessee (Poarch 1989). The smallest size-group (1.8–4.9 kg) was distributed through the lower reservoir in summer. The other three size-groups (5–14.5 kg) were attracted to cool tailwaters (<23°C) of two upstream dams. The largest size-group (11.4–14.5 kg) was consistently closest to the upstream dams in the coolest water. The three groups of progressively smaller fish sizes (8.2–11.3, 5–8.1, and 1.8–4.9 kg) were distributed at increasing distances downstream in increasingly warm water. All 45 tagged fish occupied average temperatures of 20.2–22.7°C, nearly always avoiding water above about 24°C (reservoir surface temperatures reached 32°C). In spring, movement toward tailwaters began when reservoir temperatures were near 20°C.

Lewis (1985) found differences in the distribution of gill-netted "small" (<400 mm TL) and "large" (>400 mm) striped bass in Lake Norman, North Carolina. Small fish were found in surface waters (0–5 m) in summer, whereas large fish were all collected below 5 m. Only 7 of 54 large striped bass occurred in water as warm as 27°C (just above the thermocline), whereas about half of the small fish were above 27°C. Despite stockings since 1966, only 9 of 268 striped bass collected weighed as much as 2–4 kg, and creel surveys identified 5.6% and 6.7% of catches larger than 4 kg in 1978 and 1982. The author concluded that habitat to support all large striped bass during summer and fall may not have been available at all times.

Similar responses by Gulf-strain striped bass

Striped bass from tributaries of the Gulf of Mexico appear to show influences of temperature on seasonal distribution similar to those of other stocks. Gulf stocks are restricted to riverine habitats (Long et al. 2013, this volume). There had been suggestions that the Gulf Coast-strain of striped bass might be more tolerant of high temperatures because of its long (circa 10,000 years) isolation in waters generally warmer than the East Coast stock, which was the source for reservoir stocking.

Striped bass 150–900 mm standard length in Florida's Apalachicola River were riverine and occupied coolwater refuges in summer (21°C) while the surface water temperature rose to 31°C (Wooley and Crateau 1983). Fish in these summer refuges exhibited a loss of condition, with Gulf-strain fish showing somewhat less weight loss (mean 9.4%, SD 7.0) than Atlantic-race fish (mean 15%, SD 5.1). This weight loss did not occur until fish were more than 4.5–6.7 kg in both races. In the Flint River–Lake Seminole portion of the Apalachicola River system, the fish (3.2–30 kg, average 14.1) ranged widely through the system in fall, winter, and spring (Van Den Avyle and Evans 1990). During summer, however, they occupied cool refuges almost exclusively, entering when surrounding water temperatures averaged 24.3°C (23–25°C) and remained there until October or November when ambient temperatures declined to an average of 20.1°C (17–23°C). Temperatures selected by the fish averaged 21.6°C during two summers, with an avoidance temperature of about 23°C. Weeks and Van Den Avyle (1998) characterized 22 thermal refuges in limestone springs in the Flint River and developed striped bass abundance indices for 8 of them. In the Pascagoula River, Mississippi, a river with a small and declining striped bass population, only one small summer thermal refuge (20.3–24.6°C during June–September) was occupied by tagged striped bass (Jackson et al. 2002). The similarities in thermal responses between fish from tributaries of the Gulf of Mexico and those reservoir fish of known East Coast origin are not surprising considering the mixed genetic nature of fish in the Gulf watersheds as a result of widespread stocking of East Coast fish (Long et al. 2013).

Similar responses in estuaries

The coastal stocks of striped bass from which introduced reservoir stocks were derived appear to respond similarly to those in reservoirs and rivers. Haeseker et al. (1996) studied adult striped bass in Albemarle Sound, North Carolina, in the summers of 1993–1994. Thirty-six of 78 transmitter-equipped fish (>500 mm total length, 2–3 kg) successfully roamed the Albemarle Sound in summer at temperatures above 25°C between mid-June and early September and near 29°C in July. There was high mortality of tagged fish under these conditions, however. Oxygen levels were always adequate, and no clear thermal refuges were identified except a transient one in early summer at the mouth of the Roanoke River. Fish sought the deepest (and probably coolest) depths during the warm periods. In collections of other striped bass by these authors, those 325 mm total length or longer were in relatively poor condition, with low weight in summer, and exhibited extensive red dermal lesions and external parasites. Condition worsened through the summer, but few deaths were observed. Fish smaller than 325 mm displayed better external condition during the whole period.

Pronounced differences in seasonal movements by different-sized striped bass have been documented for Atlantic coastal stocks (reviewed by Coutant 1985). Striped bass larger than 2.7 kg migrate long distances along the Atlantic coast (Chapoton and Sykes 1961; Clark 1968), whereas smaller fish usually remain near their natal rivers (Fay et al. 1983). There is a strong similarity in geographical separation of sizes in reservoirs to that seen in coastal waters. Coutant and Benson (1990) provided evidence that the seasonal movements of different size-classes of striped bass in and out of Chesapeake Bay could be related to differing responses to warming of surface waters and prevalent summer hypoxia of deeper waters in the bay similar to inland reservoirs while coastal waters are cooler in summer. Such size-dependent movement patterns are often associated with seasonal variations in temperature and food resources (Quinn and Brodeur 1991), although separation of the two variables is difficult.

Fundamental thermal niche of adult striped bass

The fundamental thermal niche (the temperature selected when all other environmental factors are favorable; Magnuson et al. 1979) of landlocked adult striped bass was estimated by Bettoli (2005). He followed 15 adult striped bass (530–1,350 mm, mean 862 mm; 3.1–13.3 kg, mean ~8.2 kg) equipped with temperature-sensing radio tags in Melton Hill Reservoir, a 92-km-long impoundment of the Clinch River, Tennessee, where a cold, hypolimnetic discharge at upstream Norris Dam and a power station's heated discharge near the reservoir's midpoint provided a wide range of temperatures year-round with high dissolved oxygen concentrations throughout. The mean temperature occupied by the fish varied seasonally and was highest in summer (17.5°C), lowest in winter (13.0°C), and intermediate in spring and fall. There was not a significant difference in selected temperatures in this size range over the year-long survey period. A fundamental thermal niche of 17.5°C (average) or 19.0°C (mode) is most relevant to the summer habitat limitations seen across water bodies with landlocked striped bass. This study suggests that temperatures occupied by adult striped bass in most field studies in summer (the realized thermal niche; e.g., about 18–24°C in most cases but occasionally higher) are somewhat above the fundamental thermal niche. This reservoir has yielded some of the largest trophy striped bass in the region, often exceeding 25 kg, likely as a result of having near optimal thermal conditions year-round with abundant dissolved oxygen and plentiful forage fish.

Exposure duration and bioenergetics

Recurring observations of striped bass in water warmer than preferred included two other variables: the duration of exposure to warm temperatures and level of nutrition (food availability) during the warm exposure. This strongly suggests that survival is a function of time-dependent, bioenergetic balance (Thompson and Rice 2013). This was particularly apparent in studies of Keystone Reservoir. Zale et al. (1990) monitored habitat selection by adult striped bass (>500 mm total length) using gill nets in 10,600-ha Keystone Reservoir, Oklahoma, where summer mortalities were common. An earlier study there (Combs and Pelz 1982) had identified movement of tagged adult striped bass to a midwater zone near the dam in summer. During summer of the Zale et al. (1990) study, striped bass were restricted to a thin layer (1–2 m) in the stratified water column consisting of the coolest water available where oxygen exceeded 2 mg/L, typical of several other reservoir studies. Based on a decline in angler catch the authors concluded that the fish stopped feeding when minimum oxygenated temperatures reached 27°C (although continued feeding has been documented at similar temperatures in other reservoirs; Thompson et al. 2010). Lochmiller et al. (1989) concluded that these fish suffered high temperature stress and had nutritional problems. No mortality occurred in 1986 when temperatures in the oxygenated layer reached 28°C and were above 27°C for 1 month. Mortality did occur in 1987 when temperatures peaked at 29.3°C and remained above 27°C for 1 month and in 1988 when temperatures peaked at 28.3°C and remained above 27°C for about 7 weeks. Mortalities continued after suitable water quality conditions returned. The authors suggested that adult striped bass can tolerate exposure to 27–28°C for about 1 month but die (probably of malnutrition) when exposed to higher temperatures for a similar period or when exposure to 28°C is prolonged. Although the fish studied by Zale et al. (1990) were adults, they were small in relation to sizes attained in cooler waters. This study is notable for explicitly considering the duration of summer exposure to high temperatures and hypothesizing a temperature-dependent nutritional point of no return that signals mortality, either immediate or delayed.

The Keystone Reservoir analysis suggests that the high temperatures apparently tolerated elsewhere may have been too short a duration or food intake sufficiently high that mortalities did not occur, even when conditions were bioenergetically detrimental. This is a plausible explanation for survival in Lake Whitney, Texas (Farquhar and Gutreuter 1989), although the duration was similar to that seen in Keystone by Zale et al. (1990). A negative bioenergetic balance is also the likely cause of the poor condition factor of striped bass seen in Lake Gaston by Jackson and Hightower (2001) when striped bass spent the summer at temperatures reaching 27–28°C but did not die. Neither Farquhar and Gutreuter (1989) nor Jackson and Hightower (2001) critically addressed why their populations did not show mortality under conditions similar to those of Zale et al. (1990).

Limited evidence suggests that the ability of striped bass to tolerate bioenergetic thermal stress decreases with increasing size. Water temperatures above 27°C were recorded for 17 adult striped bass 4–9.1 kg fitted with transmitters and tracked in the Ohio River (McAlpine Pool) for 1 year by Henley (1993). Temperatures were above 27°C in the nonstratified pool from 11 July until 15 August. Four of the largest tagged fish (5.8–9.1 kg) died of apparent thermal stress following this period, whereas other fish were either still active or otherwise accounted for (caught, emigrated), suggestive of a size-related physiological stress. Also, bioenergetic (nutritional) problems at higher temperatures were seen in striped bass functionally landlocked in the St. Johns River, Florida, which overall were in poorer condition in summer refuges than in winter (McDaniel et al. 1993). In summer, at refuge temperatures of 22.2–25.8°C, there were greater weight reductions at larger sizes: from the largest (>1.7–8.9 kg) striped bass being reduced by an average of 22.6%, medium (0.44–1.4 kg) striped bass reduced by 17.4%, and the smallest (0.14–0.33 kg) reduced by 4.9%. This situation was attributed to thermal stress because food was available, in accord with observations by Coutant (1985) and others.

If bioenergetic balance at temperatures higher than preferred is a main determinant of survival or death, rather than thermal tolerance per se, then more attention should be given to prey availability and feeding rates. Thompson and Rice (2013) have made important contributions to this issue. Higher survival and growth of smaller adult striped bass in warm water may indicate changing physiological functions with size or age that are related to bioenergetic balance, a subject that is ripe for further study.

Predictions Using Temperature–Oxygen Habitat Suitability

One value of a tested and generally confirmed hypothesis is whether it can be applied usefully to inform different situations. This one has, for problems beyond traditional fisheries management in reservoirs. Under the assumption that the thermal preferences of striped bass seen in freshwater reservoirs (about <25°C and >2 mg/L DO) reflected basic attributes of the species, studies in the late 1980s used the thermal–oxygen habitat criteria for several applications. A colleague and I carefully examined the striped bass in Chesapeake Bay in relation to the bay's seasonal water quality (Coutant and Benson 1990). Chesapeake Bay has traditionally been a focus of the East Coast striped bass population, with much of the stock reproducing in its tributaries. The stock experienced a major decline in the 1970s, and we suspected that declining water quality documented in the bay (especially high summer water temperatures and accelerating deepwater deoxygenation) might be a contributing factor. We found clear parallels between summer heating and deoxygenation in Chesapeake Bay and those processes in freshwater reservoirs, although the Chesapeake Bay is a much larger system. Seasonal movements of striped bass of various sizes in the Chesapeake Bay (as noted in historical data) matched what one would predict based on the temperature and oxygen requirements we had deduced from numerous reservoir studies. For example, young (up to 3 years old) striped bass remain in the bay year-round, whereas older, larger fish migrate into coastal waters in summer, which are cooler. In retrospect, high fishing mortality was probably the main cause of population decline at that time. However, with populations since 2000 at a high level as a result of 15–20 years of restricted harvest, the symptoms of habitat restriction typical of some reservoirs are now appearing in the bay (emaciated fish, body sores, summer mortalities, and reduced growth rates). Other investigators have followed our lead and used the temperature–oxygen niche concept for understanding striped bass population dynamics and migration in the Atlantic coast estuarine systems (Price et al. 1985; Ng et al. 2007; Costantini et al. 2008; Secor and Kraus 2012).

The habitat criteria have also been used in industrial impact assessment (Grimes 1987). He evaluated historical temperature and dissolved oxygen data on Lake Anna, Virginia before and after initiation of operation of the North Anna Nuclear Power Plant to quantify the loss of suitable habitat. There was considerable loss of suitable habitat in years when the plant operated and discharged heated water.

Two studies evaluated possible effects on striped bass habitat due to temperature rise from climate change, as estimated by atmospheric general circulation models. A reservoir water quality model for Douglas Reservoir, Tennessee was used to compare suitable habitat in 1974 (a normal year as defined by actual atmospheric and hydrological data), with water quality conditions predicted to result from double the carbon dioxide content of the air (Chang et al. 1992). A less quantitative analysis, using more simplifying assumptions, was used to estimate changed habitat in Watts Bar and Cherokee reservoirs and Chesapeake Bay (Coutant 1990). In both studies, warmer future conditions were predicted to cause overall declines in available summer habitat for adult striped bass. The Chang et al. (1992) study was generically valuable for showing that reservoir water quality models can be coupled with numerical habitat suitability criteria (temperature, dissolved oxygen) to quantitatively estimate changes in available suitable habitat (and, conversely, habitat limitations) for a particular water body. This approach has been further developed by Ruane et al. (2013) and Taylor et al. (2013).

Hybrid striped bass

There is a general perception that palmetto bass (male white bass × female striped bass) have higher temperature preferences and tolerances than pure striped bass, but the evidence is slim. On purely theoretical grounds, it would be reasonable to expect hybrids between the warmwater white bass and the apparently coolwater striped bass (at least as adults) to be intermediate. Stocking programs for enhanced fisheries have tended to shift to use of hybrids in waters where striped bass have encountered, and suffered from, limited summer habitat because of high temperature and low dissolved oxygen (Axon and Whitehurst 1985; Windham 1986). Hybrids do not seem to show the loss of condition or mortalities in summer that afflict pure striped bass, but information is anecdotal.

The clearest evidence for more thermal affinity by hybrids comes from juveniles tested in the laboratory. Effects of temperature on juvenile growth rates of both striped bass and hybrid striped bass have been studied and illustrate the differences in temperature responses between them. With ample food, maximum growth of juvenile striped bass occurred near 24°C and zero growth occurred at 33.5°C (Cox and Coutant 1981), whereas maximum growth of hybrid bass occurred at 27.9°C and growth ceased above 34.5°C (Woiwode and Adelman 1991). Hybrids continued to feed at a lower temperature (6.5°C) than striped bass (10°C) in these studies, further supporting the wider general temperature tolerance (eurythermy) of the hybrids. The comparative ecology of juveniles was studied in Claytor Lake, Virginia (Rash and Ney 2013, this volume), where they preferred similar habitats.

Differences in thermal preferences by adults in the field have not been as thoroughly addressed and, when studied, have been confounded by avoidance of hypoxia. As cooler water is often hypoxic in summer, the fish have had little choice. Like striped bass that have been shown to exhibit the behavior of occupying the coolest water available with at least 2 mg/L DO, the hybrids may have preferred somewhat cooler temperatures than where they were found. Windham (1986) used telemetry to follow 31 hybrids of about 0.9–1.9 kg in Clarks Hill (J. Strom Thurmond) Reservoir, Savannah River. These fish selected temperatures of 20.5–29.3°C with a mean of 26.3°C in July–September. There was strong thermal stratification and low oxygen in deeper waters. Tagged fish tended to reside just above the thermocline in summer, but there were no deaths reported. Twelve hybrid striped bass greater than 445 mm (1.19–2.57 kg) were tracked from April to September by Douglas and Jahn (1987) in Spring Lake, Illinois, to determine temperature and oxygen preferences. They behaved similarly to striped bass by concentrating in the deeper section of the lake during summer stratification. During this time, they were confined to a narrow layer of water just above the zone of oxygen depletion (<2 mg/L) where temperatures were 2–3°C cooler than at the surface but as high as 27°C. They dispersed from this area when reservoir temperatures declined to 25°C. Hybrids thrived in both of these situations. Recently, Hoffman et al. (2013, this volume) tracked hybrid striped bass in Monroe Lake, Indiana, finding a seasonal pattern of dispersal in spring, concentration in midwater near the dam in summer, and redispersal in the fall when temperatures decreased from 25°C to 20°C.

Evidence that preference for higher temperatures persists in hybrid adults has been shown in Claytor Lake, Virginia, where there was an opportunity to compare habitat selection and general comparative ecology of striped bass and their hybrids in the same system (Kilpatrick and Ney 2013). Spring data from this study showed higher preferred temperatures by hybrids in the short time interval before selection was constrained by hypoxia.

The apparent advantages of more thermally tolerant adult hybrids that have nearly the same angler appeal as pure striped bass need to be more thoroughly based on research data. This is important considering the genetic risks and ethical considerations of mixing hybrids with native, pure stocks of striped bass (Harrell 2013, this volume).

Discussion

The basic tenets of the hypothesis that adult striped bass select cool temperatures below about 25°C, if available, in contrast to affinities for warmer water by juveniles, and avoid dissolved oxygen levels near 2 mg/L and below (Coutant 1985) have been amply supported by more than 20 years of field studies. These studies were carried out primarily in Southeast U.S. reservoirs, rivers, and estuaries following the decade during which I developed the hypothesis. Summer habitat limitations for adult striped bass due to temperature and oxygen are real and amply documented. The strong interaction with dissolved oxygen in thermally stratified reservoirs creates a three-dimensional temperature–oxygen squeeze (perhaps better stated as a thermal niche–dissolved oxygen squeeze) in summer that restricts adults to the most suitable habitat available (realized thermal niche), which is often warmer than they would otherwise select. In general, larger adult striped bass select cooler habitat than smaller adults. If there is oxygenated water in springs or tributaries that is cooler than about 24–25°C when the main water body exceeds this general range, adult striped bass will congregate in these thermal refuges. Both occupancy of warmer water and small refuges can lead to mortality, either directly or through malnutrition or disease.

Nonetheless, the conceptual details have evolved as more studies of habitat occupancy have been carried out. There are several notable modifications. Some striped bass adults, particularly small adults less than about 2 kg, successfully occupy higher temperatures than originally thought of as suitable when forced there by avoidance of hypoxia. Where this occurs, large trophy-sized striped bass are rare or not present. Vertical and horizontal squeezes between high surface temperatures above about 25°C and deepwater low dissolved oxygen below about 2–3 mg/L have taken many forms. Thermal refuges, such as deeper water near dams, groundwater springs, cool tributaries, or cool dam tailwaters of upstream dams, are not always available throughout a summer season. The temperatures occupied differed somewhat among the various studies, reflecting the local physical geography, food availability, seasonal progression of temperature rise, and oxygen depletion. Temperatures at which most adult, landlocked striped bass are found in summer appear to be above their preferred temperature (fundamental thermal niche in summer of about 17.5–19.0°C; Bettoli 2005) and reflect the best combination of lowest temperature and sufficient oxygen that they can find. Available forage appears to influence how long striped bass adults can maintain themselves without significant weight loss or mortalities.

Tolerance by some striped bass of temperatures higher than 25°C in summer was common, at least for some periods of time. For example, the studies cited earlier by Braschler et al. (1989) in the Santee–Cooper lakes showed striped bass occupying temperatures in summer exceeding 28°C with no apparent mortalities. Similarly, Matthews et al. (1985) found striped bass in Lake Texoma in surface waters about 28°C in summer (reaching 30°C at times) without mortalities. In Lake Whitney, Texas, Farquhar and Gutreuter (1989) located striped bass up to 8.9 kg in the metalimnion near the dam at 28–30°C without mortalities seen the year of their study, although there had been mortalities there in prior years.

Tolerance of warm water appears strongly related to striped bass size, age, or both. The studies by Matthews et al. (1989) and Poarch (1989), discussed above, presented firm evidence. Reservoirs with thriving striped bass in warm surface waters in summer seem to be dominated by 2–4-kg fish (e.g., Thompson and Rice 2013), although this has not been thoroughly documented among reservoirs. Waters with trophy-sized fish (>~10 kg) all seem to be ones with abundant cool temperatures in summer (Cheek et al. 1985; Bettoli 2005). Most information on size effects suggests a continuum of change toward more sensitivity to warm water by larger fish, as indicated in our early studies (Figure 2). Drawing conclusions only from observations in waters dominated by small adult striped bass appears to be one basis for disagreement with my (1985) hypothesized upper tolerance limit of about 25°C. This apparent size effect on tolerance for warm water above preferred temperatures warrants further study.

Preference and tolerance have been shown to differ. Although preference for cool temperatures by adult striped bass is clear from seasonal field studies in well-oxygenated waters, many of these studies also showed that striped bass, especially smaller adults, could *tolerate* temperatures warmer than preferred if forced to do so by hypolimnetic hypoxia and a lack of oxygenated thermal refuges. This calls for partially modifying the original hypothesis of an upper tolerance limit of 25°C. This temperature could more properly be called an upper *preference* limit. There is ample evidence, however, that the original hypothesized upper tolerance limit applies to fish greater than ~6–7 kg, but not necessarily for smaller fish. Further evaluation of previous studies for a size effect and focused studies comparing preference and tolerance responses by sizes would be valuable.

Duration of exposure to temperatures above those preferred seems key to survival of larger striped bass. A succession of studies have increased our knowledge of the condition under which adult striped bass tolerate high temperatures (25–29°C, well above preferred) and low dissolved oxygen levels for periods in summer without mortalities, especially when there is abundant food to counteract the apparent high metabolic rates in warm water. Thus, the *duration* of the temperature–oxygen habitat restrictions and the quantity of food available are likely important covariates with temperature and dissolved oxygen levels for striped bass health and survival over summer. Several studies suggested durations of about a month were tolerable at temperatures as high as 29°C (e.g., Zale et al. 1990). Unfortunately, studies showing survival at high temperatures have not critically addressed why their populations do not experience mortalities often seen in similar water bodies (but see Thompson and Rice 2013, which does address this question). Problems with mortalities were an important motivation for the initial studies leading to the temperature–oxygen squeeze hypothesis, but it is equally important to recognize the limitations of the original hypothesis and the evolution in thinking when seeking to understand the effects of poor summer habitat on striped bass.

The modest modification of the original hypothesis largely concerns successful occupancy of water warmer than about 25°C by predominantly smaller adults, when there are no true thermal refuges available, and the effect of duration of such marginal habitat on survival. Reservoirs with marginal to unsuitable temperature–oxygen habitat for striped bass above about 10 kg, which may be lethal, can still have productive fisheries for the smaller sizes of fish (predominantly 2–4 kg with maximum sizes of 6–9 kg), but do not provide the oxygenated, coolwater summer conditions to allow fish to attain trophy sizes.

Seasonal movements of striped bass stocked in lakes and reservoirs appear to be a response to environmental temperature and dissolved oxygen that probably evolved in the ancestral stocks in Chesapeake Bay and East Coast estuaries. As discussed by Coutant and Benson (1990), there is a predominant tendency of younger fish to remain in the bay year-round while older and larger fish migrate out of the bay in summer to cooler, coastal waters. The same pattern is observed for striped bass in the Hudson River. In addition to the physiological energetic benefits this migration would afford, it may also offer survival benefits for the species by separating these large, voracious predators from their vulnerable progeny, and it may also increase access to food resources. Temperature may be the key environmental cue for this otherwise beneficial behavior, and it has taken the smaller-scale bioassay of freshwater reservoirs to clearly demonstrate it.

While the construct I used in this paper was to test the hypothesis, few of the posthypothesis reports were specifically designed to test the details of the hypothesis rigorously. Different reservoirs and other habitats occupied by striped bass were studied by means similar to ours (temperature telemetry) or by other means (e.g., gill netting). Most study reports included discussion of their results in comparison to ours and discussed the original hypothesis. Each water body studied by others had characteristics that differed from our study sites (Cherokee and Watts Bar reservoirs), but general replication in the patterns observed seems adequate to verify most aspects of the original hypothesis. At a more rigorous level, the results were consistent for the expression of preferred temperatures, avoidance of low dissolved oxygen concentrations, general behavior and distribution, size effects, and evidence of bioenergetic stress (such as loss of condition) and sometimes mortality. Some parts of the initial hypothesis, particularly the survival component (compounded by diseases, e.g., red sores, fungal infections), remain incompletely addressed (but see Rice et al. 2013; Thompson and Rice 2013).

We have an incomplete knowledge of the thermal requirements of striped bass beyond juveniles. A general decreasing fundamental thermal niche with increasing size or age is clearly evident in the studies presented, however. We do not know the "incipient lethal temperature" of adult striped bass in the context of thermal requirements of fishes developed by F. E. J. Fry and colleagues (e.g., Fry 1967) because such bioassays are nearly impossible with such large fish.

We are fortunate, however, in being able to estimate the fundamental thermal niche (Magnuson et al. 1979) of adult striped bass in summer. A fundamental niche of 17.5°C (mean) or 19°C (mode) could be estimated because of the peculiar thermal dynamics of Melton Hill Reservoir, Tennessee (Bettoli 2005). Within Bettoli's size range (3.1–13.3 kg) there was no apparent size-related change in thermal niche: the mean temperatures selected by the five smallest fish he tracked (3.1–5.1 kg) were just as cool (actually a degree cooler on average) as the mean temperatures selected by the five largest fish he tracked (6.5–13.3 kg). In light of other studies of size differences in habitat selection, this might warrant some re-evaluation. In our telemetry studies, temperatures selected by 430–680-mm (0.9–3.5 kg) striped bass in well-oxygenated, thermally stratified quarry lakes were near 22°C (Coutant and Carroll 1980), which may indicate the fundamental thermal niche for fish of this size, which is higher than for the generally larger adults studied by Bettoli. Our study could be considered as much of a fundamental thermal niche study as Bettoli's—the fish had unconstrained access to a wide range of oxygenated temperatures. The smaller fish tracked by Bettoli overlapped a small amount with the size range of those tracked in the Coutant and Carroll study, but most of the fish in the Coutant and Carroll study were less than 2 kg. Thus, that study provides evidence of a fundamental thermal niche near 22°C for fish from about 1 to 2.5 kg. Such a large shift in fundamental thermal niche between 1- and 2.5-kg fish and larger fish seems surprising, however, and warrants further study. These two studies provide good support for a decreasing thermal niche between the upper and lower end of the size spectrum, but they do not fully support a gradient with size in between the extremes, as is typically seen in the realized niches of different sizes of fish (e.g., Poarch 1989).

There is a similarity of selected temperatures and optimum growth temperatures for juveniles (0.04–0.3 kg) in the laboratory near 24°C (Cox and Coutant 1981; Coutant et al. 1984) that suggests that this might be the fundamental thermal niche for this age-group. There was a demonstrated physiological disadvantage for growth in juveniles at temperatures much above preferred, which is probably true also in larger fish. The angler reports of "skinny fish" and demonstrations of low condition factor often reported for adult striped bass in warm reservoirs in summer attest to a bioenergetic deficit. The dynamic, bioenergetic balance between temperature-dependent metabolism, feeding rates, and food availability that was only hinted at in our original Cherokee Reservoir studies by isolation of striped bass from prey while in thermal refuges has been the focus of recent research (see Thompson and Rice 2013), but further work is warranted to address this issue both generally and for specific water bodies.

Reduction of successful reproduction by large striped bass from waters having severe temperature–oxygen habitat limitations (Coutant 1987a) would be important for natural population maintenance. It would seem to provide further evidence of a summer bioenergetic deficit. This consequence is negated by stocking in most inland systems but may be a contributor to chronically low natural populations in warm habitats at the southern extent of the species' range. For example, Long et al. (2013) describe the historically low population levels of striped bass in rivers tributary to the Gulf of Mexico prior to remedial stocking. These rivers are chronically warm in summer and striped bass are known to seek thermal refuges (e.g., Van Den Avyle and Evans 1990; Weeks and Van Den Avyle 1998). Whether confined to refuges in summer with low prey availability or exposed outside refuges to high river temperatures, the largest and usually most fecund fish could have experienced sufficient bioenergetic deficit to affect egg maturation and reduce successful reproduction as seen for our fish from Cherokee Reservoir. Smaller and less fecund adults, more able to survive the warm temperatures, may have had insufficient numbers of eggs to propagate a large population. A key question is how the maturity schedule of striped bass relates to their preference (and likely bioenergetic need) for cooler water. If sufficient egg production can only be achieved by access of large fish to thermal refuges in summer, then that habitat would be essential for the population. The available evidence is suggestive and worthy of additional study.

What we have seen for striped bass—thermal sensitivity of larger body sizes, thermal stress affecting reproduction, populations in warmer waters dominated by smaller sized fish, and declining natural populations in warm rivers—has been shown to have a physiological basis in metabolic interactions of temperature and dissolved oxygen (Pörtner and Knust 2007). Several fish species have been shown to have low overall abundance and high mortality rates of older specimens at elevated temperatures where oxygen limitation was also a factor. Aerobic scope (Fry 1967) is lost at lower temperatures in larger fish such that they do not cope beyond oxygen-dependent size limits set by temperature. The limitations are expressed over a duration that depends on the severity of the high temperature and oxygen supply. Matching thresholds in field and laboratory data reviewed by Pörtner and Knust (2007) highlight what they call "oxygen- and capacity-limited thermal tolerance" involving aerobic metabolism and oxygen supply capacity. Those with an interest in the physiological mechanisms could follow this rich and developing literature.

Whether strong thermal preferences and selection of cool thermal refuges in summer are assets or liabilities for inland striped bass (Coutant 1987b) depends on context. Attraction to large thermal refuges with abundant prey appears to be a real asset for development of a population with abundant striped bass larger than about 6 kg and consistent production of trophy (>10 kg) fish as seen in Watts Bar and Melton Hill reservoirs, Tennessee (Cheek et al. 1985; Bettoli 2005; respectively). In contrast, crowding into numerous small thermal refuges formed by springs and small, cold tributaries can be a liability due to malnutrition and disease as seen in our studies of Cherokee Reservoir, Tennessee (Waddle et al. 1980; Schaich and Coutant 1980). When the fish are crowded into a thin layer of oxygenated water in the metalimnion or upper hypolimnion, the gradually thinning layer in summer can be an asset if it persists or a liability if it disappears and causes mortalities (see Rice et al. 2013).

Although speculative, a restriction of young adults to temperatures higher than optimum in a thin middepth layer or warm epilimnion could logically reduce summer growth rates such that especially large adults do not develop. This would be a liability if availability of large, trophy-sized striped bass is the management objective, but less problematic if a population of small adults is acceptable. It would also be a liability for naturally reproducing populations where most egg production comes from the largest females (Long et al. 2013). Although also speculative, persistently warm temperatures in summer may have a selective effect on reproducing populations favoring sexual maturity at a small size leading to spawning populations of small individuals 2–3 kg or less, as seen in the Albemarle Sound–Roanoke River spawning population (Haeseker et al. 1996).

While the preference of larger striped bass for cool water in summer is well demonstrated, the effects of not finding it, or finding it in limited and transient spaces, is less fully recognized. Our work, and others, helps understand the consequences for large striped bass confined to cool water in limited spaces. The work of Thompson et al. (2010) and others (e.g., Rice et al. 2013) provides understanding of the limiting effects on maximum size (and potentially survival) for fish that cannot find cool water in summer, due to the net energetic constraints resulting from interactions of food availability and higher metabolic costs due to high temperatures. But further study and evaluation is certainly warranted.

There are, and have been, implications for fisheries management from the realization of water quality habitat preferences and limitations for striped bass of various sizes. The limitations are known to be most stringent for the largest adults, especially those larger than 10 kg, but also for moderate-size fish in the 4–10-kg category. The suitability of some lakes for striped bass has been evaluated using these habitat criteria along with fish tracking to (and survival in) thermal refuges (e.g., Grimes 1987; Baker and Jennings 2005). Stocking rates have been reduced where availability of thermal refuges, including middepth layers with suitable temperature and oxygen habitat, is limiting. In some cases, stocking has been reduced to very low levels to try to maintain a limited trophy fishery for fish that are able to find good habitat. Hybrids have been substituted for pure striped bass to provide a large recreational fish more tolerant of warm water. Forage species with temperature preferences more similar to striped bass (e.g., blueback herring *Alosa aestivalis* and alewife *A. pseudoharengus*) have been introduced to avoid summer predator–prey separation (but may also cause major problems in some systems; see Rice et al. 2013). Some refuge areas where striped bass aggregate have been closed to fishing during summer (e.g., Flint River and portions of Cherokee Reservoir). Catch and release from known refugia and from warm reservoir waters has been curtailed in summer due to heavy mortalities. For example, the Santee–Cooper lakes and most of the inflowing tributaries are currently closed to fishing from 1 June through 30 September, although the large tailwater thermal refuge (Saluda River) is currently open (J. Bulak, South Carolina Department of Natural Resources, personal communication). In some cases, fishery management has concentrated on promoting harvest of large numbers of relatively small fish because of the lack of conditions suitable for producing large fish (e.g., Lake Texoma). These fishery-management actions are difficult to document in open literature publications but nonetheless have been occurring, based on communications with fishery managers. Also, climate change has implications for management of large striped bass and striped bass populations and their fisheries because of the species' thermal–oxygen habitat limitations (Chang et al. 1992; Coutant 1990).

There is a developing capability for managing the habitat to alleviate the water quality constraints for striped bass as well as the usual tailoring of fish management to existing habitat conditions. Reservoir water quality models can be used constructively to identify limiting habitats as seasonal warming and oxygen depletion progress (Chang et al. 1992; Ruane et al. 2013; Taylor et al. 2013). These models can be used to identify the causes of habitat limitations, such as nutrient enrichment or reservoir circulation (including the location of water withdrawals). The habitat limitations can be lessened or eliminated with engineering approaches such as oxygenation (Mobley et al. 2013). Although expensive, such approaches can be feasible when combined with broader priorities for water quality improvement, such as increased oxygen concentrations in reservoir tailwaters (Ruane et al. 2013). Fishery and water-resource managers could explore more use of these capabilities.

A hypothesis that was highly controversial in the late 1970s and early 1980s has had its basic tenets well supported by abundant and mostly peer-reviewed scientific evidence, with incremental modifications in accord with new information. It has provided a useful conceptual framework for understanding population dynamics and movements of striped bass in different ecosystems. Nonetheless, there is much to be learned about the effects of temperature–oxygen habitat limitations on striped bass and hybrid striped bass populations, both natural and augmented by stocking. It is more difficult to track management approaches than scientific papers, but there is every indication that management practice has embraced the cool thermal niche and temperature–oxygen squeeze ideas as a fundamental basis for striped bass management, particularly in southeastern U.S. waters. Details of site geography, seasonal temperature and oxygen dynamics, and more recently prey availability are either recognized as important for managing striped bass in particular waters or there is movement in that direction. As other papers in this volume show, there has been continued effort to refine understanding of the habitat limitations for specific waters and fisheries, especially in relation to food availability, durations of exposure, and the dynamic temperature–oxygen conditions in stratified lakes and reservoirs. Using site-specific information to develop appropriate fisheries for trophy fish, abundant small striped bass, or hybrid striped bass, we can manage inland waters for satisfied striped bass and hybrid striped bass anglers.

Acknowledgments

In addition to the Striped Bass Committee of the American Fisheries Society's (AFS) Southern Division, which developed a symposium for the 2009 AFS Annual Meeting in Nashville, Tennessee, on which this book is based, several groups stimulated this retrospective, including the U.S. Army Corps of Engineers, the U.S. Fish and Wildlife Service, and the Gulf States Marine Fisheries Commission, for whom I prepared literature summaries over several years. I thank M. Van Den Avyle, J. Isely, and an anonymous reviewer for helpful comments on earlier drafts. Extensive discussions with my coeditors have been especially helpful. Although I am now retired, much of the research and review was conducted at Oak Ridge National Laboratory, which is managed by UT-Battelle, LLC, under contract DE-AC05–00OR22725 with the U.S. Department of Energy.

References

Axon, J. R., and D. K. Whitehurst. 1985. Striped bass management in lakes with emphasis on management problems. Transactions of the American Fisheries Society 114:8–11.

Baker, T. L., and C. A. Jennings. 2005. Striped bass survival in Lake Blackshear, Georgia during drought conditions: implications for restoration efforts in Gulf of Mexico drainages. Environmental Biology of Fishes 72:73–84.

Bettinger, J. 2008. Seasonal movements of adult striped bass in the Santee–Cooper drainage. South Carolina Department of Natural Resources, Study Completion Report F-63, Columbia.

Bettoli, P. W. 2005. The fundamental thermal niche of adult landlocked striped bass. Transactions of the American Fisheries Society 134:305–314.

Bjorgo, K. A., J. J. Isely, and C. S. Thompson. 2000. Seasonal movement and habitat use by striped bass in the Combahee River, South Carolina. Transactions of the American Fisheries Society 129:1281–1287.

Braschler, D. W., M. G. White, and J. W. Folz. 1989. Movements and habitat selection of striped bass in the Santee–Cooper Reservoirs. Proceedings of the Annual Conference Southeastern Association of Fish and Wildlife Agencies 42(1988):27–34.

Chang, L. H., S. F. Railsback, and R. T. Brown. 1992. Use of a reservoir water quality model to simulate global climate change effects on fish habitat. Climatic Change 20:277–296.

Chapoton, R. B., and J. E. Sykes. 1961. Atlantic coast migration of large striped bass as evidenced by fisheries and tagging. Transactions of the American Fisheries Society 90:13–20.

Cheek, T. E., M. J. Van Den Avyle, and C. C. Coutant. 1985. Influences of water quality on distribution of striped bass in a Tennessee River impoundment. Transactions of the American Fisheries Society 114:67–76.

Clark, J. R. 1968. Seasonal movements of striped bass contingents of Long Island Sound and the New York bight. Transactions of the American Fisheries Society 97:320–343.

Combs, D. L., and L. R. Pelz. 1982. Seasonal distribution of striped bass in Keystone Reservoir, Oklahoma. North American Journal of Fisheries Management 2:66–73.

Costantini, M., S. A. Ludsin, D. M. Mason, X. S. Zhang, W. C. Boicourt, and S. B. Brandt. 2008. Effect of hypoxia on habitat quality of striped bass (*Morone saxatilis*) in Chesapeake Bay. Canadian Journal of Fisheries and Aquatic Sciences 65:989–1002.

Coutant, C. C. 1978. A working hypothesis to explain mortalities of striped bass, *Morone saxatilis*, in Cherokee Reservoir. Oak Ridge National Laboratory, ORNL/TM-6534, Oak Ridge, Tennessee.

Coutant, C. C. 1980. Environmental quality for striped bass. Pages 179–187 *in* H. Clepper, editor. Marine recreational fisheries, volume 5. Sport Fishing Institute, Washington, D.C.

Coutant, C. C. 1985. Striped bass, temperature, and dissolved oxygen: a speculative hypothesis for environmental risk. Transactions of the American Fisheries Society 114:31–61.

Coutant, C. C. 1986. Thermal niches of striped bass. Scientific American 254(8):98–104.

Coutant, C. C. 1987a. Poor reproductive success of striped bass from a reservoir with reduced summer habitat. Transactions of the American Fisheries Society 116:154–160.

Coutant, C. C. 1987b. Thermal preference: when does an asset become a liability? Environmental Biology of Fishes 18:161–172.

Coutant, C. C. 1990. Temperature–oxygen habitat for freshwater and coastal striped bass in a changing climate. Transactions of the American Fisheries Society 119:240–253.

Coutant, C. C., and D. L. Benson. 1990. Summer habitat suitability for striped bass in Chesapeake Bay: reflections on a population decline. Transactions of the American Fisheries Society 119:757–778.

Coutant, C. C., and D. S. Carroll. 1980. Temperatures occupied by ten ultrasonic-tagged striped bass in freshwater lakes. Transactions of the American Fisheries Society 109:195–202.

Coutant, C. C., K. L. Zachmann, D. K. Cox, and B. L. Pearman. 1984. Temperature selection by juvenile striped bass in laboratory and field. Transactions of the American Fisheries Society 113:666–671.

Cox, D. K., and C. C. Coutant. 1981. Growth dynamics of juvenile striped bass as functions of temperature and ration. Transactions of the American Fisheries Society 226–238.

Douglas, D. R., and L. A. Jahn. 1987. Radiotracking hybrid striped bass in Spring Lake, Illinois, to determine temperature and oxygen preferences. North American Journal of Fisheries Management 7:531–534.

Fay, C. W., R. J. Neves, and G. B. Pardue. 1983. Species profiles: life histories and environmental requirements of coastal fishes and invertebrates (mid-Atlantic)—striped bass. U.S. Fish and Wildlife Service, FWS/OBS-82/11.8, Washington, D.C.

Farquhar, B. W., and S. Gutreuter. 1989. Distribution and migration of adult striped bass in Lake Whitney, Texas. Transactions of the American Fisheries Society 118:523–532.

Fry, F. E. J. 1967. Responses of vertebrate poikilotherms to temperature. Pages 375–409 *in* A. H. Rose, editor. Thermobiology. Academic Press, London.

Grimes, D. V. 1987. Effects of power generation on adult striped bass in Lake Anna, Virginia. Virginia Power, North Anna Environmental Laboratory, Mineral, Virginia.

Grimes, D. V. 1993. Vitellolipid and vitelloprotein profiles of environmentally stressed and nonstressed populations of striped bass. Transactions of the American Fisheries Society 122:636–641.

Haeseker, S. L., J. T. Carmichael, and J. E. Hightower. 1996. Summer distribution and condition of striped bass within Albemarle Sound, North Carolina. Transactions of the American Fisheries Society 125:690–704.

Hampton, K. E., T. L. Wenke, and B. A. Zamrzla. 1988. Movements of adult striped bass tracked in Wilson Reservoir, Kansas. Prairie Naturalist 20:113–125.

Harrell, R. M. 2013. Releasing hybrid *Morone* in natural waters with congeneric species: implications and ethics. Pages 531–549 *in* J. S. Bulak, C. C. Coutant, and J. A. Rice, editors. Biology and management of inland striped bass and hybrid striped bass. American Fisheries Society, Symposium 80, Bethesda, Maryland.

Henley, D. T. 1993. Seasonal movement and distribution of striped bass in the Ohio River. Proceedings of the Annual Conference Southeastern Association of Fish and Wildlife Agencies 45(1991):370–384.

Hess, B. J., C. Jackson, and C. A. Jennings. 1999. Use of cool-water springs as thermal refuge by striped bass in the Ogeechee River, Georgia. Georgia Academy of Science 57:123–130.

Hoffman, K. J., D. S. Kittaka, and B. M. Schoenung. 2013. Evaluation and management of hybrid striped bass in Monroe Lake, Indiana. Pages 313–332 *in* J. S. Bulak, C. C. Coutant, and J. A. Rice, editors. Biology and management of inland striped bass and hybrid striped bass. American Fisheries Society, Symposium 80, Bethesda, Maryland.

Jackson, D. C., E. D. Dibble, and J. F. Mareska. 2002. Location of thermal refuge for striped bass in the Pascagoula River. Journal of the Mississippi Academy of Sciences 47:106–112.

Jackson, J. R., and J. E. Hightower. 2001. Reservoir striped bass movements and site fidelity in relation to seasonal patterns in habitat quality. North American Journal of Fisheries Management 21:34–45.

Jobling, M. 1981. Temperature tolerance and the final preferendum—rapid methods for the assessment of optimum growth temperatures. Journal of Fish Biology 19:439–455.

Kilpatrick, J. M., and J. J. Ney. 2013. Temperature and dissolved oxygen habitat use by striped bass and hybrid striped bass in Claytor Lake, Virginia. Pages 147–159 *in* J. S. Bulak, C. C. Coutant, and J. A. Rice, editors. Biology and management of inland striped bass and hybrid striped bass. American Fisheries Society, Symposium 80, Bethesda, Maryland.

Lamprecht, S. D., and W. L. Shelton. 1988. Spatial and temporal movements of striped bass in the upper Alabama River. Proceedings of the Annual Conference Southeastern Association of Fish and Wildlife Agencies 40(1986):266–274.

Lewis, R. E. 1985. Temperature selection and vertical distribution of striped bass during lake stratification. Proceedings of the Annual Conference Southeastern Association of Fish and Wildlife Agencies 37(1983):276–286.

Lewis, R. E., J. R. Siler, and L. L. Olmsted. 1979. Prevalence and intensity of *Epistylis* on fishes from two North Carolina reservoirs. Proceedings of the Annual Conference Southeastern Association of Fish and Wildlife Agencies 31(1977):372–379.

Lochmiller, R. L., J. D. Wiechman, and A. V. Zale. 1989. Hematological assessment of temperature and oxygen stress in a reservoir population of striped bass (*Morone saxatilis*). Comparative Biochemistry and Physiology A 93:535–541.

Long, E. A., C. L. Messing, K. J. Herrington, R. R. Weller, and I. I. Wirgin. 2013. Restoration of Gulf striped bass: lessons and management implications. Pages 25–63 *in* J. S. Bulak, C. C. Coutant, and J. A. Rice, editors. Biology and management of inland striped bass and hybrid striped bass. American Fisheries Society, Symposium 80, Bethesda, Maryland.

Magnuson, J. J., L. B. Crowder, and P. A. Medvick. 1979. Temperature as an ecological resource. American Zoologist 19:331–343.

Matthews, W. J. 1985. Summer mortality of striped bass in reservoirs of the United States. Transactions of the American Fisheries Society 114:62–66.

Matthews, W. J., L. G. Hill, and S. M. Schellhaas. 1985. Depth distribution of striped bass and other fish in Lake Texoma (Oklahoma–Texas) during summer stratification. Transactions of the American Fisheries Society 114:84–91.

Matthews, W. J., L. G. Hill, D. R. Edds, and F. P. Gelwick. 1989. Influence of water quality and season on habitat use by striped bass in a large southwestern reservoir. Transactions of the American Fisheries Society 118:243–250.

McDaniel, C. K., L. E. Snyder, and L. L. Connor. 1993. Impacts of thermal stress on the condition of striped bass. Proceedings of the Annual Conference Southeastern Association of Fish and Wildlife Agencies 45(1991):361–369.

Mobley, M., E. Shallenberger, M. W. Beutel, P. Gantzer, and B. Sak. 2013. Oxygen diffusers to create and maintain summer fish habitat. Pages 197–208 *in* J. S. Bulak, C. C. Coutant, and J. A. Rice, editors. Biology and management of inland striped bass and hybrid striped bass. American Fisheries Society, Symposium 80, Bethesda, Maryland.

Moss, J. L. 1985. Summer selection of thermal refuges by striped bass in Alabama reservoirs and tailwaters. Transactions of the American Fisheries Society 114:77–83.

Ng, C. L., K. W. Able, and T. M. Grothues. 2007. Habitat use, site fidelity, and movement of adult striped bass in a southern New Jersey estuary based on mobile acoustic telemetry. Transactions of the American Fisheries Society 136:1344–1355.

Poarch, S. M. 1989. A biotelemetry study of striped bass in Old Hickory Reservoir, Cumberland River, Tennessee. Master's thesis. Tennessee Technological University, Cookeville.

Pörtner, H. O., and R. Knust. 2007. Climate change affects marine fishes through oxygen limitation of thermal tolerance. Science 315:95–97.

Price, K. S., D. A. Flemer, J. L. Taft, G. B. Mackiernan, W. Nehlsen, R. B. Biggs, N. H. Burger, and D. A. Blaylock. 1985. Nutrient enrichment of Chesapeake Bay and its impact on the habitat of striped bass: a speculative hypothesis. Transactions of the American Fisheries Society 114:97–106.

Quinn, T. P., and R. D. Brodeur. 1991. Intra-specific variations in the movement patterns of marine animals. American Zoologist 31:231–241.

Rash, J. M., and J. J. Ney. 2013. Comparative ecology of juvenile striped bass and juvenile hybrid striped bass in Claytor Lake, Virginia. Pages 229–247 *in* J. S. Bulak, C. C. Coutant, and J. A. Rice, editors. Biology and management of inland striped bass and hybrid striped bass. American Fisheries Society, Symposium 80, Bethesda, Maryland.

Rice, J. A., J. S. Thompson, J. A. Sykes, and C. T. Waters. 2013. The role of metalimnetic hypoxia in striped bass summer kills: consequences and management implications. Pages 121–145 *in* J. S. Bulak, C. C. Coutant, and J. A. Rice, editors. Biology and management of inland striped bass and hybrid striped bass. American Fisheries Society, Symposium 80, Bethesda, Maryland.

Ruane, R. J., G. E. Hauser, and A. F. Sawyer. 2013. Water quality modeling as a management tool for striped bass and hybrid striped bass. Pages 161–179 *in* J. S. Bulak, C. C. Coutant, and J. A. Rice, editors. Biology and management of inland striped bass and hybrid striped bass. American Fisheries Society, Symposium 80, Bethesda, Maryland.

Schaffler, J. J., J. J. Isely, and W. E. Hayes. 2002. Habitat use by striped bass in relation to seasonal changes in water quality in a southern reservoir. Transactions of the American Fisheries Society 131:817–827.

Schaich, B. A., and C. C. Coutant. 1980. A biotelemetry study of spring and summer habitat selection by striped bass in Cherokee Reservoir, Tennessee, 1978. Oak Ridge National Laboratory, ORNL/TM-7127, Oak Ridge, Tennessee.

Scruggs, G. D., Jr., and J. C. Fuller, Jr. 1955. Indications of a freshwater population of striped bass, *Rocus saxatilis* (Walbaum), in Santee-Cooper reservoirs. Proceedings of the Annual Conference Southeastern Association of Game and Fish Commissioners 8(1954):64–69.

Secor, D. H., and R. Kraus. 2012. Modeling the thermal niche-oxygen squeeze for Chesapeake Bay striped bass. University of Maryland Center for Environmental Science, Chesapeake Biological Laboratory, Final Report, Solomons.

Setzler, E. M., W. R. Boynton, K. V. Wood, H. H. Zion, L. Lubbers, N. K. Mountford, P. Frere, L. Tucker, and J. A. Mihursky. 1980. Synopsis of biological data on striped bass, *Morone saxatilis* (Walbaum). NOAA (National Oceanic and Atmospheric Administration) Technical Report NMFS (National Marine Fisheries Service) Circular 433. Also: FAO (Food and Agriculture Organization of the United Nations) Fisheries Synopsis 121.

Stooksbury, S. W. 1977. A biotelemetry study of striped bass, *Morone saxatilis* (Walbaum), in J. Percy Priest Reservoir, Tennessee. Master's thesis. Tennessee Technological University, Cookeville.

Summers, G. L. 1982. Texoma striped bass biotelemetry. Oklahoma Department of Wildlife Conservation, Federal Aid in Fish Restoration Project F-29-R, Final Report, Oklahoma City.

Taylor, B. E., J. S. Bulak, and H. N. McKellar. 2013. Response of summer habitat for striped bass in Lake Greenwood, South Carolina to weather and phosphorus loading. Pages 181–196 *in* J. S. Bulak, C. C. Coutant, and J. A. Rice, editors. Biology and management of inland striped bass and hybrid striped bass. American Fisheries Society, Symposium 80, Bethesda, Maryland.

Thompson, J. S., and J. A. Rice. 2013. The relative influence of thermal experience and forage availability on growth of age 1–5 striped bass in two southeastern reservoirs. Pages 93–120 *in* J. S. Bulak, C. C. Coutant, and J. A. Rice, editors. Biology and management of inland striped bass and hybrid striped bass. American Fisheries Society, Symposium 80, Bethesda, Maryland.

Thompson, J. S., J. A. Rice, and D. S. Waters. 2010. Striped bass habitat selection rules in reservoirs without suitable summer habitat offer insight into consequences for growth. Transactions of the American Fisheries Society 139:1450–1464.

Van Den Avyle, M. J., and J. W. Evans. 1990. Temperature selection by striped bass in a Gulf of Mexico coastal river system. North American Journal of Fisheries Management 10:58–66.

Van Horn, S. L. 2013. A brief history of inland striped bass management. Pages 1–13 *in* J. S. Bulak, C. C. Coutant, and J. A. Rice, editors. Biology and management of inland striped bass and hybrid striped bass. American Fisheries Society, Symposium 80, Bethesda, Maryland.

Waddle, H. R., C. C. Coutant, and J. L. Wilson. 1980. Summer habitat selection by striped bass, *Morone saxatilis*, in Cherokee Reservoir, Tennessee, 1977. Oak Ridge National Laboratory, ORNL/TM-6927, Oak Ridge, Tennessee.

Weeks, G. N., and M. J. Van Den Avyle. 1998. Selection of thermal refuges by striped bass in a Gulf of Mexico coastal river. Proceedings of the Annual Conference Southeastern Association of Fish and Wildlife Commissioners 50(1996):51–61.

Wilkerson, M. L., and W. L. Fisher. 1997. Striped bass distribution, movements, and site fidelity in Robert S. Kerr Reservoir, Oklahoma. North American Journal of Fisheries Management 17:677–686.

Windham, W. T., Jr. 1986. Summer temperature selection of striped bass × white bass hybrids in a southeastern reservoir. Master's thesis. University of Georgia, Athens.

Woiwode, J. G., and I. R. Adelman. 1991. Effects of temperature, photoperiod, and ration size on growth of hybrid striped bass × white bass. Transactions of the American Fisheries Society 120:217–229.

Wooley, C. M., and E. J. Crateau. 1983. Biology, population estimates, and movement of native and introduced striped bass, Apalachicola River, Florida. North American Journal of Fisheries Management 3:383–394.

Young, S. P., and J. J. Isely. 2002. Striped bass annual site fidelity and habitat utilization in J. Strom Thurmond Reservoir, South Carolina–Georgia. Transactions of the American Fisheries Society 131:828–837.

Young, S. P., and J. J. Isely. 2004. Temporal and

spatial estimates of adult striped bass mortality from telemetry and transmitter return data. North American Journal of Fisheries Management 24:1112–1119.

Zale, A. V., J. D. Wiechman, R. L. Lochmiller, and J. Burroughs. 1990. Limnological conditions associated with summer mortality of striped bass in Keystone Reservoir, Oklahoma. Transactions of the American Fisheries Society 119:72–76.

American Fisheries Society Symposium 80:93–120, 2013

The Relative Influence of Thermal Experience and Forage Availability on Growth of Age 1–5 Striped Bass in Two Southeastern Reservoirs

JESSICA S. THOMPSON*[,1] AND JAMES A. RICE
Department of Biology, Campus Box 7617, North Carolina State University Raleigh, North Carolina 27695, USA

Abstract.—Warm epilimnetic temperatures and hypolimnetic hypoxia during summer stratification have been linked to poor growth and condition of inland striped bass *Morone saxatilis*. Contrary to expectations, however, growth occurs in some reservoirs with intense temperature–oxygen stratification in which hypoxia forces striped bass into temperatures well above their preferred range. One potential explanation for this apparent contradiction is that high forage availability may mediate the energetic costs of exposure to warm summer temperatures in some systems. To test this hypothesis, we assessed the relative influence of thermal experience and food consumption on growth of striped bass in Badin Lake and Lake Norman, North Carolina, using bioenergetics model simulations. Badin Lake is eutrophic with striped bass restricted by hypoxia to warm temperatures in the summer, but striped bass experience modest positive growth over the summer and substantial annual growth. Lake Norman is oligotrophic and striped bass are restricted to warm temperatures by hypoxia for a shorter period, but they experience almost no summer growth and minimal annual growth after age 3. Model simulations showed that Badin Lake striped bass ages 1–4 achieved high food consumption rates during the summer that continued into the fall as temperatures cooled, allowing for rapid fall growth. Lake Norman striped bass ages 1–5 experienced lower consumption rates over the summer and fall. Consumption was not sufficient to allow larger striped bass to allocate energy to growth over the summer, and these fish did not experience any season with a combination of cool temperatures and high food consumption. Habitat exchange simulations modeled how much the growth of a particular size fish in one reservoir might change if it had experienced the temperatures or food consumption of a similar sized fish in the other reservoir. These simulations showed that the relative effect of food consumption on striped bass growth was three times that of exposure temperature in the first year of the study and 37 times that of temperature in the second year. Poor striped bass growth and condition is not, therefore, linked solely to poor physical habitat. Rather, management of reservoir striped bass populations will be improved by balancing demand for and availability of prey resources for striped bass, and this balance will be especially important in reservoirs where summer hypoxia forces fish into warm temperatures that increase metabolic costs.

* Corresponding author: jessica.thompson@cnu.edu
[1] Present address: Department of Organismal and Environmental Biology, Christopher Newport University, 1 Avenue of the Arts, Newport News, Virginia 23606, USA

Introduction

Striped bass *Morone saxatilis* have been stocked into numerous reservoirs in the United States, often with the intention of establishing recreational fisheries for a large pelagic fish while controlling overabundant shad *Dorosoma* spp. populations (Axon and Whitehurst 1985; Van Horn 2013, this volume). These objectives have been achieved in some systems, but in others, growth and condition of striped bass have not met the expectations of managers and anglers (Matthews 1985). The traditional explanation for slow growth and poor condition of reservoir striped bass has focused on the role of unsuitable physical conditions during summer stratification (Coutant 1985), when the development of hypolimnetic hypoxia and warm epilimnetic temperatures can restrict larger striped bass to isolated, cool thermal refuges such as springs and tributaries or midlevel depth strata between warm temperatures and low dissolved oxygen concentrations (Cheek et al. 1985; Coutant 1985, 1986; Moss 1985; Van Den Avyle and Evans 1990; Wilkerson and Fisher 1997; Schaffler et al. 2002; Young and Isely 2002).

In the absence of thermal refuges, hypolimnetic hypoxia can force inland striped bass into epilimnetic water with temperatures as warm as 27–30°C (Matthews et al. 1985; Farquhar and Gutreuter 1989; Zale et al. 1990; Van Horn et al. 1996; Jackson and Hightower 2001; Thompson et al. 2010), well above their preferred range of 20–23°C. Coutant (1985) hypothesized that such habitat constraints would lead to poor growth and condition due to thermal stress. However, the consequences of these habitat limitations have been quite variable. Poor condition (Jackson and Hightower 2001) and apparent cessation of feeding (Zale et al. 1990) have been observed in some cases, but populations in productive reservoirs with unsuitable summer habitat do not routinely experience these problems (Matthews et al. 1985; Farquhar and Gutreuter 1989; Davias 2006; Thompson et al. 2010). Thompson et al. (2010) found that severe oxygen stratification with the development of hypoxia in both the hypolimnion and metalimnion constrained striped bass (425–804 mm total length [TL]; 0.9–6.8 kg wet weight) in Badin Lake, North Carolina, to depths just above the oxycline during the warmest months of the summer. The shallow depths occupied by striped bass were also those with the highest biomass of warmwater prey, so the severity of the temperature–oxygen structure in this system forced striped bass to overlap spatially with their prey (Thompson et al. 2010). Laboratory studies of striped bass also demonstrate that sustained feeding by fish up to 2.94 kg is possible at temperatures up to 29°C (Hartman and Brandt 1995a).

These patterns suggest that high forage availability may offset some of the high metabolic costs experienced by striped bass over the summer in reservoirs where severe oxygen stratification forces them to occupy warm epilimnetic temperatures, allowing striped bass in these systems to attain positive annual growth. Although the development of trophy striped bass fisheries will not be feasible in systems with highly unsuitable thermal conditions, many of these reservoirs do, or have the potential to, support popular fisheries for smaller-sized striped bass. A better understanding of the relative importance of exposure temperature and forage availability in determining growth of inland striped bass will, therefore, aid in optimal management of these populations.

Growth represents the physiological synthesis of an individual's entire environmental history, making it difficult to distinguish the relative influence of temperature and forage availability on fish growth based solely on observed growth patterns. Fortunately, bioenergetics modeling provides a framework for separating and analyzing the influence of biotic and abiotic conditions on growth in a way not possible from observed patterns alone (Kitchell et al. 1977; Rice et al. 1983; Railsback and Rose 1999; Petersen and Paukert 2005; Johnson et al. 2006). Bioenergetics models use a balanced energy budget in which consumed energy is set equal to growth plus metabolic costs and waste losses (Kitchell et al. 1977), allowing food consumption to be estimated based on observed growth and experienced temperatures. A striped bass bioenergetics model has been developed by Hartman and Brandt (1995a) and used to estimate consumption by coastal striped bass (Hartman and Brandt 1995b) and reservoir populations (Cyterski et al. 2002; Raborn et al. 2002; Vatland et al. 2008).

We used the striped bass bioenergetics model to determine the relative influence of food availability and thermal experience on growth of striped bass in two North Carolina reservoirs, Badin Lake and Lake Norman, which differ in forage fish abundance, temperature–oxygen conditions that determine temperatures occupied by striped bass, and observed striped bass growth and condition. Badin Lake is eutrophic (NCDENR 2002, 2007) with intense thermal-oxygen stratification leading to poor summer habitat, but striped bass still experience modest summer growth and substantial annual growth (Thompson 2006). Lake Norman is oligotrophic (NCDENR 2003) with suitable habitat available for a longer time in the summer, but striped bass grow more slowly than in Badin Lake and minimal annual weight gain is observed after age 3 (Thompson 2006). These systems were selected for this study based on input from state fisheries biologists, who were perplexed that striped bass were performing better in Badin Lake than in Lake Norman given that the dominant emphasis in the literature has been that temperature and dissolved oxygen conditions determine striped bass success.

We first conducted baseline simulations to estimate seasonal food consumption by striped bass in both reservoirs based on observed growth and thermal experience. However, even when comparing such disparate systems, simply estimating food consumption does not indicate whether food consumption or exposure temperature was more important in driving the observed growth pattern of striped bass in each reservoir. To determine the relative influence of food availability and temperature on growth, we asked how growth of a striped bass in Badin Lake would change if it experienced the food consumption of a similar-sized Lake Norman striped bass, and how a Badin Lake striped bass would grow if it occupied the temperatures experienced by a Lake Norman fish. Likewise, how would growth of a Lake Norman striped bass change if it experienced the food consumption or temperatures of a similar-sized striped bass in Badin Lake?

To answer these questions, we conducted additional bioenergetics model simulations in which we let the model estimate growth based on conditions from one reservoir with the exception that either the fish's thermal experience or estimated food consumption was taken from the other reservoir. Comparing differences between the original observed growth pattern and growth estimated in these habitat exchange simulations allowed us to assess the relative importance of temperature and food consumption to the original growth pattern. This approach builds on previous applications of bioenergetics modeling that have used these models as a tool to test hypotheses regarding the effects of variation in temperature and food availability on fish growth (Kitchell et al. 1977; Rice et al. 1983; Railsback and Rose 1999; Munch and Conover 2002) by directly comparing simulation results between systems. We discuss the implications of our results for understanding striped bass growth and improving management of stocked populations.

Study Sites

Badin Lake is a 2,165-ha, eutrophic reservoir located on the Yadkin River, a major tributary of the Pee Dee River (Figure 1). It was impounded in 1917 and is one of a series of reservoirs on the Yadkin River. Badin Lake has abundant forage dominated by threadfin shad *Dorosoma petenense*, with a minor contribution from blueback herring *Alosa aestivalis* and gizzard shad *D. cepedianum* (Thompson 2006). During the development of stratification in the early summer, striped bass can continue to ac-

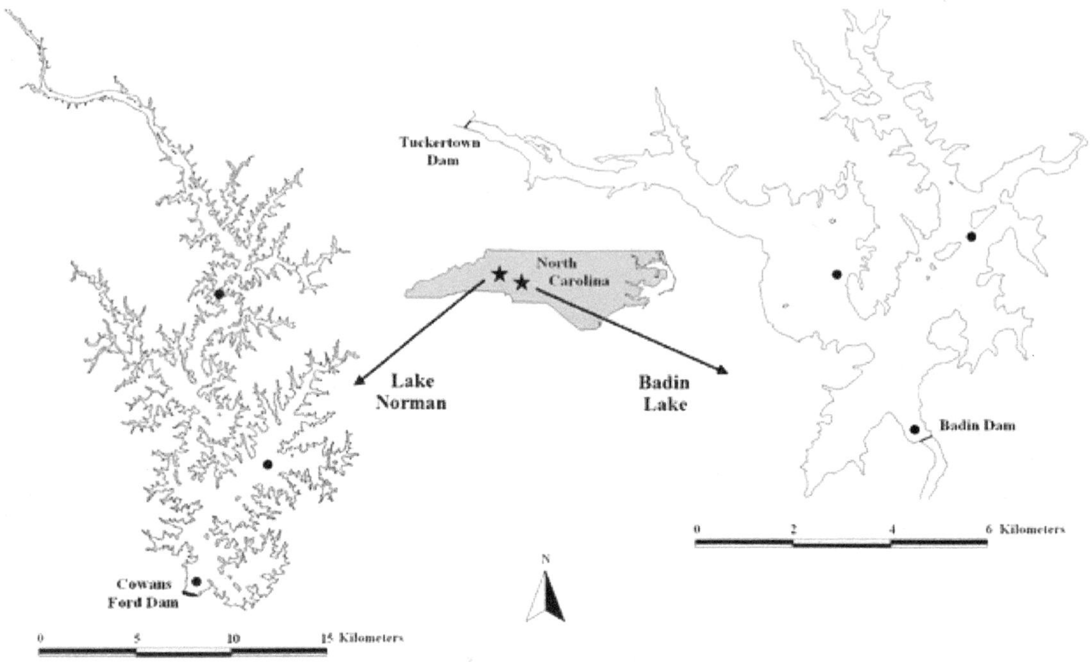

FIGURE 1. Maps of Lake Norman (left) and Badin Lake (right) showing sites of temperature and dissolved oxygen profiles (black dots) used in estimating the thermal experience of striped bass.

cess preferred temperatures for 1 to 2 weeks by occupying an oxygenated zone between hypoxic layers in the metalimnion and deeper hypolimnion (see Rice et al. 2013, this volume). However, once dissolved oxygen levels in these mid-level depth strata drop below 2 mg/L, hypolimnetic and metalimnetic hypoxia forces striped bass into shallower, epilimnetic water with temperatures above 27°C and up to 30°C for about 2 months each summer (Thompson et al. 2010). Despite occupying summer habitat considered unsuitable based on the thermal preferences of adult striped bass (Coutant 1985), all Badin Lake striped bass (ages 1–8, ~0.3–6 kg) experience positive annual growth (Figure 2) with relative weights (Anderson and Neumann 1996) in the range of 80–100 throughout the year (Thompson 2006). The population does not, however, contain many larger (>650 mm TL; Figure 3) or older (>age 4) fish (Thompson 2006) due to high fishing mortality rates (~50%/year; Thompson et al. 2007). Striped bass fishery regulations are fairly liberal on Badin Lake, with a minimum size limit of 406 mm TL (16 in) and a daily creel limit of eight fish.

Lake Norman is a 12,634-ha, oligotrophic reservoir on the Catawba River, part of the Wateree River drainage (Figure 1), and was impounded in 1963. The pelagic forage base in Lake Norman is composed primarily of threadfin shad, with a small contribution from alewife *A. pseudoharengus* and gizzard shad. Forage fish biomass in Lake Norman is approximately one-sixth that in Badin Lake, based on purse-seine catch per unit effort (Thompson 2006). As in Badin Lake, striped bass in Lake Norman can initially occupy preferred temperatures in the metalimnion during the development of summer stratification. Adequate oxygen persists in this zone for a longer time than in Badin Lake, allowing striped bass to occupy preferred temperatures for 2 to 4 weeks longer than in Badin Lake. Once dissolved oxygen drops below 2 mg/L in the metalimnion, Lake Norman striped bass must move into warmer than

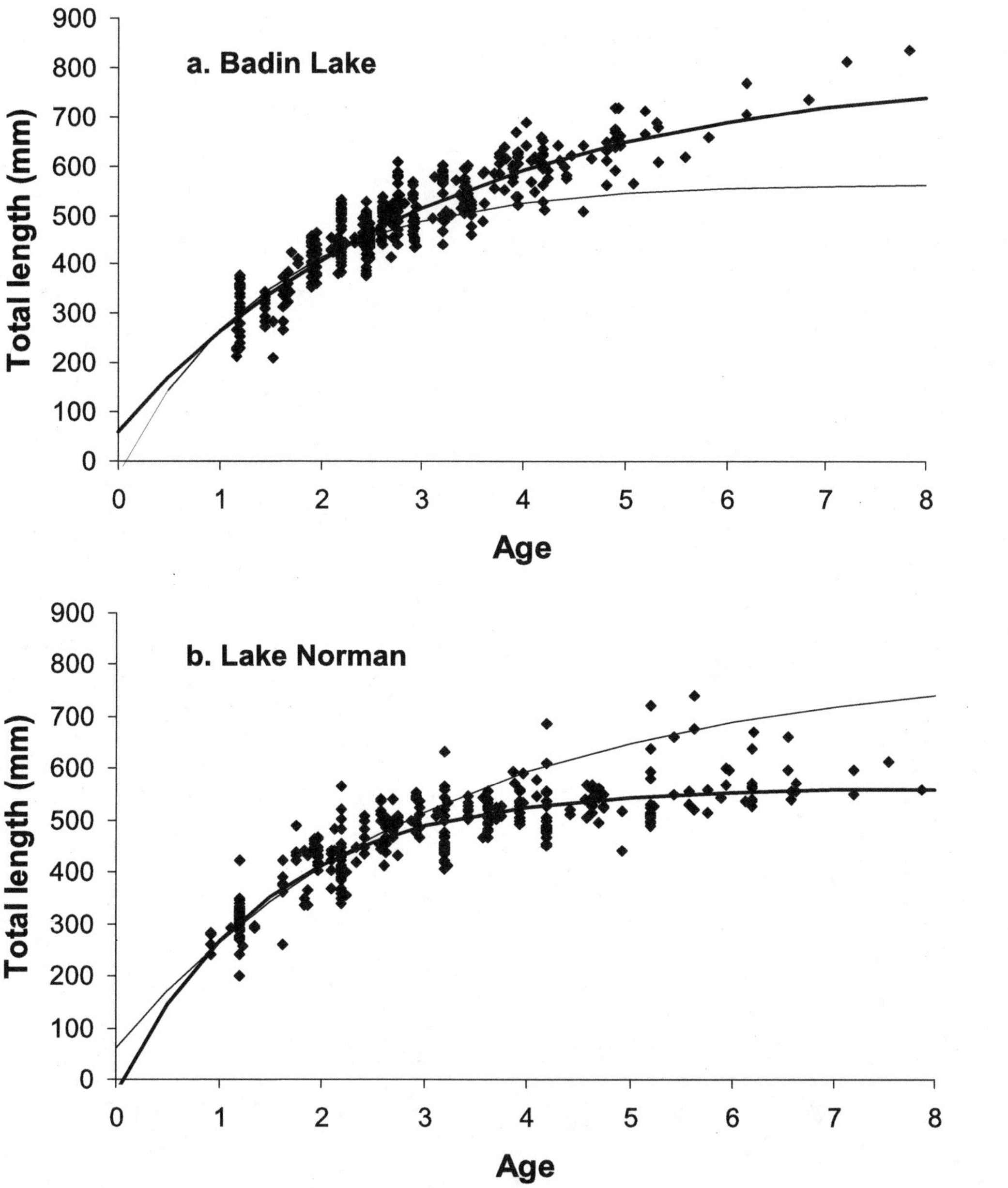

FIGURE 2. Von Bertalanffy growth curves (solid lines) fit to observed length and age at capture (points) of striped bass in (a) Badin Lake and (b) Lake Norman collected from 2000 through 2002. Dotted lines show the growth curve for the alternate reservoir for reference. Ages were determined for 347 striped bass in Badin Lake and 223 striped bass in Lake Norman.

preferred temperatures in the epilimnion or into cooler than preferred temperatures in the hypolimnion. While some fish choose to occupy cooler, hypolimnetic habitat that may also contain coolwater prey (alewife), this oxygenated refuge becomes hypoxic by midsummer, forcing any fish in it to move into oxygenated surface waters or die (Rice et al. 2013). For the remainder of the summer, all Lake Norman striped bass occupy epilimnetic temperatures of

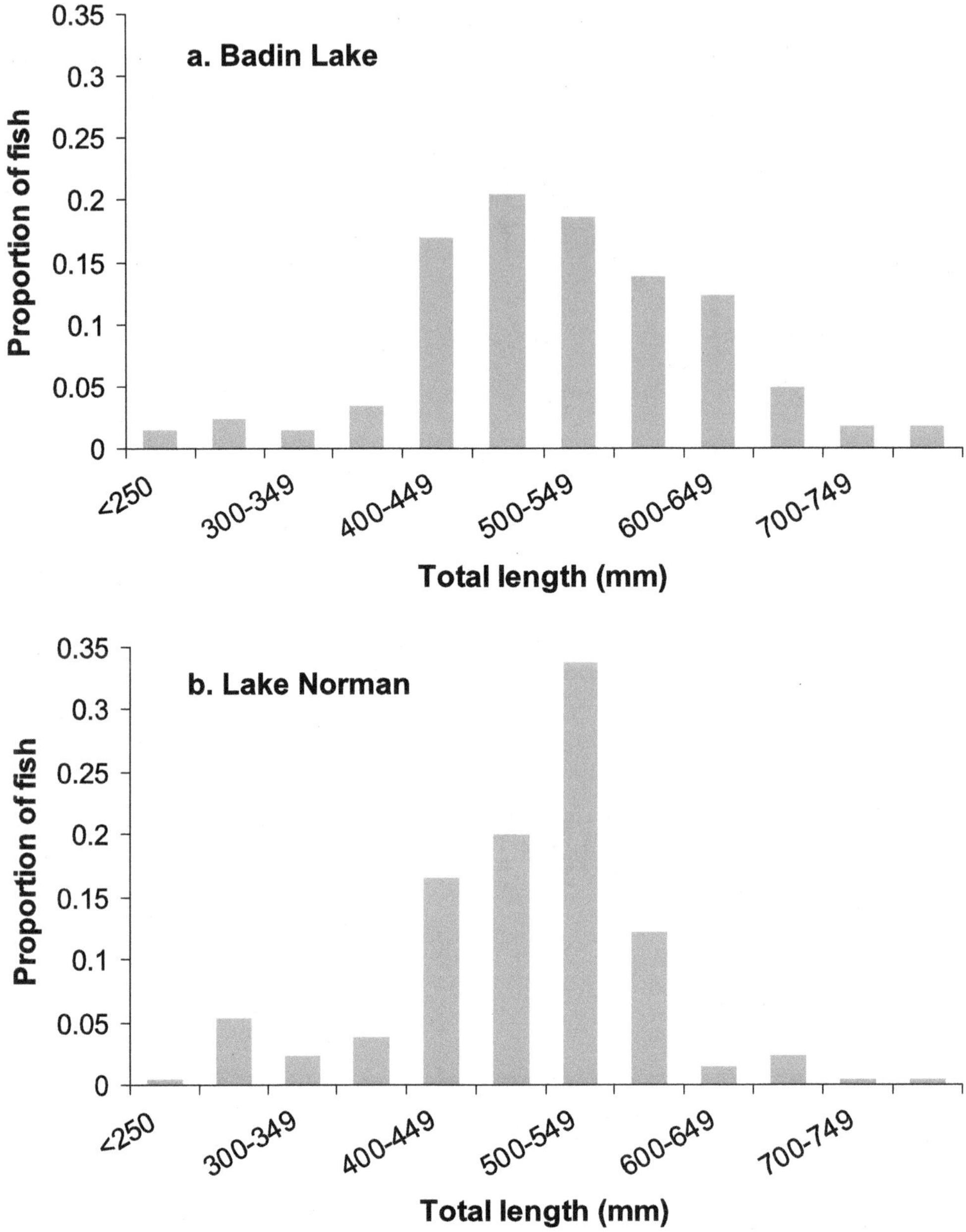

FIGURE 3. Size structure of striped bass population in (a) Badin Lake and (b) Lake Norman, based on 347 and 223 striped bass collected in Badin Lake and Lake Norman, respectively, in 2000 through 2002.

a similar range (27–30°C) to those experienced by fish in Badin Lake (Thompson 2006).

Although striped bass in Lake Norman are able to occupy cool, oxygenated habitat for a longer time in the summer than those in Badin Lake, they represent a more typical "problem" population. Their growth slows substantially after age 3 (~490 mm TL, ~1.25 kg) and essentially ceases by age 4 (~525 mm TL, ~1.45 kg; Figure 2; Thompson 2006). Therefore, although we collected a larger number of older fish in Lake Norman, the most abundant size-classes are similar in both reservoirs and fewer fish ≥600 mm TL were collected in Lake Norman (Figure 3). Condition is poor throughout the year for most striped bass in Lake Norman (relative weights in the range of 70–90) and declines significantly with increasing fish size (Thompson 2006). At the time of this study, striped bass fishery regulations were fairly restrictive on Lake Norman, with a minimum size limit of 508 mm TL (20 in) and a daily creel limit of four fish.

Methods

Bioenergetics model format

We used the Wisconsin bioenergetics model (Kitchell et al. 1977), as packaged in the software program Fish Bioenergetics 3.0 (Hanson et al. 1997), in our analysis of striped bass in Badin Lake and Lake Norman. For our purpose, model simulations were used to estimate the food consumption needed to achieve an observed pattern of weight gain or, alternatively, the weight gain that would be expected based on a certain level of consumption. The bioenergetics model partitions the energy that an individual consumes into energy put towards growth, metabolic costs, and wastes. Inputs and outputs of the model are measured and expressed in various units (e.g., grams of prey consumed or grams of growth), but the model is balanced internally in terms of energy, as explained further below. A simple mass balance equation is the basis of the model

$$C = G_S + G_G + R + S + F + U,$$

where C is consumption (g prey · g fish^{-1} · d^{-1}), G_S is somatic growth (g growth · g fish^{-1} · d^{-1}), G_G is gonadal growth (g growth · g fish^{-1} · d^{-1}), R is respiration plus metabolism associated with activity (g O_2 · g fish^{-1} · d^{-1}), S is the metabolic cost of digestion (g prey · g fish^{-1} · d^{-1}; constant proportion of C), F is egestion (g prey · g fish^{-1} · d^{-1}; constant proportion of C), and U is excretion (g prey · g fish^{-1} · d^{-1}; constant proportion of $[C - F]$). The energy densities of prey and predator are used to convert consumption to growth in energetically equivalent terms, and the energy density of the predator and an oxycalorific conversion factor of 13.6 kJ/g O_2 (Hartman and Brandt 1995a) are used to convert grams O_2 respired to grams fish lost due to respiration.

Food consumption and respiration are each modeled by an additional set of equations that incorporates species-specific physiological parameters, temperature, and fish weight (Hanson et al. 1997). The relationship used to model consumption is particularly relevant to our use of the striped bass bioenergetics model. Consumption is modeled as a proportion of the maximum feeding rate (C_{max}; g prey · g fish^{-1} · d^{-1}) that could be attained based on the fish's weight and temperature. This proportion (or P-value) accounts for ecological constraints on consumption, and when the bioenergetics model is used to estimate consumption based on weight gain, the model solves for the P-value that will adjust consumption to the proportion of C_{max} necessary to result in the observed growth pattern.

Bioenergetics model simulations to estimate food consumption require data on thermal experience, observed patterns of weight gain, diet composition, and the energy densities of the predator and the prey. The collection and analysis of these system-specific data are described below. In addition, the amount of weight lost to spawning (G_G) can be specified in the model (Hanson et al. 1997). We chose not to utilize this option in our simulations because gonadal development, as indicated by changes in gonadosomatic index (GSI), was low and variable across all ages modeled in both populations; mean spring GSI in these systems was 1.8–3.5% for males and 1.1–2.4% for females (Thompson 2006). In addition, gonadal energy density was lower than somatic energy density for most individuals (Thompson 2006), so if any spawning occurred (though none has been observed in either reservoir), the energy lost would be disproportionately low.

Bioenergetics model simulations

For each reservoir, baseline bioenergetics model simulations were used to estimate per-capita consumption (g prey consumed/d) based on observed growth and estimated exposure temperatures in 2001 and 2002, years with fairly typical environmental conditions when compared with available data from other years (Thompson 2006). For Badin Lake, consumption was estimated separately for age-1, age-2, age-3, and age-4 fish. For Lake Norman, per-capita estimates were obtained for each of these four age-classes and also age-5 fish. These age-classes were those captured with sufficient frequency to adequately determine their growth rates and made up 97% and 94% of the striped bass age 1 and older collected during the study in Badin Lake and Lake Norman, respectively (Thompson 2006). We used physiological parameters for age-1, age-2, and age-3+ striped bass (Hartman and Brandt 1995a) in our analysis of individuals in the corresponding age-classes. Modeling simulations divided the year into three seasons: spring (1 January to 15 June), summer (16 June to 15 September), and fall (16 September to 31 December), which differed in their typical thermal conditions and forage availability. Daily consumption was summed to obtain the total grams of prey consumed by an individual of each age-class in each season.

Habitat exchange simulations were then used to assess the relative impact of forage availability (as indicated by realized cumulative seasonal consumption [g prey consumed/season] estimated in the baseline simulations) and exposure temperature on growth of striped bass in each reservoir. Fish size has a large effect on rates in the bioenergetics model, so it was important to start with fish of the same size in these habitat exchange simulations. We used age-3 fish from Badin Lake and age-5 fish from Lake Norman in these simulations because they were approximately the same weight at the beginning of each year. In 2001, both age-3 Badin Lake fish and age-5 Lake Norman fish began the year at 1,729 g (equivalent to 537 and 560 mm TL in Badin Lake and Lake Norman, respectively). In 2002, age-3 Badin Lake fish began the year at 1,482 g (510 mm TL) while age-5 Lake Norman fish began the year at 1,514 g (534 mm TL; Thompson 2006).

The first consumption exchange simulation used habitat conditions experienced by an age-3 Badin Lake striped bass (including thermal experience, diet composition, and energy densities of predator and prey) but applied the cumulative seasonal food consumption originally estimated for an age-5 Lake Norman fish. The second consumption exchange simulation used habitat conditions experienced by an age-5 Lake Norman fish but substituted the cumulative seasonal food consumption estimated for an age-3 Badin Lake fish. In each consumption exchange simulation, the cumulative consumption used will represent a new proportion (*P*-value) of C_{max} because C_{max} varies with the temperatures the fish experience, which differ between the two reservoirs. Though C_{max} varies with temperature, the amount of prey the fish actually consumes depends on prey availability, which is dictated by prey density, responses of prey fish to habitat constraints, and competitive interactions. We assume that given the same prey availability, fish experiencing different temperatures will still eat the same amount (within the upper limit set by C_{max}) but will grow more or less depending on temperature. Differences in thermal conditions between the two reservoirs should not impact the ability of striped bass to access prey resources spatially. In the early part of the summer, some striped bass in both Badin Lake and Lake Norman may occupy oxygenated metalimnetic or hypolimnetic habitat that contains coolwater prey (alewife in Lake Norman and blueback herring in Badin Lake; Rice et al. 2013), but once this oxygen is depleted, hypolimnetic and metalimnetic hypoxia will constrain all striped bass to the same depths as warmwater prey in both reservoirs (Thompson et al. 2010). Therefore, by simulating growth of a striped bass in one system using the cumulative seasonal consumption estimated for a fish in the other system, we are allowing it to "experience" the forage regime of the other system while maintaining the thermal regime of the original baseline simulation. This approach allows us to isolate the effect on growth of differences in food availability between systems from the effect of differences in thermal experience.

The first temperature exchange simulation then used habitat conditions experienced by an age-3 Badin Lake striped bass (including cumulative seasonal food consumption, diet composition, and energy densities of predator and prey) but substituted the thermal experience of striped bass from Lake Norman. The second temperature exchange simulation used habitat conditions experienced by an age-5 Lake Norman fish but applied temperatures experienced by striped bass in Badin Lake. As in the consumption exchange simulations, C_{max} in each temperature exchange simulation will differ from the original baseline simulation for that system because the fish is "experiencing" a new set of temperatures. The original cumulative consumption will, therefore, represent a different proportion of this new C_{max}. However, by keeping cumulative seasonal consumption constant, we effectively maintained the food limitations in a given system while determining how much growth would change given the experienced temperatures (as dictated by thermal and dissolved oxygen conditions) of fish in the other system. Comparing the differences between simulated and observed growth due to exchanging temperature with the differences due to exchanging food consumption allowed us to quantify and compare the relative influence of each on growth of striped bass in Lake Norman and Badin Lake.

System-specific data sources

To determine the temperatures experienced by striped bass in Badin Lake and Lake Norman, we used thermal selection rules developed by Thompson et al. (2010) for striped bass in reservoirs with unsuitable summer habitat conditions. These rules were based on temperatures and dissolved oxygen levels occupied by Badin Lake striped bass tagged with temperature-sensing transmitters in 2002 and 2003. The rules state that striped bass select the warmest water available up to 20°C in the winter and spring. As the water column stratifies, striped bass remain at 20°C as long as water of this temperature is available with at least 2 mg/L dissolved oxygen. Once the dissolved oxygen level at 20°C drops below 2 mg/L, fish move up into warmer epilimnetic water and occupy the temperature at the top of the oxycline, defined as the depth just above the largest decline in dissolved oxygen occurring over a 1-m change in depth. Fish remain at the top of the oxycline until the water temperature at that depth drops to 20°C, at which point they again occupy the warmest water up to 20°C (Thompson et al. 2010). We applied these thermal selection rules to temperature and dissolved oxygen profiles collected at 1-m intervals at three sites in each reservoir (Figure 1) and averaged the resulting temperatures to obtain the temperature input for each profile date. Profiles were conducted at seasonally appropriate intervals ranging from every 3 to 4 weeks in the winter to weekly in the summer. Temperatures were linearly interpolated between profile dates to complete the seasonal pattern of temperatures experienced by striped bass in each reservoir (Figure 4).

The growth data required to estimate consumption in the baseline bioenergetics model simulations were the changes in weight for each age of fish over each seasonal time period. Von Bertalanffy growth models were used to determine the length of fish at the beginning and end of each model period; these models were fit to observed length and age at capture and back-calculation of length-at-annulus formation using sagittal otoliths from fish of each cohort (n = 36–159 observations per cohort in Badin Lake; n = 24–60 observations per cohort in Lake Norman). Striped bass length was then converted to weight using significant system-specific regression models relating weight to length and powers of day of the year (Badin Lake: analysis of variance (ANOVA), $p < 0.0001$, $R^2 = 0.97$; Lake Norman: ANOVA, $p < 0.001$, $R^2 = 0.98$; Thompson 2006). Striped bass used in the anal-

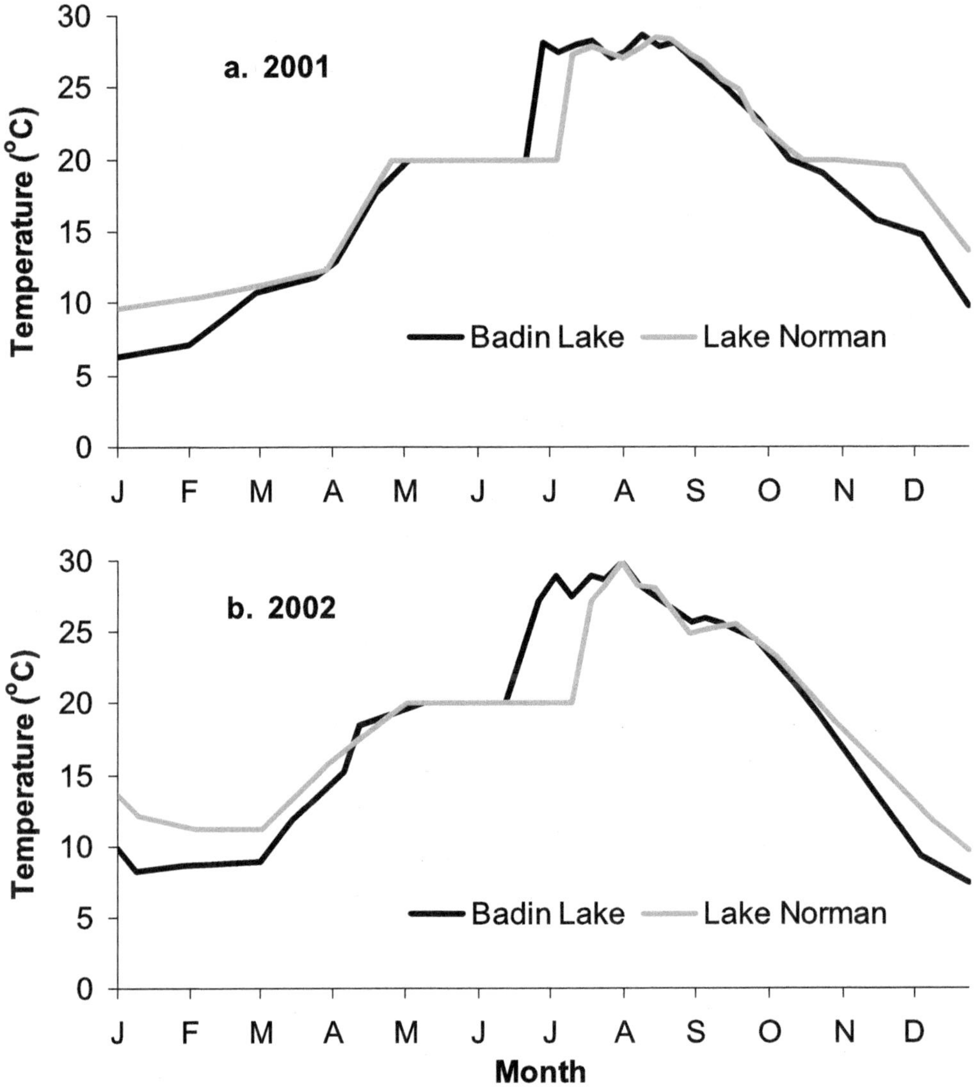

FIGURE 4. Exposure temperatures for striped bass input to striped bass bioenergetics models for Badin Lake (black lines) and Lake Norman (gray lines) in (a) 2001 and (b) 2002. Temperatures are based on applying thermal selection rules to temperature and dissolved oxygen profile data, as described in the text.

ysis of growth, energy density, and predator diet included large samples collected by gill net (51-mm and 76-mm bar mesh) in June or July, September, and December of 2000 through 2002, as well as small samples collected by a variety of methods about every 6 weeks between large samples (Thompson 2006). A total of 347 and 223 striped bass were collected in Badin Lake and Lake Norman, respectively.

Seasonal and size-specific striped bass energy density inputs to the bioenergetics models were estimated by linearly interpolating between the mean energy densities observed in each size category on each sample date in each reservoir (Figure 5). Size categories were 50-mm or 100-mm intervals depending on the observed variation in energy density (Figure 5). Energy densities were estimated by first di-

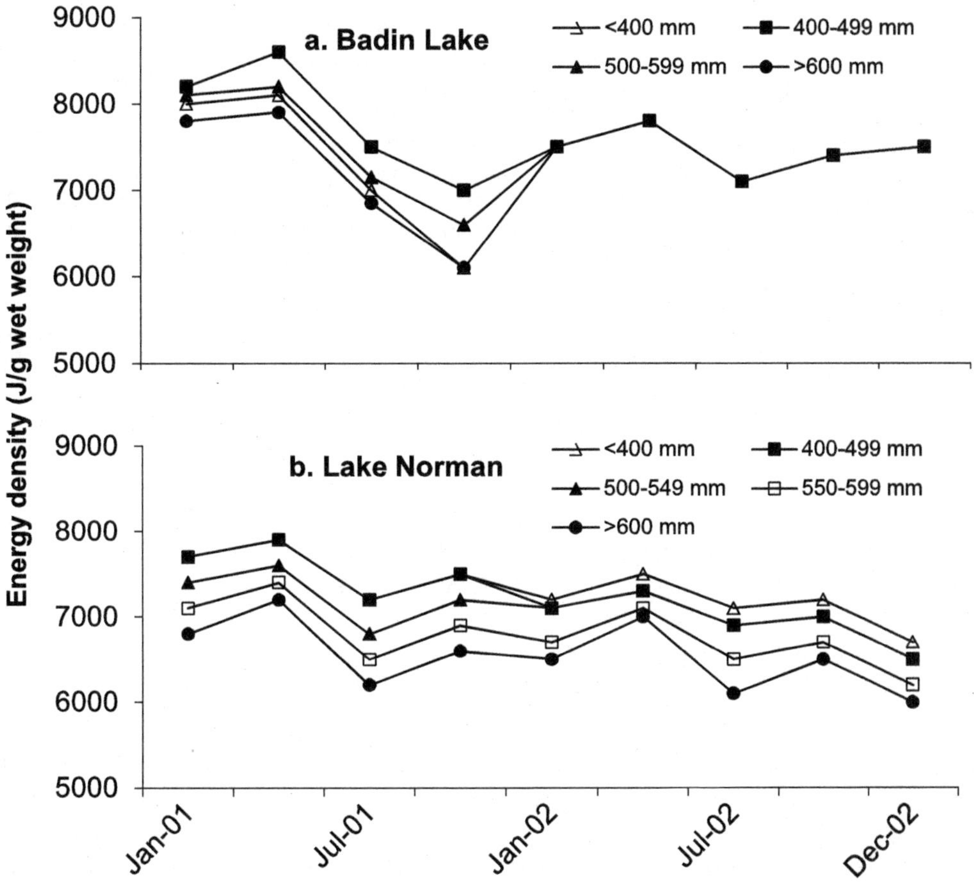

FIGURE 5. Seasonal energy density (J/g wet weight; somatic plus gonadal) of size-classes of striped bass from <400 mm total length (TL) to ≥600 mm TL in (a) Badin Lake and (b) Lake Norman, from January 2001 through December 2002. Energy density was the same for all size-classes of striped bass in Badin Lake in 2002, so data points for this year overlap. Energy density was estimated for 347 striped bass in Badin Lake and 223 striped bass in Lake Norman; sample sizes for each size-class and date ranged from 6 to 21 fish (Thompson 2006).

rectly measuring the energy densities of about 30 striped bass from each system by calorimetry (Thompson 2006) to model the relationship between energy density and the natural log of percent dry weight (DW) of the sample (Table 1). A homogeneous subsample of each remaining striped bass collected during the study was dried to a constant weight, and DW of the sample was used to determine energy density. Energy density was initially determined separately for somatic and gonadal tissue using relationships between DW and energy density of gonadal tissue for each sex (Table 1), but these values were combined into a total energy density, based on the proportionate weights of each component, for input to the bioenergetics model. No significant differences in total energy density were found between males, females, and immature fish (Badin Lake: ANOVA, p = 0.16; Lake Norman: ANOVA, p = 0.18), so all sexes were analyzed together.

Seasonal, size-specific energy densities were also determined for each species of pelagic forage fish identified in striped bass stomachs, including threadfin shad and gizzard shad in both reservoirs, blueback herring in Badin

TABLE 1. The relation between energy density (ED; J/g wet weight) and percent dry weight (DW) for striped bass somatic tissue, male gonads, female gonads, and prey species. Separate equations are presented if a significant effect of reservoir was found (analysis of variance $\alpha = 0.05$); otherwise, a single equation is presented for both Badin Lake and Lake Norman. Blueback herring were collected from Badin Lake, whereas alewives were collected from Lake Norman. The equation for striped bass somatic tissue uses $\log_e$ transformed DW values; Hartman and Brandt (1995c) used a linear relationship to relate DW and energy density of striped bass but only included data in the lower range of our DW values (approximately 21–34 DW; Figure 2 in Hartman and Brandt 1995c). Thompson (2006) also observed linearity through this lower range but $\log_e$ transformation of DW was necessary to fit data at the upper end of the DW values, which tended to have lower energy densities than would have been predicted by a linear relationship.

	Equation	R^2	Sample size	DW range
Somatic tissue	$ED = -22{,}742 + 8{,}720.70 \cdot \log_e(DW)$	0.92	62	22.1–39.2
Male gonads	$ED = -1504.22 + 295.35 \cdot DW$	0.94	46	16.4–32.9
Female gonads				
Badin Lake	$ED = -1{,}893.22 + 302.35 \cdot DW$	0.81	20	17.8–26.7
Lake Norman	$ED = -2{,}128.12 + 324.51 \cdot DW$	0.99	26	17.9–37.2
Threadfin shad	$ED = -1{,}860.03 + 287.12 \cdot DW$	0.99	34	15.8–39.8
Gizzard shad	$ED = -2{,}034.09 + 292.74 \cdot DW$	0.96	33	18.4–29.6
Blueback herring	$ED = -2{,}144.06 + 313.44 \cdot DW$	0.97	21	19.3–27.9
Alewife	$ED = -1{,}057.35 + 263.20 \cdot DW$	0.99	16	17.9–28.6

Lake, and alewife in Lake Norman (Table 2). The energy densities of 16–20 samples of each forage fish species from each reservoir, spanning the full size range encountered for each species, were determined directly by calorimetry, and these data were used to generate linear regression models relating energy density to percent DW (Table 1). For all species in each subsequent forage fish sample (i.e., a single date within a single reservoir), at least two fish from each 5-mm size-class collected were dried to determine energy density using these relationships. Pelagic forage fish were collected by purse seine (9 m deep by 118 m long net, 4.8-mm mesh) at two to three sites in each reservoir on the dates of the large gill-net striped bass samples, and smaller samples were collected by various sampling methods about every 6 weeks between large samples (Thompson 2006).

The vast majority of striped bass prey were clupeids, but invertebrates and nonclupeid fish were occasionally found in striped bass stomachs, primarily in winter and spring when clupeid prey were least abundant (Thompson 2006). Most invertebrates were Ephemeroptera, so a constant energy density of 4,705 J/g wet weight, the mean for Ephemeroptera provided in the prey energy densities in Fish Bioenergetics 3.0 (Cummins and Wuycheck 1971 in Hanson et al. 1997; Driver et al. 1974 in Hanson et al. 1997), was used for all invertebrate prey in all simulations. Nonclupeid prey fish were found in 4.8% of striped bass stomachs and included bluegill *Lepomis macrochirus*, white perch *M. americana*, and black crappie *Pomoxis nigromaculatus* (Thompson 2006). The energy density of bluegill reported in Fish Bioenergetics 3.0, 4,186 J/g wet weight (Kitchell et al. 1974 in Hanson et al. 1997), was used for all nonclupeid prey fish in all simulations.

In order for the overall energy density of the striped bass diet to be determined from these various prey energy densities, seasonal and size-specific diet composition by weight was also specified in each model simulation. These diet composition data were determined by identifying all prey items in the stomachs of striped bass captured during the study to the lowest taxonomic level possible using standard keys (Jenkins and Burkhead 1993; Voshell 2002).

TABLE 2. Size-specific energy density (J/g wet weight; thousands) of pelagic prey species from Badin Lake and Lake Norman used in the striped bass bioenergetics model. A single value is given if energy density was constant across seasons; the mean value and range are given if energy density varied seasonally. For all fish exhibiting seasonal variation, energy density was highest in the summer and lowest in the winter, except for Lake Norman alewife, which had the highest energy density in the fall (Thompson 2006). Size categories are listed by the initial size of a 10 mm total length (TL) interval, except for the final category which includes all fish ≥105 mm TL, and reflect sizes of each species found during analysis of striped bass stomach contents; not all size categories were needed for each prey species. na = not applicable.

	Size-class (mm TL)							
	35 mm	45 mm	55 mm	65 mm	75 mm	85 mm	95 mm	≥105 mm
Badin Lake								
Threadfin shad	3.2	3.4	3.7	4.0	4.3	4.5	5.0	6.0 (5.3–6.6)
Gizzard shad	na	na	na	na	na	3.6	3.7	3.9
Blueback herring	na	na	4.9 (3.8–5.8)	5.1 (3.9–6.2)	5.5 (4.0–7.0)	6.0 (4.5–7.3)	6.3 (4.7–7.7)	6.6 (5.0–8.0)
Lake Norman								
Threadfin shad	3.3	3.6	3.9	4.0	4.1	4.2	4.6 (4.2–5.2)	4.9 (4.2–5.3)
Gizzard shad	na	na	na	na	na	4.4	4.5	4.5
Alewife	na	na	4.4 (4.1–4.8)	4.6 (4.2–5.2)	4.9 (4.3–5.8)	5.2 (4.7–6.1)	5.4 (4.9–6.3)	5.7 (5.0–6.8)

For prey fish, backbone lengths were converted to total lengths and then to wet weights using species-specific regression models (Thompson 2006). Diet data were tabulated separately for three size-classes of striped bass in Badin Lake (<425 mm TL, 425 to 600 mm TL, and >600 mm TL) and two size-classes in Lake Norman (<475 mm TL and ≥475 mm TL) to account for slight differences in the most common prey items for striped bass of each size. The most common prey size was the same for all sizes of striped bass in each season, but larger striped bass increasingly included larger clupeid prey and occasional large nonclupeid prey in their diet (Thompson 2006). For each seasonal simulation, we used the data for the size-class in which a particular age striped bass started the simulation. Daily diet composition proportions were linearly interpolated between the dates for which diet data were specified in the model input.

Model Simulation Results and Interpretation

Seasonal consumption and growth

Seasonal patterns in the amount of food consumed were evident for all age-classes of striped bass in both reservoirs (Tables 3 and 4; Figures 6 and 7), although consumption by striped bass in Lake Norman was more similar among seasons than in Badin Lake. In both reservoirs and years, consumption rates were lowest during the spring, particularly among older fish (Tables 3 and 4). Growth rates were also lowest in the spring, with age-1 and age-2 fish gaining weight slowly and older fish remaining at almost constant weight or losing weight (Table 5; Figures 6 and 7). This growth pattern probably reflected a combination of cool spring temperatures (Figure 4), which would lower both metabolic rate and maximum consumption rate for a given size fish, and limited consumption due to

TABLE 3. Seasonal per-capita cumulative consumption (total g) and mean per-capita daily consumption (g/d), mean percent body weight consumed per day (%/d), and proportion of maximum consumption (*P*) attained by age-1, age-2, age-3, and age-4 Badin Lake striped bass in 2001 and 2002 estimated in bioenergetics simulations based on observed seasonal growth of each age-class. Spring refers to 1 January–15 June, summer refers to 16 June–15 September, and fall refers to 16 September–31 December. na = not applicable.

	Age 1			Age 2			Age 3			Age 4		
	Total g (g/d)	%/d	*P*	Total g (g/d)	%/d	*P*	Total g (g/d)	%/d	*P*	Total g (g/d)	%/d	*P*
2001												
Spring	1,205 (7.3)	2.3	0.46	2,940 (17.7)	1.8	0.48	3,538 (21.3)	1.2	0.36	3,977 (24.0)	0.9	0.30
Summer	2,370 (25.8)	5.2	0.84	4,044 (44.0)	3.7	0.74	6,105 (66.4)	3.5	0.78	7,114 (77.3)	3.0	0.73
Fall	2,200 (20.6)	3.3	0.59	3,184 (29.8)	2.4	0.51	5,033 (47.0)	2.3	0.53	5,541 (51.8)	1.9	0.49
Annual	5,775	3.3	na	10,168	2.4	na	14,676	2.1	na	16,632	1.7	na
2002												
Spring	1,279 (7.7)	2.4	0.47	2,479 (14.9)	1.6	0.40	3,007 (18.1)	1.2	0.32	4,096 (24.7)	1.0	0.30
Summer	2,328 (25.3)	5.0	0.83	3,536 (38.4)	3.5	0.68	5,372 (58.4)	3.7	0.76	5,900 (64.1)	2.6	0.63
Fall	2,420 (22.6)	3.4	0.72	3,185 (29.8)	2.1	0.53	4,567 (42.7)	1.9	0.52	4,803 (44.9)	1.6	0.46
Annual	6,027	3.4	na	9,233	2.2	na	12,946	2.0	na	14,799	1.6	na

TABLE 4. Seasonal per-capita cumulative consumption (total g) and mean per-capita daily consumption (g/d), mean percent body weight consumed per day (%/d), and proportion of maximum consumption (*P*) attained by age-1, age-2, age-3, age-4, and age-5 Lake Norman striped bass in 2001 and 2002 estimated in bioenergetics simulations based on observed seasonal growth of each age-class. Spring refers to 1 January–15 June, summer refers to 16 June–15 September, and fall refers to 16 September–31 December. na = not applicable.

	Age 1			Age 2			Age 3			Age 4			Age 5		
	Total g (g/d)	%/d	*P*	Total g (g/d)	%/d	*P*	Total g (g/d)	%/d	*P*	Total g (g/d)	%/d	*P*	Total g (g/d)	%/d	*P*
2001															
Spring	1,507 (9.1)	2.7	0.49	2,706 (16.3)	1.8	0.41	3,254 (19.6)	1.5	0.34	3,226 (19.4)	1.3	0.32	3,883 (23.4)	1.3	0.32
Summer	2,261 (24.6)	4.5	0.75	3,646 (39.6)	3.6	0.70	4,073 (44.3)	3.0	0.63	4,073 (44.3)	2.9	0.62	4,933 (53.6)	2.8	0.63
Fall	2,174 (20.3)	3.3	0.57	3,104 (29.0)	2.6	0.52	3,437 (32.1)	2.3	0.48	3,386 (31.6)	2.3	0.47	4,108 (38.4)	2.2	0.48
Annual	5,942	3.3	na	9,456	2.5	na	10,764	2.1	na	10,685	2.0	na	12,924	1.9	na
2002															
Spring	1,796 (10.8)	3.3	0.55	2,877 (17.3)	1.9	0.40	3,486 (21.0)	1.5	0.33	3,599 (21.7)	1.4	0.31	3,447 (20.8)	1.3	0.30
Summer	2,348 (25.5)	4.4	0.74	3,014 (32.8)	2.9	0.57	3,472 (37.7)	2.5	0.54	3,590 (39.0)	2.4	0.52	3,457 (37.6)	2.4	0.52
Fall	2,082 (19.5)	2.8	0.54	2,819 (26.3)	2.2	0.47	2,948 (27.6)	1.9	0.42	2,933 (27.4)	1.8	0.40	2,788 (26.1)	1.8	0.40
Annual	6,226	3.4	na	8,710	2.2	na	9,906	1.9	na	10,122	1.8	na	9,692	1.7	na

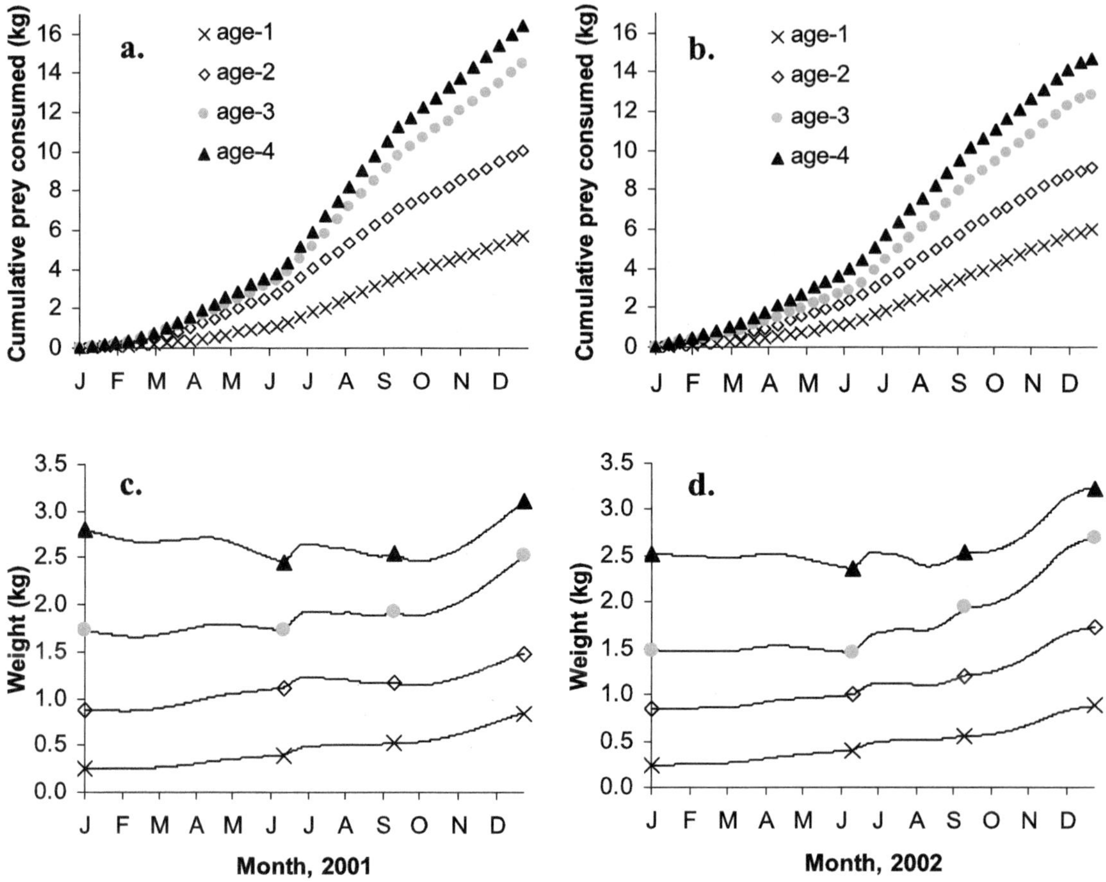

FIGURE 6. Estimates of (a, b) cumulative consumption and (c, d) weight of individual age-1 through age-4 striped bass in Badin Lake in (a, c) 2001 and (b, d) 2002. Points in panels c and d indicate observed weights input to the model, whereas lines indicate model simulations of growth.

low prey availability prior to shad spawning. The proportion of maximum consumption attained by each age-class, which will be dictated primarily by food availability, was similar between the two reservoirs in the spring and was lower than in other seasons in both systems (Tables 3 and 4). Presumed limits on food availability due to the scarcity of appropriately sized clupeid forage fish before young-of-the-year fish become available are supported by the low occurrence of clupeids in the stomachs of striped bass collected during the spring and very early summer (23.9% and 36.1% of striped bass stomachs contained clupeids in Badin Lake and Lake Norman, respectively, during this time). Instead, striped bass collected from both reservoirs during this period frequently had empty stomachs or contained invertebrates (among Lake Norman striped bass <475 mm TL) or single large nonclupeid prey fish (among striped bass >500 mm TL in both systems; Thompson 2006).

Consumption rates were highest over the summer for all age-classes in both Badin Lake and Lake Norman (Tables 3 and 4; Figures 6 and 7). In Badin Lake, growth for all age-classes was greater during the summer than during the spring but was lower than during the fall (Table 5; Figure 6), even though consumption rates were lower in the fall (see below). This result indicated that a large proportion of the energy consumed by Badin Lake striped bass during the summer was used to meet metabolic costs

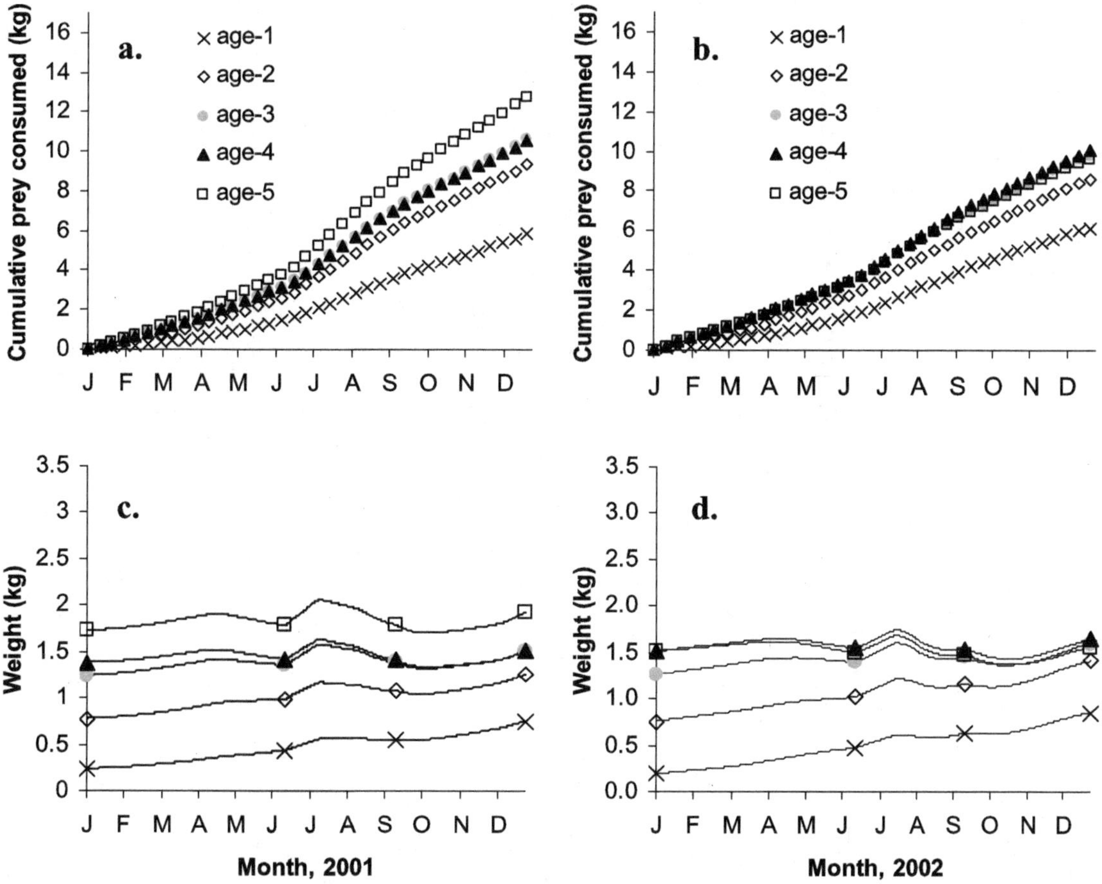

FIGURE 7. Estimates of (a, b) cumulative consumption and (c, d) weight of individual age-1 through age-5 striped bass in Lake Norman in (a, c) 2001 and (b, d) 2002. Points in panels c and d indicate observed weights input to the model, whereas lines indicate model simulations of growth.

associated with warm temperatures, rather than being allocated to growth. All ages of Badin Lake striped bass consumed a high proportion of C_{max} during the summer (Table 3), suggesting that these fish were able to meet their metabolic costs and achieve moderate growth during this period due to high food availability. Stomachs of Badin Lake striped bass collected during the late summer were full of young-of-the-year clupeids (primarily threadfin shad; Thompson 2006), showing that fish of all sizes were taking advantage of this seasonally abundant food resource. In Lake Norman, younger fish achieved some moderate growth over the summer, but growth of age-3 and older fish was negligible (Table 5; Figure 7), indicating that almost all of the energy consumed by older Lake Norman striped bass during the summer was used to meet the metabolic demands associated with warm summer temperatures. Although striped bass in Lake Norman attained their highest proportions of C_{max} during the summer (Table 4), these proportions were lower and had greater interannual variability than in Badin Lake. Lake Norman striped bass found at the top of the oxycline, as predicted by our habitat selection rules, would overlap spatially with forage fish, which Schael et al. (1995) observed from the top of the oxycline to the surface in July in Lake Norman. Thus, while seasonal availability of prey fish was highest during the summer in Lake Norman and striped bass should be able

TABLE 5. Seasonal weights (g) and growth rates (g/d) of striped bass in Badin Lake and Lake Norman in 2001 and 2002. Values under each date are the weight (g) of each age-class on that day, whereas each growth rate represents the mean rate between the dates to either side of the given value.

	2001							2002						
	1 Jan.	Growth rate (g/d)	16 June	Growth rate (g/d)	16 Sept.	Growth rate (g/d)	31 Dec.	1 Jan.	Growth rate (g/d)	16 June	Growth rate (g/d)	16 Sept.	Growth rate (g/d)	31 Dec.
Badin Lake														
Age 1	256	0.85	396	1.49	533	2.89	842	239	0.99	403	1.62	552	3.04	877
Age 2	878	1.44	1,116	0.59	1,170	2.92	1,482	842	0.96	1,000	2.27	1,209	4.83	1,726
Age 3	1,729	−0.01	1,728	2.12	1,923	5.57	2,519	1,482	−0.16	1,456	5.24	1,938	7.00	2,687
Age 4	2,803	−2.17	2,445	1.09	2,545	5.21	3,103	2,519	−0.99	2,356	2.04	2,544	6.42	3,231
Lake Norman														
Age 1	240	1.19	436	1.32	557	1.87	757	191	1.72	474	1.70	630	2.10	855
Age 2	776	1.27	986	1.01	1,079	1.70	1,261	757	1.59	1,019	1.48	1,155	2.35	1,406
Age 3	1,244	0.68	1,357	0.33	1,387	1.23	1,519	1,261	0.82	1,397	0.53	1,446	1.46	1,602
Age 4	1,385	0.20	1,418	−0.09	1,410	0.97	1,514	1,519	0.16	1,545	−0.02	1,543	0.95	1,645
Age 5	1,729	0.33	1,784	0.09	1,792	1.32	1,933	1,514	−0.11	1,496	−0.18	1,479	0.76	1,560

to access these prey, these resources were not sufficient to support substantial growth during this season because they coincide with the warmest temperatures of the year.

Model simulations of striped bass in both lakes showed a pattern of weight gain early in the summer followed by weight loss over the remainder of the summer that became more pronounced in progressively older fish, particularly in Lake Norman (Figures 6 and 7). These simulated intraseasonal growth patterns should be interpreted with caution, as they are based on the assumption that a fish consumes the same proportion of C_{max} every day of the summer. Fitting a constant *P*-value over a growth interval provides a robust estimate of the cumulative consumption required to achieve the observed net growth given experienced temperatures, regardless of how that consumption is actually distributed over the growth interval (Cochran and Rice 1982). However, if daily consumption rates vary markedly over the interval due to intraseasonal variation in forage availability, the actual pattern of growth resulting from the same cumulative consumption can be quite varied (Rice and Cochran 1984). Based on purse-seine catch per unit effort, biomass of prey fish in Badin Lake increased from early to late summer (Thompson 2006), such that the actual consumption attained by Badin Lake striped bass likely increased over the summer. Thus, growth simulated using a constant *P*-value may be overestimated in the early summer and underestimated in the late summer. Determining the actual intraseasonal pattern of growth and the factors driving it would require shorter simulation intervals between more frequent measures of striped bass size (preferably accompanied by estimates of forage availability to aid in interpretation of seasonal variation in estimated consumption rates). Regardless of the true growth trajectory of striped bass over the summer, comparisons of summer consumption estimates between the two systems (Tables 3 and 4; Figures 6 and 7) support the conclusion that abundant food in Badin Lake is essential to minimizing any potential weight loss in the late summer and allowing for positive net growth over the season. In Lake Norman, on the other hand, more limited food prevents striped bass from attaining a net increase in weight over the summer, even with a potential, simulated weight gain early in the season (Figure 7).

Fall consumption rates fell between the spring and summer rates in both reservoirs (Tables 3 and 4; Figures 6 and 7). In combination with cooler fall temperatures, which would increase the amount of growth that could be achieved for a given amount of food, these moderate consumption levels were sufficient for Badin Lake striped bass to attain high growth rates. Fall conditions in Badin Lake appeared to be the most conducive to rapid growth of striped bass as growth rates in the fall were greater than in any other seasonal period for all age-classes (Table 5; Figure 6). The proportions of C_{max} attained by each age-class of striped bass in both systems in the fall were lower than those attained in the summer period (Tables 3 and 4), suggesting that food resources, particularly young-of-the-year clupeids, became somewhat reduced as the year progressed. While these resources were still sufficient to support substantial growth in Badin Lake, fall *P*-values were 0.05–0.1 lower for age-3 fish and 0.02–0.06 lower for age-4 fish in Lake Norman (Tables 3 and 4), and growth of Lake Norman striped bass was low over the fall, especially among the older fish (Table 5; Figure 7). These results indicate that by the time temperatures began to cool, reducing metabolic costs, forage availability had become too limited in Lake Norman to support quality striped bass growth. Lake Norman striped bass of all size-classes did not, therefore, experience any season with the combination of conditions conducive to substantial net growth.

Both the mean percent body weight consumed per day and the proportion of maximum consumption attained decreased with increasing age of striped bass in all seasons in both reservoirs (Tables 3 and 4). The majority of the diet of all sizes of striped bass was composed of similar size prey items (Thompson 2006);

because smaller striped bass require less energy to meet their daily metabolic requirements than larger fish, they can satisfy this demand with fewer prey, giving them a growth advantage over larger individuals. However, it is important to note that the oldest striped bass modeled in Badin Lake (age 4) still achieved a substantial annual growth increment (Table 5; Figure 6), suggesting that while smaller fish attain a higher proportion of C_{max} in all seasons, larger fish (up to ~3.25 kg) are still able to consume sufficient resources to maintain positive annual growth. This trend of positive annual growth at older ages appears to continue for fish up to age 8 (~800 mm TL, 6 kg), the maximum age of striped bass collected in Badin Lake (Figure 2). In Lake Norman, on the other hand, a substantial decline in annual growth was observed among the older fish modeled (age 3 to age 5; ~1.25–2 kg; Table 5; Figure 7). Because older Lake Norman striped bass were considerably smaller than Badin Lake fish of the same age, the observed decline in percent body weight consumed per day represented a greater relative reduction among older Lake Norman fish. This result suggests that older fish in Lake Norman had substantially more difficulty consuming enough food to allocate much energy to growth compared to younger fish in the same system and that this contrast between age-classes was greater than in Badin Lake. Prey limitation in Lake Norman would also cause the costs associated with searching for and capturing prey compared to the benefit received from a prey item to be proportionally greater for larger fish.

Habitat exchange simulations

Results of the habitat exchange simulations showed that differences in forage availability had a larger relative influence on growth of striped bass in Badin Lake and Lake Norman than did differences in thermal experience. In 2001, annual growth of simulated fish in all habitat exchange simulations most closely resembled growth of fish in the system from which food consumption was taken (Table 6). In the temperature exchange simulations, simulated annual growth was relatively unchanged, with growth of Badin Lake fish diminishing by 164 g at Lake Norman temperatures and growth of Lake Norman fish increasing by 148 g at Badin Lake temperatures (Table 6; Figure 8). These differences may seem in the opposite direction of the expected result, but temperatures were warmer in Lake Norman over the cooler months of the year, thereby lowering potential growth for a given level of food consumption, while Badin Lake was only warmer over several weeks in early summer (Figure 4). Differences in annual growth in the consumption exchange simulations, however, were about three times those seen in the temperatures exchange simulations, with growth of Badin Lake fish diminishing by 461 g when given Lake Norman consumption levels and growth of Lake Norman fish increasing by 453 g when given Badin Lake consumption levels (Table 6; Figure 8).

The impact on simulated striped bass growth of exchanging experienced temperatures or food consumption levels between lakes varied seasonally. In the spring and fall of 2001, the effects of exchanging thermal experience and food consumption levels were fairly similar in magnitude (Figure 8). During the summer, however, differences in growth due to changes in consumption level were substantially greater than differences due to changes in temperature (Figure 8). While temperatures experienced by fish in both systems were similar once fish were forced into warm epilimnetic water by hypolimnetic and metalimnetic hypoxia, fish in Lake Norman were able to remain in cooler metalimnetic water for an additional 2 weeks in 2001 (Figure 4). We would expect this delay in exposure to warm summer temperatures to have positive energetic consequences, and indeed, Badin Lake fish experiencing Lake Norman temperatures grew an additional 134 g over the summer (Table 6; Figure 8). However, it is clear from these results that the difference in food consumption levels achieved by striped bass in the two systems had even greater repercus-

TABLE 6. Results of 2001 baseline and habitat exchange simulations using an age-3 Badin Lake striped bass (initial weight 1,729 g) and an age-5 Lake Norman striped bass (initial weight 1,729 g). Cumulative consumption is given for each of the three seasonal periods modeled and for the entire year, with the corresponding estimated proportion of maximum consumption (*P*-value) in parentheses. Net growth is given for each seasonal period modeled and for the entire year. Values in plain type are those input to the model; values in bold are model estimates.

	Consumption (g)				Growth (g)			
	Spring	Summer	Fall	Annual	Spring	Summer	Fall	Annual
Baseline simulations								
All Badin conditions	**3,538 (0.36)**	**6,105 (0.79)**	**5,033 (0.53)**	**14,676**	−1	195	596	790
All Norman conditions	**3,883 (0.32)**	**4,993 (0.63)**	**4,108 (0.47)**	**12,985**	55	8	141	204
Habitat exchange simulations								
Badin conditions, except Norman temperatures	3,538 **(0.30)**	6,105 **(0.78)**	5,033 **(0.52)**	14,676	**−94**	**326**	**394**	**626**
Norman conditions, except Badin temperatures	3,883 **(0.38)**	4,993 **(0.63)**	4,108 **(0.48)**	12,985	**170**	**−147**	**328**	**352**
Badin conditions, except Norman consumption	3,883 **(0.39)**	4,993 **(0.67)**	4,108 **(0.50)**	12,985	**99**	**−172**	**401**	**329**
Norman conditions, except Badin consumption	3,538 **(0.30)**	6,105 **(0.74)**	5,033 **(0.51)**	14,676	**−53**	**426**	**284**	**657**

sions for growth, with Badin Lake fish experiencing Lake Norman consumption showing a 367 g reduction in summer growth and Lake Norman fish experiencing Badin Lake consumption showing a 418 g increase in summer growth (Table 6; Figure 8).

The results of habitat exchange simulations based on 2002 conditions showed an even more pronounced effect of consumption on striped bass growth relative to the effect of temperature. Annual growth only diminished by 27 g for Badin Lake fish given Lake Norman temperatures and only increased by 17 g for Lake Norman fish given Badin Lake temperatures (Table 7; Figure 8). However, using food consumption estimated in the other reservoir produced a 1,023 g reduction in growth of Badin Lake striped bass and a 974 g increase in growth of Lake Norman striped bass over the year (Table 7; Figure 8). As in 2001, differences in growth due to changes in temperature versus consumption were similar in magnitude in the spring, but changes in consumption had a much greater effect on growth over the summer and, in this year, over the fall as well (Table 7; Figure 8). In the summer of 2002, Lake Norman striped bass were able to remain in cooler metalimnetic water for an additional 4 weeks beyond the date when Badin Lake fish were forced into warmer water (Figure 4). This difference again set up the potential for substantial positive temperature effects on growth during this season, but the effect of consumption was even more pronounced. Badin Lake fish experienced a 710 g reduction in growth with Lake Norman consumption levels versus a 217 g in-

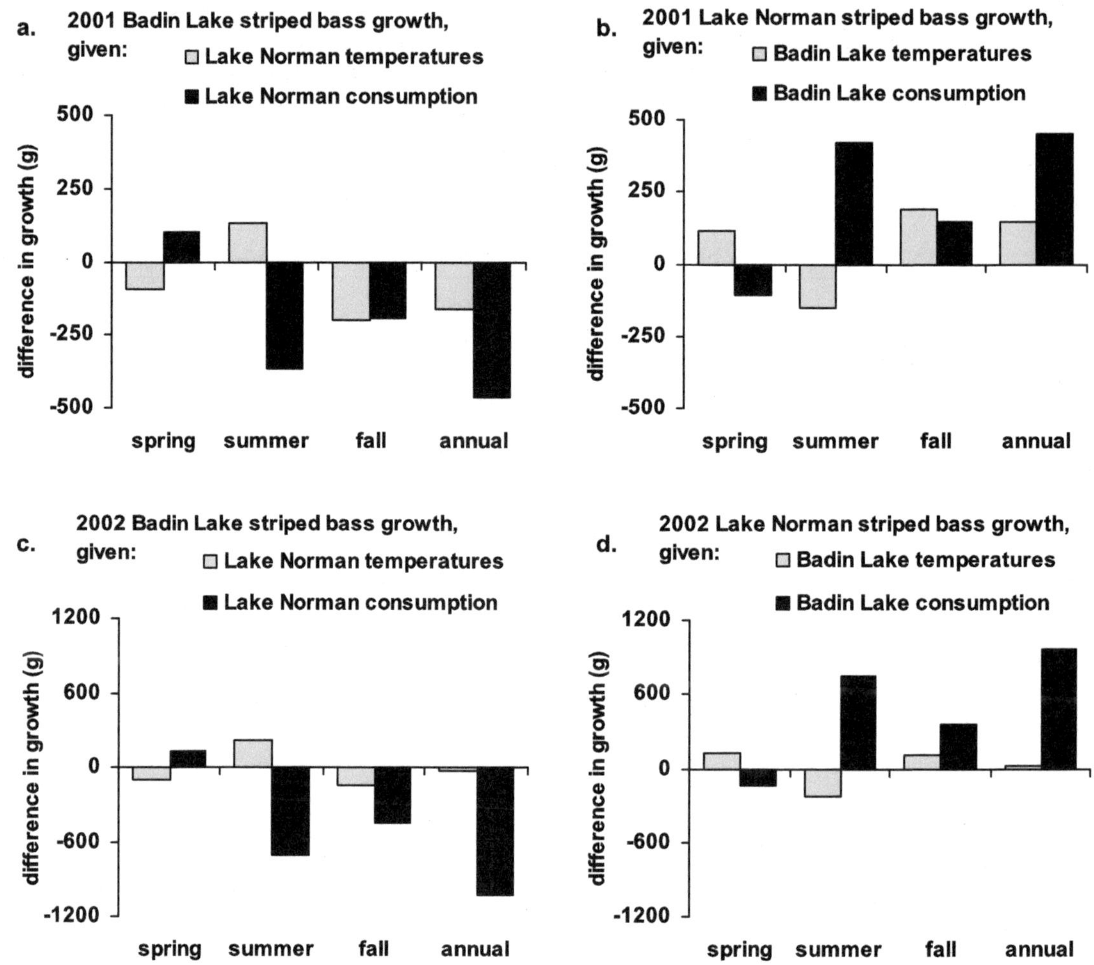

FIGURE 8. Differences between the weight gain simulated in temperature exchange simulations and the original observed weight gain (gray bars) and the weight gain simulated in consumption exchange simulations and the original observed weight gain (black bars) of (a, c) Badin Lake and (b, d) Lake Norman striped bass in (a, b) 2001 and (c, d) 2002. Temperature exchange simulations used all conditions from one reservoir but substituted the thermal regime of the other reservoir, whereas consumption exchange simulations used all conditions from one reservoir but substituted the cumulative seasonal food consumption estimated for a similar-sized fish in the other reservoir.

crease in growth with Lake Norman temperatures over the summer, while Lake Norman fish showed a 746 g increase in growth with Badin Lake consumption levels versus a 220 g decrease with Badin Lake temperatures over this time period (Table 7; Figure 8). While temperature and consumption effects were both somewhat reduced in the fall (Figure 8), the temperature effect was only about 30% of the consumption effect in both systems over this season, similar to the percentage difference observed over the summer.

Discussion

Our findings suggest that patterns of striped bass growth that seem inconsistent with habitat restrictions due to thermal and dissolved oxygen structure may be explained by differences in forage availability, and management of res-

TABLE 7. Results of 2002 baseline and habitat exchange simulations using an age-3 Badin Lake striped bass (initial weight 1,482 g) and an age-5 Lake Norman striped bass (initial weight 1,514 g). Cumulative consumption is given for each of the three seasonal periods modeled and for the entire year, with the corresponding estimated proportion of maximum consumption (*P*-value) in parentheses. Net growth is given for each seasonal period modeled, and for the entire year. Values in plain type are those input to the model; values in bold are model estimates.

	Consumption (g)				Growth (g)			
	Spring	Summer	Fall	Annual	Spring	Summer	Fall	Annual
Baseline conditions								
All Badin conditions	**3,007 (0.32)**	**5,373 (0.77)**	**4,567 (0.52)**	**12,946**	–26	482	749	1,205
All Norman conditions	**3,447 (0.30)**	**3,457 (0.52)**	**2,789 (0.40)**	**9,692**	–18	–17	81	46
Habitat exchange simulations								
Badin conditions, except Norman temperatures	3,007 **(0.27)**	5,373 **(0.74)**	4,567 **(0.47)**	12,946	**–127**	**699**	**606**	**1,178**
Norman conditions, except Badin temperatures	3,447 **(0.35)**	3,457 **(0.53)**	2,789 **(0.45)**	9,692	**99**	**–237**	**201**	**63**
Badin conditions, except Norman consumption	3,447 **(0.36)**	3,457 **(0.53)**	2,789 **(0.44)**	9,692	**111**	**–228**	**298**	**182**
Norman conditions, except Badin consumption	3,007 **(0.27)**	5,373 **(0.73)**	4,567 **(0.47)**	12,946	**–153**	**729**	**443**	**1,020**

ervoir striped bass populations should consider the important influence of forage availability on striped bass growth and condition. Highly productive reservoirs need not be automatically excluded from consideration for stocking striped bass because they lack summer habitat meeting Coutant's (1985) suggested suitability criteria of temperatures less than 25°C and dissolved oxygen levels greater than 2–3 mg/L. Clearly growth supported by a given forage base will be reduced as the summer temperatures that fish must occupy increase, so bioenergetic constraints will prevent such systems from supporting the trophy fisheries that may develop in systems with abundant prey resources and ideal temperature and dissolved oxygen conditions. However, highly productive reservoirs with severe thermal and dissolved oxygen stratification can support productive and popular fisheries for striped bass in the 1–6 kg range, with maximum sizes up to 9 kg in some systems. Thompson et al. (2010) suggested that striped bass in highly productive systems that are forced into warm epilimnetic waters by hypolimnetic and metalimnetic hypoxia may actually benefit energetically if those physical conditions increase the spatial overlap between striped bass and shallow, warmwater prey. Our results support this conclusion by demonstrating that Badin Lake striped bass experience high food consumption rates and attain high proportions of their physiological maximum consumption rate in the summer, indicating that their prey are both abundant and spatially available. These high consumption rates allow Badin Lake striped bass to achieve modest net growth over the summer, despite prolonged exposure to highly unsuitable thermal conditions. Therefore, the lack of isolated thermal refuges or an oxygenated metalimnion with preferred temperatures counterintuitively allows striped bass in systems such as Badin Lake to continue consuming prey through the warmest months of the summer, rather than potentially becoming isolated from their prey as in some other systems (Coutant 1985).

Summer conditions are also only one component of the annual thermal and forage regimes that determine striped bass growth. The continuation of high forage availability into the fall as temperatures become cooler may allow striped bass in productive systems to experience ideal conditions for rapid growth, as was seen with striped bass in Badin Lake. Such conditions in the fall may allow for substantial positive annual growth even in systems with little growth, or even negative growth, over the summer.

The importance of forage availability to patterns of striped bass growth and condition also indicates that poor growth and condition are not solely dictated by temperature and dissolved oxygen conditions outside the influence of fishery managers. Rather, growth and condition will reflect the ability of the prey community to support the predatory demand of striped bass. This predatory demand will increase as summer temperatures increase, so balancing resource supply and demand will be particularly important for the success of striped bass fisheries in southern reservoirs. Low natural mortality and high fishing mortality have been estimated for several reservoir striped bass populations (Hightower et al. 2001; Young and Isely 2004; Thompson et al. 2007), suggesting that adjusting harvest regulations or stocking rates, or both, should provide managers with an effective means of manipulating the structure of these populations and improving growth and condition.

Our study provides strong evidence that poor growth and condition of larger striped bass in Lake Norman are the direct result of insufficient prey resources to support quality growth given the number and size of fish in the system. At the time of our study, harvest regulations did not allow fish to be removed from the population until they reached age 3. Because almost all consumption by fish age 3 and older was used for maintenance rather than growth, most of the forage resources consumed by striped bass in Lake Norman were either being used by fish before they reached harvestable size or for merely keeping fish above that size alive. In Lake Norman age-3 to age-5 striped bass each eat about 160% as much forage each year as an age-1 fish, and about 120% as much as an age-2 fish, while achieving little or no growth. With such high predatory demand and low productivity, Lake Norman fish do not experience any season with the combination of cooler temperatures and abundant prey resources necessary for rapid growth. Fostering higher harvest levels of smaller fish is, therefore, the most appropriate management strategy for this reservoir. Based on the results of our study, in 2006, the minimum length limit for striped bass in Lake Norman was lowered from 508 to 406 mm TL from 1 October to 31 May, with no size restriction from 1 June to 30 September. These regulations should allow anglers to harvest fish before the size at which growth and condition declines, and with the more rapid removal of striped bass, more food should be left for those fish remaining in the system. Changes in growth rates will give useful feedback for managers when implementing such modifications to striped bass fisheries. In this context, growth rates provide a biological synthesis of information on the availability of food resources for each segment of the population in relation to the energetic costs imposed by the physical environment.

Manipulation of the striped bass population in Badin Lake may also be possible to take greater advantage of the forage resources in this system. Rapid growth in the fall, positive annual growth, and high consumption rates of Badin Lake striped bass indicate that forage availability is currently high. Therefore, increasing stocking rates may allow the system to support a larger number of fish and improve angler catch rates, or alternatively, reducing the harvest rate or increasing the size limit could increase the number of somewhat larger fish (~6–8 kg). Our bioenergetics analysis indicates that the paucity of such larger fish in Badin Lake is not due to limits on consumption and growth, and a yield-per-recruit model of this population demonstrated that older fish could be produced with a reduction in the fishing mortality rate (Thompson et al. 2007). Although summer habitat constraints will prevent striped bass in Badin Lake from reaching larger trophy sizes found in other systems with more ideal conditions, shifting the population structure to include a greater number of individuals in the larger end of the size range currently found in the system (thereby increasing the average size of harvested fish) should be possible if desired by managers and anglers. With any change in management strategies, condition and annual growth rates of each age-class should be monitored closely to ensure that management decisions do not have unexpected consequences. In Badin Lake, for example, older striped bass currently experience some weight loss during January–June, and this loss may be exacerbated by increased numbers or sizes of striped bass if growth is density-dependent during this period.

The habitat exchange simulation approach used in this study provides a powerful means for comparing the relative importance of two or more habitat conditions on growth of fish in multiple populations. In the context of understanding growth of striped bass in southern reservoirs, habitat exchange simulations using Badin Lake and Lake Norman conditions showed that the relative effect of forage availability on annual striped bass growth was about three times greater than that of temperature in 2001 and about 37 times greater in 2002. Applying habitat exchange simulations to additional reservoirs with a broader range of thermal constraints and productivities should help us determine how the relative influence of physical conditions and forage availability on striped bass growth changes as physical habitat quality varies. This broader analysis will allow us to continue to refine our understanding of the abiotic and biotic conditions required for successful reservoir striped bass fisheries.

Acknowledgments

We thank Scott Waters, Lawrence Dorsey, Bob Barwick, Scott Van Horn, Dave Coughlan, Kim Baker, Hugh Barwick, Mark Rash, Duane Harrell, and Bob Doby for field assistance. Bill Foris supplied all of the data on Lake Norman physical conditions used in this study. Numerous undergraduate technicians at North Carolina State University assisted with processing fish samples. Chuck Coutant, Jim Bulak, Mark Bevelhimer, and one anonymous reviewer provided helpful comments for improving this manuscript. Funding was provided by a North Carolina Wildlife Resources Commission grant (Federal Aid in Sport Fish Restoration Project F-68–04) to J. A. Rice and a Robert M. Jenkins Memorial Reservoir Research Scholarship from the Reservoir Committee of the Southern Division of the American Fisheries Society and a National Science Foundation Graduate Research Fellowship to J. S. Thompson.

References

Anderson, R. O., and R. M. Neumann. 1996. Length, weight, and associated structural indices. Pages 447–481 *in* B. R. Murphy and D. W. Willis, editors. Fisheries techniques, 2nd edition. American Fisheries Society, Bethesda, Maryland.

Axon, J. R., and D. K. Whitehurst. 1985. Striped bass management in lakes with emphasis on management problems. Transactions of the American Fisheries Society 114:8–11.

Cheek, T. E., M. J. Van Den Avyle, and C. C. Coutant. 1985. Influences of water quality on distribution of striped bass in a Tennessee River impoundment. Transactions of the American Fisheries Society 114:67–76.

Cochran, P. A., and J. A. Rice. 1982. A comparison of bioenergetics and direct field estimates of cumulative seasonal food consumption by largemouth bass (*Micropterus salmoides*). Pages 88–96 *in* G. Cailliet and C. Simenstad, editors. Gutshop '81: fish food habits studies. Washington Sea Grant, Seattle.

Coutant, C. C. 1985. Striped bass, temperature, and dissolved oxygen: a speculative hypothesis for environmental risk. Transactions of the American Fisheries Society 114:31–61.

Coutant, C. C. 1986. Thermal niches of striped bass. Scientific American 254:98–104.

Cummins, K. W., and J. C. Wuycheck. 1971. Caloric equivalents for investigations in ecological energetics. Mitteilung Internationale Vereinigung fuer Theoretische und Amgewandte Limnologie 18:1–151.

Cyterski, M., J. Ney, and M. Duval. 2002. Predator demand for clupeid prey in Smith Mountain Lake, Virginia. Fisheries Research 59:1–16.

Davias, L. A. 2006. A bioenergetics assessment of temperature and food consumption effects on growth of reservoir striped bass. Master's thesis. North Carolina State University, Raleigh.

Driver, E. A., L. G. Sugden, and R. J. Kovach. 1974. Calorific, chemical, and physical values of potential duck foods. Freshwater Biology 4:281–292.

Farquhar, B. W., and S. Gutreuter. 1989. Distribution and migration of adult striped bass in Lake Whitney, Texas. Transactions of the American Fisheries Society 118:523–532.

Hanson, P. C., T. B. Johnson, D. E. Schindler, and J. F. Kitchell. 1997. Fish Bioenergetics 3.0. University of Wisconsin Sea Grant Institute, Madison, Wisconsin.

Hartman, K. J., and S. B. Brandt. 1995a. Comparative energetics and the development of bioenergetics models for sympatric estuarine piscivores. Canadian Journal of Fisheries and Aquatic Sciences 52:1647–1666.

Hartman, K. J., and S. B. Brandt. 1995b. Predatory demand and impact of striped bass, bluefish, and weakfish in the Chesapeake Bay: applications of bioenergetics models. Canadian Journal of Fisheries and Aquatic Sciences 52:1667–1687.

Hartman, K. J., and S. B. Brandt. 1995c. Estimating energy density of fish. Transactions of the American Fisheries Society 124:347–355.

Hightower, J. E., J. R. Jackson, and K. H. Pollock. 2001. Use of telemetry methods to estimate natural and fishing mortality of striped bass in Lake Gaston, North Carolina. Transactions of the American Fisheries Society 130:557–567.

Jackson, J. R., and J. E. Hightower. 2001. Reservoir striped bass movements and site fidelity in relation to seasonal patterns in habitat quality. North American Journal of Fisheries Management 21:34–45.

Jenkins, R. E., and N. M. Burkhead. 1993. Freshwater fishes of Virginia. American Fisheries Society, Bethesda, Maryland.

Johnson, R. L., S. C. Blumenshine, and S. M. Coghlan. 2006. A bioenergetics analysis of factors limiting brown trout growth in an Ozark tailwater river. Environmental Biology of Fishes 77:121–132.

Kitchell, J. F., J. F. Koone, R. V. O'Neill, H. H. Shugart, Jr., J. J. Magnuson, and R. S. Booth. 1974. Model of fish biomass dynamics. Transactions of the American Fisheries Society 103:786–798.

Kitchell, J. F., D. J. Stewart, and D. Weininger. 1977. Applications of a bioenergetics model to yellow perch (*Perca flavescens*) and walleye (*Stizostedion vitreum vitreum*). Journal of the Fisheries Research Board of Canada 34:1922–1935.

Matthews, W. J. 1985. Summer mortality of striped bass in reservoirs of the United States. Transactions of the American Fisheries Society 114:62–66.

Matthews, W. J., L. G. Hill, and S. M. Schellhaass. 1985. Depth distribution of striped bass and other fish in Lake Texoma (Oklahoma–Texas) during summer stratification. Transactions of the American Fisheries Society 114:84–91.

Moss, J. L. 1985. Summer selection of thermal refuges by striped bass in Alabama reservoirs and tailwaters. Transactions of the American Fisheries Society 114:77–83.

Munch, S. B., and D. O. Conover. 2002. Accounting for local physiological adaptation in bioenergetic models: testing hypotheses for growth rate evolution by virtual transplant experiments. Canadian Journal of Fisheries and Aquatic Sciences 59:393–403.

NCDENR (North Carolina Department of Environment and Natural Resources). 2002.

Basinwide assessment report Yadkin River basin. NCDENR, Raleigh.

NCDENR (North Carolina Department of Environment and Natural Resources). 2003. Basinwide assessment report Catawba River basin. NCDENR, Raleigh.

NCDENR (North Carolina Department of Environment and Natural Resources). 2007. Lake and reservoir assessments—Yadkin-Pee Dee River basin. NCDENR, Raleigh.

Petersen, J. H., and C. P. Paukert. 2005. Development of a bioenergetics model for humpback chub and evaluation of water temperature changes in the Grand Canyon, Colorado River. Transactions of the American Fisheries Society 134:960–974.

Raborn, S. W., L. E. Miranda, and M. T. Driscoll. 2002. Effects of simulated removal of striped bass from a southeastern reservoir. North American Journal of Fisheries Management 22:406–417.

Railsback, S. F., and K. A. Rose. 1999. Bioenergetics modeling of stream trout growth: temperature and food consumption effects. Transactions of the American Fisheries Society 128:241–256.

Rice, J. A., J. E. Breck, S. M. Bartell, and J. F. Kitchell. 1983. Evaluating the constraints of temperature, activity, and consumption on growth of largemouth bass. Environmental Biology of Fishes 9:263–275.

Rice, J. A., and P. A. Cochran. 1984. Independent evaluation of a bioenergetics model for largemouth bass. Ecology 65:732–739.

Rice, J. A., J. S. Thompson, J. A. Sykes, and C. T. Waters. 2013. The role of metalimnetic hyposia in striped bass summer kills: consequences and management implications. Pages 121–145 *in* J. S. Bulak, C. C. Coutant, and J. A. Rice, editors. Biology and management of inland striped bass and hybrid striped bass. American Fisheries Society, Symposium 80, Bethesda, Maryland.

Schael, D. M., J. A. Rice, and D. J. Degan. 1995. Spatial and temporal distribution of threadfin shad in a southeastern reservoir. Transactions of the American Fisheries Society 124:804–812.

Schaffler, J. J., J. J. Isely, and W. E. Hayes. 2002. Habitat use by striped bass in relation to seasonal changes in water quality in a southern reservoir. Transactions of the American Fisheries Society 131:817–827.

Thompson, J. S. 2006. The influence of temperature and forage availability on growth and habitat selection of a pelagic piscivore. Doctoral dissertation. North Carolina State University, Raleigh.

Thompson, J. S., J. A. Rice, and D. S. Waters. 2010. Striped bass habitat selection rules in reservoirs without suitable summer habitat offer insight into consequences for growth. Transactions of the American Fisheries Society 139:1450–1464.

Thompson, J. S., D. S. Waters, J. A. Rice, and J. E. Hightower. 2007. Seasonal fishing and natural mortality of striped bass in a southeastern reservoir. North American Journal of Fisheries Management 27:681–694.

Van Den Avyle, M. J., and J. W. Evans. 1990. Temperature selection by striped bass in a Gulf of Mexico coastal river system. North American Journal of Fisheries Management 10:58–66.

Van Horn, S. L. 2013. A brief history of inland striped bass management. Pages 1–13 *in* J. S. Bulak, C. C. Coutant, and J. A. Rice, editors. Biology and management of inland striped bass and hybrid striped bass. American Fisheries Society, Symposium 80, Bethesda, Maryland.

Van Horn, S. L., J. R. Finke, and D. Degan. 1996. Summer habitat selection of striped bass in Lake Norman. Proceedings of the Annual Conference Association of Fish and Wildlife Agencies 50:91–97.

Vatland, S., P. Budy, and G. P. Thiede. 2008. A bioenergetics approach to modeling striped bass and threadfin shad predator-prey dynamics in Lake Powell, Utah–Arizona Transactions of the American Fisheries Society 137:262–277.

Voshell, Jr., J. R. 2002. A guide to common freshwater invertebrates of North America. McDonald and Woodward Publishing Company, Granville, Ohio.

Wilkerson, M. L., and W. L. Fisher. 1997. Striped bass distribution, movements, and site fidelity in Robert S. Kerr Reservoir, Oklahoma. North American Journal of Fisheries Management 17:677–686.

Young, S. P., and J. J. Isely. 2002. Striped bass annual site fidelity and habitat utilization in J. Strom Thurmond Reservoir, South Carolina–Georgia. Transactions of the American Fisheries Society 131:828–837.

Young, S. P., and J. J. Isely. 2004. Temporal and spatial estimates of adult striped bass mortality from telemetry and transmitter return data. North American Journal of Fisheries Management 24:1112–1119.

Zale, A. V., J. D. Wiechman, R. L. Lochmiller, and J. Burroughs. 1990. Limnological conditions associated with summer mortality of striped bass in Keystone Reservoir, Oklahoma Transactions of the American Fisheries Society 119:72–76.

American Fisheries Society Symposium 80:121–145, 2013

The Role of Metalimnetic Hypoxia in Striped Bass Summer Kills: Consequences and Management Implications

JAMES A. RICE*
Department of Biology, North Carolina State University
Campus Box 7617, Raleigh, North Carolina 27695, USA

JESSICA S. THOMPSON
Department of Organismal and Environmental Biology, Christopher Newport University
1 Avenue of the Arts, Newport News, Virginia 23606, USA

JAMIE A. SYKES
U.S. Army Corps of Engineers
4144 Russell Dam Drive, Elberton, Georgia 30635, USA

CHRISTIAN T. WATERS
North Carolina Wildlife Resources Commission
1751 Varsity Drive, Raleigh, North Carolina 27606, USA

Abstract.—Historically, striped bass *Morone saxatilis* summer kills have been attributed to two mechanisms: stressors associated with crowding when striped bass are confined in isolated, cool refuges; and thermal stress or energetic deficit when hypolimnetic hypoxia (dissolved oxygen less than 2 mg/L) forces them into high-temperature surface waters. Here, we present observations suggesting that a third mechanism may account for some of these striped bass mortality events. During summer stratification, many relatively deep southeastern reservoirs develop hypoxia in the metalimnion, as well as near the bottom, isolating a layer of oxygenated hypolimnetic water between them. As these hypoxic zones expand in thickness and severity of oxygen depletion, the oxygenated layer between them shrinks both horizontally and vertically, and its oxygen content declines. Evidence suggests that striped bass summer kills can occur when fish are trapped in this isolated layer and its oxygen concentration declines below 2 mg/L or disappears altogether. The presence of coolwater forage fish such as alewife *Alosa pseudoharengus* or blueback herring *A. aestivalis* may increase the risk of striped bass kills by attracting them into the hypolimnetic oxygenated layer where they may become trapped. We draw upon examples from two southeastern reservoirs to illustrate this phenomenon, and discuss its implications for reservoir fisheries management, as well as possible approaches to minimize or avoid impacts.

Introduction

Over the past 50 years striped bass *Morone saxatilis* fisheries have been established in hundreds of reservoirs throughout much of the United States (Van Horn 2013; Gustaveson and Blommer 2013; both this volume). However, the success of these fisheries has not been uniform; high variability in growth among sys-

* Corresponding author: jim_rice@ncsu.edu

tems (Wilson et al. 2013, this volume) and the occurrence of summer mortality events in many reservoirs (Matthews 1985) have posed vexing problems for fishery managers. These puzzling observations led Coutant (1985) to develop and publish the "temperature–oxygen squeeze" hypothesis, suggesting that many otherwise paradoxical observations concerning growth and survival of striped bass in different systems could be explained by interactions between habitat conditions and striped bass physiological requirements. Coutant (1985) defined suitable summer habitat for striped bass as having water temperatures of 18–25°C and dissolved oxygen (DO) concentrations greater than 2–3 mg/L, and suggested that when this habitat becomes limited or unavailable it can result in stress, reduced growth, and mortality.

In many reservoirs, striped bass avoid unsuitable habitat that develops due to warming of surface waters and deoxygenation of hypolimnetic waters as summer progresses by crowding into confined refuge areas such as springs, cool stream inflows, or uplake tailraces. If these refuges persist throughout the summer, are large enough, and contain sufficient food, they can benefit striped bass by providing suitable habitat for growth and survival (e.g., Watts Bar Reservoir, Tennessee, Cheek et al. 1985; Lake Seminole, Georgia–Florida, Van Den Avyle and Evans 1990; Robert S. Kerr Reservoir, Oklahoma, Wilkerson and Fisher 1997; Lake Murray, South Carolina, Schaffler et al. 2002). However, Coutant (1978, 1985) pointed out that even with the existence of a persistent refuge, striped bass mortality could still occur due to depletion of or segregation from food resources, increased harvest vulnerability, disease, and other stressors related to crowding. He documented examples of such stress and mortality in Cherokee Reservoir, Tennessee (Coutant 1978, 1985). Thus, issues associated with confinement in refuge habitats represent one mechanism known to account for some striped bass summer mortality events.

Coutant formulated his hypothesis based on telemetry observations of striped bass behavior in systems where suitable habitat, at least in refuges, was available throughout the summer (Coutant 1985 and 2013, this volume). However, striped bass populations have been established in many reservoirs where such habitat does not exist for extended periods during the summer. Coutant (1985) inferred from his observations that under these circumstances, striped bass would choose to occupy warmer water to avoid DO less than 2 mg/L, a prediction amply supported by subsequent studies (Matthews et al. 1985; Farquhar and Gutreuter 1989; Thompson et al. 2010; Coutant 2013). He suggested that prolonged exposure to warmer water would likely cause mortality of adult striped bass, either directly due to thermal or respiratory stress or indirectly by causing them to stop feeding and die of starvation.

Over the past 25 years, however, studies in reservoirs where habitat traditionally considered suitable is not present for part of the summer have revealed a wider range of responses in growth, condition, and survival of striped bass than predicted by Coutant's (1985) original hypothesis. Thompson et al. (2010) demonstrated that when hypoxia (DO less than 2 mg/L) forced striped bass into warmer water, they consistently moved to the top of the oxycline, sometimes occupying relatively high temperatures (27–30°C) for several weeks to several months. Yet fish in many of these reservoirs continue to survive, feed, and grow, although maximum sizes attained are limited (Matthews et al. 1985; Farquhar and Gutreuter 1989; Davias 2006; Thompson et al. 2010; Thompson and Rice 2013, this volume). However, significant striped bass mortality events are occasional or even common occurrences in other reservoirs with similar summer conditions (Matthews 1985).

Recent evidence suggests that these inconsistent growth and mortality patterns among reservoirs lacking summer thermal refuges are related to differences in food availability (Davias 2006; Thompson 2006; Thompson and Rice 2013). Striped bass forced into warm water overlap spatially with warmwater prey such as threadfin shad *Dorosoma petenense* and gizzard shad *D. cepedianum*, which commonly dominate the forage base in such systems (Schael et al. 1995; Thompson et al. 2010). Although exposure to high summer temperatures will increase metabolic costs (Hartman and Brandt 1995), the net effect of metabolic costs and food consumption rates can span a wide continuum depending on food availability, ranging from good growth if consumption rates are sufficient to compensate for high costs to poor growth and condition or even mortality if low food consumption results in an energy deficit.

Some striped bass summer mortality events do appear to be due to starvation or bioenergetic deficit. For example, Zale et al. (1990) found that common summer mortality events in Keystone Reservoir, Oklahoma occurred when hypoxia forced prolonged exposure of striped bass to 28–29°C; based on a sharp decline in angling success once temperatures reached 27°C, they inferred that mortality was caused by cessation of feeding and subsequent malnutrition. Dead striped bass from summer mortality events are rarely collected or necropsied, but in several cases, examinations have revealed enlarged and darkened gallbladders, indicating that the fish had not been feeding recently (Matthews 1985). Thus, in reservoirs without any thermal refuge, a severe energetic deficit resulting from the combined effects of limited food availability and elevated metabolic demands constitutes a second mechanism accounting for some summer mortality events.

However, we have observed some striped bass summer mortality events that cannot be adequately explained by complications of crowding in persistent thermal refuges or energetic deficits in warm water. The fish that die do not exhibit unusually poor condition or other signs of malnutrition or disease. These kills typically include the full size range of adult striped bass in the population, whereas many other die-offs are dominated by large (5–9+ kg) individuals (Matthews 1985), which are more likely to experience energetic constraints (Kitchell 1983; Hartman and Brandt 1995). Though many mortality events tend to occur later in the summer (mid-August through September; Matthews 1985; Zale et al. 1990), these kills generally occur earlier in the summer before temperature and DO conditions throughout the reservoir are at their worst. They do not continue, even though habitat conditions remain poor for the rest of the summer, and they also tend to occur in a restricted area of the reservoir, usually in the deepest area near the dam.

In this paper we make the case for a third mechanism, associated with the formation of a hypoxic or anoxic zone in the metalimnion, which may account for these striped bass summer mortality events. We draw upon observations from two reservoirs to illustrate this phenomenon, and discuss its implications for managing striped bass populations as well as possible approaches to minimize or avoid its impacts.

Metalimnetic Oxygen Minimum Formation and Implications for Fish

During summer stratification, many reservoirs not only develop a hypoxic zone at the bottom of the hypolimnion that expands upward, but also form a metalimnetic oxygen minimum, or hypoxic zone in the metalimnion, that expands and becomes more severe over time (Figure 1). This phenomenon is a common occurrence in relatively deep reservoirs (Cole and Hannon 1990; Wetzel 2001). It is caused by a variety of mechanisms (often in combination), such as biological oxygen demand from decaying organic material accumulating in the density gradient of the metalimnion, inflow of water with high oxygen demand, timing of hydrologic events in relation to the onset of thermal stratification, or dynamics of midlevel water withdrawals (Cole and Hannon 1990; Ruane et al. 2013, this volume). This hypoxic zone in the metalimnion overlays a hypolimnetic oxygen maximum, or isolated layer of higher oxygen concentrations in the upper hypolimnion (Figure 1). As the hypoxic zones on the bottom and in the met-

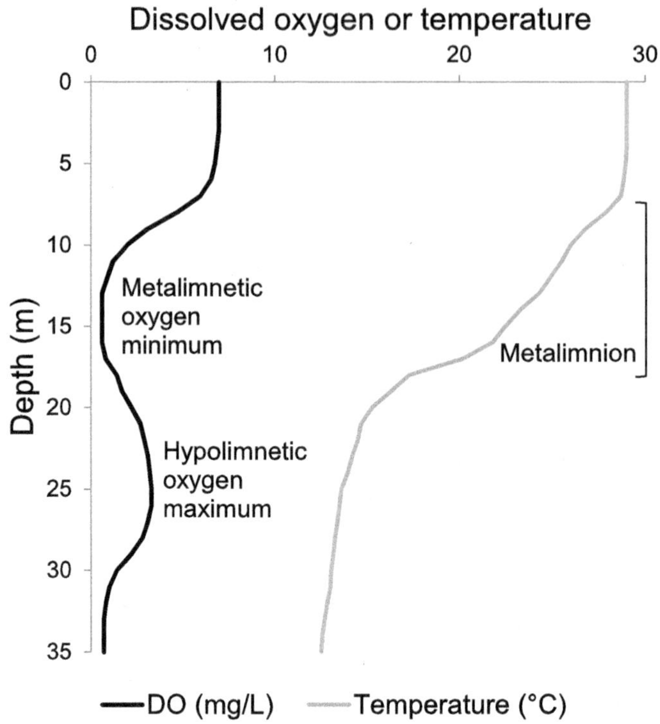

FIGURE 1. Temperature and dissolved oxygen profiles illustrating a metalimnetic oxygen minimum and a hypolimnetic oxygen maximum, which form in some deep reservoirs during summer stratification.

alimnion expand in thickness and become more severe over time, the hypolimnetic oxygenated layer between them shrinks both horizontally and vertically as its oxygen content declines. The rate of decline in oxygen concentration will depend on system-specific conditions, but we have observed rates of 0.06–0.1 mg/L/d in some southeastern U.S. reservoirs.

The impact on striped bass of the development of hypoxia in the metalimnion will depend on whether their preferred temperatures occur within the metalimnion or in the oxygenated layer below it. If their preferred temperatures are in the hypolimnetic oxygenated layer but the forage community consists only of species such as threadfin or gizzard shad that prefer warm temperatures, then striped bass must choose between occupying preferred temperatures with no food below the hypoxic metalimnion or warmer-than-preferred temperatures, with food, above the hypoxic metalimnion. Telemetry studies of striped bass indicate that when habitat with preferred temperatures and adequate DO is available, this combination apparently takes precedence over food availability (Coutant 1985). Thus, faced with this choice, striped bass are likely to occupy the oxygenated layer, even if it isolates them from their prey. However, if coolwater prey such as alewife *Alosa pseudoharengus* or blueback herring *A. aestivalis* are present in sufficient numbers along with preferred temperatures and adequate DO (as they are in some reservoirs), the hypolimnetic layer constitutes ideal habitat as described by Coutant's (1985) original hypothesis. Its long-term value will depend on whether or not these conditions persist until destratification occurs, or if fish can escape through the hypoxic metalimnion if DO in the hypolimnion drops below 2 mg/L.

In many reservoirs, formation of hypoxia in the metalimnion directly overlaps the range of temperatures preferred by adult striped bass. When DO in this zone declines below 2 mg/L, striped bass will be forced to move into water that is warmer or colder than preferred to remain at sufficient oxygen levels. The distribution of forage fish may influence that decision. In systems with only warmwater forage species, warmer epilimnetic water with available food may be more attractive than colder water with no food. If coolwater forage species like alewife or blueback herring are present in the oxygenated hypolimnetic layer, their presence may entice some striped bass to occupy this layer as well, even though it is cooler than they prefer.

Regardless of why fish choose to occupy the oxygenated layer below the hypoxic zone in the metalimnion, this behavior can be risky given the ephemeral nature of this habitat in many reservoirs. If the severity or thickness of metalimnetic hypoxia poses a sufficient barrier to movement, fish could become trapped in this isolated zone, resulting in striped bass summer kills when DO in this zone declines below critical levels. Below, we present evidence from two southeastern U.S reservoirs to support the premise that formation of a metalimnetic oxygen minimum constitutes an important, but previously overlooked, factor contributing to some striped bass mortality events not adequately explained by other mechanisms.

Lake Norman

Lake Norman, North Carolina is a 12,634-ha reservoir on the Catawba River (Figure 2). Impounded by Cowans Ford Dam in 1963, it has a mean depth of 10.2 m, has a maximum depth of 36.6 m, and has been classified as oligotrophic since the late 1970s (NCDENR 2008). It is operated by Duke Energy and is used to provide cooling water for two steam electric power generation plants, as well as for hydroelectric power generation. Striped bass have been stocked into Lake Norman by the North Carolina Wildlife Resources Commission (NCWRC) since the late 1960s; the annual stocking rate averaged about 7.8 fingerlings/ha until 1988, when it was increased to about 12.4 fish/ha (Waters and McRae 2008). Historically, the forage base and diets of striped bass have been dominated by threadfin shad, with a small contribution from gizzard shad (Lewis 1985; Thompson 2006). Striped bass in Lake Norman grow fairly well through age 3, reaching an average size of about 490 mm total length (TL) and 1.25 kg, but older fish exhibit slow growth and relatively poor condition due to habitat and forage availability limitations (Van Horn et al. 2001; Thompson 2006; Thompson and Rice 2013). Only 6% of the population is age 6 or older (Thompson et al. 2005), likely due to high angling mortality rates of striped bass in southeastern reservoirs (Hightower et al. 2001; Young and Isely 2004; Thompson et al. 2007).

Extensive long-term limnological and fisheries monitoring has been conducted on Lake Norman. From 1997 to 2010, Duke Energy personnel monitored water quality about once a month from September to June and about once a week from July to August at 15 stations (11 along the main channel) throughout Lake Norman (Figure 2), with additional sampling during striped bass mortality events. Some water quality monitoring was conducted in years prior to 1997, but the frequency of data collection varied substantially among years. Temperature, DO, and several other variables are measured with a Hydrolab data sonde at each location, starting at the surface (0.3 m) and continuing at 1-m intervals to 1 m above the bottom.

Long-term data have also been collected on the number of summer striped bass mortalities in Lake Norman. From 1983 to 2010, Duke Energy personnel conducted weekly roving boat surveys to search for floating, dead striped bass. Observers searched all open water and main shoreline areas throughout the main body and major tributary arms of the reservoir from the dam to station 15.0 (Figure 2). All dead striped bass encountered were collected, counted, and their location noted. Surveys were conducted more frequently (typically daily) during striped bass mortality events.

Each September from 1993 to 2010, Duke Energy personnel sampled pelagic forage fish in Lake Norman using a purse seine (9 m deep,

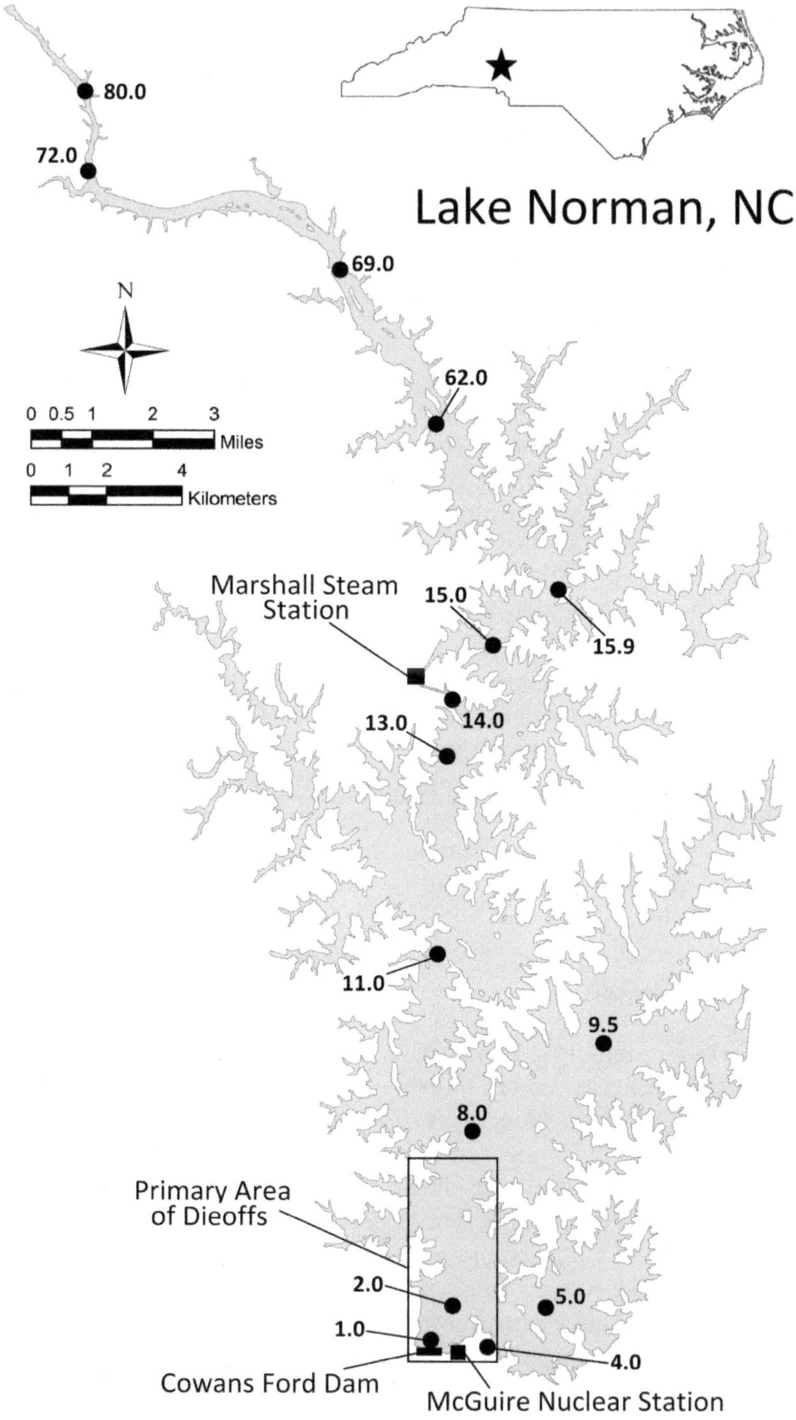

FIGURE 2. Map of Lake Norman, North Carolina, showing locations of water quality monitoring stations (numbered dots) and primary area of striped bass kills.

118 m long, 4.8 mm mesh), typically set in three main regions of the lake. These samples provided data on the species and size composition of the pelagic prey community in the epilimnion over time.

2004 striped bass mortality event

A major striped bass die-off occurred on Lake Norman during the summer of 2004. From July 22 through August 13, Duke Energy and NCWRC personnel collected nearly 2,500 dead striped bass (Figure 3). In contrast, over the previous 20 years, the total number of dead striped bass found in July and August during weekly surveys of the lake averaged 8.75 fish/year (range 0–43). All but about 40 of the fish collected during the 2004 die-off were found within 6 km of the dam (Figure 2).

In 2004, DO began to decline in the metalimnion by early June and metalimnetic hypoxia was well established throughout the main body of the lake by late June, overlaying an oxygenated layer in the hypolimnion in the lower 15 km of the lake (Figure 4). Throughout July, the metalimnion became increasingly hypoxic while the oxygenated area in the hypolimnion declined in thickness, receded down-lake toward the dam, and steadily declined in maximum DO concentration. When the kill began in earnest in late July, only a small, cold (13–14°C) vestige of oxygenated hypolimnion remained with DO slightly above 2 mg/L, separated from the warm (28–30°C), oxygenated epilimnion by a hypoxic layer in the metalimnion 14 m thick, with 7 m less than 0.5 mg/L and 4 m at 0.1 mg/L (Figure 5). Daily collections of dead striped bass increased substantially at the beginning of August as maximum hypolimnetic DO dropped below 2 mg/L, and ended as DO dropped below 0.9 mg/L (Figures 3 and 5). The timing of the mortality event in relation to the decline in hypolimnetic DO concentration immediately behind the dam (Figure 5) suggests that the kill was precipitated by loss of this oxygenated refuge.

Several other lines of evidence also support this hypothesis. Nearly all the dead striped bass were found in a restricted area near the dam, coinciding with the location of the shrinking

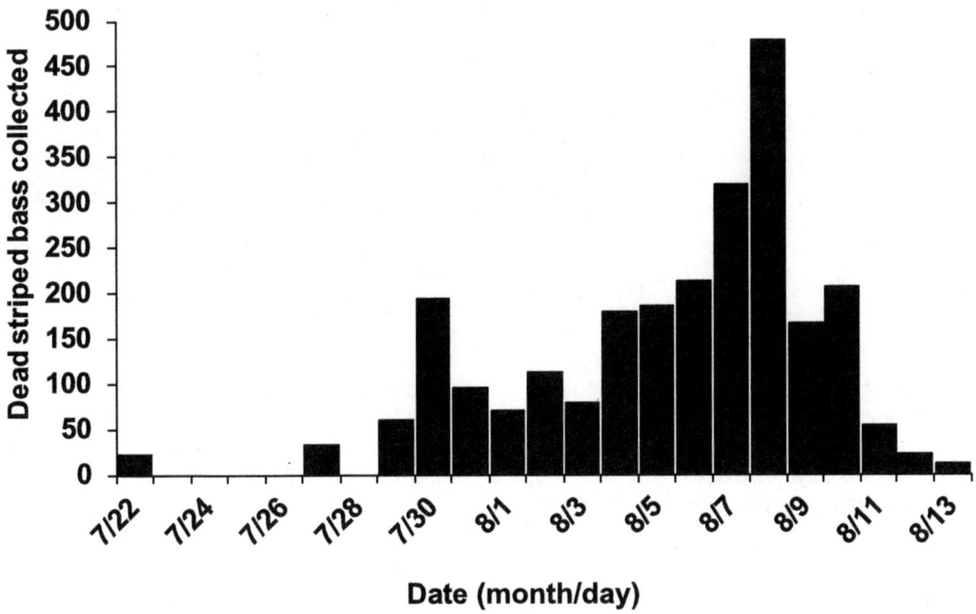

FIGURE 3. Number of dead striped bass collected on Lake Norman during the 2004 summer kill. No surveys were conducted on dates without a histogram bar. A total of 17 additional fish were collected prior to the period shown in surveys on 1, 6, 9 and 15 July, and 11 fish were collected in late August.

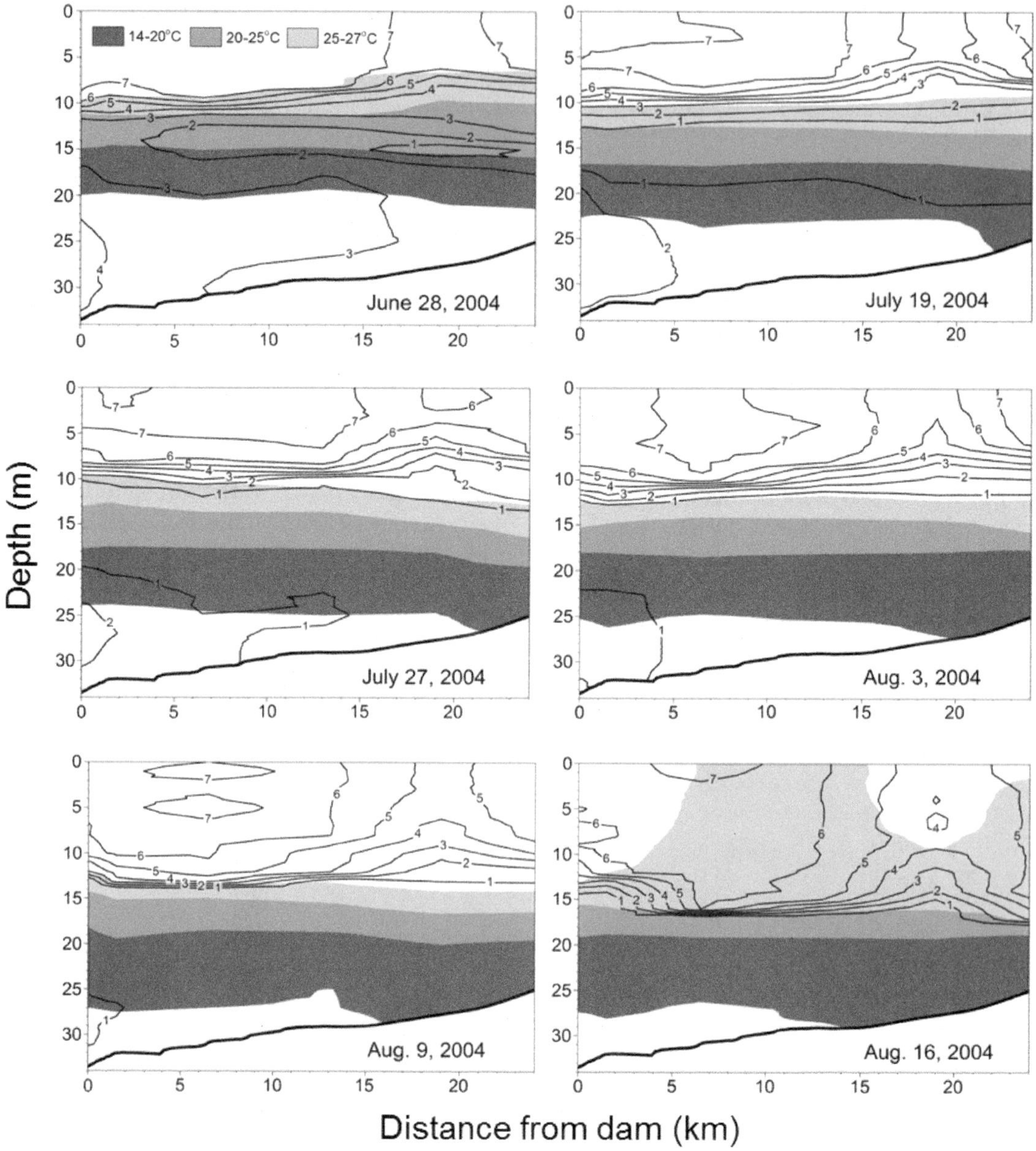

FIGURE 4. Vertical distribution of dissolved oxygen concentrations (contour lines, mg/L) and temperature (shading) along the main channel of Lake Norman from Cowans Ford Dam upstream 25 km on six dates from 28 June to 16 August 2004. Dissolved oxygen and temperature isopleths were estimated in SigmaPlot 12.0 (Systat Software, San Jose, California) using a running average with 0.1 sampling proportion and nearest neighbors bandwidth method.

oxygenated habitat in the hypolimnion. Anglers were catching striped bass in this area at depths just above and below the layer of metalimnetic hypoxia prior to and during the die-off event (B. K. Baker, Duke Energy, personal communication), indicating that at least some striped bass were occupying the oxygenated zone in the hypolimnion. Many of the dead striped bass

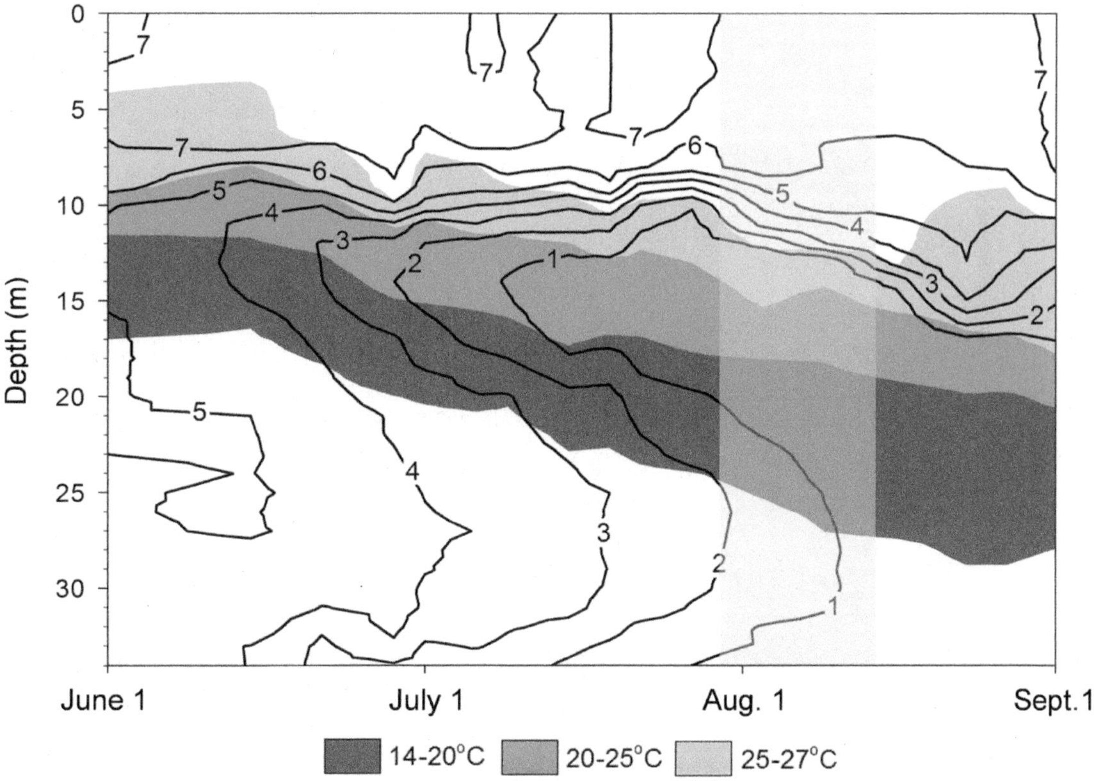

FIGURE 5. Change in depth distribution of dissolved oxygen concentrations (contour lines, mg/L) and temperature (shading) in Lake Norman at station 1.0, 320 m in front of Cowans Ford Dam, over the course of summer 2004. Shaded date range indicates the main period of dead striped bass collections. Dissolved oxygen and temperature isopleths were estimated in SigmaPlot 12.0 (Systat Software, San Jose, California) using a running average with 0.1 sampling proportion and nearest neighbors bandwidth method.

were already in an advanced state of decomposition when they appeared at the surface (fish were removed as they were found), consistent with fish dying deep in the water column and floating to the surface after several days.

Had this mortality event been due to starvation or malnutrition, we would have expected disproportionate mortality of the largest individuals in the population because total maintenance costs increase with body size (Kitchell 1983; Hartman and Brandt 1995). However, the size distribution of fish that died did not differ greatly from that of striped bass collected during annual gill-net monitoring (equal lengths of 38-, 51-, 64-, and 76-mm bar mesh) conducted the previous December by Duke Energy and NCWRC biologists (Figure 6) or from similar samples collected 2000–2002 (Thompson et al. 2005). Based on selectivity of the mesh sizes used, fish smaller and larger than 600 mm would have been underrepresented in these samples (McRae et al. 2013, this volume). Yet the fish that died included proportionally fewer smaller and larger fish than in the gill-net sample (Figure 6). Given the higher preferred temperatures of small striped bass (Coutant 1985), these fish may have been more likely to occupy the epilimnion, but if larger fish were more vulnerable to the cause of mortality, they should have constituted a higher percentage of the dead fish.

2009 striped bass mortality event

Another striped bass die-off occurred on Lake Norman during the summer of 2009 that was

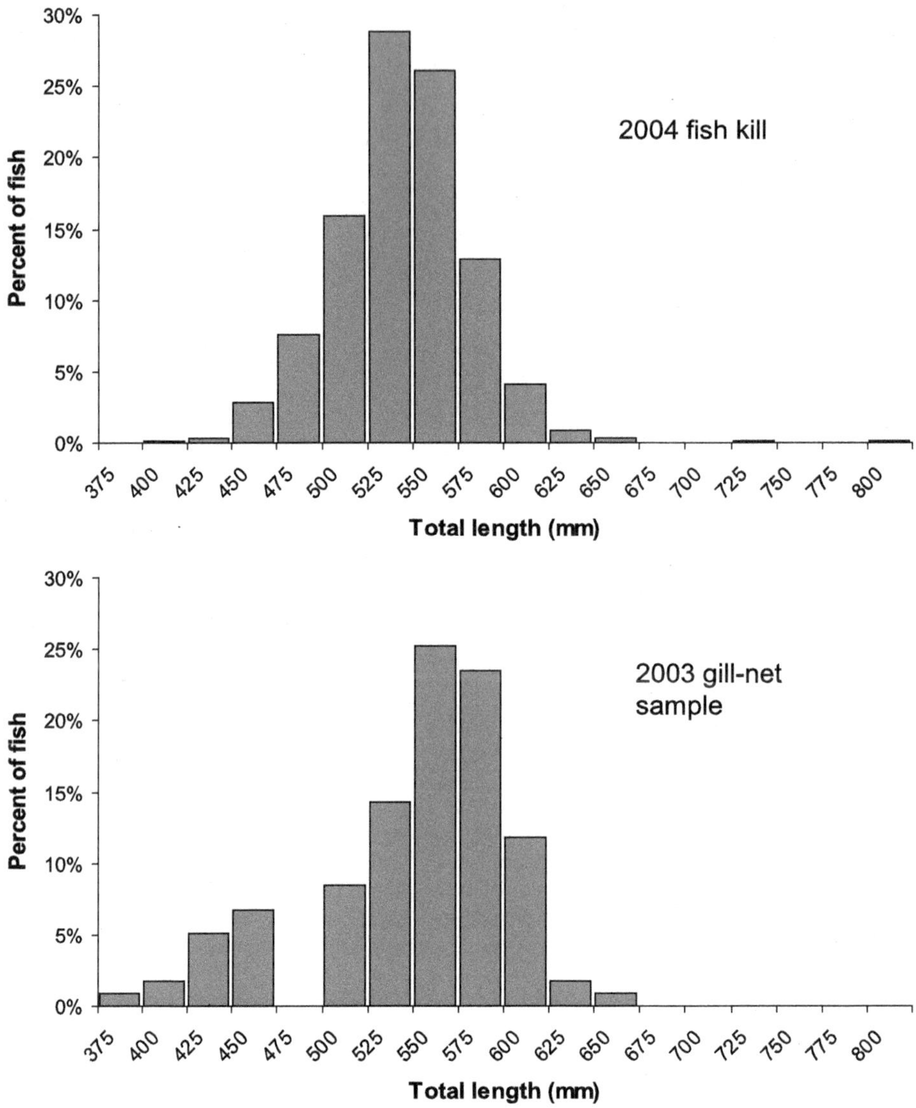

FIGURE 6. Size distributions of a subsample (n = 1,083) of dead striped bass collected in Lake Norman during 22 July–13 August 2004 (top panel) and collected in gill-net samples 22 December 2003 (n = 119; bottom panel).

smaller but otherwise quite similar to the event in 2004. From 3 through 8 August, 358 dead striped bass were collected close to the dam in the same region of the lake as the 2004 kill. Water quality conditions were comparable to 2004, with a pronounced metalimnetic hypoxic zone above an oxygenated hypolimnetic zone that decreased in size and DO concentration through time (Figure 7). Dead striped bass were first observed on Monday, 3 August, during the normal weekly survey, when maximum DO in the hypolimnion was just below 1.0 mg/L (Figure 7); more than half of the dead striped bass in this mortality event were collected on 3 August, so mortality may have started several days earlier. The physical condition and size distribution of fish that died in 2009 were nearly identical to those collected in 2004.

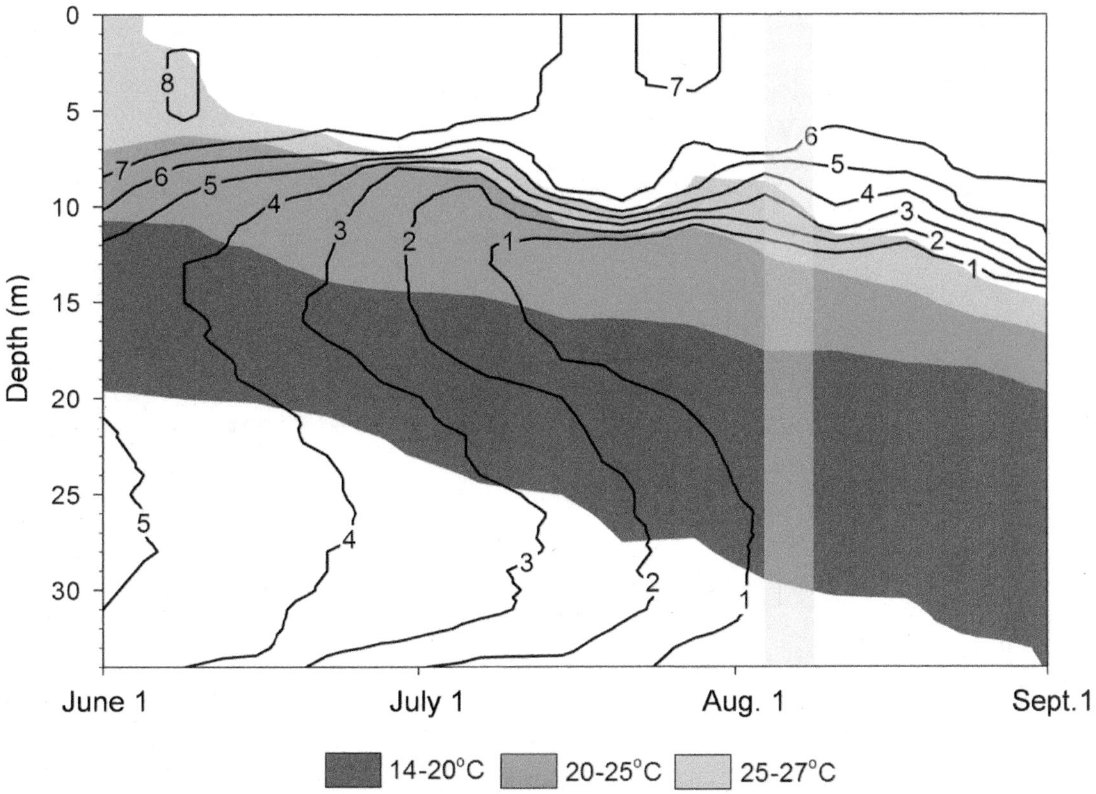

FIGURE 7. Change in depth distribution of dissolved oxygen concentrations (contour lines, mg/L) and temperature (shading) in Lake Norman at station 1.0, 320 m in front of Cowans Ford Dam, over the course of summer 2009. Shaded date range indicates the period of dead striped bass collections. Dissolved oxygen and temperature isopleths were estimated as in Figure 5.

2010 striped bass mortality event

During 2010, another major striped bass summer kill occurred on Lake Norman; from 14 July through 9 August, 6,996 dead striped bass were collected in the same region of the lake as during previous die-offs. Hypoxia had already begun to develop in the metalimnion by late May–early June, and by June 28 extended more than halfway up the main channel of the reservoir with at least 4 m of water below 2.0 mg/L DO. By the time the first nine dead striped bass were collected on 14 July, hypoxia in the metalimnion was 9 m thick with 4 m below 0.5 mg/L (Figure 8). However, unlike previous kills, mortality began while maximum DO in the hypolimnion was still 3.0–3.2 mg/L in the lower few kilometers of the reservoir. Four percent of all mortalities were collected in the first 11 d, before maximum DO in the hypolimnion declined to 2.0 mg/L. Biologists monitoring the kill suspected that these early mortalities resulted from catch-and-release mortality due to high angling pressure in the area near the dam where striped bass were concentrated. The kill accelerated on 25 July as DO in the hypolimnion dropped below 2.0 mg/L, overlain by metalimnetic hypoxia 13 m thick with 7 m at or below 0.5 mg/L and 4 m at 0.1 mg/L. Over the next 12 d, daily mortality collections ranged from 148 to 884 fish as DO in the hypolimnion declined to less than 1.0 mg/L and metalimnetic hypoxia continued to expand in thickness (Figure 8), with 7 m at 0.1 mg/L. Collections of dead striped bass drew to a close as DO in the hypolimnion dropped below 0.6 mg/L.

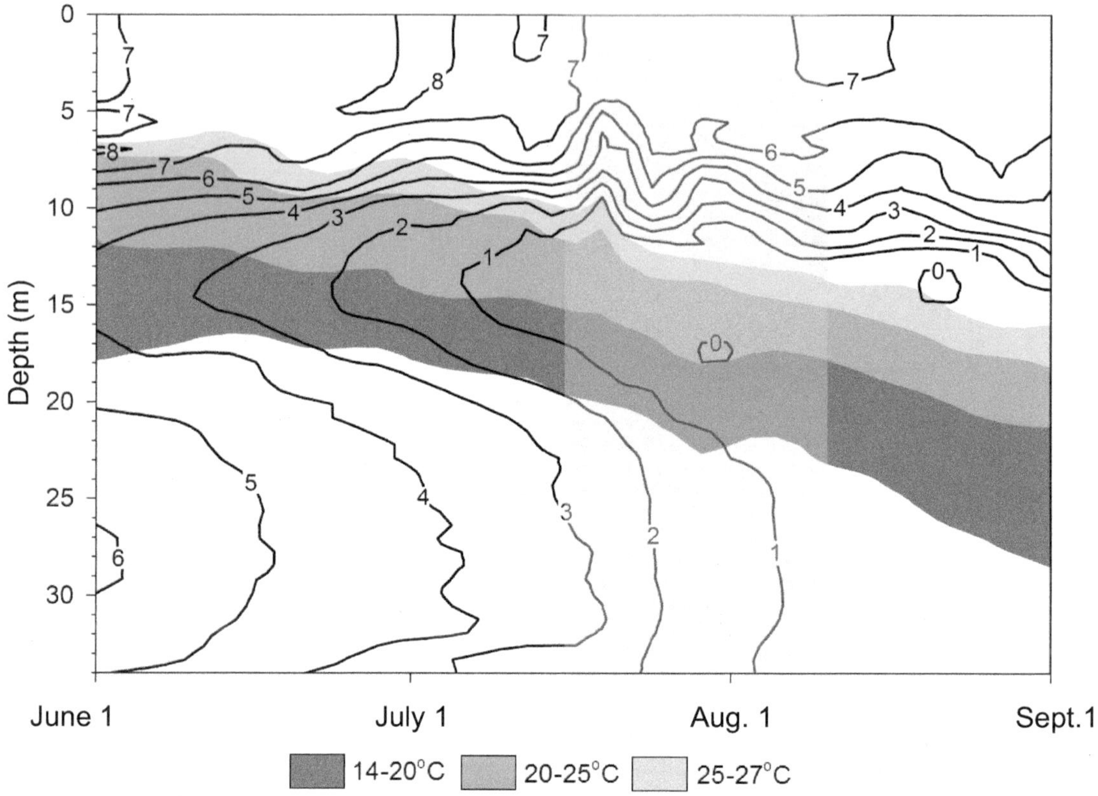

FIGURE 8. Change in depth distribution of dissolved oxygen concentrations (contour lines, mg/L) and temperature (shading) in Lake Norman at station 1.0, 320 m in front of Cowans Ford Dam, over the course of summer 2010. Shaded date range indicates the period of dead striped bass collections. Dissolved oxygen and temperature isopleths were estimated as in Figure 5.

In order to comply with permitted maximum temperature discharge requirements, Duke Energy withdrew cold water through the lower level intake (LLI) for the McGuire Nuclear Station for 7 d in July 2010 (14 July and 19–24 July). Unlike previous years, this use of the LLI occurred while there was still oxygenated habitat in the hypolimnion. Depth of the LLI (26–30 m) overlaps the depths of maximum DO concentration in the hypolimnion. To determine what impact, if any, water withdrawal through the LLI had on the loss of oxygenated habitat in the hypolimnion, we compared the rate of decline in maximum hypolimnetic DO from mid-June through mid-August in 2010 with the rate of decline over the same period in 2004 and 2009, both years with striped bass die-offs, and 2008, a year with no die-off. In 2004, 2008, and 2009, the average rate of decline in maximum hypolimnetic DO was 0.07 mg/L/d. In 2010, the average rate of decline was also 0.07 mg/L/d, except during the 6-d period of LLI use (19–24 July) when maximum DO in the hypolimnion declined from 3.0 to 2.0 mg/L (0.17 mg/L/d), 0.6 mg/L more than it would have at the average rate. Thus, the effect of withdrawing water through the LLI in 2010 was to cause maximum hypolimnetic DO to become hypoxic (and subsequently nearly anoxic) about 8 d earlier than it otherwise would have. However, because DO levels would have become too low to sustain striped bass anyway, use of the LLI only had a modest effect on when the die-off occurred, not if it would occur.

Differences in conditions among years with and without striped bass kills

Given that a metalimnetic oxygen minimum forms in Lake Norman every year but striped bass kills do not occur every year, it appears that fish occupying the hypolimnetic oxygen maximum are able to escape as it becomes hypoxic in some years, but not in others. We examined water quality data from 2000 to 2010 to see if there were any particular characteristics that consistently differentiated the 3 years with summer striped bass die-offs during that period from those without. We focused on conditions at station 1.0 (nearest the dam) on the last sample date when DO in the hypolimnion was still at least 2.0 mg/L, and the first sample date (~1 week later) when maximum DO in the hypolimnion was less than 2.0 mg/L. We quantified a variety of metrics to characterize habitat conditions, including the thickness (m) of hypoxia in the metalimnion at various levels of severity (DO <2.0, <1.5, <1.0, <0.5, and ≤0.1 mg/L), the maximum DO concentration in the hypolimnion, and the thickness (m) of the hypolimnetic oxygen maximum with DO exceeding various concentrations (Table 1). We conducted *t*-tests to determine if the thickness of metalimnetic hypoxia at each of the five levels of severity listed above differed significantly between years with and without striped bass kills in the period from 2000 to 2010. We used Bonferroni's adjusted alpha for multiple comparisons (Miller 1981) to constrain the overall alpha to 0.05 across the resulting 10 t-tests.

The extent of severe hypoxia in the metalimnion clearly differentiated conditions during summers with and without striped bass kills in the 11 years we analyzed. On the last sample date when DO in the hypolimnion still exceeded 2.0 mg/L, years with mortality events had significantly thicker zones of DO less than 0.5 mg/L (5–8 m) and at or below 0.1 mg/L (3–5 m) than in years without such events ($P < 0.05$; Table 1). None of the 8 years without a striped bass die-off had any water in the metalimnion with DO of 0.1 mg/L or less, and only one had any water (3 m) below 0.5 mg/L. In 2004, 2009, and 2010, portions of the metalimnion (3, 1, and 1 m, respectively) were actually below 0.5 mg/L DO for 2 weeks before DO in the hypolimnion declined below 2.0 mg/L. In contrast to the sharp differences in severe metalimnetic hypoxia between years with and without striped bass mortality, total thickness of metalimnetic hypoxia was not a factor. In all the years we analyzed, DO less than 2.0 mg/L formed a thick zone (7–18 m) in the metalimnion before maximum DO in the hypolimnion dropped below 2.0 mg/L (Table 1), but its thickness was not related to striped bass mortality.

These same two metrics also distinguished years with and without striped bass mortality on the first sample date after maximum hypolimnetic DO had declined below 2.0 mg/L. Metalimnetic hypoxia below 0.5 mg/L was present in five of eight nonkill years, but it was significantly thicker (7 m) in years with mortality events than in years without (0–6 m, mean = 2.6 m) ($P < 0.05$; Table 1). The metalimnetic zone at or below 0.1 mg/L was much thicker in years with mortality (6–7 m) than in nonkill years (0–3 m) ($P < 0.05$; Table 1). A week later, metalimnetic DO conditions were no longer markedly different between kill and nonkill years, as hypoxia became more severe in both the metalimnion and hypolimnion.

Potential role of forage fish in Lake Norman mortality events

A change in the Lake Norman forage fish community may have increased the likelihood of striped bass mortality events in recent years. Historically, the catch in annual epilimnetic September purse-seine samples of the pelagic prey community by Duke Energy personnel (Table 2), as well as the diets of Lake Norman striped bass (Lewis 1985; Thompson 2006), have consisted almost entirely of threadfin shad, with a few gizzard shad. These warmwater species occupy the epilimnion and upper metalimnion almost exclusively while the lake is stratified (Schael et al. 1995; Thompson et al. 2010). In the late 1990s, members of a striped bass angling group intentionally stocked alewives

TABLE 1. Characteristics of the metalimnetic oxygen minimum (M_{min}) and hypolimnetic oxygen maximum (H_{max}) at station 1.0 in Lake Norman on the last sample date (month/day) when dissolved oxygen (DO) in the H_{max} was still at least 2.0 mg/L, and the first sample date (month/day) when DO in the H_{max} had declined below 2.0 mg/L, in 2000–2010. Values indicate thickness (m) of the M_{min} and H_{max} within given DO ranges, maximum temperature and maximum DO values in the H_{max}, and coolest temperature in water above the M_{min} with at least 2.0 mg/L DO. Years with striped bass mortality events are shaded. Bold indicates those levels of metalimnetic hypoxia that were significantly thicker ($P < 0.05$, Bonferroni adjusted alpha for multiple comparisons) in years with mortality events than in those without.

	2000	2001	2002	2003	2004	2005	2006	2007	2008	2009	2010
Last sample date with DO in H_{max} ≥2.0 mg/L	7/24	8/6	7/23	7/21	7/30	7/25	7/11	7/16	7/28	7/20	7/23
M_{min} <2.0 mg/L DO	13	13	7	18	14	14	10	11	9	12	13
M_{min} <1.5 mg/L DO	11	6	0	10	11	9	0	5	5	10	10
M_{min} <1.0 mg/L DO	6	3	0	6	10	6	0	3	0	7	8
M_{min} <0.5 mg/L DO	3	0	0	0	**8**	0	0	0	0	**5**	**5**
M_{min} ≤0.1 mg/L DO	0	0	0	0	**5**	0	0	0	0	**3**	**4**
Max T in H_{max} (°C)	15.0	13.7	16.2	13.6	14.2	15.1	15.6	15.6	15.7	14.6	12.4
Max DO in Hmax (mg/L)	2.1	2.9	2.7	2.0	2.4	2.2	2.4	2.5	2.3	2.4	2.1
H_{max} ≥2 mg/L DO	2	5	12	1	8	3	10	11	8	8	8
H_{max} ≥1.5 mg/L DO	4	12	12	11	10	11	17	16	16	12	10
First sample date with DO in H_{max} <2.0 mg/L	7/31	8/10	7/29	7/28	8/3	8/1	7/19	7/23	8/4	7/27	7/26
Coolest T above M_{min} with ≥2 mg/L DO (°C)	24.7	25.8	26.4	25.2	28.2	26.0	25.5	27.0	26.7	26.3	28.7
M_{min} <2.0 mg/L DO	9	11	13	17	14	15	13	11	11	17	15
M_{min} <1.5 mg/L DO	9	11	12	15	12	13	8	7	10	14	12
M_{min} <1.0 mg/L DO	7	8	4	11	9	8	3	5	7	10	9
M_{min} <0.5 mg/L DO	4	6	0	5	**7**	4	0	2	0	**7**	**7**
M_{min} ≤0.1 mg/L DO	3	2	0	0	**6**	2	0	0	0	**7**	**6**
Max DO in Hmax (mg/L)	1.4	1.2	1.5	1.6	1.7	1.7	1.9	1.8	1.5	1.7	1.8
H_{max} ≥1.5 mg/L DO	0	0	1	1	8	3	9	11	3	7	9
H_{max} ≥1.0 mg/L DO	4	8	13	10	10	11	15	14	9	12	9

TABLE 2. Species composition (percent by number) of clupeids collected in annual September purse-seine samples in Lake Norman, 1993–2010. The purse seine sampled the upper 9 m of the water column.

	Purse-seine species composition (%)		
Year	Threadfin shad	Gizzard shad	Alewife
1993	100.0	0.0	0.0
1994	99.9	0.1	0.0
1995	99.9	0.1	0.0
1996	100.0	0.0	0.0
1997	>99.9	<0.1	0.0
1998	99.9	0.1	0.0
1999	99.3	0.3	0.5
2000	87.4	0.2	12.4
2001	76.5	0.0	23.5
2002	75.0	0.0	25.0
2003	82.6	0.1	17.3
2004	86.6	0.2	13.2
2005	98.1	0.0	1.9
2006	94.9	0.0	5.1
2007	98.3	0.0	1.7
2008	95.6	0.0	4.4
2009	91.7	0.0	8.3
2010	95.4	0.4	4.3

and blueback herring based on the notion that diversifying the prey species community would increase the forage base for striped bass. Alewives were first detected in annual purse-seine samples in 1999 and rapidly became established as a modest component of the forage fish community (Table 2) and of striped bass diets (Thompson 2006).

Hydroacoustic observations in mid-July 2004 showed fish beginning to concentrate in the upper meters of the oxygenated zone in the hypolimnion. To determine their identity, variable mesh gill nets were set (10-, 13-, 19-, 25-, 32-, and 38-mm bar mesh) overnight on 20 and 21 July (just prior to the start of striped bass mortality) in this depth zone (18.3–21.0 m deep) at station 1.0 (Figure 2). The catch confirmed the presence of both alewives (n = 44, mean TL = 119 mm) and blueback herring (n = 4, mean TL = 120 mm), as well as striped bass (n = 2, 545 and 564 mm TL). Other fish collected included 14 juvenile and adult blue catfish *Ictalurus furcatus* (101–464 mm TL), one white perch *Morone americana* (224 mm TL), and one gizzard shad (127 mm TL).

Weekly hydroacoustic surveys from the dam to about 3.2 km up the main channel confirmed that fish became increasingly concentrated in the layer of oxygenated hypolimnetic habitat as it declined in DO concentration and receded toward the dam. No fish were observed at these depths in hydroacoustic surveys once DO in the hypolimnion dropped below 0.5 mg/L. Dead alewives were not observed on the surface, but during the 2004 striped bass kill some were collected on water intake screens located just below the water surface at the McGuire Nuclear Station.

During summer 2009, weekly surveys documented a similar pattern of fish concentration in the receding oxygenated habitat in the hypolimnion. On 4 August, during the striped bass mortality event, we suspended two small-mesh gill nets (4.9 × 15.4 m panels of 6-, 10-, and 13-mm bar mesh) and three larger-mesh nets (25-, 51-, and 64-mm bar mesh, each 1.8 × 100 m) overnight in the remnant of the hypolimnetic oxygen maximum (0.6–0.8 mg/L DO, 21–26 m deep) near Cowans Ford Dam. The small-mesh nets caught 18 alewives (mean TL 155 mm), and 19 blue catfish (428–634 mm TL) were caught in the two largest-mesh nets. Two blue catfish had empty stomachs, but all the others contained alewives (mean 4.3/stomach). No striped bass were caught, but 71% of the documented striped bass mortalities had already been collected by the time these nets were set, so few, if any, striped bass may have remained in the hypolimnion. On 10 August, when maximum DO in the hypolimnion was 0.5 mg/L and 2 d after the last dead striped bass were collected on the surface, a hydroacoustic survey showed no fish in the hypolimnion. The same pattern was observed in 2010, with weekly hydroacoustic surveys documenting an increasing concentration of fish in the shrinking oxygenated hypolimnetic habitat that disappeared when DO levels dropped below about 0.5 mg/L (coinciding with the end of striped bass mortality in 2010). Since alewives became established in Lake Norman, hydroacoustic surveys in years without striped bass mortality have also shown the same pattern of hypolimnetic fish distribution in relation to DO dynamics (D. Coughlan, Duke Energy, personal communication).

During the middle of the 2010 die-off a number of dead striped bass that were collected were dissected and their stomach contents examined. Many of these contained multiple alewives about 150 mm TL (B. McRae, North Carolina Wildlife Resources Commission, unpublished data). No hypolimnetic gill netting was conducted in 2010, but the size and species of prey in these fish were consistent with the size and species of the dominant forage fish collected in the oxygenated portion of the hypolimnion during gill netting in 2004 and 2009. In contrast, summer diets of striped bass collected above the metalimnion are dominated by clupeids less than 90 mm TL (primarily threadfin shad with a small proportion of alewives; Thompson 2006).

Taken together, these observations indicate that introduction of alewives established a coolwater forage species in Lake Norman that consistently inhabited the oxygenated hypolimnetic habitat and provided a food source for striped bass also occupying that zone. Thus, it seems plausible that availability of coolwater prey may have increased the likelihood that some striped bass would choose to occupy this habitat when their preferred temperatures in the metalimnion became hypoxic.

Though the increased frequency of striped bass mortality events over the past decade is consistent with that hypothesis, it by no means confirms it; these events may still have occurred if striped bass would have occupied the hypolimnion even in the absence of coolwater forage fish. To further explore this issue, we examined all water quality data available from Lake Norman prior to 2000 to identify those years with sufficiently detailed data for us to characterize summer habitat conditions using the same metrics as for 2000–2010. Seven years met these criteria; three of them exhibited characteristics of severe metalimnetic hypoxia similar to those that typified years with striped bass mortality events during the 2000s (Table 3). In 1983, the thicknesses of metalimnetic DO less than 0.5 mg/L (7 m) and less than or equal to 0.1 mg/L (4 m) on the last sample date when DO in the hypolimnion was still at least 2.0 mg/L were in the same range as those observed in 2004, 2009, and 2010 (Table 1). In 1993, these metrics were even more extreme, with DO less than 0.5 mg/L 9 m thick, and 7 m at or below 0.1 mg/L DO. A week later, when DO in the hypolimnion had dropped below 2.0 mg/L, 12 m of the metalimnion contained less than 0.5 mg/L DO. In 1998, the extent of severe hypoxia in the hypolimnion was modest on the last date before the hypolimnion became hypoxic (2 m < 0.5 mg/L), but a week later, when DO in the hypolimnion had declined to 1.9 mg/L, this

TABLE 3. Characteristics of the metalimnetic oxygen minimum (M_{min}) and hypolimnetic oxygen maximum (H_{max}) at station 1.0 in Lake Norman on the last sample date (month/day) when dissolved oxygen (DO) in the H_{max} was still at least 2.0 mg/L, and the first sample date (month/day) when DO in the H_{max} was less than 2.0 mg/L, for years prior to 2000 with sufficient data. Values indicate thickness (m) of the M_{min} and H_{max} within given DO ranges, maximum temperature and maximum DO values in the H_{max}, and coolest temperature in water above the M_{min} with at least 2.0 mg/L DO. Shading indicates that a striped bass mortality event occurred. Values in bold indicate thickness of metalimnetic hypoxia consistent with those that differentiated years with striped bass mortality events from those without during 2000–2010 (see Table 2).

	1983	1992	1993	1996	1997	1998	1999
Last sample date with DO in H_{max} ≥2.0 mg/L	8/1	7/23	7/28	7/18	7/09	7/14	8/09
M_{min} <2.0 mg/L DO	17	14	17	10	0	16	11
M_{min} <1.5 mg/L DO	13	3	13	6	0	8	7
M_{min} <1.0 mg/L DO	10	0	11	3	0	5	5
M_{min} <0.5 mg/L DO	**7**	0	**9**	0	0	2	0
M_{min} ≤0.1 mg/L DO	**4**	0	**7**	0	0	0	0
Max T in H_{max} (°C)	14.4	14.1	13.8	13.8	15.4	14.2	14.5
Max DO in H_{max} (mg/L)	2.6	2.2	2.0	2.8	2.3	2.3	2.2
H_{max} ≥2 mg/L DO	6	3	1	8	4	7	2
H_{max} ≥1.5 mg/L DO	11	6	1	8	6	7	5
First sample date with DO in H_{max} <2.0 mg/L	9/7[a]	7/30	8/04	7/29	7/17	7/21	8/16
Coolest T above M_{min} with ≥2 mg/L DO (°C)	26.4	25.6	28.5	24.9	19.1	26.6	25.9
M_{min} <2.0 mg/L DO	–	16	17	11	8	14	12
M_{min} <1.5 mg/L DO	–	9	16	11	0	11	10
M_{min} <1.0 mg/L DO	–	3	14	7	0	8	8
M_{min} <0.5 mg/L DO	–	0	**12**	4	0	7	4
M_{min} ≤0.1 mg/L DO	–	0	**3**[b]	3	0	4	2
max DO in H_{max} (mg/L)	–	1.9	1.5	1.6	1.9	1.9	1.3
H_{max} ≥1.5 mg/L DO	–	2	4	2	4	6	0
H_{max} ≥1.0 mg/L DO	–	4	7	6	7	6	2

[a] By this date, the metalimnion and hypolimnion were completely anoxic.
[b] 10 m ≤ 0.2 mg/L.

layer of severe hypoxia had expanded to 7 m thick, as in years in the 2000s with striped bass mortality.

Of the 3 years in which some or all metrics of severe metalimnetic hypoxia were comparable to those associated with striped bass mortality in the 2000s, 1983 was the only year in which a mortality event occurred (the only striped bass kill of note in Lake Norman prior to 2004).

Between 15 and 23 August 1983, 163 dead fish were collected near the dam. Although alewives and blueback herring were not present, there was another coolwater forage fish in the reservoir at that time. In the late 1970s and early 1980s, Lake Norman had an abundant population of small yellow perch *Perca flavescens*, which were preyed upon by striped bass and were a preferred bait of striped bass anglers. A major die-off of these small yellow perch (~5,000 fish) in the same area of the lake near the dam preceded the striped bass kill by about 7–10 d. Though these observations are anecdotal, the timing and location of these mortalities suggests that both the dead yellow perch and striped bass may have been trapped in the hypolimnion as it became hypoxic.

In contrast, metalimnetic hypoxia was more severe in 1993 than in every other year we were able to evaluate (Tables 1 and 3). Had striped bass been present in the hypolimnion, the patterns observed in other years suggest that a striped bass mortality event would have occurred. By 1993, yellow perch were no longer common in Lake Norman, and alewives and blueback herring had not yet been introduced. It is possible that in the absence of coolwater prey, striped bass may have chosen not to occupy this habitat when their preferred temperatures in the metalimnion became hypoxic, accounting for the lack of striped bass mortality when oxygenated habitat in the hypolimnion disappeared.

Badin Lake

Though not nearly as comprehensive as data available from Lake Norman, observations from Badin Lake, North Carolina provide relevant information regarding striped bass behavior and mortality events in relation to formation of a metalimnetic oxygen minimum. Badin Lake is a deep (mean depth 21 m, maximum depth 54 m), 2,165-ha, eutrophic reservoir (NCDENR 2002, 2007) on the Yadkin River, stocked annually with 29.6 striped bass fingerlings/ha. The forage base is composed primarily of threadfin shad, but also includes blueback herring and gizzard shad (Thompson 2006). Three years of temperature and DO data collected during 2001–2003 show annual formation of a hypoxic zone in the metalimnion beginning in late May–early June, with DO in the hypolimnion dropping below 2.0 mg/L around late June to mid-July and disappearing shortly thereafter (Thompson et al. 2005).

Occasional die-offs of otherwise healthy-looking striped bass have been documented in Badin Lake with mortality concentrated in the lower lake near the dam, in the same area and about the same time as DO in the hypolimnion dropped below 2.0 mg/L, as shown in 2001–2003 water-quality monitoring (~400–500 fish, 13–17 July 1997; ~400–500 fish, 2–5 July 2001; ~750 fish, 20–29 June 2004; M. Hale, North Carolina Department of Environment and Natural Resources, and L. Dorsey, North Carolina Wildlife Resources Commission, unpublished data). The 2001 kill is the only one for which any limnological data are available. Sampling was not frequent enough to characterize conditions immediately before DO in the hypolimnion dropped below 2.0 mg/L, but a limnological profile was conducted in the dam forebay on 3 July, shortly after the mortality event began. On that date, 7 m of the metalimnion had DO below 0.5 mg/L and 6 m were less than 0.3 mg/L, while DO in the hypolimnetic oxygen maximum ranged from 0.9 to 1.9 mg/L. These conditions are consistent with those that differentiated years with striped bass mortality events in Lake Norman from those without such events during 2000–2010.

A striped bass telemetry study initiated in Badin Lake in 2002 provided further insight regarding striped bass behavioral responses to changing temperature and DO conditions (Thompson et al. 2010). As the reservoir warmed, striped bass moved deeper to remain in oxygenated water with temperatures of 20–21°C; by mid-June, tagged fish were concentrated in the lower portion of the lake (where previous kills had occurred) in the hypolimnetic oxygen maximum, which contained about 4 m of water with 2.0–2.5 mg/L DO at 20–21°C (Thompson et al. 2010). Thickness and severity of hypoxic conditions in the metalimnion were mild compared to 2001, never exceeding 6 m of water with DO less than 2.0 mg/L and 2 m below 0.5 mg/L. When maximum DO concentration in the hypolimnion dropped below 2.0 mg/L in late June, the tagged fish immediately moved through the metalimnetic hypoxia into surface waters with temperatures above 26°C and 4–6 mg/L DO and dispersed throughout the lake with no mortality (Thompson et al. 2010).

Discussion and Management Implications

Formation of a metalimnetic oxygen minimum overlaying a hypolimnetic oxygen maximum is a common phenomenon in many relatively deep reservoirs. The examples we have given indicate that changing habitat conditions caused by this limnological phenomenon may be more important in explaining some striped bass summer kills than previously recognized. Understanding the interactions between behavioral responses of fish and the spatial and temporal dynamics of DO and temperature in these systems may help us modify management strategies to alleviate, avoid, or minimize negative impacts on fish populations and better address concerns raised by the public about such events.

Striped bass mortality

Fortunately the extensive limnological record from Lake Norman allowed us to identify patterns in metalimnetic and hypolimnetic DO that appear to differentiate conditions likely to result in striped bass mortality from those that do not. The first prerequisite is isolation of oxygenated habitat in the hypolimnion (occupied by striped bass) that does not persist with adequate DO until fall turnover. If this habitat persists the whole summer, then the mechanism we have proposed obviously will not cause mortality (though other mechanisms could potentially cause stress, loss of condition, or mortality in such refuges (Coutant 1985, 2013). The second apparent prerequisite is formation of a thick layer of severe hypoxia in the metalimnion before DO in the hypolimnion falls below critical levels for striped bass survival. In Lake Norman, development of extensive moderate hypoxia (<2.0 mg/L) in the metalimnion was the norm (typically >10 m thick) and bore no relation to the occurrence of striped bass mortality. In contrast, years with striped bass mortality events were all characterized by 5–8 m of severe metalimnetic hypoxia (<0.5 mg/L) before the hypolimnion became hypoxic.

The patterns we observed in Lake Norman suggest that striped bass are able to leave the hypolimnion when DO declines to intolerable levels, if hypoxia in the metalimnion is relatively mild. Telemetry observations in Badin Lake confirmed that striped bass will move up through mild metalimnetic hypoxia to escape low DO when the hypolimnion becomes hypoxic (Thompson et al. 2010). Although we do not yet have definitive confirmation that thick, severe metalimnetic hypoxia will "trap" striped bass in the hypolimnion resulting in mortality as DO declines, as might be obtained from telemetered striped bass during such a mortality event, we believe the circumstantial evidence is strong, and the timing and location of numerous die-offs are consistent with this explanation.

Temperature preference is likely to influence habitat selection by striped bass as hypoxia begins to exclude them from the metalimnion. When a wide range of temperatures with oxygenated water is available, adult striped bass in the size ranges found in Lake Norman and Badin Lake usually select temperatures in the 20–24°C range, most often about 20–22°C (Coutant and Carroll 1980; Thompson et al. 2010; Coutant 2013). As metalimnetic hypoxia first begins to form in Lake Norman, these temperatures almost always occur in the developing metalimnetic hypoxic zone and in the water just above it. In contrast, the warmest oxygenated water below this hypoxic zone is typically 15–16.5°C, well below the temperatures typically selected by the size ranges of adult striped bass in the reservoirs we studied and even lower than the fundamental thermal niche of adult striped bass reported by Bettoli (2005). Thus, if habitat selection by striped bass avoiding the metalimnetic hypoxic zone was based solely on available oxygenated temperatures, it seems more likely that they would move above the hypoxic layer where preferred temperatures were still available, rather than into colder than preferred temperatures below it.

Collectively, these observations support the notion that the presence of coolwater forage fish may increase the likelihood of striped bass mortality events associated with severe metalimnetic hypoxia. Coolwater forage fish were present in the hypolimnetic habitat of Lake Norman in each year that striped bass die-offs occurred. Yet no mortality event occurred in 1993, when coolwater forage fish likely to inhabit the hypolimnion were absent or rare, even though the thickness of severe metalimnetic hypoxia was substantially greater in that year than any other.

Until concrete evidence is available to show that during these mortality events striped bass remain in the hypolimnion and succumb as DO declines to lethal levels, we cannot rule out the possibility that some striped bass may move through even a severely hypoxic metalimnion to escape. Hydroacoustic observations made near the dam in Lake Norman have documented some large targets moving up and down through severely hypoxic conditions in the metalimnion. These fish may be hypoxia-tolerant blue catfish, but if future work documents that at least some of them are striped bass, it would beg the question, why don't they all leave the hypolimnion when it becomes intolerable, even if hypoxia in the metalimnion is severe? If the fish were moving up to the lower epilimnion, mortality could possibly be due to thermal shock. However, the thermal change they would experience in years with severe hypoxia in the metalimnion would be the same as in years with mild hypoxia. In addition, if mortality occurred after fish reached the epilimnion we would expect most floating striped bass to initially look relatively fresh, whereas most looked like they had already been dead for several days when first collected during daily mortality surveys.

Potential impacts on other coolwater or coldwater species

The limnological dynamics we have described involving metalimnetic hypoxia and subsequent depletion of hypolimnetic oxygen may have important implications for other coolwater or coldwater species. Conditions of severe metalimnetic hypoxia similar to those that coincided with striped bass mortality events in Lake Norman and Badin Lake have also resulted in rainbow trout *Oncorhynchus mykiss* mortality in Bull Shoals Reservoir, Arkansas–Missouri (Love 1980). Such conditions may also contribute to entrainment or mortality of coolwater forage fish in some reservoir systems. As we observed for alewives in Lake Norman, blueback herring in J. Strom Thurmond (JST) Reservoir, Georgia–South Carolina can become concentrated in oxygenated hypolimnetic habitat isolated by extensive metalimnetic hypoxia. Hydroacoustic surveys have shown that as this oxygenated habitat shrinks the fish are crowded deeper and closer to the JST Dam penstock. In 2004, 2006, 2007, and 2008, major blueback herring entrainment events occurred at JST dam as hypolimnetic DO in the dam forebay declined below about 1.3–2.4 mg/L, and ended as it dropped to 0.7–0.8 mg/L. Visual observations indicated that in addition to entrainment through the dam, these dynamics can also result in mortality within the reservoir. Observations reported by a scuba diver following the 2004 entrainment event and photographic or video evidence obtained following the 2007 and 2008 events documented many dead blueback herring on the bottom in the dam forebay.

Limnological dynamics similar to those resulting in blueback herring entrainment through JST Dam have also caused blueback herring entrainment events at Hartwell Dam, Georgia–South Carolina in about one-third of the years since blueback herring became abundant in the mid-1980s. Using temperature and DO data, as well as fish distribution data from vertical gill netting and hydroacoustic surveys, Alexander et al. (1991) showed that entrainment occurred in years when hypolimnetic hypoxia developed after severe metalimnetic hypoxia was already in place. Entrainment events did not occur in years without this combination of limnological conditions. The potential effects of metalimnetic hypoxia and subsequent loss of oxygenated hypolimnetic habitat on entrainment and mortality of coolwater forage fishes warrant further detailed evaluation.

Management implications

The spatial and temporal dynamics of temperature and oxygen we have described are driven by interactions among multiple physical, chemical, and environmental factors (e.g., bathymetry, nutrient loading, residence time, hydrological and meteorological conditions, and reservoir operations; Cole and Wells 2002; Ruane et al. 2013). Here, we have focused on the consequences of these dynamics rather than their causes, but hydrodynamic and water quality models such as CE_QUAL-W2 (Cole and Wells 2002) have been developed that are very effective in characterizing the spatial and temporal distribution of DO and temperature throughout reservoir systems based on physical, chemical, and meteorological data that are often monitored by various public or private organizations. These models can be powerful tools for simulating water quality under various environmental conditions or management scenarios (e.g., Ruane et al. 2013; Taylor et al. 2013; this volume). Even routine empirical monitoring of water quality parameters, especially when coupled with hydroacoustic surveys of fish distributions, can help identify risks, predict impending periods of entrainment or mortality, and evaluate management options such as whether or not to stock striped bass or coolwater prey species in a particular system.

In some systems, these problems may be manageable despite the constraints of logistical limitations and competing priorities for dam operations. In reservoirs where the oxygenated hypolimnetic refuge is not going to persist all summer, hypolimnetic water withdrawals can sometimes be used to prevent its formation or remove it before fish are trapped. For example, in Lake Hartwell where blueback herring entrainment is a recurring problem, U.S. Army Corps of Engineers personnel who operate the dam have found that they can prevent herring aggregations near the penstocks by withdrawing water early in the season to eliminate oxygenated habitat in the hypolimnion before hypoxia in the metalimnion is well established. However, this management strategy can only be implemented in years with sufficient water availability. This approach may be a realistic option in some reservoirs operated primarily for hydroelectric power generation. However, in systems like Lake Norman where providing cooling water for steam electric power stations is the top priority, it is unlikely to be feasible. In Lake Norman, withdrawal of cold water from the hypolimnion is often necessary later in the summer to keep the temperature of cooling water discharged from McGuire nuclear power station below maximum permitted levels; withdrawal of hypolimnetic water early in the season might not leave sufficient cold water to meet this need.

Alternatively, in some systems particular patterns of water withdrawal may actually create problems by eliminating oxygenated hypolimnetic habitat that would otherwise persist all summer long. For example, in Lake Murray, good habitat below the hypoxic zone in the metalimnion usually persists during the summer if withdrawal from that depth is minimized. In 2005, one intake located at the same depth as the oxygenated hypolimnetic habitat was used extensively in June and July in response to high inflows, reducing this habitat, and a large striped bass kill started the last week of July when hypolimnetic DO fell below 2 mg/L (Ruane et al. 2013). Based on the results of model simulations, timing and relative use of metalimnetic and hypolimnetic water intakes in Lake Murray have since been altered to increase suitable summer habitat for striped bass (Ruane et al. 2013).

Other habitat manipulations may allow good hypolimnetic habitat to persist when it otherwise would not. Oxygen diffuser systems have been installed in more than 20 reservoirs and lakes across the United States (Mobley et al. 2013, this volume). While the primary purpose for many of these systems is to improve downstream water quality, some are specifically intended to create and maintain usable coolwater habitat for striped bass and other species. For example, in 2011, installation of a DO injection system was completed in JST Reservoir. Located 8 km upstream from the JST Dam, this system is creating a large area of well-oxygenated, cool summer habitat for striped bass (and blueback herring) to mitigate the loss of cool, oxygenated habitat in the tailrace of Richard B. Russell Dam when its reversible turbines are used to operate the dam as a pumped-storage hydropower facility (Mobley et al. 2013).

In some systems, perhaps small-scale oxygenation could be used to create refuges of adequate size to enhance oversummer survival of striped bass in reservoirs with persistent water quality issues. However, if the oxygenated area is relatively small, special regulations may be required to protect striped bass that aggregate there during the summer. For example, oxygen diffusers installed in Cherokee Reservoir to improve DO conditions downstream of the dam created a relatively small zone of ideal striped bass habitat. Anglers heavily targeted striped bass aggregated in this area, leading the Tennessee Wildlife Resources Agency to close the area to fishing 15 July–15 September to halt substantial catch and release mortality that was occurring (Mobley et al. 2013).

Decisions regarding whether or not oxygen injection or other water management changes should be implemented will depend on operational priorities, economic considerations, and other factors unique to each system. Even if a striped bass fishery's economic and recreational value alone is not sufficient to justify such changes, the fishery may benefit if installation of an oxygen injection system or other modification is the most cost-effective method to meet other priorities, such as satisfying regulatory requirements for minimum DO concentrations in tailwater releases or providing mitigation for other habitat loss or fish mortality. Modeling analyses can be useful in determining if striped bass habitat enhancement can be incorporated into such operational changes without compromising primary objectives (Ruane et al. 2013). However, both fisheries managers and anglers should recognize that in some systems with marginal habitat and recurring summer mortalities, discontinuation of striped bass stocking is a management option that may warrant consideration.

The observations we have reported here identify a significant issue affecting striped bass management in many reservoirs. We need detailed information from more systems and a wider range of conditions to determine whether or not the patterns we observed are general, and to elucidate exactly how striped bass occupying the hypolimnion below a hypoxic metalimnion are responding behaviorally to progressive loss of oxygenated habitat. Likewise, additional information is needed to clarify the fate of coolwater forage fish occupying this habitat and to assess the impacts on both striped bass and coolwater forage fish populations. Awareness of these dynamics can help us understand the constraints they impose on fisheries management and identify possible approaches to minimize or avoid their impact.

Even when management actions to alleviate striped bass summer mortalities are not possible, a clear understanding of the processes responsible for these events can still be a valuable tool for addressing questions and dispelling misinformation about their causes. Fish kills, especially of sport fish, can generate great concern among anglers and the public. When fish kills are associated with formation of metalimnetic hypoxia and the subsequent dynamics of the oxygenated layer below it, managers may be better able to address public concerns and misconceptions by conveying how limnological dynamics inherent to the reservoir system interact with habitat preferences of affected species (and with dam operations, if they are a factor) to cause such kills.

Acknowledgments

Many individuals from the North Carolina Wildlife Resources Commission, Duke Energy, South Carolina Department of Natural Resources, U.S. Army Corps of Engineers,

North Carolina Division of Water Quality and NC State University were instrumental in conducting the field work and processing the data on which much of this study depended, as well as offering insights and information that strengthened the paper. We especially thank David Coughlan, D. Hugh Barwick, and William Foris; this paper would not have been possible without the insights, advice, information, support, and editorial comments they provided during its development. We also thank Brian McRae, Lawrence Dorsey, Scott Van Horn, Troy Thompson, Nick Jeffers, Heather Dendy, Bryan Richardson, B. Kim Baker, Mike Abney, Duane Harrell, Gene Vaughan, Mark Rash, Brad Garner, Bob Doby, Bryan Kalb, Chase Fulk, Josh Quinn, Mark Auten, Sabrina Baker, Keith Finley, Ben Lastra, Glenn Long, Todd Lynn, John Shaw, Wade Bales, Hal Beard, Ron Ahle, Mark Hale, Scott Waters, Marybeth Brey, Sally Petre, Miranda Wood, J. Casey, D. Williams, and Benjamin Kornegay. This research was funded by the North Carolina Wildlife Resources Commission through Federal Aid in Sport Fish Restoration grant F-68–8.

References

Alexander, C. M., S. J. Bennett, J. H. Carroll, and R. L. Kasul. 1991. Entrainment of blueback herring (*Alosa aestivalis*) at Hartwell Dam, South Carolina and Georgia: a case study. U.S. Army Engineer, Savannah District, Savannah, Georgia.

Bettoli, P. W. 2005. The fundamental thermal niche of adult landlocked striped bass. Transactions of the American Fisheries Society 134:305–314.

Cheek, T. E., M. J. Van Den Avyle, and C. C. Coutant. 1985. Influences of water quality on distribution of striped bass in a Tennessee River impoundment. Transactions of the American Fisheries Society 114:67–76.

Cole, T. M., and H. H. Hannon. 1990. Dissolved oxygen dynamics. Pages 71–107 *in* K. W. Thornton, B. L. Kimmel, and F. E. Payne, editors. Reservoir limnology: ecological perspectives. Wiley, New York.

Cole, T. M., and S. A. Wells. 2002. CE_QUAL-W2: a two-dimensional, laterally averaged hydrodynamic and water quality model, version 3.1. User manual. U.S. Army Corps of Engineers, Instruction Report EL-02–1, Washington, D.C.

Coutant, C. C. 1978. A working hypothesis to explain mortalities of striped bass, *Morone saxatilis*, in Cherokee Reservoir. Oak Ridge National Laboratory, ORNL/TM-6534, Oak Ridge, Tennessee.

Coutant, C. C. 1985. Striped bass, temperature, and dissolved oxygen: a speculative hypothesis for environmental risk. Transactions of the American Fisheries Society 114:31–61.

Coutant, C. C. 2013. When is habitat limiting for striped bass? Three decades of testing the temperature–oxygen squeeze hypothesis. Pages 65–91 *in* J. S. Bulak, C. C. Coutant, and J. A. Rice, editors. Biology and management of inland striped bass and hybrid striped bass. American Fisheries Society, Symposium 80, Bethesda, Maryland.

Coutant, C. C., and D. S. Carroll. 1980. Temperatures occupied by ten ultrasonic-tagged striped bass in freshwater lakes. Transactions of the American Fisheries Society 109:195–202.

Davias, L. A. 2006. A bioenergetics assessment of temperature and food consumption effects on growth of reservoir striped bass. Master's thesis. North Carolina State University, Raleigh.

Farquhar, B. W., and S. Gutreuter. 1989. Distribution and migration of adult striped bass in Lake Whitney, Texas. Transactions of the American Fisheries Society 118:523–532.

Gustaveson, W., and G. Blommer. 2013. History of striped bass management in the Colorado River. Pages 15–23 *in* J. S. Bulak, C.C. Coutant, and J. A. Rice, editors. Biology and management of inland striped bass and hybrid striped bass. American Fisheries Society, Symposium 80, Bethesda, Maryland.

Hartman, K. J., and S. B. Brandt. 1995. Comparative energetics and the development of bioenergetics models for sympatric estuarine piscivores. Canadian Journal of Fisheries and Aquatic Sciences 52:1647–1666.

Hightower, J. E., J. R. Jackson, and K. H. Pollock. 2001. Use of telemetry methods to estimate natural and fishing mortality of striped bass in Lake Gaston, North Carolina. Transactions of the American Fisheries Society 130:557–567.

Kitchell, J. F. 1983. Energetics. Pages 313–338 *in* P. W. Webb and D. Weihs, editors. Fish biomechanics. Praeger, New York.

Lewis, R. E. 1985. Temperature selection and vertical distribution of striped bass during lake stratification. Proceedings of the Annual Conference Southeastern Association of Fish and Wildlife Agencies 37(1983):276–286.

Love, W. T. 1980. Effects of a high rainfall year on trout habitat in Bull Shoals Reservoir. Proceedings of the Annual Conference Southeastern Association of Fish and Wildlife Agencies 33(1979):331–343.

Matthews, W. J. 1985. Summer mortality of striped bass in reservoirs of the United States. Transactions of the American Fisheries Society 114:62–66.

Matthews, W. J., L. G. Hill, and S. M. Schellhaass. 1985. Depth distribution of striped bass and other fish in Lake Texoma (Oklahoma–Texas) during summer stratification. Transactions of the American Fisheries Society 114:84–91.

McRae, B. J., J. S. Bulak, B. E. Taylor, and C. T. Waters. 2013. Evaluation of the use of gill nets for monitoring reservoir striped bass fisheries. Pages 263–278 *in* J. S. Bulak, C.C. Coutant, and J. A. Rice, editors. Biology and management of inland striped bass and hybrid striped bass. American Fisheries Society, Symposium 80, Bethesda, Maryland.

Miller, R. G., Jr. 1981. Simultaneous statistical inference, 2nd edition. Springer-Verlag, Berlin.

Mobley, M., E. Shallenberger, M. W. Beutel, P. Gantzer, and B. Sak. 2013. Oxygen diffusers to create and maintain summer fish habitat. Pages 197–208 *in* J. S. Bulak, C.C. Coutant, and J. A. Rice, editors. Biology and management of inland striped bass and hybrid striped bass. American Fisheries Society, Symposium 80, Bethesda, Maryland.

NCDENR (North Carolina Department of Environment and Natural Resources). 2002. Basinwide assessment report: Yadkin River basin. NCDENR, Raleigh, North Carolina.

NCDENR (North Carolina Department of Environment and Natural Resources). 2007. Lake and reservoir assessments: Yadkin–Pee Dee River basin. NCDENR, Raleigh, North Carolina.

NCDENR (North Carolina Department of Environment and Natural Resources). 2008. Lake and reservoir assessments: Catawba River basin. NCDENR, Raleigh, North Carolina.

Ruane, R. J., G. E. Hauser, and A. F. Sawyer. 2013. Water quality modeling as a management tool for striped bass and hybrid striped bass. Pages 161–179 *in* J. S. Bulak, C. C. Coutant, and J. A. Rice, editors. Biology and management of inland striped bass and hybrid striped bass. American Fisheries Society, Symposium 80, Bethesda, Maryland.

Schael, D. M., J. A. Rice, and D. J. Degan. 1995. Spatial and temporal distribution of threadfin shad in a southeastern reservoir. Transactions of the American Fisheries Society 124:804–812.

Schaffler, J. J., J. J. Isely, and W. E. Hayes. 2002. Habitat use by striped bass in relation to seasonal changes in water quality in a southern reservoir. Transactions of the American Fisheries Society 131:817–827.

Taylor, B. E., J. S. Bulak, and H. N. McKellar. 2013. Response of summer habitat for striped bass in Lake Greenwood, South Carolina to weather and phosphorus loading. Pages 181–196 *in* J. S. Bulak, C. C. Coutant, and J. A. Rice, editors. Biology and management of inland striped bass and hybrid striped bass. American Fisheries Society, Symposium 80, Bethesda, Maryland.

Thompson, J. S. 2006. The influence of temperature and forage availability on growth and habitat selection of a pelagic piscivore. Doctoral dissertation. North Carolina State University, Raleigh.

Thompson, J. S., and J. A. Rice. 2013. The relative influence of thermal experience and forage availability on growth of age 1–5 striped bass in two southeastern reservoirs. Pages 93–120 *in* J. S. Bulak, C. C. Coutant, and J. A. Rice, editors. Biology and management of inland striped bass and hybrid striped bass. American Fisheries Society, Symposium 80, Bethesda, Maryland.

Thompson, J. S., J. A. Rice, and D. S. Waters. 2010. Striped bass habitat selection rules in reservoirs without suitable summer habitat offer insight into consequences for growth. Transactions of the American Fisheries Society 139:1450–1464.

Thompson, J. S., J. A. Rice, D. S. Waters, J. C. Taylor, J. E. Hightower, L. A. Davias, and P. S. Rand. 2005. Energetics of reservoir striped bass populations: phase 1. North Carolina Wildlife Resources Commission, Federal Aid in Sport Fish Restoration, Project F-68–04, Final Report, Raleigh.

Thompson, J. S., D. S. Waters, J. A. Rice, and J. E. Hightower. 2007. Seasonal fishing and natural mortality of striped bass in a southeastern reservoir. North American Journal of Fisheries Management 27:681–694.

Van Den Avyle, M. J., and J. W. Evans. 1990. Temperature selection by striped bass in a Gulf of Mexico coastal river system. North American Journal of Fisheries Management 10:58–66.

Van Horn, S. L. 2013. A brief history of inland striped bass management. Pages 1–13 *in* J. S. Bulak, C. C. Coutant, and J. A. Rice, editors. Biology and management of inland striped bass and hybrid striped bass. American Fisheries Society, Symposium 80, Bethesda, Maryland.

Van Horn, S. L., B. K. Baker, and M. Rash. 2001. Growth and condition response of Lake Norman striped bass to increased stocking rates and more restrictive harvest regulations. Proceedings of the Annual Conference Southeastern Association of Fish and Wildlife Agencies 53(1999):193–199.

Waters, C. T., and B. J. McRae. 2008. Communicating with anglers in the face of conflict: a Lake Norman case history. Pages 207–217 *in* M. S. Allen, S. Sammons, and M. J. Maceina, editors. Balancing fisheries management and water uses for impounded river systems. American Fisheries Society, Symposium 62, Bethesda, Maryland.

Wetzel, R. G. 2001. Limnology: lake and river systems, 3rd edition. Academic Press (Elsevier USA), San Diego, California.

Wilkerson, M. L., and W. L. Fisher. 1997. Striped bass distribution, movements, and site fidelity in Robert S. Kerr Reservoir, Oklahoma. North American Journal of Fisheries Management 17:677–686.

Wilson, D. M., V. J. DiCenzo, and J. Odenkirk. 2013. Comparisons of reservoir characteristics with striped bass growth and relative weight in the southeastern United States. Pages 209–218 *in* J. S. Bulak, C. C. Coutant, and J. A. Rice, editors. Biology and management of inland striped bass and hybrid striped bass. American Fisheries Society, Symposium 80, Bethesda, Maryland.

Young, S. P., and J. J. Isely. 2004. Temporal and spatial estimates of adult striped bass mortality from telemetry and transmitter return data. North American Journal of Fisheries Management 24:1112–1119.

Zale, A. V., J. D. Wiechman, R. L. Lochmiller, and J. Burroughs. 1990. Limnological conditions associated with summer mortality of striped bass in Keystone Reservoir, Oklahoma. Transactions of the American Fisheries Society 119:72–76.

American Fisheries Society Symposium 80:147–159, 2013

Temperature and Dissolved Oxygen Habitat Use by Striped Bass and Hybrid Striped Bass in Claytor Lake, Virginia

JOHN M. KILPATRICK*
Georgia Department of Natural Resources
2024 Newton Road, Albany, Georgia 31701, USA

JOHN J. NEY
Virginia Polytechnic Institute and State University
Blacksburg, Virginia 24062, USA

Abstract.—We compared the habitat selection (temperature and dissolved oxygen) of striped bass (STB) *Morone saxatilis* and hybrid striped bass (HSB; palmetto bass [male white bass *M. chrysops* × female striped bass]) in Claytor Lake, Virginia. Striped bass (n = 28) and HSB (n = 33) were implanted with temperature-sensitive radio tags and tracked biweekly for a period of 19 months to record position and habitat use. Striped bass averaged 715 mm total length (range 533–940) whereas HSB averaged 547 mm total length (range 460–659). Habitat separation was most apparent in the warm season months of June through August. During this period, when deoxygenation of the metalimnion and hypolimnion was not limiting, STB and HSB segregated according to apparent temperature preferences, with HSB selecting warmer temperatures. As deoxygenation progressed, HSB and especially STB were forced into warmer strata than preferred. Results from this study suggest that if favorable dissolved oxygen levels are maintained, STB and HSB will vertically segregate during summer stratification. Temperature selection and qualitative observations of fish mortality in summer indicated that adult HSB would be less vulnerable than adult STB to a temperature–dissolved oxygen squeeze in summer and its detrimental effects.

Introduction

Many southeastern reservoirs (e.g., Norfolk Reservoir, Arkansas; Cherokee Reservoir, Tennessee; Lake Seminole, Georgia; and Claytor Lake, Virginia) have been stocked with both striped bass (STB) *Morone saxatilis* and hybrid striped bass (HSB; palmetto bass [male white bass *M. chrysops* × female striped bass]). Stocking objectives for these reservoirs have included establishment of a recreational fishery, development of a trophy fishery, and the utilization of clupeid forage fishes (Bailey 1975; Combs and Peltz 1982; Ott and Malvestuto 1984; Murphy and Valentine 1985; Neal et al. 1999). Food habits have been found to overlap between adult STB and HSB (Bonds 2000) providing evidence for potential competition. However, there are no published reports on the temperature and dissolved oxygen levels (DO) used by populations of STB and HSB that occur together in the same water body.

Physical habitat conditions for inland populations of STB are most often defined as the combination of acceptable temperature and DO (Coutant 1985). The utilization of summer temperature refugia has been documented by many authors (Waddle 1979; Schaich and Coutant 1980; Bjorgo et al. 2000; Schaffler et al.

* Corresponding author: john.kilpatrick@dnr.state.ga.us

2002; Young and Isely 2002; Bettoli 2005). Temperature preference during summer stratification for STB varies with size, with larger fish occupying cooler temperatures if available (Schaich and Coutant 1980). Preferred temperature for adult STB is most often cited between 18°C and 24°C (Coutant and Carroll 1980; Crance 1984; Cheek et al. 1985) but may be as low as 15–16°C (Schaich and Coutant 1980). Coutant and Carroll (1980) suggested a fundamental niche for STB of 19–23°C. However, Bettoli (2005), conducting a study in a reservoir where cool, oxygenated water was readily available, found that the average temperature occupied by STB during the summer stratification period was 17.5°C. Optimal DO level for STB is usually considered at or above 5 mg/L while tolerable or suboptimal levels are from 2 or 3 mg/L to 5 mg/L (Waddle 1979; Coutant 1985, 1990). Levels of DO less than 2 mg/L are considered as poor or uninhabitable (Crance 1984). During summer stratification in southeastern reservoirs, STB may simply select the coolest water available with DO levels above 2–3 mg/L (Coutant and Benson 1990).

Similar to STB, the combination of temperature and dissolved oxygen is considered of primary importance for physical HSB habitat (DeMauro 1990). Habitat needs and preferences for HSB are not as well defined as for STB, but it is believed that HSB are more able to tolerate warm summertime conditions than STB (Coutant 1985; Moss 1985; Muncy et al. 1990). The exact temperature preference for HSB is unknown but may be intermediate to the parental STB and white bass (Barans and Tubb 1973; Gammon 1973). Based on findings from previous telemetry studies, HSB prefer 21–27°C water with dissolved oxygen greater than 4.5 mg/L (Douglas and Jahn 1987; Muncy et al. 1990). Like STB, they are intolerant of low (<2 mg/L) levels of DO. For example, Piner (1993) found that HSB in Fairfield Reservoir, Texas inhabited water up to 32°C if well oxygenated (>5mg/L), even though cooler water with DO of 2–4 mg/L was available. Anecdotal evidence for differences in summertime habitat selection between STB and HSB indicates that HSB may only rarely be found in close proximity with adult STB (Coutant 1985; Moss 1985).

Coutant (1985) pointed out that a scarcity of larger STB (>6 kg) observed by Kohler (1980), coupled with the temperature–DO squeeze identified by Boaze (1972), was evidence of marginal habitat in Claytor Lake, Virginia. Claytor Lake has been stocked with both STB and HSB since 1993, and fisheries for both have become well established. This dual stocking provided an opportunity to investigate the degree of habitat overlap between STB and HSB. While both STB and HSB utilized gizzard shad *Dorosoma cepedianum* and alewife *Alosa pseudoharengus* as primary forage in Claytor Lake (Bonds 2000), spatial occupancy patterns may heighten or lessen the potential for competitive habitat overlaps. Defining spatial occupancy patterns of STB and HSB will help to identify whether these species are complementary to one another, thus both diversifying the fishery and increasing potential moronid biomass available to anglers.

Although there is an abundance of literature on landlocked STB, and to a lesser degree HSB, their behavior when found in the same water body has not been described. The objective of this study was to compare seasonal habitat use and overlap versus habitat availability for adult STB and HSB in Claytor Lake.

Study Site

Claytor Lake (Figure 1), located in the New River Valley in southwest Virginia, was impounded in 1939 by the Appalachian Power Company (Rosebery 1951). It has a surface area of 1,810 ha (4,500 acres) at normal pool and is 34 km long with a 35-m maximum depth. The average retention time is a relatively short 30 d (Copeland 1999). It is dimictic, experiencing both a summer and winter stratification (Kohler et al. 1986). The reservoir maintains a relatively stable water elevation, only fluctuating about 1.6 m annually.

Striped bass were first introduced into Claytor Lake in 1969, and HSB were first introduced in 1993 (J. R. Copeland, Virginia

FIGURE 1. Collection sites for implanting radio-tags in striped bass (circles) and hybrid striped bass (triangles) from spring 2001 through spring 2002 in Claytor Lake, Virginia. Points may represent multiple captures at one location.

Department of Game and Inland Fisheries, personal communication). Both STB and HSB were good stocking candidates, as Claytor has limited littoral habitat (Kohler et al. 1986). Striped bass were stocked at a rate of about 70,000 fingerlings per year (37 fingerlings/ha) from 1969 until 1993, when stocking rates were halved to about 35,000 fingerlings per year (18.5/ha) to accommodate new HSB stockings (also 18.5 fingerlings/ha). In 1998 and 2001–2009, the STB stocking rate was again increased to 70,000 fingerlings annually in an attempt to boost STB abundance, while HSB stocking numbers remain at 35,000/ha.

Methods

Fish habitat use was assessed by tagging adult (≥460 mm total length) STB and HSB with implanted radio tags (Table 1). Both STB and HSB were captured by shoreline electrofishing, angling, or gill nets from various locations throughout the reservoir (Figure 1). Shoreline electrofishing was conducted using a boom electrofisher with pulsed DC current during spring of 2001 and 2002. Angling was conducted during all seasons using live and artificial baits. Gill nets were used during the winter of 2001–2002, when surface water temperature was below

TABLE 1. Number and sizes of striped bass and hybrid striped bass caught and tagged with internal implant radio tags. TL = total length.

	Striped bass				Hybrid striped bass			
Time period	*N*	Mean TL	Min TL	Max TL	*N*	Mean TL	Min TL	Max TL
Apr–Jun 01	8	657	533	864	10	531	460	648
Oct–Dec 01	8	716	641	775	9	547	505	659
Jan–Mar 02	6	784	645	940	0	–	–	–
Apr–Jun 02	6	738	673	899	14	565	483	638
Overall	28	718	533	940	33	550	460	659

10°C. Sinking gill nets, measuring from 30.5 to 61 m in length and 2.4 to 3.0 m in height with bar-mesh sizes of 64, 89, and 114 mm, were used. Gill nets were placed in high fish-density areas identified from boat-mounted sonar readings, they were checked frequently (<30 min), and fish were quickly tagged and released.

Following capture, a temperature-sensitive internal radio tag (Advanced Telemetry Systems, Model 5955, Isanti, Minnesota) was surgically implanted into the fish using absorbable sutures (Hart and Summerfelt 1975; Walsh et al. 2000). Individual surgical implantation required 105–150 s to complete. Each cylindrical transmitter measured 17 mm diameter, 51 mm in length, had an approximately 25-cm trailing antenna, and weighed 20 g in air. Each transmitter had a 2/3 AA lithium battery, individually coded radio frequency, duty cycle (4 d on/3 d off), and a 0–40°C temperature sensor with pulse indicator. Radio tags were calibrated at temperatures ranging from 4°C to 30°C using water baths prior to use; linear calibration curves for each tag were established.

Beginning in May 2001 and continuing until December 2002, transmitted fish were located at least twice per month throughout the period of thermal stratification (May–October) and at least once monthly otherwise. Fish were tracked via a boat-mounted loop-style antenna (ATS low-frequency loop, Isanti, Minnesota) and a scanning radio receiver (ATS R2000, Isanti, Minnesota). The entire reservoir was searched in a loop fashion so that both sides and all coves were searched during each tracking event. Tracking was completed both during daylight and nighttime hours. It took from 8 to 20 h to complete a tracking assessment of the reservoir.

Each time a telemetered fish was located, the temperature of the fish was estimated by the timed interval of the temperature-sensitive tags (tag pulse rate varied linearly with temperature). Once temperature was determined, a temperature/DO probe (YSI model 58, Yellow Springs, Ohio) was lowered to the depth at which that temperature was found to estimate DO. Thus, temperature, depth, and DO of fish occupancy could be estimated when the reservoir was thermally stratified.

When fish temperatures, as read from the pulse rate of the radio tag, were not reasonable, boat-mounted sonar was used to estimate fish depth. This was especially problematic given that small errors in temperature could result in large errors in estimated DO in the occupied habitat because DO sometimes varied sharply over 1–2 m depth. Boat-mounted sonar (Lowrance LCX-15 MT with a 12°-cone-angle transducer, Tulsa, Oklahoma) provided a second method for estimating fish depth and thus temperature and DO occupancy of each fish. Often, large schooling fish (presumably STB or HSB) were located in the immediate vicinity of a radio-tagged fish. In cases where temperature differed by more than 1°C or DO differed by more than 1 mg/L between radio-tag and sonar-predicted estimates, the sonar-predicted values were used when they could be validated by the locations of other fish. If temperature readings differed significantly (more 1°C) from the environment but no other radio-tagged fish were located within the same school as any fish in question, that data point was not utilized. Because the positive vertical identification of a specific temperature was only possible when the reservoir was stratified, the boat-mounted sonar was also used for estimating fish depth during isothermal conditions.

Available habitat was determined by taking vertical profiles of temperature and dissolved oxygen every 2 m from surface to bottom at 3.2-km (2 mi) intervals along the main channel of the impounded New River and along Peak Creek, resulting in 13 sample sites. Profiles were taken between 1000 and 1400 hours, within 1–2 d of each tracking event, and from the center of the reservoir or at the deepest spot within each transect. A YSI model 58 dissolved oxygen meter was used for all measurements. Readings were taken twice monthly, from July 2001 through December 2002. Temperature and DO were recorded to the nearest 0.1°C and 0.1 mg/L.

We calculated and graphed the mean monthly temperature, DO level, and depth occupied by STB and HSB; 95% confidence intervals for the means were calculated. Also, for June through August, 2002, we graphically compared temperature and dissolved oxygen selection by HSB and STB to available habitat conditions in the reservoir. Analysis of variance was used to test for differences ($\alpha = 0.05$) in the temperature, DO, and depth selection both between and within each species. Fisher's least significant difference was used to identify significant differences within comparable time periods. The Kolmogorov-Smirnov goodness of fit test ($\alpha = 0.05$) for continuous data (Zar 1999) was used to evaluate if fish occupancy of habitat was distributed similar to available lake habitat.

Results

Twenty-eight STB and 33 HSB were implanted with radio tags. The average length of STB in this study was 718 mm total length (TL) and ranged from 533 to 940 mm TL, whereas the average length of HSB was 550 mm TL and ranged from 460 to 659 mm TL (Table 1). Telemetered fish were detected from June 2001 through December 2002. Gill nets proved the most efficient collecting technique for both STB and HSB. Hook-and-line sampling was moderately effective but much less efficient. Survival of gill-netted, radio-tagged fish was greater than 90%.

Twenty-one HSB and 18 STB were located a minimum of 10 times and were included in our analysis; fish located less than 10 times were excluded. Of the fish that were located fewer than 10 times, five HSB and four STB were confirmed as dead or harvested during the study. An additional seven HSB and five STB were missing at the conclusion of the study. One fish was only located six times, although it was alive at the end of the study. It is notable that some of the transmitters (4 of the first 20 received) either went off duty-cycle or ceased transmitting radio signals within a few months of receiving them. It is possible that some tags were put in fish and then never located again due to transmitter malfunction.

Habitat quality, as determined by the amount of available habitat, was diminished during the summer of 2002 compared to 2001 because of a largely anoxic metalimnion and hypolimnion (Figure 2). Stratification during August of 2001 provided coolwater refuge areas (<25°C) with DO levels of greater than 5.0 mg/L, whereas stratification during August of 2002 left water that was less than 25°C devoid of DO and thus uninhabitable (Figure 2). As a result, in 2002, the only thermal habitat that was available from late July to early September was greater than 24.0°C. Anoxic water (<1.0 mg/L) during late August of 2001 represented only about 25% of Claytor Lake, whereas more than 45% of Claytor Lake contained less than 1.0 mg/L in August of 2002.

The radio-telemetry equipment utilized during this study was unable to consistently locate fish that were deeper than about 17 m. Therefore, fish that inhabited the summer, coolwater refuges available in 2001 were not reliably located for up to 3 months. During this time period, STB became increasingly difficult to find, but readings from the boat-mounted sonar and reports from fishermen supported the idea that STB were located near the bottom of the metalimnion at a depth of approximately 20 m, where temperatures were 19–20°C and DO levels were at 4–5 mg/L during summer stratification in 2001 (Figure 2). During this same time, HSB occupied shallower (5–15 m) and warmer (21–26°C) water and were vertically separated from STB. For instance, during the period of June–August 2001, STB were located 44.8% of the time (30 of 67 opportunities). During that same period, HSB were located 85.1% of the time (80 of 94 opportunities). In contrast, during the period of September–November 2001, STB were located 100% of the time (25/25) and HSB were located 95.6% (43/45). However, in 2002, due to the DO deficiency and subsequent loss of useable habitat (DO > 2.0 mg/L), both STB and HSB were located in depths shallower than 10 m and were reliably located.

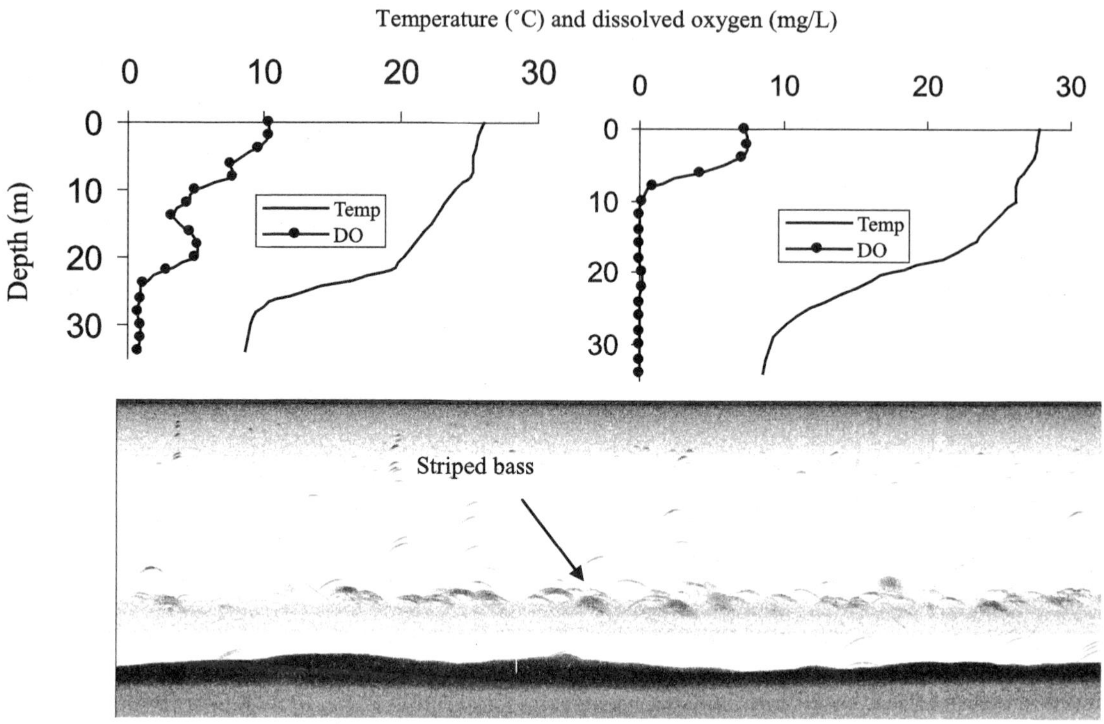

FIGURE 2. Vertical water quality profiles from August 2001 (left) and August 2002 (right) near Claytor Lake dam. Bottom image illustrates a school of striped bass, confirmed by angling efforts, that was located 20 m in depth during August 2001.

Throughout most of the fall, winter, and early spring periods, STB and HSB were in similar habitats, usually at depths less than 2.0 m apart, temperatures less than 1.5°C apart, and DO less than 1 mg/L apart (Figure 3). Monthly habitat use for STB and HSB was quite variable (Figure 3). Striped bass occupied temperatures ranging from 3.7°C to 27.0°C, DO from 0.8 to 11.7mg/L, and depths from 0.5 to 26.0 m; HSB occupied temperatures ranging from 3.4°C to 27.4°C, DO from 0.2 to 12.0 mg/L, and depths from 0.5 to 28.8 m.

Striped bass and HSB differed in their habitat use during the warm-season stratification of both 2001 and 2002 (Figure 3). Hybrid striped bass were found in warmer average water temperatures (25.5 versus 24.0; $P < 0.05$) than STB during July–September of 2002. Striped bass were found in significantly deeper average water depths during this period in 2001 (12.5 versus 11.3 m; $P < 0.05$) and 2002 (9.1 versus 6.0 m; $P < 0.05$). Additionally, because of the inability to detect fish that were below about 17 m and our belief that STB were located at depths below 18 m during the summer of 2001, STB likely were at cooler temperatures and greater depths than documented. During July–September, striped bass were found in water with higher average DO than HSB during 2001 (4.9 versus 4.0 mg/L; $P < 0.05$), but HSB occupied water with higher average DO levels in 2002 (5.8 versus 4.4 mg/L; $P < 0.05$).

June through August 2002 provided a good example of habitat progression (Figure 4). During June 2002, STB occupied cooler (mean 19.3°C) water than HSB (mean 23.2°C; $P < 0.01$). While DO use differed between STB and HSB ($P < 0.05$), all fish were found in concentrations higher than 3 mg/L. July 2002 also found STB in cooler (mean 22.4°C) water than HSB (mean 25.8°C; $P < 0.01$). Dissolved oxygen habitat occupied by STB (mean

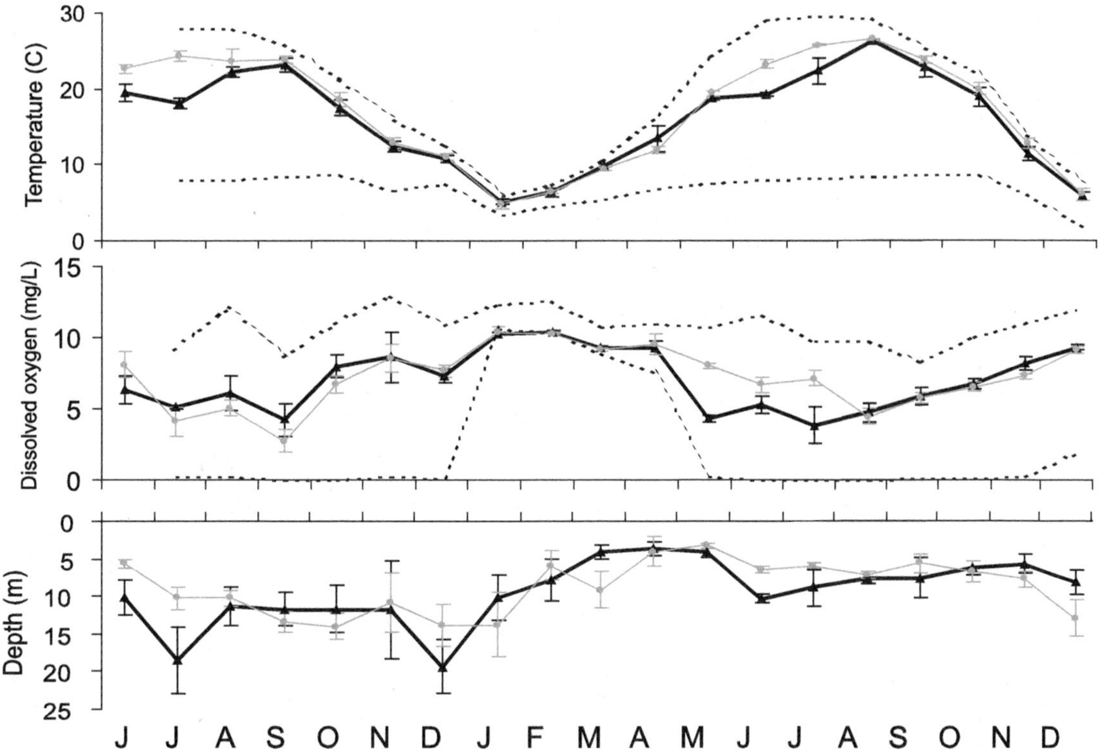

FIGURE 3. Temperature, dissolved oxygen, and depth selection by striped bass (black lines) and hybrid striped bass (gray lines) in Claytor Lake, Virginia, June 2001–December 2002. Dashed lines represent the maximum and minimum value available. Error bars represent 95% confidence intervals.

3.4 mg/L) was significantly lower than those used by HSB (mean 7.1 mg/L; $P < 0.01$). During August, the majority (66%) of Claytor Lake was ≥24°C, whereas every located STB and HSB were in this upper stratum as all temperatures cooler than 24°C had DO levels less than 2 mg/L. During August, STB occupied similar ($P > 0.05$) water temperatures (mean 26.4°C) and DO (4.7 mg/L) as HSB (mean 26.5°C; 4.4 mg/L). Throughout this 3-month period, both STB and HSB were in similar horizontal locations (Figure 4) but were only vertically separated during June and July. Therefore, when refugia habitat (preferred temperature, suboptimal to optimal DO) existed for both STB (15–20°C and DO > 2.5 mg/L) and HSB (21.5–25.5°C and DO > 2.0 mg/L), they occupied two "stories" of Claytor Lake. Habitat occupancy during June through August of 2001 also differed between STB and HSB (Figure 3), but results were tempered by the inability to detect fish that were deeper than 17 m.

Habitat occupancy was different in summer (i.e. July–September) 2001 than summer 2002. During summer of 2001, STB utilized cooler (mean 22.3°C) and deeper (mean 12.5 m) water than in 2002, where they were found in warmer (mean 24.0°C) and shallower (mean 9.1 m) water; DO at which STB were located was similar between the two summers (mean 4.9 mg/L in 2001 versus 4.4 mg/L in 2002). Similar to STB, tagged HSB were found in cooler (mean 23.8°C) and deeper (mean 11.3m) water in 2001 than in 2002, where they were found in warmer (mean 25.5°C) and shallower (mean 6.0 m) water during summer stratification. However, DO levels where HSB were found differed between the summer of

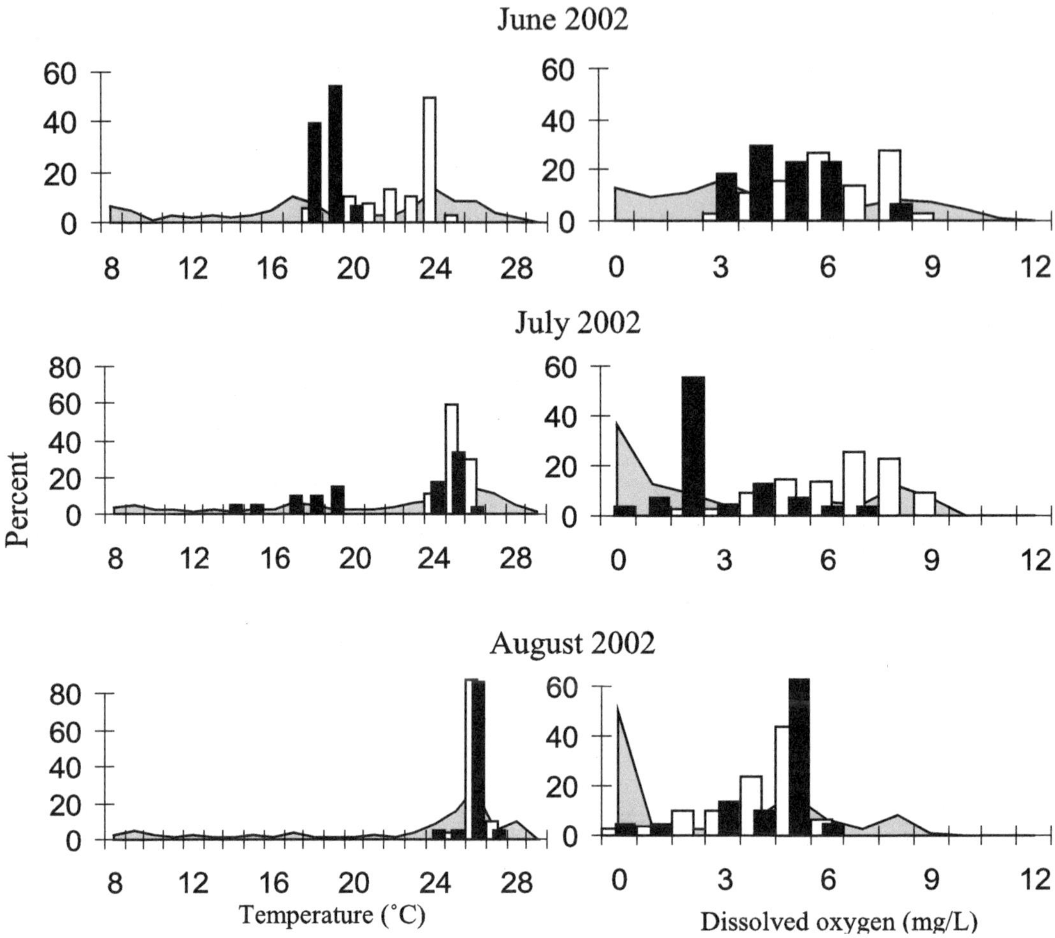

FIGURE 4. Habitat usage and habitat availability in Claytor Lake during stratification in 2002. Habitat availability is weighted by reservoir volume. Gray area represents available habitat, black bars represent striped bass habitat usage, and white bars represent hybrid striped bass usage.

2001 (mean 4.0 mg/L) and 2002 (mean 5.8 mg/L).

There were large differences in the amount of Claytor Lake volume that was suitable moronid habitat between the summers of 2001 and 2002 due to differences in stratification patterns. For instance, anoxic water (<1.0 mg/L) during late August of 2001 represented only about 25% of Claytor Lake, whereas more than 45% of Claytor Lake contained less than 1.0 mg/L in August of 2002. Habitat quality, as determined by the amount of refuge habitat, was diminished in 2002 because of a largely anoxic metalimnion and hypolimnion (Figure 2). As a result, in 2002, the only thermal habitat that was available from late July to early September was greater than 24.0°C.

Discussion

Striped bass and hybrid striped bass exhibited unique temperature preferences in June–August when oxygen levels did not limit access to cool water in either the metalimnion or hypolimnion. Additionally, habitat use for both STB and HSB was disproportional to its availability in Claytor Lake. In 2001, STB were trapped in a classic temperature–DO squeeze, similar to that documented by Coutant (1985), whereas HSB occupied shallower, warmer water than STB within the upper metalimnion and lower epilimnion. In mid-summer stratification of 2002, however, the entire portion of the reservoir below the epilimnion was anoxic, and both STB and HSB were forced to occupy the shallower but warmer (>25°C) habitat. In both years, STB occupied water that, on average, was at least 1.5°C cooler in July–September than HSB.

Coutant and Carroll (1980) suggested a temperature preference for STB centered around 22°C, and Crance (1984) modeled optimal habitat at 20–25°C. However, Young and Isely (2002) found that STB during summer stratification would tolerate suboptimal DO (3.0–4.3 mg/L) in order to inhabit water at 16.5–19.4°C, and Schaffler et al. (2002) found STB inhabited water of 16–18°C even when DO was less than 5 mg/L. Striped bass in Claytor Lake were found in cooler temperatures (15–20°C) as long as this water had adequate DO (2–3 mg/L). For instance, in late June of 2002, oxygenated water was available from 17°C to 29°C; STB were found at 18–19°C while HSB selected water temperatures of 23–25°C. At this time, fish were in the same longitudinal locations within the reservoir. In early summer of 2001, STB were found in 18–20°C water, when at the same time and approximate same longitudinal locations, HSB were found in 21–25°C water. At no other times of the year were wide ranges of oxygenated temperatures available to fish, which led to our finding of vertical habitat overlap except in June through August when oxygenated (≥2.5 mg/L) water 15–25°C was available.

Average summertime temperatures occupied by STB in Claytor Lake (22.3–24.9°C) were cooler than STB used in Lake Whitney, Texas (27–29°C; Farquhar and Gutreuter 1989) and Lake Norman, North Carolina (22–28°C; Van Horn et al. 1998) but warmer than Melton Hill Reservoir, Tennessee (14.1–16.7°C; Bettoli 2005) and Lake Murray, South Carolina (16–18°C; Schaffler et al. 2002). However, similar to the findings from Van Horn et al. (1998), STB in Claytor were found to occupy temperatures up to 27°C during the warmest parts of summer. Average summer DO use by STB in Claytor Lake (4.7–4.9 mg/L) was higher than that found from transmitted fish in Lake Norman (2.3–4.7 mg/L; Van Horn et al. 1998) but similar to that found with fish from J. Strom Thurmond Reservoir, South Carolina (5–6 mg/L; Young and Isely 2002) and Watts Bar Reservoir (>4.0 mg/L; Cheek et al. 1985). However, summer temperature selection by reservoir STB is highly influenced by availability of dissolved oxygen.

Average summer temperatures occupied by HSB in Claytor Lake (23.8–25.5°C) were cooler than temperatures occupied by telemetered HSB in Clarks Hill Reservoir, Georgia–South Carolina (25–28°C; Windham 1986); Ross Barnett Reservoir, Mississippi (27–29°C; Muncy et al. 1990); and Lake Fairfield, Texas (30–33°C; Piner 1993) but similar to HSB from Spring Lake, Illinois (24–25°C; Douglas and Jahn 1987), largely due to the amount of available habitat present during stratification in these reservoirs. Average summer DO habitat used by HSB in Claytor Lake (4.0–5.8 mg/L) was similar to those in Clarks Hill Reservoir, Georgia–South Carolina (>4 mg/L; Windham 1986); Ross Barnett Reservoir, Mississippi (>4 mg/L; Muncy et al. 1990); and Lake Fairfield, Texas (>5.0 mg/L; Piner 1993) but higher than fish from Spring Lake, Illinois (2–5 mg/L; Douglas and Jahn 1987).

During summer stratification and similar to the findings of Piner (1993), the vertical distribution of STB was determined more by DO than by temperature if DO was less than 2–3 mg/L. Temperature was more likely to determine the vertical distribution of STB if DO levels were greater than 3 mg/L. Therefore, it cannot be said that STB preferred summer temperatures occupied in this study, but rather used these temperatures to avoid low DO levels. During summer stratification, STB were consistently found at the lowest temperatures where acceptable DO levels were available

(≥2.5 mg/L). However, if DO levels dropped below 2.5 mg/L, STB simply selected the coolest possible water with DO ≥ 2.5 mg/L. In many systems, STB may use a coolwater refuge area, such as a tailrace receiving cooler water than found in the main reservoir. However, no coolwater input thermal refuge areas were identified in this study and both STB and HSB were forced into water in excess of 27°C during summer stratification in 2002 to avoid water less than 2.5 mg/L DO.

Striped bass have been suggested to prefer DO levels of greater than 5 mg/L (Crance 1984; Coutant 1985; Young and Isely 2002), whereas HSB have been found to select DO levels greater than 4.5 mg/L (Douglas and Jahn 1987; DeMauro 1990; Muncy et al. 1990; Piner 1993). Our study suggests that when 15–25°C water is available, STB will utilize the coolest water available as long as there is ≥2.5 mg/L of dissolved oxygen. In August of 2002, STB were in water greater than 25.0°C, but most were also found at DO levels of greater than 5 mg/L. Cooler water was available but primarily only contained less than 2 mg/L DO. Hybrid striped bass in this study also showed a preference for DO levels of greater than 4.5 mg/L. For example, in August of 2002, HSB were found in temperatures greater than 25.5°C, but most fish were found at DO levels of 4–5 mg/L in order to avoid cooler water that had less oxygen. If adequate DO levels existed for HSB (>2.0 mg/L), they sought out a temperature of 21.5–25.5°C, while STB sought a temperature of 15–20°C. Similar to STB, if DO levels dropped below 2 mg/L, HSB inhabited water in excess of 27°C.

Based on the habitats occupied by telemetered fish during the 17 months of this study under different combinations of temperature and DO, several habitat generalizations can be made. In order to inhabit temperatures less than 20°C, STB tolerated DO levels as low as 2.5 mg/L. However, to avoid DO levels of less than 2.0 mg/L, STB tolerated water temperatures in excess of 25°C. Hybrid striped bass tolerated water temperatures in excess of 26°C to avoid DO levels of less than 2.0 mg/L. They would, however, tolerate DO levels as low as 2–3 mg/L to inhabit water temperatures less than 26°C.

There was a substantial decline of summer moronid habitat in 2002 as compared to 2001. Ruane et al. (2013, this volume) conducted water quality modeling on Claytor Lake and suggested that system morphometry and hydrology were the primary factors influencing available summer habitat. The flow regime in Claytor differed between the 2 years. Although daily mean March–June flows in 2001 (76.3 m^3/s) were similar to those in 2002 (61.5 m^3/s), 2 d in spring of 2002 produced flow averaging about 566.3 m^3/s, and a large amount of eroded organic matter was deposited into Claytor (G. J. Devlin, Virginia Department of Environmental Quality, personal communication). As these organics gradually settled into the deeper layers of Claytor Lake, the lake began to stratify. By May 2002, the reservoir had fully stratified, but flows remained low. The organics remaining in the hypolimnion coupled with the low degree of mixing from incoming water currents rendered the deeper portions of Claytor anoxic in 2002 and led to almost a complete loss of STB habitat (<25°C and ≥2.5 mg/L) during the warmest months of 2002 (late July–August). Flows were below average during March–April 2001, but flows during the warmer periods of May–August of 2001 remained approximately average. Therefore, in 2001, water turnover rate was relatively high in the metalimnion and hypolimnion, not as many organics were introduced in the spring, and STB habitat persisted throughout the year. Therefore, drought years may be especially detrimental to the maintenance of coolwater habitat during summer stratification, especially if high rain events during early stratification deposit large amounts of organic debris into Claytor Lake and a high water retention rate ensues. Managers must consider the range of summer conditions a reservoir may experience, as severe loss of habitat could result in a major fish mortality event. Evaluating habitat dynamics with limnological measurements and modeling is a potentially fruitful field for refining *Morone* sp. management.

Qualitative observations of summer fish mortality suggested that HSB were more tolerant of restricted habitat conditions than STB. Although there was not a definitive pattern in radio-tagged fish mortality, there was an increase in sightings of dead, floating striped bass in the summer of 2002. Dead fish were located during every radio-tracking event during the summer of 2002 but only rarely observed during 2001. The observed fish mortalities, whether from natural or hooking mortality, were primarily larger (>5 kg) STB, especially in 2002. Few HSB mortalities were noted during this study, even during times of apparent high concurrent STB mortality. Striped bass mortality from loss of habitat has been documented in many other reservoirs, with most occurring in southeastern reservoirs during warmer months. Larger STB may have less temperature tolerance (Coutant 1985) and are likely more susceptible to summer mortality (Matthews 1985). This observation is consistent with a higher temperature tolerance of HSB, presumably subjecting them to a less physiological stress during the summers of 2001 and 2002.

These observations suggest that low abundance of large (>6 kg) STB in Claytor Lake may be due to inadequate summer habitat, especially for these fish. In contrast, HSB did not seem to be limited by habitat in Claytor Lake. This may have contributed to the increased relative abundance of HSB over STB reported by Bonds (2000).

This study augments the limited, published literature concerning the relative behavior of STB and HSB when they coexist in a reservoir. Striped bass and HSB occupied similar vertical (depth) habitats in Claytor Lake throughout most of the study period. Fish were also found in similar horizontal positions, thus supporting Bonds (2000) findings that the two species are trophic equivalents and are potential competitors. However, the two species occupied different vertical habitats during early stratification of 2001 and 2002 when preferred temperatures were available and DO levels were ≥ 2.5 mg/L. During the summer of 2001, STB also were in deeper and cooler water than HSB. While we defined summer temperature preferences, the comparative ability of STB and HSB to physiologically tolerate warm, summertime conditions was not defined and demands additional investigation.

Acknowledgments

Funding was provided by American Electric Power and the Virginia Department of Game and Inland Fisheries. Project support was provided through staff at the Department of Fisheries and Wildlife of Virginia Tech University. Field support was provided by Jake Rash, Marcy Anderson, John Harris, Jamie Roberts, Dennis Brown, and other personnel from Virginia Tech University and John Copeland, Dan Wilson, and other personnel from the Virginia Department of Game and Inland Fisheries. Jim Peterson helped this manuscript with review and data analysis support.

References

Bailey, W. M. 1975. An evaluation of striped bass introductions in the southeastern United States. Proceedings of the Annual Conference Southeastern Association of Game and Fish Commissioners 28(1974):54–68.

Barans, C. A., and R. A. Tubb. 1973. Temperatures selected seasonally by four fishes from western Lake Erie. Journal of the Fisheries Research Board of Canada 30:1697–1703.

Bettoli, P. W. 2005. The fundamental thermal niche of adult landlocked striped bass. Transactions of the American Fisheries Society 134:305–314.

Bjorgo, K. A., J. J. Isely, and C. S. Thomason. 2000. Seasonal movement and habitat use by striped bass in the Combahee River, South Carolina. Transactions of the American Fisheries Society 129:1281–1287.

Boaze, J. L. 1972. Effects of landlocked alewife introduction on white bass and walleye population, Claytor Lake, Virginia. Master's thesis. Virginia Polytechnic Institution and State University, Blacksburg.

Bonds, C. C. 2000. Assessment of the response of

piscivorous sportfishes to the establishment of gizzard shad in Claytor Lake, Virginia. Master's thesis. Virginia Polytechnic Institution and State University, Blacksburg.

Cheek, T. E., M. J. Van Den Avyle, and C. C. Coutant. 1985. Influences of water quality on distribution of striped bass in a Tennessee River impoundment. Transactions of the American Fisheries Society 114:67–76.

Combs, D. L., and L. R. Peltz. 1982. Seasonal distribution of striped bass in Keystone Reservoir, Oklahoma. North American Journal of Fisheries Management 2:66–73.

Copeland, J. R. 1999. Claytor Lake management report 1998. Virginia Department of Game and Inland Fisheries, F111R7, Final Report, Richmond.

Coutant, C. C. 1985. Striped bass temperature and dissolved oxygen: a speculative hypothesis for environment risk. Transactions of the American Fisheries Society 114:31–61.

Coutant, C. C., and D. S. Carroll. 1980. Temperatures occupied by ten ultrasonic-tagged striped bass in freshwater lakes. Transactions of the American Fisheries Society 109:195–202.

Coutant, C. C. 1990. Temperature-oxygen habitat for freshwater and coastal striped bass in a changing climate. Transactions of the American Fisheries Society 119:240–253.

Coutant, C. C., and D. L. Benson. 1990. Summer habitat suitability for striped bass in Chesapeake Bay: reflections on a population decline. Transactions of the American Fisheries Society 119:757–778.

Crance, J. H. 1984. Habitat suitability index models and instream flow suitability curves: inland stocks of striped bass. U.S. Fish and Wildlife Service, Washington, D.C.

DeMauro, R. A. 1990. Movements, habitat associations, and food of hybrid striped bass in Ross Barnett Reservoir, Mississippi. Master's thesis. Mississippi State University, Mississippi State.

Douglas, D. R., and L. A. Jahn. 1987. Radiotracking hybrid striped bass in Spring Lake, Illinois, to determine temperature and oxygen preferences. North American Journal of Fisheries Management 7:531–534.

Farquhar, B. W., and S. Gutreuter. 1989. Distribution and migration of adult striped bass in Lake Whitney, Texas. Transactions of the American Fisheries Society 118:523–532.

Gammon, J. R. 1973. The response of fish populations in the Wabash River to heated effluents. Pages 513–523 *in* D. J. Nelson, editor. Proceedings of third national symposium on radioecology, Chapman and Hall, London.

Hart, L. G., and R. C. Summerfelt. 1975. Surgical procedures for implanting ultrasonic transmitters into flathead catfish (*Pylodictis olivaris*). Transactions of the American Fisheries Society 1:56–59.

Kohler, C. C. 1980. Trophic ecology of an introduced landlocked alewife (*Alosa pseudoharengus*) population and assessment of alewife impact on resident sportfish and crustacean zooplankton communities in Claytor Lake, Virginia. Doctoral dissertation. Virginia Polytechnic Institute and State University, Blacksburg.

Kohler, C. C., J. J. Ney, and W. E. Kelso. 1986. Filling the void: development of a pelagic fishery and its consequences to littoral fishes in a Virginia mainstream reservoir. Pages 166–177 *in* G. E. Hall and M. J. Van Den Avyle, editors. Reservoir fisheries management: strategies for the 80's. American Fisheries Society, Southern Division, Reservoir Committee, Bethesda, Maryland.

Matthews, W. J. 1985. Summer mortality of striped bass in reservoirs of the United States. Transactions of the American Fisheries Society 114:62–66.

Moss, J. L. 1985. Summer selection of thermal refuges by striped bass in Alabama reservoirs and tailwaters. Transactions of the American Fisheries Society 114:77–83.

Muncy, R. J., R. A. DeMauro, L. E. Miranda, and P. S. Phalen. 1990. Habitats used by radio-tagged adult hybrids in Ross Barnett Reservoir. Mississippi Cooperative Fish and Wildlife Research Unit, Completion Report, Federal Aid to Fisheries Project F-82-1, Mississippi State.

Murphy, B. R., and G. Valentine. 1985. Hybrid striped bass for control of stunted white crappie. Pages 10–11 *in* Annual proceedings of the Texas Chapter of the American Fisheries Society, volume 25. American Fisheries Society, Texas Chapter, Bethesda, Maryland.

Neal, J. W., R. L. Noble, and J. A. Rice. 1999. Fish community response to hybrid striped bass introduction in small warmwater impoundments. North American Journal of Fisheries Management 19:1044–1053.

Ott, R. A., and S. P Malvestuto. 1984. The striped bass × white bass hybrid in West Point Reservoir. Proceedings of the Annual Conference Southeastern Association of Fish and Wildlife Agencies 35(1981):641–646.

Piner, K. R. 1993. Seasonal distribution of hybrid striped bass in a power-plant cooling reservoir and a comparison to red drum. Master's thesis. Texas A&M University, Galveston.

Rosebery, D. A. 1951. Fisheries management of Claytor Lake, and impoundment on the New River in Virginia. Transactions of the American Fisheries Society 80:194–209.

Ruane, R. J., G. E. Hauser, and A. F. Sawyer. 2013. Water quality modeling as a management tool for striped bass and hybrid striped bass. Pages 161–179 *in* J. S. Bulak, C. C. Coutant, and J. A. Rice, editors. Biology and management of inland striped bass and hybrid striped bass. American Fisheries Society, Symposium 80, Bethesda, Maryland.

Schaffler, J. J., J. J. Isely, and W. E. Hayes. 2002. Habitat use by striped bass in relation to seasonal changes in water quality in a southern reservoir. Transactions of the American Fisheries Society 131:817–827.

Schaich, B. A., and C. C. Coutant. 1980. A biotelemetry study of spring and summer habitat selection by striped bass in Cherokee Reservoir, Tennessee, 1978. Oak Ridge National Laboratory, ORNL/TM-7127, Oak Ridge, Tennessee.

Van Horn, S. L., J. R. Finke, and D. Degan. 1998. Summer habitat selection of striped bass in Lake Norman. Proceedings of the Annual Conference Southeastern Association of Fish and Wildlife Agencies 50(1996):91–97.

Waddle, H. R. 1979. Summer habitat selection by striped bass, *Morone saxatilis*, in Cherokee Reservoir, Tennessee, 1977. Master's thesis. University of Tennessee, Knoxville.

Walsh, M. G., K. A. Bjorgo, and J. J. Isely. 2000. Effects of implantation method and temperature on mortality and loss of simulated transmitters in hybrid striped bass. Transactions of the American Fisheries Society 129:539–544.

Windham, W. T. 1986. Summer temperature selection of striped bass × white bass hybrids in a southeastern reservoir. Master's thesis. University of Georgia, Athens.

Young, S. P., and J. J. Isely. 2002. Striped bass annual site fidelity and habitat utilization in J. Strom Thurmond Reservoir, South Carolina–Georgia. Transactions of the American Fisheries Society 131:828–837.

Zar, J. H. 1999. Biostatistical analysis, fourth edition. Prentice Hall, Upper Saddle River, New Jersey.

American Fisheries Society Symposium 80:161–179, 2013

Water Quality Modeling as a Management Tool for Striped Bass and Hybrid Striped Bass

Richard J. Ruane*
Reservoir Environmental Management, Inc.
900-5 Vine Street, Chattanooga, Tennessee 37403, USA

Gary E. Hauser
Loginetics, Inc.
6009 Mont Richer Avenue, Knoxville, Tennessee 37918, USA

Andrew F. Sawyer
Reservoir Environmental Management, Inc.
900-5 Vine Street, Chattanooga, Tennessee 37403, USA

Abstract.—Striped bass *Morone saxatilis* habitat in water bodies is affected by many factors such as hydrological and meteorological conditions, eutrophication, reservoir operations, dam outlet levels, lake characteristics, and watershed characteristics. The CE-QUAL-W2 water quality model is a tool that can integrate the effects of all these factors on striped bass habitat. Once a baseline model is calibrated, it can be used to diagnose constraints to striped bass habitat, identify potential enhancement measures, and evaluate ways to alleviate the impacts of conflicting water uses. Importantly, the model integrates the best available information within the best available scientific method of evaluating water quality or habitat issues. Centering the discussion around an agreed-upon scientific tool helps to ensure that the subjective concerns expressed by stakeholders are objectively evaluated. In the three case studies explored in this paper, a change of hydropower operation was agreed to for Lake Murray, South Carolina that would help maintain summer habitat for striped bass; simulations indicated that hydropower operations were not a major factor affecting striped bass habitat in Claytor Lake, Virginia; and an efficient oxygen injection system was designed for J. Strom Thurmond Reservoir, South Carolina and Georgia to mitigate for habitat loss associated with a change in hydropower operation. Water quality modeling is an important tool for objectively evaluating the maintenance or enhancement of striped bass and hybrid striped bass (white bass *M. chrysops* × striped bass) habitat in reservoirs.

Introduction

Striped bass *Morone saxatilis* are dependent on summer habitat with acceptable levels of temperature, dissolved oxygen (DO), and available forage (Coutant 1985 and 2013, this volume; Thompson and Rice 2013, this volume). Lack of suitable habitat can lead to poor condition during summer (McDaniel et al. 1991) or even large-scale mortality events if habitat becomes unacceptable or fish are trapped in water of poor quality (Matthews 1985; Rice et al. 2013, this volume). Both poor condition and mortal-

* Corresponding author: jimruane@comcast.net

ity events in reservoirs are generally associated with a deterioration of habitat as summer progresses and reservoir stratification patterns are altered (Zale et al. 1990).

The summer thermal stratification of a reservoir and associated variations in oxygen concentration with depth are complex processes that vary among reservoirs as a function of physical and chemical factors. These processes are described in limnology texts such as Wetzel (2001) and model reference documents such as Cole and Wells (2002). In a stratified reservoir during summer, acceptable cool, well-oxygenated water may be found in either the hypolimnion or the metalimnion, if it does not become excessively consumed by in-lake DO demands. The metalimnion serves as a transition zone between the epilimnion and the hypolimnion in reservoirs that is transient over time, especially at hydropower projects that involve relatively high water flows passing through a reservoir. The dynamics of thermal stratification and DO concentrations are affected by many factors, including nutrient levels, residence time of water in the habitat, outlet level(s) for the reservoir releases, nominal residence time for water in the lake, meteorological conditions, inflow temperatures, annual hydrologic variations, and other water uses of the lake (e.g., flood control, lake level management, tailwater needs, thermal effluents, and water and wastewater treatment plants).

Generally, adequate physical, chemical, and meteorological data are available to model the dynamics of summer stratification in a reservoir as these parameters are often monitored by a variety of public and private entities. Once a baseline model is configured and calibrated, it can be used to determine if the reservoir is likely to provide suitable habitat under usual, as well as extreme, environmental conditions and then, if needed, evaluate those controllable factors that can influence the quality and quantity of summer striped bass habitat. Model simulations can allow managers to assess whether habitat improvements can be achieved and, if so, at what impact relative to other water uses (e.g., hydropower, water supplies, wastewater assimilation, recreation, and land owners) of the reservoir. For example, holding pool elevations higher during the summer and autumn periods might benefit other lake users, tailrace temperature, and water supplies; however, holding pool elevation higher may impact power generation. Another example is that lake oxygenation on a water supply reservoir might benefit striped bass habitat as well as the water treatment plant.

The modeling efforts reported herein are based upon state-of-the-art approaches that are logical, sound extensions of well-founded research and studies conducted over the past half century. With any use of models, it should be recognized that modeling results provide a general indicator of what is likely to occur under given sets of conditions. As is the case in all aquatic environments, natural conditions are more complex than models, so the models tend to reproduce the major patterns that are observed in the field but will lack resolution, inputs, or formulations to reproduce all the minor patterns that are observed. Models are internally consistent and based on rigorous governing equations, so they can often help explain apparent discrepancies in field observations.

The objective of this paper is to use data from three case studies to demonstrate to fishery managers how water quality modeling was able to (1) define baseline habitat conditions, and (2) define water management or engineering strategies that would enhance striped bass habitat in three southeastern reservoirs, Lake Murray, South Carolina; Claytor Lake, Virginia; and J. Strom Thurmond, South Carolina–Georgia. The model results presented in this chapter were used for decision-making purposes for determining ways of enhancing striped bass habitat.

Methods

We used the CE-QUAL-W2 water quality model to simulate reservoir water quality. CE-QUAL-W2 is a two-dimensional, hydrodynamic and water quality model for reservoirs and rivers, and its use and formulation are described by Cole and Wells (2002). The CE-QUAL-W2 model is deterministic (i.e., mechanistic), not stochastic. Modeled temperatures within lakes are driven by boundary conditions, including inflows, outflows and their withdrawal zones, inflow temperatures, and other forcing functions such as heat loadings and atmospheric heat exchange driven by meteorology. Modeled water quality within lakes is driven by inflow water quality (especially temperature, organic matter, nutrients, turbidity, etc.), transport of water through the lake, solar radiation and wind, algal production and death, bacterial decomposition, and sediment–water interactions. Calibration and application of the model to lake water quality requires interdisciplinary knowledge of hydrodynamics, heat transfer, physical/chemical/biological processes and stoichiometry, limnological processes, meteorology, numerical methods, computerized data assembly and analyses, lake sediment processes and sediment–water interactions, stream hydrologic and water quality processes, power plant operations, wastewater treatment plants, and statistics.

Calibration of a model with actual data from the reservoir is an essential component of model building. Model calibration is the process whereby the modeler applies the selected model to site-specific conditions. The W2 model requires about 200 settings to be used to run the model, but default values usually can be used for about 75% of these values, leaving about 30–40 critical parameters to be adjusted to calibrate the model to temperature and water quality conditions. A trial-and-error process is used to develop the final adjustments to attain a calibration of the model that yields an acceptable level of fit between observed data and model simulations. We required tight calibrations for temperature and DO as a function of inflow, as well as in-lake data on temperature, nutrients, and organic matter. In planning mode (looking back and comparing model results with actual effects of various operations), historical measurements are typically used as boundary conditions. In forecast mode (projecting into the future), boundary conditions are unknown so the user must take care to provide meaningful boundary-condition projections. Since forecasts of future hydrologic conditions are not certain, projecting boundary conditions often involves use of analogous historical years or sensitivity simulations covering a range of possible future conditions. In the course of calibrating and applying the CE-QUAL-W2 model for use in evaluating striped bass habitat, we found that the following provisions within version 3 would improve the performance of the model for meeting our specific objectives:

1. Provide for the phosphorus and nitrogen content of organic matter to be different for labile and refractory organic matter. This was desired since refractory matter accounts for much of the organic matter but has very little phosphorus and nitrogen content. This modification allowed a more effective calibration to the data through more direct control over mass of phosphorus and nitrogen in the system, which influences algal growth and oxygen depletion.
2. Allow the release of organic matter from the sediments under hypoxic conditions. This was desired since this organic matter exerts an additional DO demand on the water column, and it allows inclusion of this source of organic matter, providing more effective calibrations to measured data.
3. Use the Wuest wind drag coefficient so that a higher level of mixing could be induced under low wind speed conditions.
4. Program the CE-QUAL-W2 to conserve phosphorus when the phosphorus content in algal assemblages is not equal to the phosphorus content in organic matter.
5. Precipitate phosphorus from the water column to help account for the effects of finely suspended clay on phosphorus sorption and settling.

Documentation for the version of CE-QUAL-W2 we used is provided in the user manual authored by Cole and Wells (2002), and upgrades are reported in Sawyer and Ruane (2006), available at www.loginetics.com/w2/docs.

Lake Murray, South Carolina

Lake Murray is a 20,200-ha hydroelectric reservoir located in central South Carolina. Mean annual retention time of this reservoir is 400 d. Striped bass are stocked annually to provide a sport fishery (Schaffler et al. 2002), though summer mortalities of striped bass adults occur intermittently (Sawyer and Ruane 2007). During this work, the Lake Murray Project, operated by South Carolina Electric and Gas, was under relicensing consideration by the Federal Energy Regulatory Commission. The Water Quality Technical Working Committee from the relicensing process identified the following issues to be addressed using the CE-QUAL-W2 model:

- Determine the causes of striped bass fish kills during summer, especially factors related to hydropower operations.
- Determine the effects of operating turbine 5, outlet depth at 24 m, on striped bass habitat and entrainment of blueback herring *Alosa aestivalis*; turbines 1–4 have an outlet depth at 53 m.
- Determine operational changes and pool level management alternatives that would increase habitat for striped bass and blueback herring.
- Assess benefit of reducing nutrient loading to coolwater species.
- Track any impacts that could occur to the tailwater trout fishery due to potential operational changes in the reservoir.

Hydrological and environmental conditions for the years 1992, 1996, and 1997 were selected to calibrate the model. After developing calibrated models for all 3 years, it became apparent that one model could be developed for all 3 years if sediment oxygen demand was adjusted for each year. Water balance for specific calibration years was derived using daily hydropower releases and reservoir storage changes to back-calculate total daily inflows. Measured inflows were subtracted from total inflow, and the remainder of total inflow was apportioned by drainage area among the local inflows. Water quality data collected by the South Carolina Department of Health and Environmental Control was used to develop model inputs. The model was calibrated to pool elevation, total phosphorous, chlorophyll *a*, temperature, and DO data obtained during the calibration years. Calibration of the model is covered in much more detail in Sawyer and Ruane (2006). Once the model was calibrated, we defined acceptable summer striped bass habitat as water less than 27°C and DO greater than 2.5 mg/L.

We used the model and historical data to estimate the amount and timing of acceptable striped bass habitat that was available in eight representative years (Figure 1). Table 1 summarizes flow conditions experienced by Lake Murray for these 8 years. The mean flow for the months April through July had the highest correlation with availability of striped bass habitat in Lake Murray. Of those 8 years, 1991, 1998, and 2005 were high-flow years that experienced summers with major striped bass fish kills. Modeled volume of habitat is shown in Figure 1, and this plot shows that habitat declined most rapidly in these 3 years and 1997. The year 1997 did not experience a fish kill probably because DO was about 1–2 mg/L at minimum and fish "escaped" or survived in water that was about 20°C. Habitat was most favorable in 1992, 2000, and 2001. Historical records indicated that 1992 experienced cool meteorological conditions while 2000 and 2001 were low flow years. It appeared that 1996 might have been an acceptable habitat year, but a special, significant drawdown of the pool level starting in late August and extending into September probably exacerbated the loss of habitat. Based on these initial simulations, it appeared that high flow through the system increased the probability of the loss of striped bass habitat.

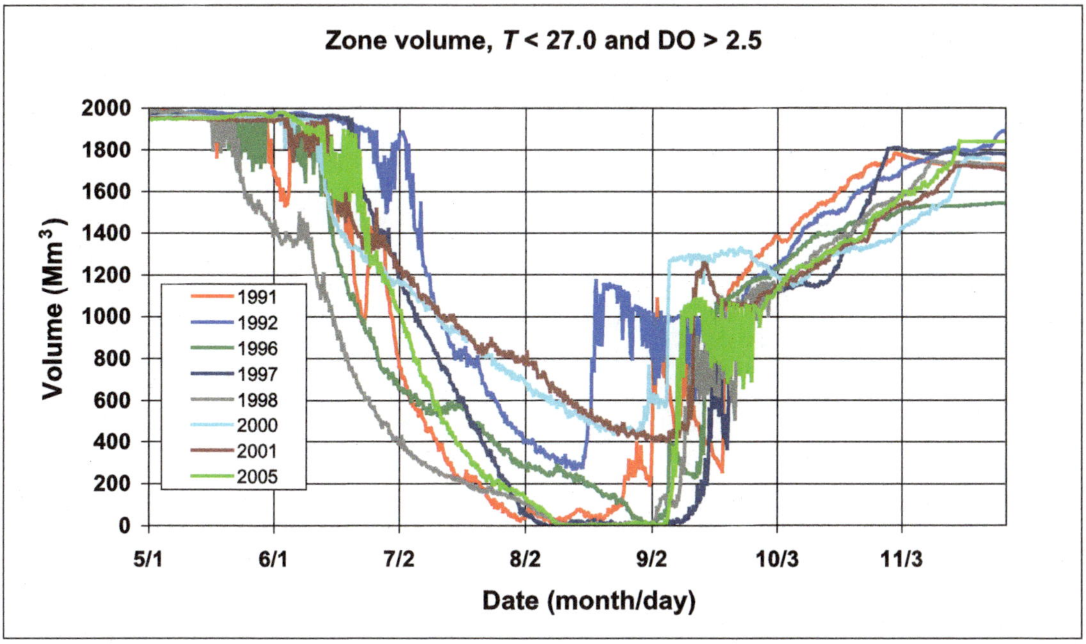

FIGURE 1. Volume of water (Mm^3) with suitable striped bass habitat (<27°C and >2.5 mg/L dissolved oxygen) for eight modeled years for Lake Murray, South Carolina.

We used the model to assess the effect of turbine 5 operation on available habitat in the lake and in the tailwater. Three operational options for turbine 5 in relation to other turbines were considered. The operational scenarios for turbine 5 were as follows: (1) Unit 5 was the first turbine turned on for all flow levels up to 170 m^3/s from 1 November through 15 June, (2) Settlement Agreement Unit 5 on first until 1 August—Unit 5 was the first turbine turned on when flow levels were between 42.5 and 170 m^3/s until 1 August, and (3) Settlement Agreement Unit 5 on last—involved operating Unit 5 when Units 1–4 were already on and the total flow level from all turbines was between 272 and 442 m^3/s. Three years were modeled that represented a range of hydrologic conditions. The model was set up for a normal flow year (1997), a high flow year (1998), and a low flow year (2000). In both the 1997 and 1998 model runs, the hypolimnion of Lake Murray and the tailwater did not warm as much as they did under actual conditions (Table 2). While this result was true for both the "Unit 5 on first" and the "Unit 5 on last" operations of Unit 5, the "Unit 5 on first" operations resulted in cooler water in the bottom of Lake Murray as well as in the tailwater. With cooler water preserved in the bottom of Lake Murray, DO in the release was higher and there was an increase in

TABLE 1. Mean flow (m^3/s) of Lake Murray, South Carolina hydropower releases in all modeled years.

	Mean flow	
Year	Annual	April–July
1991	74.8	93.6
1992	74.6	55.4
1996	83.0	60.7
1997	71.0	63.6
1998	109.2	103.9
2000	40.2	19.2
2001	27.0	19.3
2005	66.1	81.3

TABLE 2. Actual and modeled water temperatures in hydropower releases at Lake Murray Dam, South Carolina. Full descriptions of model scenarios are found in text. SA = settlement agreement.

Scenarios	Temperature °C		
	1 August	1 September	1 October
1997			
Actual	16.6	18.2	19.0
Unit 5 on first	14.5	15.7	16.4
SA Unit 5 on first	15.8	16.8	17.4
SA Unit 5 on last	16.4	17.5	18.0
1998			
Actual	15.7	16.6	17.3
Unit 5 on first	14.1	15.0	15.8
SA Unit 5 on first	14.7	15.5	16.2
SA Unit 5 on last	15.1	15.9	16.6
2000			
Actual	14.1	15.0	15.6
Unit 5 on first	13.3	14.3	15.0
SA Unit 5 on first	14.6	15.3	15.9
SA Unit 5 on last	15.1	15.7	16.2

the availability of tolerable striped bass habitat in Lake Murray (Figure 2). In the 2000 model run, the bottom of Lake Murray and the tailwater warmed more than they did under actual conditions, especially for the case with "Unit 5 on last" operations. However, the temperatures were still significantly cooler than in normal and high flow years. Under all conditions in 2000, the release temperatures were relatively cool and were not seen as a significant issue for trout in the downstream tailwater. For all years modeled, the Unit 5 last-on, first-off scenario resulted in warmer temperatures in Lake Murray and warmer hydropower releases than the Unit 5 first-on, last-off scenario. Therefore, model simulations indicated that "first-on" operation of turbine 5 from November through July was the preferred operational scenario. Simulations also showed that the benefit in habitat was dependent on hydrological conditions in a given year (Figure 2).

Maintaining the target pool level for April through August at 109.1 m above mean sea level either increased or had no effect on striped bass habitat. Of the 8 years modeled, there was some improvement in the volume of striped bass habitat in 4 years: 1992, 1996, 1998, and 2001.

The current phosphorous load to Lake Murray had a major effect on the amount of summer habitat for striped bass (Figure 3). Under current loading, there was essentially no acceptable striped bass habitat in most of August in 1998 and 2005, the years selected for simulations. When phosphorous loading was reduced by 60%, there was a substantial increase in the availability of striped bass habitat. Actual increase in habitat would be substantial but less than indicated in Figure 3 because when phosphorus is reduced, the ratio of organic matter to phosphorous can increase and cause additional oxygen demand, which was not considered within the model.

Weather conditions exerted a substantial effect on striped bass habitat (Figure 4). We incorporated 1992 cooler than normal meteorological data within the 2005 CE-QUAL-W2 model that had high-flow hydrology. Striped

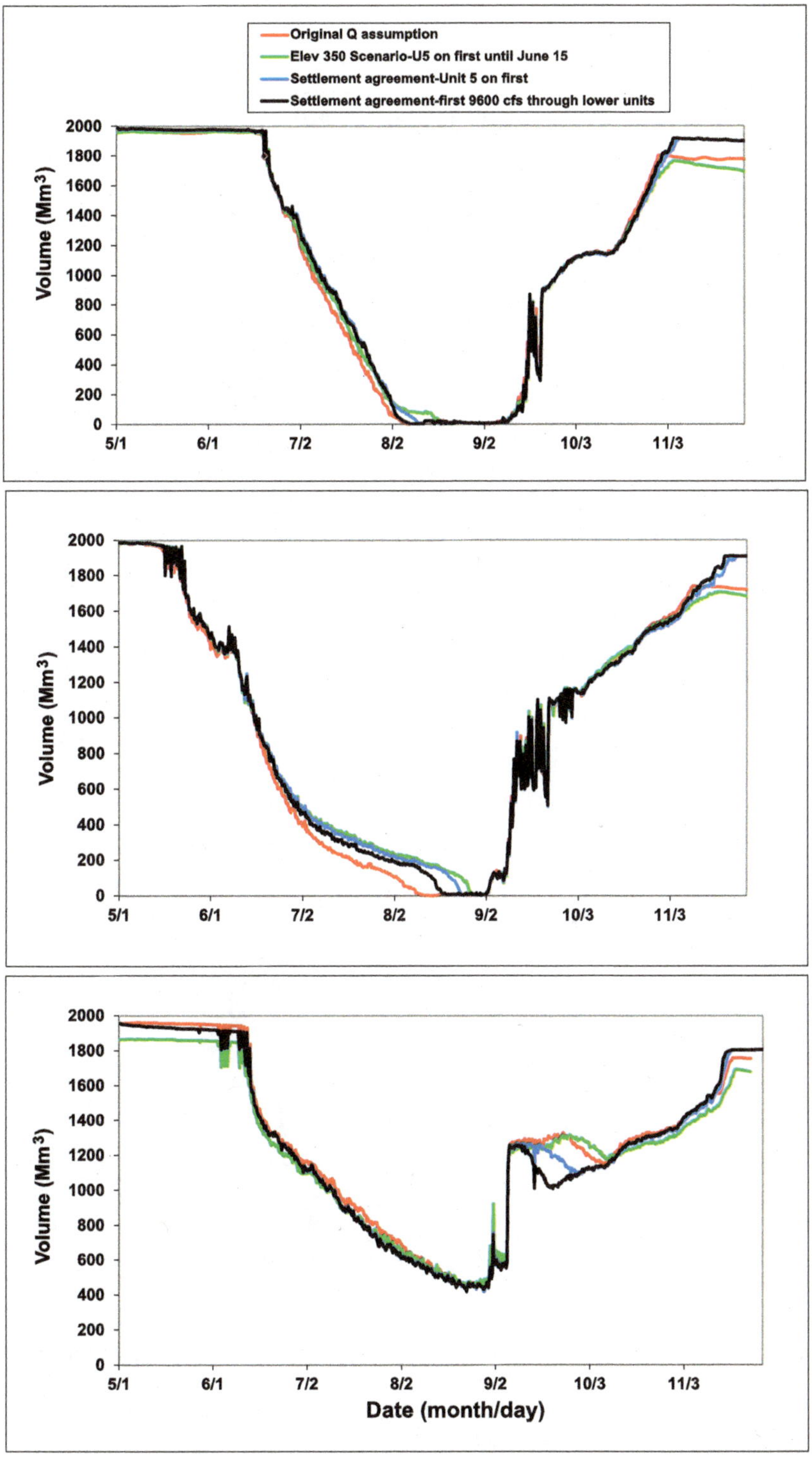

Figure 2. Comparison of the projected volume of striped bass habitat in Lake Murray, South Carolina, in 1997, 1998, and 2000 (top to bottom) under four operational scenarios (fully defined in text) using a calibrated CE-QUAL-W2 water-quality model. Striped bass habitat was defined as having temperatures less than 27°C and dissolved oxygen greater than 2.5 mg/L.

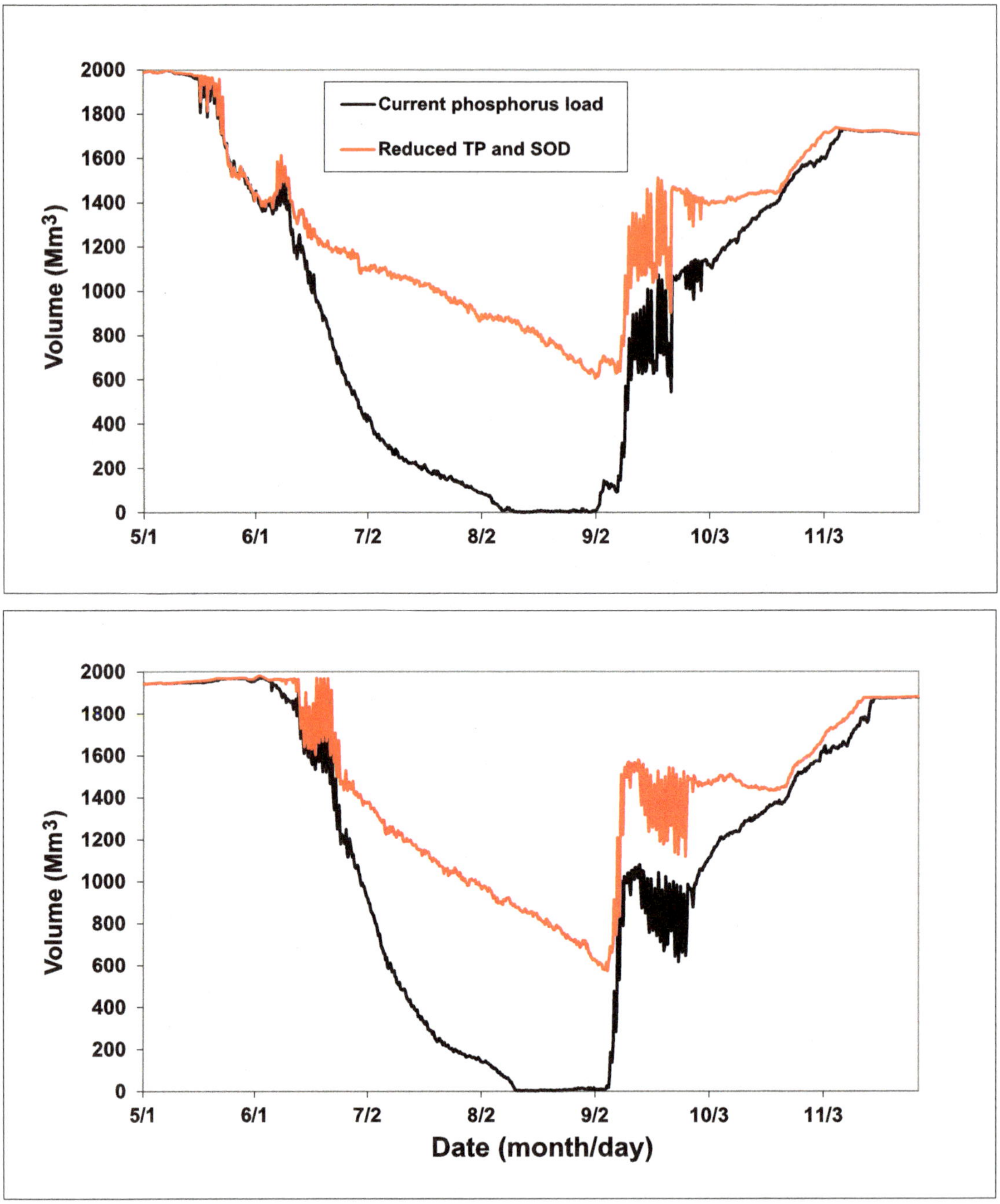

FIGURE 3. A comparison using a calibrated CE-QUAL-W2 water-quality model of the volume of striped bass habitat in Lake Murray, South Carolina, in 1998 (top) and 2005 (bottom), under current conditions and a 60% reduction in phosphorus loading. Striped bass habitat had temperatures less than 27°C and dissolved oxygen greater than 2.5 mg/L.

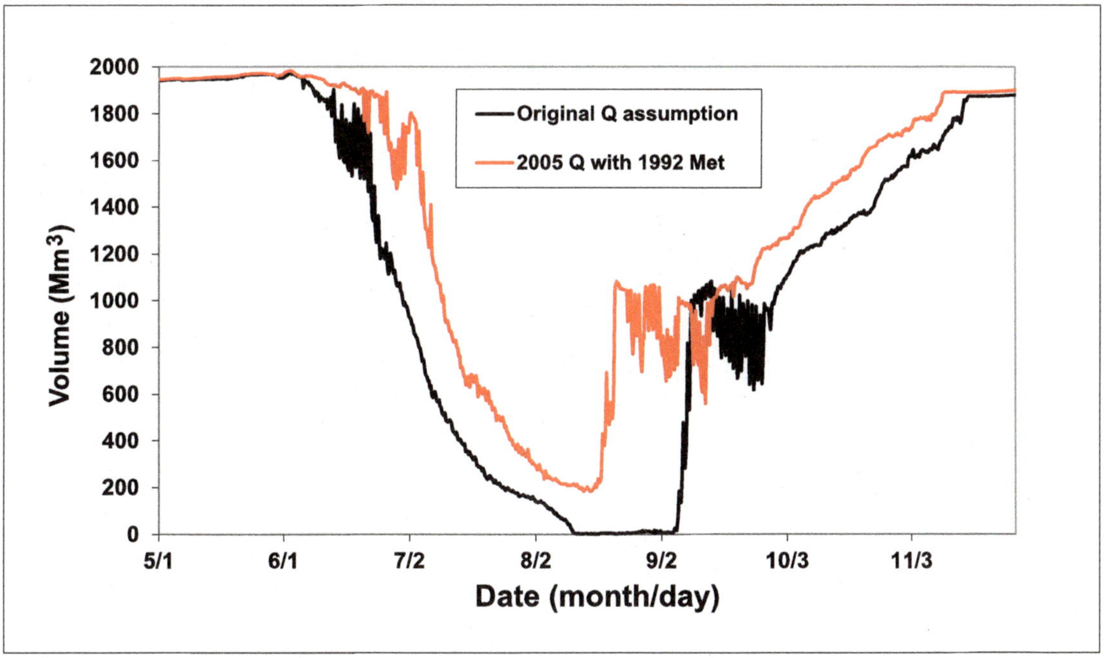

FIGURE 4. A comparison of striped bass habitat in Lake Murray, South Carolina, in 2005, with projected habitat using 2005 flows ('Q') and 1992 meteorological conditions ('Met'). A CE-QUAL-W2 water-quality model was used. Striped bass habitat was defined as temperatures less than 27°C and dissolved oxygen greater than 2.5 mg/L.

bass summer mortalities were reported in 2005 but probably would not have occurred in 2005 if 1992 meteorological conditions had occurred. In 2005 model runs, the DO was less than 2 mg/L in August at depths greater than 6–8 m, as well as in the upper metalimnion where temperature was about 28°C. When the 1992 meteorological data were used in the 2005 model, the results yielded DO levels of about 3 mg/L at the mid-depth range of 20–45 m in the first 11 km upstream from the dam and greater than 5 mg/L in the upper 14 m of the lake where temperature was about 26–27°C.

In summary, water quality modeling was used to help define controllable and noncontrollable factors that affected striped bass habitat in Lake Murray. High annual inflows and outflows, especially during April–July, were associated with fish kills under current nutrient loads to Lake Murray. Higher annual outflows cause the bottom of the lake to warm, and lower DO levels are associated with this warmer water. Annual meteorological conditions, hydrology, and nutrient load were factors that exerted a substantial effect on habitat but were not under control of the dam operator. However, lake elevation and turbine use were under the control of the dam operator. In the draft license agreement, the dam operator indicated that it would maximize efforts to fill Lake Murray to full pool by 1 March rather than 1 June (as was previously done) and change the operation of turbine 5 so that it is operated on a first-on basis for flows greater than 457.2 m^3/s up to 31 July of each year. Both of these actions will increase the possibility of increased summer habitat for striped bass without impacting a downstream tailwater fishery. An added benefit of using the model during the relicensing process was that it provided a central, best-available source of information upon which most stakeholders could agree. Additionally, all stakeholders had the opportunity to suggest and view the effect of various operational modifications through different model runs.

Claytor Lake, Virginia

Claytor Lake is a 1,820-ha hydroelectric reservoir operated by American Electric Power, and located in southwestern Virginia. Mean annual retention time of this reservoir is 33 d. Striped bass and hybrid striped bass (white bass *M. chrysops* × striped bass) are stocked annually and provide a major sport fishery (Kilpatrick and Ney 2013; Rash and Ney 2013; both this volume). Major striped bass summer mortalities have occurred intermittently over the years.

Claytor Lake stratifies during summer months (June–August), relegating coolwater fish habitat to the hypolimnetic waters below the thermocline. According to a study conducted by Kilpatrick (2003) on Claytor Lake, striped bass and hybrid striped bass preferred a narrow band of habitat near the thermocline at 15–21 m below the water's surface. These areas are often relatively low in DO due to anaerobic conditions at or near the lake bottom. This creates a "summer temperature–oxygen squeeze" (Coutant 1985) that is known to limit the growth and survival of coolwater fish species such as striped bass.

The FERC relicensing of Claytor Lake Hydroelectric Project called for the use of a dynamic stratification model for the fisheries assessment, documenting the extent and location of available coolwater habitat during summer months. The model would also be used to discern how hydropower operations might influence habitat. The CE-QUAL-W2 water quality model was used for the dynamic stratification model. This model was developed based on data inputs from the collected bathymetry and other relicensing studies such as water quality, sediment, and fish entrainment and impingement (Sawyer and Ruane 2009). The calibrated CE-QUAL-W2 model of Claytor Lake characterized the distribution and concentrations of major water quality constituents (temperature, DO, and nutrients) within the reservoir under a variety of project operations, including, but not limited to, lake level, inflow and outflow regimes, and meteorological extremes.

The objectives of the modeling effort were to (1) simulate effects of hydropower project operations on thermal stratification of the lake and the summer habitat for striped bass, (2) conduct an assessment of project effects on primary and secondary productivity under a variety of project operations, (3) assess whether phytoplankton and zooplankton communities relate to striped bass and hybrid striped bass summer habitat, and (4) investigate the causes of summer mortalities of striped bass and hybrid striped bass.

Considering data availability and hydrologic conditions, the years selected for calibration were 2002, 2004, and 2007. Study year 2007 was chosen as the primary calibration year with the intention of applying the same coefficients and inflow water quality to 2002 and 2004 conditions to check model robustness. Water balance for specific calibration years was derived using Claytor Lake levels and discharges, as well as flow data from U.S. Geological Survey gauges. Water-quality data collected in the Claytor Lake watershed were used to develop model inputs. Data collected in Claytor Lake and the releases from Claytor Hydro by were used to calibrate the model.

To characterize the availability of striped bass habitat in Claytor Lake, four types of habitat were defined: optimal, suboptimal temperature and optimal DO, optimal temperature and suboptimal DO, and tolerable. The criteria for these habitat types were defined based on criteria developed by Kilpatrick (2003). Optimal habitat was defined as water temperature between 15.0°C and 20.0°C and DO exceeding 5.0 mg/L. The first suboptimal category was defined as temperatures between 10.0°C and 15.0°C or between 20.0°C and 25.0°C and DO concentrations exceeding 5.0 mg/L. The second suboptimal category was defined as temperatures between 15.0°C and 20.0°C and DO concentrations between 2.5 and 5.0 mg/L. Tolerable habitat was defined as all other habitat having temperatures less than 27°C and DO concentrations greater than 2.5 mg/L.

The calibrated model showed that optimal habitat was not available in Claytor Lake during the summer months, but in the 3 years modeled, there was always some amount of volume that met the tolerable definition (Figure 5). For the normal flow year, 2004, habitat for suboptimal temperature and suboptimal DO was available during the summer months. This habitat occurred at the mid-depths of the reservoir and was attributed to higher inflows in 2004 compared to the low-flow years of 2002 and 2007.

The calibrated model was used to investigate a summer mortality event of striped bass in 2007. In that year, an isolated pocket of oxygenated, cool water near the dam was lost, possibly trapping striped bass in anoxic water (Figure 6). Rice et al. (2013) described

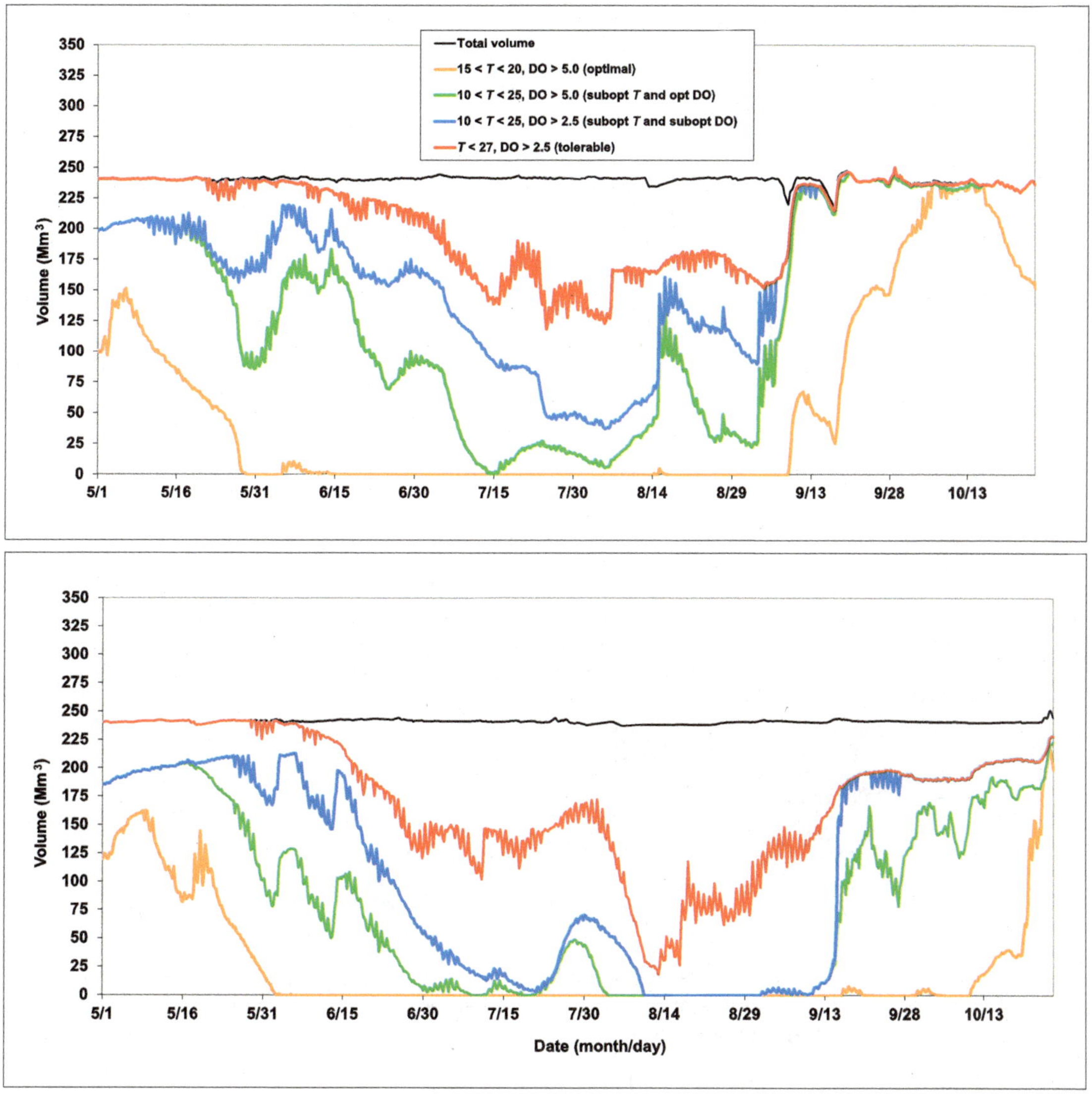

FIGURE 5. Striped bass habitat in the New River Arm of Claytor Lake, Virginia, in 2004 (top) and 2007 (bottom). Water temperature (T, °C) and dissolved oxygen (DO, mg/L) combinations defined habitat, which was estimated using a CE-QUAL-W2 water-quality model.

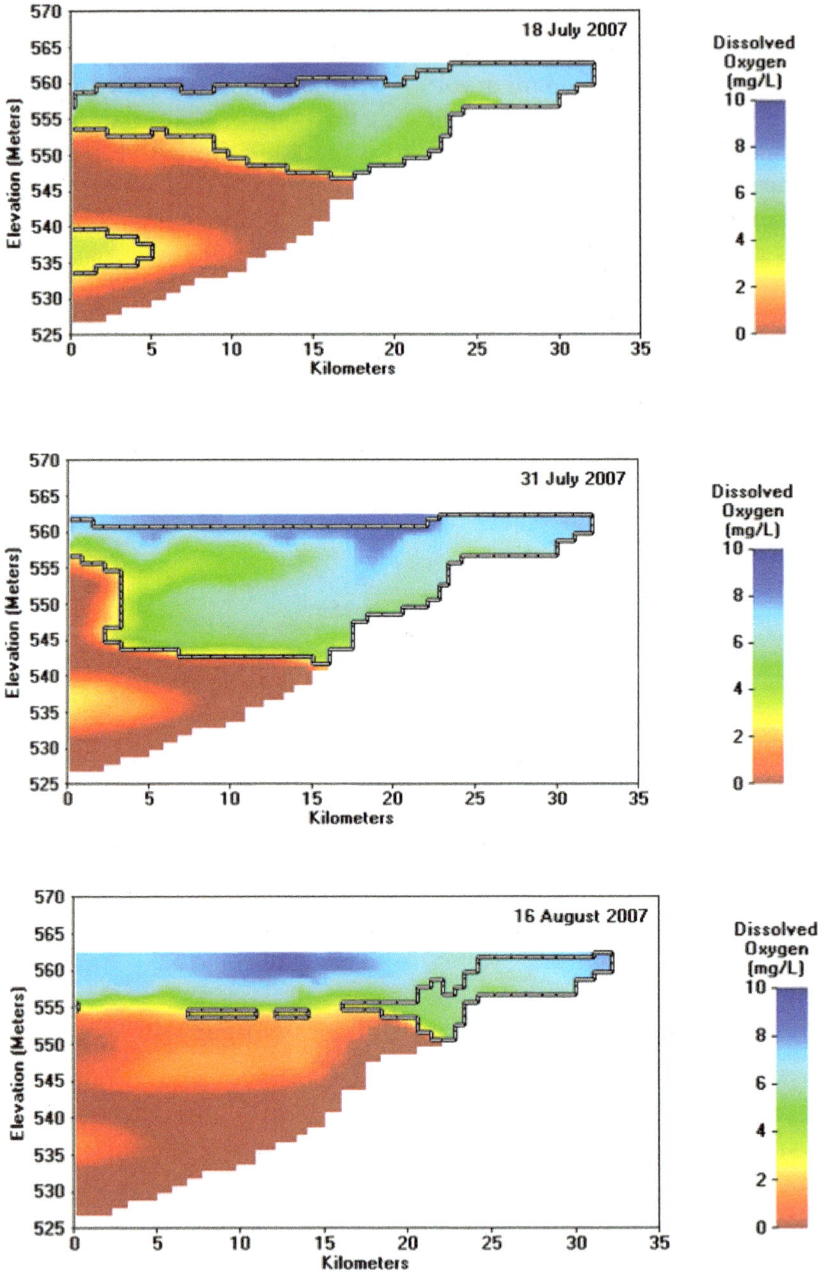

FIGURE 6. Change in oxygen profile of Claytor Lake, Virginia, between 18 July and 16 August 2007, a year of summer mortality of adult striped bass. Dashed lines outline zones of tolerable striped bass habitat (<27°C and >2.5 mg/L dissolved oxygen).

the dynamics of striped bass mortality events related to loss of hypolimnetic and metalimnetic oxygen.

The calibrated CE-QUAL-W2 two-dimensional water quality model was used for the 2002 and 2004 water years to evaluate habitat availability under four alternative operational scenarios: (1) existing conditions, (2) existing conditions without winter drawdown, (3) peaking conditions without winter drawdown, and (4) run of the river. The pool levels and release flows from Claytor Dam for the scenarios were provided by American Electric Power. Although these alternative operational scenarios were important to hydropower operations, they involved relatively minor variations in pool elevation over the course of the period April through October, not differing by more than about 30 cm over a narrow range of 46 cm.

Water quality modeling evaluations showed that striped bass habitat (Figure 7) and algal production were generally insensitive to the reservoir operating scenarios that were considered. This suggested that system morphometry and hydrology were the primary factors influencing summer habitat and algal production, over and above the influences of hydropower operations.

In summary, the CE-QUAL-W2 water quality model was used successfully on Claytor Lake to estimate the volume of striped bass habitat, investigate a summer mortality event of striped bass, and investigate whether operational changes could substantially increase summer habitat. Results indicated that system hydrology, residence time, and system morphometry dominated the production of acceptable summer habitat for striped bass. Summer mortality was related to exhaustion of an isolated pocket of oxygenated habitat in the hypolimnion. Thus, in Claytor Lake, hydropower operational changes had minimal effect on summer habitat availability.

J. Strom Thurmond Reservoir, South Carolina–Georgia Border

J. Strom Thurmond (JST) Reservoir is a 28,400-ha reservoir operated by the U.S. Army Corps of Engineers (Corps). The reservoir is stocked and supports a popular fishery for striped bass. The Richard B. Russell Dam tailwater at the upstream end of JST Reservoir provides cool, oxygenated water during summer for striped bass, including large, trophy-sized fish. In recent years, the Corps expressed an interest in changing operations at Richard B. Russell Dam from conventional to pump-back hydropower. Water-quality modeling simulations indicated that this change in operation would diminish the amount of cool, oxygenated water available in summer months for striped bass in the tailwater. As part of a legal settlement, the Corps agreed to install an oxygen injection system in the main body of JST Reservoir to mitigate perceived fishery impacts associated with the expected change in operation (Mobley et al. 2013, this volume). The design criteria for the 18–24°C thermal layer of the reservoir in July through September was to (1) oxygenate to 5 mg/L for 1 mi (1.6 km) anywhere in the lower 8 mi (12.9 km) of the reservoir, (2) oxygenate to 4 mg/L for 4 mi (6.4 km) in the lower 8 mi (12.9 km) of the reservoir, and (3) oxygenate to 3 mg/L just upstream from the turbine intakes, prior to turbine aeration. Also, to avoid damage to diffusers from surface recreation, the upper diffuser level was limited to no higher than 16.76 m below normal pool. Water quality modeling was used to design and assess the magnitude of habitat improvement associated with the oxygen injection system.

A version of the CE-QUAL-W2 two-dimensional water quality model with an embedded bubble-plume (BP) submodel was applied to this project. The BP submodel calculates hydrodynamic and water quality variables for a plume comprised of a bubble-water inner core and an annulus of entrained water from the ambient reservoir. For a range of diffuser types, the BP submodel integrates the governing equations for seven fluxes (water, momentum, heat, DO, dissolved nitrogen, gaseous oxygen, and gaseous nitrogen) and five equations of state (pressure, water density, bubble-water mixture density, gas volume, and bubble radius) based on Wuest et al. (1992). Bubble plume simulates upwelling associated with air or oxygen plumes in a thermally stratified ambient environment based on time-variant gas flow inputs and initial bubble size. It includes bubble-size changes

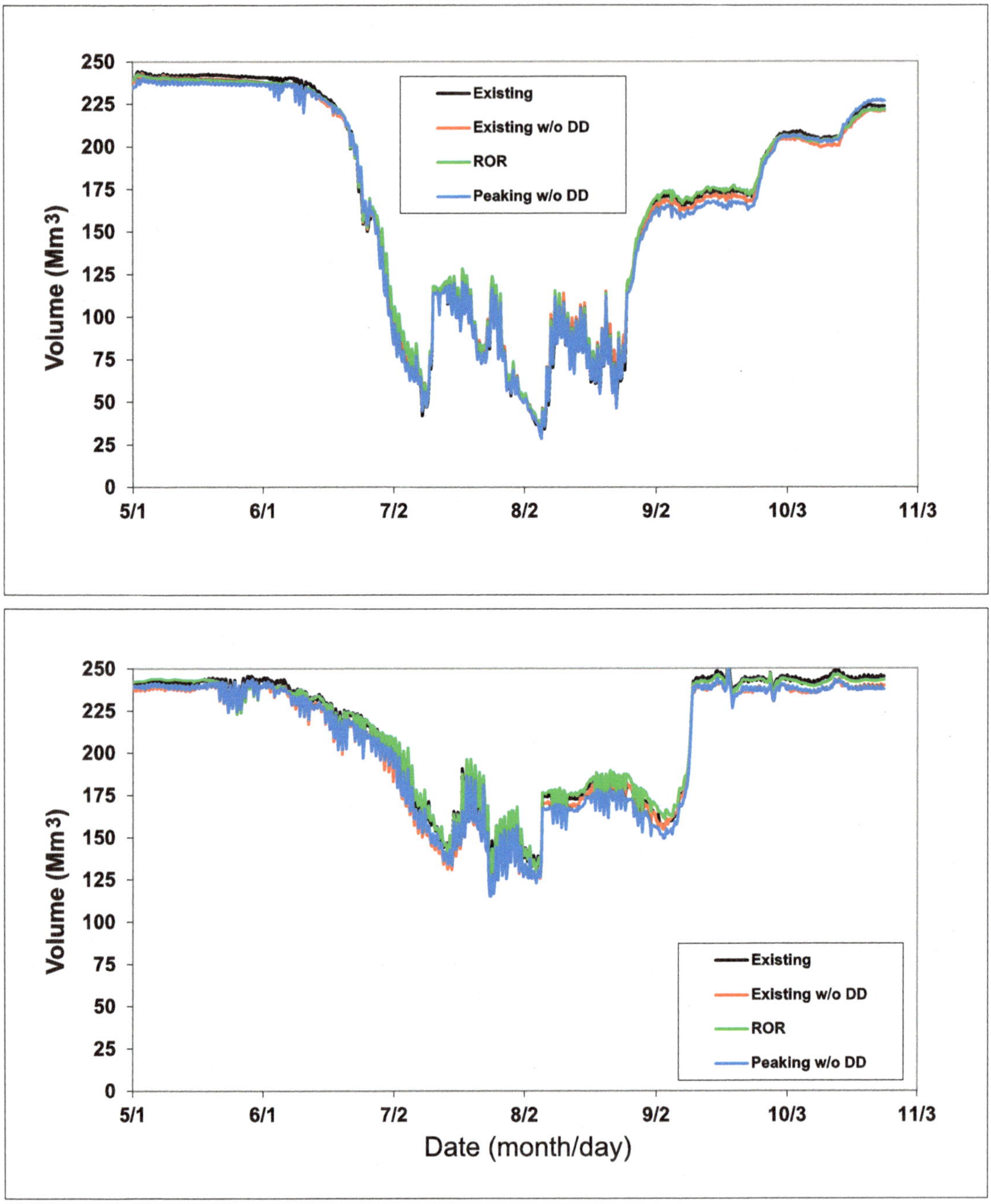

FIGURE 7. A comparison of tolerable striped bass habitat in Claytor Lake, Virginia, in 2002 (top) and 2004 (bottom), under four distinct reservoir operational strategies. Habitat was estimated using a CE-QUAL-W2 water-quality model. Tolerable striped bass habitat was defined as water having a temperature less than 27°C and dissolved oxygen greater than 2.5 mg/L.

from decompression and gas transfer as the bubbles rise and exchange gases with entrained water. When the plume has entrained enough ambient water to destroy its momentum, the bubble plume continues to rise but separates from the oxygenated entrained water that falls back into the water column and accrues in water layers having similar density. Bubbles break from the detraining plume and continue to rise at that point, starting a new plume. This process is repeated until the final plume reaches the surface or the bubbles dissolve completely. In CE-QUAL-W2 terminology, entrainment of ambient water along the length of the plume is treated as a withdrawal from the reservoir and each plume detrainment is treated as tributary injection back into the reservoir. Bubble plume has been verified in field tests (G. E. Hauser, unpublished data).

After review of pertinent hydrologic and water quality conditions at JST Reservoir, the CE-QUAL2-W2-BP model was calibrated for temperature and water quality using three hydrologic years—2003 (wet), 2004 (moderate), and 2007 (dry). Model calibration was based on weekly temperature and DO profiles for these 3 years, along with all available additional water quality data for organic matter, nutrients, algae, pH, and other variables. To address withdrawal zone issues, alternative calibrations using altered bathymetry were also developed for 2003 and 2007 (Ruane et al. 2008).

Diffuser elevations and capacity

After review of the JST Reservoir diffuser system goals and JST Reservoir water quality, conceptual diffuser designs based on a stand-alone version of bubble plume were developed and compared. A high-flux oxygen diffuser was proposed to (1) reduce overall diffuser length and cost, and (2) enable oxygen placement high in the water column with deeper diffusers. The calibrated CE-QUAL-W2-BP model was used to quantify the maximum oxygen injection capacity required in each of the three calibration years for a range of candidate diffuser configurations.

Original scoping for the high-flux diffuser using the standalone BP submodel called for a system of five diffusers (200, 400, 600, 1,000, 1,000 m lengths) placed at the same elevation in a region near Modoc, South Carolina that spanned three model segments. Model runs using the CE-QUAL-W2-BP model with the conceptual diffuser system showed that the diffusers needed to be at different elevations to be able to (1) place the oxygen plume in the 18–24°C target temperature zone under all situations across all model calibration years, and (2) provide a thicker layer of striped bass habitat.

With the diffusers at different levels, the upper diffusers were turned on first in the model, followed by deeper diffusers as the top and bottom of the 18–24°C temperature zone became deeper over the season. In these simulations, it became apparent that shorter diffusers were needed to provide more flexibility to incrementally add oxygen at each level, without having to suddenly turn on a 1,000-m diffuser at two of the elevations. It was also determined that when low-flow operations occur through JST Reservoir and DO is only marginally less than the goals, shorter diffusers were needed to provide smaller oxygen increments at target elevations for cost-effective use of the liquid oxygen supply.

Simulations showed that the deeper the oxygen is injected in the 18–24°C zone, the greater the resulting oxygen concentrations for the same gas flow. This occurs due to higher oxygen transfer efficiency at higher water pressures at depth and less physical reservoir volume at deeper depths.

Based on model simulations, the final diffuser configuration consisted of nine 400-m diffusers (Table 3). The diffuser elevations chosen provide the greatest operational flexibility for targeted vertical placement of oxygen (Hauser 2010). According to the model, having four levels also allows independent use of the lower levels to supplement oxygen goals in JST Reservoir releases, regardless of the oxygen flux required in the other two diffusers. The lower levels also provide better oxygen transfer effi-

Table 3. Oxygen diffuser depth configuration in J. Strom Thurmond Reservoir, South Carolina and Georgia. Full pool elevation is 100.6 m.

Diffuser(s)	Depth (m) at full pool
1 and 4	17.1
2, 5, and 7	20.7
3, 6, and 8	24.4
9	28.1

ciencies. The very lowest level allows flexibility for late in September and for years with low pool levels, when the bottom of the 18–24°C zone drops significantly in elevation. These diffuser levels also covered the range of possible plume behaviors resulting from uncertainty in BP submodel inputs. Sensitivity simulations showed that minor changes in diffuser position in the vicinity of the diffuser field would not significantly affect performance, nor would any anticipated variations in diffuser level and flux along the length of each diffuser.

Diffuser operation

To help anticipate operational issues, a series of simulations were made with the calibrated models to emulate actual operations. Oxygen injection rate was varied in 10–20-d increments as needed to just meet goals. Simulations demonstrated the designed system would meet project goals (Figure 8). These simulations showed that oxygen fluxes required in actual operations may range from 50% to 100% of the maximum estimated injection capacities over the summer.

Simulations also showed that wind on the reservoir would increase operational challenges. The model showed that surface winds produce a secondary current in the opposite direction in the oxygenated zone. The 4 and 5 mg/L DO plumes can migrate 1.5 mi (2.4 km) upstream or downstream due to modest but sustained winds lasting 1–2 d. In one simulation, a 3-mi (4.8 km) plume excursion occurred over 2 d in response to a sudden reversal in the prevailing wind direction.

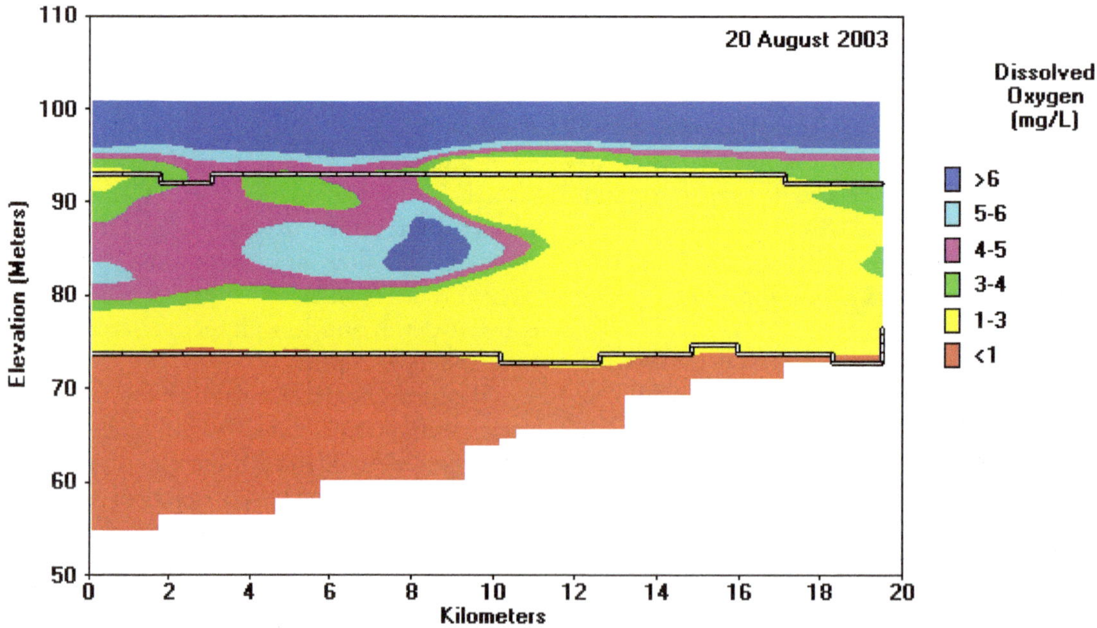

Figure 8. Projected dissolved oxygen concentrations in J. Strom Thurmond Reservoir, South Carolina and Georgia, obtained with an injection of 71 tons/d of liquid oxygen and using environmental characteristics of 20 August 2003. Marked zone is the 18–24°F temperature range. Dashed lines bound the 18–24°C range.

Modeling results indicated that the oxygen system would provide approximately 150–200 Mm^3 of additional striped bass habitat in the reservoir for most of the summer and fall (Figure 9). Striped bass habitat was defined as water between 18°C and 24°C with at least 3 mg/L of DO. This volume can be visualized as a habitat zone about 25 m thick by 1,000 m wide by 7 km long.

In summary, water quality modeling was used to design and evaluate an oxygen injection system for JST Reservoir. A final design of shorter diffusers at varying depths provided maximum efficiency for achieving the desired oxygen enhancement in the reservoir. The oxygen injection system is expected to provide approximately 150–200 Mm^3 of additional habitat in the lower 8 mi (12.9 km) of JST Reservoir, mitigating for the expected loss of habitat as a result of a change in operations from conventional, one-directional hydropower to cycling pump-back hydropower.

Discussion

The CE-QUAL-W2 model is well suited for reservoirs with striped bass fisheries due to the habitat requirements related to temperature and DO conditions that are affected by a wide range

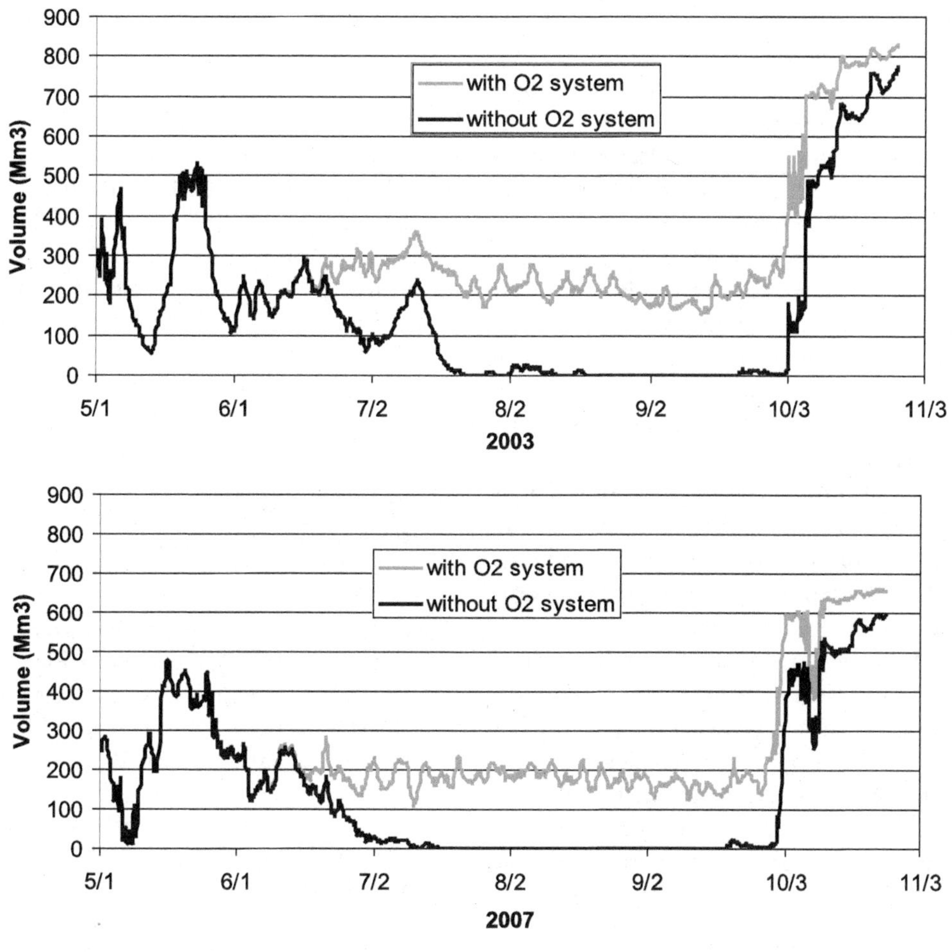

FIGURE 9. A CE-QUAL-W2 water quality model comparison of striped bass habitat in J. Strom Thurmond Reservoir, with and without an oxygen injection system. Striped bass habitat was defined as water between 18°C and 24°C with at least 3 mg/L of dissolved oxygen.

and complexity of hydrologic, meteorological, and limnological conditions. This model is especially useful for reservoirs like Lake Murray and J. Strom Thurmond that have relatively long residence times where water quality is significantly affected by limnological processes.

Water quality in reservoirs, especially striped bass habitat, is affected by numerous factors that are site-specific and include external as well as internal processes. The CE-QUAL-W2 model is an excellent tool for reconciling all available information and data (or that can be obtained) and combining these inputs with scientific understanding that is encompassed by the model. CE-QUAL-W2 modeling is based upon state-of-the-art approaches that are logical, sound extensions of well-founded research and studies conducted over the past half century. Modeling results provide a general indicator of what is likely to occur under given sets of conditions. As is the case in all aquatic environments, natural conditions are more complex than models, so models tend to reproduce major patterns that are observed in the field, but lack resolution, inputs, or formulations to reproduce all the minor patterns that occur. Models are internally consistent and based on rigorous governing equations, so they can often help explain apparent discrepancies in field observations as well as be used diagnostically to identify key factors affecting modeling results. Models often are the most comprehensive approach for integrating and analyzing field data collected on reservoirs; however, good data as well as an understanding of the data are crucial for developing a good model.

While water-quality modeling for striped bass habitat can be difficult, one can take comfort in knowing that others have shown that models can be successfully applied to challenging situations. In the case studies presented, important insights relevant to management were found for three southeastern reservoirs that stratify during summer and contain striped bass populations.

Acknowledgements

The authors (as water quality modelers) appreciate significant assistance received from many others involved in the three model applications presented herein that address striped bass habitat. The South Carolina Department of Natural Resources, American Rivers, and Midlands Striper Club provided significant multi-year information on striped bass in Lake Murray and the rivers downstream from Saluda Hydro. This work was performed for South Carolina Electric and Gas as part of their FERC relicensing process for Saluda Hydro that was guided by Kleinschmidt Associates. The Virginia Department of Game and Inland Fisheries reviewed and provided comments on the results of the Claytor model. This work was performed for Appalachian Power Company as part of their FERC relicensing process for the Claytor Project that was guided by Devine Tarbell & Associates, part of HDR. Modeling for the oxygen diffuser system at J. Strom Thurmond Reservoir was guided by Jamie Sykes (Savannah District, Army Corps of Engineers) and part of the modeling was conducted through Mobley Engineering, Incorporated. We appreciate the assistance of the editors, especially in condensing the results of three water quality model applications so that a range of model applications could be presented.

References

Cole, T. M., and S. A. Wells. 2002. CE_QUAL-W2: a two-dimensional, laterally averaged hydrodynamic and water quality model, version 3.1. User manual. U.S. Army Corps of Engineers, Instruction Report EL-02-1, Washington, D.C.

Coutant, C. C. 1985. Striped bass, temperature, and dissolved oxygen: a speculative hypothesis for environmental risk. Transactions of the American Fisheries Society 114:31–61.

Coutant, C. C. 2013. When is habitat limiting for striped bass? Three decades of testing the temperature–oxygen squeeze hypothesis. Pages 65–91 *in* J. S. Bulak, C. C. Coutant, and J. A. Rice, editors. Biology and management of inland striped bass and hybrid striped bass. American Fisheries Society, Symposium 80, Bethesda, Maryland.

Hauser, G. E. 2010. J. Strom Thurmond Reservoir—modeling to support diffuser design/

placement for initial field implementation. Report prepared for U.S. Army Corps of Engineers, Savannah District, Savannah, Georgia.

Kilpatrick, J. M. 2003. Habitat use, movements, and exploitation of striped bass and hybrid striped bass in Claytor Lake, Virginia. Master's thesis. Virginia Polytechnic Institute and State University, Blacksburg.

Kilpatrick, J. M., and J. J. Ney. 2013. Temperature and dissolved oxygen habitat use by striped bass and hybrid striped bass in Claytor Lake, Virginia. Pages 147–159 *in* J. S. Bulak, C. C. Coutant, and J. A. Rice, editors. Biology and management of inland striped bass and hybrid striped bass. American Fisheries Society, Symposium 80, Bethesda, Maryland.

Matthews, W. J. 1985. Summer mortality of striped bass in reservoirs of the United States. Transactions of the American Fisheries Society 114:62–66.

McDaniel, C. K., L. E. Snyder, and L. L. Connor. 1991. Impacts of thermal stress on the condition of striped bass. Proceedings of the Southeastern Association of Fish and Wildlife Agencies 1991:361–369.

Mobley, M., E. Shallenberger, M. W. Beutel, P. Gantzer, and B. Sak. 2013. Oxygen diffusers to create and maintain summer fish habitat. Pages 197–208 *in* J. S. Bulak, C. C. Coutant, and J. A. Rice, editors. Biology and management of inland striped bass and hybrid striped bass. American Fisheries Society, Symposium 80, Bethesda, Maryland.

Rash, J. M., and J. J. Ney. 2013. Comparative ecology of juvenile striped bass and hybrid striped bass in Claytor Lake, Virginia. Pages 229–247 *in* J. S. Bulak, C. C. Coutant, and J. A. Rice, editors. Biology and management of inland striped bass and hybrid striped bass. American Fisheries Society, Symposium 80, Bethesda, Maryland.

Rice, J. A., J. S. Thompson, J. A. Sykes, and C. T. Waters. 2013. The role of metalimnetic hypoxia in striped bass summer kills: consequences and management implications. Pages 121–145 *in* J. S. Bulak, C. C. Coutant, and J. A. Rice, editors. Biology and management of inland striped bass and hybrid striped bass. American Fisheries Society, Symposium 80, Bethesda, Maryland.

Ruane, R. J., G. E. Hauser, D. F. McGinnis, M. H. Mobley, and A. F. Sawyer. 2008. J. Strom Thurmond dissolved oxygen system—CE-QUAL-W2 modeling for oxygen diffuser placement. Report prepared for U.S. Army Corps of Engineers, Savannah District, Savannah, Georgia.

Sawyer, A. F., and R. J. Ruane. 2006. Calibration of the CE-QUAL-W2 model for Lake Murray. Report prepared for SCE&G, Columbia, South Carolina.

Sawyer, A. F., and R. J. Ruane. 2007. Applications of the CE-QUAL-W2 model for Lake Murray relicensing issues. Report prepared for SCE&G, Columbia, South Carolina.

Sawyer, A. F., and R. J. Ruane. 2009. Development and application of the CE-QUAL-W2 model for the Claytor Lake aquatic resources assessment. Report prepared for Appalachian Power Company, Roanoke, Virginia.

Schaffler, J. A., J. J. Isely, and W. E. Hayes. 2002. Habitat use by striped bass in relation to seasonal changes in water quality in a southeastern reservoir. Transactions of the American Fisheries Society 131:817–827.

Thompson, J. S., and J. A. Rice. 2013. The relative influence of thermal experience and forage availability on growth of age 1–5 striped bass in two southeastern reservoirs. Pages 93–120 *in* J. S. Bulak, C. C. Coutant, and J. A. Rice, editors. Biology and management of inland striped bass and hybrid striped bass. American Fisheries Society, Symposium 80, Bethesda, Maryland.

Wetzel, R. G. 2001. Limnology: lake and river ecosystems, 3rd edition. Academic Press (Elsevier USA), San Diego, California.

Wuest, A., N. Brooks, and D. Imboden. 1992. Bubble plume modeling for lake restoration. Water Resources Research 28:3235–3250.

Zale, A. V., J. D. Wiechman, R. L. Lochmiller, and J. Burroughs. 1990. Limnological conditions associated with summer mortality of striped bass in Keystone Reservoir, Oklahoma. Transactions of the American Fisheries Society 119:72–76.

American Fisheries Society Symposium 80:181–196, 2013

Response of Summer Habitat for Striped Bass in Lake Greenwood, South Carolina to Weather and Phosphorus Loading

Barbara E. Taylor*, James S. Bulak, and Henry N. McKellar
South Carolina Department of Natural Resources
Freshwater Fisheries Research Laboratory
1921 VanBoklen Road, Eastover, South Carolina 29044, USA

Abstract.—A CE-QUAL-W2 water quality model was used to characterize the availability of striped bass *Morone saxatilis* habitat in Lake Greenwood, South Carolina, during 2004 and 2005. Although the lake has a productive fishery, water quality and aquatic habitat are affected by nutrient loading, algal blooms, and extensive oxygen depletion in the bottom waters. The main objectives were to characterize habitat availability and predict the implications of a change in phosphorus loading from the Saluda and Reedy rivers. The baseline scenario of the model showed that habitat was most critical during July and August, when as little of 5% of the reservoir contained tolerable habitat (temperature <28°C and dissolved oxygen >2 mg/L). Favorable habitat (temperature <25°C and dissolved oxygen >2 mg/L) was usually absent for most of July and August. Pulses of higher inflow or freshets produced short-term increases in tolerable habitat, especially in the upper end of the reservoir. Phosphorus-loading scenarios predicted that large reductions (50% or more) would be required to improve habitat substantially during midsummer. For the manager of a striped bass fishery, water quality models can be useful tools for evaluating habitat, especially under marginal conditions, and for predicting the impact of altered water management practices.

Introduction

Whether a particular reservoir provides suitable habitat for striped bass *Morone saxatilis* is a question often faced by fishery managers. How the available habitat may change temporally or spatially due to anthropogenic, climatic, and other factors is often another question of interest, as destruction of habitat can lead to fish kills and loss of a fishery (Wilson 2013, this volume). Van Horn (2013, this volume) documents the history of the establishment of reservoir fisheries for striped bass.

Striped bass have been intermittently stocked into Lake Greenwood, South Carolina, for more than a half century, producing a modest fishery. The first stocking took place in 1955 when 293 adult fish were released (Stevens 1958). Beginning in 1968, the reservoir was stocked either annually or biennially with fingerlings. From 2008 to 2012, an annual average of 167,000 fingerlings was stocked during the spring. Hayes (2004) conducted a creel survey of Lake Greenwood and reported that largemouth bass *Micropterus salmoides* and black crappie *Pomoxis nigromaculatus* were the species of greatest interest; striped bass accounted for 2% of angler-hours and 5% of the total biomass harvest.

Lake Greenwood is a major impoundment on the Saluda River in the Piedmont region of

* Corresponding author: taylorb@dnr.sc.gov

South Carolina (Figure 1). The reservoir has an average retention time of 78 d (Inabinet and Pearse 1981), a maximum depth of 21 m, an average depth of 7 m, and a total surface area of 4,614 ha (SCDHEC 2004). According to the U.S. Geological Survey (USGS) Geographic Information Retrieval and Analysis System (GIRAS) 2001 land use classification, more than half of the watershed was forested, about a quarter was in agricultural use, and about a tenth was developed (Taylor et al. 2008). The reservoir is formed by the junction of the Saluda and Reedy rivers, which supply 62% and 22% of the total reservoir inflow. Lake Greenwood receives high concentrations of nutrients from the Saluda and Reedy rivers. In 2004–2005, total phosphorus concentrations in the Saluda and Reedy Arms were typically near 0.06 mg/L (McKellar and Bulak 2008). Greenwood County, South Carolina, currently operates the Buzzard Roost hydroelectric station, located at Lake Greenwood dam. Oxygen injection lines

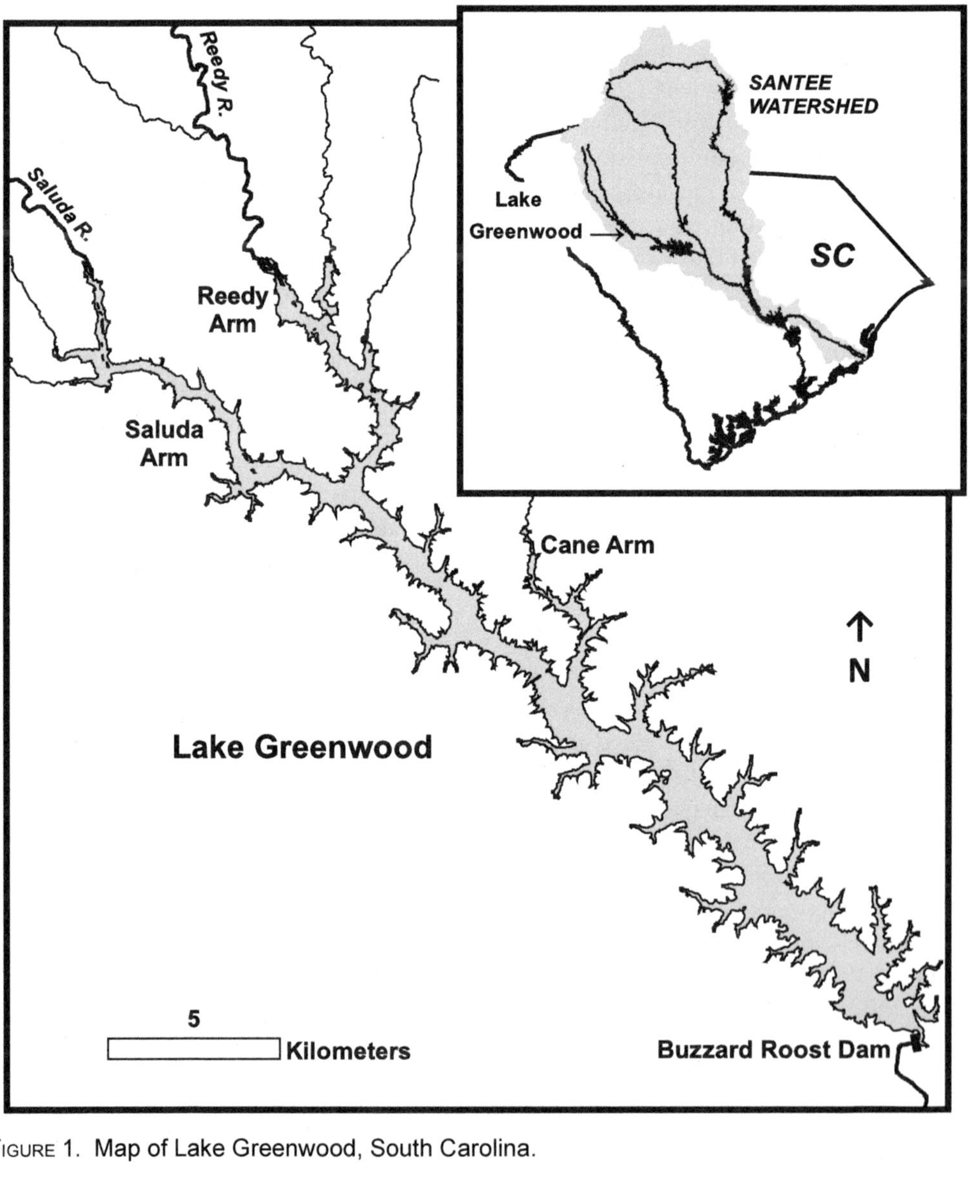

FIGURE 1. Map of Lake Greenwood, South Carolina.

extend from the dam for approximately 0.9 km into the reservoir to maintain dam discharge within state dissolved oxygen standards; this oxygen injection provides a small volume of improved habitat for striped bass during the summer months. Overall, fish biomass (330 kg/ha from cove rotenone sampling in 1986–1987) and harvest (31.6 kg/ha/year, excluding catfish, in 1985–1987) ranked in the upper third among 17 reservoirs in the Piedmont region of the Carolinas (Rodriguez and Olmsted 1995).

High levels of nutrient loading may impair water quality and habitat for striped bass in lakes and reservoirs. In springtime, rapid depletion of dissolved oxygen below the thermocline creates a large volume of hypoxic or anoxic bottom water that persists through the summer and into the fall. Although some oxygen depletion below the thermocline is a natural process in stratified lakes (Wetzel 1983; Kalff 2002), the rate of depletion, and the extent and duration of hypoxic or anoxic conditions can be affected by nutrient loading, subsequent algal blooms, and oxygen consumption of excess organic matter (Lee and Jones 1984).

The stocking and management of striped bass in reservoirs such as Lake Greenwood presents a special problem because adult striped bass seek cool, well-oxygenated waters, which may be limited during the summer. Summer die-offs of striped bass occur intermittently in Lake Greenwood. For example, Hayes (2003) reported a significant summer die-off of striped bass in August 2002 brought about by high temperatures and low dissolved oxygen in thermal refuges near the dam. Snoots (1993) tracked 12 transmitter-equipped adult striped bass and concluded that Lake Greenwood possessed little or no suitable habitat (water temperature <25°C and dissolved oxygen >4 mg/L) for striped bass during the summer months. He concluded that fish survived by using a tributary stream and the main channel of the reservoir, especially at the lower end, during summer.

For striped bass, a habitat "squeeze" occurs when the depth range of favorable or tolerable conditions is narrowed because of high temperatures in the surface waters and low dissolved oxygen in the cooler, deeper waters (Coutant 1985 and 2013, this volume). The habitat requirements are not precisely known. Coutant (1985) first defined habitat for adult striped bass as temperature less than 25°C and dissolved oxygen levels of 3–4 mg/L. Subsequent evaluations revealed that striped bass could tolerate higher temperatures and lower dissolved oxygen than initially hypothesized; these studies are reviewed and discussed by Coutant (2013). Suitable habitat may change as striped bass grow and age (Coutant et al. 1984). Thompson and Rice (2013, this volume) have shown that striped bass adults can overcome poor habitat conditions during summer if adequate prey is readily available, helping to offset the bioenergetic stress associated with poor habitat.

Our objective was to use a water quality model to explore the effects of phosphorus loading and weather on habitat for striped bass. For this exercise, we defined favorable habitat as water less than 25°C and greater than 2 mg/L of dissolved oxygen; tolerable habitat was defined as water less than 28°C and greater than 2 mg/L of dissolved oxygen. We use the results to develop management expectations for the striped bass fishery and evaluate the potential for improving striped bass habitat by reducing phosphorus loading to the reservoir.

Methods

The Lake Greenwood water quality model we employed was a simulation in CE-QUAL-W2 (Cole and Wells 2002), using W2i pre- and postprocessing software from Loginetics, Inc. (Knoxville, Tennessee). The model was built and calibrated by McKellar et al. (2008). The model simulates water quality in Lake Greenwood during 2004 and 2005.

The main sources of spatial variability in long, narrow reservoirs such as Lake Greenwood are longitudinal (upstream–downstream) and vertical (surface–bottom). Longitudinal variability arises largely from the transition between a riverine-like environment at the upstream end (typically the source of hydrologic inputs) and a lacustrine environment at the downstream end (typically near the dam with its hydrologic releases) (Thornton et al. 1990). The vertical variability arises mainly from seasonal thermal stratification between warm upper layers (epilimnion) and the cooler bottom layers (hypolimnion) (Wetzel 1983). The U.S. Army Corps of Engineers lake and reservoir model (CE-QUAL-W2) is a laterally averaged, two-dimensional (longitudinal and vertical) hydrodynamic and water quality model that is well suited for long, narrow water bodies (Cole and Wells 2002).

Lake Greenwood was represented in CE-QUAL-W2 by a series of cells representing longitudinal segments and vertical layers. There were 48 longitudinal segments: 34 in the Saluda Arm and main lake (Branch 1), 8 in the Reedy Arm (Branch 2), and 6 in the Cane Creek Arm (Branch 3). The segments were subdivided into 1-m vertical layers. The simulations were driven by time series of data specifying water volumes entering and exiting the lake, water quality constituents of the inflowing water, and weather. The data sources for lake bathymetry, hydrology, water quality constituents, and meteorology are described in Table 1.

The baseline scenario was calibrated to reproduce the observed vertical distributions of temperature, as well as concentrations of dissolved oxygen, total phosphorus, and chlorophyll α along the longitudinal axes of the main lake and the modeled tributaries (McKellar et al. 2008). The overall absolute mean errors for the 2-year simulations were 0.9°C for temperature, 1.1 mg/L for dissolved oxygen, 0.05 mg/L for total phosphorus, and 6.8 μg/L for chlorophyll α.

Flows in the Saluda River at Ware Shoals (USGS gaging station 02163500) during the period of the simulation were compared to records for 1940–2007. Records of similar duration were not available for the Reedy River.

To investigate the effects of hydrology and weather on habitat for striped bass, we exchanged meteorological data for May and June of the 2 years, producing a simulation combining low flow in those months with high temperature in the first year and high flow and low temperature in the second year. The entire array of hourly meteorological data, including temperature and wind speed, was exchanged, as well as water temperatures for all tributary and distributed inflows.

To investigate the effects of phosphorus loading on habitat for striped bass, we simulated a 25% increase and 25, 50, 75, and 90% decreases in total phosphorus inputs from the Saluda and Reedy rivers. Total phosphorus was computed as the sum of bioavailable phosphorus and the phosphorus content of organic matter in the water (algal biomass, labile and refractory dissolved, and particulate organic matter). We modified all of these constituents proportionately. The scenarios are similar to those reported by McKellar et al. (2008), except that (1) initial values for water quality constituents throughout the lake were set to the values for input from the Saluda River on the first day of the simulation for each scenario, and (2) algal constituents were included in the load reductions.

To account for longer-term effects of altered phosphorus loads, we also adjusted sediment oxygen demand (SOD) in each of the phosphorus load-change scenarios, following Ruane and Hauser (2006). The SOD rate for each model segment was

$$\mathrm{SOD} = \mathrm{SOD}_c\,(L/L_c)^{0.5},$$

where SOD is the segment-specific SOD (g/m^2/d) for the selected scenario, SOD_c is the SOD in the baseline calibration, L is the phosphorus load for the selected scenario, and L_c is the phosphorus load of the baseline calibration.

McKellar et al. (2008) noted uncertainty about the time scale and magnitude of the response of sediment oxygen demand to altered phosphorus loads. Heeding their recommendation for sensitivity analysis, we compared simulations with load-adjusted SOD parameters to otherwise identical simulations with baseline SOD parameters.

TABLE 1. Main components of the Lake Greenwood model, with sources of data used in model construction, calibration, and simulations.

Component	Sources
Bathymetry	Map of 1-m contours developed from more than 11,000 georeferenced depths (McKellar et al. 2008)
Hydrology	
Tributary inflows (gaged)	Daily stream flow at the Saluda River at Ware Shoals (USGS 02163500) and the Reedy River at Ware Shoals (USGS 02165000); records for the Reedy River after storm damage to the gage in September 2004 were estimated from a 2003 regression of Reedy River discharge with Saluda River discharge
Distributed inflows (estimated)	Back-calculated from gaged inflows and outflows to give water levels approximating Federal Energy Regulatory Commission rule curve for water level in Lake Greenwood
Drinking water withdrawal (gaged)	W.R. Wise Water Treatment Plant (D. Tuck, W. R. Wise Water Treatment Plant, personal communication)
Outflow at dam (gaged)	Daily stream flow in the tailrace canal (USGS 02166501)
Water quality constituents	
Includes algae (three taxonomic groups), bioavailable phosphorus (dissolved and adsorbed), ammonium, nitrate and nitrite, organic matter (labile and refractory, dissolved and particulate), dissolved oxygen, and temperature	
Inflowing water	Monthly samples from South Carolina Department of Health and Environmental Control, supplemented with intervals of more frequent samples by Clemson University (Reedy River near Ware Shoals: DHEC S-021 and Clemson RR3; Saluda River near Ware Shoals, DHEC S-125 and Clemson SR2)
Lake	Monthly and more frequent samples at 11 stations in Lake Greenwood in 2004–2005, including vertical profiles at 1-m intervals for temperature and dissolved oxygen (McKellar and Bulak 2008)
Weather	Hourly air temperature, dew point, wind speed and direction, and cloud cover at Greenwood County airport (Southeastern Regional Climate Center, Columbia, South Carolina)

Results

There were substantial differences in hydrology and weather between 2004 and 2005 (Figure 2). Monthly flow from Saluda River in 2004 was near normal from January through August. The flow in September, 2004, was exceptionally high—the 68-year record for the month—due to tropical storms, which produced daily discharge as high as 26×10^6 m^3. Saluda River flows for March through July of 2005 were above normal; the values were about twice as high as the values for 2004. Flow was near normal for the remainder of 2005. The Saluda River accounted for about 60% of the modeled hydrologic input to Lake Greenwood.

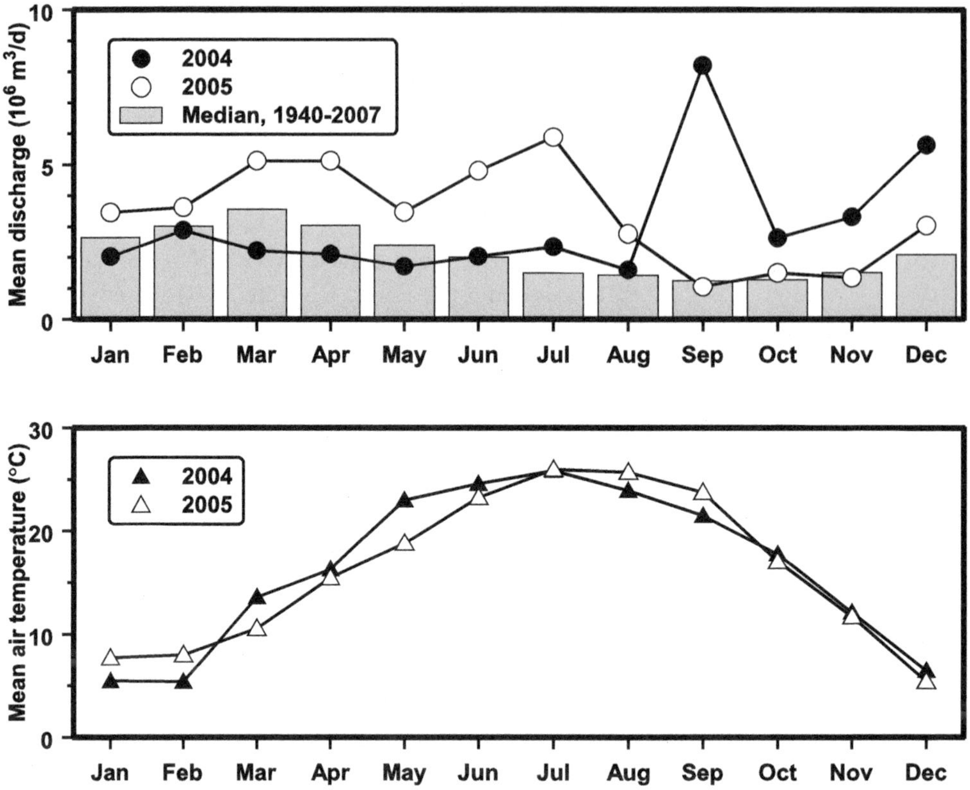

FIGURE 2. Monthly mean values for inflow from the Saluda River and air temperature in 2004–2005. Inflows are based on daily U.S. Geological Survey data for the Saluda River with extrapolations to fill missing data. Air temperatures are based on hourly data from the Greenwood County Airport.

Discharge was highly variable. In summer (June, July, and August), typical (median) daily discharge in the Saluda River was 1.5×10^6 m^3 for the years 1940–2007. Daily discharges of greater than 3×10^6 m^3 occurred at least once during those 2 months in most (51 of 68) years.

Air temperatures differed most in May, when 2004 average temperature was 4.2°C warmer than 2005 (Figure 2).

Simulated water temperatures in Lake Greenwood ranged from minima around 5°C in winter and maxima exceeding 30°C in summer (Figure 3). The main part of the lake was strongly and consistently stratified from May through October in both years.

During June, July, and August, the warmest months, tolerable habitat was continuously present, but its volume varied substantially among months and between years (Table 2). Favorable habitat was usually absent in July and August; presence and volume in June differed greatly between years.

In both years, 90% or more of the lake (by volume) did not meet the criteria for tolerable habitat for a month or more, in July and August (Figure 4). During summer, the main areas of tolerable habitat were generally located in the shallow upper reaches of the impoundment (Figure 4). During summer in the lower 20 km of the lake, tolerable habitat was generally restricted to a narrow band (<2 m) of the metalimnion at an approximate depth of 6–8 m (Figure 5).

Water temperature, rather than dissolved oxygen, was the prime cause for the lack of tolerable habitat during the summer months (Figure 6). When 90% or more of the lake did not meet the criteria for tolerable habitat, 59–72%

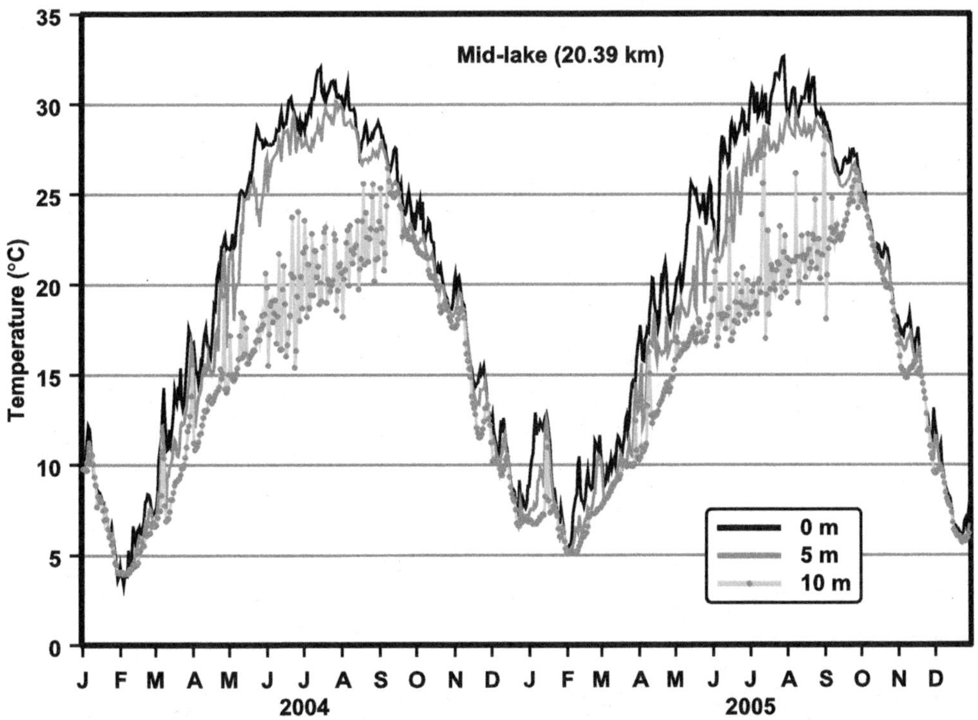

FIGURE 3. Water temperature at mid-lake in the baseline modeled scenario for Lake Greenwood.

of the total volume was intolerable because of excessively high temperature alone, 22–34% because of insufficient dissolved oxygen alone, and 0–8% because of both conditions. The dynamics of both thermal stratification and dissolved oxygen depletion differed between the 2 years of the simulation (Figure 6). By late May 2004, 40% of the lake was intolerable habitat. This extent of habitat loss was not reached until late June in 2005. The earlier loss of habitat in 2004 occurred with the warmer temperatures and lower inflows of May and June 2004.

TABLE 2. Favorable and tolerable striped bass habitat in June, July, and August 2004 and 2005 in the baseline scenario for Lake Greenwood. Favorable striped bass habitat is defined as temperature less than 25 °C and dissolved oxygen greater than 2 mg/L, tolerable habitat as temperature less than 28 °C and dissolved oxygen greater than 2 mg/L.

		Favorable habitat		Tolerable habitat	
Year	Month	Mean volume (% of total)	Days present	Mean volume (% of total)	Days present
2004	June	1.1%	18	25.6%	30
	July	<0.1%	1	5.3%	31
	August	<0.1%	5	33.8%	31
2005	June	29.9%	27	62.7%	30
	July	<0.1%	6	17.5%	31
	August	<0.1%	1	5.8%	31

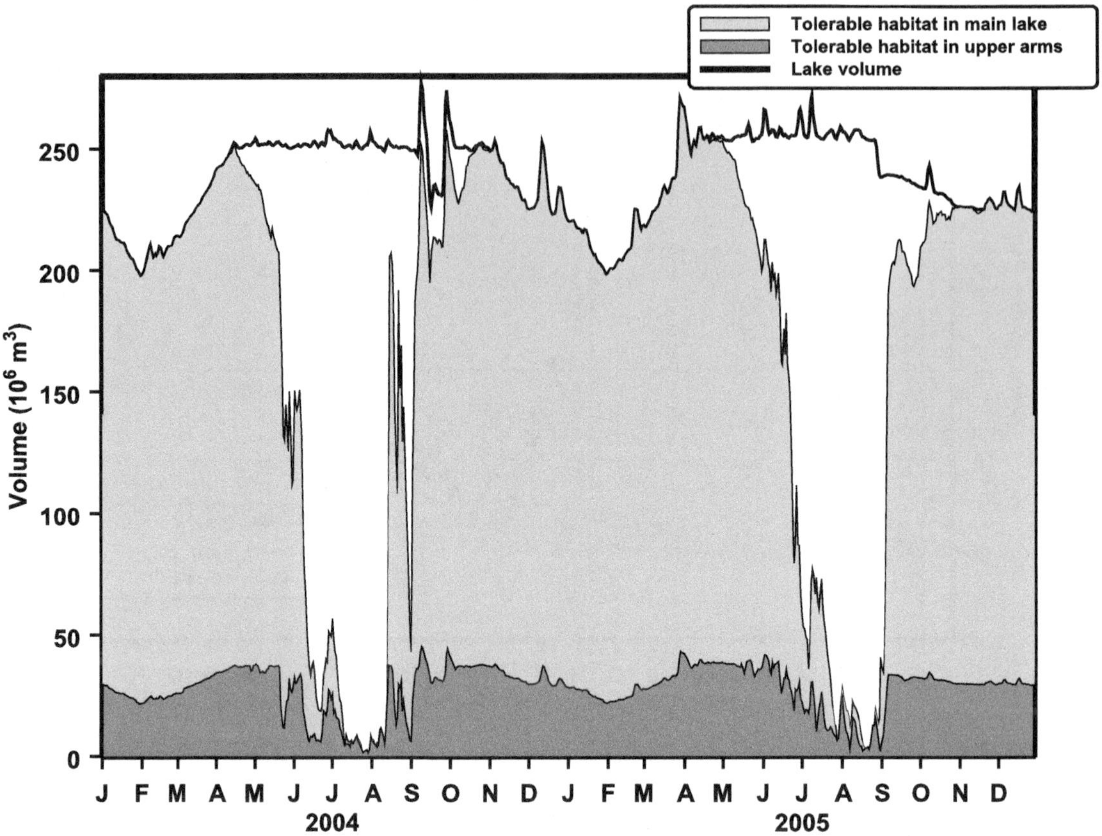

FIGURE 4. Tolerable striped bass habitat by region in the baseline modeled scenario for Lake Greenwood. Tolerable habitat is defined as temperature less than 28°C and dissolved oxygen greater than 2 mg/L. The upper arms were defined as Cane Creek arm and areas upstream of the junction of the Saluda and Reedy rivers.

Exchanging May–June weather, but not inflows, delayed habitat loss by more than a month in 2004. In 2005, habitat loss was advanced by about 3 weeks. These effects did not persist. By mid-July of both years, daily habitat volumes in the simulation with exchanged May–June weather differed from those in the baseline simulation by less than 4% of total lake volume.

Fluctuations in tolerable habitat in late June–August were associated with pulses of inflow greater than 5×10^6 m^3 for the Saluda and Reedy rivers combined or greater than 3×10^6 m^3 for the Saluda River. During these pulses, inflow increased by a factor of two or more. These inflows affected mainly the upper waters of the lake, temporarily alleviating the most extreme temperatures. However, the main part of the lake remained strongly stratified, and these pulses had little effect on hypoxia in the cooler, deeper waters. In the 68-year record for the Saluda River, such pulses of flow occurred at least once in June–August in more than 80% of years. None occurred in June–August 2002, the year of fish kills reported by Hayes (2003).

Simulated changes in phosphorus loading from the Saluda and Reedy rivers had substantial effects on algal productivity, as indicated by chlorophyll-α concentrations, and on habitat for striped bass. Average chlorophyll-α concentrations during April through August changed linearly with the phosphorus loading; the 90% reduction scenario resulted in 2.9 μg/L as compared to 11.8 μg/L for the baseline scenario. During the critical months of July and Au-

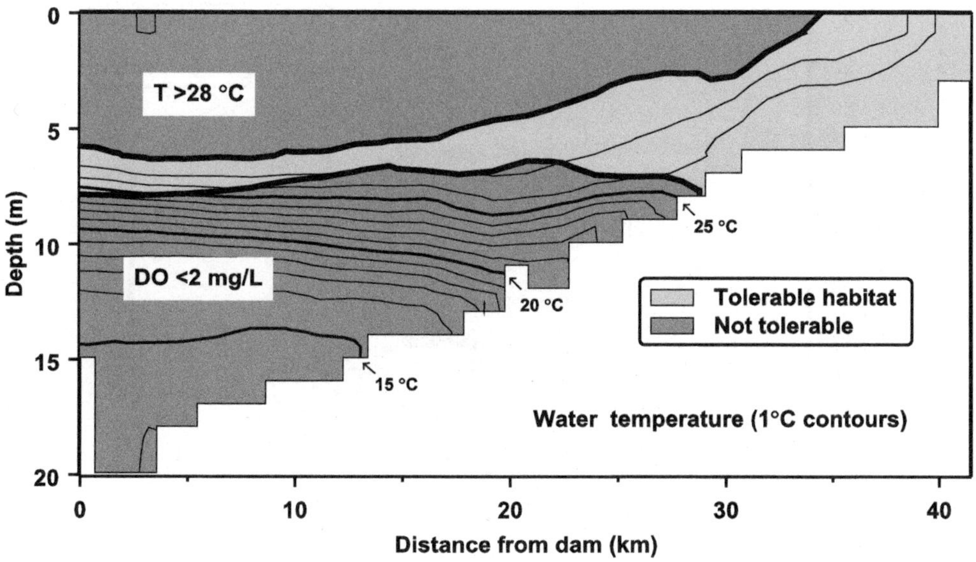

FIGURE 5. Water temperature in the main branch on 1 July 2004 in the baseline modeled scenario for Lake Greenwood. DO = dissolved oxygen.

gust, striped bass habitat increased in all of the phosphorus reduction scenarios (Table 3; Figure 7). The 90% phosphorus reduction scenario doubled average tolerable habitat and increased favorable habitat from near zero to nearly 10% in July and August. Also, in the 90% scenario, favorable habitat was present during each day of July and August. The response to loading was not linear: the 25% changes in loading had disproportionately small effects on habitat for striped bass.

The extent of extreme hypoxic conditions was sensitive to the sediment oxygen demand parameters, especially in the higher phosphorus reduction scenarios (Figure 8). The load-adjusted parameters produced a 61% decrease in the extent of extreme hypoxia in the 90% load decrease scenario as compared to the baseline parameters. Metrics such as mean total phosphorus concentration released at the dam and average chlorophyll-α concentration were insensitive (<5% change) to the parameters.

Discussion

The CE-QUAL-W2 water quality model for Lake Greenwood enabled us to describe the temporal and spatial change in striped bass habitat in Lake Greenwood in detail over a 2-year period, to evaluate influences of weather in spring and summer, and to predict effects of changes in phosphorus loads from the main tributaries.

The calibrated model closely reproduced the dynamics of temperature and dissolved oxygen in the lake over two annual cycles with differing weather. We judge that it should accurately estimate striped bass habitat for baseline and near-baseline scenarios. The phosphorus loading scenarios included adjustments of the parameters controlling sediment oxygen demand. The adjustments are plausible but uncalibrated for Lake Greenwood. The adjustments enhanced the response of dissolved oxygen to the changes in phosphorus loading, with disproportionately greater effect under the more extreme scenarios. We thus are less confident that the model accurately estimates striped bass habitat for the more extreme load-change scenarios.

Habitat for striped bass in Lake Greenwood becomes severely limited in summer. For both years of the simulation, favorable habitat (temperature <25°C, dissolved oxygen >2 mg/L) was usually absent in July and August; tolerable

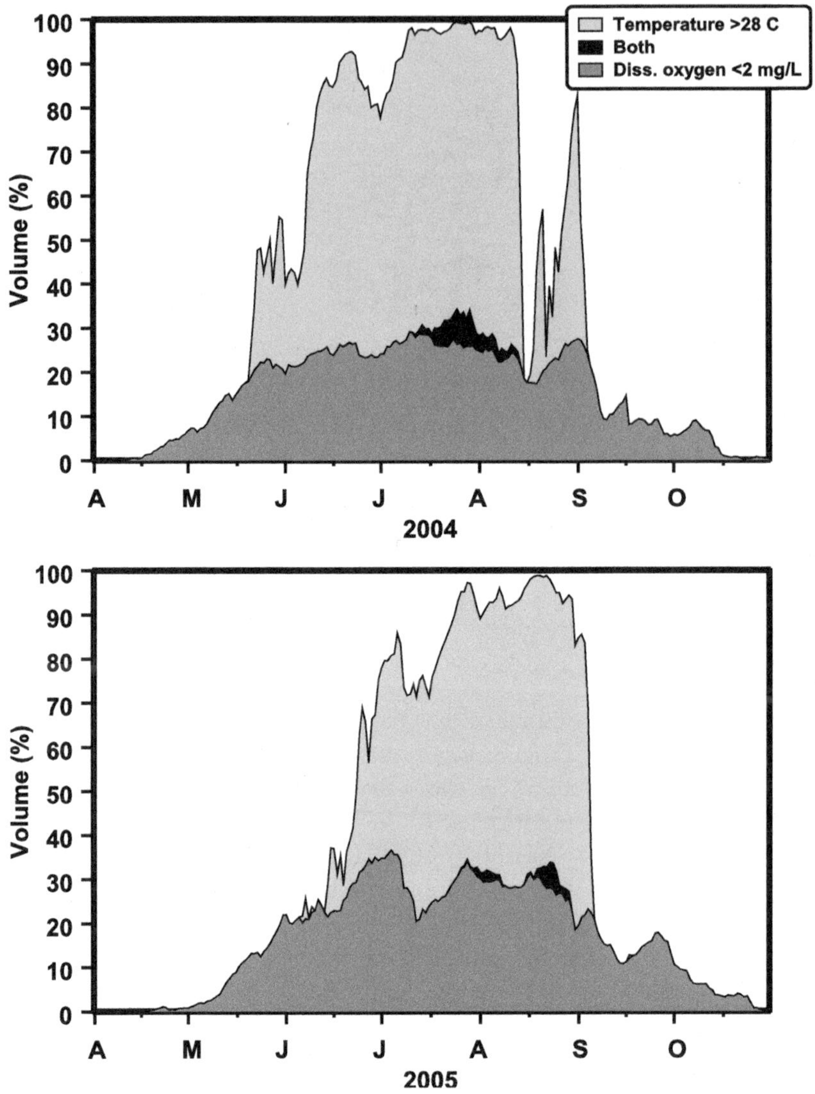

FIGURE 6. Lake volume failing to meet criteria for tolerable striped bass habitat in the baseline modeled scenario for Lake Greenwood. Tolerable habitat is defined as temperature less than 28°C and dissolved oxygen greater than 2 mg/L. Volume is expressed as percentage of the total lake volume for each day.

habitat (temperature <28°C, dissolved oxygen >2 mg/L) was always present but sometimes limited to small volumes in the upper arms of the lake. Such conditions seem unsuitable to production of trophy-sized (>9 kg) striped bass. Telemetry studies in reservoirs that do support trophy striped bass showed the availability of relatively cool (<25°C), oxygenated areas (Cheek et al. 1985; Young and Isely 2004; Bettoli 2005, for example), while similar studies in reservoirs where trophy striped bass were rare showed summer temperatures approaching and sometimes exceeding 28°C (Matthews et al. 1985; Hampton et al. 1988; Zale et al. 1990, for example).

In both years of the simulation, summer pulses of inflow or freshets expanded habitat for striped bass, mainly in the upper arms of the

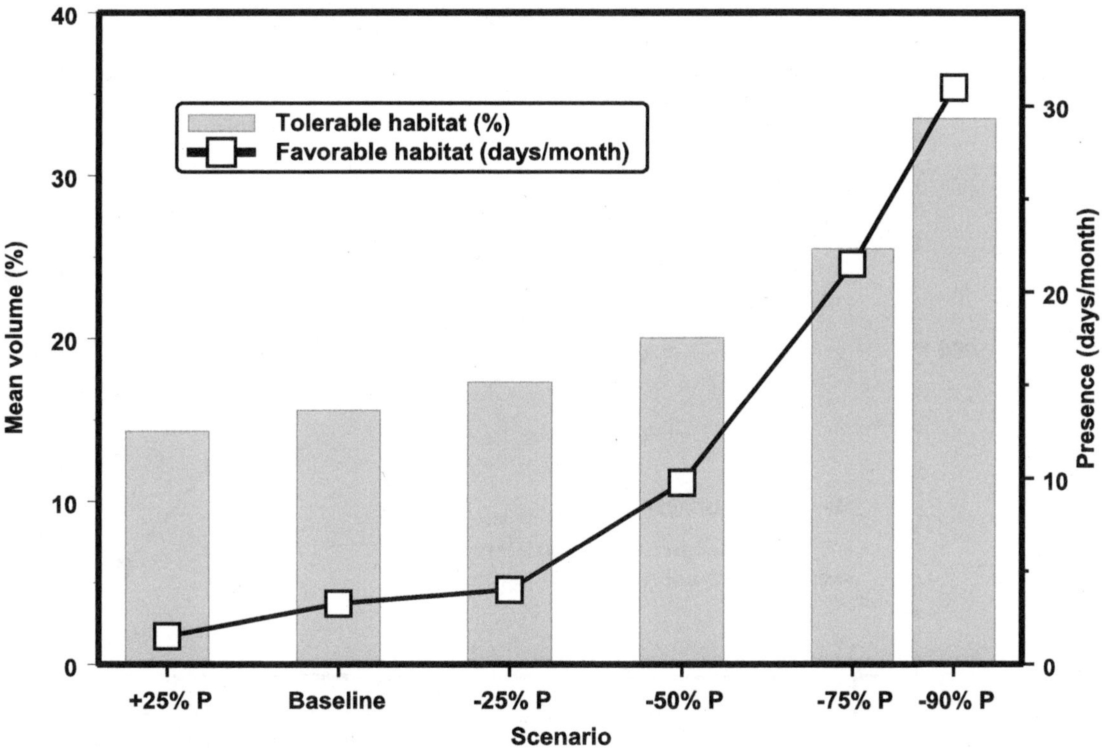

FIGURE 7. Favorable and tolerable striped bass habitat in July and August in the baseline and phosphorus (P) loading scenarios for Lake Greenwood. Favorable habitat is defined as temperature less than 25°C and dissolved oxygen greater than 2 mg/L, tolerable habitat as temperature less than 28°C and dissolved oxygen greater than 2 mg/L. Mean volume is the mean of the daily percentage of the total lake volume.

TABLE 3. Favorable and tolerable striped bass habitat in July–August in the baseline and phosphorus loading scenarios for Lake Greenwood. Favorable habitat is defined as temperature less than 25 °C and dissolved oxygen greater than 2 mg/L, tolerable habitat as temperature less than 28 °C and dissolved oxygen greater than 2 mg/L.

	Favorable habitat		Tolerable habitat	
Scenario	Mean volume (% of total)	Mean days present (per month)	Mean volume (% of total)	Mean days present (per month)
+25% P	<0.1%	1.50	14.3%	31
Baseline	<0.1%	3.25	15.6%	31
–25% P	<0.1%	4.00	17.4%	31
–50% P	0.3%	9.75	20.1%	31
–75% P	2.3%	21.50	25.5%	31
–90% P	8.5%	31.00	33.5%	31

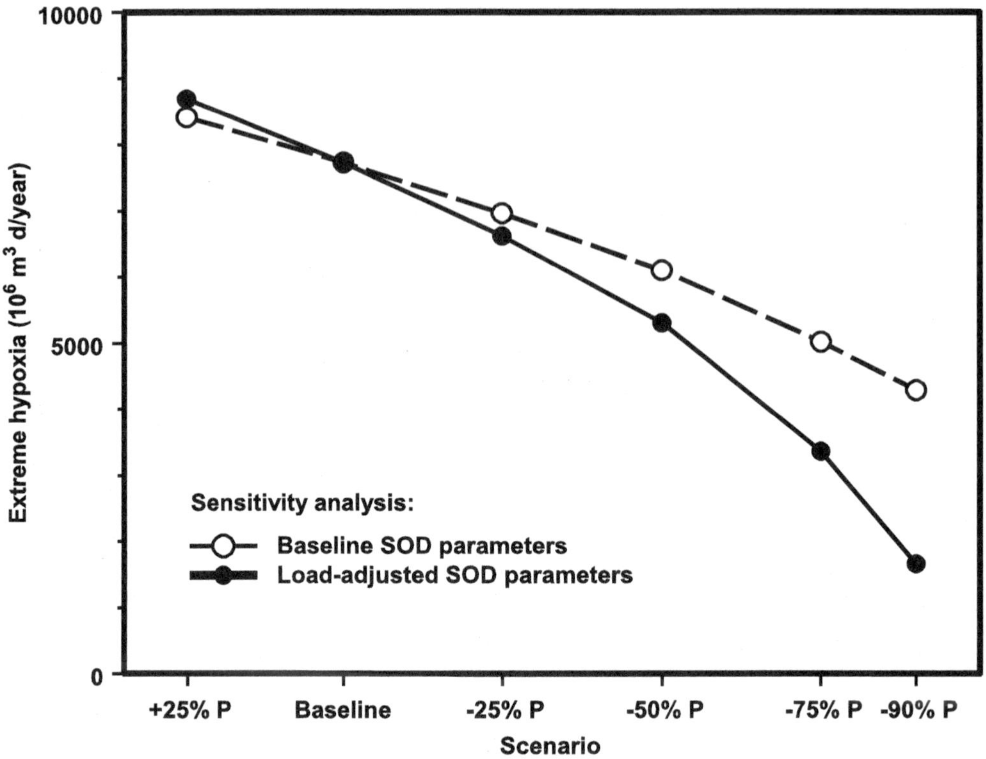

FIGURE 8. Sensitivity of extreme hypoxia in the sediment oxygen demand (SOD) parameters in the baseline and phosphorus-loading modeled scenarios for Lake Greenwood. The extent of extreme hypoxia was measured by the cumulative volume of lake water with dissolved oxygen less than 1 mg/L.

reservoirs. These events were driven by rainfall, including tropical storms. The summer telemetry study at Lake Greenwood showed that striped bass used refuges in both the main lake and the Reedy Arm (Snoots 1993), but further study is needed to learn how striped bass react to summer freshets. Grimes (1987) observed extensive movement of striped bass into habitat expanded by cool summer rains in Lake Anna, Virginia.

The longer-term record for the Saluda River indicates that some years (about one in five) do not have the summer freshets observed in 2004 and 2005 during June–August. It seems plausible that more extreme habitat limitation could occur in those years; the 2002 fish kill occurred in such a year.

Differences in May and June weather between the 2 years of the simulation affected the onset of habitat loss in the scenario, as revealed by the scenario with exchanged May–June weather. These influences did not persist. In July and August, they were overwhelmed by the influence of the summer freshets in these months.

Abundant forage may partially offset the absence of favorable summer habitat for striped bass in Lake Greenwood. Thompson and Rice (2013) showed that abundant forage in Badin Lake, North Carolina during summer allowed age-1 to age-5 striped bass to tolerate summer habitat stress better than similar striped bass in Lake Norman, North Carolina, where forage was poor; they suggested that forage-rich reservoirs that lacked optimal habitat could support a striped bass fishery, though not a fishery for trophy-sized fish. As reported by Rodriguez and Olmsted (1995), forage in Lake Greenwood (334 kg/ha) was about twice as high as in Lake Norman (118 kg/ha), although not quite so high in Badin Lake (449 kg/ha).

Phosphorus loading affects algal productivity, which in turn affects striped bass habitat through depletion of oxygen in deeper, cooler water during summer. Although algal productivity responded linearly to phosphorus loading in the simulations, small changes in loading had disproportionately weak effects on habitat for striped bass. With a 50% reduction in loading, tolerable habitat increased only from 16% to 20% of the lake in July–August, and favorable habitat from less than 0.1% to 0.3%. These effects seem small; their benefit would depend on the ability of the striped bass to use the additional habitat. The effects of the greater load reductions were substantial, although there is also more uncertainty about their accuracy. The 90% reduction scenario, for example, predicted continuous occurrence of favorable habitat in summer, raising the possibility that the lake could support a trophy fishery.

These greater phosphorus load reductions would not be easily achieved. Taylor and Bulak (2009) estimated that 50% (median for 1999–2006; range 40–70%) of the phosphorus load to Lake Greenwood from the Saluda and Reedy rivers derived from point sources, specifically wastewater discharge. The 50% load-reduction scenario would thus entail eliminating the phosphorus from all of the point sources or all of the nonpoint sources (or some combination) in the Saluda and Reedy watersheds; the 90% load reduction scenario would entail eliminating most of the phosphorus from both point and nonpoint sources.

Phosphorus reduction would, however, need to be evaluated in terms of the potential reduction in forage productivity and its influence on the total fishery, as well as its cost. Several researchers have demonstrated strong relationships among various combinations of phosphorus concentration, algal biomass, forage fish biomass, and sport fish harvest (Yurk and Ney 1989; Ney et al. 1990; Rodriguez and Olmsted 1995; Maceina and Bayne 2001), although the biomass of forage and sport fish tends to decline in hyper-eutrophic conditions (Kautz 1982). For example, for 14 Piedmont reservoirs in the Carolinas, Rodriguez and Olmsted's log–log regression predicts forage fish biomass ranging from 70 to 800 kg/ha for chlorophyll-α ranging from 2 to 40 μg/L. Improving habitat for striped bass at Lake Greenwood would likely reduce the abundance of forage fish that presently may help support persistence of striped bass through the summer periods of stressful habitat.

For striped bass in Lake Greenwood, our interpretation of the simulation results suggests that

1. striped bass persist through habitat restrictions in midsummer due to their ability to find and use small volumes of tolerable (or near tolerable) habitat;
2. abundant forage may partially offset the stress of high temperatures, but the prolonged absence of favorable habitat precludes development of a trophy fishery;
3. in summers unrelieved by freshets of cooler inflowing water, more severe and prolonged habitat loss may cause kills of striped bass, such as that which occurred in 2002; and
4. large reductions in phosphorus loading would be required to improve habitat substantially for adult striped bass during midsummer.

This example demonstrates how a water-quality model can be used to develop expectations for a striped bass fishery under baseline conditions and to assess the potential effects of environmental change. The combination of water quality modeling with analyses of longer-term meteorological records offers fisheries managers a basis for estimating frequencies of conditions that might severely limit adult striped bass habitat, producing summer fish kills and causing population bottlenecks.

Other applications of water quality models demonstrate their potential utility in evaluating and managing habitat for striped bass. Chang et al. (1992) illustrated changes in striped bass habitat in Douglas Reservoir, Tennessee that could occur with predicted temperature rise during climate change. Ruane et al. (2013, this volume) assessed the effects of turbine withdrawals from different depths, nutrient reduction, and an oxygenation system. Nestler et al. (2002) predicted the in-reservoir, summer movements of blueback herring *Alosa aestivalis* in relation to changing habitat conditions in a stratified reservoir; this same approach could be used to help predict summer movements of striped bass and hybrid striped bass. For the future, we envision the merging of telemetry (Hightower et al. 2013, this volume) and bioenergetic models (Thompson and Rice 2013) with water quality models to further improve our understanding of habitat preferences and utilization by striped bass and hybrid striped bass.

Acknowledgments

The original studies were supported by the Buzzard's Roost Fisheries Mitigation Trust Fund (administered by Duke Energy, U.S. Fish and Wildlife Service, and South Carolina Department of Natural Resources), a grant from the V. Kann Rasmussen Foundation to the Saluda-Reedy Watershed Consortium, and Upstate Forever. The Sport Fish Restoration Act of the U. S. Fish and Wildlife Service also provided support. This application of the model was undertaken, in part, to honor the memory of our deceased colleague Hank McKellar, who developed the initial model and guided us and many others through the fun and rigors of model development. We are very grateful to C. Coutant and an anonymous reviewer for their comments on the manuscript.

References

Bettoli, P. W. 2005. The fundamental thermal niche of adult landlocked striped bass. Transactions of the American Fisheries Society 134:305–314.

Chang, L. H., S. F. Railsback, and R. T. Brown. 1992. Use of a reservoir water quality model to simulate global climate change effects on fish habitat. Climatic Change 20:277–296.

Cheek, T. E., M. J. Van Den Avyle, and C. C. Coutant. 1985. Influences of water quality on distribution of striped bass in a Tennessee River impoundment. Transactions of the American Fisheries Society 114:67–76.

Cole, T. M., and S. A. Wells. 2002. CE-QUAL-W2: a two-dimensional, laterally averaged hydrodynamic and water quality model. Version 3.1 user manual. U.S. Army Corps of Engineers, Instruction Report EL-02–1, Washington, D.C.

Coutant, C. C. 1985. Striped bass, temperature, and dissolved oxygen: a speculative hypothesis for environmental risk. Transactions of the American Fisheries Society 114:31–61.

Coutant, C. C. 2013. When is habitat limiting for striped bass? Three decades of testing the temperature–oxygen squeeze hypothesis. Pages 65–91 *in* J. S. Bulak, C. C. Coutant, and J. A. Rice, editors. Biology and management of inland striped bass and hybrid striped bass. American Fisheries Society, Symposium 80, Bethesda, Maryland.

Coutant, C. C., K. L. Zachmann, D. K. Cox, and B. L. Pearman. 1984. Temperature selection by juvenile striped bass in laboratory and field. Transactions of the American Fisheries Society 113:666–671.

Grimes, D. V. 1987. Effects of power generation on adult striped bass in Lake Anna, Virginia. Virginia Power, North Anna Environmental Laboratory, Virginia Power, Mineral, Virginia.

Hampton, K. E., T. L. Wenke, and B. A. Zamrzla. 1988. Movements of adult striped bass tracked in Wilson Reservoir, Kansas. Prairie Naturalist 20:113–125.

Hayes, W. E. 2003. Fisheries investigations in lakes and streams—District III. South Carolina Department of Natural Resources, Annual Progress Report, F-63–8-3, July 1, 2002–December 31, 2003. Columbia.

Hayes, W. E. 2004. Fisheries investigations in lakes and streams—District III. South Carolina Department of Natural Resources, Annual Progress Report, F-63–9-3, January 1, 2004–December 31, 2004, Columbia.

Hightower, J. E., J. C. Taylor, and D. J. Degan. 2013. Estimating abundance of adult striped bass in reservoirs using mobile hydroacoustics. Pages 279–289 *in* J. S. Bulak, C. C. Coutant, and J. A. Rice, editors. Biology and management of inland striped bass and hybrid striped bass. American Fisheries Society, Symposium 80, Bethesda, Maryland.

Inabinet, J. R., and J. G. Pearse. 1981. A nutrient assessment of Lake Greenwood, South Carolina. South Carolina Department of Health and Environmental Control, Technical Report No. 013–81, Columbia.

Kalff, J. 2002. Limnology: inland water ecosystems. Prentice Hall, Upper Saddle River, New Jersey.

Kautz, R. S. 1982. Effects of eutrophication on the fish communities of Florida lakes. Proceedings of the Annual Conference Southeastern Association of Fish and Wildlife Agencies 34(1980):67–80.

Lee, G. F., and R. A. Jones. 1984. Summary of US OECD eutrophication study: results and their application to water quality management. Internationale Vereinigung für Theoretische und Angewandte Limnologie Verhandlungen 22:261–267.

Maceina, M. J., and D. R. Bayne. 2001. Changes in the black bass community and fishery with oligotrophication in West Point reservoir, Georgia. North American Journal of Fisheries Management 21:745–755.

Matthews, W. J., L. G. Hill, and S. M. Schellhaass. 1985. Depth distribution of striped bass and other fish in Lake Texoma (Oklahoma–Texas) during summer stratification. Transactions of the American Fisheries Society 114:84–91.

McKellar, H., and J. Bulak. 2008. Appendix A: phosphorus, algae, and oxygen dynamics in Lake Greenwood. Pages 39–56 *in* H. McKellar, J. Bulak, and B. Taylor. A dynamic water quality model of Lake Greenwood, SC: development and application toward issues of phosphorus loading, algal dynamics, and oxygen depletion. South Carolina Department of Natural Resources, Freshwater Fisheries Research Laboratory, Eastover. Available: www.dnr.sc.gov/fish/fwfi/files/LakeGreenwoodmodelreportMcKellar.pdf. (December 2012).

McKellar, H., J. Bulak, and B. Taylor. 2008. A dynamic water quality model of Lake Greenwood, SC: development and application toward issues of phosphorus loading, algal dynamics, and oxygen depletion. South Carolina Department of Natural Resources, Freshwater Fisheries Research Laboratory, Eastover. Available: www.dnr.sc.gov/fish/fwfi/files/LakeGreenwoodmodelreport-McKellar.pdf. (December 2012).

Nestler, J. M., R. Goodwin, T. Cole, D. Degan, and D. Dennerline. 2002. Simulating movement patterns of blueback herring in a stratified southern impoundment. Transactions of the American Fisheries Society 131:55–69.

Ney, J. J., C. M. Moore, M. S. Tisa, J. J. Yurk, and R. J. Neves. 1990. Factors affecting the sport fishery in a multiple-use Virginia (USA) reservoir. Lake and Reservoir Management 6:21–32.

Rodriguez, M. S., and L. L. Olmsted. 1995. Variation in fish community composition, biomass, and harvest relative to environmental parameters in reservoirs of North and South Carolina. 1993 Proceedings of the Annual Conference Southeastern Association of Fish and Wildlife Agencies 47(1993):552–570.

Ruane, J., and G. Hauser. 2006. Catawba-Wateree river reservoir and tailwater models (SC) and modeling approach. Duke Power Catawba-Wateree relicensing report. Reservoir Environmental Management, Inc., Chattanooga, Tennessee and Loginetics, Inc., Knoxville Tennessee.

Ruane, R. J., G. E. Hauser, and A. F. Sawyer. 2013. Water quality modeling as a management tool for striped bass and hybrid striped bass. Pages 161–179 *in* J. S. Bulak, C. C. Coutant, and J. A. Rice, editors. Biology and management of inland striped bass and hybrid striped bass. American Fisheries Society, Symposium 80, Bethesda, Maryland.

SCDHEC (South Carolina Department of Health and Environmental Control). 2004. Watershed water quality assessment: Saluda River basin. SCDHEC, Bureau of Water, Technical Report No. 0004-04, Columbia.

Snoots, T. R. 1993. Summer distribution and availability of habitat for adult striped bass in Lake Greenwood, South Carolina. Master's thesis. Clemson University, Clemson, South Carolina.

Stevens, R. 1958. The striped bass of Santee–Cooper Reservoir. Proceedings of the Annual Conference Southeastern Association of Game and Fish Commissioners 11(1957):253–264.

Taylor, B., and J. Bulak. 2009. Assessing effects of watershed change on phosphorus loading to Lake Greenwood, South Carolina. Final report. Supplement: impact of point sources on phosphorus loading to Lake Greenwood. South Carolina Department of Natural Resources, Freshwater Fisheries Research Laboratory, Eastover. Available: www.dnr.sc.gov/fish/fwfi/files/SaludaReedywatershedsupplement.pdf. (December 2012).

Taylor, B., J. Bulak, and H. McKellar. 2008. Assessing effects of watershed change on phosphorus loading to Lake Greenwood, South Carolina. Final report. South Carolina Department of Natural Resources, Freshwater Fisheries Research Laboratory, Eastover. Available: www.dnr.sc.gov/fish/fwfi/files/SaludaReedywatershedTaylor.pdf. (December 2012).

Thompson, J. S., and J. A. Rice. 2013. The relative influence of thermal experience and forage availability on growth of age 1–5 striped bass in two southeastern reservoirs. Pages 93–120 *in* J. S. Bulak, C. C. Coutant, and J. A. Rice, editors. Biology and management of inland striped bass and hybrid striped bass. American Fisheries Society, Symposium 80, Bethesda, Maryland.

Thornton, K. W., B. L. Kimmel, and F. E. Payne. 1990. Reservoir limnology: ecological perspectives. Wiley, New York.

Van Horn, S. L. 2013. A brief history of inland striped bass management. Pages 1–13 *in* J. S. Bulak, C. C. Coutant, and J. A. Rice, editors. Biology and management of inland striped bass and hybrid striped bass. American Fisheries Society, Symposium 80, Bethesda, Maryland.

Wetzel, R. G. 1983. Limnology, 2nd edition. Saunders College Publishing, Philadelphia.

Wilson, D. M. 2013. Utilizing anglers to improve striped bass management. Pages 333–349 *in* J. S. Bulak, C. C. Coutant, and J. A. Rice, editors. Biology and management of inland striped bass and hybrid striped bass. American Fisheries Society, Symposium 80, Bethesda, Maryland.

Young, S. P., and J. J. Isely. 2004. Temporal and spatial estimates of adult striped bass mortality from telemetry and transmitter return data. North American Journal of Fisheries Management 24:1112–1119.

Yurk, J. J., and J. J. Ney. 1989. Phosphorus–fish community biomass relationships in southern Appalachian reservoirs: can lakes be too clean for fish? Lake and Reservoir Management 5:83–90.

Zale, A. V., J. D. Wiechman, R. L. Lochmiller, and J. Burroughs. 1990. Limnological conditions associated with summer mortality of striped bass in Keystone Reservoir, Oklahoma. Transactions of the American Fisheries Society 119:72–76.

American Fisheries Society Symposium 80:197–208, 2013

Oxygen Diffusers to Create and Maintain Summer Fish Habitat

MARK MOBLEY*
Mobley Engineering, Inc.
Post Office Box 600, Norris, Tennessee 37828, USA

ED SHALLENBERGER
Colville Confederated Tribes
64 Schoolhouse Loop Road, Nespelem, Washington 99155, USA

MARC W. BEUTEL
Department of Civil and Environmental Engineering, Washington State University
Post Office Box 642910, Pullman, Washington 99164, USA

PAUL GANTZER
Gantzer Water Resources Engineering, LLC
14816 119th Place NE, Kirkland, Washington 98034, USA

BRIAN SAK
Fisheries and Wildlife Section, Natural Resources and Lands Management Division
San Francisco Public Utilities Commission
8653 Calaveras Road, Sunol, California 94586, USA

Abstract.—Habitat for striped bass *Morone saxatilis*, hybrid striped bass, salmonids, and other coolwater and coldwater fishes can be limiting in stratified reservoirs during summer and early autumn as surface water temperatures increase above tolerable ranges and deeper waters are depleted of dissolved oxygen (DO). Usable habitat can be increased in these reservoirs using oxygen diffusers to increase DO concentrations in the cooler, deeper waters. Several oxygen diffuser systems are currently in operation. Some of the systems were originally designed to increase DO in hydropower reservoir releases, but have also created fish habitat as a result of the diffuser system's efficient oxygen transfer capabilities in the reservoir. Several other systems are operated to improve water quality in the reservoir for water supply, and two systems have specific fish habitat maintenance goals. Improvements in DO for fish have been obtained at Calaveras Reservoir, California by the San Francisco Public Utility Commission, and fish studies at this reservoir are currently underway. In North Twin Lake, Washington, the Colville Confederated Tribes and Washington State University have documented improved trout habitat and reduced sulfide concentrations. Oxygenation of cool, deep water is now a proven technology that can alleviate summertime thermal and oxygen stress on striped bass and hybrid striped bass and can minimize habitat-related mortalities. The technology is being implemented specifically for striped bass with a large installation

* Corresponding author: mark@mobleyengineering.com

in J. Strom Thurmond Reservoir, South Carolina, for the Savannah District of the U. S. Army Corps of Engineers.

Introduction

Deterioration of summertime habitat can be a major cause of striped bass *Morone saxatilis* mortality and was first identified in Cherokee Reservoir, Tennessee (Coutant 2013, this volume). Suitable habitat is typically lost by warming of surface waters above about 25–27°C and depletion of dissolved oxygen (DO) to 2 mg/L or less in cooler, deep waters isolated by thermal stratification. This seasonal habitat restriction for striped bass and hybrid striped bass and its detrimental consequences have been confirmed as a general feature of many freshwater bodies into which these fish have been introduced (Coutant 2013; Rice et al. 2013; Thompson and Rice 2013; Van Horn 2013; all this volume). The temperature and oxygen limitations also affect distribution and abundance of prey for these large, pelagic predators, contributing to varying degrees of tolerance for the marginal physico-chemical conditions among water bodies (Thompson and Rice 2013). Various approaches have been used to maintain fisheries for large striped bass and their hybrids in the face of summer habitat limitations, including protection and expansion of naturally oxygenated thermal refuges such as cool tributary springs (Long et al. 2013, this volume), and summer fishing restrictions. Although artificial cooling is impractical, artificial oxygenation has been more promising, especially since it has additional benefits beyond in-reservoir habitat for striped bass.

In 1992, the Tennessee Valley Authority (TVA) installed oxygen diffusers in Cherokee Reservoir to improve DO conditions in the Holston River, downstream of the dam. Cherokee was the third TVA hydropower project to utilize oxygen diffusers in the reservoir to improve the water quality of hydropower releases, but the first with abundant striped bass in the reservoir. The Cherokee oxygenation system is the largest operated by TVA and is capable of delivering more than 136,000 kg/d of pure oxygen gas to counteract extremely low summertime DO conditions in the reservoir and enhance DO levels in released waters. The diffusers extend over 3 km upstream of the dam and spread oxygen over an even larger area. Operation of the diffusers increased DO levels in the deeper, cooler waters of the reservoir (Figure 1), which included the depth range that contained optimal temperatures for adult striped bass. The improved oxygen content created additional usable fish habitat near the dam during the warm summer months that attracted striped bass. As a result, the area became so heavily used by anglers that the Tennessee Wildlife Resources Agency closed the area to fishing 15 July–15 September due to substantial catch and release mortality of striped bass (http://news.tn.gov/node/7454). Since 1992, TVA has installed oxygen diffusers at other sites and currently operates reservoir oxygen diffuser systems at nine hydropower projects. Reservoir conditions have been enhanced near the diffusers at all of these reservoirs, but none have experienced the extreme striped bass response that occurred at Cherokee.

Since 1999, Mobley Engineering, Inc. has applied state-of-the-art bubble plume modeling techniques pioneered at Eawag, the Swiss Federal Institute of Aquatic Science and Technology (Wuest et al. 1992), Virginia Tech University (McGinnis et al. 2004; Singleton and Little 2006; Singleton et al. 2007), and TVA (Brown et al. 1989) to design and build reservoir oxygen systems nationwide. A small percentage of the systems had specific requirements for creating and maintaining usable fish habitat in reservoirs and lakes. Others were concerned simply with water quality.

This paper presents a brief description of the diffuser system, discusses two recent case studies where oxygenation was used for improv-

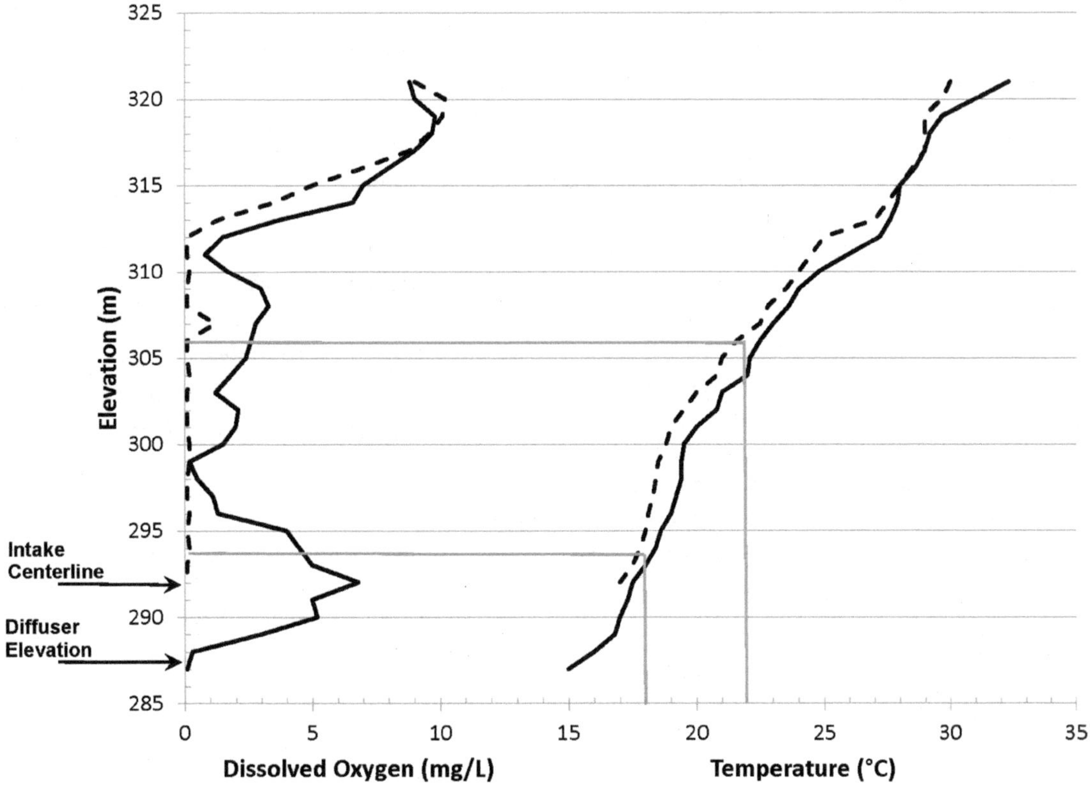

FIGURE 1. Dissolved oxygen and temperature profiles in Cherokee Reservoir, Tennessee, 14 August 1995, upstream (dashed lines) and downstream (solid lines) of oxygen diffusers (Mobley 1997). Elevations of the hydropower water intake and the oxygen diffuser are indicated. Gray lines indicate approximate depth of the estimated optimal temperature range for adult striped bass.

ing coldwater habitat for trout, and describes an ongoing project specifically designed for striped bass. In 2005, the San Francisco Public Utilities Commission installed a diffuser system to expand summer habitat for resident rainbow trout *Oncorhynchus mykiss* in Calaveras Reservoir, California. In Washington state, the Colville Confederated Tribes and Washington State University began conducting the first full year of a pilot study in 2009, operating a test diffuser system installed in 2008 to increase summertime habitat for stressed rainbow trout and brook trout *Salvelinus fontinalis* in North Twin Lake, Washington while using South Twin Lake as a study reference (Beutel et al. 2011). While these projects were specifically focused on enhancing habitat for trout with temperature preferences below 20°C (Biggs 2007), it is clear that similar reservoir oxygen systems can provide comparable habitat enhancements for other coolwater and coldwater species such as striped bass and its hybrids. A recent installation in J. Strom Thurmond Reservoir, South Carolina–Georgia is an example of a project where habitat for striped bass was the primary concern.

Reservoir Diffuser System

The line diffuser system for these applications was based on a design initially developed at TVA (Mobley 1997; Beutel and Horne 1999). This type of diffuser design is currently installed and operated at nine TVA reservoirs and at 11 reservoirs for other utilities from Maine to California. The line diffuser is a simple and economical design that spreads bubbles over a large area and is installed and retrieved for required maintenance from a 6 × 2.5 m "jon" boat without the assistance of divers. The system uses long lines of flexible porous hose (commercial "soaker hose" often used for irrigation, formed from shredded automobile tires) to avoid clogging and other maintenance problems experienced by previous designs that used ceramic diffusers. The diffuser is a two-pipe system that includes a high-density polyethylene supply pipe to distribute the gas along the length of the diffuser and a buoyancy pipe that can support the entire weight of the diffuser system when filled with air (Figure 2). Flow control orifices control the oxygen flow to independent lengths of porous hose to maintain a continuous flow of small gas bubbles along the full length of the diffuser. Concrete anchors are attached to the diffuser piping with stainless steel cable and saddle connections. The line diffuser has proven to be an economical oxygenation diffuser design that transfers oxygen efficiently to deep water, minimizes temperature destratification to preserve the coolwater habitat, and avoids sediment disruption. The three case studies (Calavares Reservoir, Twin Lakes, and J. Strom Thurmond Reservoir) illustrate its deployment and oxygenation performance.

Calaveras Reservoir

Calaveras Reservoir is part of a water supply system owned and operated by the San Francisco Public Utility Commission (SFPUC), San Francisco, California. The reservoir is located above and between Sunol and Milpitas, California and has a surface area of 581 ha, a

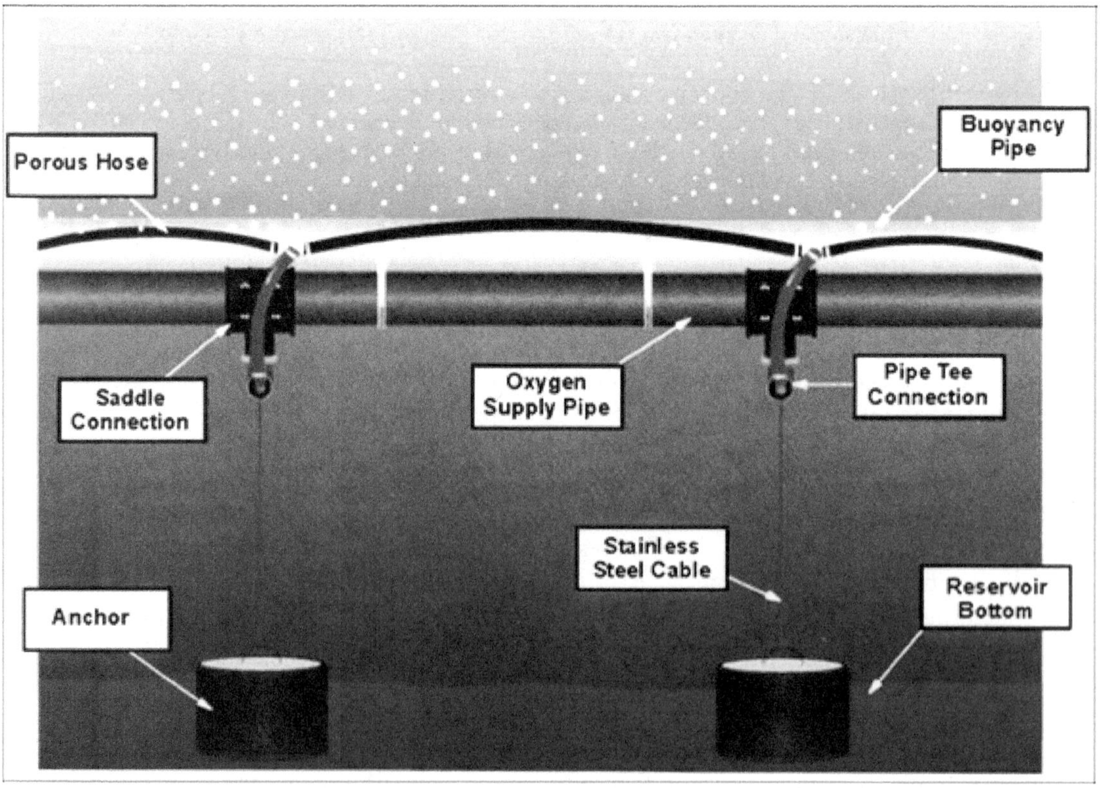

FIGURE 2. Schematic of diffuser system components.

maximum depth of 43 m, and a volume of 120 million m^3 at full pool elevation of 230 m. In 2004, the SFPUC undertook an improvement program to replace Calaveras Dam. In the interim, the California Division of Safety of Dams mandated an operational reservoir level of 215 m, reducing the water depth to less than 27 m, the surface area to 348 ha, and the storage capacity to 42 million m^3. The reduced size of the reservoir resulted in reduced fish habitat and algal blooms that caused taste and odor problems. The fish habitat issue was especially important to the SFPUC (2005), state and federal fisheries agencies, and environmental advocates because the reservoir contained a unique, self-sustaining population of rainbow trout.

In 2004, the SFPUC undertook several studies that determined that a hypolimnetic oxygenation system (HOS) was a viable means to maintain reservoir water quality and fish habitat. Dissolved oxygen objectives for the HOS were defined as a minimum of 2 mg/L in the hypolimnion and 5 mg/L in the volume of water between 18°C and 22°C for favorable trout habitat (also the preferred temperature range for large striped bass; Coutant 2013). After an extensive evaluation of alternatives, the reservoir oxygen diffuser system was recommended as the best approach to achieve the project objectives at Calaveras Reservoir (Weiss and Associates 2004).

A site study evaluated potential diffuser layouts utilizing state-of-the-art bubble plume modeling to predict oxygen placement over a range of reservoir conditions, diffuser layouts, and oxygen flow rates (Mobley 2004). To enhance key fish habitat, it was particularly important to place oxygen at the correct elevation in the reservoir water column to oxygenate waters that corresponded with optimal temperatures for fish. A diffuser layout that consisted of two diffuser lines, each 305 m long, was recommended. One diffuser was placed in the deepest part of the reservoir near the dam. The second diffuser was extended almost 1.6 km upstream of the dam to place oxygen higher in the water column and spread oxygen into an upstream cove, named Arroyo Hondo, to maintain prime fish habitat. The recommended oxygen system delivery capacity was 3,600 kg/d of oxygen to match the maximum measured oxygen demands in the reservoir. Reservoir oxygen demand is calculated from measured DO and temperature profiles in the reservoir during the onset of oxygen depletion and stratification. The reduction in DO over time during development of summer stratification multiplied by the affected reservoir volume estimates reservoir oxygen demand. In the Calaveras Reservoir design additional oxygen delivery capacity was specified to provide the capability to increase DO levels after a shutdown or failure.

The Calaveras Reservoir oxygenation system was installed in late summer of 2005 at a cost of about US$1.2 million. The diffusers at Calaveras Reservoir are supplied with oxygen from an onsite bulk oxygen storage facility. The facility includes a 41,000-L liquid oxygen tank, two vaporizers, and associated equipment to supply pressurized gaseous oxygen to the diffusers. Oxygen is stored in the tank at approximately −150°C and vaporized as needed to supply the system. Pressure caused by the vaporization is sufficient to provide the oxygen flow to the diffusers in the reservoir. When operating in the reservoir, the oxygen diffusers produce a long, straight curtain of very small bubbles that are spread out enough to create only a weak upwelling plume. Most of the bubbles dissolve before they reach the surface as oxygen is transferred to the surrounding water. Those that reach the surface create a faint bubble pattern that is visible when the water is calm. Oxygen transfer efficiency is 85% to 90% with the 20–27-m depths targeted at Calaveras Reservoir.

The Calaveras Reservoir oxygenation system was placed in operation on 6 September 2005. Initially, the system was operated at higher than normal oxygen flow rates to overcome the existing hypoxic conditions in the reservoir. The reservoir DO levels responded quickly to the oxygen input, reaching objectives within 3 weeks of start-up (Figure 3). Dissolved oxygen levels were increased well into Arroyo Hondo,

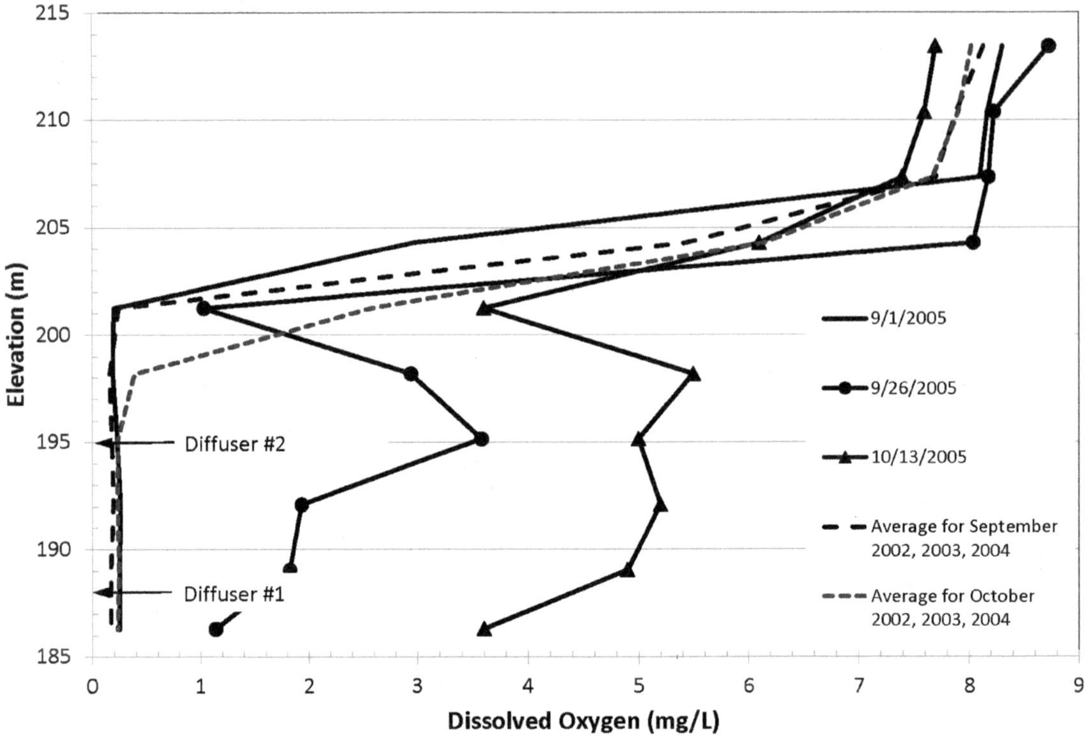

FIGURE 3. Depth distribution of dissolved oxygen (DO) 5 d before start-up of the Calaveras Reservoir oxygenation system on 6 September 2005 (9/1/2005), 3 weeks after start-up (9/26/2005) and about 5 weeks after start-up (10/13/2005). Average DO depth distributions from September and October 2002, 2003, and 2004 are plotted for comparison.

even though high sediment oxygen demand reduced the levels near the bottom. Similar DO levels have been maintained every year since the system was installed, providing suitable coldwater trout habitat. Costs for operation and maintenance have been about $50,000, annually.

The Calaveras Reservoir oxygenation system has successfully met SFPUC objectives for increased DO levels in the hypolimnion. The system has been so successful that SFPUC installed a similar hypolimnetic oxygen system in the larger San Antonio Reservoir in 2009 to achieve similar goals for water quality and fish habitat enhancement.

North Twin Lake

North and South Twin Lakes are natural lakes on the Colville Confederated Tribes (CCT) Reservation northwest of Spokane, Washington. The lakes are in the hills above Inchelium, Washington; are very similar in volume, shape, and depth; and are connected by a short channel. North Twin Lake has a surface area of 371 ha, a maximum depth of 15 m, and a volume of 36 million m^3. The lakes have historically been stocked with rainbow trout (coastal) and brook trout by the CCT Fish and Wildlife Department. The Twin Lakes support a strong tribal trout fishery and are a tourist attraction. More recently, CCT has emphasized stocking and reestablishment of Columbia basin redband trout *O. m. gairdneri*, a rainbow trout subspecies native to Northwest lakes and streams east of the Cascade Mountain Range (Clegg 2010) in the Twin Lakes.

The water quality and fisheries in the lakes have been monitored and studied by researchers at Washington State University (WSU) and

CCT (Christensen and Moore 2008, 2009, 2010). Both lakes experience low DO levels in the bottom waters that typically become anoxic below about 7 m depth in the late summer when surface temperatures reach 22°C. The lack of cool, oxygenated water has limited trout habitat in the lakes. Trout have exhibited signs of stress and reduced growth rates during the summer, and total annual mortality has exceeded 90% (Clegg 2010).

Seeking a way to improve summertime fish habitat, WSU and CCT collaborated on a proposal to the Columbia River Water Management Program to design, install, and operate an on-site pilot oxygenation system that would deliver oxygen to the bottom waters of one lake (North Twin), with the other (South Twin) serving as a reference location. The dual nature of North and South Twin lakes provides an ideal setting to scientifically evaluate the effect of lake oxygenation on trout habitat.

In the summer of 2008, an oxygen diffuser system was installed in North Twin Lake to enhance DO levels. The total installation cost, using a temporary oxygen supply facility, was about $400,000. Initial funding for the project was provided by the Columbia River Water Management Program. The system included a 745-m line diffuser that distributes oxygen near the bottom of the lake; oxygen is supplied from a 23,000-L temporary liquid oxygen facility onshore.

The oxygen delivery capacity for the North Twin Lake oxygenation system was based on lake dissolved-oxygen profiles measured about monthly in the center of the lake from 2004 through 2007. The delivery capacity included factors to account for diffuser-induced oxygen demand, recovery from a shutdown, diffuser oxygen transfer efficiency, plus an additional safety factor. The diffuser-induced oxygen demand is the increased demand exerted by the reservoir sediments when water is stirred over them. This factor can range from 2 to 4 depending on site-specific conditions. The total capacity requirement was calculated to be 4,300 kg/d, a conservative engineering estimate for the design capacity. About 1,500 kg/d was estimated to be required from May to November, starting with aerobic conditions.

Monitoring during the study was conducted by WSU and CCT and included extensive monitoring of temperature, dissolved oxygen, and fish distribution. Prior to initial testing of the system in the fall of 2008, the hypolimnion exhibited severe oxygen depletion with DO levels less than 0.5 mg/L. Within 24 h of starting oxygenation DO levels were observed to increase to 1–2 mg/L near the diffuser.

In 2009, operation of the oxygenation system was initiated as soon as DO levels began to decline in late spring. Over the course of the summer of 2009, an oxygen flow rate of 540 to 800 kg/d was sufficient to maintain high DO levels in North Twin Lake, while the hypolimnion of South Twin Lake became severely hypoxic or anoxic (Figure 4). In June, DO levels were 5–10 mg/L in the bottom waters of North Twin Lake, while bottom waters were less than 1 mg/L in South Twin Lake. Overall oxygen transfer efficiency for the 2009 oxygenation season was estimated to be 83%. Installation of the diffuser system and site preparation was about $400,000. Operating costs have run about $80,000 per year, including liquid oxygen tank rental and remote delivery surcharges. The tribe is pursuing installation of an onsite oxygen generation system to reduce operating costs.

Recent fishery studies suggest enhanced deepwater habitat for trout in North Twin Lake (Clegg 2010; Lanouette 2011; Shallenberger 2011). In 2010, horizontal gill nets were set overnight on a monthly basis in the pelagic zones of both North and South Twin lakes at depths of 2–5 m, 5–8 m, and 9–12 m, as well as along shore and bottom. The fish captured in the nets (Table 1) demonstrate that trout did not use the hypoxic hypolimnion of South Twin Lake during periods of stratification, while the oxygenated hypolimnion of North Twin Lake was regularly used. These studies were corroborated by monthly hydroacoustic surveys.

As with Calaveras Reservoir, the results were so positive that an additional deepwater

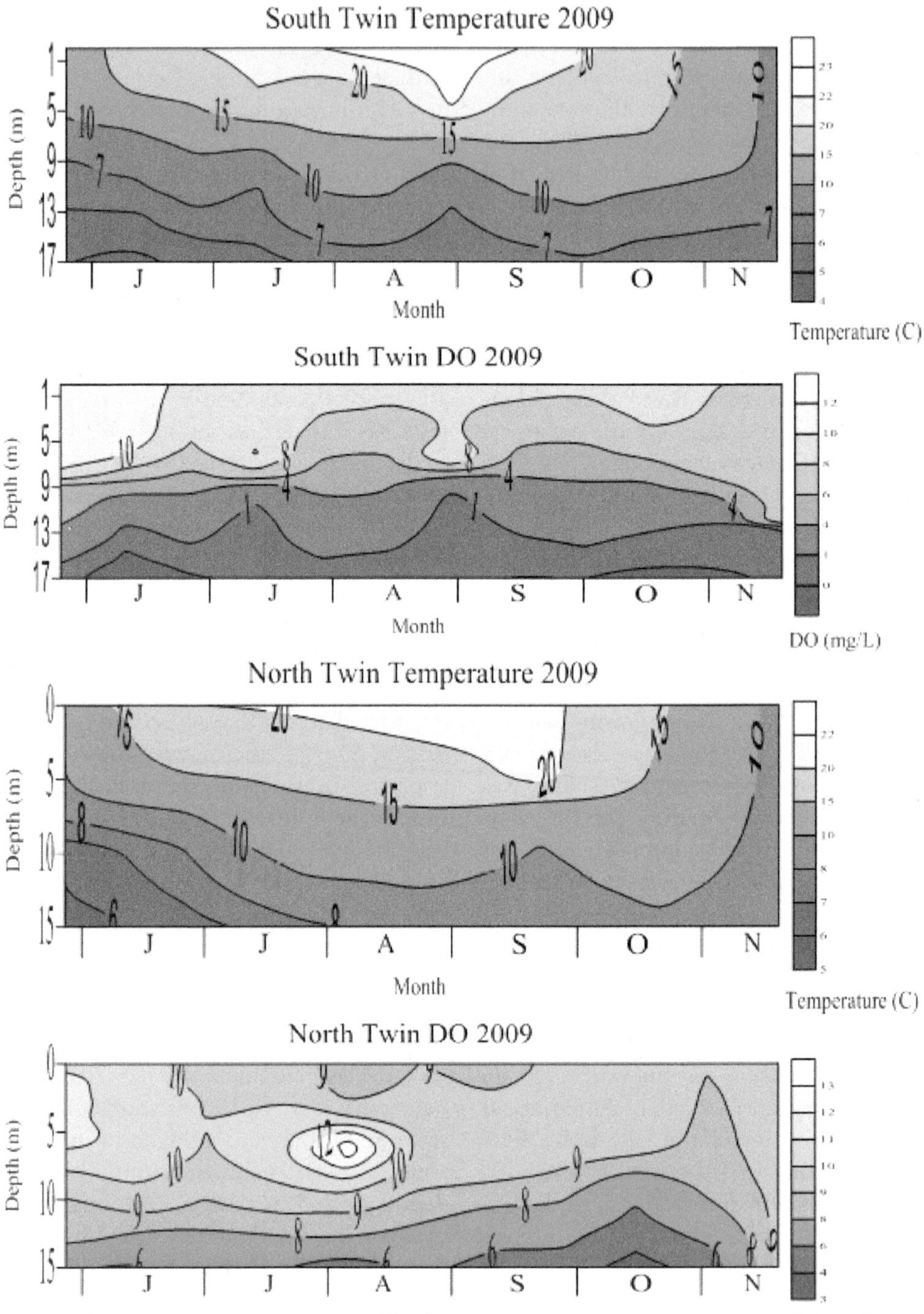

FIGURE 4. Depth distributions of temperature (°C) and dissolved oxygen (mg/L) in North and South Twin Lakes from late May through late November 2009. North Twin Lake is oxygenated by a diffuser at a depth of 14 m. Figures compliments of Dr. Barry Moore, Washington State University.

TABLE 1. Fish captured in 2010 in gill-net surveys in North and South Twin Lakes, Washington by lake, sampling stratum, and sampling date.

Stratum	6 Apr	19 Apr	6 May	9 Jun	14 Jul	11 Aug	10 Sep	21 Oct	Total
North Twin Lake									
Inshore	37			7				60	104
2–5 m		4	18	6	0	0	4	8	40
5–8 m		2	11	13	5	31	10	4	76
9–12 m		0	3	1	1	5	1	0	11
Benthic		0	0	0	0	0	1		1
South Twin Lake									
Stratum	6 Apr	19 Apr	6 May	9 Jun	14 Jul	11 Aug	10 Sep	21 Oct	Total
Inshore	39			4				14	57
2–5 m		1	5	1	0	0	0	4	11
5–8 m		2	6	9	3	4	0	0	24
9–12 m		0	0	1	0	0	0	0	1
Benthic		0	0	0	0	0	0	0	0

diffuser system with a capacity around 4,300 kg/d was installed in South Twin Lake in 2010 to improve trout habitat in that lake.

J. Strom Thurmond Reservoir

Work was recently completed by a team, including the senior author, to design, install, and test a striped bass habitat enhancement system for the U.S. Army Corps of Engineers' J. Strom Thurmond Reservoir near Augusta, Georgia, on the Savannah River. Oxygenation of the lower reservoir is intended to mitigate for loss of cool, oxygenated water in the tailrace of the upstream Richard B. Russell Dam (a popular striped bass fishery) when that dam operates as a pumped-storage hydropower facility (thus eliminating the persistent coolwater summer discharges typical of previous single-pass hydropower generation). The reservoir oxygen diffuser system was installed in the summer of 2010 and spring of 2011 and is capable of delivering up to 182,000 kg/d of oxygen. It is designed to maintain a large volume (1.6 km long) of striped bass habitat in the reservoir approximately 8 km upstream of the J. Strom Thurmond Dam. The specific design goals are 5 mg/L DO for 1.6 km (no less than 4.5 mg/L) and 4 mg/L DO for 6.4 km (no less than 3.5 mg/L) in the lower 13 km and in the 18–24°C temperature zone of the reservoir. A further goal is 3 mg/L DO at the entrance to the turbines to enhance the downstream water quality of the hydropower releases from J. Strom Thurmond Dam.

Extensive water quality modeling (Ruane et al. 2013, this volume) was done in support of a diffuser design that could place oxygen as needed to maintain striped bass habitat in the 18–24°C temperature zone of the reservoir as it broadens and deepens over the course of the summertime stratification. The final diffuser system design included nine diffusers at four elevations and more than 10 km of underwater supply piping. The system was fully operational in June of 2011 and was operated during the summer of 2011 to verify performance of contractual requirements, demonstrate flexibility, and maintain design goals. More than 400 water quality profiles were obtained to document distribution of the oxygen along the length and width of the lower 13 km of the reservoir; design goals were met or exceeded for the entire summer (Figure 5). Further monitoring of the habitat and response of the striped bass is ongoing.

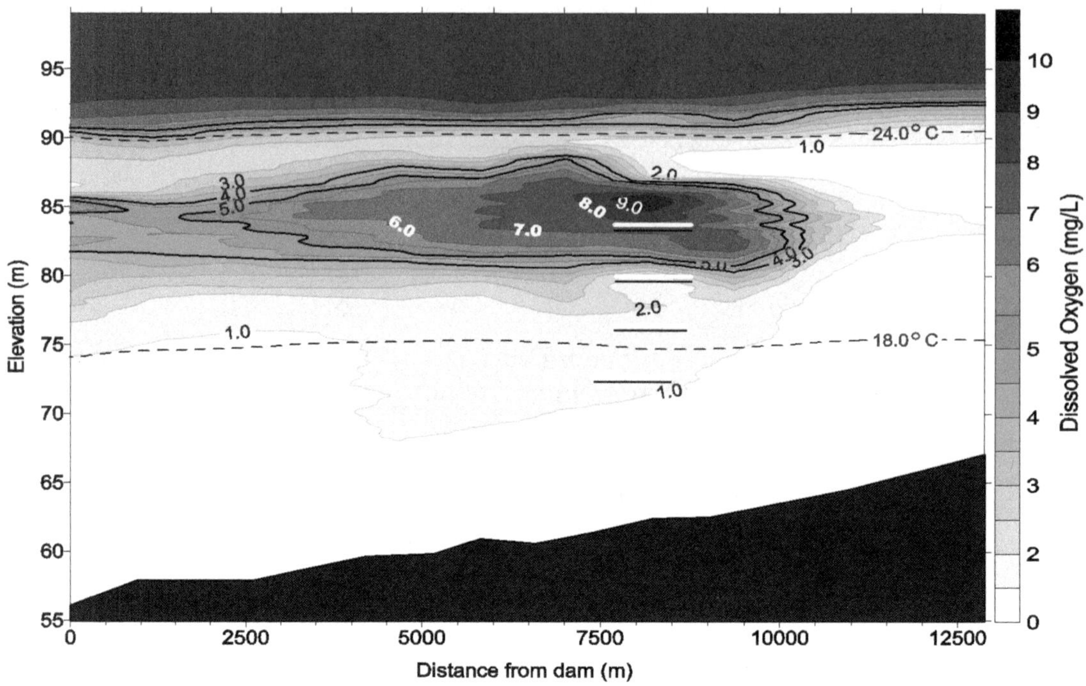

FIGURE 5. Example depth distribution of dissolved oxygen concentrations throughout the lower 13 km of J. Strom Thurmond Reservoir based on water quality profiles from 2011. Dashed lines indicate the 18°C and 24°C depth contours. Black-and-white bars indicate the four oxygen diffusers and their elevations.

Discussion

The DO and fish-habitat enhancements achieved in Calaveras Reservoir and North Twin Lake demonstrate the capability of a reservoir oxygen diffuser system to create and maintain coldwater fish habitat in diverse settings. At each of these sites, the gaseous oxygen bubble plumes placed oxygen in specific depths or temperature ranges in the lake or reservoir based on the habitat requirements of the desired fish species. Oxygen was added at rates that were determined by analyses of oxygen demand of the deep water during summer stratification. These calculations could have been improved with currently available reservoir water quality models (Ruane et al. 2013).

The examples we have described clearly demonstrate that reservoir oxygen diffuser systems can enhance habitat for coldwater and coolwater species, including striped bass and their hybrids. Striped bass habitat can be enhanced in habitat-limited reservoirs by placing oxygen in the optimal temperature regions of the reservoir (18–24°C) that may normally lack sufficient oxygen content during the summer months. Installation of a diffuser system upstream of Cherokee Dam in 1992 for tailwater oxygenation serendipitously demonstrated the attractiveness of artificially oxygenated water to striped bass in the summer. The two trout-related case studies presented here confirm that the diffuser system, which often includes thousands of feet of diffuser, can be located to distribute oxygen into specific areas and thermal strata of a reservoir as needed to optimize coolwater or coldwater fish habitat. Preliminary results from the initial year of operation of the system installed in J. Strom Thurmond Reservoir specifically to enhance striped bass habitat suggest that similar success can be achieved on a much larger scale.

As further experience is gained by deploying the system in water bodies of varying sizes and water-quality conditions, it should be technically feasible to artificially remove the temperature-DO habitat limitations for striped bass and their hybrids, which have been demonstrated in many inland waters across the country. Whether the technical feasibility is matched by economic desirability will depend on public attitudes toward striped bass and their hybrids in inland waters and the values of striped bass and hybrid fisheries (McMullin 2013; Miranda and Rayborn 2013; Whitehead 2013; all this volume). In most cases, oxygenation is likely to be too expensive to implement solely for striped bass habitat enhancement. The economic desirability increases when it is also the best option for meeting tailrace water quality standards (as in TVA reservoirs), as environmental mitigation for development projects (as at J. Strom Thurmond Reservoir), or as part of water treatment improvements at a water supply reservoir (Calaveras Reservoir). A multipurpose design that optimizes in-reservoir and downstream benefits would seem most fruitful and cost-effective. Where corollary benefits are not present, the scale of oxygenation might be scaled down to provide smaller volumes of enhanced DO in lake hypolimnia. This approach could provide DO-temperature refuges for striped bass during critical periods of summer, so long as angling for concentrated fish is controlled.

References

Beutel, M. W., and A. J. Horne. 1999. A review of the effects of hypolimnetic oxygenation on lake and reservoir water quality. Lake and Reservoir Management 15:285–297.

Beutel, M. W., B. C. Moore, P. Gantzer, and E. Shallenberger. 2011. Ease the squeeze in Twin Lakes: the Colville Confederated Tribes work to enhance a trout fishery and improve water quality in reservation lakes. Lakeline 30(4):31–37.

Biggs, M. J. 2007. Seasonal habitat use and movement by Columbia River redband trout in Twin Lakes, WA. Master's thesis. Washington State University, Pullman.

Brown, R. T., J. A. Gordon, and C. E. Bohac. 1989. Measurement of upwelling flow from air diffuser. Journal of Environmental Engineering 115:1269–1275.

Christensen, D. R., and B. C. Moore. 2008. Diet composition and overlap in a mixed warm- and coldwater fish community. Journal of Freshwater Ecology 23:195–204.

Christensen, D. R., and B. C. Moore. 2009. Using stable isotopes and a multiple mixing model to evaluate fish dietary niches in a mesotrophic lake. Lake and Reservoir Management 25:167–175.

Christensen, D. R., and B. C. Moore. 2010. Largemouth bass consumption demand on hatchery rainbow trout in two Washington lakes. Lake and Reservoir Management 26:200–211.

Clegg, E. 2010. Movement and distribution of trout following hypolimnetic oxygenation in Twin Lakes, Washington. Master's thesis. Washington State University, Pullman.

Coutant, C. C. 2013. When is habitat limiting for striped bass? Three decades of testing the temperature–oxygen squeeze hypothesis. Pages 65–91 *in* J. S. Bulak, C. C. Coutant, and J. A. Rice, editors. Biology and management of inland striped bass and hybrid striped bass. American Fisheries Society, Symposium 80, Bethesda, Maryland.

Lanouette, B.P. 2011. Changes in salmonid vertical distribution following hypolimnetic oxygenation in North Twin Lake, Washington. Master's thesis. Washington State University, Pullman.

Long, E. A., C. L. Messing, K. J. Herrington, R. R. Weller, and I. I. Wirgin. 2013. Restoration of Gulf striped bass: lessons and management implications. Pages 25–63 *in* J. S. Bulak, C. C. Coutant, and J. A. Rice, editors. Biology and management of inland striped bass and hybrid striped bass. American Fisheries Society, Symposium 80, Bethesda, Maryland.

McGinnis, D. F., A. Lorke, A. Wuest, A. Stöckli, and J. C. Little. 2004. Interaction between a bubble plume and the near field in a stratified lake. Water Resources Research 40, W10206, doi:10.1029/2004WR003038.

McMullin, S. L. 2013. Managing user conflicts on reservoirs supporting striped bass or hybrid striped bass fisheries. Pages 521–530 *in* J. S. Bulak, C. C. Coutant, and J. A. Rice, editors. Biology and management of inland striped bass and hybrid striped bass. American Fisheries Society, Symposium 80, Bethesda, Maryland.

Miranda, L. E., and S. W. Rayborn. 2013. Interactions between striped bass and other game fish in reservoirs. Pages 501–519 *in* J. S. Bulak, C. C. Coutant, and J. A. Rice, editors. Biology and management of inland striped bass and hybrid striped bass. American Fisheries Society, Symposium 80, Bethesda, Maryland.

Mobley, M. H. 1997. TVA Reservoir Aeration Diffuser System (TVA Technical Paper 97-3), ASCE WaterPower '97 Conference Proceeding Papers, August 5-8, Atlanta, GA. American Society of Civil Engineers, Reston, Virginia.

Mobley, M. H. 2004. Oxygen diffuser system for water quality and fish habitat enhancement at Calaveras Reservoir. Site study report prepared for the San Francisco Public Utility Commission, San Francisco.

Rice, J. A., J. S. Thompson, J. A. Sykes, and C. T Waters. 2013. The role of metalimnetic hypoxia in striped bass summer kills: consequences and management implications. Pages 121–145 *in* J. S. Bulak, C. C. Coutant, and J. A. Rice, editors. Biology and management of inland striped bass and hybrid striped bass. American Fisheries Society, Symposium 80, Bethesda, Maryland.

Ruane, R. J., G. E. Hauser, and A. F. Sawyer. 2013. Water quality modeling as a management tool for striped bass and hybrid striped bass. Pages 161–179 *in* J. S. Bulak, C. C. Coutant, and J. A. Rice, editors. Biology and management of inland striped bass and hybrid striped bass. American Fisheries Society, Symposium 80, Bethesda, Maryland.

SFPUC (San Francisco Public Utility Commission). 2005. Population size estimates for adult rainbow trout (*Oncorhynchus mykiss*) in San Antonio and Calaveras reservoirs. SFPUC, Water Quality Bureau, Technical Memorandum No. 2-04-006, Sunol, California.

Shallenberger, E. 2011. Twin Lakes enhancement annual report for 2010. Colville Confederated Tribes Fish and Wildlife Department, CCT/RF-2011-1, Nespelem, Washington.

Singleton, V. L., P. Gantzer, and J. C. Little. 2007. Linear bubble plume model for hypolimnetic oxygenation: full-scale validation and sensitivity analysis. Water Resources Research 43, W02405, doi:10.1029/2005WR004836.

Singleton, V. L., and J. C. Little. 2006. Designing hypolimnetic aeration and oxygenation systems—a review. Environmental Science & Technology 40:7512–7520.

Thompson, J. S., and J. A. Rice. 2013. The relative influence of thermal experience and forage availability on growth of age 1–5 striped bass in two southeastern reservoirs. Pages 93–120 *in* J. S. Bulak, C. C. Coutant, and J. A. Rice, editors. Biology and management of inland striped bass and hybrid striped bass. American Fisheries Society, Symposium 80, Bethesda, Maryland.

Van Horn, S. L. 2013. A brief history of inland striped bass management. Pages 1–13 *in* J. S. Bulak, C. C. Coutant, and J. A. Rice, editors. Biology and management of inland striped bass and hybrid striped bass. American Fisheries Society, Symposium 80, Bethesda, Maryland.

Weiss and Associates. 2004. Feasibility study for Calaveras Reservoir hypolimnetic oxygenation system. Prepared for City and County of San Francisco Public Utilities Commission, San Francisco.

Whitehead, J. C. 2013. Economics of recreational fisheries for inland striped bass and hybrid striped bass. Pages 551–558 *in* J. S. Bulak, C. C. Coutant, and J. A. Rice, editors. Biology and management of inland striped bass and hybrid striped bass. American Fisheries Society, Symposium 80, Bethesda, Maryland.

Wuest, A., N. Brooks, and D. Imboden. 1992. Bubble plume modeling for lake restoration. Water Resources Research 28:3235–3250.

American Fisheries Society Symposium 80:209–218, 2013

Comparisons of Reservoir Characteristics with Striped Bass Growth and Relative Weight in the Southeastern United States

Daniel M. Wilson* and Victor J. DiCenzo
Virginia Department of Game and Inland Fisheries
1132 Thomas Jefferson Road Forest, Virginia 24551, USA

John Odenkirk
Virginia Department of Game and Inland Fisheries
1320 Belman Road Fredericksburg, Virginia 22401, USA

Abstract.—Growth and relative weight (W_r) condition index were estimated for striped bass *Morone saxatilis* (ages 1–6) from 15 southeastern U.S. reservoirs spanning seven states. On average, striped bass reached quality length for a fishery (510 mm) in about 2.5 years. Predicted lengths up to age 6 were highly variable among reservoirs and overlapped among age-classes. Cooling degree-days above 15.5°C was negatively related with growth for ages 2–6 and, when combined with higher water level fluctuation, explained 47–56% of the variation for ages 2–3. For ages 4–6, the best models for growth included increased reservoir volume, reduced retention time, and fewer cooling degree-days, as predictors, explaining 69–74% of the variation. Correlations between growth and W_r were not significant. Relative weights for each reservoir were similar but declined with fish age. Significant relationships between W_r and reservoir age, elevation, mean depth, area, and water level fluctuation only occurred for some ages. In general, physical characteristics used in this study explained only modest amounts of the variation in growth and W_r for the reservoirs we investigated. Our results underscore the complexity of individual reservoir systems and differences among reservoirs when defining expectations for reservoir striped bass populations.

Introduction

Striped bass *Morone saxatilis* have been established in reservoirs across the United States through extensive stocking programs, particularly in southeastern U.S. systems (Van Horn 2013, this volume). Popular fisheries have resulted, producing related beneficial economic impacts (Bettoli 2013, this volume). However, physical reservoir characteristics, such as variable reservoir hydrology and basin morphology, may constrain these fisheries. Assessment of growth and relative weight (W_r) condition index is an important aspect of striped bass management, but little has been done to relate these indices to physical reservoir characteristics.

Most successful striped bass populations in reservoirs are sustained through stocking, but determining appropriate stocking rates and fishing regulations to maintain the fishery are confounded by the lack of available data describing the reservoir characteristics necessary to maintain healthy populations. Currently, condition and growth of reservoir striped bass are influenced primarily by thermal regimes, avail-

* Corresponding author: dan.wilson@dgif.virginia.gov

ability of suitable summer habitat, and forage (Coutant 1985; Mathews et al. 1985; Thompson et al. 2010; Thompson and Rice 2013, this volume). Vatland et al. (2008) found a positive relationship between striped bass condition and shad abundance in Lake Powell, and Ney (1993) highlighted the importance and challenges for fisheries managers to maintain a stable predator–prey relationship. However, due to the complexity of monitoring seasonally changing thermal regimes and prey populations and developing predator–prey relationships for specific reservoirs, a more easily measured alternative based on physical reservoir characteristics would be desirable for managers.

Identification of the physical and morphological factors related to W_r and growth may describe the potential of reservoirs to produce quality striped bass populations. Condition indices such as W_r have often been considered useful as an inexpensive and easily measured surrogate for growth and assessment of the general well-being of fish populations, but Anderson and Neumann (1996) cautioned that W_r may not be an accurate predictor of growth rate and can be constrained by season and fish size variability. Van Horn et al. (2001) were among the first to describe growth and condition of adult striped bass in one reservoir, but few studies have gone beyond this to compare other reservoir populations. Recent studies have focused on habitat selection, movement, and bioenergetics rather than comparative studies (e.g., Cyterski 1999; Jackson and Hightower 2001; Thompson et al. 2010). Some managers, however, face funding and personnel limitations to conduct similarly detailed studies on their popular and potentially controversial striped bass populations. Additionally, Ney (1993) cautioned that the extensive data demands and uncertain accuracy of bioenergetic models may constrain their utility to address many fish management questions. A less data-demanding middle ground is needed. Therefore, the objectives of this study were to (1) compare growth and W_r of reservoir striped bass in systems across the southeastern United States, (2) examine the relationships of striped bass growth and W_r to reservoir morphometric variables, and (3) determine if striped bass growth and W_r are related to each other.

Methods

We obtained data from 15 reservoirs from seven states within the southeastern United States: Virginia, North Carolina, South Carolina, Georgia, Kentucky, Oklahoma, and Texas. Biologists from additional states were contacted, but they did not have sufficient data available for inclusion in this study. Fisheries data requested for each reservoir were striped bass lengths (mm total length [TL]), ages (derived from otoliths), weights (g), and collection dates for each fish. In 14 of the reservoirs, striped bass were collected with gill nets from October to March; in Hartwell Lake (Georgia, South Carolina), data were collected with electrofishing in March, and some fish from Kerr Reservoir were collected in April–May with electrofishing.

For each reservoir, we also requested morphological data that were readily available and commonly considered important descriptors of reservoirs in other studies (e.g., DiCenzo et al. 1995; Maceina 2003), including age, elevation, area, volume (surface area × mean depth), mean annual discharge, retention time (volume/mean annual discharge), mean depth, maximum depth, and mean annual water fluctuation (Table 1). Annual accumulated cooling degree-days (CDD; a measure of how much, and for how long, air temperature was higher than a specific temperature) were determined from the Web site www.degreedays.net with a base temperature of 15.5°C at the nearest weather station of each reservoir. Cooling degree-days were determined using only days with average air temperatures greater than 15.5°C and calculated by summing, for a 12-month period, each day's average temperature minus 15.5. These data were used as a representative of warm air temperature and possible indicator of summer habitat constraints. Finally, biologists subjectively classified reservoirs as eutrophic or oligo-mesotrophic.

Only reservoirs that had data sets with a

TABLE 1. Summary statistics for abiotic variables from 15 southeastern U.S. striped bass reservoirs.

Variable	Mean	Minimum	Maximum
Reservoir age (years)	57	28	82
Elevation (m)	177	23	567
Area (ha)	16,142	1,323	69,203
Volume (ha/m)	169,054	11,248	488,053
Retention time (d)	219	8	640
Mean depth (m)	11.3	4.9	24.0
Maximum depth (m)	36.4	20.1	63.7
Annual fluctuation (m)	2.7	0.0	15.0
Cooling degree-days (d)	1,144	833	1,516

minimum of 180 striped bass collected since 2000 were included in the database, other reservoirs submitted lacked data for ages 1–6, and most age data collected prior to 2000 were derived from scales. A total of 14,927 fish were included in the study. We analyzed data from fish ages 1–6 because they were the only ages with data available for all reservoirs.

Relative weight of striped bass was described with the relative weight index (W_r), using the standard weight (W_s) equation from Anderson and Neumann (1996): $\log_{10}(W_s) = -4.924 + 3.007 \cdot \log_{10}(\text{TL})$.

To develop length-at-age data, each fish was assigned a fractional age based on the month of collection to account for variable collection periods; annulus formation for all fish was assigned to May 1. For example, fish with six annuli collected in November was assigned an age of 6.5. The equation TL = slope[$\log_e$(age)] + intercept was developed for each reservoir to describe length at age and allow comparisons between reservoirs. Stock, quality, and preferred lengths are from Gabelhouse (1984) and refer to lengths used for computing stock density indices.

We used stepwise multiple regression to characterize the relationships between mean length and W_r at age with abiotic variables. Spearmans correlation was used to assess relations between W_r of age-1 and age-6 fish. Pearson's product moment correlation analysis was used to assess relationships between mean length at age and relative weight. We used analysis of variance to test for differences in mean length at age and mean W_r between oligo-mesotrophic and eutrophic reservoirs. Statistical significance was set at $P \leq 0.10$. There was no seasonal variability considered for W_r assessments.

Results

The length-at-age relations estimated from fish ages 1–6 were generally robust with coefficients of determination ranging from 0.64 to 0.94 (Table 2). Predicted length at age varied considerably among reservoirs (Table 3). Except for Smith Mountain Lake, all reservoirs that had above average length at age 3 continued to have above average lengths through age 6. Length at age 1 was not a consistent predictor of length at ages 3–6.

Predicted lengths up to age 6 varied greatly among reservoirs, overlapped among age-classes, and became more variable with fish age (Table 3). On average, striped bass reached stock length (300 mm TL) in less than 1 year and quality length (510 mm TL) in about 2.5 years. Only one reservoir reached preferred length (760 mm TL) by age 6 (Lake Cumberland).

Mean relative weight of striped bass across all reservoirs was 96 for age-1 fish and generally decreased for each subsequent age-class (Table 4). Relative weights for age-1 fish were negatively correlated with W_r values for age-6 fish (Spearmans correlation, $r = -0.82$, $P < 0.01$). For 11 of the 15 reservoirs, mean W_r declined

TABLE 2. Growth models used to describe striped bass length-at-age data for each reservoir with the following equation y = slope($\log_e x$) + intercept; y is fish length (total length, mm) and x is fish age (years). The number of striped bass for each reservoir is denoted by N. Data were sent from the state listed.

Reservoir (state)	Equation	R^2	N
Lake Anna (Virginia)	$y = 175.68 \log_e x + 304.60$	0.85	267
Lake Buchanan (Texas)	$y = 181.25 \log_e x + 337.89$	0.85	350
Claytor Lake (Virginia)	$y = 211.86 \log_e x + 336.34$	0.91	579
Lake Cumberland (Kentucky)	$y = 237.80 \log_e x + 354.14$	0.90	731
Lake Gaston (North Carolina)	$y = 164.81 \log_e x + 348.37$	0.91	421
Hartwell Lake (Georgia)	$y = 233.27 \log_e x + 322.85$	0.93	301
Jordan Lake (North Carolina)	$y = 213.87 \log_e x + 323.80$	0.92	633
Kerr Reservoir (Virginia)	$y = 174.47 \log_e x + 367.71$	0.94	1,267
Leesville Lake (Virginia)	$y = 218.61 \log_e x + 334.20$	0.89	184
Lake Marion (South Carolina)	$y = 181.19 \log_e x + 353.31$	0.90	1,476
Lake Murray (South Carolina)	$y = 151.17 \log_e x + 339.81$	0.76	1,273
Lake Norman (North Carolina)	$y = 97.66 \log_e x + 422.78$	0.64	805
Smith Mountain Lake (Virginia)	$y = 182.01 \log_e x + 346.61$	0.88	4,167
Lake Texoma (Oklahoma)	$y = 176.34 \log_e x + 282.32$	0.76	2,183
Lake Tillery (North Carolina)	$y = 192.83 \log_e x + 349.05$	0.89	290

significantly between ages 1 and 6 (Table 4). Only two reservoirs (Marion and Murray) had Wr values of 100 or more for all six age-classes; in contrast, mean W_r values of age-3 and older fish did not exceed 90 in eight reservoirs, and in three of these, mean W_r did not exceed 90 for any age-class. Thus, in reservoirs where mean W_r was low for younger striped bass, it tended to be low for older fish as well; likewise, where W_r was high for younger fish, it generally remained relatively high for older age-classes.

Cooling degree-days was the most consistent predictor of mean length at age (Table 5). Annual water level fluctuation and cooling degree-days were significant regressors for striped bass mean length at ages 2 and 3, explaining 47% and 56% of the variation, respectively. For striped bass ages 4, 5 and 6, volume, retention time, and cooling degree-days were significant regressors and explained 74, 71, and 69% of the variation. None of the nine abiotic variables we evaluated were a significant predictor of striped bass mean length at age 1.

Stepwise regressions explained 50–79% of the variation in mean W_r by age-class. However, no environmental variable consistently predicted condition factor (Table 6).

Length at age and W_r were generally unrelated (Table 7). Mean length and W_r of age-2 striped bass were negatively correlated (r = –0.45). For all other comparisons, there was no significant correlation between growth and W_r.

Biologists classified 10 reservoirs as oligo-mesotrophic and 5 as eutrophic (Lake Anna, Jordon Lake, Kerr Reservoir, Lake Marion, and Lake Texoma). For striped bass ages 1–6, there was no difference in mean W_r between these two trophic states. Striped bass mean lengths at ages 1–6 in oligo-mesotrophic reservoirs and eutrophic reservoirs were also similar ($P > 0.10$).

Discussion

This study was the first, to our knowledge, to document striped bass growth and W_r on a regional scale. While the findings for growth and W_r were highly variable among reservoirs, the results provide fisheries managers with descriptive statistics of growth and W_r. Understanding distinctions between populations showing dif-

TABLE 3. Predicted length-at-age data for striped bass from 15 southeastern U.S. reservoirs. Standard deviation was computed from reservoir means during the months specified. Data was sent from the state listed and ranked by length at age 3. Bolded reservoirs had above average growth by age-class. A score was assigned to each reservoir based on the number of times that predicted length exceeded mean length at age.

	Length (mm) at age						
	1	2	3	4	5	6	Score
Lake Texoma[e] (Oklahoma; Feb)	282	405	476	527	566	598	0
Lake Anna[e] (Virginia; Nov)	305	426	498	548	587	618	0
Lake Murray[o] (South Carolina; Nov–Dec)	340	445	506	549	583	611	0
Lake Gaston[o] (North Carolina; Dec–Jan)	348	463	529	577	614	644	1
Lake Norman[o] (North Carolina; Dec–Jan)	423	490	530	558	580	598	2
Lake Buchanan[o] (Texas; Jan–Mar)	338	464	537	589	630	663	0
Smith Mountain Lake[o] (Virginia; Oct–Dec)	347	473	547	599	640	673	3
Lake Marion[e] (South Carolina; Dec–Feb)	353	479	552	604	645	678	6
Jordon Lake[e] (North Carolina; Dec–Jan)	324	472	559	620	668	707	5
Kerr Reservoir[e] (Virginia; April–May, Oct–Dec)	360	486	559	611	652	685	6
Lake Tillery[o] (North Carolina; Dec–Jan)	349	483	561	616	659	695	6
Claytor Lake[o] (Virginia; Oct–Dec)	336	483	569	630	677	714	5
Leesville Lake[o] (Virginia; Oct–Dec)	334	486	574	637	686	726	5
Hartwell Lake[o] (Georgia; Mar)	323	485	575	646	698	741	5
Lake Cumberland[o] (Kentucky; Nov–Dec)	354	519	615	684	737	780	6
Mean	342	471	546	600	641	675	
25% quartile	323	451	525	558	591	618	
75% quartile	349	485	568	630	686	726	
SD	30.9	27.1	33.6	41.0	47.2	52.5	

[o] Reservoir was classified as oligo-mesotrophic.

[e] Reservoir was classified as eutrophic.

ferent growth and W_r and their connection to each reservoir's individual set of conditions can lead to better fisheries management decisions, including harvest regulations and stocking rates (Van Horn et al. 2001).

Intuitively, managers realize all reservoirs are different, and even the same reservoir may produce different habitat conditions from year to year (e.g., Sammons and Bettoli 2000). Since striped bass fisheries of most southeastern U.S. reservoirs are maintained by stocking (Axon and Whitehurst 1985), recruitment mechanisms are less important than physical features that may dictate habitat conditions and fish responses to them. For example, temperature and dissolved oxygen regimes may affect adult striped bass survival, condition, and growth (Schaffler et al. 2002; Thompson et al. 2010). These and other factors such as overfishing or stocking success, combined with forage availability (Raborn et al. 2007), ultimately dictate the success or failure of a given fishery. Our attempts to correlate physical reservoir features with striped bass growth and W_r provide further insight into mechanisms impacting these fisheries, even though the standard physical characteristics examined are not fully defining.

The growth rate of an individual fish is highly pliant, species- and size-specific response to habitat conditions and food availability (Brandt

TABLE 4. Relative weights (W_r) of striped bass from 15 southeastern U.S. reservoirs. Standard deviation was computed from reservoir means during the months specified. Data were sent from the state listed and ranked by W_r at age 3. Slopes are from regression of mean W_r (dependant variable) and age (independent variable). Asterisks denote significance ($P < 0.10$).

	Mean W_r for each age						
	1	2	3	4	5	6	Slope
Lake Buchanan[o] (Texas; Jan–Mar)	96	92	81	79	81	75	–4.00*
Lake Norman[o] (North Carolina; Dec–Jan)	85	83	81	79	76	79	–1.51*
Lake Anna[e] (Virginia ; Nov)	88	89	82	76	76	74	–3.29*
Lake Cumberland[o] (Kentucky; Nov–Dec)	93	86	83	82	82	79	–2.37*
Smith Mountain Lake[o] (Virginia; Oct–Dec)	90	84	83	81	79	77	–2.34*
Leesville Lake[o] (Virginia; Oct–Dec)	92	83	85	84	86	80	–1.49
Hartwell Lake[o] (Georgia; Mar)	94	85	87	80	86	87	–0.69
Claytor Lake[o] (Virginia; Oct–Dec)	95	95	90	85	81	82	–3.20*
Kerr Reservoir[e] (Virginia; April–May, Oct–Dec)	97	88	92	91	88	87	–1.46*
Jordon Lake[e] (North Carolina; Dec–Jan)	100	97	94	94	92	91	–1.71*
Lake Tillery[o] (North Carolina; Dec–Jan)	100	98	96	93	91	89	–2.26*
Lake Gaston[o] (North Carolina; Dec–Jan)	96	99	97	99	99	89	–0.94
Lake Texoma[e] (Oklahoma; Feb)	96	101	97	97	86	88	–2.43*
Lake Marion[e] (South Carolina; Dec–Feb)	106	106	106	110	101	103	–0.74
Lake Murray[o] (South Carolina; Nov–Dec)	118	115	106	108	105	100	–3.46*
Mean	96	93	91	89	87	85	
25% quartile	91	85	83	80	81	79	
75% quartile	100	98	96	93	91	89	
SD	8.0	9.4	8.9	10.7	8.9	8.5	

[o] Reservoir was classified as oligo-mesotrophic.
[e] Reservoir was classified as eutrophic.

Table 5. Stepwise regression results for striped bass mean length at ages 1–6 and reservoir abiotic variables. Variables included in significant models were annual water fluctuation (AWF), cooling degree-days (CDD), retention time (RT), and volume (VOL). Significance was set at $P < 0.10$. TL = total length.

Age	Model	R^2
1	No significant model	–
2	TL = 3.207(AWF) – 0.049(CDD) + 536.848	0.47
3	TL = 3.640(AWF) – 0.076(CDD) + 652.646	0.56
4	TL = 0.000(VOL) – 0.089(RT) – 0.127(CDD) + 787.932	0.74
5	TL = 0.000(VOL) – 0.098(RT) – 0.146(CDD) + 856.540	0.71
6	TL = 0.000(VOL) – 0.105(RT) – 0.160 (CDD) + 910.609	0.69

TABLE 6. Stepwise regression results for striped bass mean relative weight (W_r) at ages 1–6 as a function of reservoir abiotic variables. Variables included in significant models were reservoir age (RA), elevation (ELV), mean depth (MD), area (ARE), and annual water fluctuation (AWF). Significance was set at $P < 0.10$.

Age	Model	R^2
1	W_r = 0.318(RA) – 0.007(ELV) + 82.407	0.58
2	W_r = 0.365(RA) – 0.714(MD) + 81.195	0.58
3	W_r = 0.349(RA) – 0.812(MD) + 81.165	0.58
4	W_r = 0.000(ARE) – 0.991(AWF) + 87.744	0.52
5	W_r = 0.274(RA) – 0.012(ELV) + 78.853	0.51
6	W_r = 0.000(ARE) – 0.734(MD) + 90.111	0.79

and Kirsch 1993). Our finding that cooling degree-days was a negative predictor of striped bass growth is consistent with the literature, which is replete with cases of thermal impacts limiting striped bass populations (e.g., Coutant 1985, 2013, this volume; Zale et al. 1990; McDaniel et al. 1993). Bettoli (2005) recently stated that the fundamental thermal niche of adult landlocked striped bass may be lower than previously believed, which could explain bioenergetic stresses in warmer water, especially for large fish. Increased cooling degree-days was likely related to summer water temperatures and associated with reduced striped bass growth in this study; however, adequate forage may still be the most important variable in the success of a given population with moderately acceptable thermal regimes (Thompson et al. 2010; Thompson and Rice 2013, this volume).

Reservoirs with faster growth had more dramatic water level fluctuations, greater volume, and shorter retention time combined with cooler summers and may have inherent productivity gradients or other factors (such as stratification patterns) more likely to provide optimal striped bass habitat. Cooling degree-days for ages 2–6 appeared as significant stepwise regression results reinforcing Coutant's (1985) emphasis on the importance of summer temperatures. Increased water level fluctuation was an important variable for young fish, possibly due to factors not measured in this study. Our data also suggests that older fish grew faster in reservoirs with greater volume and reduced retention time, possibly due to primary productivity and/or greater amounts of available preferred habitat. However, Matthews (1985) found some combinations of these extremes resulted in a higher frequency of adult striped bass mortality. Larger reservoirs with greater volume may have tempered the ex-

Table 7. Pearson's product-moment correlation coefficients comparing striped bass length at age with relative weight (W_r). Asterisks denote significance at $P < 0.10$.

	Mean W_r					
Length at	Age 1	Age 2	Age 3	Age 4	Age 5	Age 6
Age 1	–0.15	–0.19	–0.17	–0.15	–0.14	–0.12
Age 2	–0.26	–0.45*	–0.44	–0.41	–0.39	–0.37
Age 3	–0.16	–0.29	–0.29	–0.28	–0.26	–0.25
Age 4	–0.14	–0.21	–0.21	–0.20	–0.18	–0.17
Age 5	–0.08	–0.10	–0.10	–0.09	–0.08	–0.07
Age 6	–0.19	–0.19	–0.14	–0.10	–0.08	–0.06

tremes of these conditions responsible for reducing fish health and ultimately growth.

Just as Van Horn et al. (2001) found for Lake Norman, we found that in the southeastern United States, W_r of striped bass declined with age. While the results were not consistent for all ages, there was a suggestion that higher relative weights of striped bass may be associated with larger and older reservoirs with shallower depths located at lower elevations. Reservoir size may have been related to trophic status from increased nutrient input and were probably more likely to contain adequate forage than less productive systems. Thompson et al. (2010) recently found that warm, productive reservoirs without permanent thermal refuges can provide suitable conditions for quality striped bass growth and condition from age 1 to age 5 due to abundant and diverse forage. While reservoirs with greater mean surface areas had higher summer mortality of striped bass, Matthews (1985) cautioned that mortality and poor habitat were widespread geographically in reservoirs of the United States and highly variable.

Growth and W_r were unrelated, even though it was hypothesized that these two variables would be positively correlated. Brown and Murphy (1991) found that W_r was positively correlated with growth of juvenile striped bass in aquaria. Anderson and Neumann (1996) suggested that relative weight is most likely to be a predictor of growth when data are collected during the period of most rapid growth. However, striped bass growth occurs in spurts and the collection of W_r information may not coincide with strong growth periods (Thompson and Rice 2013, this volume). Additionally, Moss (1985) found that relative weights may be depressed during summer months and could have affected some early fall collections, but Haeseker et al. (1996) reported normal values by fall after summer lows. However, if forage is limiting, striped bass W_r may be poor even during winter months (Van Horn et al. 2001).

Relative weights computed in this study were consistently less than 100. This may indicate that the W_s equation may not accurately describe the condition and potential of a striped bass population in southeastern U.S. reservoirs. Managers should evaluate W_r data cautiously and not have preconceived notions about a given system based on general reservoir morphology but should endeavor to collect needed biological and physical data in efforts to elucidate the relationships between striped bass and habitat, as described in other chapters of this book (Coutant 2013; Rice et al. 2013; Ruane et al. 2013; Thompson and Rice 2013; all this volume).

This study and others cited indicate that use of standard reservoir morphological and trophic features to predict striped bass growth and condition has important limitations. We did show that, regionally, striped bass growth and W_r may be enhanced in larger reservoirs with characteristics including cooler summer temperatures and shorter retention time over systems having the opposite extremes of these variables. While not a revelation, the limitations of our study serve as a reminder that each system is extraordinarily complex and influenced by a host of abiotic and biotic variables—many of which were unmeasured in the current study.

Acknowledgments

We thank the biologists from Georgia, Kentucky, North Carolina, Oklahoma, South Carolina, and Texas who collected and provided data on fish and reservoirs. We also thank Dave Steffen for his statistical review and assistance. This manuscript benefited from the reviews of Scott Smith and two anonymous reviewers.

References

Anderson, R. O., and R. M. Neumann. 1996. Length, weight, and associated structural indices. Pages 447–482 *in* B. R. Murphy and D. W. Willis, editors. Fisheries techniques, 2nd edition. American Fisheries Society, Bethesda, Maryland.

Axon, J. R., and D. H. Whitehurst. 1985. Striped bass management in lakes with emphasis on management problems. Transactions of the American Fisheries Society 114:8–11.

Bettoli, P. W. 2005. The fundamental thermal niche of adult landlocked striped bass. Transactions of the American Fisheries Society 134:305–314.

Bettoli, P. W. 2013. Challenges and opportunities in studying and managing striped bass and its hybrids in the 21st century. Pages 575–588 *in* J. S. Bulak, C. C. Coutant, and J. A. Rice, editors. Biology and management of inland striped bass and hybrid striped bass. American Fisheries Society, Symposium 80, Bethesda, Maryland.

Brandt, S. B., and J. Kirsch. 1993. Spatially explicit models of striped bass growth potential in Chesapeake Bay. Transactions of the American Fisheries Society 122:845–869.

Brown, M. L., and B. R. Murphy. 1991. Relationship of relative weight (W_r) to proximate composition of juvenile striped bass and hybrid striped bass. Transactions of the American Fisheries Society 120:509–518.

Coutant, C. C. 1985. Striped bass, temperature, and dissolved oxygen: a speculative hypothesis for environmental risk. Transactions of the American Fisheries Society 114:31–61.

Coutant, C. C. 2013. When is habitat limiting for striped bass? Three decades of testing the temperature–oxygen squeeze hypothesis. Pages 65–91 *in* J. S. Bulak, C. C. Coutant, and J. A. Rice, editors. Biology and management of inland striped bass and hybrid striped bass. American Fisheries Society, Symposium 80, Bethesda, Maryland.

Cyterski, M. J. 1999. Analysis of the trophic support capacity of Smith Mountain Lake, Virginia, for piscivorous fish. Doctoral dissertation. Virginia Polytechnic Institute and State University, Blacksburg.

DiCenzo, V. J., M. J. Maceina, and W. C. Reeves. 1995. Factors related to growth and condition of the Alabama subspecies of spotted bass in reservoirs. North American Journal of Fisheries Management 15:794–798.

Gabelhouse, D. W., Jr. 1984. A length-categorization system to assess fish stocks. North American Journal of Fisheries Management 4:273–285.

Haeseker, S. L., J. T. Carmichael, and J. E. Hightower. 1996. Summer distribution and condition of striped bass within Albemarle Sound, North Carolina. Transactions of the American Fisheries Society 125:690–704.

Jackson, J. R., J. E. Hightower. 2001. Reservoir striped bass movements and site fidelity in relation to seasonal patterns in habitat quality. North American Journal of Fisheries Management 21:34–45.

Maceina, M. J. 2003. Verification of the influence of hydrologic factors on crappie recruitment in Alabama reservoirs. North American Journal of Fisheries Management 23:470–480.

Mathews, W. J. 1985. Summer mortality of striped bass in reservoirs of the United States. Transactions of the American Fisheries Society 114:62–66.

Matthews, W. J., L. G. Hill, and S. M. Schellhaass. 1985. Depth distribution of striped bass and other fish in Lake Texoma (Oklahoma–Texas) during summer stratification. Transactions of the American Fisheries Society 114:84–91.

McDaniel, C. K., L. E. Snyder, and L. L. Conner. 1993. Impacts of thermal stress on the condition of striped bass. Proceedings of the Annual Conference Southeastern Association of Fish and Wildlife Agencies 45(1991):361–369.

Moss, J. L. 1985. Summer selection of thermal refuges by striped bass in Alabama reservoirs and tailwaters. Transactions of the American Fisheries Society 114:77–83.

Ney, J. J. 1993. Bioenergetics modeling today: growing pains on the cutting edge. Transactions of the American Fisheries Society 122:736–748.

Raborn, S. W., L. E. Miranda, and M. T. Driscoll. 2007. Prey supply and predator demand in a reservoir of the southeastern United States. Transactions of the American Fisheries Society 136:12–23.

Rice, J. A., J. S. Thompson, J. A. Sykes, and C. T. Waters. 2013. The role of metalimnetic hypoxia in striped bass summer kills: consequences and management implications. Pages 121–145 *in* J. S. Bulak, C. C. Coutant, and J. A. Rice, editors. Biology and management of inland striped bass and hybrid striped bass. American Fisheries Society, Symposium 80, Bethesda, Maryland.

Ruane, R. J., G. E. Hauser, and A. F. Sawyer. 2013. Water qualtiy modeling as a management tool for striped bass and hybrid striped bass.

Pages 161–179 *in* J. S. Bulak, C. C. Coutant, and J. A. Rice, editors. Biology and management of inland striped bass and striped bass hybrids. American Fisheries Society, Symposium 80, Bethesda, Maryland.

Schaffler, J. J., J. J. Isely, and W. E. Hayes. 2002. Habitat use by striped bass in relation to seasonal changes in water quality in a southern reservoir. Transactions of the American Fisheries Society 131:817–827.

Sammons, S. M., and P. W. Bettoli. 2000. Population dynamics of a reservoir sport fish community in response to hydrology. North American Journal of Fisheries Management 20:791–800.

Thompson, J. S., and J. A. Rice. 2013. The relative influence of temperature and forage availability on growth of age 1–5 striped bass in two southeastern reservoirs. Pages 93–120 *in* J. S. Bulak, C. C. Coutant, and J. A. Rice, editors. Biology and management of inland striped bass and hybrid striped bass. American Fisheries Society, Symposium 80, Bethesda, Maryland.

Thompson, J. S., J. A. Rice, and D. S. Waters. 2010. Striped bass habitat selection rules in reservoirs without suitable summer habitat offer insight into consequences for growth. Transactions of the American Fisheries Society 139:1450–1464.

Van Horn, S. L. 2013. A brief history of inland striped bass management. Pages 1–13 *in* J. S. Bulak, C. C. Coutant, and J. A. Rice, editors. Biology and management of inland striped bass and hybrid striped bass. American Fisheries Society, Symposium 80, Bethesda, Maryland.

Van Horn, S. L., B. K. Baker, and M. Rash. 2001. Growth and condition response of Lake Norman striped bass to increase stocking rates and more restrictive harvest regulations. Proceedings of the Annual Conference Southeastern Association of Fish and Wildlife Agencies 53(1999):193–199.

Vatland, S., P. Budy, and G. P. Thiede. 2008. A bioenergetics approach to modeling striped bass and threadfin shad predator–prey dynamics in Lake Powell, Utah–Arizona. Transactions of the American Fisheries Society 137:262–277.

Zale, A. V., J. D. Wiechman, R. L. Lochmiller, and J. Burroughs. 1990. Limnological conditions associated with summer mortality of striped bass in Keystone Reservoir, Oklahoma. Transactions of the American Fisheries Society 119:72–76.

American Fisheries Society Symposium 80:219–227, 2013

Comparisons of Growth for Hybrid Striped Bass in North America

Randall D. Schultz* and Andy L. Fowler
Iowa Department of Natural Resources
24570 U.S. Highway 34, Chariton, Iowa 50049, USA

Jason M. Goeckler
Kansas Department of Wildlife, Parks and Tourism
1820 Merchant Street, Emporia, Kansas 66801, USA

Michael C. Quist
U.S. Geological Survey – Idaho Cooperative Fish and Wildlife Research Unit
Department of Fish and Wildlife Resources, University of Idaho
Moscow, Idaho 83844, USA

Abstract.—Growth of hybrid striped bass (white bass *Morone chrysops* × striped bass *M. saxatilis*) throughout North America was summarized to evaluate latitudinal differences in growth. Age was estimated from scales and otoliths (nondifferentiated) collected from 29 populations in 12 states. Hybrid striped bass populations were delineated by midwestern, southeastern, and southwestern regions. Growth among regions was compared by fitting a von Bertalanffy growth model to each population and by comparing mean length at capture (fall-sampled fish) for ages 1 and 3 and maximum ages. Midwestern populations exemplified the highest theoretical maximum length (L_∞), followed by southeastern populations, although differences were not significant among regions. Likewise, growth coefficients (K) and maximum ages did not differ among regions. Southeastern populations had greater length at age than midwestern populations but were similar to southwestern values. These results provide a framework for comparing North American hybrid striped bass populations and for managing this important sport fish in reservoir systems.

Introduction

Many states have stocked hybrid striped bass (male white bass *Morone chrysops* × female striped bass *M. saxatilis*) (also referred to as original cross or palmetto bass) due to its popularity among anglers. In particular, hybrid striped bass exhibit fast growth, are aggressive, are considered to have superior fighting abilities, and are capable of reaching large sizes and providing a trophy fishery (Jahn et al. 1987). Additionally, biologists have been interested in using this species to restructure stunted panfish populations (Layzer and Clady 1984; Jahn et al. 1987; Neal et al. 1999; Hutt et al. 2008), control invasive species (Iowa Department of Natural Resources, unpublished data), and manage abundant gizzard shad *Dorosoma cepedianum* populations (Dettmers et al. 1996). Although hybrid striped bass continue to expand their distribution through movement and additional

* Corresponding author: randy.schultz@dnr.iowa.gov

stockings, little is known about growth of hybrid striped bass.

An understanding of growth is important for properly managing sport fish and prey populations. Growth is allied with environmental conditions (e.g., temperature and water quality; Miranda and Durocher 1986; Woiwode and Adelman 1991; McInerny and Cross 1999; Quist et al. 2003; Wuellner et al. 2010), prey availability (Dettmers et al. 1996; Olson et al. 2007; Thompson and Rice 2013, this volume) and genetic factors (Avise and Van Den Avyle 1984; Leitner et al. 2002), making it a useful metric to evaluate habitat suitability, prey availability, or the effect of management strategies targeting sport fish and their prey. In addition, growth assessments are commonly used to assess potential problems (e.g., overfishing; Schultz and Dodd 2008) and management actions, such as regulations (Paukert et al. 2007), lake renovations (Heman et al. 1969), and prey stockings (Noble 1981).

Hybrid striped bass populations have been established across a range of latitudes and climatic zones, primarily in the midwestern and southern United States. Their habitat ranges from large, deep impoundments (Kilpatrick and Ney 2013, this volume) to shallow urban lakes (Schultz and Dodd 2008) and ponds (Neal et al. 1999). However, stockings have had mixed success, dependent upon habitat quality (Ruane et al. 2013, this volume) and prey size (Crandall 1979; Dennerline and Van Den Avyle 2000) and quantity (Ott and Malvestuto 1984; Dennerline and Van Den Avyle 2000). Previous research has demonstrated that fish in southern latitudes generally have greater growth but less longevity than fish in northern latitudes (Colby et al. 1979). Beverton (1987) compared walleye *Sander vitreus* populations along a latitudinal gradient from Canada to Texas and showed that southern walleye populations grew faster, matured later, but died earlier than more northern populations. Although latitudinal differences in temperature and growing season are known to influence walleye population dynamics (Quist et al. 2003), fish assemblage structure also differs with latitude (Hansen et al. 2010), confounding latitudinal effects on sport fish populations, through differences in prey availability (Santucci and Wahl 1993). For example, southern aquatic systems have higher species diversity, potentially increasing the number and abundance of available prey species (Quist et al. 2003). Likewise, latitudinal differences in growing season, temperature, and prey composition may have similar effects on hybrid striped bass populations, but this theory has not been tested.

Despite the effort and resources expended to stock hybrid striped bass, their importance to anglers, and their potential role in restructuring prey communities, little is known about how their growth compares throughout their distribution (Carlander 1997). Therefore, the objective of this study was to gain an understanding of hybrid striped bass growth across its distribution, with emphasis on latitudinal differences. Specifically, we compared regional growth by fitting a von Bertalanffy growth model to each population and comparing mean length at ages 1 and 3 and maximum ages for 29 populations in 12 states to provide a framework for comparing growth among North American hybrid striped bass populations.

Methods

Age and growth data on hybrid striped bass (fall sampled; length at age) were solicited from state natural resource agencies throughout the geographical distribution of hybrid striped bass. Growth data were based on ages derived from both scales and otoliths. Although some structures may provide biased age estimates depending on latitude or age (DeVries and Frie 1996), it was assumed that growth data provided by agencies utilized structures that provided relatively accurate estimates of age for their region (Maceina et al. 2007). Populations were not differentiated based on sex. In addition, few biologists responded with data on both palmetto bass (male white bass × female striped bass) and sunshine bass (female white bass × male striped bass). The number of sunshine bass populations (N = 4) was relatively low, and therefore this cross was not included in any analyses. Hybrid striped bass populations were separated into one of three regions: (1) midwestern populations (Illinois, Indiana, Kansas, Nebraska, and Ohio), (2) southeastern populations (Georgia, Kentucky, North Carolina, Tennessee, and Virginia), and (3) southwestern populations (Oklahoma and Texas) (Figure 1). Data were truncated to only include age-8 and younger fish because few populations contained fish older than 8 years old, but also to reduce potential errors in aging by scales and to minimize the effects that a few, long-lived, slow growing populations might have on the overall analysis. We limited the number of lakes within each region to no more than 10 to minimize the influence of any one region (Jackson et al. 2008). Lakes were chosen based on the largest sample sizes, after first ensuring that all possible states were included within a region. When growth data were provided for more than 1 year for a particular lake, the study with the largest sample size was chosen. Our objective was to compare growth of hybrid striped bass in the regions. This was done by first fitting a von Bertalanffy growth model to each population. We used the model

$$L_{age} = L_{\infty}[1 - e^{-K(age - t_0)}],$$

where L_{age} is the mean length at age, L_{∞} is the theoretical maximum length, K is the growth coefficient, and t_0 is the theoretical age when length equals zero. Growth models were fit to L_{age} data using nonlinear regression techniques (PROC NLIN) in the Statistical Analysis System (SAS; Freund and Littell 1991) following methods described in Quist et al. (2003). Analysis of variance (ANOVA) was used to determine differences in growth variables (L_{∞}, K, mean length at ages 1 and 3, and maximum age) among hybrid striped bass populations by region. Results were considered significantly different at $P < 0.05$.

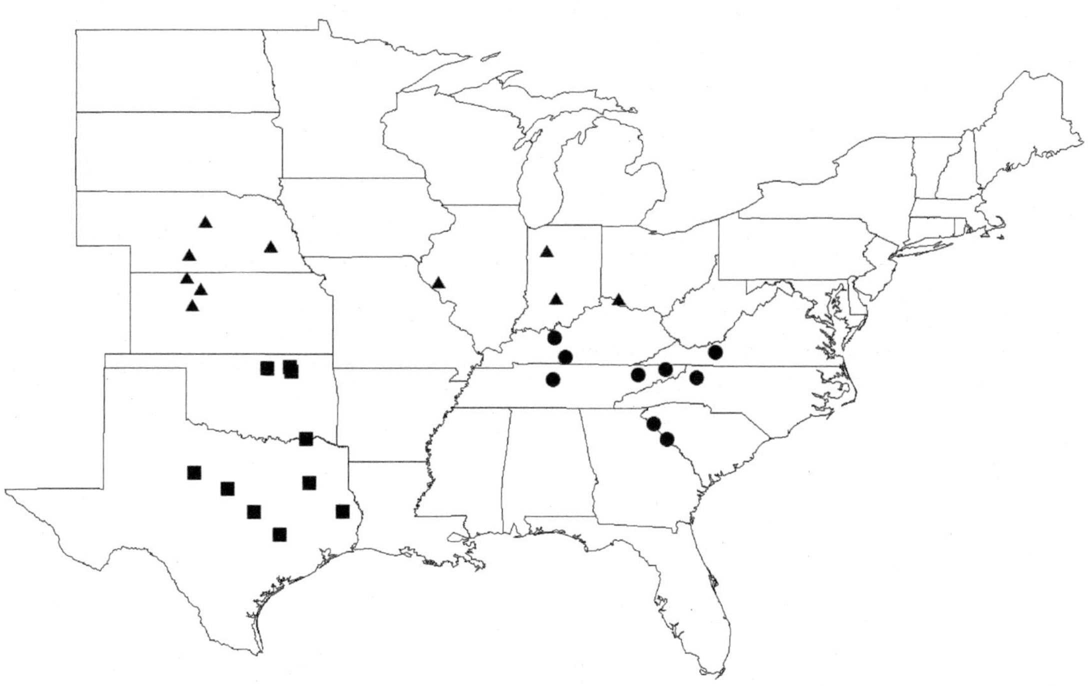

FIGURE 1. Water bodies used in current study. Midwestern lakes are identified by triangles, southeastern lakes by circles, and southwestern lakes by squares.

Results

Growth data were obtained from 29 populations representing 12 states (Table 1). The number of populations from individual states varied from 1 to 31, with Nebraska having the largest number of midwestern sites (N = 14 populations), Texas having the largest number of southwestern sites (N = 31 populations), and Tennessee contributing three sites for the southeastern region. Less than one-third of the sites selected for analyses had hybrid striped bass greater than 8 years old.

Standard growth models (Figure 2) indicated that midwestern populations had the highest L_∞, followed by southeastern populations, although differences were not significant among regions (ANOVA; F = 2.28, df = 2, 28; P = 0.12). However, growth coefficients (K) were highest in the southern regions, although differences were not significant (ANOVA; F = 1.93; df = 2, 28; P = 0.17). Southwestern populations had the oldest fish (8.4 years ± 1.6 SD), followed by southeastern populations (8.2 years ± 1.9 SD). Maximum age for midwestern populations was 7.3 years ± 1.2 SD. Maximum age did not differ among the three regions (ANOVA; F = 1.45; df = 2, 28; P = 0.25). Mean lengths at ages 1 and 3 were highest for southeastern populations (age 1 = 358 mm total length [TL]; age 3 = 530 mm TL), followed by southwestern populations (age 1 = 327 mm TL; age 3 = 497 mm TL). Midwestern populations had mean lengths at age 1 equaling 289 mm; mean lengths at age 3 were 475 mm. Significant differences for mean lengths at age occurred between southeastern and midwestern populations (ANOVA; $F > 3.52$; df = 2, 28; $P < 0.05$; Table 2).

Discussion

Age estimates of hybrid striped bass used in this study contained a mixture of scales and otoliths. Because concern exists over whether scales accurately portray maximum ages and, therefore, growth at older ages, we truncated the ages that were included in the analysis to 8 years and younger. Additionally, this allowed us to minimize the effects that a few, long-lived, slow-growing populations might have on the overall analysis. Welch et al. (1993) showed that age estimates are more precise for striped bass otoliths than scales, but precision is similar between otoliths and scales for white bass (Soupir et al. 1997). Unfortunately, precision between otoliths and scales for hybrid striped bass has not been evaluated. Even though different structures were used to obtain age estimates, this should not be problematic. For example, Jackson et al. (2008) found that growth standards did not differ among structures for nine North American species.

Several studies have shown that fish from northern latitudes demonstrate slower growth but greater longevity (Colby et al. 1979; Graham 1999; Quist et al. 2003). This phenomenon has been linked to an increased growing season, warmer water, and a higher diversity of prey in southern latitudes. Although our initial expectation was that hybrid striped bass populations would exhibit latitudinal differences in growth, our study did not fully support this generalization. We did not observe a difference in theoretical maximum length nor maximum ages between midwestern, southeastern, and southwestern populations of hybrid striped bass. However, it is known that the apparent maximum age of a population may be influenced by the extent of angler harvest, which can truncate size and age distributions (Maceina et al. 1998; Isermann and Paukert 2010). Additionally, small numbers of older fish may not be representative in samples and can alter von Bertalanffy growth models. The growth coefficient (i.e., K), however, was greatest for fish in southern populations, but differences did not exist among regions. Mean lengths at capture for both ages 1 and 3 were greatest in the southeastern region. Differences were marginally significant over that observed in the midwestern region (P = 0.044); however, differences between southeastern and midwestern regions for mean length at age 3 were highly significant (P = 0.01). Mean length at capture for both ages

TABLE 1. Waterbody, von Bertalanffy parameter estimates (theoretical maximum length [L_{∞}, mm], growth coefficient [K], theoretical age when length equals zero [t_0]) and maximum age (Age_{max}) for hybrid striped bass populations across North America.

Water body	L_{∞}	K	t_0	Age_{max}
Midwestern lakes				
Illinois				
Lake Pittsfield	685	0.394	0.471	6
Indiana				
Lake Freeman	786	0.275	−1.086	6
Monroe Reservoir	641	0.498	−0.130	9
Kansas				
Cedar Bluff Reservoir	665	0.342	−1.06	6
Norton Reservoir	731	0.274	−1.393	8
Webster Reservoir	843	0.204	−1.294	7
Nebraska				
Branched Oak Lake	658	0.323	−0.979	9
Calamus Reservoir	663	0.274	−1.086	7
Elwood Reservoir	651	0.334	−1.047	8
Ohio				
East Fork Lake	624	0.341	−0.171	7
Southeastern lakes				
Georgia				
Clarks Hill Reservoir	673	0.374	−0.830	7
Lake Hartwell	709	0.251	−2.171	7
Kentucky				
Barren River Lake	645	0.520	−0.545	9
Rough River Lake	617	0.254	−1.992	6
North Carolina				
West Kerr Scott Reservoir	684	0.636	−0.041	6
Tennessee				
Boone Reservoir	629	0.544	−0.755	10
Cherokee Reservoir	693	0.365	−0.692	8
J. Percy Priest Reservoir	670	0.435	−1.124	10
Virginia				
Claytor Lake	610	0.408	−1.203	11
Southwestern lakes				
Oklahoma				
Birch Lake	537	0.581	−0.865	8
Skiatook Lake	633	0.266	−2.268	9
Sooner Lake	502	0.630	−0.870	8
Texas				
Belton Lake	760	0.244	−1.211	8
Fort Phantom Hill Lake	666	0.347	−0.391	8
Lake Sommers	712	0.224	−2.869	6
Lake Palestine	603	0.585	−0.142	11
Pat Mayse Lake	672	0.334	−1.436	7
Proctor Lake	612	0.543	−0.182	11
Sam Rayburn Reservoir	639	0.490	0.288	8

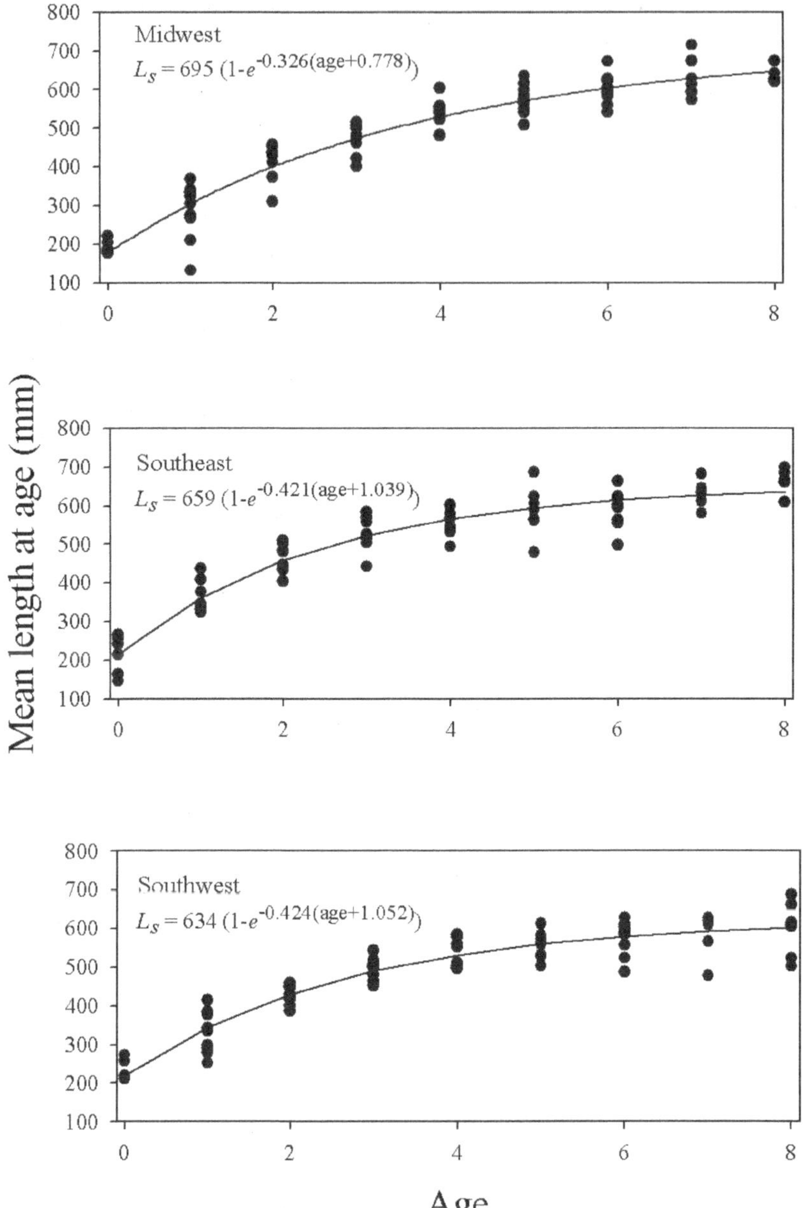

FIGURE 2. Mean length at age for hybrid striped bass populations throughout North America. The line represents the best-fit von Bertalanffy growth model (i.e., the standardized growth model) and was used to estimate age-specific standard lengths (i.e., L_s).

was intermediate in the southwestern populations. Hybrid striped bass may exhibit faster growth than either parental species (Carlander 1997). Therefore, hybrids may exemplify similar growth and age structure across a wider distribution than that which exists for either parental species. Unfortunately, studies to support or refute this supposition for hybrids are sparse. Hubert (1999) and Jackson (1999) found similar results where channel catfish *Ictalurus punctatus*

TABLE 2. Region, von Bertalanffy parameter estimates (theoretical maximum length [L_∞, mm], growth coefficient [K], theoretical age when length equals zero [t_0]), maximum age [Age_{max}], and lakes with ages > 8 years old [%] for North America hybrid striped bass populations.

Region	L_∞	K	t_0	Age_{max}	Ages > 8 years (%)
Midwestern lakes	695	0.326	–0.778	9	20
Southeastern lakes	659	0.421	–1.039	11	44
Southwestern lakes	634	0.424	–1.052	11	30

and flathead catfish *Pylodictis olivaris* did not exhibit regional differences in growth.

The results of this study are an important initial summarization of hybrid striped bass growth in North America. Latitudinal gradients in growth reveal that realistic expectations in hybrid striped bass growth are necessary for the management goals across their distributional range. Currently, little information is being reported for hybrid striped bass, especially in relation to striped bass (Bettoli 2013, this volume); this is especially true even for basic fisheries management information such as growth, recruitment and mortality. Continued development of population dynamics is needed to facilitate hybrid striped bass comparisons among populations to guide management activities. We encourage managers to evaluate and report population dynamics of hybrid striped bass they manage, utilizing this study as a starting point to better define management goals and objectives for this highly valuable sport fish species.

Acknowledgments

This study was funded through the Federal Aid in Sport Fish Restoration Program. Appreciation is extended to Gordon Schneider, Kyle Austin, Tommie Berger, Lynn Davignon, John Reinke, and Steve Price for assistance with data collection. Special appreciation is extended to biologists who shared hybrid striped bass age and growth assessments.

References

Avise, J. C., and M. J. Van Den Avyle. 1984. Genetic analysis of reproduction of hybrid white bass × striped bass in the Savannah River. Transactions of the American Fisheries Society 113:563–570.

Bettoli, P. W. 2013. Challenges and opportunities in studying and managing striped bass and its hybrids in the 21st century. Pages 575–588 *in* J. S. Bulak, C. C. Coutant, and J. A. Rice, editors. Biology and management of inland striped bass and hybrid striped bass. American Fisheries Society, Symposium 80, Bethesda, Maryland.

Beverton, R. J. H. 1987. Longevity in fish: some ecological and evolutionary considerations. Pages 161–185 *in* A. D. Woodhead and K. H. Thompson, editors. Evolution of longevity in animals: a comparative approach. Plenum, New York.

Carlander, K. D. 1997. Handbook of freshwater fishery biology, volume 3. Iowa State University Press, Ames.

Colby, P. J., R. E. McNicol, and R. A. Ryder. 1979. Synopsis of biological data on the walleye, *Stizostedion v. vitreum* (Mitchill 1818). FAO Fisheries Synopsis 119.

Crandall, P. S. 1979. Evaluation of striped bass × white bass hybrids in a heated Texas reservoir. Proceedings of the Annual Conference Southeastern Association of Fish and Wildlife Agencies 32(1978):588–598.

Dennerline, D. E., and M. J. Van Den Avyle. 2000. Sizes of prey consumed by two pelagic predators in US reservoirs: implications for quantifying biomass of available prey. Fisheries Research 45:147–154.

Dettmers, J. M., D. R. DeVries, and R. A. Stein. 1996. Quantifying responses to hybrid striped bass predation across multiple trophic levels: implications for reservoir biomanipulation. Transactions of the American Fisheries Society 125:491–504.

DeVries, D. R., and R. V. Frie. 1996. Determination of age and growth. Pages 483–512 *in* B. R. Murphy and D. W. Willis, editors. Fisheries techniques, 2nd edition. American Fisheries Society, Bethesda, Maryland.

Freund, R. J., and R. C. Littell. 1991. SAS system for regression, 2nd edition. SAS Institute, Cary, North Carolina.

Graham, K. 1999. A review of the biology and management of blue catfish. Pages 37–50 *in* E. R. Irwin, W. A. Hubert, C. F. Rabeni, H. L. Schramm, Jr., and T. Coon, editors. Catfish 2000: proceedings of the international ictalurid symposium. American Fisheries Society Symposium 24, Bethesda, Maryland.

Hansen, M. J., N. P. Lester, and C. C. Krueger. 2010. Natural lakes. Pages 449–500 *in* W. A. Hubert, and M. C. Quist, editors. Inland fisheries management in North America, 3rd edition. American Fisheries Society, Bethesda, Maryland.

Heman, M. L., L. C. Redmond, and R. S. Campbell. 1969. Manipulation of fish populations through reservoir drawdown. Transactions of the American Fisheries Society 98:293–304.

Hubert, W. A. 1999. Biology and management of channel catfish. Pages 3–22 *in* E. R. Irwin, W. A. Hubert, C. F. Rabeni, H. L. Schramm, Jr., and T. Coon, editors. Catfish 2000: proceedings of the international ictalurid symposium. American Fisheries Society Symposium 24, Bethesda, Maryland.

Hutt, C. P., J. W. Neal, and T. J. Lang. 2008. Stocking harvestable hybrid striped bass in an urban fishing program: angling success, angler satisfaction, and influence on bluegill size structure. Pages 403–412 *in* R. T. Eades, J. W. Neal, T. J. Lang, K. M. Hunt, and P. Pajak, editors. Urban and community fisheries programs: development, management, and evaluation. American Fisheries Society, Symposium 67, Bethesda, Maryland.

Isermann, D.A., and C.P. Paukert. 2010. Regulating harvest. Pages 185–212 *in* W. A. Hubert and M. C. Quist, editors. Inland fisheries management in North America, 3rd edition. American Fisheries Society, Bethesda, Maryland.

Jackson, D. C. 1999. Flathead catfish: biology, fisheries, and management. Pages 23–36 *in* E. R. Irwin, W. A. Hubert, C. F. Rabeni, H. L. Schramm, Jr., and T. Coon, editors. Catfish 2000: proceedings of the international ictalurid symposium. American Fisheries Society Symposium 24, Bethesda, Maryland.

Jackson, Z. J., M. C. Quist, and J. G. Larscheid. 2008. Growth standards for nine North American fish species. Fisheries Ecology and Management 15:107–118.

Jahn, L. A., D. R. Douglas, M. J. Terhaar, and G. W. Kruse. 1987. Effects of stocking hybrid striped bass in Spring Lake, Illinois. North American Journal of Fisheries Management 7:522–530.

Kilpatrick, J. M., and J. J. Ney. 2013. Temperature and dissolved oxygen habitat use by striped bass and hybrid striped bass in Claytor Lake, Virginia. Pages 147–159 *in* J. S. Bulak, C. C. Coutant, and J. A. Rice, editors. Biology and management of inland striped bass and hybrid striped bass. American Fisheries Society, Symposium 80, Bethesda, Maryland.

Layzer, J. B., and M. D. Clady. 1984. Evaluation of the striped bass × white hybrid for controlling stunted bluegills. Proceedings of the Annual Conference Southeastern Association of Fish and Wildlife Agencies 35(1981):297–310.

Leitner, J., J. Bulak, and R. Dunham. 2002. A comparison of first and third year growth of two strains of largemouth bass in South Carolina. Pages 365–370 *in* D. P. Philipp and M. S. Ridgway, editors. Black bass: ecology, conservation, and management. American Fisheries Society, Symposium 31, Bethesda, Maryland.

Maceina, M. J., P. W. Bettoli, S. D. Finley, and V. J. DiCenzo. 1998. Analyses of the sauger fishery with simulated effects of a minimum size limit in the Tennessee River of Alabama. North American Journal of Fisheries Management 18:66–75.

Maceina, M. J., J. Boxrucker, D. L. Buckmeier, R. Scott Gangl, D. O. Lucchesi, D. A. Isermann, J. R. Jackson, and P. J. Martinez. 2007. Current status and review of freshwater fish aging procedures used by state and provincial fisheries agencies with recommendations for future directions. Fisheries 32:329–340.

McInerny, M. C., and T. K. Cross. 1999. Effects of lake productivity, climate warming, and intraspecific density on growth and growth patterns of black crappie in southern Minnesota lakes. Journal of Freshwater Ecology 14:255–264.

Miranda, L. E., and P. P. Durocher. 1986. Effects of environmental factors on growth of largemouth bass in Texas reservoirs. Pages 115–121 *in* G. E. Hall and M. J. Van Den Avyle, editors. Reservoir fisheries management: strategies for the 80's, American Fisheries Society, Southern Division, Reservoir Committee, Bethesda, Maryland.

Neal, J. W., R. L. Noble, and J. A. Rice. 1999. Fish community response to hybrid striped bass introduction in small warmwater impoundments. North American Journal of Fisheries Management 19:1044–1053.

Noble, R. L. 1981. Management of forage fishes in impoundments of the southern United States. Transactions of the American Fisheries Society 110:738–750.

Olson, N. W., C. S. Guy, and K. D. Koupal. 2007. Interactions of three top-level predators in a polymictic Great Plains reservoir. North American Journal of Fisheries Management 27:268–278.

Ott, R. A., and S. P. Malvestuto. 1984. The striped bass × white bass hybrid in West Point Reservoir. Proceedings of the Annual Conference Southeastern Association of Fish and Wildlife Agencies 35(1981):641–646.

Paukert, C. P., M. A. McInerny, and R. D. Schultz. 2007. Historical trends in creel limits, length-based limits, and season restrictions for black basses in the United States and Canada. Fisheries 32:62–72.

Quist, M. C., C. S. Guy, R. D. Schultz, and J. L. Stephen. 2003. Latitudinal comparisons of walleye growth in North America and factors influencing growth of walleyes in Kansas reservoirs. North American Journal of Fisheries Management 23:677–692.

Ruane, R. J., G. E. Hauser, and A. F. Sawyer. 2013. Water qualtiy modeling as a mangagement tool for striped bass and hybrid striped bass. Pages 161–179 *in* J. S. Bulak, C. C. Coutant, and J. A. Rice, editors. Biology and management of inland striped bass and hybrid striped bass. American Fisheries Society, Symposium 80, Bethesda, Maryland.

Santucci, V. J., Jr., and D. H. Wahl. 1993. Factors influencing survival and growth of stocked walleye (*Stizostedion vitreum*) in a centrarchid-dominated impoundment. Canadian Journal of Fisheries and Aquatic Sciences 50:1548–1558.

Schultz, R. D., and B. J. Dodd. 2008. Growth, mortality, and harvest of walleye and hybrid striped bass in an Iowa urban lake: simulated effects of minimum size limits. Pages 413–424 *in* R. T. Eades, J. W. Neal, T. J. Lang, K. M. Hunt, and P. Pajak, editors. Urban and community fisheries programs: development, management, and evaluation. American Fisheries Society, Symposium 67, Bethesda, Maryland.

Soupir, C. A., B. B. Blackwell, and M. L. Brown. 1997. Relative precision among calcified structures for white bass age and growth assessment. Journal of Freshwater Ecology 12:531–534.

Thompson, J. S., and J. A. Rice. 2013. The relative influence of thermal experience and forage availability on growth of age 1–5 striped bass in two southeastern reservoirs. Pages 93–120 *in* J. S. Bulak, C. C. Coutant, and J. A. Rice, editors. Biology and management of inland striped bass and hybrid striped bass. American Fisheries Society, Symposium 80, Bethesda, Maryland.

Welch, T. J., M. J. Van Den Avyle, R. K. Betsill, and E. M Driebe. 1993. Precision and relative accuracy of striped bass age estimates from otoliths, scales, and anal fin rays and spines. North American Journal of Fisheries Management 13:616–620.

Woiwode, J. G., and I. R. Adelman. 1991. Effects of temperature, photoperiod, and ration size on growth of hybrid striped bass × white bass. Transactions of the American Fisheries Society 120:217–229.

Wuellner, M. R., S. R. Chipps, D. W. Willis, and W. E. Adams, Jr. 2010. Interactions between walleyes and smallmouth bass in a Missouri River reservoir with consideration of the influence of temperature and prey. North American Journal of Fisheries Management 30:445–463.

American Fisheries Society Symposium 80:229–247, 2013

Comparative Ecology of Juvenile Striped Bass and Juvenile Hybrid Striped Bass in Claytor Lake, Virginia

Jacob M. Rash*
North Carolina Wildlife Resources Commission
645 Fish Hatchery Road, Marion, North Carolina 28752, USA

John. J. Ney
Virginia Polytechnic Institute and State University
Blacksburg, Virginia 24062, USA

Abstract.—Comparisons of juvenile ecology between coexisting striped bass *Morone saxatilis* and hybrid striped bass (white bass *M. chrysops* × striped bass) that may limit recruitment to age 1 were considered in terms of trophic relationships, physiological indices of health, overwinter survival, and poststocking predation in Claytor Lake, Virginia. Both fishes preferred habitat types characterized by structure-free sand or gravel substrates, but striped bass and hybrid striped bass did not exhibit significant diet overlap during the growing season. At approximately 120 mm total length (TL), juvenile moronids shifted from a diet of zooplankton and invertebrates to a diet primarily comprised of age-0 fishes. However, piscivorous striped bass primarily preyed on age-0 alewives *Alosa pseudoharengus* while hybrid striped bass consumed age-0 sunfishes *Lepomis* spp. The delayed stocking date of hybrid striped bass appeared to contribute to the dietary difference. Striped bass achieved mean total lengths of 229 and 173 mm by the end of the 2001 and 2002 growing seasons, respectively. Mean total lengths of hybrid striped bass stocked into the reservoir 3 months later than striped bass reached 133 mm at the end of the 2002 growing season. Overwinter loss of smaller (<150 mm TL) striped bass was observed for the 2001–2002 sampling period. Predation upon stocked fingerlings was not considered significant in limiting survival; only three fingerling moronids were found in the stomach contents of 200 potential predators captured near stocking sites. It does not appear that resource competition with hybrid striped bass during the growing season resulted in overwinter mortality of juvenile striped bass in our study. Delayed hybrid striped bass stocking may lessen the potential for trophic competition between striped bass and hybrid striped bass at early life stages.

Introduction

Claytor Lake, Virginia supports fisheries for both striped bass *Morone saxatilis* and hybrid striped bass (female white bass *M. chrysops* × male striped bass). Axon and Whitehurst (1985) noted the popularity of both fishes, reporting that 34 states have established fisheries for striped bass, hybrid striped bass, or a combination of both. However, knowledge concerning the ecology of juvenile striped bass and juvenile hybrid striped bass following stocking is sparse. Through our investigation into the early life history of these two fishes in Claytor Lake, Virginia, we sought to determine if the inter-

* Corresponding author: jacob.rash@ncwildlife.org

actions of striped bass and hybrid striped bass at the juvenile life stage potentially resulted in competition and eventual mortality.

Diets of adult striped bass and hybrid striped bass in Claytor Lake have been found to overlap significantly, with each being heavily dependent on the clupeid forage base (Bonds 2000). Juvenile *Morone* in other systems have been observed to shift food habits from zooplankton and aquatic insects to larval and juvenile fishes at approximately 100 mm total length (TL; Stevens 1958; Markle and Grant 1970; Ware 1971; Axon 1979; Van Den Avyle et al. 1983; Ott and Malvestuto 1984; Saul and Wilson 1984; Kinman 1987; Matthews et al. 1992; Sutton and Ney 2001; Sutton and Ney 2002). Sutton and Ney (2001, 2002) reported that in Smith Mountain Lake, Virginia, smaller striped bass juveniles (<120 mm TL) maintained a mixed diet of invertebrates and age-0 cyprinids while larger juveniles (>150 mm TL) were strictly piscivorous. Saul and Wilson (1984) reported similar results for juvenile hybrid striped bass in Cherokee Reservoir, Tennessee, with individuals less than 120 mm TL feeding primarily on invertebrates while individuals greater than 120 mm TL were largely piscivorous.

Fast growth during the initial months of life can be critical for overwinter survival of piscivorous fishes in temperate climates (Forney 1966; Chevalier 1973; Nielsen 1980; Craig 1982; Minton and McLean 1982; Sutton 1997; Sutton and Ney 2001). Both striped bass and hybrid striped bass are characterized as having rapid first-year growth. Hybrid striped bass do not reach the same ultimate size as striped bass but do grow more rapidly during the first 2 years of life than either white bass or striped bass (Tuncer et al. 1990; Jenkins and Burkhead 1993). In controlled environments, hybrid striped bass fry can exhibit first-year growth rates of up to 1.8 g/d (Zhang et al. 1994), ultimately becoming 23.0% longer than striped bass fry over a 71-d period (Logan 1968). In natural systems, hybrid striped bass can achieve lengths greater than 200 mm TL during the first year of growth (Crandall 1979; Gilliland and Clady 1984; Layzer and Clady 1984; Saul and Wilson 1984; Austin and Hurley 1987).

During the growing season, age-0 striped bass tend to frequent shallow, structure-free littoral areas over sand or gravel substrate (Mensinger 1971; Van Den Avyle and Higginbotham 1980; Van Den Avyle et al. 1983; Matthews et al. 1992; Sutton 1997; Eggleton et al. 2005). In East Fork Lake, Ohio, Austin and Hurley (1987) found that juvenile hybrid striped bass exhibited preference for these habitats types as well. Matthews et al. (1992) found a significant overlap in habitat use by juvenile white bass and juvenile striped bass in Lake Texoma, Oklahoma–Texas, but little information is available concerning the spatial distribution of sympatric juvenile striped bass and juvenile hybrid striped bass. Additionally, age-0 hybrid striped bass exhibit optimal growth and peak conversion efficiency at 26.8°C (Woiwode and Adelman 1991) while juvenile striped bass have a preferred thermal niche of 24–26°C (Coutant 1985). Thus, during the growing season, juvenile hybrid striped bass and juvenile striped bass were expected to occupy similar thermal habitats in Claytor Lake.

First-year survival of stocked striped bass fingerlings in inland water bodies is often less than 20% (Bailey 1975; Moore et al. 1991; Michaelson et al. 2001; Sutton and Ney 2001; Sutton et al. 2013, this volume), but causes of this early mortality have received little attention until recently. Michaelson et al. (2001) quantified losses of stocked striped bass by predation to largemouth bass *Micropterus salmoides* (the primary littoral piscivore) in Smith Mountain Lake, Virginia. Only 14 striped bass were recovered in 1,147 largemouth bass stomachs over the 2-year study period, resulting in a 0.1–3.0% estimated loss of fingerlings to predation. McGovern and Olney (1988) noted similar results when they examined 235 stomachs of 14 species of fishes collected in the Pamunkey River, Virginia, and provided no evidence of predation on striped bass eggs or larvae. Austin and Hurley (1987) reported predation by black basses and yearling striped bass on hybrid striped bass fingerlings in East Fork Lake, Ohio, to be minimal. Therefore, mortality of *Morone* fingerlings due to predation was anticipated to be insignificant in the regulation of first-year survival in Claytor Lake, where the principal littoral piscivores are *Micropterus* spp.

Low survival of juvenile *Morone* is more likely the result of size-dependent starvation (Austin and Hurley 1987; Sutton 1997; Sutton and Ney 2001). Likelihood of overwinter survival for juvenile *Morone* is a function of the size of the individual at the end of the growing season (Hurst and Conover 1998; Sutton and Ney 2001) and is not species-specific (Harrell et al. 1988). Larger individuals are able to accumulate greater lipid reserves than smaller individuals, and smaller fishes are more vulnerable to overwinter mortality (Post and Evans 1989; Johnson and Evans 1990; Hurst and Conover 1998; Sutton and Ney 2001).

The ability of an individual to obtain adequate energy reserves is critical for its survival. If both juvenile striped bass and juvenile hybrid striped bass eat similar food items and occupy the same habitat types, the potential exists for competition between the two species to limit growth and ultimately survival. Therefore, we (1) quantified and compared first-year growth of juvenile striped bass and juvenile hybrid striped bass, (2) determined diet overlap of age-0 striped bass and age-0 hybrid striped bass, (3) compared habitat associations of juvenile striped bass and juvenile hybrid striped bass, and (4) assessed the impacts of predation and starvation upon survival of stocked fingerlings of striped bass and hybrid striped bass. By identifying aspects of juvenile striped bass and juvenile hybrid striped bass ecology, this study provides information to enable managers to design and implement stocking programs to enhance the recruitment of juvenile *Morone* into their fisheries.

Study Area

Claytor Lake is a 1,820-ha impoundment created in 1939 by construction of a dam on New River in Pulaski County, Virginia (Figure 1). The impoundment has 161 km of shoreline, a 15-m average depth, and a maximum depth of 37.5 m (Kohler et al. 1986; Copeland 1999; Palmer 1999). Its steep-gradient shape results in a limited littoral zone best developed in a small number of bays and the major tributary, Peak Creek.

Initial stocking programs featured centrarchid species and walleye *Sander vitreus* (Kohler et al. 1986; Copeland 1999). Since 1960, stocking programs have been directed toward the development of a pelagic fishery (Copeland 1999). Preliminary attempts to establish threadfin shad *Dorosoma petenense*, rainbow trout *Oncorhynchus mykiss*, and brown trout *Salmo trutta* failed, and in 1968, striped bass fingerlings were introduced into the reservoir to help fill the pelagic niche (Kohler et al. 1986; Copeland 1999). In 1993, sympatric stockings of striped bass and hybrid striped bass were initiated. Alewife *Alosa pseudoharengus*, gizzard shad *D. cepedianum*, and sunfishes *Lepomis* spp. serve as principal forage fishes (Bonds 2000).

Methods

Striped bass and hybrid striped bass stockings

In 2001, approximately 70,000 and 35,000 striped bass and hybrid striped bass fingerlings, respectively, were evenly distributed to three stocking sites along the longitudinal axis of the reservoir: Claytor Lake State Park, Dehaven Park, and Lighthouse Marina (Table 1; Figure 1). In 2002, half of the approximately 70,000 striped bass fingerlings were stocked at Claytor Lake State Park and the other half at Lighthouse Marina, while roughly 33,500 hybrid striped bass were again equally distributed among the three stocking sites (Table 1; Figure 1). Striped bass fingerlings stocked on 6 June 2001 and 14 June 2002 averaged approximately 40 mm TL and less than 1.0 g (Table 1). Hybrid striped bass fingerlings, averaging 68 mm TL and 3.2 g, were stocked in September 2001, a stocking date 3 months later than striped bass.

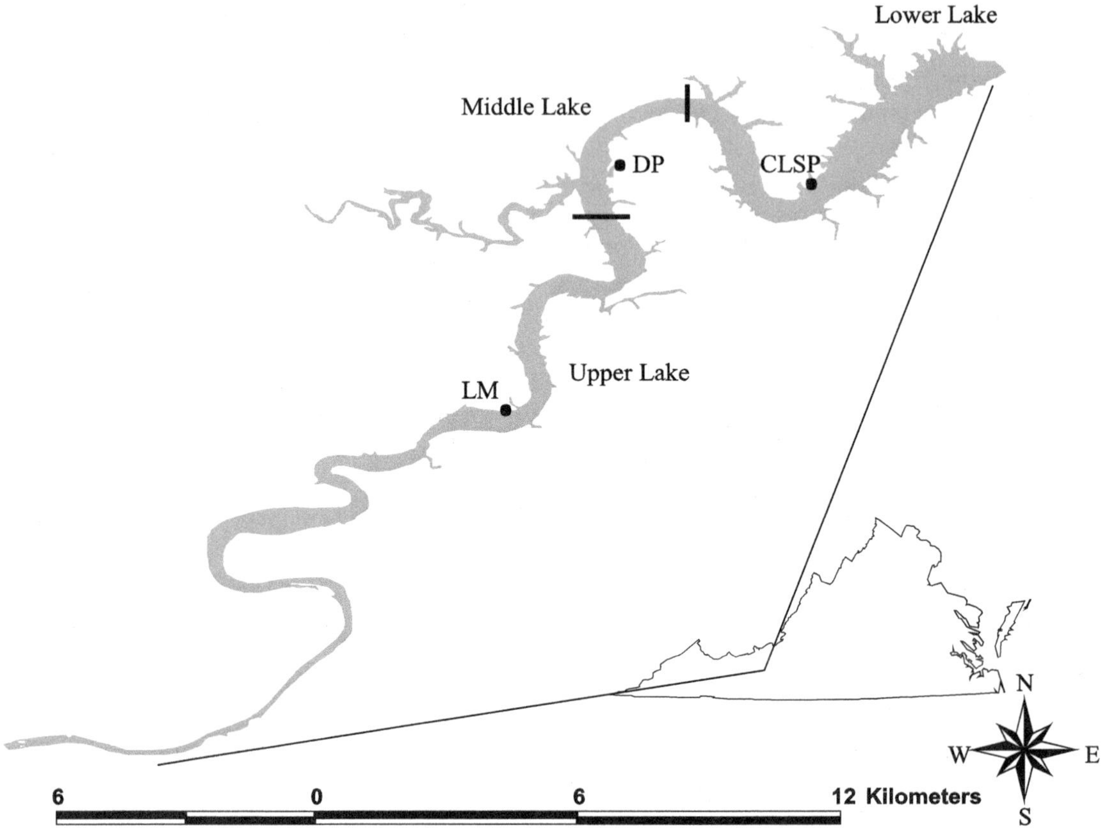

FIGURE 1. Map of Claytor Lake, Virginia, with sampling partitions (Lower Lake = Claytor Lake Dam to Twin Coves, Middle Lake = Twin Coves to Peak Creek, and Upper Lake = upstream of Peak Creek) and stocking locations (CLSP = Claytor Lake State Park, DP = Dehaven Park, and LM = Lighthouse Marina).

Hybrid striped bass were stocked in August 2002 at an average length of 94 mm TL and average weight of 11.0 g, a significantly larger size than hybrid striped bass stocked in 2001 (Table 1). All striped bass fingerlings originated from the Brookneal, Virginia hatchery operated by the Virginia Department of Game and Inland Fisheries (VDGIF). Hybrid striped bass fingerlings were purchased for stocking by VDGIF from Keo Fish Farm in Arkansas (2001) and Southland Fisheries Corporation in South Carolina (2002).

Field collections

Age-0 striped bass and age-0 hybrid striped bass were sought weekly from stocking dates until the second week of December 2001 and 2002 (the approximate end of the growing season). The reservoir was partitioned into three sampling regions based upon longitudinal distribution (Figure 1). Sampling efforts were concentrated upon areas where juveniles were consistently found (sand or gravel substrates without cover) in a single region each week to maximize capture success. Additional habitats in each region were also monitored to quantify habitat usage. Fishes were again sought during the postwinter, pregrowing season period in March and April 2002 and 2003. Lack of fishes in initial 2003 postwinter samples (March and April) required additional collections in May 2003.

To obtain representative samples of age-0 *Morone* over the size range present in the lake, a variety of sampling gears were used. Elec-

TABLE 1. Summary of juvenile striped bass (SB) and juvenile hybrid striped bass (SBH) stockings during 2001 and 2002 at Claytor Lake, Virginia. Mean total lengths and wet weights are given with range of values in parentheses. CLSP = Claytor Lake State Park; DP = Dehaven Park; and LM = Lighthouse Marina.

Year	Stock	Approximate number stocked	Stocking sites	Rearing site	Total length (mm)	Wet weight (g)
2001	SB	70,000	CLSP DP LM	Brookneal Fish Hatchery—Virginia Department of Game and Inland Fisheries (Brookneal, VA)	41 (31–57)	0.86 (0.33–2.25)
	SBH	33,500	CLSP DP LM	Keo Fish Farm (Keo, AR)	68 (55–78)	3.22 (1.05–5.39)
2002	SB	70,000	CLSP LM	Brookneal Fish Hatchery—Virginia Department of Game and Inland Fisheries (Brookneal, VA)	41 (33–49)	0.65 (0.35–1.04)
	SBH	33,500	CLSP DP LM	Southland Fisheries Corp. (Hopkins, SC)	94 (64–110)	10.77 (3.21–16.80)

trofishing was conducted one night per week throughout each region to target fish in the littoral zone using a boom-type electrofisher with pulsed DC. Horizontal gill nets were also deployed each week September–December 2001 and 2002, March–April 2002 and 2003, and May 2003 to target fish not susceptible to electrofishing. Three bipanel monofilament gill nets (30.5 m long and 1.8 m deep, consisting of two 15.2-m-long panels, with bar mesh sizes of 25 and 19 mm, 19 and 13 mm, and 25 and 13 mm) were set perpendicular to the shoreline and anchored to the bottom. One of each net type was deployed weekly; nets were set at dusk and retrieved at dawn the following day.

We sought to collect at least 30 juvenile striped bass and 30 juvenile hybrid striped bass in total using all gear types employed during each weekly sampling period. All striped bass and hybrid striped bass collected were immediately placed on ice. Upon return from field collection, all fish were blotted dry and weighed to the nearest 0.01 g, measured to the nearest 1 mm TL, and frozen whole for future stomach-content analysis.

First-year growth

Subsamples of juvenile *Morone* fingerlings were obtained from hatchery vehicles at stocking sites or by the immediate sampling of stocked fingerlings to provide baseline size information on stocked *Morone* (Table 1). Individuals collected during weekly samples were pooled by month to increase sample sizes for evaluation of first-year growth. Temporal patterns in growth were described by calculating a monthly mean length and weight of juveniles. One-way analysis of variance (ANOVA) was used to test differences in means; analyses were considered significant at $P < 0.05$ for type I error. Monthly length-frequency histograms for 2001–2002 and 2002–2003 were interpreted to describe patterns in growth and size distribution of juvenile striped bass and juvenile hybrid striped bass (Sutton and Ney 2001).

Postwinter survival

Overwinter survival was evaluated by comparing length-frequency histograms from late fall and early spring. Juvenile striped bass have been reported to exhibit little to no growth overwinter when water temperatures are below 10°C (Cox and Coutant 1981; Kerby et al. 1987; Woiwode and Adelman 1991; Hurst and Conover 1998; Sutton and Ney 2001). Sutton and Ney (2001) noted that determination of size-dependent overwinter mortality from length-frequency data requires that (1) mean length of fish increased over the winter while the variability in length decreased, and (2) the decrease in variability should be due to an upward shift in the minimum but not the maximum length. Because failure to attain some minimum size prior to the winter period may result in the inability of an individual to obtain the energy reserves required to survive the overwinter period (Sutton and Ney 2001), we evaluated health of juvenile moronids in this study via assessment of Fulton condition factor (Anderson and Neuman 1996), relative weight (Brown and Murphy 1991; Anderson and Neuman 1996), and lipid index values (Sutton 1997). Rash (2003) provides further information regarding the calculation and interpretation of these indices of health values for juvenile striped bass and juvenile hybrid striped bass in Claytor Lake.

Food habits

To assess diet composition, fish were thawed, and stomachs, from the base of the esophagus to the anterior portion of the intestine, were removed and preserved in 10% formalin for subsequent analysis. Stomach contents were identified to the lowest possible taxon (order or family for invertebrates and genus or species for fish). Prey items were blotted dry and weighed to the nearest 0.01 g.

The percent contribution by weight of each prey type (Hyslop 1980) was calculated using the following equation:

$$\%\mathrm{WTP}_i = \left[\sum(\mathrm{WTP}_i / \mathrm{WT}) / N\right] \times 100,$$

where $\%\mathrm{WTP}_i$ is the average percent contribution by weight for prey type i, WTP_i is the weight of prey i consumed by an individual predator, WT is the total weight of all prey types consumed by that predator, and N is the total number of predators of that type sampled that contained food (Sutton 1997). Percent contribution by weight allows food types to be quantified in directly comparable mass units so that the relative importance of these diet items can be estimated in terms of approximate nutrition gained by the predator (Bowen 1996; Sutton 1997).

Diet overlap was calculated between juvenile striped bass and hybrid striped bass by month during the 2002 growing season and the 2003 postwinter periods to provide an index of potential interspecific trophic competition (Sutton 1997). Overlap was evaluated by using diet composition weight percentages and Schoener's (1970) overlap index, which is defined as

$$C_{\mathrm{xy}} = 1.0 - 0.5\sum_{i=1}^{n}\left|p_{\mathrm{x}i} - p_{\mathrm{y}i}\right|,$$

where C_{xy} is the overlap index, $p_{\mathrm{x}i}$ is the proportion of food type i used by species x (striped bass), $p_{\mathrm{y}i}$ is the proportion of food type i used by species y (hybrid striped bass), and n is the total number of food categories (Crowder 1990). Overlap values of 0.60 or greater are considered indicative of potential trophic competition (Crowder 1990).

Habitat usage

During both sampling seasons, while collecting fish for growth and diet analysis, the habitat usage of the species in the littoral zone was documented. The dominant substrate, sand (particle size 0.06–1 mm), gravel (2–15 mm), cobble (64–256 mm), or boulder (>256 mm) (Bain 1999) and percent and type of cover (woody debris and vegetation) over which each individual was captured were documented. Dominant substrate was determined by visual estimation of the largest percent of substrate type by area.

During the 2002 growing season, we searched for 100-m units of shoreline in the lower region of the reservoir that represented one of the possible habitat categories (four substrates, with and without cover). Due to the lack of "gravel with cover" habitat, this investigation resulted in a total of seven categories of 100-m units (four substrate types, with and without cover, exclusive of "gravel with cover"). Once per month, each habitat unit was electrofished for a continuous 30-min time interval to investigate for the presence of age-0 striped bass and age-0 hybrid striped bass.

Monthly habitat usage overlap by juvenile striped bass and juvenile hybrid striped bass then were calculated via Schoener's (1970) overlap index, which is defined as

$$C_{xy} = 1.0 - 0.5\sum_{i=1}^{n}\left|p_{xi} - p_{yi}\right|,$$

where C_{xy} is the overlap index, p_{xi} is the proportion of habitat type i used by species x (striped bass), p_{yi} is the proportion of habitat type i used by species y (hybrid striped bass), and n is the total number of habitat categories. Habitat electivity was calculated by using the Strauss linear selection index:

$$L_i = r_i - p_i,$$

where r_i is the usage proportion of habitat i by species, and p_i is the proportion of habitat i in the environment (Crowder 1990). Positive values indicate preference for a habitat type and negative values indicate avoidance of a habitat type.

Predation

In 2002, electrofishing was conducted along the entire shorelines of the stocking coves (approximately 460 and 75 m at Claytor Lake State Park and Dehaven Park, respectively, and 410 m in the cove immediately downstream of Lighthouse Marina) and approximately 100 m above and below each stocking cove on the date of stocking, 1 d poststocking, and 3 d poststocking to capture potential predators (all omnivorous and piscivorous species). Electrofishing was then conducted weekly (100 m above and below and in stocking coves) for 1 month thereafter to target *Micropterus* spp., the primary littoral piscivores of Claytor Lake. To evaluate predation, predator stomach contents were evacuated with clear acrylic tubes and predators then were released (Michaelson et al. 2001). All stomach contents were preserved on ice and identified to the lowest possible taxon upon return from field collection.

Intensity of predation was assessed via frequency of occurrence of juvenile striped bass and juvenile hybrid striped bass in the diets of predators. When all specimens had been examined, the proportion of the potential predators that contained one or more juvenile striped bass or juvenile hybrid striped bass was calculated as the frequency of occurrence for that juvenile *Morone* as a food type (Bowen 1996).

Results

Field collections

A total of 1,628 age-0 moronids (N = 1,507 for striped bass; N = 121 for hybrid striped bass) were collected from Claytor Lake during the 2001–2002 and 2002–2003 sampling seasons. An additional 285 fingerlings (N = 153 for striped bass; N = 132 for hybrid striped bass) were obtained from hatchery trucks on the stocking date. More striped bass were collected during 2001–2002 (N = 842) than during 2002–2003 (N = 665). Significantly greater numbers of hybrid striped bass were collected in 2002–2003 (N = 119) than in 2001–2002 (N = 3); for that reason, inter- and intrastock comparisons using juvenile hybrid striped bass were limited to just one season (2002–2003).

First-year growth

Age-0 Morone consistently increased in size over each growing season (Figure 2). By December 2001, striped bass juveniles ranged from 174 to 278 mm TL (mean = 229 mm TL, SE = 4.31). No juvenile Morone were captured in December 2002, but juvenile striped bass and hybrid striped bass collected in November 2002 ranged from 112 to 227 mm TL (mean = 173 mm TL, SE = 4.45) and 98–157 mm TL (mean = 133 mm TL, SE = 13.49), respectively. Additionally, observed size distributions of each

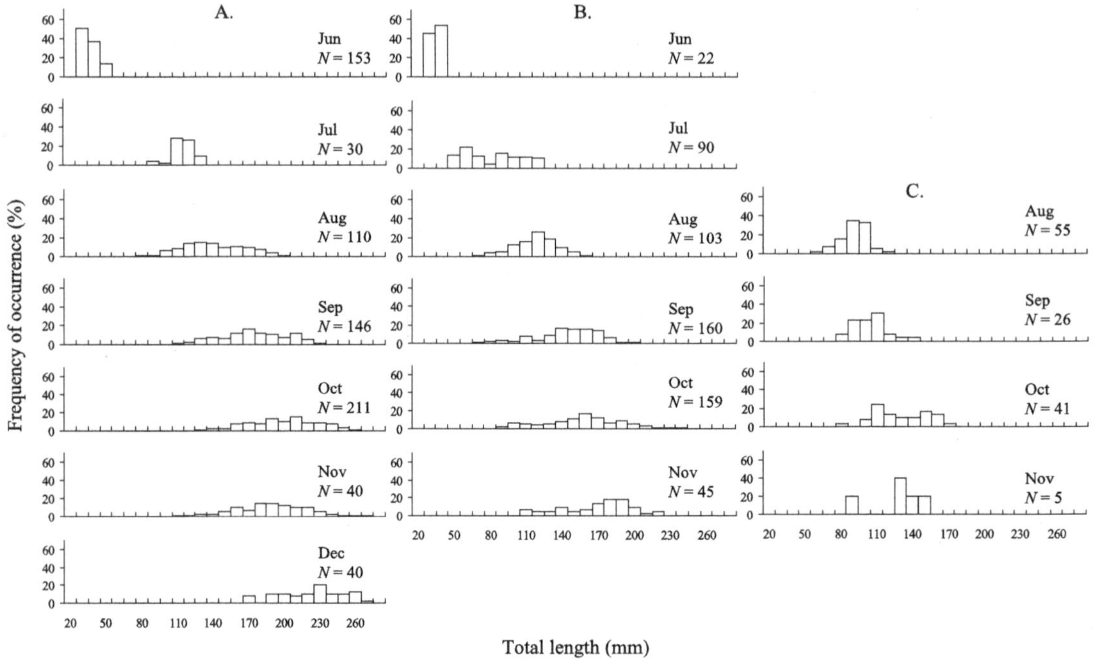

FIGURE 2. Monthly length-frequency distributions over the 2001 and 2002 growing seasons in Claytor Lake, Virginia, by (A) juvenile striped bass in 2001, (B) juvenile striped bass in 2002, and (C) juvenile hybrid striped bass in 2002.

juvenile moronid failed to exhibit a clear divergence in size through the progression of either growing season (Figure 2).

Postwinter survival

Condition factor, relative weight, and lipid index values were indicators of poor health over the growing seasons for each stock (Rash 2003). Mean juvenile striped bass total lengths differed significantly between fall (November and December) 2001 and spring (March and April) 2002; spring individuals were 9% larger (ANOVA, F = 28.69, df = 343, P < 0.001). Length-frequency distributions before and after winter do not exhibit clear bimodality (Figure 3). However, there was an overall shift in the population mean to 219 mm TL (SE = 2.35) and the resulting loss of small (<150 mm TL) individuals (5% < 150 mm TL in the fall versus 1% < 150 mm TL in the spring), with no increase in the maximum total length (278 mm TL). No juvenile hybrid striped bass were collected during the 2002 postwinter period.

In 2003, we observed a substantial upward shift in size distribution for striped bass and hybrid striped bass following the overwinter period. Age-0 striped bass mean total lengths differed significantly between fall (October and November) 2002 and spring 2003; mean total length of 238 mm TL (SE = 4.56) was 45% larger in spring (ANOVA, F = 263.87, df = 288, P < 0.001), coupled with a 105-mm increase in the maximum length to 332 mm TL. Juvenile striped bass failed to display a bimodal size distribution before the winter (Figure 4), but once more, striped bass did exhibit a shift to larger-mode individuals in the spring distribution (32% < 150 mm TL in the fall versus 1% < 150 mm TL in the spring).

Juvenile hybrid striped bass also had significantly different mean total lengths between fall 2002 and spring 2003 periods; spring individuals had a mean total length of 264 mm (SE = 8.12) and were 97% larger (ANOVA, F = 236.78, df = 84, P < 0.001). Again, it is difficult to detect the presence of bimodality upon

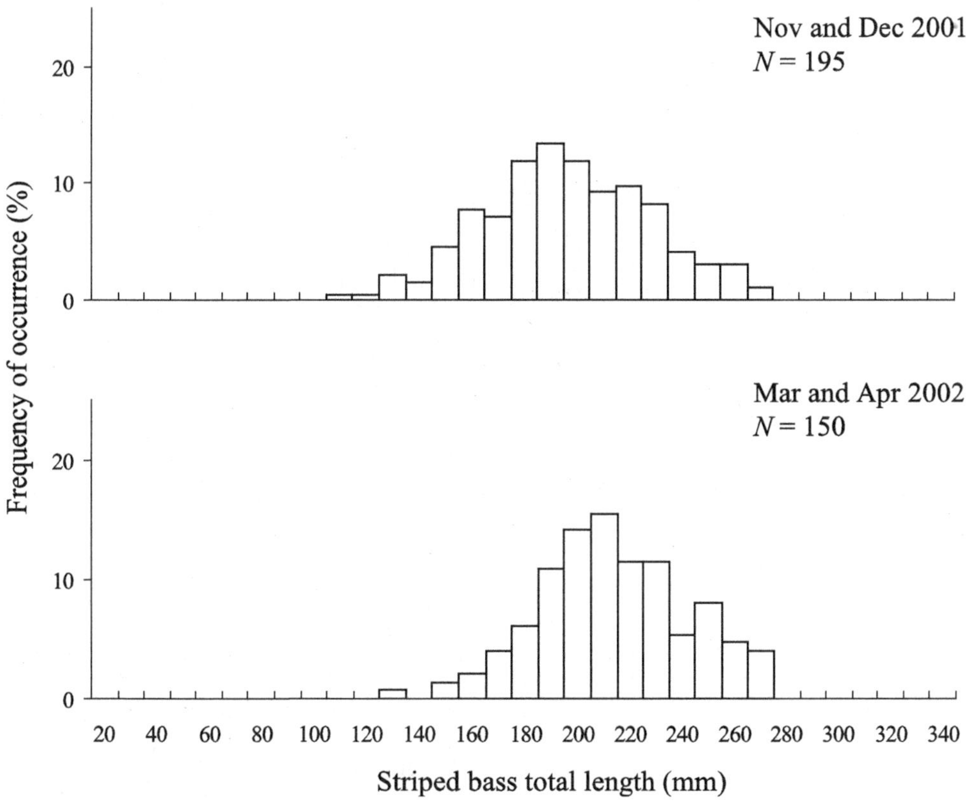

FIGURE 3. Length-frequency distributions of juvenile striped bass before and after overwintering in Claytor Lake, Virginia, during the 2001–2002 sampling season.

visual observation of the length-frequency distribution for hybrid striped bass (Figure 4). In late fall, 70% of hybrid striped bass collected were less than 150 mm TL versus only 5% in the spring. Spring hybrid striped bass mean total lengths were 11% greater than mean striped bass total lengths (ANOVA, $F = 9.62$, df = 124, $P = 0.002$).

Food habits

Juvenile striped bass exhibited a size-dependent shift in food habits over both the 2001 and 2002 growing seasons (Figure 5). Stomach contents of individuals greater than 120 mm TL contained progressively greater percentages of fishes, with age-0 alewives serving as the principal prey item (Figure 5). Unlike juvenile striped bass, hybrid striped bass stomachs in 2002 did not contain alewives (Figure 5). Juvenile hybrid striped bass exhibited a similar shift to piscivory at approximately 120 mm TL, but individuals greater than 120 mm TL had 100% of their diet by weight comprised of age-0 sunfishes. During the 2002 growing season, juvenile striped bass and juvenile hybrid striped bass did not exhibit diet overlap values greater than 0.60 for any month. However, the juvenile *Morone* did exhibit an overlap value of 0.80 for spring 2003, when diets were predominately comprised of alewives.

Habitat usage

Positive habitat electivity values reflect the preference of both juvenile striped bass and juvenile hybrid striped bass for sand and gravel substrates without cover in 2002 (Table 2). Yet, habitat overlap values greater than 0.60 were not detected for any individual month of the

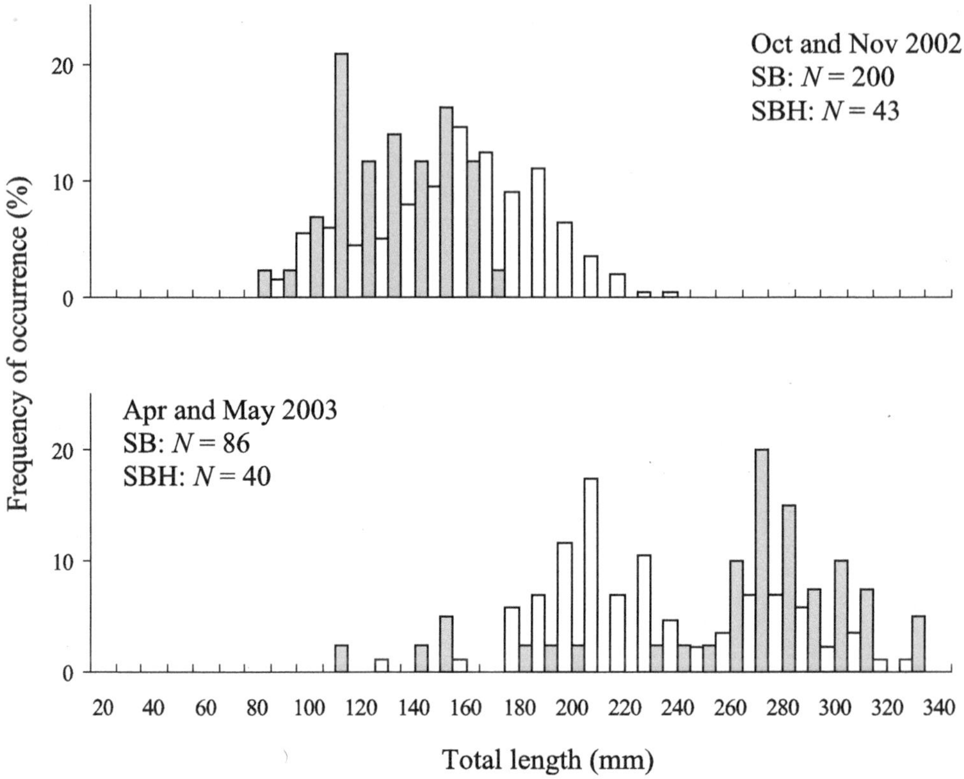

FIGURE 4. Length-frequency distributions of juvenile striped bass (SB) and juvenile hybrid striped bass (SBH; shaded bars) before and after overwintering in Claytor Lake, Virginia, during the 2002–2003 sampling season.

growing season or for all months combined. It appears that both fishes exhibited preference for the same habitat types but failed to simultaneously occupy these sampling units during our monthly surveys.

Predation

Frequency of occurrence of juvenile striped bass and juvenile hybrid striped bass fingerlings in the diets of potential predators was minimal. Predators evaluated after striped bass stocking in 2002 consisted of 55 largemouth bass, 20 smallmouth bass *M. dolomieu*, 17 spotted bass *M. punctulatus*, seven yellow perch *Perca flavescens*, three striped bass, two rock bass *Ambloplites rupestris*, and one flathead catfish *Pylodictis olivaris*. None of the 105 predators sampled following striped bass introductions contained an age-0 striped bass. Predators investigated for the consumption of fingerling hybrid striped bass were 40 spotted bass, 29 smallmouth bass, 12 largemouth bass, six flathead catfish, five yellow perch, two yellow bullhead *Ameiurus natalis*, and one rock bass. A largemouth bass, smallmouth bass, and spotted bass each contained a fingerling hybrid striped bass on the night of stocking; therefore, only 3 of the 95 predators sampled following hybrid striped bass stocking in 2002 contained a hybrid striped bass fingerling.

Discussion

Our efforts to investigate the comparative ecology of age-0 striped bass and age-0 hybrid striped bass throughout this study were impaired by the failure to collect juvenile hybrid striped bass during the 2001–2002 sampling period. In 2001, hybrid striped bass were stocked a month later, at approximately 72%

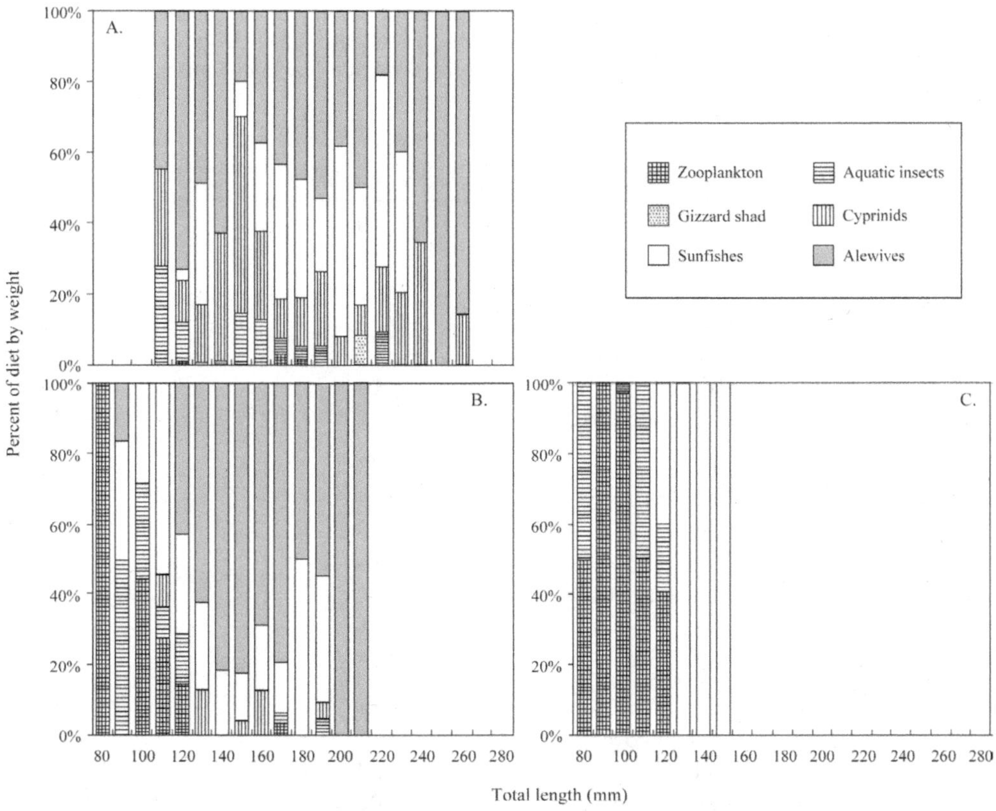

FIGURE 5. Size-dependent patterns of diet composition over the 2001 and 2002 growing seasons in Claytor Lake, Virginia, by (A) juvenile striped bass in 2001, (B) juvenile striped bass in 2002, and (C) juvenile hybrid striped bass in 2002.

of the mean total length and in poorer condition than hybrid striped bass stocked in 2002 (Rash 2003). As a result, relatively poor survival of these less-fit, 2001 hybrid striped bass fingerlings might be expected, especially because they were stocked late in the growing season. Capture of only three juvenile hybrid striped bass during the weeks following their stocking in 2001 and the failure to capture any juvenile hybrid striped bass following the postwinter period in 2002 supports the hypothesis of poor survival of 2001 hybrid striped bass.

Growth of striped bass in Claytor Lake is consistent with the descriptions of first-year growth reported for age-0 striped bass in other southeastern reservoirs, with some specimens achieving total lengths greater than 200 mm (Stevens 1958; Mensinger 1971; Ware 1971; Erickson et al. 1972; Axon 1979; Van Den Avyle and Higginbotham 1980; Dey 1981; Van Den Avyle et al. 1983; Moss and Lawson 1985; Sutton and Ney 2001). In contrast, age-0 hybrid striped bass (maximum TL = 157 mm) did not display total lengths as great as those reported in the literature for first-year growth (Crandall 1979; Ott and Malvestuto 1984; Saul and Wilson 1984; Moss and Lawson 1985; Austin and Hurley 1987; Kinman 1987). In addition, juvenile hybrid striped bass did not exhibit the more rapid growth rate than striped bass as reported for these fishes in other systems (Bishop 1968; Logan 1968; Tuncer et al. 1990; Jenkins and Burkhead 1993), perhaps because hybrid striped bass were stocked 2 months later than striped bass.

Variation in striped bass growth between years of this study may be related to habitat alterations resulting from drought conditions

TABLE 2. Temporal patterns in linear selection habitat electivity values for juvenile striped bass and juvenile hybrid striped bass during the 2002 growing season in Claytor Lake, Virginia. The total electivity values reflect the pooling of all individuals from each month. Positive electivity values are in bold. Hybrid striped bass were not captured in October.

	Habitat unit	Striped bass electivity value	Hybrid striped bass electivity value
September	Sand	**0.19**	**0.86**
	Sand with cover	–0.14	–0.14
	Gravel	**0.52**	–0.14
	Cobble	–0.14	–0.14
	Cobble with cover	–0.14	–0.14
	Boulder	–0.14	–0.14
	Boulder with cover	–0.14	–0.14
October	Sand	**0.23**	–
	Sand with cover	–0.14	–
	Gravel	**0.38**	–
	Cobble	–0.14	–
	Cobble with cover	–0.14	–
	Boulder	–0.14	–
	Boulder with cover	–0.14	–
November	Sand	**0.52**	–0.14
	Sand with cover	–0.14	–0.14
	Gravel	**0.02**	**0.86**
	Cobble	**0.02**	–0.14
	Cobble with cover	–0.14	–0.14
	Boulder	–0.14	–0.14
	Boulder with cover	–0.14	–0.14
Total	Sand	**0.25**	**0.69**
	Sand with cover	–0.14	–0.14
	Gravel	**0.39**	**0.02**
	Cobble	–0.07	–0.14
	Cobble with cover	–0.14	–0.14
	Boulder	–0.14	–0.14
	Boulder with cover	–0.14	–0.14

(VWRRC 2002) during the 2002 growing season. Reduced pool elevation resulted in the loss of preferred juvenile habitat in the littoral zone (sand and gravel substrates without cover) during summer 2002. Consequently, foraging success of recently stocked fingerlings was likely decreased as fishes were forced to seek prey outside of their preferred habitats. These juveniles may have been forced to reallocate energy from initial growth to storage of energy reserves during this period of environmental stress (Post and Parkinson 2001). Additionally, variation in ration has been found to impact moronid populations (Miranda and Raborn 2013, this volume), and it may account for the yearly difference in growth for juvenile striped bass in Claytor Lake, but determination of forage availability was beyond the scope of this study.

Analysis of the differences in length-frequency distributions between fall and spring sampling periods for striped bass in 2001–2002 suggests that size-dependent mortality over the winter period potentially limits the recruitment of juveniles in Claytor Lake. Large fish typically accumulate greater energy reserves (i.e., higher percentage of body composition in lipids) than smaller fish, suggesting that overwinter survival, a period when fish feed very little, could be critical for juvenile fishes that do not attain some minimum size prior to their first overwintering period (Sutton and Ney 2001). The observation of smaller individuals experiencing disproportionately higher overwinter mortality than larger individuals is reported in numerous studies for a variety of age-0 fishes. Sutton and Ney (2001) reported substantial mortality of smaller (<150 mm TL) age-0 striped bass during the overwinter period in Smith Mountain Lake, Virginia. The loss of small-mode (<150 mm TL) individuals from the age-0 cohort also was observed for hybrid striped bass in East Fork Lake, Ohio (Austin and Hurley 1987). Similar size-dependent overwinter mortality has been documented for largemouth bass (Shelton et al. 1979; Garvey et al. 1998; Fullerton et al. 2000), white perch *Morone americana* (Johnson and Evans 1990), yellow perch (Toneys and Coble 1980; Post and Evans 1989), Colorado pikeminnow *Ptychocheilus lucius* (Thompson et al. 1991), pumpkinseed *Lepomis gibbosus* (Benard and Fox 1997), freshwater drum *Aplodinotus grunniens* (Braaten and Guy 2004), Atlantic cod *Gadus morhua* (Gotceitas et al. 1999), and windowpane *Scophthalmus aquosus* (Neuman and Able 2009).

A clear bimodal size distribution never developed for either striped bass or hybrid striped bass juveniles during this study. Thus, greater attention was placed upon detecting the presence of an upward shift in the mean, without an increase in the maximum total length of the population, to evaluate overwinter mortality (Sutton and Ney 2001). Examination of length-frequency distributions did identify the loss of smaller individuals (<150 mm TL) overwinter in the 2001–2002 sampling season. By spring, the mean total length had increased 9% from fall while the maximum total length remained at approximately 270 mm TL. In addition, only 1% of individuals in the spring were less than 150 mm TL, compared with the 5% that had entered into the winter.

In late fall 2002, juvenile striped bass were smaller (32% < 150 mm TL versus 5% < 150 mm TL in 2001) and exhibited lipid reserves lower than individuals of late fall 2001 (Rash 2003). Therefore, with striped bass individuals entering the harsh winter of 2002 at smaller sizes and at a poorer condition in comparison to juvenile striped bass in 2001, the potential existed for overwinter mortality in the 2002–2003 sampling season. However, it was not possible to accurately detect size-dependent overwinter mortality in 2002–2003 fish because the upward shift in the minimum and mean lengths of striped bass and hybrid striped bass in the spring 2003 was accompanied by an increase in the maximum total length for both fishes of 105 and 173 mm, respectively. The increase in the maximum size indicates that overwinter disappearance of small individuals may simply be due to their growth beyond that size range. Increases in total lengths from fall 2002 to spring 2003 are likely due to resumed growth by fishes in the spring period before becoming susceptible to sampling gear. Sampling began in March for both the 2002 and 2003 spring sampling periods. In 2002, the majority of individuals were caught during March and the first 2 weeks of April, but in 2003, individuals were not collected until the last 2 weeks of April and the first 2 weeks of May. In addition, by the time both striped bass and hybrid striped bass were collected in 2003, both fishes had established diets composed of sunfishes, crayfishes, and predominately alewives. Striped bass collected in spring 2002 were found to have a mixed diet of invertebrates and fishes. The additional time to grow coupled with the switch to an energetically profitable diet gave juvenile striped bass and juvenile hybrid striped bass the opportunity to grow before becoming susceptible to the sampling gear deployed in spring 2003.

Both juvenile striped bass and juvenile hybrid striped bass exhibited size-dependent shifts in food habits at approximately 120 mm TL. Moronids 80–120 mm TL principally consumed aquatic insects and zooplankton while individuals over 120 mm TL mostly preyed upon age-0 fishes. Comparable alterations in food habits for individuals greater than 100 mm TL have been reported for other reservoir populations of age-0 striped bass (Ware 1971; Axon 1979; Van Den Avyle et al. 1983; Matthews et al. 1992; Sutton and Ney 2001; Sutton and Ney 2002) and age-0 hybrid striped bass (Ott and Malvestuto 1984; Saul and Wilson 1984; Austin and Hurley 1987; Kinman 1987).

Even though the two moronids were primarily piscivorous, they did not concentrate on the same prey items during the growing season, resulting in low diet overlap. Examination of stomach contents for juvenile striped bass and juvenile hybrid striped bass revealed temporal feeding patterns. Given the early June stocking date for fingerling striped bass, they were able to forage on age-0 alewives by July 2001 and 2002. In contrast, juvenile hybrid striped bass (stocked 2 months later in the growing season at a mean of approximately 94 mm TL) did not feed upon age-0 alewives as did juvenile striped bass, which in August averaged approximately 122 mm TL. Instead, juvenile hybrid striped bass preyed upon age-0 sunfishes and cladocerans throughout the growing season. Delayed stocking of hybrid striped bass in Claytor Lake may be limiting the ability of juvenile hybrid striped bass to utilize age-0 alewives as a continuous food source throughout the growing season.

Nigro (1980) found that by August, age-0 alewives in Claytor Lake exhibited mean total lengths of approximately 110 and 90 mm in 1978 and 1979, respectively. In addition, Tisa (1988) reported that age-0 alewives in Smith Mountain Lake, Virginia reached a mean total length of approximately 70 mm by early August. In 2002, hybrid striped bass fingerlings were stocked into Claytor Lake in August at a mean total length of 93.6 mm. Based on the relationship between predator mouth gape and prey body depth (Sutton 1997), age-0 striped bass 90 mm TL can only ingest alewife less than or equal to 30 mm TL. If the maximum total length ingestibility limit values developed for Smith Mountain Lake are applied to the juvenile hybrid striped bass entering Claytor Lake in August, hybrid striped bass could not ingest age-0 alewives. Alewives possess the highest gross energy values among clupeid forage fishes (Strange and Pelton 1987). The inability of juvenile hybrid striped bass to consume age-0 alewives during the growing season has two potentially significant impacts upon the trophic interactions in Claytor Lake. First, it allows the age-0 alewife forage base to be utilized more heavily by other lake piscivores (especially age-0 striped bass). Second, the inability to take advantage of alewives as a source of prey may force juvenile hybrid striped bass to seek alternative prey items, such as cyprinids, age-0 gizzard shad, and age-0 sunfishes.

Although low overlap was observed during the growing season, juvenile striped bass and juvenile hybrid striped bass exhibited a significant diet overlap in the postwinter period (0.80). By spring 2003, hybrid striped bass were large enough to overcome their apparent initial inability to ingest alewives, forcing higher overlap values indicative of potential competition, but these values were only observed after the "critical period" of winter. Thus, it appears that due to the late stocking of hybrid striped bass, it is unlikely (based on 2002–2003) that juvenile hybrid striped bass are causing overwinter starvation mortality in juvenile striped bass. However, because we were unable to capture striped bass in the spring of 2003 prior to the onset of new growth, we could not compare late fall versus spring length-frequency distributions for evidence of size-dependent mortality.

Habitat electivity values indicated preference by juvenile striped bass and juvenile hybrid striped bass for similar habitat units during the months of September through November 2002. The preference of both species for sand or gravel substrates without cover is consistent for these fishes (Mesinger 1970; Van Den Avyle and Higginbotham 1980; Van Den Avyle et al. 1983; Austin and Hurley 1987; Matthews et al. 1992; Sutton 1997; Eggleton et al. 2005). Habitat overlap values as calculated in this study may have been low due to striped bass and hybrid striped bass being temporally separated even as they utilized similar habitats. Perhaps, increased monitoring of habitat overlap (more than once a month) would have revealed higher overlap values. During weekly specimen collection for the remaining objectives of the study, juvenile hybrid striped bass were always collected in the presence of juvenile striped bass. Consequently, it appears that habitat overlap values in this study were unrepresentative of the actual habitat use of age-0 striped bass and age-0 hybrid striped bass, while habitat electivity values were accurate in identifying the habitat preference for these fishes in Claytor Lake as sand and gravel substrates lacking cover.

Effects of predation upon the first-year survival of juvenile striped bass and juvenile hybrid striped bass in Claytor Lake were minimal. None of the predators sampled following striped bass stocking were found to have consumed a fingerling striped bass, and only 3 of 95 predator stomachs examined after the stocking of hybrid striped bass contained a hybrid striped bass fingerling. The low level of predation on juvenile *Morone* in Claytor Lake is consistent with that reported in other studies (Austin and Hurley 1987; McGovern and Olney 1988; Michaelson et al. 2001). If the total percentages of fingerlings lost to predation, as estimated by Michaelson et al. (2001), were applied to fingerlings stocked in Claytor Lake, loss of fingerling hybrid striped bass would only range from 35 to 420 individuals. In addition, during this period of increased fingerling vulnerability following stocking, *Micropterus* spp. (the principal littoral predator in Claytor Lake) were found to primarily consume bluegills *Lepomis macrochirus* and crayfish in Claytor Lake (Bonds 2000).

Results of our study indicate that the delayed stocking of hybrid striped bass appears to help alleviate potential competitive interactions with age-0 striped bass in Claytor Lake. Although both fishes were found to prefer the same habitat types, they did not eat the same prey items during the growing season, and both fishes had equivalent indices of health values (Rash 2003). If competition is to occur, it is likely going to cause a reduction in biomass at later life stages in Claytor Lake. We found high diet overlap among striped bass and hybrid striped bass once we were able to capture them in late April 2003, when both fishes primarily consumed alewives. If both striped bass and hybrid striped bass rely upon alewives as forage following winter, potential trophic competition could occur at this period. This is especially important because energy reserves of juvenile moronids are significantly reduced over the winter, and the immediate acquisition of food following winter is required to maintain metabolic processes and preferred prey (alewives) may become scarce if overwinter die-offs occur.

In conclusion, the delayed stocking date of hybrid striped bass does not appear to limit their recruitment to the adult life stage in Claytor Lake (J. Copeland, Virginia Department of Game and Inland Fisheries, personal communication). It is likely that their condition at the time of stocking allows hybrid striped bass to enter the reservoir with sufficient energy reserves to avert significant overwinter mortality. To avoid substantial diet overlap between striped bass and hybrid striped bass, hybrid striped bass should continue to be stocked later in the growing season and at larger size than striped bass in Claytor Lake. Stocking hybrid striped bass fingerlings in August appears to allow striped bass (stocked in June) to use age-0 alewives as prey throughout the growing season while not having to share the resource with age-0 hybrid striped bass. Furthermore, given our inability to assess trophic compatibility across all seasons for these fishes during their first year of life, further investigation into the understanding of the mechanisms influencing postwinter survival is warranted.

Acknowledgments

We thank American Electric Power, the Virginia Department of Game and Inland Fisheries, and the Department of Fisheries and Wildlife Sciences at the Virginia Polytechnic Institute and State University for their financial support. We are also appreciative of J. Duty, J. Harris, and J. Kilpatrick for assisting with data collection, and we are indebted to J. Copeland, E. Hallerman, and K. Finne for sharing their expertise. J. Bulak, C. Coutant, J. Kilpatrick, J. Rice, T. Sutton, and D. Yow graciously provided comments to improve this manuscript.

References

Anderson, R. O., and R. M. Neuman. 1996. Length, weight, and associated structural indices. Pages 447–482 *in* B. R. Murphy and D. W. Willis, editors. Fisheries techniques, 2nd edition. American Fisheries Society, Bethesda, Maryland.

Austin, M. R., and S. T. Hurley. 1987. Evaluation of a striped bass (*Morone saxatilis*) × white bass (*Morone chrysops*) hybrid introduction in East Fork Lake, Ohio. Ohio Division of Wildlife, Federal Aid in Fish Restoration Project F-29-R, Study 20, Final Report, Columbus.

Axon, J. R. 1979. An evaluation of striped bass introductions in Herrington Lake. Kentucky Department of Fish and Wildlife Resources, Fisheries Bulletin 63, Frankfort.

Axon, J. R., and D. K. Whitehurst. 1985. Striped bass management in lakes with emphasis on management problems. Transactions of the American Fisheries Society 114:8–11.

Bailey, W. M. 1975. An evaluation of striped bass introductions in the southeastern United States. Proceedings of the Annual Conference Southeastern Association of Game and Fish and Commissioners 28(1974):54–68.

Bain, B. B. 1999. Substrate. Pages 95–104 *in* M. B. Bain and N. J. Stevenson, editors. Aquatic habitat assessment. American Fisheries Society, Bethesda, Maryland.

Benard, G., and M. G. Fox. 1997. Effects of body size and population density on overwinter survival of age-0 pumpkinseeds. North American Journal of Fisheries Management 17:581–590.

Bishop, D. R. 1968. Evaluation of the striped bass (*Roccus saxatilis*) and white bass (*R. chrysops*) hybrids after two years. Proceedings of the Annual Conference Southeastern Association of Game and Fish Commissioners 21(1967):245–254.

Bonds, C. C. 2000. Assessment of the response of piscivorous sportfishes to the establishment of gizzard shad in Claytor Lake, Virginia. Master's thesis. Virginia Polytechnic Institute and State University, Blacksburg.

Bowen, S. H. 1996. Quantitative description of the diet. Pages 513–532 *in* B. R. Murphy and D. W. Willis, editors. Fisheries techniques, 2nd edition. American Fisheries Society, Bethesda, Maryland.

Braaten, P. J., and C. S. Guy. 2004. First-year growth, condition, and size-selective winter mortality of freshwater drum in the lower Missouri River. Transactions of the American Fisheries Society 133:385–398.

Brown, M. L., and B. R. Murphy. 1991. Standard weights (W_s) for striped bass, white bass, and hybrid striped bass. North American Journal of Fisheries Management 11:451–467.

Chevalier, J. R. 1973. Cannibalism as a factor in first year survival of walleye in Oneida Lake. Transactions of the American Fisheries Society 102:739–744.

Copeland, J. R. 1999. Claytor Lake management report 1998. F111R7, Final Report. Virginia Department of Game and Inland Fisheries, Richmond.

Coutant, C. C. 1985. Striped bass, temperature, and dissolved oxygen: a speculative hypothesis for environmental risk. Transactions of the American Fisheries Society 114:31–61.

Cox, D. K., and C. C. Coutant. 1981. Growth dynamics of juvenile striped bass as functions of temperature and ration. Transactions of the American Fisheries Society 110:226–238.

Craig, J. F. 1982. Population dynamics of Windermere perch. Report of the Freshwater Biological Association 50:49–59.

Crandall, P. S. 1979. Evaluation of striped bass × white bass hybrids in a heated Texas reservoir. Proceedings of the Annual Conference

Southeastern Association of Fish and Wildlife Agencies 32(1978):588–598.

Crowder, L. B. 1990. Community ecology. Pages 609–632 *in* C. B. Schreck and P. B. Moyle, editors. Methods for fish biology. American Fisheries Society, Bethesda, Maryland.

Dey, W. P. 1981. Mortality and growth of young-of-the-year striped bass in the Hudson River estuary. Transactions of the American Fisheries Society 110:151–157.

Eggleton, M. A., R. Ramirez, C. W. Hargrave, K. B. Gido, J. R. Masoner, G. D. Schnell, and W. J. Matthews. 2005. Predictability of littoral-zone fish communities through ontogeny in Lake Texoma, Oklahoma–Texas, USA. Environmental Biology of Fishes 73:21–36.

Erickson, K. E., J. Harper, G. C. Mesinger, and D. Hicks. 1972. Status and artificial reproduction of striped bass from Keystone Reservoir, Oklahoma. Proceedings of the Annual Conference Southeastern Association of Game and Fish Commissioners 25(1971):513–522.

Forney, J. L. 1966. Factors affecting first-year growth of walleyes in Oneida Lake, New York. New York Fish and Game Journal 13:146–167.

Fullerton, A. H., J. E. Garvey, R. A. Wright, and R. A. Stein. 2000. Overwinter growth and survival of largemouth bass: interactions among size, food, origin, and winter severity. Transactions of the American Fisheries Society 129:1–12.

Garvey, J. E., A. Wright, and R. A. Stein. 1998. Overwinter growth and survival of age-0 largemouth bass (*Micropterus salmoides*): revisiting the role of body size. Canadian Journal of Fisheries and Aquatic Sciences 55:2414–2424.

Gilliland, E. R., and M. D. Clady. 1984. Diet overlap of striped bass × white bass hybrids and largemouth bass in Sooner Lake, Oklahoma. Proceedings of the Annual Conference Southeastern Association of Fish and Wildlife Agencies 35(1981):317–330.

Gotceitas, V., D. A. Methven, S. Fraser, and J. Brown. 1999. Effects of body size and food ration on over-winter survival and growth of age-0 Atlantic cod, *Gadus morhua*. Environmental Biology of Fishes 54:413–420.

Harrell, R. M., D. W. Meritt, J. N. Hochheimer, D. W. Webster, and W. D. Miller. 1988. Overwintering success of striped bass and hybrid striped bass held in cages in Maryland. The Progressive Fish-Culturist 50:120–121.

Hurst, T. P., and D. O. Conover. 1998. Winter mortality of young-of-the-year Hudson River striped bass (*Morone saxatilis*): size-dependent patterns and effects on recruitment. Canadian Journal of Fisheries and Aquatic Sciences 55:112–1130.

Hyslop, E. J. 1980. Stomach contents analysis—a review of methods and their application. Journal of Fish Biology 17:411–429.

Jenkins, R. E., and N. M. Burkhead. 1993. Freshwater fishes of Virginia. American Fisheries Society, Bethesda, Maryland.

Johnson, T. B., and D. O. Evans. 1990. Size dependent winter mortality of young-of-the-year white perch: climate warming and invasion of the Laurentian Great Lakes. Transactions of the American Fisheries Society 119:301–313.

Kerby, J. H., J. M. Hinshaw, and M. T. Huish. 1987. Increased growth and production of striped bass × white bass hybrids in earthen ponds. Journal of the World Aquaculture Society 18:35–43.

Kinman, B. T. 1987. Evaluation of hybrid striped bass introductions in Herrington Lake. Kentucky Department of Fish and Wildlife Resources, Fisheries Bulletin 82, Frankfort.

Kohler, C. C., J. J. Ney, and W. E. Kelso. 1986. Filling the void: development of a pelagic fishery and its consequences to littoral fishes in a Virginia mainstream reservoir. Pages 166–177 *in* G. E. Hall and M. J. Van Den Avyle, editors. Reservoir fisheries management: strategies for the 80s. American Fisheries Society, Southern Division, Reservoir Committee, Bethesda, Maryland.

Layzer, J. B., and M. D. Clady. 1984. Evaluation of the striped bass × white bass hybrid for controlling stunted bluegills. Proceedings of the Annual Conference Southeastern Association of Fish and Wildlife Agencies 35(1981):297–310.

Logan, H. J. 1968. Comparison of growth and survival rates of striped bass and striped bass × white bass hybrids under controlled environments. Proceedings of the Annual Conference Southeastern Association of Game and Fish Commissioners 21(1967):260–263.

Markle, D. F., and G. C. Grant. 1970. The summer food habits of young-of-the-year striped bass in three Virginia rivers. Chesapeake Science 11:50–54.

Matthews, W. J., F. P. Gelwick, and J. J. Hoover. 1992. Food and habitat use by juveniles of species of *Micropterus* and *Morone* in a southwestern reservoir. Transactions of the American Fisheries Society 121:54–66.

McGovern, J. C., and J. E. Olney. 1988. Potential predation by fish and invertebrates on early life history stages of striped bass in Pamunkey River, Virginia. Transactions of the American Fisheries Society 117:152–161.

Mensinger, G. C. 1971. Observations on the striped bass, *Morone saxatilis*, in Keystone Reservoir, Oklahoma. Proceedings of the Annual Conference Southeastern Association of Game and Fish Commissioners 24(1970):447–463.

Michaelson, D. P., J. J. Ney, and T. M. Sutton. 2001. Largemouth bass predation on stocked striped bass in Smith Mountain Lake, Virginia. Transactions of the American Fisheries Society 21:326–332.

Minton, J. W., and R. B. McLean. 1982. Measurements of growth and consumption of sauger (*Stizostedion canadense*): implication for fish energetics studies. Canadian Journal of Fisheries and Aquatic Sciences 39:1396–1403.

Miranda, L. E., and S. W. Raborn. 2013. Interactions between striped bass and other gamefish in reservoirs. Pages 501–519 *in* J. S. Bulak, C. C. Coutant, and J. A. Rice, editors. Biology and management of inland striped bass and hybrid striped bass. American Fisheries Society, Symposium 80, Bethesda, Maryland.

Moore, C. M., R. J. Neves, and J. J. Ney. 1991. Survival and abundance of stocked striped bass in Smith Mountain Lake, Virginia. North American Journal of Fisheries Management 11:393–399.

Moss, L. J., and C. S. Lawson. 1985. Evaluation of striped bass and hybrid striped bass stockings in eight Alabama public fishing lakes. Proceedings of the Annual Conference Southeastern Association of Fish and Wildlife Agencies 36(1982):33–41.

Neuman, M. J., and K. W. Able. 2009. Overwinter growth and mortality of young-of-the-year windowpanes: cohort-specific responses. Marine and Coastal Fisheries: Dynamics, Management, and Ecosystem Science 1:133–142.

Nielsen, L. A. 1980. Effect of walleye (*Stizostedion vitreum vitreum*) predation on juvenile mortality and recruitment of yellow perch (*Perca flavescens*) in Oneida Lake, New York. Canadian Journal of Fisheries and Aquatic Sciences 37:11–19.

Nigro, A. A. 1980. Reproduction and early life history of landlocked population of alewife (*Alosa pseudoharengus*) in Claytor Lake, Virginia. Master's thesis. Virginia Polytechnic Institute and State University, Blacksburg.

Ott, R. A., Jr., and S. P. Malvestuto. 1984. The striped bass × white bass hybrid in West Point Reservoir. Proceedings of the Annual Conference Southeastern Association of Fish and Wildlife Agencies 35(1981):641–646.

Palmer, G. C. 1999. Genetic characterization and intermixing of walleye stocks in Claytor Lake and the upper New River, Virginia. Master's thesis. Virginia Polytechnic Institute and State University, Blacksburg.

Post, J. R., and D. O. Evans. 1989. Size-dependent overwinter mortality of young-of-the-year yellow perch (*Perca flavescens*): laboratory, in situ enclosure, and field experiments. Canadian Journal of Fisheries and Aquatic Sciences 46:1958–1968.

Post, J. R., and E. A. Parkinson. 2001. Energy allocation strategy in young fish: allometry and survival. Ecology 82:1040–1051.

Rash, J. M. 2003. Comparative ecology of juvenile striped bass and juvenile hybrid striped bass in Claytor Lake, Virginia. Master's thesis. Virginia Polytechnic Institute and State University, Blacksburg.

Saul, B. M., and J. L. Wilson. 1984. Food habits and growth of young-of-the-year white bass × striped bass hybrids in Cherokee Reservoir, Tennessee. Proceedings of the Annual Conference Southeastern Association of Fish and Wildlife Agencies 35(1981):311–316.

Schoener, T. W. 1970. Non-synchronous spatial overlap of lizards in patchy habitats. Ecology 51:408–418.

Shelton, W. L., W. D. Davies, T. A. King, and T.

J. Timmons. 1979. Variation in the growth of the initial year class of largemouth bass in West Point Reservoir, Alabama and Georgia. Transactions of the American Fisheries Society 108:142–149.

Stevens, R. E. 1958. The striped bass of the Santee–Cooper Reservoir. Proceedings of the Annual Conference Southeastern Association of Game and Fish Commissioners 11(1957):253–264.

Strange, R. J., and J. C. Pelton. 1987. Nutrient content of clupeid forage fishes. Transactions of the American Fisheries Society 116:60–66.

Sutton, T. M. 1997. Early life history dynamics of a stocked striped bass (*Morone saxatilis*) population and assessment of strategies for improving stocking success in Smith Mountain Lake, Virginia. Doctoral dissertation. Virginia Polytechnic Institute and State University, Blacksburg.

Sutton, T. M., and J. J. Ney. 2001. Size-dependent mechanisms influencing first-year growth and winter survival of stocked striped bass in a Virginia mainstream reservoir. Transactions of the American Fisheries Society 130:1–17.

Sutton, T. M., and J. J. Ney. 2002. Trophic resource overlap between age-0 striped bass and largemouth bass in Smith Mountain Lake, Virginia. North American Journal of Fisheries Management 22:1250–1259.

Sutton, T. M., D. M. Wilson, and J. J. Ney. 2013. Biotic and abiotic determinants of stocking success for striped bass in inland waters. Pages 365–382 *in* J. S. Bulak, C. C. Coutant, and J. A. Rice, editors. Biology and management of inland striped bass and hybrid striped bass. American Fisheries Society, Symposium 80, Bethesda, Maryland.

Thompson, J. M., E. P. Bergersen, C. A. Carlson, and L. R. Kaeding. 1991. Role of size, condition, and lipid content in the overwinter survival of age-0 Colorado squawfish. Transactions of the American Fisheries Society 120:346–353.

Tisa, M. S. 1988. Compatibility and complementarity of alewife *Alosa pseudoharengus* and gizzard shad *Dorosoma cepedianum* as forage fish in Smith Mountain Lake, Virginia. Doctoral dissertation. Virginia Polytechnic Institute and State University, Blacksburg.

Toneys, M. L., and D. W. Coble. 1980. Mortality, hematocrit, osmolality, electrolyte regulation, and fat depletion of young-of-the-year freshwater fishes under simulated water conditions. Canadian Journal of Fisheries and Aquatic Sciences 37:225–232.

Tuncer, H., R. M. Harrell, and E. D. Houde. 1990. Comparative energetics of striped bass (*Morone saxatilis*) and hybrids (*M. saxatilis* × *M. chrysops*) juveniles. Aquaculture 86:387–400.

Van Den Avyle, M. J., and B. J. Higginbotham. 1980. Growth, survival, and distribution of striped bass stocked into Watts Bar Reservoir, Tennessee. Proceedings of the Annual Conference Southeastern Association of Fish and Wildlife Agencies 33(1979):361–370.

Van Den Avyle, M. J., B. J. Higginbotham, B. T. James, and F. J. Bulow. 1983. Habitat preferences and food habits of young-of-the-year striped bass, white bass, and yellow bass in Watts Bar Reservoir, Tennessee. North American Journal of Fisheries Management 3:163–170.

VWRRC (Virginia Water Resources Research Center). 2002. Virginia Water Central October 2002. VWRRC, Newsletter 23, Blacksburg.

Ware, F. J. 1971. Some early life history of Florida's inland striped bass, *Morone saxatilis*. Proceedings of the Annual Conference Southeastern Association of Game and Fish Commissioners 24(1970):439–447.

Woiwode, J. G., and I. R. Adelman. 1991. Effects of temperature, photoperiod, and ration size on growth of hybrid striped bass × white bass. Transactions of the American Fisheries Society 120:217–229.

Zhang, Q., R. C. Reigh, and W. R. Wolters. 1994. Growth and body composition of pond-raised hybrid striped basses, *Morone saxatilis* × *M. chrysops* and *M. saxatilis* × *M. mississippiensis*, fed low and moderate levels of dietary lipid. Aquaculture 125:119–129.

American Fisheries Society Symposium 80:249–262, 2013

Tagging Methods for Estimating Population Size and Mortality Rates of Inland Striped Bass Populations

JOSEPH E. HIGHTOWER*
U.S. Geological Survey
North Carolina Cooperative Fish and Wildlife Research Unit
Department of Biology, North Carolina State University
Raleigh, North Carolina 27695, USA

KENNETH H. POLLOCK
Departments of Biology, Biomathematics, and Statistics
North Carolina State University
Raleigh, North Carolina 27695, USA

Abstract.—Striped bass *Morone saxatilis* in inland reservoirs play an important role ecologically and in supporting recreational fishing. To manage these populations, biologists need information about abundance and mortality. Abundance estimates can be used to assess the effectiveness of stocking programs that maintain most reservoir striped bass populations. Mortality estimates can indicate the relative impact of fishing versus natural mortality and the need for harvest regulation. The purpose of this chapter is to evaluate tagging studies as a way of obtaining information about abundance and mortality. These approaches can be grouped into three broad categories: tag recapture, tag return, and telemetry. Tag-recapture methods are typically used to estimate population size and other demographic parameters but are often difficult to apply in large systems. A fishing tournament can be an effective way of generating tagging or recapture effort in large systems, compared to using research sampling only. Tag-return methods that rely on angler harvest and catch and release can be used to estimate fishing (F) and natural (M) mortality rates and are a practical approach in large reservoirs. The key to success in tag-return studies is to build in auxiliary studies to estimate short-term tagging mortality, short- and long-term tag loss, reporting rate, and mortality associated with catch and release. F and M can also be estimated using telemetry tags. Advantages of this approach are that angler nonreporting does not bias estimates and fish with transmitters provide useful ecological data. Cost can be a disadvantage of telemetry studies; thus, combining telemetry tags with conventional tag returns in an integrated analysis is often the optimal approach. In summary, tagging methods can be a powerful tool for assessing the effectiveness of inland striped bass stocking programs and the relative impact of fishing versus natural mortality.

* Corresponding author: jhightower@ncsu.edu

Introduction

Striped bass *Morone saxatilis* have been widely introduced into inland reservoirs, not only to provide recreational fishing opportunities, but also to fill an ecological role as a top-level predator. These introductions have generally been successful, but most populations are maintained by stocking because, with few exceptions, striped bass do not reproduce in reservoirs. Management of these stocked populations is generally limited to combinations of size and creel limits to control harvest. Information regarding population size or mortality rates would be helpful in developing or assessing regulations and stocking rates, but striped bass populations are difficult to survey because of their pelagic distribution (Moore et al. 1991).

Prior studies have shown that a variety of data sources can be combined to estimate approximate population size of striped bass in large reservoirs. For example, Moore et al. (1991) used age-composition data from standardized gill netting to generate a catch curve (Ricker 1975) estimate of the total mortality rate for striped bass in Smith Mountain Lake, Virginia. Natural mortality was thought to be low, so the exploitation rate was assumed to be equal to the total mortality rate. They then obtained an estimate of age-4 abundance by dividing a creel survey estimate of 1977 age-4 harvest by the assumed exploitation rate. That abundance estimate was used to scale up gill-net catch-per-unit-effort data in order to estimate abundance for other ages and years. Cyterski et al. (2002) produced additional estimates of abundance for Smith Mountain Lake striped bass. They used the average number of stocked juveniles and survival estimates from gill-net data to generate a population vector. For the reproducing striped bass population in the Santee–Cooper system in South Carolina, Bulak et al. (1995) estimated population size from estimates of egg production in tributary rivers, a mark–recapture estimate of juvenile abundance, a published estimate of juvenile mortality, and catch-curve estimates of mortality for fish age 1 and older.

A large-scale tagging study can provide direct estimates of population size and mortality that would be useful for management. For example, tagging estimates of population size can be used to evaluate stocking success or to adjust stocking rates. Population estimates can also be used in bioenergetic models to examine the balance between predators and prey resources (Cyterski et al. 2002). Fishing and natural mortality estimates from a tagging study can be used to construct yield models in order to examine the potential benefits of different fishing mortality rates or size limits (Thompson et al. 2007). Information about fishing versus natural mortality is also helpful in determining whether a stocking program is meeting its goal or whether other factors (e.g., poor habitat quality) limit the program's potential (Hightower et al. 2001).

The purpose of this chapter is to examine various tagging methods (tag recapture, tag return, telemetry) to determine what types of estimates each method can provide and then suggest under what conditions the method would be practical in the field. We also include guidance about the reliability of tag-return estimates as a function of study design and mortality rates. This review of tagging methods is directed towards striped bass researchers but should be equally relevant for fisheries biologists studying hybrid striped bass and other fish species in large reservoirs.

Tag-Recapture Models

Closed tag-recapture models are used to estimate population size. The simplest version is the Petersen model for two sampling occasions (Ricker 1975; Seber 1982). It is applicable in short-term studies where the population is closed (no changes due to recruitment, mortality, emigration, or immigration). We provide below an overview of the Peterson model and then consider more general models for three or more sampling occasions. There are versions of these more general models for situations where the animals may have unequal capture probabilities due to heterogeneity (variation among individuals) or trap response (where being captured affects the odds of being caught again). Open models allowing for recruitment, mortality, emigration, and immigration are used to estimate population sizes and other demographic parameters.

Petersen two-sample model

The Petersen closed population model has a long history and can be applied whether tagging and recapture are done solely by the researcher or whether angler catches provide fish for tagging or recapture. This approach assumes that the population is closed (no mortality and no recruitment or migration in or out), so sampling would need to occur over a month or two, given the typical annual survival rate of 40% for adult striped bass (e.g., Moore et al. 1991; Bulak et al. 1995; Hightower et al. 2001; Young and Isely 2004; Thompson et al. 2007).

For inland striped bass, a two-sample population estimate could be obtained by research gill netting. A practical concern about relying on research collections for tagging and recapture is that for a large and widely dispersed reservoir population, it is difficult to achieve a suitable capture probability (probability of capturing an individual on a sampling occasion). For a reservoir population estimate, capture probability can also be thought of as the fraction of the population sampled on each occasion. As an example, consider a 4,000-ha reservoir that could have 65,200 age-1 and older striped bass (based on the estimate of 16.3/ha from Moore et al. 1991). Guidelines from Robson and Regier (1964) suggest that a two-sample Petersen estimate should be based on capture of roughly 2,000 fish in each of the two samples in order to be within 25% of the true population size (i.e., between 48,900 and 81,500). Capturing 2,000 fish on each occasion from a population of 65,200 is a capture probability of about 0.03, which is low but would still be difficult to achieve in practice. The highest average gill-net catch per unit effort for Smith Mountain Lake was 10.06 (age 1) striped bass per overnight set of three gill nets (Wilson 2004), which implies that 200 three-net sets would be needed to achieve a sample size of 2,000 fish. Another practical issue is that mortality associated with gill-net capture would need to be measured and accounted for in the population estimate.

An alternate approach would be to take advantage of the sampling effort generated by recreational anglers. For example, Baldwin et al. (2003) estimated the size of a walleye *Sander vitreus* (formerly *Stizostedion vitreum*) population by tagging fish caught in a fishing tournament and obtaining recaptures through short-term gill netting and electrofishing. This approach of using different gears for tagging and recapture was recommended by Ricker (1975), since it would be unlikely for the same biases to be present in both gears. One concern about using tournament catches for the tagging sample is the potential for hooking (tagging) mortality in striped bass. Edwards et al. (2004) tagged largemouth bass *Micropterus salmoides* by electrofishing and obtained recaptures through fishing tournaments on two Connecticut lakes. If a fishing tournament is used to obtain recaptures for a two-sample Petersen estimate, the second sample should be the entire catch of the tournament, with all the fish examined for tags. If fish released during the day ("culled") provide tag returns, it would also be necessary to include the number of culled fish in the second sample. Tag returns from anglers (nontournament fishing) could be used as the recaptures, but this approach is subject to much greater uncertainty compared to a tournament because the size of the second sample (total harvest) would need to be estimated through a creel survey. Nonreporting of tags is a potentially large bias in using fishery tag returns as recaptures. We postpone further discussion of reporting rate to when we discuss tag-return models to estimate fishing and natural mortality.

General tag-recapture models

Tag recapture typically refers to the situation where capture, tagging, release, and one or more live recaptures are made by the researcher. This is distinctly different from tag-return studies (discussed later) that utilize angler harvested tag returns to obtain one recapture. Tag-recapture models are generally categorized as either closed, with the population totally static during the short study, or open, where allowance is made for changes due to births, deaths, emigration, and immigration.

If the focus is on estimating population size and not mortality rates, a short-term study is recommended so that the population can be assumed to be closed (Pine et al. 2003). The Petersen model for two samples is the simplest closed model and has been discussed in the previous section. More general closed population models for more than two samples have a long history, going back to Schnabel (1938). The Schnabel method is very simple and requires the strong assumption of equal capture probabilities of all fish on each sampling occasion (no heterogeneity and no trap response). This approach can be used on a large lake, although it is difficult in practice to obtain large enough samples for tagging and recapture. Hansen et al. (2008) used multiple occasion sampling and a Schnabel model to generate population estimates of lake trout *Salvelinus namaycush* in a 38,300-ha lake. Large trap nets were used to estimate abundance of mature fish during fall spawning periods in 2 years, and gill nets were used in a third year to generate an estimate applicable to all sizes (not just mature fish). Although the estimates required substantial effort (709–1,039 trap-net-nights in the first 2 years and 137 gill-net-nights in the third season), they provided valuable guidance to fishery managers about the current and future effects of this introduced species on native fauna.

The original Schnabel model has been generalized to allow for heterogeneity and trap response to influence capture probabilities (Otis et al. 1978). Those models are available through the software programs CAPTURE and MARK (Otis et al. 1978; White and Burnham 1999; Pine et al. 2003) and would be very useful in situations where abundance was the parameter of interest. In most cases, however, the primary focus for inland striped bass has been on estimating mortality, so we do not discuss them in detail here.

Open tag-recapture models (allowing births, deaths, emigration, and immigration) are the appropriate choice for longer-term studies. They provide information about abundance and mortality, but they have higher data requirements because there are more parameters to estimate. In addition to population size, there are parameters related to losses (apparent mortality, which is usually confounded with emigration) and gains (apparent recruitment, which is usually confounded with immigration). We have found very few published examples of these models being applied in lakes or reservoirs (Hightower and Gilbert 1984; Fabrizio et al. 1997; Mills et al. 2002; Pollock et al. 2007). For a large population, preliminary estimates of population size and survival (within 50% of the true value) can be obtained from a Jolly–Seber model with six sampling occasions and capture probabilities in the range of 0.02–0.04 (Hightower and Gilbert 1984). For the striped bass example considered above (65,200 fish age 1 and older), this would imply a sample size of 1,304–2,608 fish on each occasion. Such a large sample size would be difficult to achieve, even if sampling could be done at a time when the fish are more aggregated (e.g., during a spring spawning migration).

Tag-Return Models for Mortality Estimation

Tag-return models are commonly used when the focus of the study is on estimating mortality rates rather than population size. These so-called "Brownie models" were originally developed for migratory waterfowl (Brownie et al. 1985); thus, they are an effective approach for large or widely dispersed populations that are harvested, such as inland striped bass in reservoirs. These models are generally used to estimate the total mortality rate, but if information is available about the tag-reporting rate (probability that the tag on a harvested animal is reported), total mortality can be partitioned into harvest and natural death components (Pollock et al. 1991; Hoenig et al. 1998b). These estimates greatly enhance the value of tagging studies as they can be used to determine the need for harvest regulations or to determine if the current harvest rate is at or below a target level.

Tag-return studies are often conducted for a single year in order to get a point estimate of the exploitation rate. The multi-year approach developed by Brownie et al. (1985) produces estimates of annual total mortality, and this can be split into fishing and natural mortality if reporting rate can be estimated. The multi-year approach has higher cost but provides much more information about the demographics of the population. To obtain acceptable results, the study should span at least 3 to 5 years with a release of at least 300 tagged fish per year (Pine et al. 2003). These recommended sample sizes assume no mortality due to capture, handling, or tagging so they should be increased to compensate for any estimated tagging mortality. Tagging is preferably done at a time when fish are aggregated and readily collected for tagging. For example, tagging of the Roanoke River striped bass population occurs each spring when the fish are concentrated on the spawning grounds (Thomas et al. 2008). Tagging of the coastal migratory striped bass stock occurs during winter when the fish are aggregated off the coasts of North Carolina and Virginia (Welsh et al. 2007). In both cases, tag returns occur throughout the year, from recreational and commercial fisheries and from across the entire geographic range of the population. This wide temporal and spatial coverage would not be practical if dependent on research sampling for recaptures.

Assumptions of tag-return studies include (1) that tagged fish are representative of the entire population, (2) that there is homogeneity of the mortality processes over all fish considered in the model, (3) that the fates of tagged fish are independent, and (4) that tag returns are correctly reported by year (or period, if using a time step other than annual). There are quite a few potential fates of tagged fish (Figure 1), and the study should include auxiliary studies to account for potential biases such as tag loss and mortality due to the tagging process. Information about those factors could be taken from the literature, but every study is unique (e.g., skill level for taggers, water temperatures when tagging done) so using published values introduces an unknown bias. It is far preferable to estimate the rate of tag loss by double-tagging a proportion of fish and noting how many recoveries from double-tagged fish include one versus two tags. Mortality associated with tagging and immediate tag shedding can be estimated through short-term cage (holding pen) studies. Holding pen studies must be carefully designed to avoid bias (Pollock and Pine 2007). For example, mortality due to the holding pen can be accounted for by comparing survival of tagged and control fish, with the latter group treated identically except for the tagging process.

A critical step in conducting a tag-return study is to build in a method for estimating the tag reporting rate so that total mortality can be partitioned into fishing and natural mortality components. A common approach for estimating the reporting rate is through the use of high-reward tags (Pollock et al. 2001). The value of each high-reward tag is assumed to be high enough (e.g., $100) to ensure 100% reporting so the reporting rate can be estimated by comparing the tag-return rate for high-reward versus standard tags (Pollock et al. 2001). High-reward tagging should be done every year; otherwise, one must also assume that reporting rate is constant over years. Other approaches for estimating the tag reporting rate are through the use of planted tags (secretly planting tags in fishers' catches) or getting recoveries through an observer or creel sampling program (Hearn et al. 1999, 2003).

Another important step is to ensure that tagged fish are well mixed within the population as a whole (Hoenig et al. 1998a). This can be difficult to achieve if tagging is done on an aggregated subset of the population (e.g., spawners below an upstream dam) because

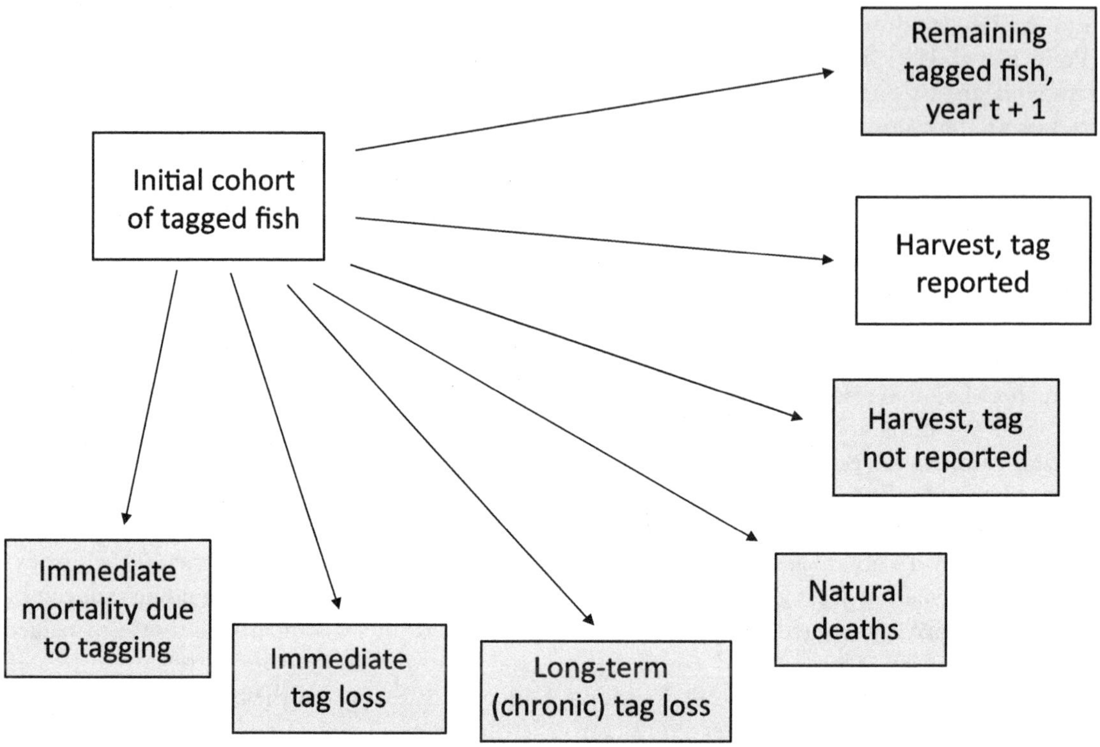

FIGURE 1. Possible fates of tagged fish in a tag-return study. Unfilled boxes are observed (initial number released, reported harvest); boxes with gray shading are fates that are not observed.

those aggregations attract fishing activity. This can result in many tag returns after only a short period at large and may overstate the true intensity of fishing. One analytical approach for dealing with nonmixing is to estimate separate mortality rates for newly tagged and previously tagged fish (Hoenig et al. 1998a). However, this reduces the sample size for the parameters of interest: those that apply to the untagged population. To avoid this consequence, the best approach (if possible) is to spread tagging effort widely across time and space. Information about the spatial distribution of striped bass (such as from telemetry) would be useful for allocating tagging effort.

Another consideration is whether the model needs to account for fish age (size). This might be necessary if vulnerability to harvest (or catch and release) varied substantially by size, for example, due to a minimum size regulation. If only fish above a minimum size limit are tagged, then the standard instantaneous-rates tag return model for adult fish (Hoenig et al. 1998a) should suffice. If some tagged fish are below the minimum harvestable size, the most practical approach is to use an age-structured model (Jiang et al. 2007) and set up the age-specific vulnerabilities to account for size effects.

The tag-return model must also account for tag returns from catch and release as well as harvest. If the tag is clipped when a fish is caught and released, it reduces the pool of tagged fish at risk but does not contribute to fishing mortality (except for catch-and-release mortality, discussed below). This can be accounted for analytically by modeling the population of tags at risk, rather than tagged fish (Jiang et al. 2007). An additional fishing mortality rate F' is introduced to account for the "mortality" of clipped tags. The population of tags declines due to fishing (F), clipping of tags for fish caught and released (F'), and natural mortality (M), so the total instantaneous mortality rate (Z) for tags is $F + F' + M$. Z for fish is $F + \delta F' + M$, where δ is a value between 0 and 1, accounting for mortality due to catch and release. The probability of catch-and-release mortality has to be estimated from some kind of auxiliary study (Bettinger and Wilde 2013, this volume). Reported tags from fish that are caught and released with the tag left intact are difficult to account for analytically. The situation is somewhat similar to the tag-recapture models discussed earlier, but in this case, recaptures by anglers occur at random times (rather than on a set schedule of research sampling occasions). More work is needed on this topic because very valuable information could be gained if fish were recaptured (caught and released) multiple times. Until new modeling approaches can accommodate these data, two practical options are to ignore those reported tags (Jiang et al. 2007) or to treat those tags as if they had been clipped (ignoring any later reports of those tags; Bacheler et al. 2008).

The precision of tag-return estimates depends on the number of returned tags, which depends on the number of tagged fish, rates of fishing and natural mortality, the reporting rate, and the number of years of recoveries. A biologist planning a tag-return study can use simulation with assumed rates to determine how many fish to tag in order to obtain useful results. Appendix 1 contains simulation code for program SURVIV, which can be run using a Web browser (www.mbr-pwrc.usgs.gov/software/survive.html).

The example is based on a standard Brownie model with 3 years of releases (at 200 fish per year) and 5 years of tag returns. The relative standard errors (SE/estimate × 100) show that precision is excellent for the survival rate (assumed constant over the experiment) and tag-return rates for years 1–3 (Table 1). The precision of survival estimates would be substantially reduced if estimated separately for each year. Precision of recovery rate estimates decreases considerably for the final 2 years, when no additional tags are released. The example code can be modified to explore how changes in study design or the assumed rates affect the expected precision (e.g., change number of released tags to 100 per year).

Tag-return methods are designed to estimate mortality rates, but it would be possible to estimate population size as well if a creel survey estimate of harvest was available. The tag-return estimates of F and M would be used to calculate an exploitation rate: $u = F/Z \cdot [1 - \exp(-Z)]$. Dividing the creel survey estimate of harvest by the exploitation rate would provide an estimate of abundance for fish within the exploited size range. There would be considerable uncertainty associated with an estimate of this type because it would include variance components due to the tag-return estimates of F and M and due to the creel survey estimate of harvest. Before planning a field study that would use this approach, it would be wise to simulate the tagging experiment (as in Appendix A) and to obtain at least a rough estimate of the variance for the creel survey estimate of harvest. An approximate variance for the population estimate can be obtained using the delta method (similar to equation (12.2) in Williams et al. 2002).

Telemetry-Based Approaches for Estimating Mortality Rates

Telemetry-based methods have proven effective for short- and long-term studies to estimate mortality rates. They differ from the previously described tag-recapture and tag-return studies in that fish with transmitters are rarely captured again after tagging. Instead information comes from "virtual" recaptures, when the telemetered fish is detected and some indication of its status is determined (e.g., its current position or change in position since the last detection).

Short-term telemetry studies have addressed management issues such as catch-and-release (hooking) mortality (Bendock and Alexandersdottir 1993; Lee and Bergersen 1996; Bettoli and Osborne 1998) or mortality associated with downstream passage through turbines (Skalski et al. 2002). These studies typically rely on transmitters that are attached

TABLE 1. Estimated rates, SEs, and relative SEs (SE/estimate × 100) from a simulated tag-return experiment (Appendix 1) with 3 years of releases (200 tagged fish per year) and 5 years of returns, using 100 replicates. Survival was 0.4 for all 5 years and was estimated as a single parameter; true annual recovery rates were 0.4, 0.5, 0.4, 0.5, and 0.4 for years 1–5.

Parameter	Mean estimate	Mean SE	Relative SE (%)
Survival rate	0.404	0.030	7.47
Recovery rate, year 1	0.396	0.032	8.19
Recovery rate, year 2	0.499	0.031	6.25
Recovery rate, year 3	0.399	0.030	7.40
Recovery rate, year 4	0.499	0.073	14.67
Recovery rate, year 5	0.397	0.109	27.39

quickly, either externally or by insertion into the stomach. This can be a better approach for studying catch-and-release mortality of striped bass than cage studies because of the potential for stress or mortality when held in cages (Haeseker et al. 1996; Bettoli and Osborne 1998). On the other hand, the concern for short-term telemetry studies is stress or possible mortality associated with the transmitter. Bettoli and Osborne (1998) noted that their short-term mortality estimates were low except during summer, which argued for a seasonal effect related to catch and release rather than due to transmitter attachment.

Improvements in transmitter and receiver technology have dramatically increased our capability to conduct longer-term studies that are the focus of this article. These studies can provide monthly, quarterly, seasonal, or annual estimates of fishing and natural mortality. The methods and models were first developed for terrestrial populations (Trent and Rongstad 1974; Pollock et al. 1989a, 1989b; White and Garrott 1990; Pollock et al. 1995), then modified for fish because the status of a telemetered fish is generally determined indirectly, through movement (Hightower et al. 2001). The approach requires that the entire study area be searched on a regular basis (e.g., monthly). Telemetered fish that change locations between search occasions are assumed to be alive, those that remain stationary are eventually classified as a natural mortality (or catch-and-release mortality), and those that disappear are classified as harvested. If fish can emigrate from the study area, fixed receiver arrays can be used to detect out-migration so that those fish can be censored from the analysis (and not incorrectly attributed to harvest).

This approach accounts for relocations of live fish and natural mortalities and allows for relocation probabilities less than one (Hightower et al. 2001). Model assumptions are as follows: (1) every marked animal present in the study area at time i (whether alive or dead due to natural causes) has the same probability of being relocated on the ith search occasion; (2) every marked animal alive and in the study area at time i has the same probability of surviving to the next search occasion; (3) the probability of transmitter failure or of a transmitter being shed is negligible; (4) all animals behave independently with respect to relocation and survival probabilities; (5) fish that leave the study area are either harvested or, if emigrating, are detected and censored; (6) a fish located repeatedly at the same site represents a natural or catch-and-release mortality; (7) natural mortality occurs immediately prior to the first relocation at the final site occupied by that fish; and (8) marked and unmarked animals are assumed to have equal survival rates.

Telemetry estimates of mortality rates have been produced for three reservoir populations of striped bass (Hightower et al. 2001; Young and Isely 2004; Thompson et al. 2007). The mortality estimates were generally precise and varied strongly among seasons. The seasonality of fishing can be important because of the high catch-and-release mortality observed when water temperatures are high (Wilde et al. 2000; Bettinger et al. 2005, Bettinger and Wilde 2013). A great benefit of these studies is the estimation of natural mortality, which is a parameter very hard to estimate any other way. Information about F versus M is helpful for understanding which source of mortality has a greater impact on abundance and the age or size distribution of the population. In these three examples, fishing mortality was substantially higher than natural mortality, suggesting that limits on harvest and catch-and-release mortality would produce more trophy-sized fish. Telemetry results can provide the information needed for yield modeling, for example, to examine different minimum size limits (Thompson et al. 2007).

Conventional tags only provide information at the times of tagging and recapture, whereas telemetry tags can provide information on multiple search occasions. Seasonal locations of sonic- or radio-tagged fish provide insights into habitat use (including summer refugia), spawning sites, and areas used for foraging. Telemetry locations can also aid in understanding habitat-related natural mortalities and spatial patterns in fishing. For example, all natural mortalities of telemetered striped bass in J. Strom Thurmond Reservoir occurred in mid to late summer in an area of unsuitable summer habitat (Young and Isely 2004). Fishing mortalities were concentrated in late spring and late summer in and around the tailrace of the next upstream dam (Young and Isely 2004).

A limitation of the telemetry approach is that it is not possible to distinguish between natural mortality and other sources of mortality that result in a stationary transmitter within the reservoir. The most likely cause would be mortality due to catch-and-release fishing, but another possibility would be for a fish to be filleted in the field, with the carcass (and transmitter) discarded in the water. These potential biases have not been a practical concern to date because the estimated Ms in previous telemetry studies of reservoir striped bass have been quite low (suggesting that other causes of stationary transmitters were negligible). It may be worth examining the spatial pattern of natural mortalities in order to assess the potential for these sources of bias. If apparent natural mortalities of telemetered fish are concentrated in areas of high fishing activity, then catch-and-release mortality or discarding of carcasses may be a nonnegligible fraction of the estimate of M.

One design issue to consider is whether a telemetered fish should also be given an external tag. External tags can provide some additional biological data and might be necessary in order to recover archival tags (Thompson et al. 2007). For estimating mortality, a reported external tag can clarify the fate of a fish that disappeared from the study area, for example, whether the transmitter failed, the fish was harvested, or the fish emigrated undetected and was caught (whether released or harvested) outside the study area. This information is useful but difficult to use in a quantitative way because of the typically high rates of underreporting. Hightower et al. (2001) recommended against use of an external tag because it might affect an angler's decision about whether to harvest or release a fish.

Combining Conventional Tags and Telemetry

We believe that the most cost effective strategy for a tag-return study would be a combination of conventional tags (e.g., internal anchor) and telemetry. Conventional tags have the advantage of allowing large sample sizes because of the low cost of tags. Another advantage is that relatively little effort is required other than at times when tagging occurs. These tags provide direct information about fishing mortality (through the return of tags) but only indirect information about natural mortality. In contrast, telemetry tags are expensive so most studies are carried out with small sample sizes. Considerable effort is required to search the entire study area during each period (e.g., monthly), but as noted above, this provides reliable seasonal information about mortality in addition to ecological information. Telemetry provides direct information about natural mortality (telemetered fish that stop moving) but only indirect information about fishing (through the disappearance of transmitters).

The combined method requires the independent release of fish with conventional tags (e.g., internal anchor) and transmitters. As the two tag types have complementary strengths, the combined analysis may produce better estimates of fishing and natural mortality than either method alone (Pollock et al. 2004). It also provides a reliable estimate of the reporting rate for conventional tags without the need for auxiliary methods such as high-reward or planted tags. Another advantage is that combined estimates can be compared to estimates based only on conventional tags or only on telemetry. This is helpful in determining whether the two tag types are providing consistent information. One possible complication is that the conventionally tagged fish and the telemetry tagged fish may not always be monitored over the same spatial scales, so they may not be subjected to the same mortality forces over the whole course of the study, leading to inconsistent estimates. However, even in such cases, the two different estimates of natural mortality could provide very useful information. There could also be violations of assumptions for one or both tag types (e.g., if there was unexpected shedding of conventional tags, or transmitter failure). Bacheler et al. (2009) used the combined method on estuarine red drum *Sciaenops ocellatus* and found similar temporal patterns in fishing mortality but moderate differences in magnitude. Conventional tags in that study were applied over a much larger region than the telemetry study area, so differences in fishing or natural mortality could have been due to spatial differences in mortality rates. As in the simulations by Pollock et al. (2004), the results for red drum showed that information on natural mortality was obtained mostly from telemetry, whereas conventional tags provided most of the information about fishing mortality.

Conclusions

Much of the information needed for effective management of inland striped bass can be obtained through tagging studies. Abundance can best be estimated through tag-recapture methods, either using research sampling alone or in combination with angler catches. A short-term study is preferable so that population closure can be assumed. It can be difficult to generate adequate sample sizes in the large systems where inland striped bass generally occur, so it is valuable to conduct simulation studies prior to the start of field work to determine if expected sample sizes will provide acceptably precise results.

A tag-return study is a practical approach for estimating mortality rates of reservoir striped bass. It is critical to build in auxiliary studies to estimate potential biases such as tag loss, mortality due to tagging, and tag-reporting rate. Simulation studies done prior to the start of field work can assess the likely precision of study results, given planned sample sizes. The simulations could include the potential impact of factors such as nonreporting or tag loss, both of which reduce the sample size of returned tags. The best approach for separating total mortality into component rates F and M is to combine release of a large annual sample of conventional tags with a smaller release of telemetry-tagged fish. Tag-return estimates of F and M can also be used, in conjunction with a creel survey estimate of harvest, to estimate population size for harvestable fish.

An estimate of adult striped bass population size, whether generated through a tag-recapture or tag-return study, would be helpful in assessing the effectiveness of a stocking program. For populations dependent on stocked juveniles, the estimate of adult abundance could also be used to estimate survival over the difficult-to-study stage between juvenile and adult. Other management questions such as optimal stocking rate or striped bass impacts on prey species (e.g., Cyterski et al. 2002) could be answered effectively with the information produced through a tagging study.

Acknowledgments

We thank M. Allen, J. Bulak, and J. Rice for comments on previous versions of the manuscript. The Cooperative Fish and Wildlife Research Unit is jointly supported by North Carolina State University, North Carolina Wildlife Resources Commission, U.S. Geological Survey, U.S. Fish and Wildlife Service, and Wildlife Management Institute. Any use of trade, product, or firm names is for descriptive purposes only and does not imply endorsement by the U.S. Government.

References

Bacheler, N. M., J. A. Buckel, J. E. Hightower, L. M. Paramore, and K. H. Pollock. 2009. A combined telemetry: tag return approach to estimate fishing and natural mortality rates of an estuarine fish. Canadian Journal of Fisheries and Aquatic Sciences 66:1230–1244.

Bacheler, N. M., J. E. Hightower, L. M. Paramore, J. A. Buckel, and K. H. Pollock. 2008. An age-dependent tag return model for estimating mortality and selectivity of an estuarine-dependent fish with high rates of catch and release. Transactions of the American Fisheries Society 137:1422–1432.

Baldwin, C. M., J. G. McLellan, M. C. Polacek, and K. Underwood. 2003. Walleye predation on hatchery releases of kokanees and rainbow trout in Lake Roosevelt, Washington. North American Journal of Fisheries Management 23:660–676.

Bendock, T., and M. Alexandersdottir. 1993. Hooking mortality of Chinook salmon released in the Kenai River, Alaska. North American Journal of Fisheries Management 13:540–549.

Bettinger, J. M., J. R. Tomasso, and J. J. Isely. 2005. Hooking mortality and physiological responses of striped bass angled in freshwater and held in live-release tubes. North American Journal of Fisheries Management 25:1273–1280.

Bettinger, J. M., and G. R. Wilde. 2013. Catch-and-release mortality of inland striped bass and hybrid striped bass. Pages 473–499 *in* J. S. Bulak, C. C. Coutant, and J. A. Rice, editors. Biology and management of inland striped bass and hybrid striped bass. American Fisheries Society, Symposium 80, Bethesda, Maryland.

Bettoli, P. W., and R. S. Osborne. 1998. Hooking mortality and behavior of striped bass following catch and release angling. North American Journal of Fisheries Management 18:609–615.

Brownie, C., D. R. Anderson, K. P. Burnham, and D. S. Robson. 1985. Statistical inference from band recovery data: a handbook. U.S. Fish and Wildlife Service, Resource Publication 156, Washington, D.C.

Bulak, J. S., D. S. Wethey, and M. G. White III. 1995. Evaluation of management options for a reproducing striped bass population in the Santee–Cooper system, South Carolina. North American Journal of Fisheries Management 15:84–94.

Cyterski, M., J. Ney, and M. Duval. 2002. Predator demand for clupeid prey in Smith Mountain Lake, Virginia. Fisheries Research 59:1–16.

Edwards, G. P., Jr., R. M. Neumann, R. P. Jacobs, and E. B. O'Donnell. 2004. Impacts of small club tournaments on black bass populations in Connecticut and the effects of regulation exemptions. North American Journal of Fisheries Management 24:811–821.

Fabrizio, M. C., M. E. Holey, P. C. McKee, and M. L. Toneys. 1997. Survival rates of adult lake trout in northwestern Lake Michigan, 1983–1993. North American Journal of Fisheries Management 17:413–428.

Haeseker, S. L., J. T. Carmichael, and J. E. Hightower. 1996. Summer distribution and condition of striped bass within Albemarle Sound, North Carolina. Transactions of the American Fisheries Society 125:690–704.

Hansen, M. J., N. J. Horner, M. Liter, M. P. Peterson, and M. A. Maiolie. 2008. Dynamics of an increasing lake trout population in Lake Pend Oreille, Idaho. North American Journal of Fisheries Management 28:1160–1171.

Hearn, W. S., J. M. Hoenig, K. H. Pollock, and D. A. Hepworth. 2003. Tag reporting rate estimation: 3. Use of planted tags in one component of a multiple-component fishery. North American Journal of Fisheries Management 23:66–77.

Hearn, W. S., T. Polacheck, K. H. Pollock, and W. Whitelaw. 1999. Estimation of tag reporting rates in age-structured multicomponent fisheries where one component has observers. Canadian Journal of Fisheries and Aquatic Sciences 56:1255–1265.

Hightower, J. E., and R. J. Gilbert. 1984. Using the Jolly-Seber model to estimate population size, mortality, and recruitment for a reservoir fish population. Transactions of the American Fisheries Society 113:633–641.

Hightower, J. E., J. R. Jackson, and K. H. Pollock. 2001. Use of telemetry methods to estimate natural and fishing mortality of striped bass in Lake Gaston, North Carolina. Transactions of the American Fisheries Society 130:557–567.

Hoenig, J. M., N. J. Barrowman, K. H. Pollock, E. N. Brooks, W. S. Hearn, and T. Polacheck. 1998a. Models for tagging data that allow for incomplete mixing of newly tagged animals. Canadian Journal of Fisheries and Aquatic Sciences 55:1477–1483.

Hoenig, J. M., N. J. Barrowman, W. S. Hearn, and K. H. Pollock. 1998b. Multiyear tagging studies incorporating fishing effort data. Canadian Journal of Fisheries and Aquatic Sciences 55:1466–1476.

Jiang, H. H., K. H. Pollock, C. Brownie, J. M. Hoenig, R. J. Latour, B. K. Wells, and J. E. Hightower. 2007. Tag return models allowing for harvest and catch and release: evidence of environmental and management impacts on striped bass fishing and natural mortality rates. North American Journal of Fisheries Management 27:387–396.

Lee, W. C., and E. P. Bergersen. 1996. Influence of thermal and oxygen stratification on lake trout hooking mortality. North American Journal of Fisheries Management 16:175–181.

Mills, K. H., S. M. Chalanchuka, and D. J. Allan. 2002. Abundance, annual survival, and recruitment of unexploited and exploited lake charr, *Salvelinus namaycush*, populations at the Experimental Lakes Area, northwestern Ontario. Environmental Biology of Fishes 64:281–292.

Moore, C. M., R. J. Neves, and J. J. Ney. 1991. Survival and abundance of stocked striped bass in Smith Mountain Lake, Virginia. North American Journal of Fisheries Management 11:393–399.

Otis, D. L., K. P. Burnham, G. C. White, and D. R. Anderson. 1978. Statistical inference from capture data on closed populations. Wildlife Monographs 62.

Pine, W. E., K. H. Pollock, J. E. Hightower, T. J. Kwak, and J. A. Rice. 2003. A review of tagging methods for estimating fish population size and components of mortality. Fisheries 28(10):10–23.

Pollock, K. H., and W. E. Pine. 2007. The design and analysis of field studies to estimate catch-and-release mortality. Fisheries Management and Ecology 14:123–130.

Pollock, K. H., J. Yoshizaki, M. C. Fabrizio, and S. T. Schram. 2007. Survival rates of lake trout estimated by a mark-recapture study in Lake Superior, 1969–1996. Transactions of the American Fisheries Society 136:185–194.

Pollock, K. H., C. M. Bunck, S. R. Winterstein, and C.-L. Chen. 1995. A capture–recapture survival analysis model for radio-tagged animals. Journal of Applied Statistics 22:661–672.

Pollock, K. H., H. Jiang, and J. E. Hightower. 2004. Combining telemetry and fisheries tagging models to estimate fishing and natural mortality rates. Transactions of the American Fisheries Society 133:639–648.

Pollock, K. H., J. M. Hoenig, and C. M. Jones. 1991. Estimation of fishing and natural mortality when a tagging study is combined with a creel survey or port sampling. Pages 423–434 *in* D. Guthrie, J. M. Hoenig, M. Holliday, C. M. Jones, M. J. Mills, S. A. Moberly, K. H. Pollock, and D. R. Talhelm, editors. Creel and angler surveys in fisheries management. American Fisheries Society, Symposium 12, Bethesda, Maryland.

Pollock, K. H., J. M. Hoenig, W. S. Hearn, and B. Calingaert. 2001. Tag reporting rate estimation: 1. An evaluation of the high reward tagging method. North American Journal of Fisheries Management 21:521–532.

Pollock, K. H., S. R. Winterstein, and M. J. Conroy. 1989a. Estimation and analysis of survival distributions for radio-tagged animals. Biometrics 45:99–109.

Pollock, K. H., S. R. Winterstein, C. M. Bunck, and P. D. Curtis. 1989b. Survival analysis in telemetry studies: the staggered entry design. Journal of Wildlife Management 53:7–15.

Ricker, W. E. 1975. Computation and interpretation of biological statistics of fish populations. Fisheries Research Board of Canada Bulletin 191.

Robson, D. S., and H. A. Regier. 1964. Sample size in Petersen mark-recapture experiments. Transactions of the American Fisheries Society 93:215–226.

Schnabel, Z. E. 1938. The estimation of the total fish population of a lake. The American Mathematical Monthly 45:348–352.

Seber, G. A. F. 1982. The estimation of animal abundance and related parameters, 2nd edition. Charles Griffin, London.

Skalski, J. R. R. Townsend, J. Lady, A. E. Giorgi, J. R. Stevenson, and R. D. McDonald. 2002. Estimating route-specific passage and survival probabilities at a hydroelectric project from smolt radiotelemetry studies. Canadian Journal of Fisheries and Aquatic Sciences 59:1385–1393.

Thomas, C. D., K. J. Dockendorf, and J. W. McCargo. 2008. Review of striped bass fisheries and monitoring programs in the Roanoke River, North Carolina—2007. Report to the Atlantic States Marine Fisheries Commission, Striped Bass Technical Committee. North Carolina Wildlife Resources Commission, Division of Inland Fisheries, Raleigh.

Thompson, J. S., D. S. Waters, J. A. Rice, and J. E. Hightower. 2007. Seasonal natural and fishing mortality of striped bass in a Southeastern reservoir. North American Journal of Fisheries Management 27:681–694.

Trent, T. T., and O. J. Rongstad. 1974. Home range and survival of cottontail rabbits in southwestern Wisconsin. Journal of Wildlife Management 38:459–472.

Welsh, S. A., D. R. Smith, R. W. Laney, and R. C. Tipton. 2007. Tag-based estimates of annual fishing mortality of a mixed Atlantic coastal stock of striped bass. Transactions of the American Fisheries Society 136:34–42.

White, G. C. and K. P. Burnham. 1999. Program MARK: Survival estimation from populations of marked animals. Bird Study 46(Supplement):120–138.

White, G. C., and R. A. Garrott. 1990. Analysis of wildlife radio-tracking data. Academic Press, San Diego, California.

Wilde, G. R., M. I. Muoneke, P. W. Bettoli, K. L. Nelson, and B. T. Hysmith. 2000. Bait and temperature effects on striped bass hooking mortality in freshwater. North American Journal of Fisheries Management 20:810–815.

Williams, B. K., J. D. Nichols, and M. J. Conroy. 2002. Analysis and management of animal populations: modeling, estimation, and decision making. Academic Press, San Diego, California.

Wilson, D. M. 2004. Effects of expanded stocking locations on striped bass survival. Proceedings of the Annual Conference Southeastern Association of Fish and Wildlife Agencies 56(2002):79–85.

Young, S. P., and J. J. Isely. 2004. Temporal and spatial estimates of adult striped bass mortality from telemetry and transmitter return data. North American Journal of Fisheries Management 24:1112–1119.

APPENDIX A. SURVIV code for simulating a tag-return experiment, with 3 years of releases of 200 fish per year and 5 years of returns. Cell probabilities are obtained from a standard Brownie model (model estimates survival and tag-recovery rates), with survival assumed to be constant for simplicity. Tag recovery values (cell frequencies) are determined in the simulations and are set at 0 except for the final cell, which is set equal to the number released.

```
PROC MODEL NPAR=6 ADDCELL /* From Brownie et al. (1978), p 21 */ ;
 COHORT = 200        /* Tagged yr 1 */ ;
   0:S(2)            /* Tags recovered yr 1 */ ;
   0:S(1)*S(3)       /* Tags recovered yr 2 */ ;
   0:S(1)**2*S(4)    /* Tags recovered yr 3 */ ;
   0:S(1)**3*S(5)    /* Tags recovered yr 4 */ ;
 200:S(1)**4*S(6)    /* Tags recovered yr 5 */ ;
 COHORT = 200        /* Tagged yr 2 */ ;
   0:s(3)            /* Tags recovered yr 2 */ ;
   0:s(1)*s(4)       /* Tags recovered yr 3 */ ;
   0:s(1)**2*s(5)    /* Tags recovered yr 4 */ ;
 200:s(1)**3*s(6)    /* Tags recovered yr 5 */ ;
 COHORT = 200        /* Tagged yr 3 */ ;
   0:s(4)            /* Tags recovered yr 3 */ ;
   0:s(1)*s(5)       /* Tags recovered yr 4 */ ;
 200:s(1)**2*s(6)    /* Tags recovered yr 5 */ ;
 LABELS;
   s(1)=Survival;
   s(2)=Tag-rec_rate_1;
   s(3)=Tag-rec_rate_2;
   s(4)=Tag-rec_rate_3;
   s(5)=Tag-rec rate_4;
   s(6)=Tag-rec_rate_5;
PROC SIMULATE NSIM=100 SEED=4567555 ERROR TEST ;
INITIAL /* Statements that follow fix the true values for simulation */;
  s(1)=0.4;
  s(2)=0.4;
  s(3)=0.5;
  s(4)=0.4;
  s(5)=0.5;
  s(6)=0.4;
PROC ESTIMATE;
INITIAL /* Statements that follow fix the starting values for estimation */;
  s(1)=0.5;
  s(2)=0.5;
  s(3)=0.5;
  s(4)=0.5;
  s(5)=0.5;
  s(6)=0.5;
```

American Fisheries Society Symposium 80:263–278, 2013

Evaluation of the Use of Gill Nets for Monitoring Reservoir Striped Bass Fisheries

Brian J. McRae*
North Carolina Wildlife Resources Commission
1751 Varsity Drive, Raleigh, North Carolina 27606, USA

James S. Bulak and Barbara E. Taylor
South Carolina Department of Natural Resources
1921 Van Boklen Road, Eastover, South Carolina 29044, USA

Christian T. Waters
North Carolina Wildlife Resources Commission
1751 Varsity Drive, Raleigh, North Carolina 27606, USA

Abstract.—Gill nets are commonly used to evaluate recreational and commercial striped bass *Morone saxatilis* fisheries. Using historical gill-net data for striped bass from four southeastern reservoirs, we fitted selection curves to evaluate size selectivity of the nets, assessed the effect of net selectivity on stock assessment metrics, defined the variability associated with two commonly used survey metrics, catch per unit effort (CPUE) and length at age, and evaluated the likelihood of observing a true change in these metrics between two time periods. Selection curves fitted to data from an experimental gill net with bar mesh sizes of 25, 38, 51, 64, and 76 mm indicated that the net was most efficient at capturing striped bass around 600 mm in length; efficiency of the net declined sharply for fish less than 300 mm in length. Adjusting catch data for net selectivity did not have a significant effect on either mortality rates from age 1 to age 3 or mean length at age. Variability associated with length-at-age data was relatively small while variability in CPUE data was relatively large. We found that historical sample sizes of age-1 and age-2 striped bass would be adequate to detect changes in mean length at age as small as 15 mm between two 5-year periods with a power of 0.8. However, in most instances, the historical number of net sets per year needed to be substantially increased to detect either a 30% or 50% change in CPUE between the two periods with a power of 0.8. In three of the four reservoirs, the historical number of nets set per year was sufficient to detect a 100% change in CPUE between the two periods with a power greater than 0.8. Variance associated with abundance-based metrics will generally be more difficult to overcome than that associated with fish-specific data or cohort-specific data, such as mean length at age. We suggest that managers carefully evaluate the population-specific variability associated with the metrics of interest and consider the acceptable level of risk for failing to detect a change.

* Corresponding author: brian.mcrae@ncwildlife.org

Introduction

Gill nets are commonly used to evaluate recreational and commercial fisheries (Diana et al. 2006; Isermann 2007; Adlerstein et al. 2008; Lauer et al. 2008). Waters (2005) surveyed reservoir managers from fish and wildlife agencies in the southeastern United States and found that of the agencies that managed reservoir striped bass *Morone saxatilis* fisheries, 9 of the 10 respondents indicated that gill nets were used to monitor striped bass populations. Thus, gill nets appeared to be the principal gear used to assess reservoir striped bass populations in southeastern reservoirs.

Waters (2005) further indicated that annual sampling effort and frequency of sampling varied greatly. Most often, a survey design based upon fixed sites sampled over time was employed, while a few agencies used either a random or targeted approach. Also, net configurations varied from experimental nets (i.e., a net with multiple panels of differing meshes) with up to eight panels to individual, single-mesh nets; mesh sizes ranged from 13- to 102-mm bar mesh. Most agencies sampled in late fall and winter, between October and February.

The survey of southern agencies indicated that data analyses employed were similar with almost all of the respondents evaluating some measure of condition, relative abundance, and growth (Waters 2005). In addition, all but one of the agencies indicated that they used gill-net data to develop length- or age-frequency distributions or estimates of mortality rates. Since the same analyses are routinely used despite substantial differences in sampling methodology and gill-net configuration, there is a need to assess the variability and bias associated with gill-net data and to evaluate the robustness of the stock assessment metrics typically used by managers.

Success at capturing fish in gill nets is influenced by rates of encounter with and retention by the net (Rudstam et al. 1984; Anderson 1998). Accounting for these factors is critical during the experimental design and data analysis phases of a sampling program.

Encounter rate (i.e., the probability of a fish coming into contact with a gill net) is directly related to the fish's behavior. Therefore, the efficiency of a gill net is a function of fish movement (Hubert 1996), and movement is influenced by species-specific life history traits and environmental factors. However, encounter probabilities can also differ within species. Rudstam et al. (1984) suggested that large fish have a higher probability of encountering gill nets than small fish because they swim faster and travel farther. Encounter probabilities and catch rates also depend on the length of time the nets soak in the water (Hansen et al. 1998). Catch rates decline with increasing soak time as a result of net saturation and the loss of available mesh or because fish avoid the net as they observe other dead or struggling fish captured in the net (Kennedy 1951; Minns and Hurley 1988; Olin et al. 2004). Because striped bass are highly mobile, particularly in the fall and winter period when most gill-net surveys are conducted, measuring encounter rates in a large reservoir is difficult.

The retention of fishes in a gill net is dependent on fish becoming wedged, gilled, or tangled by the mesh once a net is encountered (Hubert 1996). Mesh size is the most important factor determining whether a fish is retained in a net. The efficiency with which particular mesh sizes capture fish of different lengths is referred to as "net selectivity." Net selectivity has been evaluated in many species (Henderson and Wong 1991; Kraft and Johnson 1992; Pierce et al. 1994; Wilde 1995; Anderson and Neumann 2000). Baranov (1948) suggested that the majority of fish caught in a gill net will be within ±20% of the optimum capture length associated with a particular mesh. Hamley (1980) indicated that the optimum capture girth is about 1.25 times the mesh perimeter. For male striped bass in the Roanoke River, North Carolina, Trent and Hassler (1968) reported that mean fork length (in inches) of fish captured in a specific mesh was 2.24 + 3.41 × the stretched mesh (in inches). Although reservoir striped bass are often sampled by gill net, net selectivity has not been evaluated for these populations.

Using historical gill-net data for reservoir striped bass, our objectives were to (1) define the selection curves for reservoir striped bass; (2) assess the effect of net selectivity on stock assessment metrics; (3) define the variability associated with two commonly used survey metrics, catch per unit effort (CPUE) and length at age; and (4) evaluate the likelihood of observing a true change in these metrics between two time periods.

Methods

Historical gill-net data

We used gill-net data from four reservoirs located in the southeastern United States, B. Everett Jordan Reservoir (2004–2008), Lake Norman (2003–2007), Smith Mountain Lake (2004–2008), and the Santee–Cooper system (1983–2004). B. Everett Jordan Reservoir is a 5,270-ha flood control reservoir that was impounded in 1982 and is located near Raleigh, North Carolina. Lake Norman is a 13,156-ha impoundment located north of Charlotte, North Carolina, constructed in 1963. Also impounded in 1963, Smith Mountain Lake is an 8,337-ha reservoir located southeast of Roanoke, Virginia. The Santee–Cooper system, constructed in 1941, is located northwest of Charleston, South Carolina and consists of Lake Marion (40,469 ha) and Lake Moultrie (22,622 ha).

Striped bass were collected annually from each reservoir for the time periods reported. However, effort and net configuration varied among the four systems. At B. Everett Jordan Reservoir, bottom-set, single-panel gill nets that were 61 m long and 2.4 m deep were used to target striped bass. Approximately eight net-nights of 51- and 76-mm bar mesh were set in December each of the 5 years. In 2006, 64-mm mesh was added and approximately four net-nights were fished in each of the remaining 3 years. In 2007, 89-mm mesh nets were added and six net-nights were fished in 2007 and 2008. At Lake Norman, suspended, experimental nets that were 76.2 m long and 6.1 m deep were used. Two nets with 38.1-m panels of 38- and 51-mm bar mesh and two nets with 64- and 76-mm bar mesh were set for one night in December in most years. At Smith Mountain Lake, surface set, single-panel nets that were 61 m long and 2.4 m deep were set at six locations in October, November, and December of each year. On each sampling occasion, two 32-, 51-, and 76-mm bar mesh nets were set. At the Santee–Cooper system, experimental gill nets were set at four locations in both lakes Moultrie and Marion during December, January, and February. Each experimental net was 2.4 m deep and constructed of five 15.2-m panels with bar mesh sizes of 25, 38, 51, 64, and 76 mm. On each sampling occasion, one sinking net and one floating net were set overnight.

For each data set, total length (mm) and weight (g) were measured for each striped bass captured. In addition, mesh of capture was noted except at Lake Norman in 2003 and 2004. Sagittal otoliths were removed for later age determination. Since sample collections straddled 31 December, which is commonly used as the point at which age changes, we defined 1 May as the birth date of all striped bass.

Net selectivity

Striped bass data from the Santee–Cooper system were compiled and analyzed for the period 1983–2004. Analyses included length and age distributions from each mesh (25, 38, 51, 64, and 76 mm), mean length at age, and mortality as calculated by catch curve analysis (Ricker 1975). Catch curves were calculated using data for the 1997–2001 cohorts only. To investigate the possible effects of net selectivity on these metrics and indices, we used an indirect method to estimate a type B size selection curve for the series of mesh sizes fished at the Santee–Cooper system (Regier and Robson 1966; Hamley 1975). The selection curve for the net estimates the relative probability that a fish of a given length is captured if it encounters the sampling gear.

Hamley (1975) provided a thorough review of the development of indirect analytical methods. Millar (2000) discussed the statistical pros and cons of three popular analytical strategies that have been used recently. We used the maximum likelihood estimation method of Millar (www.stat.auckland.ac.nz/~millar/selectware/code.html). For each of the five meshes, we scaled selectivity relative to the size-class with the highest probability of capture, which was assigned a value of 1.0. We evaluated the lognormal model, the normal model with equal spread, the normal model with unequal spread, and the gamma model.

We compared the selection curves to estimate the size range for which each mesh had the highest fitted selectivity. From the field data, we estimated the size range for which each mesh had the highest capture rate. We also estimated optimal capture length using the equation of Trent and Hassler (1968).

We summed the selectivities by size-class across meshes to define the selection curve for the five-panel experimental net. For each size-class i, we calculated the adjusted catch a_i using the equation

$$a_i \frac{\sum_{j=1}^{5} c_{ij}}{\sum_{j=1}^{5} s_{ij}},$$

where c = observed catch, s = fitted selectivity, and j = mesh size-class.

We used the observed and adjusted gill-net data to calculate mean length at age and a mortality (Z) estimate using cohort-specific, catch curve analysis (Miranda and Bettoli 2007). Analysis of covariance (ANCOVA) was used to test (P = 0.05) whether adjusted and observed slopes were different.

Variability in length at age and catch per unit effort

Historical data were pooled within reservoirs and 5-year means and associated SD for length at age 1, length at age 2, and catch per net-night at age 1 (CPUE) were calculated (Table 1). Data were not corrected for net selectivity. For length-at-age comparisons, catch data from all meshes were used on B. Everett Jordan Reservoir (2004–2008), Lake Norman (2003–2007), Smith Mountain Lake (2004–2008), and the Santee–Cooper system (2000–2004). The cohort of interest for CPUE evaluation was age-1 fish. Cohort CPUE was evaluated from experimental gill nets at Lake Norman and the Santee–Cooper system, whereas only fish captured from 51-mm bar mesh gill nets were evaluated at B. Everett Jordan Reservoir (2004–2008) and only fish from 32- and 51-mm bar mesh gill nets were evaluated at Smith Mountain Lake (2004–2008). Data from 2000 to 2004 were analyzed for the Santee–Cooper system, and CPUE for age-1 striped bass from Lake Norman were only calculated from the 3-year period of 2005–2007. In addition to these metrics, average annual sample size (i.e., number of striped bass for length at age and number of net-nights for CPUE) was calculated.

Power analysis was used to identify the sample sizes needed to detect specific changes in the metrics over time if two-sample t-tests were used (Brown and Guy 2007). The metrics that were evaluated were length at age 1, length at age 2, and age-1 CPUE. For the length-at-age metrics, we investigated changes from the historical means of 5, 10, 15, 20, and 25 mm. For CPUE, changes from the historical means of 30%, 50%, and 100% were evaluated.

We calculated the total sample size needed during a 5-year period to obtain a power ($1 - \beta$) of 0.8 with α = 0.05. For the length-at-age metrics, total sample size was defined as the number of striped bass collected. For CPUE, total sample size was the number of net-nights set. To obtain an estimate of annual sample size for a given metric, the total sample size required for the 5-year period was divided by five.

Power of the two-sample t-test was analyzed with a SAS (SAS 2002) program (Appendix A). Sample standard deviations were

TABLE 1. Five-year means (SD) of length at age 1 and length at age 2 from all striped bass collected and 5-year means (SD) of age-1 striped bass catch per unit effort (CPUE) from all nets sets from B. Everett Jordan Reservoir (2004–2008), Lake Norman (2003–2007), Smith Mountain Lake (2004–2008), and the Santee–Cooper system (2000–2004). Mean (SD) annual sample sizes among years are also identified. Cohort CPUE was evaluated from experimental gill nets at Lake Norman and the Santee–Cooper system, whereas only fish captured from 51-mm bar mesh gill nets were evaluated at B. Everett Jordan Reservoir and only fish from 32- and 51-mm bar mesh gill nets were evaluated at Smith Mountain Lake. Estimates were not corrected for net selectivity.

	B. Everett Jordan	Norman	Smith Mountain	Santee–Cooper
5-year mean length (mm) at age 1	413 (20)	456 (46)	428 (28)	433 (38)
Mean sample size age-1 fish/year	37 (10)	28 (13)	171 (104)	118 (89)
5-year mean length (mm) at age 2	530 (24)	526 (34)	526 (31)	539 (37)
Mean sample size age-2 fish/year	37 (37)	34 (21)	39 (21)	73 (81)
5-year mean age-1 CPUE[a]	5 (5)	6 (8)	14 (14)	5 (9)
Mean net-nights/year	8 (5)	5 (2)	12 (0)	24 (1)

[a] Historical cohort CPUE from Lake Norman was only calculated for the 3-year period 2005–2007.

obtained from the historical data sets (Table 1) and assumed to be equal for both periods. We also assumed that the two-sample *t*-test was robust against data that were not normally distributed. Similarly, we assumed that among-site variability was not significant for length-at-age data. For the power analysis, the sample size for the historical period was held constant and equal to the historical sample size (Table 1).

Results

Net selectivity

More than 5,500 striped bass were captured from lakes Marion and Moultrie in gill nets during the study period (1983–2004). As expected, length of captured fish increased as mesh size increased (Figure 1). Shape of the distributions also changed as mesh size increased.

For estimating selectivity of the five-panel gill nets, the lognormal model (Figure 2) gave the best fit (model deviance of 1299 on 250 degrees of freedom), compared with normal with equal spread, normal with proportional spread, and gamma models. The normal model with equal spread gave a result (model deviance of 1466 on 250 degrees of freedom) that was close to the lognormal model. However, the lognormal model better represented the pattern in size distributions of captured fish.

The experimental, multi-mesh net was most efficient at capturing fish around 600 mm in length (Figure 2). Efficiency declined sharply for fish less than 300 mm. The mean length at age for age-0 fish (285 mm for the observed data, Table 2) lies within this range of sharply reduced efficiency while the mean lengths for fish ages 1–3 (433–655 mm for the observed data) lie within the range of similar, relatively high efficiency.

The selection curves reproduced the observed ranges of the relative capture rates better for the smaller meshes than for the larger meshes (Table 3). The selection curves were dis-

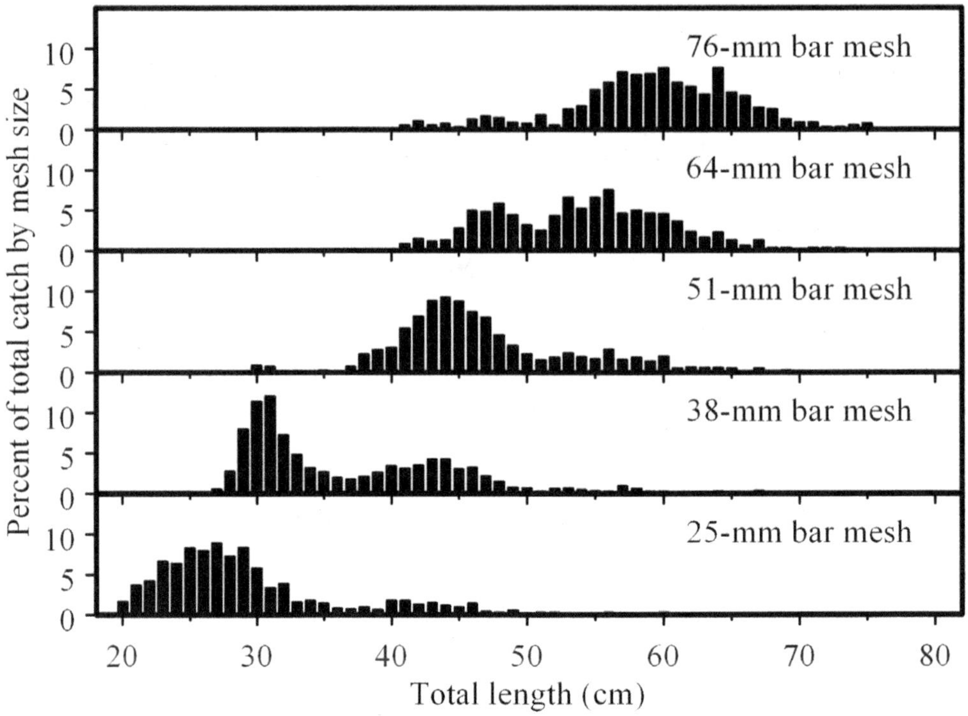

FIGURE 1. Length-frequency distributions of striped bass collected with experimental gill nets from the Santee–Cooper system, South Carolina, deployed December–February of each year from 1984 to 1985 through 2003–2004.

placed upward for the larger meshes. The modes of the curves lay at the edge or above the ranges of the highest observed relative capture rates for the 51- and 64-mm meshes. The optima, as predicted by Trent and Hassler (1968), and the midpoints of the ranges of the highest observed relative capture rates were more similar, differing by 5–22.5 mm (middle three meshes only).

Estimated instantaneous mortality (Z) from age 1 to age 3 was –0.83 (SE = 0.29) and –0.92 (SE = 0.29) for the observed and selectivity-adjusted data. Numbers of fish of ages 4 and older were too small for inclusion in this analysis. ANCOVA indicated that slopes of the observed and adjusted catch curves (Figure 3) were not statistically different (interaction effect: $F(1,26) = 0.04$, $p > 0.05$). Because age-0 fish were probably undersampled by the mesh configuration (see Figure 2), they were omitted from the comparison of slopes.

Mean lengths at age calculated using the observed and adjusted data sets differed by less than 5 mm for age-classes 0 through 4 (Table 2).

Variability in length at age and catch per unit effort

Length at age.—Mean lengths for all age-1 and age-2 striped bass over the 5-year historical period were generated for the four study reservoirs (Table 1). The data demonstrated similar variability around the means for both age-1 and age-2 striped bass, varying by less than 10%.

Historical average sample sizes of age-1 striped bass (Table 1) were adequate to detect a 25-mm change in total length during the future 5-year period for each reservoir if a power of at least 0.8 was desired. Even for a change as small as 15 mm, a power greater than 0.78 was attained for each reservoir if the historical sample sizes were obtained. However, to detect a 5-mm change between the two periods with

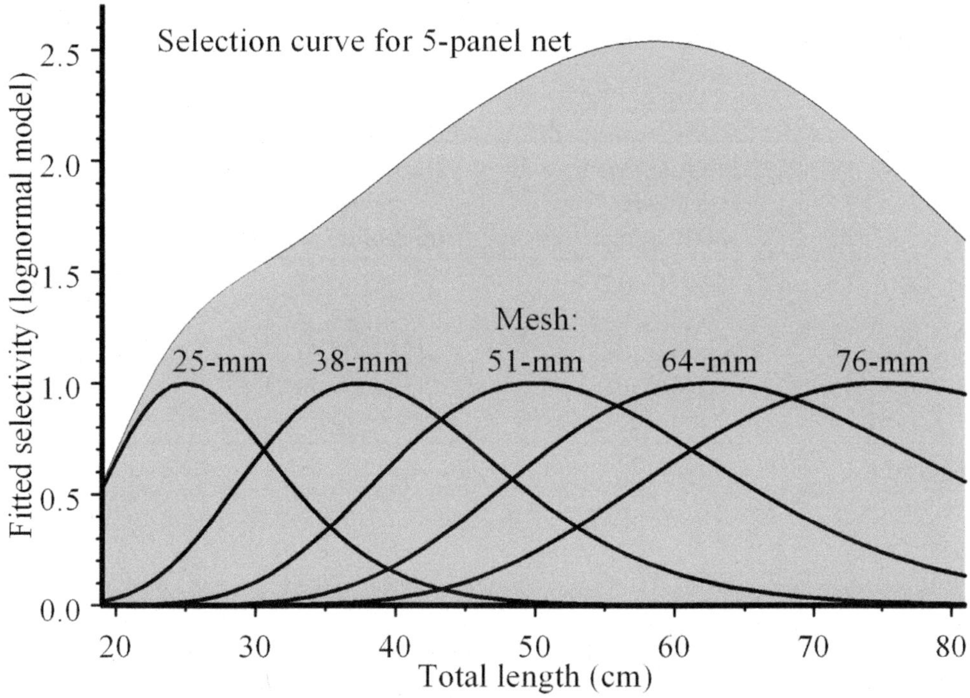

FIGURE 2. Fitted selectivity curves for striped bass for each individual mesh size in the five-panel experimental gill nets consistently deployed at the Santee–Cooper system, South Carolina 1984–1985 through 2003–2004, and the cumulative selection curve for all mesh sizes combined.

TABLE 2. Mean total length (mm) at age for striped bass from the Santee-Cooper system, South Carolina 1984–1985 through 2003–2004, as determined via experimental gill netting in December–February of each year. Top section shows observed data. Bottom section shows data corrected for gill-net selectivity. Values in parentheses indicate sample size. Estimates in bold indicate the largest sample for each age-class.

	Bar mesh (mm)					
Age	25	38	51	64	76	All mesh
0	**264 (912)**	307 (815)	301 (39)	–	–	285 (1766)
1	404 (203)	417 (677)	**440 (913)**	465 (270)	464 (55)	433 (2118)
2	529 (26)	545 (82)	539 (225)	**550 (409)**	574 (222)	552 (964)
3	602 (12)	600 (22)	608 (63)	602(151)	**619 (156)**	610 (404)
4		651 (11)	649 (23)	655 (44)	**657 (61)**	655 (139)
	Bar mesh (mm)					
Age	25	38	51	64	76	All mesh
0	**257 (912)**	306 (815)	298 (39)	–	–	281 (1766)
1	399 (203)	413 (677)	**438 (913)**	463 (270)	462 (55)	430 (2118)
2	529 (26)	544 (82)	538 (225)	**550 (409)**	574 (222)	552 (964)
3	601 (12)	600 (22)	608 (63)	603 (151)	**620 (156)**	610 (404)
4		652 (11)	651 (23)	658 (44)	**658 (61)**	656 (139)

TABLE 3. Comparison of fitted selectivities and observed capture rates of gill-net meshes for striped bass from the Santee–Cooper system. Estimate of Trent and Hassler (1968) is included for comparison.

	Length of fish (mm)		
Bar mesh (mm)	Highest fitted selectivity—mode (range)	Highest observed capture—midpoint (range)	Trent and Hassler optimum
25	250 (*, 305)	* (*, 285)	250
38	380 (305, 445)	345 (285, 405)	340
51	500 (445, 565)	453 (405, 500)	430
64	620 (565, 685)	553 (500, 605)	530
76	750 (685, *)	* (595, *)	620

* Undetermined.

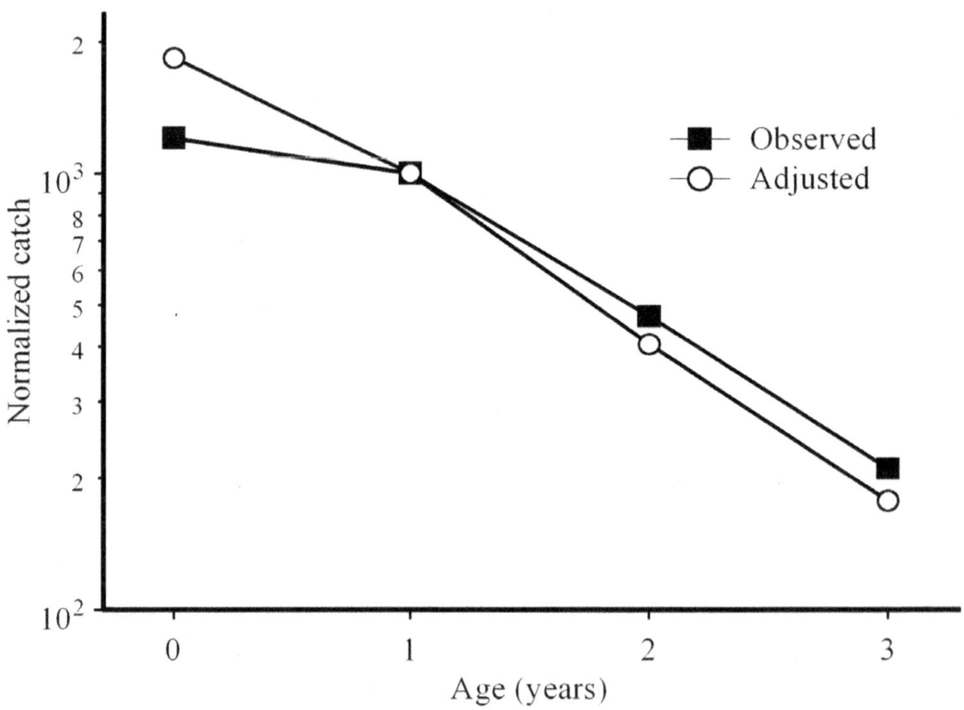

FIGURE 3. Ricker catch curves for the 1997–2001 striped bass cohorts in the Santee–Cooper system, South Carolina. Relative population sizes are based on combined gill-net data for all sampled years. Observed data (■) and data adjusted for net selectivity (○) are shown. To facilitate visual comparison, values were normalized to a catch of 1,000 at age 1.

a power of 0.8, only the historical sample size at Smith Mountain Lake was adequate (Figure 4). To detect a 5-mm change, the sample size of age-1 fish needed to increase from 37 to 81 fish at B. Everett Jordan Reservoir and from 118 to 400 fish at the Santee–Cooper system; a power of 0.8 was not attained at Lake Norman even if the historical sample size were increased fivefold. We caution that detection of differences on the order of 5–10 mm or less may be confounded by effects of net selectivity on estimates of average size.

Similar results were observed when changes in length at age for age-2 fish were evaluated. To detect a 25-mm change in total length, a power greater than 0.8 was attained if historical sample sizes of age-2 fish were collected from each reservoir. Similarly, at a power of at least 0.8, historical sample sizes from all four reservoirs were adequate to detect a 15-mm change. However, to detect a 5-mm change between the two periods with a power of 0.8, greater than 5 times the historical sample size for age-2 striped bass was needed at all four reservoirs (Figure 4). Again, we caution that detection of small differences may be confounded by effects of net selectivity.

Catch per unit effort.—Age-1 CPUE values from all nets set during the 5-year historical period varied among reservoirs (Table 1). For each reservoir, the catches also varied substantially per net set with standard deviations generally equal to or greater than mean CPUE (Table 1).

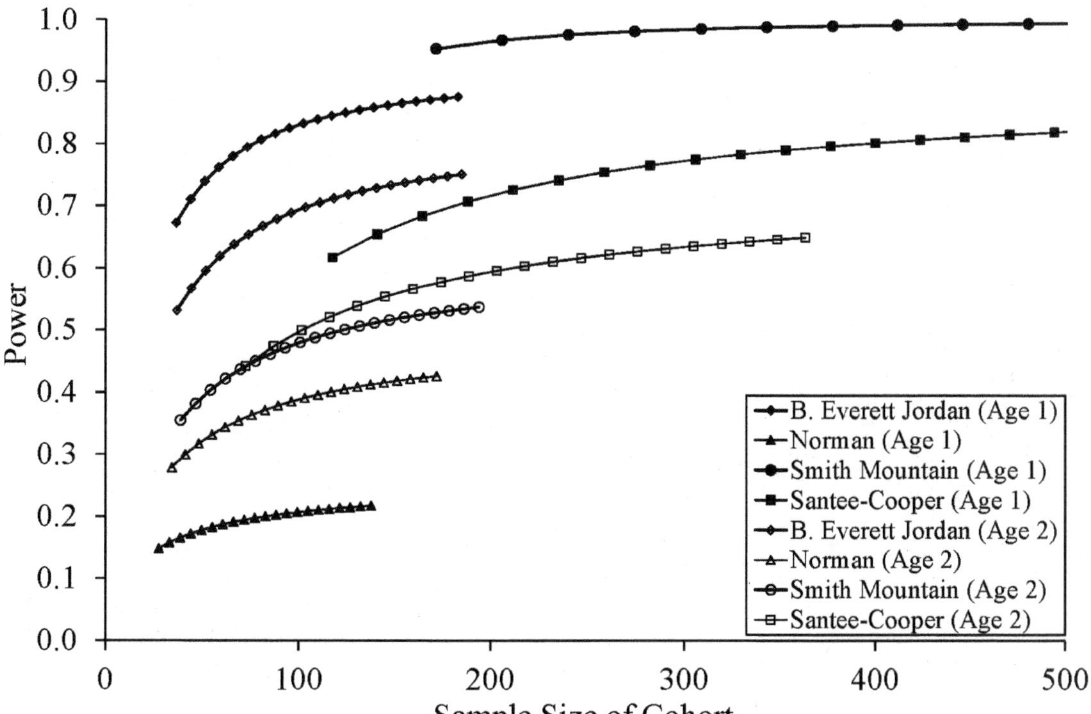

FIGURE 4. Power curves for two-sample *t*-tests when a 5-mm change from historical values in length at age 1 and length at age 2 was specified for striped bass from B. Everett Jordan Reservoir (2004–2008), Lake Norman (2005–2007), Smith Mountain Lake (2004–2008), and the Santee–Cooper system (2000–2004). The sample size range for each curve extends from the historical average catch (Table 1) to five times that catch (curves for age-1 fish in Smith Mountain Lake and the Santee–Cooper system are truncated at 500 fish).

Given the observed variability in catch among net sets, a considerable increase in number of net-nights was needed for each reservoir to detect a 30% change from historical CPUE values at a power of 0.8 (Figure 5). Increasing the number of net-nights 10 times as compared to the mean number of net-nights set during the 5-year historical period only attained power that ranged from 0.18 at Lake Norman to 0.62 at Smith Mountain Lake.

Smith Mountain Lake was the only system where the historical number of net-nights was sufficient to detect a 50% change in CPUE of age-1 fish with a power of 0.8 (Figure 5). In the Santee–Cooper system, a tripling of net-nights generated a power of 0.7 to detect a similar change, but power of 0.8 could only be achieved with a sevenfold increase in net-nights. However, at B. Everett Jordan Reservoir, even a 10-fold increase in the number of gill nets set per year only resulted in a power of 0.7 to detect a 50% change in CPUE of age-1 fish. In Lake Norman, increasing the number of net-nights by 10 times the historical number of sets only attained a power of 0.4 to detect a 50% change in CPUE (Figure 5).

Historical sample sizes at three out of the four reservoirs produced a power greater than 0.9 to detect a 100% change from the historical CPUE values. At Lake Norman, a power of only 0.7 was attained when the historical sample size was increased from 5 net-nights per year to 25 net-nights per year. Further, there was a negligible increase in power if the sample

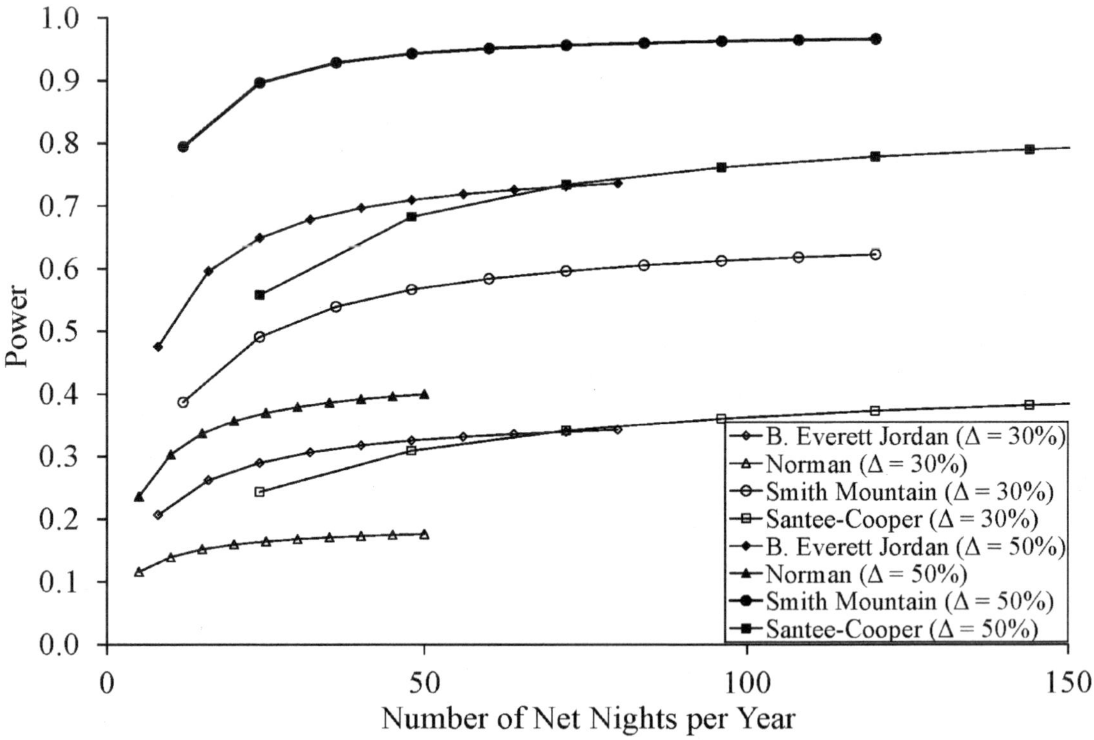

FIGURE 5. Statistical power of a two-sample *t*-test to detect a 30% or 50% change from historical values in catch per unit effort, in relation to the number of gill nets set per year during a future 5-year period, for age-1 striped bass from B. Everett Jordan Reservoir (2004–2008), Lake Norman (2005–2007), Smith Mountain Lake (2004–2008), and the Santee–Cooper system (2000–2004). The sample size range for each curve extends from the historical average number of net-nights per year (Table 1) to 10 times that number of net-nights (curves for the Santee-Cooper system are truncated at 150 net-nights).

size at Lake Norman was increased to 50 net-nights per year.

Discussion

Gill nets are a useful tool for evaluating reservoir striped bass populations and fisheries. However, to appropriately use gill nets, one must clearly define the question or hypothesis of interest and then appropriately consider the (1) size selectivity of various meshes, (2) magnitude and variance of daily gill-net catch numbers, (3) variance in annual cohort abundance, and (4) desired statistical certainty associated with a monitoring or study result.

Net selectivity for reservoir striped bass has not been previously documented, using either direct or indirect methods. The direct method requires either knowledge of the actual size structure of the population or comparison to catch in a gear with known selectivity. For example, Hamley and Regier (1973) tagged walleye *Sander vitreus* in a small lake and recaptured them using several mesh sizes to estimate selectivity. Trent and Hassler (1968) compared the catch of striped bass in experimental gill nets to the catch in bow nets, which were assumed to catch male striped bass randomly in relation to the length range of interest. The application of the direct method within a large reservoir would require a substantial expenditure of effort.

Therefore, in this study, we employed an indirect method to estimate net selectivity for striped bass in the Santee–Cooper system. It is important to understand selectivity when evaluating a fishery because indices calculated from gill-net data might otherwise lead to inappropriate management decisions (Finstad et al. 2000; Finstad and Berg 2004).

A selection curve (such as Figure 2) for a set of meshes can be used to determine the relative efficiency associated with sampling age- or size-classes of interest. Catch estimates can then be corrected for gill-net selectivity (Spangler and Collins 1992; Hovgard 1996; Anderson 1998; Helser et al. 1998; Irwin et al. 2008).

For the Santee–Cooper striped bass, adjusting for net selectivity did not substantially affect the estimate of mortality for ages 1–3 or estimates of mean size for ages 0–4, although it would have affected the mortality estimate obtained from ages 0–3. However, discrepancies between the observed and fitted size selectivities raise concerns about performance of the lognormal model for larger net meshes. This concern is reinforced by comparison with Trent and Hassler's (1968) estimates, which are more similar to the observed selectivities. We speculate that the poor fit of the model for the larger meshes is related to the broad overlap in size-classes captured by these meshes.

In any case, the net selection curve indicates important general properties of the gill-net data, particularly reduced selectivities toward the extremes of the sampled size range. An age-class with its median size near one of the tails will be underrepresented in calculations based on observed data, rather than on adjusted data. Given that the correction is evidently imperfect, a more robust alternative may be to use the estimates of selectivity simply to identify size ranges that are likely to have been undersampled.

Although mortality estimates were corrected for net selectivity, the variance in the abundance-at-age estimates used in the catch curve may still be considerable. Additionally, the standard error associated with the regression estimate of Z was substantial. Thus, while gill-net data were used to produce a statistically significant estimate of mortality, managers should exercise caution and use a range of mortality values when evaluating potential management actions.

Gill nets are appropriate gear when the objective is to evaluate changes in growth, as long as the appropriate meshes are selected. Growth of striped bass has been evaluated in many projects where the fish have been collected using gill nets (Schramm et al. 2001; Van Horn et al. 2001). For the four southeastern reservoirs in our study, we determined that the historical sample sizes would be adequate to detect small changes in mean length at age with relatively high confidence between 5-year periods. Differences in catch rates among nets do not influence the variation of lengths found within an age-class. Although the biases introduced by size selectivity appeared to be small in the data we analyzed, we recommend some caution in interpreting results for age-classes at the extremes of the sizes captured by the gear, such as age-0 fish from the Santee–Cooper system in this study. We conclude that the variability in metrics calculated from fish-specific data, such as length at age or presumably condition, is small enough to allow detection of change.

Unlike the variance in length-at-age data, we found that the variance associated with CPUE was very large for each of the four reservoirs. In all cases, mean standard deviation was at least 100% of the mean age-1 CPUE. Unlike fish- or cohort-specific data, there is no limit to the amount of variance that is possible when using gill nets to monitor relative abundance, as catches can range from zero to the full capacity of each net.

The variance in catch data obtained from gill nets will affect the level of change in CPUE that a manager can expect to detect. What level of change in the population is significant is a basic question to consider when using gill nets to monitor changes in relative abundance. It is likely that annual variation in abundance is a function of a specific reservoir. Lamprecht et al. (2013, this volume) show long-term monitoring data for the Santee–Cooper striped bass population; their data suggest that abundance can double from a period of weak recruitment to a period of strong recruitment. Thus, for this population, a 100% change in relative abundance for two 5-year periods may be significant. However, if changes in population abundance are less severe in a particular reservoir, then the ability to detect a 100% change may not be useful.

We found that most sampling designs were adequate to detect a 100% change in CPUE between two 5-year periods. However, we determined that the variability in daily catch was generally too large to detect a 50% change between two 5-year periods. Only at Smith Mountain Lake would the historic gill-net sampling strategy allow the statistical detection of a 50% change in CPUE of age-1 striped bass with a power of 0.8. Although a power of 0.8 is a common choice for comparisons (Brown and Guy 2007), managers could alternatively accept a lower power and higher risk of a type II statistical error when evaluating a given change of interest. For example, at the Santee–Cooper system, we found that if the net-nights were increased from 24 to 48 per year, a 50% change in CPUE would be detected with a power of 0.7. Similarly, Wagner et al. (2009) found that gill-net surveys had relatively low power to detect small temporal changes in walleye CPUE in Lake Erie.

We acknowledge that a two-sample *t*-test would not be the most powerful test to evaluate changes in CPUE. However, our hypothetical test revealed that managers should be cautious when comparing CPUE obtained from gill nets. We also acknowledge that growth and CPUE have among-year variance. However, we pooled historical data within reservoirs to negate the among-year variance in an effort to evaluate variability in commonly used metrics on a large scale to assist reservoir managers with sampling designs.

Variability associated with catch and data collected on a population scale from gill nets can mainly be attributed to either encounter probability or retention rates (Rudstam et al. 1984; Anderson 1998). Encounter rates are particularly problematic as they are difficult to estimate and address. However, if encounter probabilities can be increased, a corresponding decrease in variance could be observed. Historical gill-net data or telemetry data could be used to set nets in habitats where striped bass are likely to occur. For example, telemetry studies have shown that striped bass congregate during spring spawning runs and in the epilimnion or in refuge areas in the thermocline and hypolimnion during summer stratification (see Coutant 2013, this volume). This approach would increase the encounter probability and could decrease the associated variance, which is needed for a more meaningful comparison of some metrics. However, for this directed sampling approach to be valid for condition or length-at-age data, the manager would need to assume that the metrics in question are not spatially dependent. Similarly, for CPUE data, the manager would need to assume that mean CPUE is related to population size and that changes in CPUE over time are not related to changes in habitat use. Also, standard errors associated with abundance-based sampling could be decreased by reallocating sampling effort so that the number of samples was increased.

An a priori determination of the desired statistical certainty associated with monitoring or hypothesis testing is needed when using gill nets. Many studies acknowledge the variability inherent in catch data obtained from gill nets (Hansen et al. 1998; Olin et al. 2004). It is the magnitude of that variability in combination with a manager's willingness to risk failing to detect change (type II statistical error) that ultimately determines which stock assessment metrics are appropriate to use. In our analyses of gill-net data for striped bass from four reservoirs in the southeastern United States, variability among length at age was relatively small while variability in CPUE was relatively large. Our results suggest that the variance associated with abundance-based metrics, such as relative abundance and mortality, will generally be more difficult to overcome than that associated with fish-specific data, such as condition, or cohort-specific data, such as mean length at age.

We conclude that encounter probability is the main limiting factor when evaluating catch of striped bass with gill nets. However, if encounter probabilities can be increased such as by making use of knowledge of seasonal aggregation behaviors, variability in catch among nets would likely decrease, thereby making comparisons more valid. Our results suggested that we would need to substantially increase the number of gill nets fished per year to have moderate confidence in catch-related comparisons in the four study reservoirs. While selection curves as presented here can be utilized to refine catch data for size-selective bias, this does not address the variability associated with encounter rates. Selection curves can help managers focus the mesh sizes employed to maximize the catch of the size- and age-classes of interest. However, we caution that the range of meshes employed must extend beyond the size-classes of interest to prevent undersampling at the tails of the curves. Managers must still be aware that the needed effort to detect a given change, depending upon the level of acceptable risk of a type II statistical error, might still be too large to be practical.

Acknowledgments

We thank N. C. Oakley from the North Carolina Wildlife Resources Commission, S. D. Lamprecht from the South Carolina Department of Natural Resources, D. M. Wilson from the Virginia Department of Game and Inland Fisheries, and M. A. Abney from Duke Energy Company for providing the data used in this manuscript. We also thank John Grego from the University of South Carolina Statistics Laboratory who provided assistance with the initial use of Millar's gill net selectivity model and the power analyses.

References

Adlerstein, S. A., E. S. Rutherford, R. M. Claramunt, D. F. Clapp, and J. A. Clevenger. 2008. Seasonal movements of Chinook salmon in Lake Michigan based on tag recoveries from recreational fisheries and catch rates in gill-net assessments. Transactions of the American Fisheries Society 137:736–750.

Anderson, C. S. 1998. Partitioning total size selectivity of gill nets for walleye (*Stizostedion vitreum*) into encounter, contact, and retention components. Canadian Journal of Fisheries and Aquatic Sciences 55:1854–1863.

Anderson, M. R., and R. M. Neumann. 2000. Indirect estimate of gill net mesh size selectivity for landlocked alewives. Journal of Freshwater Ecology 15:491–496.

Baranov, F. I. 1948. The theory and assessment of fishing gear. Pishchepromisdat, Moscow.

Brown, M. L., and C. S. Guy. 2007. Science and statistics in fisheries research. Pages 1–29 *in* C. S. Guy and M. L. Brown, editors. Analysis and interpretation of freshwater fisheries data. American Fisheries Society, Bethesda, Maryland.

Coutant, C. C. 2013. When is habitat limiting for striped bass? Three decades of testing the temperature–oxygen squeeze hypothesis. Pages 65–91 *in* J. S. Bulak, C. C. Coutant, and J. A. Rice, editors. Biology and management of inland striped bass and hybrid striped bass. American Fisheries Society, Symposium 80, Bethesda, Maryland.

Diana, C. M., J. L. Jonas, R. M. Claramunt, J. D. Fitzsimons, and J. E. Marsden. 2006. A comparison of methods for sampling round goby in rocky littoral areas. North American Journal of Fisheries Management 26:514–522.

Finstad, A. G., and O. K. Berg. 2004. Bimodal population size distributions and biased gillnet sampling. Canadian Journal of Fisheries and Aquatic Sciences 61:2151–2157.

Finstad, A. G., P. A. Jansen, and A. Langeland. 2000. Gillnet selectivity and size and age structure of an alpine Arctic char (*Salvelinus alpinus*) population. Canadian Journal of Fisheries and Aquatic Sciences 57:1718–1727.

Hamley, J. M. 1975. Review of gillnet selectivity. Journal of the Fisheries Research Board of Canada 32:1943–1969.

Hamley, J. M. 1980. Sampling with gill nets. Pages 37–53 *in* T. Backiel and R. Welcomme, editors. Guidelines for sampling fish in fresh water. Food and Agriculture Organization of the United Nations, European Inland Fisheries Advisory Committee Technical Paper 33, Rome.

Hamley, J. M., and H. A. Regier. 1973. Direct estimates of gillnet selectivity to walleye (*Stizostedion vitreum vitreum*). Journal of the Fisheries Research Board of Canada 30:817–830.

Hansen, M. J., R. G. Schorfhaar, and J. H. Selgeby. 1998. Gill-net saturation by lake trout in Michigan waters of Lake Superior. North American Journal of Fisheries Management 18:847–853.

Helser, T. E., J. P. Geaghan, and R. E. Condrey. 1998. Estimating gillnet selectivity using nonlinear response surface regression. Canadian Journal of Fisheries and Aquatic Sciences 55:1328–1337.

Henderson, B. A., and J. L. Wong. 1991. A method for estimating gillnet selectivity of walleye (*Stizostedion vitreum vitreum*) in multimesh multifilament gill nets in Lake Erie, and its application. Canadian Journal of Fisheries and Aquatic Sciences 48:2420–2428.

Hovgard, H. 1996. A two-step approach to estimating selectivity and fishing power of research gill nets used in Greenland waters. Canadian Journal of Fisheries and Aquatic Sciences 53:1007–1013.

Hubert, W. A. 1996. Passive capture techniques. Pages 157–192 *in* B. R. Murphy and D. W. Willis, editors. Fisheries techniques. American Fisheries Society, Bethesda, Maryland.

Irwin, B. J., T. J. Treska, L. G. Rudstam, P. J. Sullivan, J. R. Jackson, A. J. VanDeValk, and J. L. Forney. 2008. Estimating walleye (*Sander vitreus*) density, gear catchability, and mortality using three fishery independent data sets for Oneida Lake, New York. Canadian Journal of Fisheries and Aquatic Sciences 65:1366–1378.

Isermann, D. A. 2007. Evaluating walleye length limits in the face of population variability: case histories from western Minnesota. North American Journal of Fisheries Management 27:551–568.

Kennedy, W. A. 1951. The relationship of fishing effort by gill nets to the interval between lifts. Journal of the Fisheries Research Board of Canada 8:264–274.

Kraft, C. E., and B. L. Johnson. 1992. Fyke-net and gill-net size selectivities for yellow perch in Green Bay, Lake Michigan. North American Journal of Fisheries Management 12:230–236.

Lamprecht, S. D., M. D. Mauck, and V. J. DiCenzo. 2013. Natural reproduction: how it has affected striped bass management of John H. Kerr and Santee–Cooper reservoirs and Lake Texoma. Pages 413–430 *in* J. S. Bulak, C. C.

Coutant, and J. A. Rice, editors. Biology and management of inland striped bass and hybrid striped bass. American Fisheries Society, Symposium 80, Bethesda, Maryland.

Lauer, T. E., J. C. Doll, P. J. Allen, B. Breidert, and J. Palla. 2008. Changes in yellow perch length frequencies and sex ratios following closure of the commercial fishery and reduction in sport bag limits in southern Lake Michigan. Fisheries Management and Ecology 15:39–47.

Millar, R. B. 2000. Untangling the confusion surrounding the estimation of gillnet selectivity. Canadian Journal of Fisheries and Aquatic Sciences 57:507–511.

Minns, C. K., and D. A. Hurley. 1988. Effects of net length and set time on fish catches in gill nets. North American Journal of Fisheries Management 8:216–223.

Miranda, L. E., and P. W. Bettoli. 2007. Mortality. Pages 229–277 *in* C. S. Guy and M. L. Brown, editors. Analysis and interpretation of freshwater fisheries data. American Fisheries Society, Bethesda, Maryland.

Olin, M., M. Kurkilahti, P. Peitola, and J. Ruuhijarvi. 2004. The effects of fish accumulation on the catchability of multimesh gillnet. Fisheries Research 68:135–147.

Pierce, R. B., C. M. Tomcko, and T. D. Kolander. 1994. Indirect and direct estimates of gill-net selectivity for northern pike. North American Journal of Fisheries Management 14:170–177.

Regier, H. A., and D. S. Robson. 1966. Selectivity of gill nets, especially to lake whitefish. Journal of the Fisheries Research Board of Canada 23:423–454.

Ricker, W. E. 1975. Computation and interpretation of biological statistics of fish populations. Fisheries Research Board of Canada Bulletin 191.

Rudstam, L. G., J. J. Magnuson, and W. M. Tonn. 1984. Size selectivity of passive fishing gear: a correction for encounter probability applied to gill nets. Canadian Journal of Fisheries and Aquatic Sciences 41:1252–1255.

Trent, L., and W. W. Hassler. 1968. Gill net selection, migration, size and age composition, sex ratio, harvest efficiency, and management of striped bass in the Roanoke River, North Carolina. Chesapeake Science 9:217–232.

SAS. 2002. SAS 9.2. SAS Institute, Cary, North Carolina.

Schramm, H. L., Jr., J. E. Kraai, and C. R. Munger. 2001. Intensive stocking of striped bass to restructure a gizzard shad population in a eutrophic Texas reservoir. Proceedings of the Annual Conference Southeastern Association of Fish and Wildlife Agencies 53(1999):180–192.

Spangler, G. R., and J. J. Collins. 1992. Lake Huron fish community structure based on gill-net catches corrected for selectivity and encounter probability. North American Journal of Fisheries Management 12:585–597.

Van Horn, S. L., B. K. Baker, and M. Rash. 2001. Growth and condition response of Lake Norman striped bass to increased stocking rates and more restrictive harvest regulations. Proceedings of the Annual Conference Southeastern Association of Fish and Wildlife Agencies 53(1999):193–199.

Wagner, T., C. S. Vandergoot, and J. Tyson. 2009. Evaluating the power to detect temporal trends in fishery-independent surveys: a case study based on gill nets set in the Ohio waters of Lake Erie for walleyes. North American Journal of Fisheries Management 29:805–816.

Waters, C. T. 2005. A summary of striped bass sampling strategies used on southeastern reservoirs. American Fisheries Society, Southern Division, Bethesda, Maryland. Available: www.sdafs.org/sbtc/2005Symposium/Presentations/Waters2/Waters2.htm. (September 2009).

Wilde, G. R. 1995. Gill net sample size requirements for temperate basses, shads, and catfishes. Proceedings of the Annual Conference Southeastern Association of Fish and Wildlife Agencies 47(1993):588–595.

Appendix

SAS code used to conduct power analysis

```
data power;
    *specify where output file will be sent;
file 'C:\...\name.txt';
    * specify alpha;
alpha=.05;
    * a = number of years in each contrast set;
a=5;
    *nold = number of historical samples per year;
nold=100;
    * sig =SD;
sig=37.96;
    * L = change from historic mean you wish to detect;
    *nfac = the multiples of the base data set you want to evaluate:
do L=25 to 50 by 2;
do nfac=1 to 5 by 0.2;
nc=nold*a*nfac*L*L/(sig*sig*(1+nfac));
nma=a*(nold-1)+a*(nold*nfac-1);
fcrit=finv(1-alpha,1,nma);
power=1-probf(fcrit,1,nma,nc);
    *  nnew = new sample size;
nnew=nold*nfac;
put nnew L power;
end;
end;
run;
```

American Fisheries Society Symposium 80:279–289, 2013

Estimating Abundance of Adult Striped Bass in Reservoirs Using Mobile Hydroacoustics

Joseph E. Hightower*
U.S. Geological Survey
North Carolina Cooperative Fish and Wildlife Research Unit
Department of Biology, North Carolina State University
Raleigh, North Carolina 27695, USA

J. Christopher Taylor
National Oceanic and Atmospheric Administration
National Ocean Service, Center for Coastal Fisheries and Habitat Research
101 Pivers Island Road, Beaufort, North Carolina 28516, USA

Donald J. Degan
Aquacoustics, Inc.
29824 Birdie Haven Court, Post Office Box 1473
Sterling, Alaska 99672, USA

Abstract.— Hydroacoustic surveys have proven valuable for estimating reservoir forage fish abundance but are more challenging for adult predators such as striped bass *Morone saxatilis*. Difficulties in assessing striped bass in reservoirs include their low density and the inability to distinguish species with hydroacoustic data alone. Despite these difficulties, mobile hydroacoustic surveys have potential to provide useful data for management because of the large sample volume compared to traditional methods such as gill netting and the ability to target specific areas where striped bass are aggregated. Hydroacoustic estimates of reservoir striped bass have been made using mobile surveys, with data analysis using a threshold for target strength in order to focus on striped bass-sized targets, and auxiliary sampling with nets to obtain species composition. We provide recommendations regarding survey design, based in part on simulations that provide insight on the level of effort that would be required to achieve reasonable estimates of abundance. Future surveys may be able to incorporate telemetry or other sonar techniques such as side-scan or multibeam in order to focus survey efforts on productive habitats (within lake and vertically). However, species apportionment will likely remain the main source of error, and we see no hydroacoustic system on the horizon that will identify fish by species at the spatial and temporal scale required for most reservoir surveys. In situations where species composition can be reliably assessed using traditional gears, abundance estimates from hydroacoustic methods should be useful to fishery managers interested in developing harvest regulations, assessing survival of stocked juveniles, identifying seasonal aggregations, and examining predator–prey balance.

* Corresponding author: jhightower@ncsu.edu

Introduction

Hydroacoustic methods have proven effective for estimating reservoir forage fish abundance. These mobile surveys are cost-effective to carry out (after the initial purchase cost of equipment) and provide fine-scale information on spatial distribution (e.g., Degan and Wilson 1995; Schael et al. 1995; Taylor et al. 2005). The ability to sample a large volume, compared to other methods such as gill netting, is important for forage species such as threadfin shad *Dorosoma petenense* that often occur at high but patchy densities. Estimating forage fish abundance by species is often straightforward because the pelagic zone typically contains only a few prey species.

Applying hydroacoustic methods to adult predators such as striped bass *Morone saxatilis* has been more challenging, but prior attempts at assessments of land-locked salmonids suggest that the application may hold promise for large predators in other systems (Yule 2000; Gangl and Whaley 2004; and reviewed in Taylor and Maxwell 2007). Practical difficulties include the relatively low density of adult predators and the inability to distinguish species with hydroacoustic data alone. Effective hydroacoustic methods for adult striped bass would be welcome, however, because other assessment techniques such as capture–recapture methods are laborious (Hightower and Pollock 2013, this volume) or provide only relative indices of abundance (e.g., annual gill-net surveys).

The purposes of this chapter are to review prior hydroacoustic surveys of striped bass and discuss design considerations in planning future surveys. We also consider recent developments in hydroacoustic equipment and whether technological advances will enhance our ability to survey reservoir populations of striped bass.

Prior Hydroacoustic Studies

We found three published studies that used hydroacoustic methods to estimate abundance of adult striped bass. Two were conducted in lakes and the third was carried out in the lower Hudson River estuary. All three studies (summarized below) used scientific fishery hydroacoustic systems. This type of system consists of an echo sounder that produces the digital signal that is transmitted underwater by the transducer. Returning echoes from a fish or other object are received by the transducer and amplified by the echo sounder. The timing of the returning echo determines the range of the object from the transducer while the intensity of the returning echo (target strength) is proportional to the size of the fish target (Love 1977) and can be used to separate individuals according to expected sizes (e.g., of predators versus smaller prey). A laptop computer controls the data acquisition parameters and also stores data from the echo sounder for later analysis. Older studies used either single- or dual-beam transducers, which had limited ability to resolve positions of fishes within the acoustic beam (Ehrenberg and Torkelson 1996). More recent studies have adopted split-beam transducers, which increase the precision of positioning within the beam and also decrease the variation in the estimate of target strength of the returning echo (Ehrenberg and Torkelson 1996). Transducer beam widths are chosen to optimize the detection of fish targets while limiting acoustic noise not attributed to fish. Typically, acoustic beams are cones between 6° and 15° wide. The volume of water sampled increases with distance from the transducer (apex), with the volume near the apex poorly sampled. Two of the three studies reviewed below use a single vertically oriented transducer to sample from 2 m below the surface to near bottom. If the fishes are expected to occupy near-surface waters, an additional transducer is oriented horizontally to sample the upper water column (Figure 1). Two of the three studies used split-beam target tracking to identify individuals and estimate abundance. In target tracking, echoes showing a consistent trajectory through the acoustic beam are grouped together and counted as a single fish track (Taylor and Maxwell 2007). One study used echo integration to estimate density. This analytical approach is

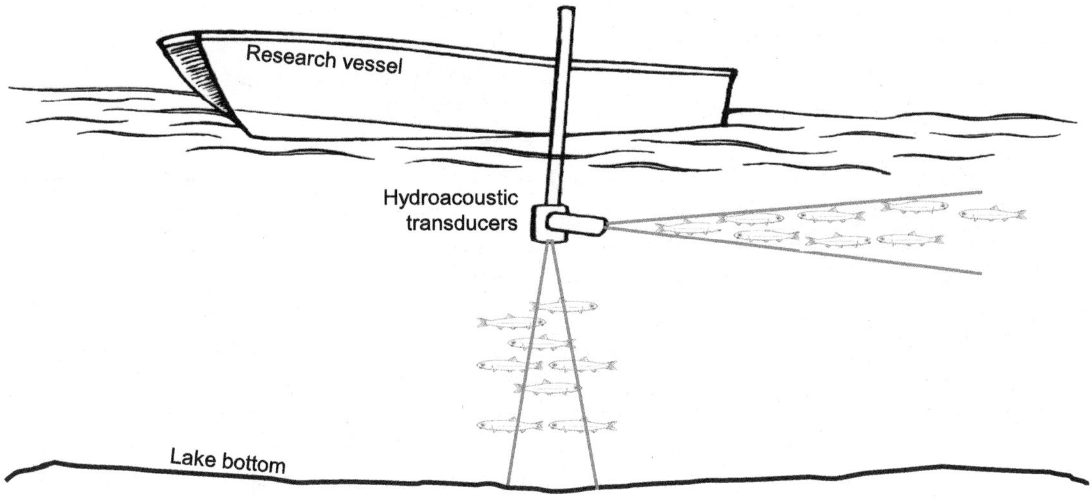

FIGURE 1. Two transducers deployed from the research vessel for simultaneous down-looking and side-looking deployments to sample fish targets throughout the water column.

appropriate when fish density is high and it is not feasible to track individual fish. It is based on the principle that the total acoustic energy returned from a sampled water volume is proportional to the number of fish in that volume (Taylor and Maxwell 2007). Information about the acoustic size (target strength) of an average individual fish is required to translate the total acoustic energy returned into an estimate of fish density. We focus below on the unique aspects of each study. Detailed information on fishery hydroacoustic theory and principles are provided elsewhere (Brandt 1996; Simmonds and MacLennan 2005).

Mueller and Horn (1999) carried out mobile hydroacoustic surveys of Lakes Mead and Powell and estimated striped bass abundance from areas of the lake greater than 20 m deep. The surveys were conducted in August of 1996 and 1997. Densities of striped-bass sized targets were relatively low, allowing individual targets to be counted to produce fish abundance estimates. They used a minimum threshold target strength of −40 dB to limit the analysis to fish estimated to be larger than 16 cm, based on studies relating target strength to fish size (Love 1977; MacLennan and Simmonds 1992). Catches from vertical gill netting (surface to depths >30 m) were used to partition the hydroacoustic estimates among species, with striped bass comprising an estimated 74–89% of the fish larger than 16 cm. The abundance estimates varied substantially between the 2 years (1.9 and 3.0 million in Lake Powell, 0.8 and 1.3 million in Lake Mead) and 90% confidence intervals were close to or included 0, so there was clearly much uncertainty associated with the estimates. The striped bass estimates along with similar estimates for prey species were used to examine predator:prey relationships and to explain the poor condition and die-offs observed for striped bass (Gustaveson and Blommer 2013, this volume).

Hartman and Nagy (2006) used mobile hydroacoustic surveys to estimate the abundance of striped bass, white perch *Morone americana* and bay anchovy *Anchoa mitchilli* in the lower Hudson River estuary. Their analysis was done from a depth of 2 to 0.5 m above bottom (with depths ranging up to 18 m), and the estimates were extrapolated to account for the entire water column. They estimated fish densities using echo integration and partitioned the density estimates into size-classes using an equation from Hartman and Nagy (2005) relating target strength to fish length. Midwater trawl catches were used to partition the size-specific density estimates among species. They also made a limited number of gill-net sets to determine whether their target species occurred within 0.5 m of the bottom. No striped bass were collected within this acoustically inaccessible zone. Their estimates of total abundance for striped bass (90 mm and larger) were 0.76 million in December 1995 and 1.47 million in December 1997. A high percentage of the estimated total in 1997 was in the size range classified as age 0.

Stewart et al. (2007) conducted mobile surveys during February using side- and down-looking transducers to estimate striped bass abundance in Lake Pleasant, Arizona. The focus of their study was to evaluate the impact of invading striped bass on an important and historically strong largemouth bass *Micropterus salmoides* fishery. Their analysis was based on individual fish tracks. Fish estimated to be larger than 125 mm (2005) or 150 mm (2006), based on Love's equation (Love 1977), were assumed to be potential predators. Striped bass comprised 25–38% of the fish in this size range, assuming that horizontal gill-nets set at the surface and on bottom obtained a representative sample of the population of interest.

Design Considerations

A primary advantage of mobile hydroacoustic surveys is the ability to sample a greater proportion of the reservoir in a sampling day compared to fixed-gear (e.g., gill net) or other direct sampling (e.g., trawls). Also, the volume sampled can be reliably estimated, whereas fixed-gear methods sample an unknown volume and catch rates depend partly on fish movement. Despite these advantages, a mobile hydroacoustic survey may still only represent a small fraction of the entire volume of the lake because of the narrow beam width of the hydroacoustic transducers. An effective survey design will consider time and resources available and sample a sufficiently representative geographic area and portion of the total volume, based on available information about density and seasonal distribution of striped bass. As an example, we calculated the number of fish that might be observed along a 1,000-m transect of a mobile hydroacoustic survey, assuming homogeneous distribution throughout a reservoir and using a striped bass population estimate of 136,227 fish age 1 and older from Smith Mountain Lake (Moore et al. 1991). The storage volume for Smith Mountain Lake (1,082,480 acre-feet) was obtained from a Federal Energy Regulatory Commission environmental impact statement document. Converting to cubic meters (and assuming a completely random striped bass distribution) generates a density estimate of 0.0001020 age-1+ striped bass per cubic meter. Unless the survey transects are predominately done at depths greater than 10 m, it is not likely that more than a single striped bass would be observed along any 1,000-m transect if using a 6° downward-oriented transducer (Table 1). Using a wider 15° transducer slightly more than doubles the expected number of fish encountered, but it also reduces the resolution, so there is less ability to separate individual fish targets, particularly near the bottom.

We then simulated a variety of spatial distributions of striped bass in a reservoir and simulated sampling of the population using a mobile acoustic survey and a range of sampling efforts. The population distribution and mobile survey were simulated using Wildlife Simulation Package (WiSP; Zucchini et al. 2007) in the statistical programming language R (R Development Core Team 2009). The simulated striped bass population was based on the assessment used above from Smith Mountain Lake, Virginia (Moore et al. 1991). The simulated reservoir had a total area of about 10,000 ha and a population of 137,000 individuals. To simplify the simulation, we assumed that all fish were located at the 10-m depth stratum and that the survey was done using a 15° conical downward-oriented transducer. This results in a higher encounter rate than depicted in Table 1 and would be similar to seasonal periods in which striped bass are concentrated in a vertical band

TABLE 1. Calculated beam width and sampling volume for a 1,000-m-long transect. Beam width calculated for two common transducer beam widths, 6° and 15°. Beam volume is calculated for 5-m depth interval centered on each stratum. Fish counted per 1,000 m assumes a system-wide density of 0.000102 fish/m³, as derived from Moore et al. (1991).

	6° beam			15° beam		
Depth stratum (m)	Beam width (m)	Beam volume (m^3)	Fish per 1,000 m	Beam width (m)	Beam volume (m^3)	Fish per 1,000 m
5	0.52	2,620	0.27	1.32	6,583	0.67
10	1.05	5,241	0.53	2.63	13,165	1.34
15	1.57	7,861	0.80	3.95	19,748	2.01
20	2.10	10,482	1.07	5.27	26,330	2.69
25	2.62	13,102	1.34	6.58	32,913	3.36
30	3.14	15,722	1.60	7.90	39,496	4.03
35	3.67	18,343	1.87	9.22	46,078	4.70
40	4.19	20,963	2.14	10.53	52,661	5.37
45	4.72	23,584	2.41	11.85	59,244	6.04
50	5.24	26,204	2.67	13.17	65,826	6.71

of suitable or preferred habitat (Coutant 2013; Rice et al. 2013; Thompson and Rice 2013; all this volume). The assumed vertical distribution determined the beam width and detection of individuals along a survey transect. We chose four types of distribution for this exercise: (1) a homogeneous and random (Poisson) distribution of individuals throughout the reservoir, (2) a distribution gradient with higher densities in one region of the reservoir as may be found during seasons of restricted thermal habitats or spawning season, (3) a patchy distribution that may be driven by the distribution of forage species or habitat preferences, and (4) a patchy distribution within a gradient (Figure 2). The effort for each survey varied from five parallel transects (approximately 40 km total survey length) to 50 transects (400 km total survey length). The first transect was randomly located and remaining transects were evenly spaced across the hypothetical reservoir. If an individual was within the detection width of the transducer beam, it was counted and the sampled area was estimated. A distance method was used to estimate the total abundance (Borchers et al. 2002). We simulated each level of effort 30 times and calculated the mean abundance and coefficient of variation for each level. The results are also presented in terms of the total distance traveled to provide guidance for designing a survey or monitoring program. Mobile surveys are usually conducted from small vessels at speeds between 1.5 and 2 m/s (5–7 km/h), which would equate to a maximum of 40–50 km survey distance per 8-h sample day.

This exercise highlighted two important considerations when designing a mobile hydroacoustic survey for fishes like striped bass in reservoirs. First, using a minimal effort level of 40 km of transects, and assuming a random and homogeneous distribution of fish in the system, coefficient of variation approached 20% (Figure 2). Second, as the population distribution departs from homogeneous, uncertainty in abundance estimates increases. In the extreme case of two large patches, coefficient of variation approached 60% for the lowest effort level. The coefficient of variation did not fall below 20% until effort was increased by 60% or another half-day of effort (Figure 2). For this 10,000-ha reservoir and the types of spatial distribution and patchiness simulated

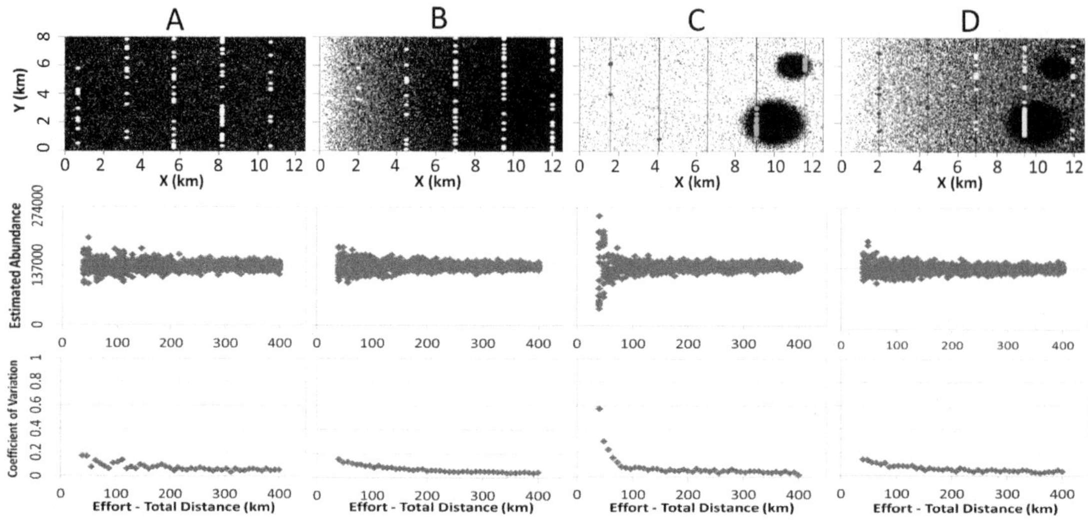

FIGURE 2. Results from a simulation survey of four types of spatial distribution of striped bass and varying level of survey effort. See text for description of system, population and survey. Top row shows the simulated 8,000 × 12,500 m rectangular reservoir with four types of striped bass distribution: (A) a homogeneous random distribution, (B) a gradient distribution varying 10-fold along the *x*-axis, (C) a patchy distribution with densities 100-fold higher in patches compared to surrounding system, and (D) two patches within a gradient distribution with densities 30- to 100-fold higher in patches compared to surrounding system. Vertical lines are shown as an example of the level of effort equivalent to 40 km and individuals that were detected along each transect (highlighted dots). Second row shows estimated abundances from 30 simulations at each level of survey effort (40 to 400 km). Bottom row shows coefficient of variation at each level of survey effort (SD/mean).

here, a survey effort level of 100 km of transects (2 d) would produce reasonable population estimates for large pelagic fishes such as striped bass. However, this exercise included some simplifying population and survey conditions: (1) single species, (2) fixed depth of all individuals, and (3) simple systematic (parallel) survey design. We encourage biologists and managers to consider this simulation approach when designing their survey for striped bass, paying particular attention to any prior knowledge about the spatial arrangement (e.g., patchiness) of the striped bass population and possible logistical constraints that may dictate a different survey design (e.g., zigzag pattern). As we discuss below, considerations must also be given to mixed-species communities and the need to apportion the hydroacoustic density estimates based on directed sampling for species identification.

Another potential design issue is that striped bass may occupy habitats that are not readily surveyed using hydroacoustics. For example, striped bass that are in shallow near-shore areas, within about 0.5 m of the bottom, within submerged standing timber, or in highly localized summer refuges such as springs or tributaries (Coutant 1985) would be difficult or impossible to survey. The best strategy here would be to determine times of the day or year when striped bass are separated from structure, bottom, and other species. For example, Mueller and Horn (1999) reported that during winter, striped bass were often closely associated with the bottom but, at night, would rise and disperse throughout the water column.

A hydroacoustic survey using a split-beam sonar provides some information about fish size based on the strength of returning echoes. If the survey is done using target tracking (which we recommend), then the average target strength for the series of echoes from an individual fish can be used to estimate fish size, using a regression equation such as that developed by Hartman and Nagy (2005). Using this approach to track individual fish, it should be possible to separate small forage fishes from larger fish such as striped bass (Figure 3). This does not eliminate the need for traditional sampling to get species composition (e.g., gill netting), but it does mean that species composition can be estimated for a specific size range of interest (e.g., the range containing adult striped bass).

Vertical gill netting can be a useful method for estimating species composition in open water (Mueller and Horn 1999). The main limiting factor for this method is that catches tend to be small, resulting in poor estimates of species composition over the larger zone of the hydroacoustic transect. Horizontal gill nets fished at the surface or bottom may be effective if large targets believed to be striped bass are distributed throughout the water column. Fishing a suspended gill net at a particular depth could be very effective in certain seasons (e.g., when striped bass are concentrated due to summer habitat squeeze), although setting the nets can be difficult and catches are often low (Stewart et al. 2007; Bergthold and Bettoli 2009).

One strategy for improving the precision of hydroacoustic surveys for striped bass would be to narrow the spatial scope of the survey. A data-driven approach for defining the spatial scope would be to implant transmitters in a subset of fish. Locations of radio- or sonic-tagged fish within the reservoir would make it possible to stratify by zone or habitat type, with proportionally more hydroacoustic sam-

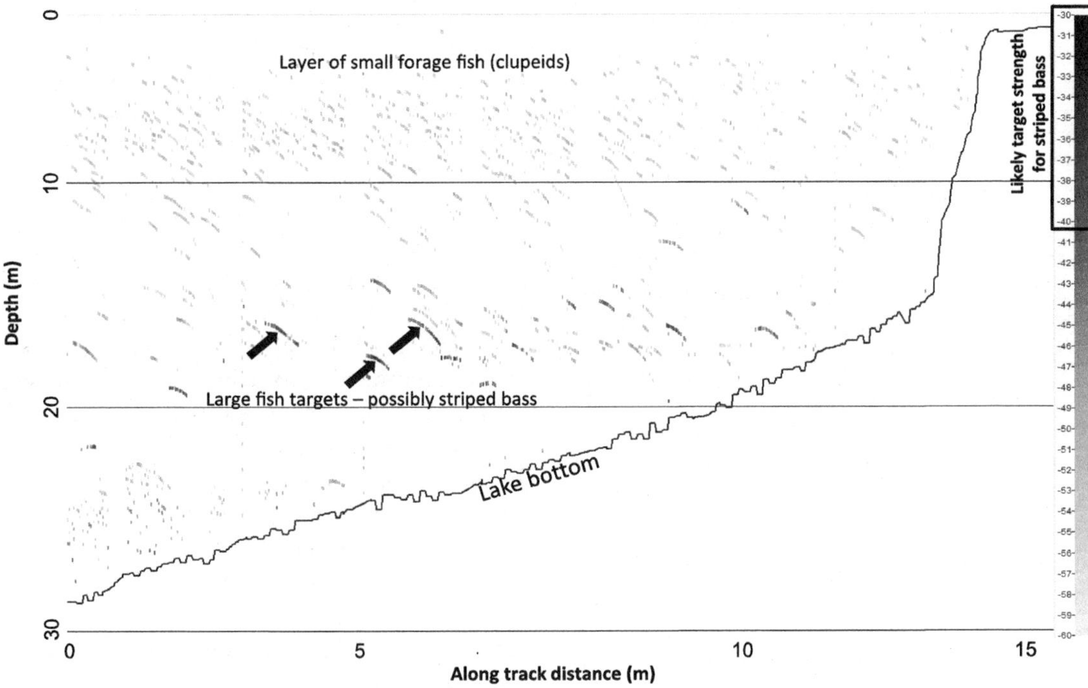

FIGURE 3. Example echogram from Badin Lake, North Carolina showing a layer of small forage fish (clupeids) within the first 10 m with larger fish targets (possibly including striped bass) between 10 and 20 m deep. The dashed box on the right side of the echogram indicates the range of target strengths (TS; –40 to –27 dB) that would be expected for striped bass larger than 150 mm, based on the relationship provided by Hartman and Nagy (2005). For clarity, only three large fish targets are identified.

pling in strata containing more striped bass. If depth- or temperature-sensing transmitters were used, the reservoir could be stratified in three dimensions. Information about the vertical distribution of striped bass should increase the efficiency of the hydroacoustic survey, simplify the analysis of hydroacoustic data, and reduce uncertainty about species composition of hydroacoustic targets. For example, Thompson (2006) used depth- and temperature-sensing tags to show that striped bass in Badin Lake, North Carolina were concentrated into a limited volume of (marginally) suitable habitat during summer due to temperature and dissolved oxygen requirements. Because of this "habitat squeeze" (Coutant 1985, 2013), summer would be a very efficient time to conduct a hydroacoustic survey for striped bass. However, one practical concern about a summer survey is that fish sampling for species composition could result in considerable mortality of striped bass.

If a data-driven approach is not feasible, initial weights for proportional sampling by stratum could be equal or based on information from anglers or area biologists. Preliminary weights could be updated as hydroacoustic data become available. For the hydroacoustic survey of Lakes Powell and Mead, Mueller and Horn (1999) limited their analysis of hydroacoustic data to areas of the lake greater than 20 m deep, based on the experience of area biologists.

A hydroacoustic survey during summer could also be useful for monitoring the decline in suitable habitat (habitat squeeze). The survey would provide three-dimensional information about where striped bass-sized targets are concentrated. Coupled with water quality data and predictions from reservoir water quality models (Ruane et al. 2013, this volume), it should lead to a greater understanding of how habitat squeeze impacts reservoir striped bass. It could also be used to assess the effectiveness of potential engineering methods intended to create better temperature-dissolved oxygen habitat (Mobley et al. 2013, this volume).

Field Trial of Simulated Survey Methodology

The mobile survey methodology described above was field tested on 2 September 2010 at Jordan Lake near Raleigh, North Carolina. The pilot survey was conducted in the Haw River arm, where summer die-offs of striped bass had been observed in prior years (N. C. Oakley, North Carolina Wildlife Resources Commission, personal communication). Four transects with a total distance of 10.1 km were done using a down-looking BioSonics DT-X split-beam system with a frequency of 430 kHz and a 7° transducer beam. Fish tracking was done using Echoview software, and tracks with an average target strength less than –35 dB were excluded. This target strength should correspond to an average size of about 35 cm, based on Love's (1971) dorsal aspect equation.

We detected six fish tracks with mean target strength greater than –35 dB. These fish were located at a depth of about 3 m, where water temperature was 29°C and dissolved oxygen concentration was about 2 mg/L. The dissolved oxygen level was less than 1 mg/L at a depth of 4 m, so we assumed the effective search volume was between our starting range of 0.5 m and the bottom or 4 m, whichever was less. This produced an estimated sampling volume for our transects of 16,204 m^3 or a fish density of 0.0004/m^3. This density is slightly higher than the estimate assumed in Table 1, but striped bass in Jordan Lake were concentrated into a limited depth range, and the six fish we detected may have included multiple species. The low number of detected targets also makes clear the difficulty of conventional sampling to get species composition. Fish sampling might be aided here by the apparently clumped spatial distribution for striped bass in this arm of the lake. All six targets we detected were less than 1 km apart and three of the six were at the same location.

Conclusions and Future Directions

Hydroacoustic surveys have several advantages over traditional methods such as gill netting.

Gear selectivity (relative vulnerability to capture across size-classes) is not an issue, whereas both gill nets and electrofishing are known to be size selective. Hydroacoustic surveys can cover large areas at fine resolution compared to the coarse resolution of traditional gears. It is also straightforward to expand hydroacoustic survey results to a whole-system abundance estimate if sufficient sampling is performed in each stratum, whereas traditional methods provide only an index of relative abundance.

Generating a lake-wide abundance estimate for striped bass-sized targets would typically require less than a week of transect effort in order to achieve reasonable precision. Relatively more effort would be needed if the distribution of striped bass is patchy, although information about the patchiness from telemetry, preliminary hydroacoustic surveys, or other data sources could result in a much more efficient and cost-effective survey.

Additional field effort would be required to partition the lake-wide abundance estimate among species. The sampling locations (areas and depths) and gears (e.g., gill-net orientation and mesh sizes) would be based on the hydroacoustic results. This should make the fish sampling more effective, as there may be relatively few species within the habitats occupied by adult striped bass. For example, striped bass comprised 74–89% of the vertical gill-net catch of fish 16 cm and larger within Lakes Powell and Mead (Mueller and Horn 1999). Similarly, mid-water trawl sampling in the lower Hudson River indicated that 100% of the fish between 30 and 100 cm were striped bass (Hartman and Nagy 2006).

A future direction for mobile hydroacoustic surveys may be to increase the sampling volume through the use of other sonar technologies. As we show in Table 1, striped bass may be rare along any given transect owing to the narrow transducer beam used. Other sonar technologies may provide information on the distribution of large fishes in reservoirs that could help guide the survey design. Multibeam echo sounders can increase the sampling swath by 10-fold, and many side-scan sonar systems are capable of sampling at least 100-m-wide swaths. Neither of these systems can currently provide quantitative information that could produce abundance estimates, but presence of large targets or groups or schools of large fishes could be used to focus effort using the split-beam echo sounder. Another approach is to orient a split-beam transducer at an oblique angle at the depth where striped bass are expected to occupy. Sample volume could be increased threefold compared to a vertically oriented transducer; however, this approach would only work if striped bass are well above the bottom. Alternatively, a horizontally aimed transducer towed at the expected depth of striped bass vertical distribution would increase the sample volume compared to surface-mounted transducers by more than sevenfold.

Abundance estimates from a hydroacoustic survey should be valuable to the fishery manager. These estimates would provide a context for interpreting creel survey results, that is, in assessing the extent to which fishing impacts the adult population. Hydroacoustic estimates can also be used to examine the effectiveness of a stocking program, not only survival from stocking to recruitment into the fishery, but also the adult population produced at different stocking rates. Estimates of population size can also be valuable for examining the predator:prey ratio. Hydroacoustic surveys often produce abundance estimates for forage fish as well as predators and would be valuable for examining whether current (or proposed) stocking levels might overwhelm the prey base. Tagging methods can also be used to estimate absolute abundance of striped bass, but it is difficult to generate the sample sizes needed to obtain precise estimates (see Hightower and Pollock 2013). Our view is that hydroacoustic surveys have the greatest potential of the available sampling gears for producing whole-lake abundance estimates that would allow for more advanced management approaches.

Acknowledgments

We thank P. Bettoli, C. Coutant, J. Rice, and an anonymous reviewer for comments on previous drafts of this manuscript and J. Hughes for assistance in the Jordan Lake pilot survey.

References

Bergthold, C. L., and P. W. Bettoli. 2009. Growth, diet, and sampling of rainbow trout in three Tennessee reservoirs. Tennessee Wildlife Resources Agency, Fisheries Report 09-04, Nashville.

Borchers, D. L., S. T. Buckland, and W. Zucchini. 2002. Estimating animal abundance: closed populations. Springer-Verlag, London.

Brandt, S. B. 1996. Acoustic assessment of fish abundance and distribution. Pages 385–432 *in* B. R. Murphy and D. W. Willis, editors. Fisheries techniques, 2nd edition. American Fisheries Society, Bethesda, Maryland.

Coutant, C. C. 1985. Striped bass, temperature, and dissolved oxygen: a speculative hypothesis for environmental risk. Transactions of the American Fisheries Society 114:31–61.

Coutant, C. C. 2013. When is habitat limiting for striped bass? Three decades of testing the temperature–oxygen squeeze hypothesis. Pages 65–91 *in* J. S. Bulak, C. C. Coutant, and J. A. Rice, editors. Biology and management of inland striped bass and hybrid striped bass. American Fisheries Society, Symposium 80, Bethesda, Maryland.

Cyterski, M., J. Ney, and M. Duval. 2003. Estimation of surplus biomass of clupeids in Smith Mountain Lake, Virginia. Transactions of the American Fisheries Society 132:361–370.

Degan, D. J., and W. Wilson. 1995. Comparison of four hydroacoustic frequencies for sampling pelagic fish populations in Lake Texoma. North American Journal of Fisheries Management 15:924–932.

Ehrenberg, J. E., and T. C. Torkelson. 1996. Application of dual-beam and split-beam target tracking in fisheries acoustics. ICES Journal of Marine Science 53:329–334.

Gangl, R. S., and R. A. Whaley. 2004. Comparison of fish density estimates from repeated hydroacoustic surveys on two Wyoming waters. North American Journal of Fisheries Management 24:1279–1287.

Gustaveson, W., and G. Blommer. 2013. History of striped bass management in the Colorado River. Pages 15–23 *in* J. S. Bulak, C. C. Coutant, and J. A. Rice, editors. Biology and management of inland striped bass and hybrid striped bass. American Fisheries Society, Symposium 80, Bethesda, Maryland.

Hartman, K. J., and B. W. Nagy. 2005. A target strength and length relationship for striped bass and white perch. Transactions of the American Fisheries Society 134:375–380.

Hartman, K. J., and B .W. Nagy. 2006. Winter distribution and abundance of Hudson River fishes using hydroacoustics. Pages 175–196 *in* J. R. Waldman, K. E. Limburg, and D. L. Strayer, editors. Hudson River fishes and their environment. American Fisheries Society, Symposium 51, Bethesda, Maryland.

Hightower, J. E., and K. H. Pollock. 2013. Tagging methods for estimating population size and mortality rates of inland striped bass populations. Pages 249–262 *in* J. S. Bulak, C. C. Coutant, and J. A. Rice, editors. Biology and management of inland striped bass and hybrid striped bass. American Fisheries Society, Symposium 80, Bethesda, Maryland.

Love, R. H. 1971. Dorsal-aspect target strength of an individual fish. Journal of the Acoustical Society of America 49:816–823.

Love, R. 1977. Target strength of an individual fish at any aspect. Journal of the Acoustical Society of America 62:1397–1403.

MacLennan, D. N., and E. J. Simmonds. 1992. Fisheries acoustics. Chapman and Hall, New York.

Mobley, M., E. Shallenberger, M. W. Beutel, P. Gantzer, and B. Sak. 2013. Oxygen diffusers to create and maintain summer fish habitat. Pages 197–208 *in* J. S. Bulak, C. C. Coutant, and J. A. Rice, editors. Biology and management of inland striped bass and hybrid striped bass. American Fisheries Society, Symposium 80, Bethesda, Maryland.

Moore, C. M., R. J. Neves, and J. J. Ney. 1991. Survival and abundance of stocked striped bass in Smith Mountain Lake, Virginia. North American Journal of Fisheries Management 11:393–399.

Mueller, G., and M. J. Horn. 1999. Description of the pelagic zooplankton and fish communities of Lakes Powell and Mead. U.S. Geological Survey, Bureau of Reclamation, Open

File Report 99-109. Available: http://pubs.usgs.gov/of/1999/0109/report.pdf (January 2012).

R Development Core Team. 2009. R: a language and environment for statistical computing. R Foundation for Statistical Computing, Vienna. Available: www.r-project.org. (January 2012)

Rice, J. A., J. S. Thompson, J. A., Sykes, and C. T. Waters. 2013. The role of metalimnetic hyposia in striped bass summer kills: consequences and management implications. Pages 121–145 *in* J. S. Bulak, C. C. Coutant, and J. A. Rice, editors. Biology and management of inland striped bass and hybrid striped bass. American Fisheries Society, Symposium 80, Bethesda, Maryland.

Ruane, R. J., G. E. Hauser, and A. F. Sawyer. 2013. Water qualtiy modeling as a management tool for striped bass and hybrid striped bass. Pages 161–179 *in* J. S. Bulak, C. C. Coutant, and J. A. Rice, editors. Biology and management of inland striped bass and hybrid striped bass. American Fisheries Society, Symposium 80, Bethesda, Maryland.

Schael, D. M., J. A. Rice, and D. J. Degan. 1995. Spatial and temporal distribution of threadfin shad in a southeastern reservoir. Transactions of the American Fisheries Society 124:804–812.

Simmonds, J., and D. MacLennan. 2005. Fishery acoustic theory and practice. Blackwell Scientific Publications, Oxford, UK.

Stewart, W. T., M. M. Meding, and D. R. Rogers. 2007. Lake Pleasant striped bass. Arizona Game and Fish Department, Research Branch, Technical Guidance Bulletin No. 11, Phoenix.

Taylor, J. C., and S. W. Maxwell. 2007. Hydroacoustics: lakes and reservoirs. Pages 153–172 *in* D. H. Johnson, B. M. Shrier, J. S. O'Neal, J. A. Knutzen, X. Augerot, T. A. O'Neill, and T. N. Pearsons. Salmonid field protocols handbook: techniques for assessing status and trends in salmon and trout populations. American Fisheries Society, Bethesda, Maryland.

Taylor, J. C., J. S. Thompson, P. S. Rand, and M. Fuentes. 2005. Sampling and statistical considerations for hydroacoustic surveys used in estimating abundance of forage fishes in reservoirs. North American Journal of Fisheries Management 25:73–85.

Thompson, J. S. 2006. The influence of temperature and forage availability on growth and habitat selection of a pelagic piscivore. Doctoral dissertation. North Carolina State University, Raleigh.

Thompson, J. S., and J. A. Rice. 2013. The relative influence of thermal experience and forage availability on growth of age 1–5 striped bass in two southeastern reservoirs. Pages 93–120 *in* J. S. Bulak, C. C. Coutant, and J. A. Rice, editors. Biology and management of inland striped bass and hybrid striped bass. American Fisheries Society, Symposium 80, Bethesda, Maryland.

Yule, D. 2000. Comparison of horizontal acoustic and purse-seine estimates of salmonid densities and sizes in eleven Wyoming waters. North American Journal of Fisheries Management 20:759–775.

Zucchini, W., D. L. Borchers, M. Erdelmeier, E. Rexstad, and J. Bishop. 2007. A Wildlife Simulation Package, WiSP 1.2.4. Institut fur Statistik und Okonometrie, Geror-August-Universitat Gottingen, Platz der Gottinger Seiben 5, Gottingen, Germany.

American Fisheries Society Symposium 80:291–312, 2013

Options for Estimating Striped Bass Catch and Harvest: Effectiveness of Creel Surveys

J. Warren Schlechte* and Nathan G. Smith
Heart of the Hills Fisheries Science Center
Texas Parks and Wildlife Department
Inland Fisheries Division - Research
5103 Junction Highway, Mt. Home, Texas 78058, USA

John B. Taylor
Texas Parks and Wildlife Department
Inland Fisheries Division - Headquarters
4200 Smith School Road, Austin, Texas 78744

Abstract.—Striped bass *Morone saxatilis* provide important commercial and recreational fisheries in many Atlantic coast states, in addition to providing popular recreational reservoir fisheries in numerous inland states. Measurements of fishery-dependent data such as harvest, catch, and effort are essential to determining whether management actions are effective. Many states, including Texas, use creel-survey methods for measuring striped bass catch and effort statistics for recreational anglers. However, a systematic overview of the performance of the various types of creel surveys, or creel surveys in general, for measuring striped bass catch and effort has not been performed previously. In many states, both roving and access creel surveys are used, and effort may be allocated using unequal probabilities. In fisheries with low harvests of *Morone* spp., survey statistics for these fish typically have high relative standard errors (RSE), from 30% to 115%. As directed effort increases, the RSE typically decreases; in systems with high angling effort directed at striped bass, RSE is much lower (e.g., 18% in Lake Texoma, Texas–Oklahoma). We used these and other data from striped bass creel surveys throughout the United States to assess the reliability of estimates from various methods. We recommend making survey sampling effort coincide with the fishing effort. While it may be possible to improve the precision by increasing the amount of manpower devoted to the current creel survey, stratifying or using unequal spatial and temporal probabilities are techniques more likely to improve precision in a cost-effective manner. Once the data have been collected, we show how using model-based estimation, such as using Kalman filters or empirical Bayesian estimation, could also prove advantageous.

Introduction

Striped bass *Morone saxatilis* and other *Morone* species provide important commercial and recreational fisheries in many Atlantic coast states (Koo 1970; U.S. Department of the Interior, Fish and Wildlife Service, and U.S. Department of Commerce, U.S. Census Bureau 2006) in addition to providing popular recreational reser-

* Corresponding author: warren.schlechte@tpwd.state.tx.us

voir fisheries in many inland reservoirs (Matthews 1985; Van Horn 2013, this volume). To document the importance of particular fisheries, as well as to evaluate management activities and regulations, agencies often utilize fishery-dependent data (Hudy and Berry 1983; U.S. Department of the Interior, Fish and Wildlife Service, and U.S. Department of Commerce, U.S. Census Bureau 2006), including directed effort, catch (all fish captured, whether retained or released), and harvest (only fish that are retained). Because moronids (i.e., striped bass, hybrid striped bass, and white bass *M. chrysops*) can provide important recreational fisheries, many states are interested in measuring catch and effort statistics of their recreational anglers.

Fisheries managers throughout the United States use creel surveys to estimate catch, harvest, and effort within recreational fisheries (Pollock et al. 1994; Malvestuto 1996; Allen et al. 2003; Moczygemba et al. 2006). In our discussions with biologists throughout the southeastern United States, we were told that rarely are these surveys directed toward striped bass fisheries, as very few reservoirs have striped bass fisheries of sufficient importance to warrant dedicated creel surveys. Instead, biologists use creel surveys to estimate statistics for all species being sought within their reservoirs and moronid angling statistics are ancillary. There have been discussions about whether the current approach is sufficient or if changes in creel-survey designs need to be considered and implemented to improve our knowledge of moronid fisheries. While it is well understood that effective fisheries management requires understanding the fishery, creel surveys can be expensive to conduct. Hence, the costs of the current survey methods and the potential costs of changes to those methods should be balanced with the benefit of improved data quality.

Our objectives were to examine the reliability of catch and effort statistics gathered from creel surveys for striped bass and similar species. We also examined the literature to see if available data might suggest biases in current designs and what steps could be taken to address those biases. Finally, using the available information, we provide some alternatives to consider for future collection of fishery-dependent data.

Current Creel-Survey Methods

Most states use both access-point and roving creel surveys to interview their anglers (for general coverage of these types of surveys, along with advantages and disadvantages, see Pollock et al. 1994). Access-point creel surveys are generally used when a water body has few access points or when only a few of the potential access points have the majority of the effort (Hayne 1991). Under these circumstances, it is advantageous for the biologist to stay in one place and conduct interviews as anglers leave the water body. When several access points exist, a bus-route design might be considered (Jones and Robson 1991). A bus-route design allows the biologist to visit several adjacent access points and is used to increase the number of interviews obtained. Care must be taken in the design of the bus-route survey or estimates could be biased (Chen and Woolcock 1999). An advantage to an access-point survey is that all data are collected at the end of the fishing trip, so no extrapolation is needed to estimate that angler's catch, harvest or effort.

Roving creel surveys are generally used when there are numerous public or private access points, such that remaining at a fixed location would result in only a few interviews (Hayne 1991; Robson 1991). Roving creel surveys can also be useful when bank fishing is important but ingress and egress points are poorly defined. Under these circumstances, the biologist actively searches for anglers (roves) and interviews anglers while they are fishing. In this way, more interviews may be conducted for each unit of time the biologist is at the reservoir (Pollock et al. 1994). Although roving creel surveys are statistically defensible and efficient, use of the data requires more assumptions. For example, because trips are not complete at the time of the interview, the biologist must assume that the catch rate in the completed portion of the trip is the same as the catch rate in the remaining portion. Comparisons between catch rates of incomplete and completed trips show that this assumption may not always be valid (Keefe et al. 2009). This suggests that one may be ill-advised to compare catch and harvest statistics from surveys using these different designs. Effort is estimated using either an instantaneous counting method or a progressive method. If the latter is used, care must be taken to avoid introducing bias (Pollock et al. 1994). Roving creels also result in a nonrandom sample of anglers—anglers who stay longer are more likely to be interviewed—which leads to a length-of-stay bias (Robson 1991; Wade et al. 1991; Pollock et al. 1994). However, the roving creel method can provide good estimates and may be the best choice for some systems.

Creel surveys are designed and conducted very similarly throughout the southeastern U.S. Texas' creel survey provides one example of how a typical creel survey is designed and conducted. In Texas, biologists typically design creel surveys that sample a given water body on 36 randomly chosen days throughout a year-long period. These 36 d are divided into four seasonal strata (sampling 9 d per seasonal stratum) and further divided into two day-type strata: a weekday stratum (5 d) and a weekend-holiday stratum (4 d). Stratification is a statistical technique in which portions of the population that are relatively homogeneous (e.g., summer weekdays) are sampled as a unit; therefore, catch and effort values are more alike within a given season and day type but may be quite different outside these groupings. Seasons and day type are stratified to improve the overall precision of the estimates. The amount of stratification within the creel survey design should be a balance between the precision increase associated with adding more strata and the cost increase associated with adding more samples to represent the new strata adequately. Meredith and Malvestuto (1991) demonstrated that while some stratification was useful, increasingly complex stratification did little to improve precision. Days may be further divided into two or more intervals, often 4–8 h long. Typically, days are subdivided into intervals for one or more of the following reasons: (1) the fishing day is longer than the biologist's workday, (2) fishing effort is not uniform throughout the day, (3) multiple within-day counts are desired, (4) the biologist wishes to sample multiple sites within a day, or (5) after accounting for travel time, only a fraction of the day can be sampled. While days may be subdivided into subday intervals, these subday intervals are not usually treated as strata, but instead are treated as subsampling units (see Pollock et al. 1994). Sample allocation for the creel survey is random within the constraints imposed by the stratification. Where prior information on angler-use patterns exists, biologists are encouraged to use nonuniform probabilities that reflect those patterns in apportioning their sampling effort (Hayne 1991). All creel surveys have these basic elements. What differs among them are the details, such as the designation of strata, sampling probabilities, and the levels of survey effort.

To estimate angling effort, anglers are counted, although in some designs, surrogates (e.g., boat trailers that are parked at launching ramps) are counted. The general approach for estimating angling effort with a creel survey is to count all anglers encountered during the sampling period. For an access-point creel survey, not only are all encountered anglers counted, but anglers are also asked about the length of their completed trip. Total angler-hours for an access-point creel are estimated by multiplying the total number of anglers by the average trip length and by a factor that accounts for the days that were not sampled within the period of interest. When all anglers are interviewed (i.e., a census is obtained), total angler-hours can be estimated by simple summation of the data for that subsample. Getting a census of the anglers within a subsample can be useful because it can reduce the total variability of the estimates. For a roving creel survey, a slightly different method is used. Angler counts within a subsample are expanded by the probability associated with sampling that area, and further expanded to account for unobserved days when sampling did not occur. Angler-hours are estimated by multiplying the estimated number of anglers by the length of the fishing day. The length of the fishing day is typically defined by the biologist as daylight hours. In rare instances, angling effort may be estimated using an aerial survey. To estimate angler effort from an aerial survey, the angler counts are handled like the instantaneous counts from a roving-creel survey. For more specific details on these types of surveys and the mathematics used in calculations, see Pollock et al. (1994) and Malvestuto (1996).

Catch and harvest are also dealt with slightly differently depending upon whether an angler's trip is complete when the angler is interviewed. If the trip is complete (as is typical with an access-point creel survey), then catch and harvest statistics are handled in the same fashion as the effort data: average counts (or harvest) per hour are expanded by the total angler-hours. When all anglers are interviewed, total catch and harvest are estimated by simple summation of the data. For completed trips, the ratio-of-means approach (Malvestuto 1996; Hoenig et al. 1997) is recommended. This estimator is calculated by summing all the catch (or harvest) observed during the creel survey and placing that sum in the numerator, then dividing that by the total of hours fished for all completed trips observed during the creel survey. In this approach, each angler-hour contributes equally to the estimator. For two trips with the same catch rate, the longer trip gets more weight than the shorter trip because more fish were caught and more effort was expended.

If the trip is not completed when the angler is interviewed (as is typical of a roving creel survey), then the data are often adjusted to account for the unobserved remainder of the trip. Typically, this is done by assuming that the angler's catch per unit effort (CPUE) will be the same across the entire trip, and therefore the CPUE reported for the completed portion of the trip is applied to the entire day. If the fishing day is subsampled, CPUE estimates are applied to that subsample, and the daily estimate would be the weighted average CPUE of the subsamples. As Malvestuto (1996) reports, studies that have tested this assumption have conflicting results. Thus, it may be worth testing within each fishery. While the ratio-of-means approach could be used to estimate the average catch rate, because anglers are encountered at different times within their fishing day, the mean-of-ratios approach is suggested (Pollock et al. 1994; Malvestuto 1996; Hoenig et al. 1997). In this approach, each angler contributes equally to the estimator, regardless of how long they have been fishing. For two trips with the same catch rate, the longer trip gets the same weight as the shorter trip because it is assumed that the catch rate is time invariant.

Study Methodology

We reviewed the designs of, and the results from, numerous creel surveys from several states throughout the southeastern United States to evaluate the effectiveness of currently used creel-survey methods, emphasizing striped bass and other moronids (i.e., hybrid striped bass and white bass) fisheries. We examined catch and effort statistics, estimates of precision, the proportion of the fishery represented by moronids, and, where published, the number of fish observed within the creel.

In many water bodies, it is common for biologists to pool data on all *Morone* species, as moronids may comprise a small fraction of the fishery and identification of the species can be difficult (Williams 1976; Storey et al. 2002). The parentage of the hybrids is undefined within some state creel reports. For this report, we treated all hybrids as a single category. Most hybrids are likely palmetto bass (male white bass × female striped bass) or sunshine bass (female white bass × male striped bass). Several authors have worked on quantifying characteristics that can be used to discriminate the moronids (Harrell and Dean 1988; Muoneke et al. 1991; Storey et al. 2002); however, no set of characteristics has proven 100% discriminatory. Depending upon the situation and the objectives, absolute discrimination may not be necessary. We decided to examine the creel survey results for all moronids in cases where data were pooled, as well as examine results for specific moronids where data were available. In addition to collecting creel-survey data from these states, we also asked the biologists involved if they were satisfied with the results from their creel surveys.

How Well Do Current Creel Surveys Work: Precision and Bias

In this review, we noted issues that could contribute to poor precision and issues that may bias estimates. To improve precision, one can either increase the amount of sampling effort or reallocate the sampling effort using known mathematical relationships between sampling effort and precision. To eliminate bias, the survey design must be constructed in such a way that all members of the population have a known probability of selection. When fractions of the population are over- or underrepresented, estimates will be biased. Increasing sampling effort will not decrease the bias of the estimates; bias is a result of either methodology or implementation. Accuracy is the combined effect of precision and bias and is typically measured using a quantity known as mean squared error (Pollock et al. 1994). Because quantifying bias is often difficult, we typically focus on estimating precision (e.g., standard error).

Table 1 shows results from various states' creel surveys; Table 2 shows similar data from a creel survey conducted at Lake Texoma. We found that in most states, striped bass frequently represented a small fraction of the overall fishery (Table 1); in contrast, striped bass represented a large fraction of the Lake Texoma fishery (Table 2). In most instances, the creel surveys used to estimate striped bass (or the moronid complex) catch and harvest were designed to sample all anglers regardless of the species of fish they were targeting. This means much of the sampling effort was only useful for confirming that most anglers were not fishing for striped bass and that no catch or harvest of striped bass was occurring. The resultant zeros in the catch and effort data (i.e., the biologist did not observe any catch or targeted effort for striped bass in the creel survey) have the effect of increasing the variance of the resulting striped bass estimates. Although it can be useful to confirm that anglers are neither targeting nor catching striped bass, if these designs were being used to get information on striped bass fisheries, we suggest that biologists' time and energy might have been better allocated so most of the sampling occurred when anglers actually were targeting and catching striped bass.

Estimates of overall (i.e., all species combined) angling effort from the surveys we reviewed were often quite precise (Table 1), suggesting that the surveys were well designed and that sufficient sampling effort was expended. Where available, catch and harvest statistics were often less precise because of the additional variability associated with fishing success, but they too seemed quite satisfactory based on generally accepted guidelines of precision (Robson and Regier 1964). Unfortunately, in many instances, measures of precision for the moronid portion of the creel were not given, making it difficult to judge the value of the data. In instances where precision for the moronid portion of the creel was given (or could be calculated), we found that estimates were often quite imprecise (Table 1). Despite this, the biologists suggested the results were sufficient for their current management. As expected, as more effort was directed towards the moronid fishery and more fish were harvested, the estimates became more precise (Figures 1 and 2). Most likely, this was because the data became more homogeneous.

Despite overall estimates being quite precise, there are elements that could be improved if better estimates for moronid fisheries are desired. Underrepresentation of nighttime fishing is one example that can lead to both poor precision and biased estimates. Most states' creel

TABLE 1. Results from creel surveys throughout the Southeastern United States reporting data on moronid fisheries. S = striped bass, W = white bass, and H = hybrids; R = roving and A = access.

State	Year	Water body	Design	Species	Effort [angler-hours for all fish] (RSE[a])	Effort [*Morone* angler-hours] (RSE[a])	Percent of total effort directed at *Morone*	*Morone* harvest (RSE[a])	Percent of total harvest	References
SC	2000	Lake Moultrie	R	S	266,000 (8.2%)	19,000 (8–15%)	7%	8,000	7%	[b]
SC	2001	Lake Moultrie	R	S	348,000 (6.6%)	27,000	8%	8,500	5.4%	[b]
SC	1998	Lake Marion	R	S	883,000 (7%)	48,000	5%	18,000	3.5%	[b]
SC	1999	Lake Marion	R	S	467,000 (10%)	38,000	8%	10,000	4.3%	[b]
KY	2005	Rough River Lake	R	H	147,400 (3%)	20,000	13%	3,000	5.2%	[c]
KY	2008	Lake Cumberland (lower lake)	R	S	360,000 (2%)	150,000	42%	18,000	14.8%	[c]
KY	2008	Lake Cumberland (upper lake)	R	S	190,000 (3%)	18,000	9%	2,000	2%	[c]
KY	2004	Herrington Lake	R	W and H	73,000 (3%)	14,000	19%	2,400 and 3,100	20%	[c]
KY	2006	Fishtrap Lake	R	H	24,000 (4%)	1,100	5%	343	2.8%	[c]
KY	2007	Barren River	R	H and W	128,000 (3%)	15,000	12%	13,000 and 529	15%	[c]
KY	2006	Lake Taylorsville	R	H and W	142,000 (3%)	7,200	5%	8,100 and 4,000	3%	[c]
GA	1971	Lake Nottely	R	S	26,000 (NA)	0	0%	0	0%	[d]
GA	1972	Lake Nottely	R	S	33,000 (NA)	300	0.9%	43	0.3%	[d]
GA	1973	Lake Nottely	A	S	54,000 (12%)	100	0.2%	11 (73%)	0.2%	[d]

TABLE 1. Continued.

State	Year	Water body	Design	Species	Effort [angler-hours for all fish] (RSE[a])	Effort [*Morone* angler-hours] (RSE[a])	Percent of total effort directed at *Morone*	*Morone* harvest (RSE[a])	Percent of total harvest	References
GA	1975	Lake Nottely	A	S	33,000 (10%)	NA	NA	19 (105%)	0.2%	d
GA	1984	Lake Nottely	R	S and W	64,000 (NA)	5,600 and NA	9% and NA	358 and 9	0.9% and 0%	e
GA	1979	Lake Lanier	R	S and W	1,400,000 (9%)	36,400 and 9,800	2.6% and 0.7%	316 and 8,000	0.1% and 1.3%	f
GA	1980	Lake Lanier	R	S and W	1,000,000 (6%)	11,000 and 4,000	1.1% and 0.4%	40 and 4,000	<0.1% and 0.7%	f
GA	1981	Lake Lanier	R	S and W	800,000 (6%)	8,800 and 8,000	1.1% and 1%	31 and 6,000	<0.1% and 1.3%	f
GA	1980	Lake Oconee	R and A	S and W	294,000 (NA)	0 and 3,400	0 and 1.1%	523 and 25,000	0.1% and 8%	g
GA	1981	Lake Oconee	R and A	S and W	573,000 (NA)	0 and 719	0 and 0.1%	1,200 and 7,000	0.2% and 1%	g
GA	1982	Lake Oconee	R and A	S and W	515,000 (NA)	2,400 and 77	0.5% and <0.1%	800 and 1,700	0.2% and 0.4%	g
GA	1977	Lake Walter F. George	R	H and W	177,000 (9.6%)	74 (84%) and 19 (74%)	<0.1% and <0.1%	1,800 (52%) and 3,500 (50%)	1.5% and 3%	h
GA	1978	Lake Walter F. George	R	H and W	79,000 (9%)	616 (66%) and 375 (79%)	0.8% and 0.5%	1,100 (56%) and 1,200 (55%)	4% and 4%	h
GA	1978	Lake Seminole	Effort from aerial survey; catch from roving	S, H, and W	550,000 (6.5%)	0, 793, and 16,891	0%, 0.1%, and 3.1%	726 (33%), 5,000 (40%), and 15,000 (18%)	0.1%, 0.7%, and 2%	i
GA	1979	Lake Seminole	Effort from aerial survey; catch from roving	S, H, and W	625,000 (6.1%)	0, 5,900, and 5,400	0%, 0.9%, and 0.9%	7 (57%), 5,500 (24%), and 9,000 (24%)	<0.1%, 0.6%, and 1.1%	i
GA	1978–1979	Bartlett's Ferry	A	H and W	33,000 (36%)	560 and NA	1.7% and NA	133 (104%) and 111 (105%)	0.8% and 0.7%	j

Table 1. Continued.

State	Year	Water body	Design	Species	Effort [angler-hours for all fish] (RSE[a])	Effort [*Morone* angler-hours] (RSE[a])	Percent of total effort directed at *Morone*	*Morone* harvest (RSE[a])	Percent of total harvest	References
GA	1978–1979	Goat Rock	A	W	21,000 (13%)	0	0%	100 (104%)		j
GA	1978–1979	Lake Oliver	A	H and W	22,000 (33%)	NA and 0	NA and 0%	63 (105%) and 1,100 (102%)	1% and 25%	j
GA	1982	Lake Blackshear	R	H and W	123,000 (6%)	2,400 (32%) and 14,000 (12%)	1.9% and 11%	2,400 (17%) and 9,200 (18%)	1.7% and 6.8%	k
GA	1983	Bartlett's Ferry	R	H and W	37,000 (NA)	238 and 0	0.7% and 0%	129 (31%) and 13 (77%)	0.6% and 0.1%	l
GA	1984	Lake Walter F. George	R	H and W	400,000 (NA)	7,400 and 2,100	1.9% and 0.5%	456 (48%) and 533 (74%)	0.4% and 0.5%	l
GA	1985	Lake Seminole	Effort from aerial survey; catch from roving	H and W	334,000 (NA)	1,000 and 0	2.9% and 0%	5,100 (21%) and 1,200 (34%)	3% and 0.7%	l
GA	1986	Lake Blackshear	Roving	H and W	58,000 (NA)	144 and 4,100	0.2% and 7.1%	1,000 (20%) and 6,000 (41%)	1.2% and 7.3%	l
GA	1976–1978	Ocmulgee River – Section I	A	S	95,000/year (NA)	192/year	0.2%	148/year (95%)	0.1%	m
GA	1976–1978	Ocmulgee River – Section II	A	S	183,000/year (NA)	500/year	0.3%	490/year (39%)	0.1%	m
GA	1980–1982	Altamaha River	A	S	186,000/year (5%)	5,100/year	2.7%	2,400/year (27%)	0.6%	n
GA	1980–1982	Savannah River – Freshwater Section	A	S and H	97,000/year	4,600 (32%) and 4,300 (46%)	4.7% and 4.4%	633/year (39%) and 843/year (24%)	0.2% and 0.3%	o
GA	1980–1982	Savannah River – Estuarine Section	A	S and H	21,000/year	6,600/year (18%) and 30/year (82%)	32% and 0.2%	1,100 (24%) and 353 (23%)	2.9% and 0.9%	o
GA	1986	Ogeechee River	A	S	159,000 (14%)	1,000 (63%)	0.6%	113 (79%)	0.1%	p

Table 1. Continued.

State	Year	Water body	Design	Species	Effort [angler-hours for all fish] (RSE[a])	Effort [*Morone* angler-hours] (RSE[a])	Percent of total effort directed at *Morone*	*Morone* harvest (RSE[a])	Percent of total harvest	References
GA	1979 (Jul–Dec)	Jackson Lake	R	W, S, and H	50,000 (18%)	1,900 (37%), 200 (24%), and 0 (0%)	3.8%, 0.4%, and 0%	1,300 (42%), 6 (83%), and 0 (0%)	3.1%, <0.1%, and <0.1%	q
GA	1980	Jackson Lake	R	W, S, and H	123,000 (24%)	1,300 (35%), 200 (58%), and 0 (0%)	1.1%, 0.2%, and 0%	2,500 (30%), 24 (54%), and 0 (0%)	2.5%, <0.1%, and 0%	q
GA	1981	Jackson Lake	R	W, S, and H	78,000 (14%)	79 (110%), 0 (0%), and 0 (0%)	0.1%, 0%, and 0%	115 (44%), 0 (0%), and 0 (0%)	0.2%, 0%, and 0%	q
GA	1982	Jackson Lake	R	W, S, and H	75,000 (13%)	1,100 (26%), 0 (0%), and 20 (80%)	1.5%, 0%, and <0.1%	570 (36%), 12 (108%), and 94 (95%)	0.6%, 0%, and 0.1%	q
GA	1987	Lake Sinclair	R	W, S, and H	400,000 (NA)	9,300 (35%) for all *Morone* combined	2%	3,200 (24%), 507 (64%), and 5,200 (58%)	1%, 0.2%, and 2%	r
GA	1980	Allatoona Reservoir	R	W, S, and H	294,000 (7%)	NA	NA	2,400 (34%), 0 (0%), and 0 (0%)	0.8%, 0%, and 0%	s
GA	1983	Allatoona Reservoir	R	W, S, and H	374,000 (8%)	NA	NA	1,400 (47%), 951 (42%), and 565 (51%)	0.9%, 0.6%, and 0.4%	s
GA	1986	Allatoona Reservoir	R	W, S, and H	456,000 (6%)	NA	NA	3,500 (51%), 3,200 (18%), and 763 (20%)	2.1%, 2%, and 0.5%	s
NC	2007	Lake Norman	R	S	310,000	46,000 (9%)	15%	7,000 (18%)	5%	t
NC	2007	Lake Norman	A	S	218,000	33,000 (21%)	15%	3,000 (46%)	5%	t

TABLE 1. Continued.

State	Year	Water body	Design	Species	Effort [angler-hours for all fish] (RSE[a])	Effort [*Morone* angler-hours] (RSE[a])	Percent of total effort directed at *Morone*	*Morone* harvest (RSE[a])	Percent of total harvest	References
NC	2007	Lake Gastron	R	S	320,000	16,000 (18%)	5%	1,200 (28%)	5%	t
TX	2008–2009	Tawakoni	A	All moronids	114,000 (18%)	41,000 (25%)	36%	9,300 (52%)	NA	u
TX	2005–2006	Toledo Bend	R	All moronids	650,000 (12%)	17,000 (29%)	3%	92,000 (50%)	14%	v
TX	1999	Canyon	R	S		12,000 (33%)		1,600 (53%)	NA	w

[a] Relative standard error where reported.
[b] White and Lamprecht 2002.
[c] J. Ross, Kentucky Department of Fish and Wildlife Resources, personal communication.
[d] England 1977.
[e] Weaver 1985.
[f] Weaver and England 1982.
[g] Martin and Hess 1984.
[h] Keefer 1980.
[i] Keefer 1981a.
[j] Keefer 1981b.
[k] Keefer 1984.
[l] Keefer 1988.
[m] Coomer and Holder 1980.
[n] Hottell et al. 1983.
[o] Schmitt and Hornsby 1985.
[p] Schmitt 1988.
[q] Ellis 1983.
[r] Evans 1989.
[s] Beisser 1989.
[t] B. McRae, North Carolina Wildlife Resources Commission, personal communication.
[u] K. Storey, Texas Parks and Wildlife Department, personal communication.
[v] T. Driscoll, Texas Parks and Wildlife Department, personal communication.
[w] C. Munger, Texas Parks and Wildlife Department, personal communication.

TABLE 2. Creel survey results from Lake Texoma, where the primary species sought is striped bass. The Bayesian estimate uses the previous data for the prior, and updates using the current data each year. RSE = relative standard error. R = roving; S = Striped bass.

State	Year	Design	Species	Effort [*Morone* angler-hours] (RSE)	Percent of total effort	Harvest (RSE)	Reference	Bayesian estimate of harvest (RSE)
TX	1997	R	S	939,000 (17%)	68%	903,000 (18%)	a	
TX	1998	R	S	801,000 (15%)	68%	819,000 (18%)	a	856,912 (13%)
TX	1999	R	S	1,146,000 (15%)	66%	1,233,000 (17%)	a	937,191 (10%)
TX	2000 (Dec–Feb)	R	S	71,000 (22%)	68%	94,000 (23%)	a	

[a] B. Hysmith, Texas Parks and Wildlife Department, personal communication.

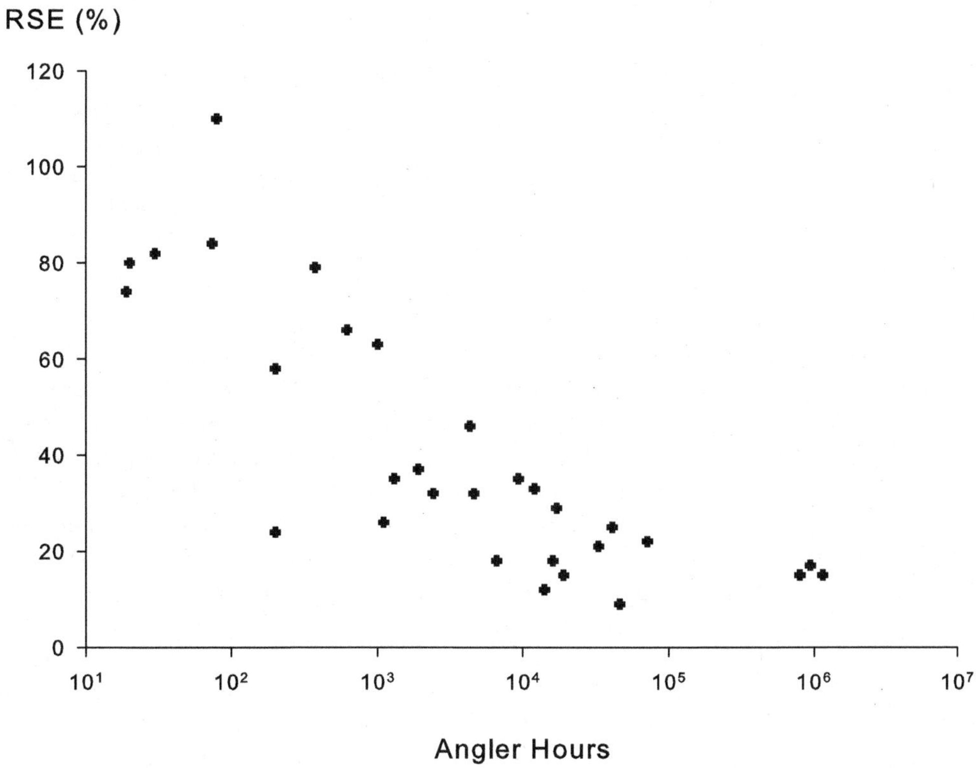

FIGURE 1. The relationship between relative standard error (RSE) and the estimated angler-hours from moronid fishery creel surveys. Note, as the number of angler-hours increases, the RSE decreases. Notice that the horizontal axis (the abscissa) is plotted on a log-scale. The points to the far right of the graph represent data from Lake Texoma. Data are from Tables 1 and 2.

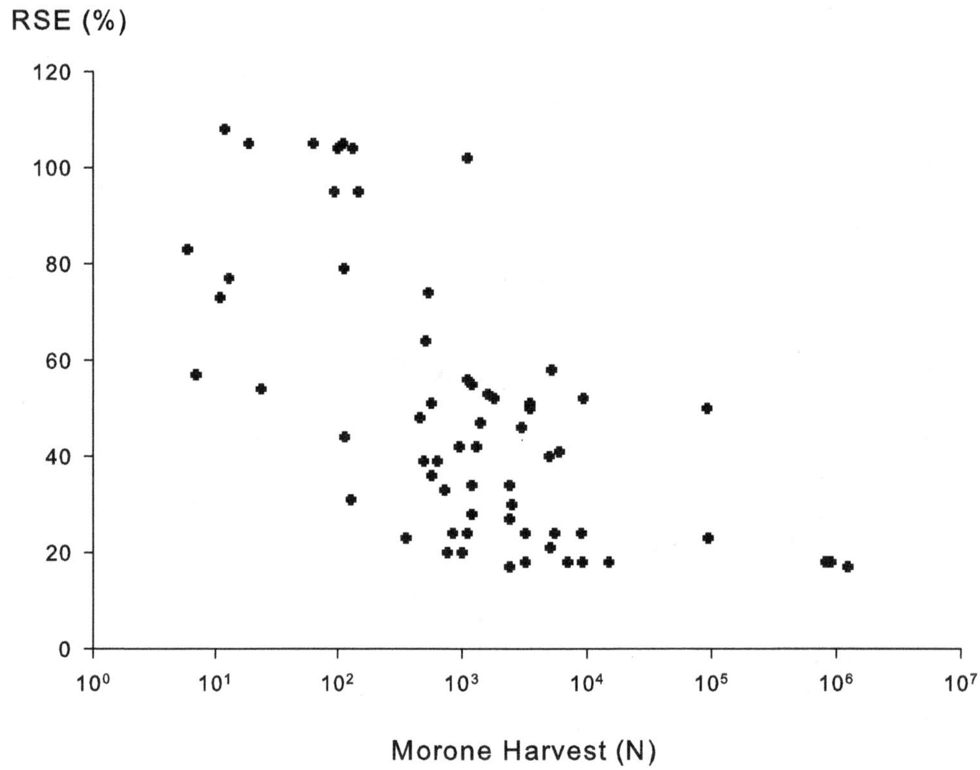

FIGURE 2. The relationship between relative standard error (RSE) and the estimated harvest of moronids. Note, as the harvest increases, the RSE decreases; however, the relationship is more variable than the relationship between angler-hours and RSE. Notice that the horizontal axis (the abscissa) is plotted on a log scale. The points to the far right of the graph represent data from Lake Texoma. Data from Tables 1 and 2.

surveys are typically conducted only during the daylight hours, primarily due to safety concerns. A tag-recapture study done concurrently with the creel survey in South Carolina (White and Lamprecht 2002) suggested that much of the fishery may be underrepresented and underreported because fishing does occur outside of the survey hours. Twenty percent of the recaptured fish came from night fishing, suggesting that the night-fishing component was substantial, both in effort and catch.

Mismatch between the seasonal timing of the creel survey effort and the moronid angling effort can result in imprecise estimates. However, estimates should be unbiased because each day has a known probability of selection. Not surprisingly, peaks of striped bass angling effort were frequently (but not always) associated with striped bass migrations; other moronid fisheries had other seasonal patterns. From studies that tracked seasonal harvest (Table 3), two patterns became apparent. Summer and fall seasons frequently had the highest reported harvest rates. However, while this pattern generally held true, the seasonal pattern could be quite different in different water bodies for a number of reasons. Foremost, effort and catch are not 100% correlated. Hence, angling effort may peak at a time that differs from angler catch. Second, because there are no set definitions of the seasons, one agency's spring may be another agency's summer. Third, because of the geographic extent of these species, spawning runs likely occur at different times of the year. And fourth, different water bodies have different moronid fisheries. For instance, harvest in Lake Norman, North

TABLE 3. Timing of greatest *Morone* spp. harvest as reported in creel survey reports. R = roving and A = access; S = striped bass, W = white bass, and H = hybrids.

State	Year	Water body	Design	Species	Period of greatest harvest	Period of second greatest harvest	Period of third greatest harvest	Reference
SC	1998	Lake Moultrie	R	S	May	June	July	a
KY	2005	Rough River Lake	R	H	August	June	July	b
KY	2008	Lake Cumberland (lower lake)	R	S	June	November	July	b
KY	2008	Lake Cumberland (upper lake)	R	S	April	November	July	b
KY	2004	Herrington Lake	R	W and H	May	June	March	b
KY	2006	Fishtrap Lake	R	H	August	June	July	b
KY	2007	Barren River Lake	R	H and W	July	August	June	b
KY	2006	Taylorsville	R	H and W	July	May	June	b
GA	1972	Lake Nottely	R	S	July	October	–	c
GA	1973	Lake Nottely	A	S	June	–	–	c
GA	1975	Lake Nottely	A	S	June	–	–	c
GA	1984	Lake Nottely	R	S and W	July	March	October	d
NC	2007	Lake Norman	R	S	Winter	Fall	Spring	e
NC	2007	Lake Norman	A	S	Fall	Winter	Summer	e
NC	2007	Lake Gaston	R	S	Spring	Summer	Fall	e
TX	2008–2009 (Jun–Feb)	Tawakoni	A	All moronids	Sep–Nov	Dec–Feb	Jun–Aug	f
TX	2005–2006	Toledo Bend	R	All moronids	Jun–Aug	Sep–Nov	Mar–May	g

[a] White and Lamprecht 2002.
[b] J. Ross, Kentucky Department of Fish and Wildlife Resources, personal communication.
[c] England 1977.
[d] Weaver 1985.
[e] B. McRae, North Carolina Wildlife Resources Commission, personal communication.
[f] K. Storey, Texas Parks and Wildlife Department, personal communication.
[g] T Driscoll, Texas Parks and Wildlife Department, personal communication.

Carolina appeared to be highest in fall or winter, whereas at Lake Gaston, North Carolina, harvest was highest in the spring. Most creel surveys do not use this information in their design, which leads to inefficient allocation of effort when precise estimates are desired for the moronid fisheries. To improve the precision of a statistic, more effort in the creel design should be allocated to those times with high fishing effort and high variability (Hayne 1991).

Most creel surveys also ignore subpopulations of the anglers. Estimates of angler success assume that surveyed anglers represent the average angler when, in fact, some angler subpopulations may differ substantially from the typically surveyed angler in their catch rates or the sizes of fish they harvest. A creel study on Lake Texoma (Moczygemba et al. 2006) indicated that 60% of the directed effort for striped bass was by guided anglers. Further, guided anglers were also more successful than unguided anglers (Moczygemba et al. 2006). Likewise, a pair of tag-recapture studies showed that anglers with a professional guide caught 11% of total recaptures in Lake Marion and 29% in Lake Moultrie (both in South Carolina) while only representing 5% and 17% of the anglers, respectively (Whiute and Lamprecht 2002). Ignoring angler subpopulation characteristics such as these and combining all anglers into a single estimate can decrease precision. If characteristics of these subpopulations lead to them being sampled in an unrepresentative manner, estimates can also be biased.

Our review of these studies suggests that merely increasing the amount of effort within the current creel-survey format may be less effective at improving assessments of moronid fisheries than redesigning creel surveys to match local fishing conditions and angling populations. When specific sections of a fishery are missed, the resulting estimates will be biased, as the full population of anglers has not been included in the sampling frame. Understanding the nature of specific moronid fisheries is integral to designing a creel-sampling process that will both improve the precision of the estimates and ensure that estimates are unbiased.

Options for Improving Effectiveness of Creel Surveys

Striped bass creel-survey results for most of the water bodies we evaluated (Table 1) indicated that striped bass made up a low percentage of total fishing effort and harvest. Although the RSE was generally not reported, the available estimates of RSE suggested that estimates for moronids were frequently imprecise. Robson and Regier (1964) suggest that for preliminary management surveys, an RSE of 50% is sufficient; for more accurate management work, an RSE of 25% may be acceptable; and for careful research, biologists should attempt to achieve an RSE of 10%. It should also be noted that it is unclear whether Robson and Regier (1964) were prescribing these guidelines for all studies or for just mark–recapture studies. Regardless, these guidelines have been repeated in various papers (Malvestuto 1996; Dumont and Schlechte 2004; Logsdon et al. 2009; and many others); it is rarely stated why the reported precision is appropriate for the given objective. Without fully understanding how the data will be used, the levels of precision promoted by Robson and Regier (1964) and others must be viewed as general guidelines. In the case of moronid creel surveys, whose main objective seems to be to broadly characterize the use of these fisheries alongside other, often more important fisheries, the Robson and Regier (1964) guidelines often appear inappropriately restrictive.

Before a creel survey is undertaken, it is incumbent upon the biologist to define the objective of that specific creel survey and the level of precision appropriate for the given measurements (Malvestuto 1996). This objective-based sampling approach will help to ensure that the appropriate amount of sampling effort is expended and that sampling effort is allocated with regard to the temporal and spatial extents of the fishery. Hence, while it may seem intuitive to want to minimize variability, improving the precision of a number for a small-scale fishery is often neither necessary nor cost-effective.

For example, in Georgia, at Lake Nottely in 1973 and 1975, RSEs for striped bass harvest were reported as 73% and 105%, respectively. These are obviously substantially larger than Robson and Regier (1964) suggested for even preliminary management studies. However, overall harvest was estimated to be 11 and 19 fish, respectively. If one were to assume that a normal confidence interval around the estimate was appropriate, the upper 95th confidence interval would estimate that at most 27 and 58 fish were harvested in 1973 and 1975, respectively. Hence, even with these high RSEs, the overall information coming from the creel survey was that very few striped bass were being harvested. Increasing the effort in these creel surveys to reduce the estimated RSEs would likely take much effort, yet would be of little benefit if the manager just wants to note the general level of catch.

In the example above, the lower confidence bound extends below zero. Since some fish were observed and there cannot be negative fish, the normal approximation may be inappropriate. If this is concerning, one can create asymmetric confidence intervals. One technique to create an asymmetric bound is to use a bootstrap technique (Efron 1987; DiCiccio and Efron 1996; Schnute and Haigh 2003). A score-statistic confidence interval is another approach (Cadigan 2011) that creates a confidence interval with good coverage and disallows a lower bound that is negative. While these approaches create more accurate confidence intervals, they are computer-intensive. An easily implemented approximation is to set the lower bound at the number of fish actually observed within the creel survey and extend the upper bound of the confidence interval until the interval contains 95%.

When striped bass are much more important to the fishery (Figures 1 and 2; Table 2), current creel survey designs can provide reasonably precise estimates of effort and harvest. In these instances, because effort and harvest are more consistent, each subsample looks more similar, so the RSE is reduced. Therefore, we feel an important question to ask is whether improved precision of estimates is needed. To answer this, the manager should consider the scale of the fishery and the relative importance of the objectives of the creel survey.

There are times when precise data may be needed. Population modeling often requires accurate (i.e., precise and unbiased) measures of current fishing mortality, as well as age and growth statistics. In cases where several alternative models exist, being able to choose between alternatives may require accurate estimates of these vital rates. In turn, models with these well-defined parameters can be used to better inform managers (Allen and Gwinn 2013, this volume). Specific modeling needs may alter how managers design and implement their creel surveys; improved designs could be of great benefit for meeting management objectives. Communication between the management agencies and researchers (when different) can help the manager decide if common objectives exist and if more effort or a new design might be better suited to collect the data.

When the decision is made that more precise estimates of moronid fishing effort and harvest are needed, the first reaction may be to simply increase the manpower devoted to the current creel survey design. However, given that moronid fishing effort and harvest are typically very low relative to other species, simply increasing the overall manpower devoted to a typical creel survey design may not provide the desired effect. There are two reasons that may reduce the effectiveness of this approach. First, assuming that moronid fisheries are characterized by small clusters of high catch and effort, one result of increased sampling may be to actually increase the variability. While in theory, increases in sample size result in more precise estimates, such considerations only apply on average. For any given sample, the variability may increase. The reasoning behind this is that the biologist would be more likely to encounter groups of successful anglers, but much of the survey effort could still occur on days when little or no fishing for moronids was occurring.

Thus, the range of the observed catch and effort could increase, leading to an overall increase in the variance, yet much of the data would only confirm that fishing effort for moronids was low or nonexistent. The second process that could reduce the effectiveness of simply increasing survey effort is the mathematical relationship between sample size and precision. The calculation of the relative standard error is a function of the square root of the sample size. Thus, even if one assumes the observed variance will remain constant, to reduce the RSE by half requires a fourfold increase in the sample size. Therefore, survey effort would have to be increased substantially to achieve any improvement in precision of RSE.

An alternative to simply increasing the overall manpower devoted to standard surveys would be to focus sampling where the fishing effort is allocated, spatially and temporally. The use of sampling strategies that correspond to the anglers' effort would improve spatial and temporal coverage (Sztramko 1991; Newman et al. 1997; Betsill and Pitman 2002) and should improve the precision. For instance, Malvestuto (1996) and Malvestuto and Knight (1991) noted that increasing the within-day sampling effort is a more cost-effective way of improving the estimates than adding more sampling days. A relatively cost-effective option for adding sampling days, hence improving precision of effort estimates from access point creel data, is to use remote sensing devices (e.g., trail cameras, boat trailer counts) to increase coverage (Steffe et al. 2008). Access creels are easily supplemented in this manner considering the increasing availability and low cost of remote monitoring equipment. Unmanned aerial vehicles are a remote-sensing device that could be used for estimating effort for roving creels. Unfortunately, while these would likely improve the overall precision for the estimates of effort, variability for the striped bass portions of effort and harvest may remain higher than desired.

In a creel design without strata, creating strata could be useful. To create strata, divide the entire fishery into a set of mutually exclusive and totally exhaustive subpopulations (Cochran 1977). This means each element of the entire fishery should be within one and only one stratum, yet together, these subpopulations comprise the entire fishery. The general rules for creating useful strata are that within a stratum, the data elements should be homogenous (i.e., low variability between data elements), whereas between strata, the data elements should be quite heterogenous (i.e., the mean values of different strata should be as disparate as possible). For a striped bass creel, this would mean that it would be best to have all the high-catch and high-effort activity in one stratum and all the low-catch and low-effort activity in a separate stratum. Some good bases for strata for striped bass creels would be season, time of day, and location.

In a creel design that already has strata (temporal or spatial), one way to improve precision is to allocate additional sampling effort to those strata (Cochran 1977) in which you expect the most moronid activity and the highest variability. Because many of the striped bass fisheries are seasonal in nature; precision could be greatly improved simply by focusing more survey effort when striped bass anglers are fishing. Typically, each stratum is allocated at least two samples, so a mean and variance can be estimated for that stratum. However, for extremely seasonal fisheries, all the effort could be placed in a single stratum and the remainder of the fishery could be assumed to have no catch and no effort. This is a strong assumption, so biologists may want to check the validity of this assumption on a regular interval. Due to variability in timing of striped bass harvest across the range of populations surveyed (Table 3), seasonal timing of additional creel days to capture striped bass anglers is likely to be site-specific. In addition, by using unequal sampling probabilities (Hayne 1991), one could focus more of the biologist's effort on times and areas when striped bass anglers are more likely to be encountered. Another way to improve precision is to simply pool data across strata. Pooling data across strata only allows one to achieve better precision for the pooled estimate. To achieve better precision both within and across strata, one must allocate additional sampling effort to those strata in which one expects the most moronid activity and the highest variability.

When striped bass fisheries are minor relative to total fishing effort, the probability of encountering a striped bass angler in most of the current creel-survey designs is low. There are options for designing creel surveys that can increase the encounter rate of these "rare events." If we assume that striped bass anglers are clustered spatially or temporally, a two-stage sampling design could be adopted (Cochran 1977). This design employs a rule-based adaptive sampling framework where certain outcomes of the basic sampling protocol trigger additional or auxiliary sampling. For example, assume that one was using one of the current creel-survey designs, but wanted to improve the precision of estimates from a striped bass fishery. Once a striped bass angling event was encountered within the current creel survey design, the surveying effort would be temporarily increased in order to sample more striped bass anglers. For this approach to work well, striped bass anglers must be tightly clustered in either time or space. Fortunately, given the nature of striped bass angling and the data that exist on seasonality, this assumption is expected to hold in some fisheries. Alternatively, if striped bass angling effort or harvest was correlated with a more commonly measured variable, a model-based sampling design may be appropriate (Edwards et al. 2005). Although we do not know of a good model-based survey design for striped bass in the literature, a design based on hydrology, water temperature, and date might prove useful for many riverine striped bass fisheries.

Ultimately, the levels of imprecision within striped bass fisheries are associated with poor spatial and temporal coverage of current creel surveys combined with low rates of catch and harvest of striped bass. This stands in contrast to other species such as largemouth bass, which are quite frequently the largest fraction of a fishery. Because the largemouth bass fishery is both large-scale and seasonally consistent, Texas' creel surveys, as an example, routinely have RSEs for largemouth bass effort of 15% and for largemouth bass catch of 30%. Surveys throughout the Southeast report extremely low relative standard errors for fish caught most frequently, with higher RSEs for rarer fish species.

Relatively few anglers often make up a large proportion of the effort and harvest of rarer species. In such cases, alternative survey methods such as angler diaries or logbooks can be more useful than standard roving or access-point creel surveys. These techniques require little additional effort on the part of biologists and can provide extensive effort and harvest information when the participants are motivated to take part. Several options exist for creating a sampling framework for striped bass anglers. First, many striped bass lakes have active guiding services that would be a natural starting point for initiating a logbook survey. Additionally, networking with local lake associations or angling clubs could provide a sample of motivated anglers. Pollock et al. (1994) provides a detailed discussion of pros and cons of diary or logbook data. If one samples only guided trips, biologists should consider that there are likely differences in catch rates and effort between guided and unguided trips (Knapp and Goeman 2005; Moczygemba et al. 2006).

In situations where creel surveys are conducted regularly and prior information exists on effort and catch, it may be helpful to consider Bayesian or time-series analyses to improve the precision of catch and effort estimates. Often, managers may be more interested in trends or large changes from historical conditions and less interested in the actual values of effort and harvest at a specific time. The accuracy needed to detect a trend is primarily a function of the signal-to-noise ratio (McCabe and Wolock 1991). This ratio combines the two factors that control detection time of a trend: (1) magnitude of change, and (2) measured variability of the parameter, which is a combination of natural and sampling variability. The larger the change in the abundance or the longer the time series, the less precise the data need to be to observe that change. For creel-survey data, Kalman filtering, or state-space models, can be used to account for both sampling error and process error while at the same time identifying historical trends (Kendall and Ord 1990). In addition, this approach allows one to test whether newly compiled creel data fall within historical norms. Unlike a typical creel-survey approach, in which each year is estimated independently of previous years' data, state-space models assume that angler effort and harvest in the current year are not independent of the previous years' effort and harvest. This approach requires a slightly different model formulation but has the advantage of including previously collected creel survey data within new estimates. When processes follow a well-defined trend, weight is given to the previously collected data. Because more data are included in the estimation procedure, the resulting parameter estimates are more precise. When the process does not follow a well-defined trend, little to no weight is given to the previously collected data. Hence, the parameter estimate tends to be comparable to an estimate that uses only the current year's data.

We used 3 years of creel-survey data for Lake Texoma as an example of Bayesian updating of a creel survey. We assumed a basic model, that there is no trend in the data. We used the previous year's estimated mean and variance as the prior distribution and the new measurement as the newest data to update the prior data and give us the new posterior distribution (Gelman et al. 2000). Under the Bayesian framework, with normally distributed data the posterior mean is a weighted average of the prior data and the current observation, with the weights being proportional to the variance. Thus, an observation with high precision gets more weight and one with low precision gets less weight. The precision of the posterior mean is the average of the precisions. In our example, the Bayes estimates fluctuate less than each year's estimates (Table 2). While each year's RSE is quite high, by suggesting that one year's estimate is likely to be similar to the previous year's (through the Bayesian updating), the RSE drops to 13% at the first update and down to 10% with the second. This works well if the system is stable or is following a well-defined trend. If not, then this type of updating will produce a smoother and lagged signal, especially with imprecise data. This is because both techniques are using not just the current data, but the historic data as well, hence creating moving, weighted-average estimates.

Although we do not know of any other examples of Bayesian updating or Kalman filtering being applied to creel-survey data, Quinn and Deriso (1999) highlighted several examples of Kalman filtering in other fisheries applications, including spawner-recruit models, delay-difference models, and catch-age models. A breeding bird survey (Sauer et al. 2008) used spatial and temporal replicates of bird counts to estimate population status and trend using a hierarchical model within a Bayesian framework (Link and Sauer 1998; Sauer and Link 2002; Link et al. 2006). This methodology allows for increased precision of the parameters (a term called "Bayesian shrinkage"). Although data from moronid creel surveys will not be as extensive as data from the breeding bird survey, we expect that appropriate application of these techniques could prove useful.

Throughout the southeastern United States, creel surveys seem to be doing an adequate job of documenting the importance of the moronid fisheries relative to other fisheries. We have identified several potentially useful avenues for consideration by managers who desire an improvement in the estimation of creel statistics. We recommend making the survey sampling effort coincide with the fishing effort. While it may be possible to improve the precision by increasing the amount of manpower devoted to the current creel survey, stratifying or using unequal spatial and temporal probabilities are techniques more likely to improve precision in a cost-effective manner. Once the data have been collected, using model-based estimation, such as using Kalman filters or empirical Bayesian estimation, could also prove advantageous. We envision many challenges and opportunities in management of striped bass and its hybrids in the future (Bettoli 2013, this volume); understanding fishing effort, catch rates, and harvest of these fisheries is critical. Appropriately designed, conducted, and analyzed creel surveys are essential for developing this understanding.

Acknowledgments

This work would not have been possible without the cooperation of the following biologists who contributed creel data, creel reports, or opinions on data quality: John Findeisen, John Tibbs, Stephen Magnelia, Todd Driscoll, Kevin Storey, Rick Ott, John Moczygemba, and Bruce Hysmith of Texas Parks and Wildlife; Dave van Voorhees of National Marine Fisheries Service; Brian McRae of North Carolina Wildlife Resources Commission Division of Inland Fisheries; Larry Bull and Larry Pugh of Mississippi Department of Wildlife, Fisheries and Parks; Dan Rankin, Hal Beard, Weston Houck, and Scott Lamprecht of South Carolina Department of Natural Resources; Jeff Ross of the Kentucky Department of Fish and Wildlife Resources; Tim Bonvechio, Dennis Schmitt, and Elizabeth Colvin of Georgia Department of Natural Resources; Mike Siepker of Missouri Department of Conservation; Tim Churchill and Pat Black of Tennessee Wildlife Resources Agency; and Greg Summers of Oklahoma Department of Wildlife Conservation. We are especially grateful to the editors and to in-house and anonymous reviewers whose comments helped us to substantially improve our manuscript. This work was funded through Federal Aid in Sport Fish Restoration Grants SFR F-148-M and SFR F-22-D to the Texas Parks and Wildlife Department, Inland Fisheries Division.

References

Allen, M. S., and D. Gwinn. 2013. Population models for assessment and management of inland striped bass fisheries. Pages 351–364 *in* J. S. Bulak, C. C. Coutant, and J. A. Rice, editors. Biology and management of inland striped bass and hybrid striped bass. American Fisheries Society, Symposium 80, Bethesda, Maryland.

Allen, M. S., K. I. Tugend, and M. J. Mann. 2003. Largemouth bass abundance and angler catch rates following a habitat enhancement project at Lake Kissimmee, Florida. North American Journal of Fisheries Management 23:845–855.

Beisser, G. S. 1989. The fish populations and sport fishery of Allatoona Reservoir. Georgia Department of Natural Resources, Final Report, Federal Aid Project F-36–10, Atlanta.

Betsill, R. K., and V. M. Pitman. 2002. Comparison of creel statistics for river and reservoir components of a Texas white bass fishery. North American Journal of Fisheries Management 22:659–664.

Bettoli, P. W. 2013. Challenges and opportunities in studying and managing striped bass and its hybrids in the 21st century. Pages 575–588 *in* J. S. Bulak, C. C. Coutant, and J. A. Rice, editors. Biology and management of inland striped bass and hybrid striped bass. American Fisheries Society, Symposium 80, Bethesda, Maryland.

Cadigan, N. G. 2011. Confidence intervals for trawable abundance from stratified-random bottom trawl surveys. Canadian Journal of Fisheries and Aquatic Sciences 68:781–794.

Chen, S. X., and J. L. Woolcock. 1999. A condition for designing bus-route type access site surveys to estimate recreational fishing effort. Biometrics 55:799–804.

Cochran, W. G. 1977. Sampling techniques, 3rd edition. Wiley, New York.

Coomer, C., Jr., and D. R. Holder. 1980. A fisheries survey of the Ocmulgee River. Georgia Department of Natural Resources, Final Report, Federal Aid Project F-29–8, Atlanta.

DiCiccio, T. J., and B. Efron. 1996. Bootstrap confidence intervals (with Discussion). Statistical Science 11:189–228.

Dumont, S. C., and J. W. Schlechte. 2004. Use of resampling to evaluate a simple random sampling design for general monitoring of fishes in Texas reservoirs. North American Journal of Fisheries Management 24:408–416.

Edwards, T. C., Jr., D. R. Cutler, N. E. Zimmermann, L. Geiser, and J. Alegria. 2005. Model-based stratifications for enhancing the detection of rare ecological events. Ecology

86:1081–1090.
Efron, B. 1987. Better bootstrap confidence intervals (with Discussion). Journal of the American Statistical Association 82:171–200.
Ellis, F. S. 1983. A survey of the Jackson Lake fishery. Georgia Department of Natural Resources, Final Report, Federal Aid Project F-33–2, Atlanta.
England, R. H. 1977. Striped bass introductions into Lake Nottely. Georgia Department of Natural Resources, Final Report, Federal Aid Project F-25–3, Atlanta.
Evans, J. W. 1989. A survey of the largemouth bass fishery on Lake Sinclair. Georgia Department of Natural Resources, Final Report, Federal Aid Project F-33–7, Atlanta.
Gelman, A., J. B. Carlin, H. S. Stern, and D. B. Rubin. 2000. Bayesian data analysis. Chapman and Hall/CRC, New York.
Harrell, R. M., and J. M. Dean. 1988. Identification of juvenile hybrids of *Morone* based on meristics and morphometrics. Transactions of the American Fisheries Society 117:529–535.
Hayne, D. W. 1991. The access point creel survey: procedures and comparison with the roving-clerk creel survey. Pages 123–138 *in* D. Guthrie, J. M. Hoenig, M. Holliday, C. M. Jones, M. J. Mills, S. A. Moberly, K. H. Pollock, and D. R. Talhelm, editors. Creel and angler surveys in fisheries management. American Fisheries Society, Symposium 12, Bethesda, Maryland.
Hoenig, J. M., C. M. Jones, K. H. Pollock, D. S. Robson, and D. L. Wade. 1997. Calculation of catch rate and total catch in roving surveys of anglers. Biometrics 53:306–317.
Hottell, H. E., C. Coomer, Jr., and D. R. Holder. 1983. A fisheries survey of the Altamaha River. Georgia Department of Natural Resources, Final Report, Federal Aid Project F-29–9, Atlanta.
Hudy, M., and C. R. Berry. 1983. Performance of three strains of rainbow trout in a Utah reservoir. North American Journal of Fisheries Management 3:136–141.
Jones, C. M., and D. S. Robson. 1991. Improving precision in angler surveys: traditional access design versus bus route design. Pages 177–188 *in* D. Guthrie, J. M. Hoenig, M. Holliday, C. M. Jones, M. J. Mills, S. A. Moberly, K. H. Pollock, and D. R. Talhelm, editors. Creel and angler surveys in fisheries management. American Fisheries Society, Symposium 12, Bethesda, Maryland.
Keefe, D. G., R. C. Perry, and J. G. Luther. 2009. A comparison of two methodologies for estimating brook trout catch and harvest rates using incomplete and complete fishing trips. North American Journal of Fisheries Management 29:1058–1064.
Keefer, L. C. 1980. A survey of the Lake Walter F. George fishery. Georgia Department of Natural Resources, Final Report, Federal Aid Project F-28–16, Atlanta.
Keefer, L. C. 1981a. A survey of the Lake Seminole fishery with emphasis on the success of stocking *Morone* hybrids. Georgia Department of Natural Resources, Final Report, Federal Aid Project F-28–18, Atlanta.
Keefer, L. C. 1981b. Creel survey on three Chattahoochee River reservoirs. Georgia Department of Natural Resources, Final Report, Federal Aid Project F-28–21, Atlanta.
Keefer, L. C. 1984. A survey of the Lake Blackshear fishery. Georgia Department of Natural Resources, Final Report, Federal Aid Project F-28–24, Atlanta.
Keefer, L. C. 1988. A survey of the sport fisheries of four major reservoirs in southwest Georgia. Georgia Department of Natural Resources, Final Report, Federal Aid Project F-28–25, Atlanta.
Kendall. M., and J. K. Ord. 1990. Time series, 3rd edition. Oxford University Press, New York.
Knapp, M. L., and T. J. Goeman. 2005. Catch and harvest by fishing guides compared with other anglers in five central Minnesota lakes. North American Journal of Fisheries Management 25:885–889.
Koo, T. S. Y. 1970. The striped bass fishery in the Atlantic states. Chesapeake Science 11:73–93.
Link, W. A., and J. R. Sauer. 1998. Estimating population change from count data: application to the North American breeding bird survey. Ecological Applications 8:258–268.
Link, W. A., J. R. Sauer, and D. K. Niven. 2006. A hierarchical model for regional analysis of population change using Christmas bird count data, with application to the American

black duck. The Condor 108:13–24.

Logsdon, D. E., L. M. Miller, and C. S. Anderson. 2009. Evaluation of long-term retention and detection of oxytetracycline marks in walleye otoliths using genetic methodology. Transactions of the American Fisheries Society 138:872–887.

Malvestuto, S. P. 1996. Sampling the recreational creel. Pages 591–623 *in* B. R. Murphy and D. W. Willis, editors. Fisheries techniques, 2nd edition. American Fisheries Society, Bethesda, Maryland.

Malvestuto, S. P., and S. S. Knight. 1991. Evaluation of the components of variance for a stratified two-stage roving creel design with implications for sample size allocation. Pages 108–115 *in* D. Guthrie, J. M. Hoenig, M. Holliday, C. M. Jones, M. J. Mills, S. A. Moberly, K. H. Pollock, and D. R. Talhelm, editors. Creel and angler surveys in fisheries management. American Fisheries Society, Symposium 12, Bethesda, Maryland.

Martin, C. R. and T. B. Hess. 1984. Analysis of the fish population and fishery of Lake Oconee. Georgia Department of Natural Resources, Final Report, Federal Aid Project F-26–5, Atlanta.

Matthews, W. J. 1985. Summer mortality of striped bass in reservoirs of the United States. Transactions of the American Fisheries Society 114:62–66.

McCabe, G. J., and D. M. Wolock. 1991. Detectability of the effects of a hypothetical temperature increase on the Thornthwaite moisture index. Journal of Hydrology 125:25–35.

Meredith, E. K., and S. P. Malvestuto. 1991. An evaluation of survey designs for the assessment of effort, catch rate and catch for two contrasting river fisheries. Pages 223–232 *in* I. G. Cowx, editor. Catch and effort sampling strategies: their application in freshwater fisheries management. Fishing News Books, Oxford, UK.

Moczygemba, J. H., B. T. Hysmith, and J. W. Schlechte. 2006. Comparison of guided and unguided fishing trips from two Texas reservoirs. Pages 17–30 *in* C. C. Bonds, editor. Annual proceedings of the Texas Chapter, American Fisheries Society, volume 27. American Fisheries Society, Texas Chapter, Bethesda, Maryland.

Muoneke, M. I., O. E. Maughan, and M. E. Douglas. 1991. Multivariate morphometric analysis of striped bass, white bass, and striped bass × white bass hybrids. North American Journal of Fisheries Management 11:330–338.

Newman, S. P., P. W. Rasmussen, and L. M. Andrews. 1997. Comparison of a stratified, instantaneous count creel survey with a complete mandatory creel census on Escanaba Lake, Wisconsin. North American Journal of Fisheries Management 17:321–330.

Pollock, K. H., C. M. Jones, and T. L. Brown. 1994. Angler survey methods and their applications in fisheries management. American Fisheries Society, Special Publication 25, Bethesda, Maryland.

Quinn, T. J., II, and R. B. Deriso. 1999. Quantitative fish dynamics. Oxford University Press, New York.

Robson, D. S. 1991. The roving creel survey. Pages 19–24 *in* D. Guthrie, J. M. Hoenig, M. Holliday, C. M. Jones, M. J. Mills, S. A. Moberly, K. H. Pollock, and D. R. Talhelm, editors. Creel and angler surveys in fisheries management. American Fisheries Society, Symposium 12, Bethesda, Maryland.

Robson, D. S., and H. A. Regier. 1964. Sample size in Petersen mark–recapture experiments. Transactions of the American Fisheries Society 93:215–226.

Sauer, J. R., J. E. Hines, and J. Fallon. 2008. The North American breeding bird survey, results and analysis 1966–2007, version 5.15.2008. U.S. Geological Survey Patuxent Wildlife Research Center, Laurel, Maryland. Available: www.mbr-pwrc.usgs.gov/bbs/bbs.html (March 2012).

Sauer, J. R., and W. A. Link. 2002. Hierarchical modeling of population stability and species group attributes from survey data. Ecology 86:1743–1751.

Schmitt, D. N., and J. H. Hornsby. 1985. A fisheries survey of the Savannah River. Georgia Department of Natural Resources, Final Report, Federal Aid Project F-30–6, Atlanta.

Schmitt, D. N. 1988. A fisheries survey of the Ogeechee River. Georgia Department of Natural Resources, Final Report, Federal Aid

Project F-30–7, Atlanta.

Schnute, J. T., and R. Haigh. 2003. A simulation model for designing groundfish trawl surveys. Canadian Journal of Fisheries and Aquatic Sciences 60:640–656.

Steffe, A. S., J. J. Murphy, and D. D. Reid. 2008. Supplemented access point sampling designs: a cost-effective way of improving the accuracy and precision of fishing effort and harvest estimates derived from recreational fishing surveys. North American Journal of Fisheries Management 28:1001–1008.

Storey, K. W., J. W. Schlechte, and L. T. Fries. 2002. Field identification accuracy for white bass and hybrid striped bass. Proceedings of the Annual Conference Southeastern Association of Fish and Wildlife Agencies 54(2000):97–106.

Sztramko, L. K. 1991. Improving precision of roving-creel-survey estimates: implications for fisheries with a closed season. Pages 116–121 *in* D. Guthrie, J. M. Hoenig, M. Holliday, C. M. Jones, M. J. Mills, S. A. Moberly, K. H. Pollock, and D. R. Talhelm, editors. Creel and angler surveys in fisheries management. American Fisheries Society, Symposium 12, Bethesda, Maryland.

U.S. Department of the Interior, Fish and Wildlife Service, and U.S. Department of Commerce, U.S. Census Bureau. 2006. National survey of fishing, hunting, and wildlife-associated recreation. U.S. Fish and Wildlife Service, Washington, D.C.

Van Horn, S. L. 2013. A brief history of inland striped bass management. Pages 1–13 *in* J. S. Bulak, C. C. Coutant, and J. A. Rice, editors. Biology and management of inland striped bass and hybrid striped bass. American Fisheries Society, Symposium 80, Bethesda, Maryland.

Wade, D. L., C. M. Jones, D. S. Robson, and K. H. Pollock. 1991. Computer simulation techniques to assess bias in the roving-creel-survey estimator. Pages 40–46 *in* D. Guthrie, J. M. Hoenig, M. Holliday, C. M. Jones, M. J. Mills, S. A. Moberly, K. H. Pollock, and D. R. Talhelm, editors. Creel and angler surveys in fisheries management. American Fisheries Society, Symposium 12, Bethesda, Maryland.

Weaver O. R. 1985. Evaluation of the walleye population of Lake Nottely. Georgia Department of Natural Resources, Final Report, Federal Aid Project F-25–14, Atlanta.

Weaver O. R., R. H. England. 1982. Evaluation of the sportfish population of Lake Lanier. Georgia Department of Natural Resources, Final Report, Federal Aid Project F-25–9, Atlanta.

White, M. G., III, and S. D. Lamprecht. 2002. Fisheries investigations in lakes and streams. South Carolina Department of Natural Resources, Final Report, Federal Aid Project F-63–7-5, Columbia.

Williams, H. M. 1976. Characteristics for distinguishing white bass, striped bass and their hybrid (striped bass × white bass). Proceedings of the Annual Conference Southeastern Association of Game and Fish Commissioners 29(1975):168–171.

American Fisheries Society Symposium 80:313–332, 2013

Evaluation and Management of Hybrid Striped Bass in Monroe Lake, Indiana

KEVIN J. HOFFMAN*, DAVE S. KITTAKA, AND BRIAN M. SCHOENUNG
Division of Fish and Wildlife, Indiana Department of Natural Resources
Indianapolis, Indiana 46204, USA

Abstract.—Striped bass *Morone saxatilis* movements and population dynamics have been well studied in midwestern impoundments; however, a paucity of information exists on hybrid striped bass in these systems. We studied palmetto bass (male white bass *M. chrysops* × female striped bass) population dynamics and seasonal movements from 2005 to 2007 at Monroe Lake, a 4,350-ha flood control impoundment in south-central Indiana. Palmetto bass were collected, measured, and aged using otoliths to assess growth and mortality. Thirty fish were tagged with radio tags in April 2006 to determine seasonal movements and distributions for 1 year. Oxygen and temperature profiles were measured throughout the year to determine available habitat. Total annual mortality was 32% (±5%). Catch-curve residuals were modeled against the number of fish stocked to investigate recruitment variability. More than 95% of the variation in year-class strength was explained by the number of age-0 fish (25–50 mm total length) stocked in early summer. We determined that optimal stocking rates were between 12 and 23 fish/ha. Fish moved more during spring (501 m/d) than other seasons (109–220 m/d). Fish were concentrated in the lower basin of the lake during summer and in the upper basin during winter. Based on minimum habitat suitability index values for temperature and dissolved oxygen, there was no optimal habitat available in winter or spring. The volume of optimal habitat increased to 22% of the lake during summer and 47% in fall. Based on our evaluation of population dynamics, seasonal movements, available habitat, and stocking efficiency, current management strategies of hybrid striped bass in Monroe Lake are adequate to sustain a healthy fishery at similar levels of mortality.

Introduction

A paucity of information exists on movements and population dynamics of palmetto bass (male white bass *Morone chrysops* × female striped bass *M. saxatilis*) in midwestern impoundments even though striped bass have been well studied there. To rectify this situation, we studied hybrid striped bass population dynamics and seasonal movements from 2005 to 2007 at Monroe Lake, Indiana. Monroe Lake is a 4,350-ha, U.S. Army Corps of Engineers (USACE) flood control reservoir located in Brown and Monroe counties southeast of Bloomington, Indiana (Figure 1). Constructed in 1965, it is the largest inland body of water in Indiana. There are nine publicly owned boat ramps and several privately owned recreational facilities such as boat rentals, sports shops, marinas, and campgrounds. Monroe Lake is bisected by Highway 446, resulting in

* Corresponding author: khoffman@dnr.in.gov

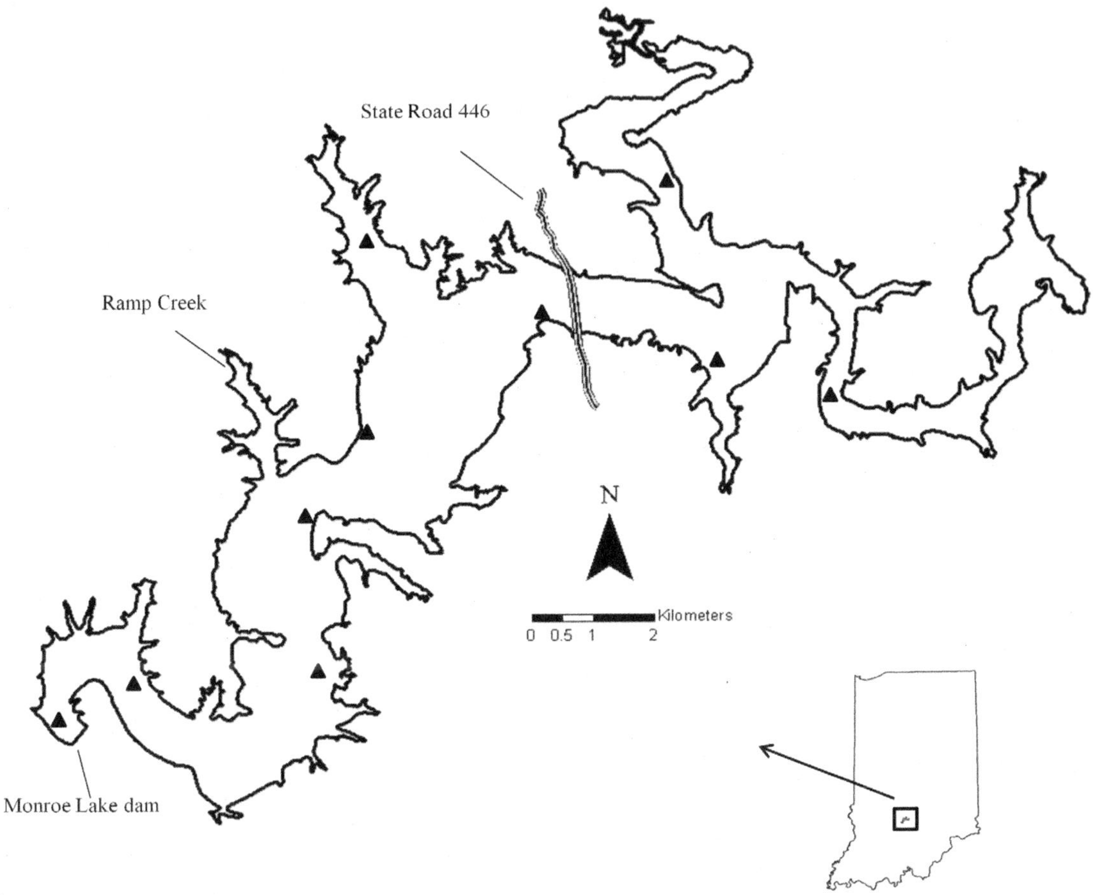

FIGURE 1. Monroe Lake, Indiana. Water flows northeast to southwest. The impoundment has two distinct basins, separated by a causeway (State Road 446). Black triangles represent the 10 water quality stations where temperature and dissolved oxygen profiles were collected. Inset indicates location in the state of Indiana, USA.

two very distinct basins. The lower basin contains approximately 80% of the total volume of water and represents approximately 64% of the surface area at normal pool (164 m above sea level). In addition to its recreational opportunities, Monroe Lake serves as a primary water supply for the city of Bloomington. The USACE maintains the lake at normal pool throughout the year and the lake is not drawn down in the winter. The USACE allows the lake's conservation pool to rise as much as 5 m before water is released during rain events.

The Monroe Lake fishery has been managed primarily for largemouth bass *Micropterus salmoides* and panfish fishing (i.e., bluegill *Lepomis macrochirus* and crappies *Pomoxis* spp.). Since about 1982, the fish community has been characterized by an overabundance of forage fish, mainly gizzard shad *Dorosoma cepedianum*, with few predator fish. Fish management activities at Monroe Lake have included several supplemental stockings of predator species. Early stockings included northern pike *Esox lucius* and walleye *Sander vitreus* but were for the most part unsuccessful. Recent stockings have included walleye and palmetto bass. The intent of these stockings has been to provide additional fishing opportunities and utilize surplus forage fish, particularly gizzard shad.

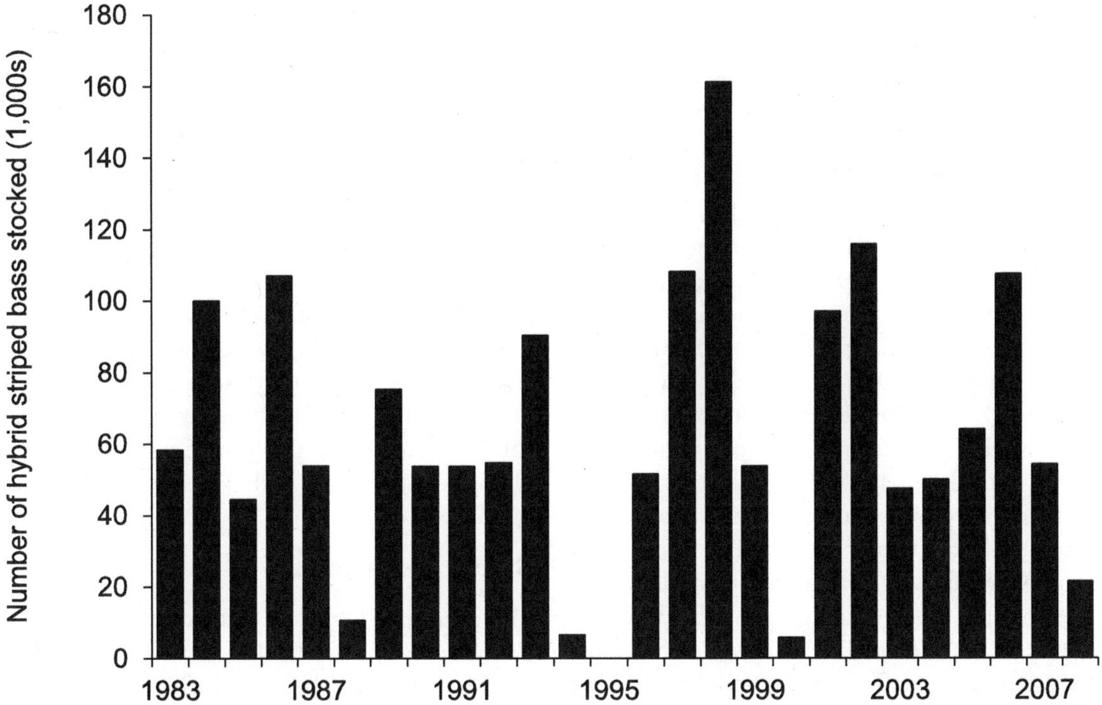

FIGURE 2. Number of palmetto bass fingerlings (25–50 mm total length, except 2004 when 75–100 mm) stocked in Monroe Lake, Indiana, from 1983 to 2008.

Palmetto bass have been stocked at Monroe Lake since 1983 with a stocking goal of 12 fish/ha or 53,750 fingerlings (usually 25–50 mm total length [TL]) annually (Figure 2). In 2004, fingerlings were approximately 75–100 mm TL. The Indiana Department of Natural Resources (IDNR) has an internal policy not to stock hybrids in a water body if white bass are present because hybrids and white bass have been known to back-cross (Avise and Van Den Avyle 1984). Although white bass are not present in Monroe Lake, yellow bass *Morone mississippiensis* are present. White bass and palmetto bass have been managed statewide with the same regulations because it was thought that anglers would not be able to distinguish between small, temperate bass species. Currently, hybrids and white bass are managed with a statewide daily creel limit of 12 fish singly or in aggregate, with no more than two fish exceeding 432 mm TL.

This paper presents a description and evaluation of the palmetto bass management program in Monroe Lake as an example that may be useful, along with other papers in this volume (see Collier et al. 2013; Kilpatrick and Ney 2013; and Rash and Ney 2013; all this volume), for managing hybrid fisheries. Several IDNR fisheries management surveys and angler creel surveys have been conducted at Monroe Lake to monitor the fish community. Results from those surveys confirmed that the hybrid fishery at Monroe Lake was well established with increased popularity among anglers. We were particularly interested in analyzing (1) population dynamics, (2) seasonal movements, and (3) available habitat by season (i.e., dissolved oxygen and water temperature by volume of water).

Methods

Population dynamics

We analyzed current population dynamics of hybrids in Monroe Lake using fish-population survey data from fall 2005, stocking records, and creel data. Hybrids were collected using experimental gill nets (300 m long; 3 m deep; bar mesh sizes = 38, 51, 64, and 76 mm) from 17 to 26 October 2005 as part of a standard monitoring program for hybrids in Indiana. All hybrids collected were measured to the nearest 0.25 cm TL and weighed. One hundred fish of representative sizes were sacrificed and otoliths removed for aging. Otoliths were viewed under a dissecting microscope in whole view, and if three or more annuli were present, otoliths were cracked through the core, placed vertically in putty with the broken plane up, submersed in water, and illuminated with a fiber-optic light, and annuli were counted by one reader. An age-length key was constructed to assign ages to nonaged fish. We calculated mean length at age from the age-length key using the midpoint method (DeVries and Frie 1996), and we fitted a von Bertalanffy growth model to the data (Busacker et al. 1990).

We estimated total annual mortality using three variations (Heinke, Robson and Chapman [1961], and linear catch curve) of the catch-curve method, as described by Miranda and Bettoli (2007). These methods were averaged to produce a final estimate of total annual mortality. Hybrids were fully recruited to gill net by age 1 in the fall and were included in the estimates of total annual mortality. Fish over age 8 were not included in the catch-curve analysis because there were fewer than three fish in those age-classes. Natural mortality and conditional natural mortality were estimated using methods described in Fishery Analyses and Simulation Tools software (FAST; Slipke and Maceina 2000). Conditional natural mortality was estimated using the Quinn and Deriso (1999) option in FAST, assuming that approximately 3% of the population of hybrids in Monroe Lake reached a maximum age of 13 years old (oldest age observed in this population sample), which resulted in a conditional natural mortality of 0.24. We used 3% as the proportion that survives to tmax because approximately 3% of our sample constituted the oldest age-classes. Slipke and Maceina (2000) also recommended using between 1% and 5% when calculating conditional natural mortality using the Quinn and Deriso (1999) method. We modeled yield-per-100 recruits using FAST software over a range of exploitation rates (0.0–0.8) and size limits (150, 350, 425, and 500 mm TL). According to the creel data collected for this project at Monroe Lake, anglers began harvesting hybrids at approximately 150 mm TL, which represented "no size limit" for this analysis. Percent change in yield-per-100 recruits was evaluated among the different size limits to investigate potential changes in population dynamics of hybrids in Monroe Lake.

Stocking evaluation

The number of stocked hybrids varied from 0 to more than 160,000 fish annually since the stocking program began in 1983. Since 1996, most fry have been obtained from interstate trade with other fish and wildlife agencies. The majority of hybrids stocked in Indiana up to 2007 were a standard cross (i.e., palmetto bass). Fry were typically transported to Indiana in May each year and cultured in earthen ponds for approximately 30 d. Fingerling hybrids were typically 25–50 mm TL when stocked in early June. Stocking rates ranged from 1 to 37 fish/ha from 1997 to 2005. Most years, the fish were stocked at multiple boat ramps to distribute the fish. Assuming no natural reproduction, we investigated variation in year-class strength in relation to number stocked. To investigate variability in year-class strength and optimal stocking rates, we plotted standardized-residuals from a catch curve against the number of hybrids stocked that year (Maceina 1997; Maceina and Bettoli 1998).

Creel survey

An angler survey was conducted at Monroe Lake from 2 April to 31 October 2007. The objectives of the survey were to measure fishing pressure and fish harvest and to evaluate angler interest in and harvest of palmetto bass, largemouth bass, and walleye. We used an "access point," two-stage cluster design, with nonuniform probabilities (Pollock et al. 1994). Ramp and shift probabilities were assigned using observed effort from previous creel survey data on this lake. From April through October, the fishing day was divided into three periods: 0600–1400, 1400–2200, and 2200–0600. Fishing activity probabilities assigned to these periods were 0.25, 0.66, and 0.09, respectively, based on prior experience. We stratified the lake by upper- and lower-basin boat ramps. One clerk was assigned the four upper-basin boat ramps (Stratum 1), and the remaining five ramps (Stratum 2) were assigned to a second clerk. Work shifts were assigned randomly based on site, day of week, and time of day probabilities. Creel clerks worked 8-h shifts, 9 d per 2-week period. Anglers were interviewed at the end of their fishing trips. Interview data included length of trip, number in party, county of residence, species sought, number and length (nearest 1.3 cm TL) of fish harvested, and number of largemouth bass, hybrids, and walleye caught and released. Weights of hybrids harvested were estimated using length-weight regressions from hybrids collected at Monroe Lake.

Tagging

Radio tags were surgically implanted in 30 hybrids (range 559–668 mm TL) on 10 and 12 April 2006 (Table 1). Radio tags, purchased from Advanced Telemetry Systems (model F1850B), operated at frequencies between 150 and 152 MHz, a pulse rate of 40 pulses per minute, and a pulse width of 20 ms. Tags were three-stage, which offers a clearer signal and double the detection range of a two-stage transmitter but, consequently, approximately 60% lower battery life than a two-stage transmitter. The duty cycle was 14 d on and 14 d off to increase battery life. Tags had a warranty battery life of 302 d and a capacity of 752 d. Tags were fitted with a mortality switch and a 254-mm trailing whip antenna. Tags weighed 25.9 g and were less than 1% of the total weight of each fish.

We collected hybrids using experimental gill nets that were set for less than 1 h at night along the face of the dam. All hybrids were immediately removed from gill nets, placed in a tank, and transported less than 100 m to the surgical station. Fish were kept in the tanks, and temperature and dissolved oxygen were monitored. They were measured, weighed, and placed in a basket that supplied a small electrical current for immobilization (Jennings and Looney 1998). We placed a small tube connected to a 1,363-L/h, 12-V bilge pump in the mouth of the fish, which supplied a constant stream of oxygenated water through the gills. A few scales were removed along the ventral line between the pelvic fins and the anus. A 50-mm incision was made, and the tag was inserted with the antenna protruding out of the incision. Monofilament nylon (#2/0) sutures were used to close the incision, and petroleum jelly antibiotic was applied to the incision to provide a temporary water barrier. Tagged fish were removed from the surgical basket and immediately placed into a 0.5% saltwater solution, allowed to equilibrate, and released. All fish were released within 100 m of the capture site after tags were implanted (Table 1).

Tracking

Based on preliminary field testing, we determined that tags near the surface could be detected at distances up to 1,000 m. Tags at approximately 7 m deep could easily be detected while the boat was in motion from approximately 200 m, or 400 m without the boat motor running. Typically, the hypolimnion in Monroe Lake was anoxic below 6 m during summer stratification. We traveled in a zigzag pattern covering the entire lake, keeping in mind the detection distances and boat speed. Tracking began 7 d postimplantation. Fish were tracked bimonthly, from a boat, approximately the same time of day during each sampling trip (0800–1800). A four-element Yagi antenna was mounted to a 2-m pole in the boat so that it was easily accessible to the driver and rotated to triangulate tag location. Another

TABLE 1. Date tagged, size (length, weight), tag number, operation time, number of relocations, and days at large for radio-tagged hybrid striped bass, Monroe Lake, Indiana, 2006–2007.

Date tagged	Total length (mm)	Weight (kg)	Tag #	Operation time (seconds)	Number of relocations	Days at large
4/10/2006	655	3.75	803	321	3	85
4/10/2006	655	4.25	812	320	3	56
4/10/2006	660	4.50	823	370	19	427
4/12/2006	612	3.00	833	354	2	33
4/10/2006	622	4.25	843	309	8	168
4/10/2006	610	3.50	852	342	17	392
4/10/2006	594	3.00	862	309	18	427
4/12/2006	640	4.25	873	458	9	146
4/12/2006	569	2.75	883	305	12	181
4/12/2006	584	2.75	892	424	8	146
4/12/2006	597	3.25	903	376	19	372
4/12/2006	617	3.50	912	328	3	34
4/12/2006	638	4.00	923	308	2	57
4/12/2006	605	3.00	933	305	20	392
4/10/2006	668	4.50	943	379	2	57
4/10/2006	622	4.00	953	367	11	252
4/12/2006	572	3.00	962	318	21	427
4/12/2006	574	3.00	972	293	6	119
4/12/2006	610	3.25	983	307	21	427
4/12/2006	602	3.25	991	343	15	343
4/10/2006	630	3.75	1,001	328	16	366
4/10/2006	605	3.25	1,013	413	9	168
4/12/2006	594	3.00	1,022	259	19	372
4/12/2006	584	3.00	1,032	278	20	427
4/12/2006	607	3.25	1,043	347	19	392
4/12/2006	658	4.00	1,053	324	18	372
4/10/2006	630	4.00	1,064	NA	2	8
4/12/2006	594	3.25	1,073	292	20	392
4/12/2006	559	3.00	1,083	278	17	343
4/10/2006	579	3.00	1,093	323	19	427

antenna, which consisted of a bare-end coaxial cable inserted into a small diameter polyvinyl chloride (PVC) tube, was attached to the back of the boat with the PVC tube inserted into the water. As a tagged hybrid was approached, the driver switched from the Yagi antenna to the submersed antenna to further pinpoint the fish's location. Upon isolating a location with the coaxial cable, a Global Positioning System (GPS) coordinate was recorded using a Trimble GPS unit. Temperature and dissolved oxygen profiles were taken at 1-m intervals at most sites where fish were located, and maximum depth was recorded.

Most (94%) of the hybrid locations were postprocessed using Pathfinder Office software and base files for GPS collected from Bloomington, Indiana. Postprocessing point locations were needed to correct fish locations because of potential errors in GPS coordinates due to, for example, atmospheric interference (Trimble Navigation Limited 2004). The remaining 6% of locations were not postprocessed because of power failures with the GPS base station during the sampling event. All locations were loaded into ArcGIS 9.3 (ESRI, Redlands, California), and fish movements were digitized manually between successive locations following the shortest route in water. Mean, minimum distance between locations (displacement) was calculated using the XTools extension in ArcGIS. Seasonal movements were analyzed using PROC MIXED in SAS (SAS Institute, Cary, North Carolina) with fish nested within season. Seasons were defined as spring (March–May), summer (June–August), fall (September–November), and winter (December–February). A midline was digitized in ArcGIS to represent the middle of the main lake and major embayments. Fish locations were "snapped" to the line at right angles using Hawth's Tools extension for ArcGIS, and distance from the dam along the midline was calculated. Mean distance from the dam was analyzed in SAS using a mixed model with fish nested within season.

We calculated utilization distributions (UD) using a fixed-kernel density estimator with the Home Range Tools extension for ArcGIS (Rodgers et al. 2007). Utilization distribution describes the distribution of an animal's position in a plane and represents a nonparametric approach for determining a home range of an animal using a probabilistic model (Worton 1989). Because we relocated fish only 2–21 times, we could not estimate a true home range, which requires a minimum of 30 relocations (Seaman et al. 1999). However, UDs were calculated for each fish for all relocations by season to get a general idea of the area each fish used. Utilization distributions were calculated using values of $\alpha = 0.50$ (UD50) and 0.95 (UD95) (i.e., a UD95 represents an area used by a fish with 95% confidence).

Diel movement of tagged fish was evaluated on two dates, 23 May and 30 August 2006. On each date, three to four fish were tracked and located approximately every hour for 24 h. Point locations were loaded into ArcGIS, and displacement was determined between consecutive points, similar to the analysis for seasonal movements. Diel periods were defined as day (0600–2000) and night (2000–0600), and pooled displacement of palmetto bass was compared between time periods.

Habitat

Dissolved oxygen and temperature profiles at 1-m intervals were collected at 10 fixed stations (Figure 1) approximately 2 km apart, generally on the same day as a tracking event. Temperature and dissolved oxygen were collected using a YSI meter. Digital contour maps were acquired from topographic maps at approximately 3-m depth intervals. The lake was split into 10 "cells" in ArcGIS, with each cell corresponding to the vicinity of each of the 10 water quality stations. Data from the water quality station were assigned to the corresponding cells. Volume of each cell was calculated using ArcGIS 3D Analyst, and temperature and dissolved oxygen data were assigned to the corresponding depth stratum. Total volume containing temperature and dissolved oxygen within certain values was summarized for the entire lake among seasons.

As suggested by Kilpatrick (2003), we defined optimal habitat for hybrids as temperatures ranging from 21.5°C to 25.5°C and dissolved oxygen greater than 4.5 mg/L; suboptimal habitat was defined as temperatures less than 21.5°C and between 25.5°C and 27°C, and dissolved oxygen between 2.0 and 4.5 mg/L; poor habitat was defined as temperatures greater than 27°C and dissolved oxygen less than 2.0 mg/L. Depth strata were assigned a value of optimal, suboptimal, or poor based on the minimum habitat suitability index score for dissolved oxygen or temperature (e.g., if dissolved oxygen received a "poor" rating and temperature received an "optimal" rating, the overall rating would be "poor."). Volumes of optimal, suboptimal, and poor habitat were summarized and compared among seasons.

Results

Population dynamics and stocking evaluation

A total of 229 hybrids was collected, ranging in total length from 104 to 650 mm. Fish up to 650 mm TL, 3.8 kg, and age 13 were collected. Mean lengths at age during fall for ages 1–8 were 275, 475, 510, 577, 607, 624, 608, and 625 mm TL, respectively. A von Bertalanffy growth model was fitted to the data; parameters for the model were $K = 0.724$, $L_{inf} = 615$, and $t_0 = 0.192$ ($P < 0.0001$; $r^2 = 0.9890$). Total annual mortality estimates of 34% (±9.8%; Heinke), 29% (95% confidence interval = 13–41%; catch curve; Figure 3), and 32% (±5%; Robson and Chapman 1961) resulted in a mean total annual mortality estimate of 32% based on the population structure in this one year of sampling. Conditional natural mortality was estimated at 24% using FAST and the method described by Quinn and Deriso (1999).

The number of hybrids stocked between 1997 and 2008 ranged from 5,732 to 161,250. Standardized residuals (i.e., year-class strength) from a catch curve were plotted against number stocked (Figure 4). A second-order polynomial was fitted to the data, and the number stocked explained more than 95% of the variation in year-class strength. In general, year-class strength increased as the number of stocked fish increased from 0 to 100,000 fish but decreased as the number of hybrids stocked increased over 100,000 fish. We determined that optimal stocking rates were between 11 and 23 fish/ha.

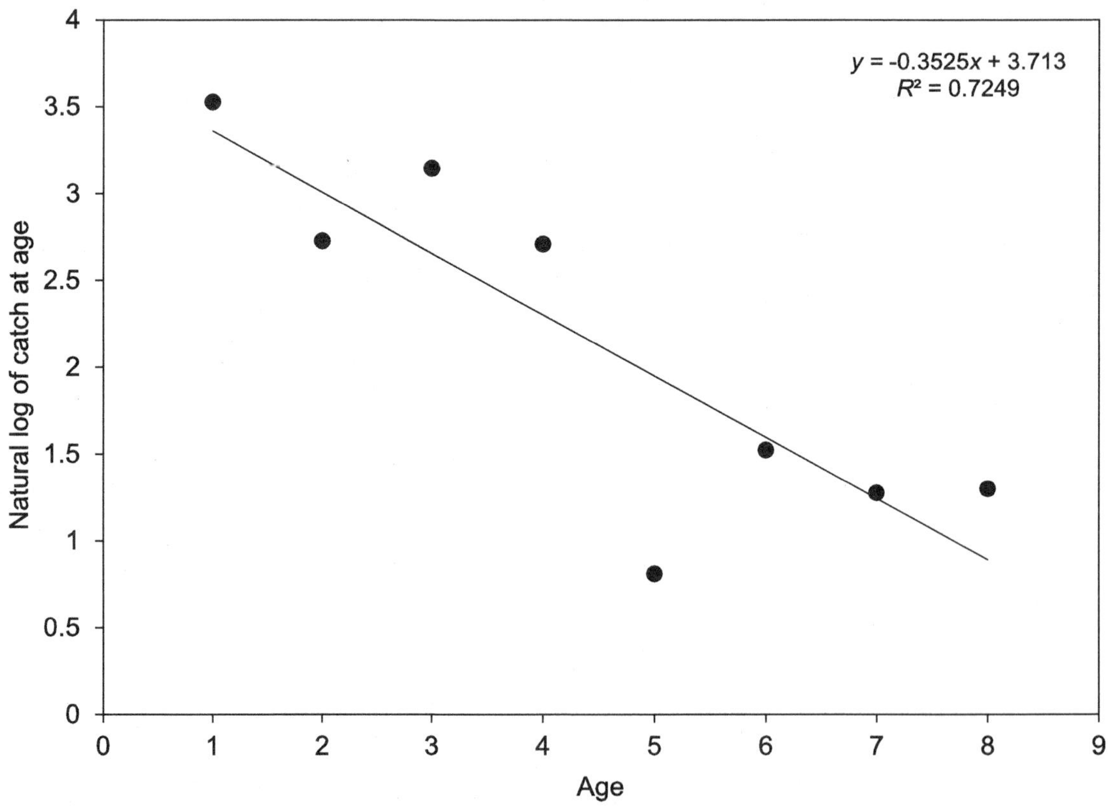

FIGURE 3. Catch-curve regression for palmetto bass in Monroe Lake, Indiana, fall 2005.

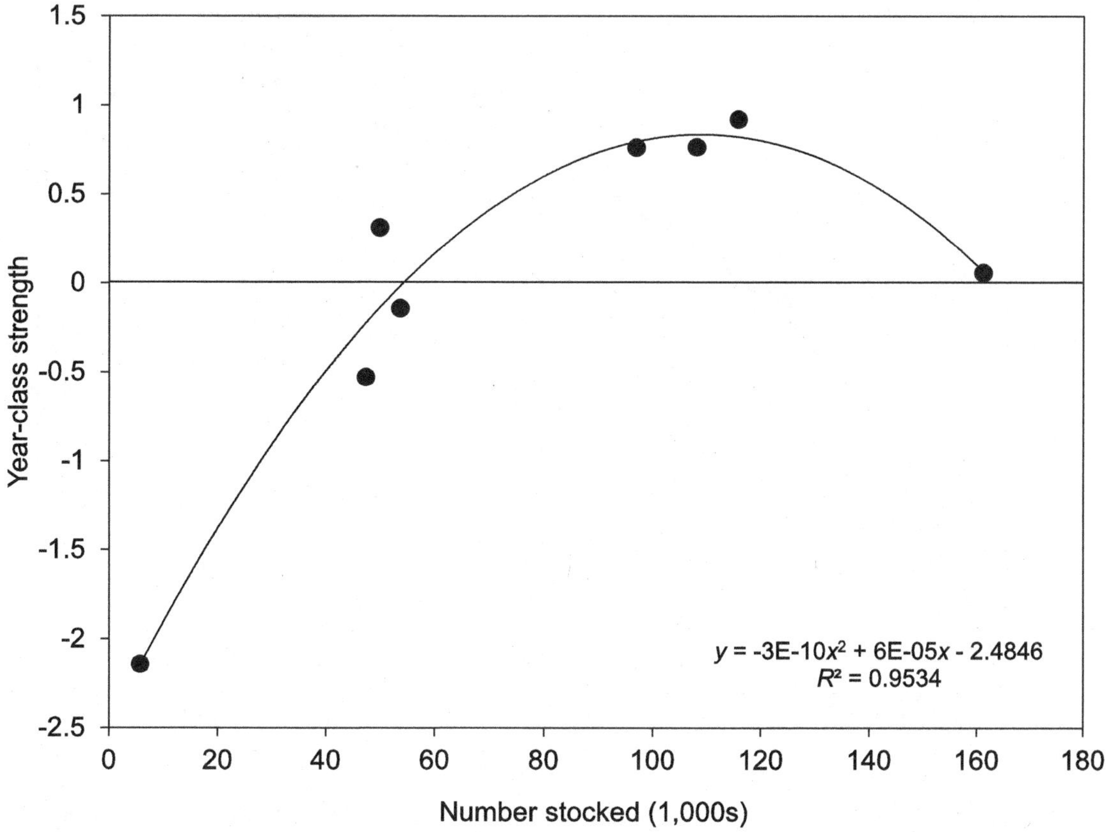

FIGURE 4. Catch-curve residuals (i.e., year-class strength) for palmetto bass modeled against the numbers stocked that summer, Monroe Lake, Indiana.

Yield-per-100 recruits did not differ among size limits from 0 to approximately 10% exploitation (Figure 5). At 150-mm-TL minimum size limit (MSL), palmetto bass would begin experiencing growth overfishing at approximately 20% exploitation. There was no difference in yield among the 350-, 425-, and 500-mm-TL size limits until approximately 30% exploitation and greater. The model predicted that yield-per-100 recruits may increase by 21% when going from a 150-mm-TL MSL to a 350-mm-TL MSL.

Movement

Tagged palmetto bass were relocated 545 times, averaging 12.6 times per fish with an average at-large period of 260 d. Although wind parameters were not recorded, fish were often located on windswept shorelines in shallow (<3 m) water. If sustained wind was not present, fish were not as concentrated, unless we saw them actively feeding on gizzard shad.

Locations and amount of movement differed among seasons through the 2006–2007 tracking period. During spring 2006, tagged fish were located throughout the lake, often in shallow water in flooded brush (Figure 6). Greatest movement between locations occurred in spring (Table 2). One palmetto bass traveled more than 19 km from the release point up to the upper basin after 7 d (or less). Mean distance from the dam differed among seasons ($F = 72.66$; df = 3; $P < 0.0001$). In spring, fish were spread out, with many in mid-lake, with a mean distance from the dam

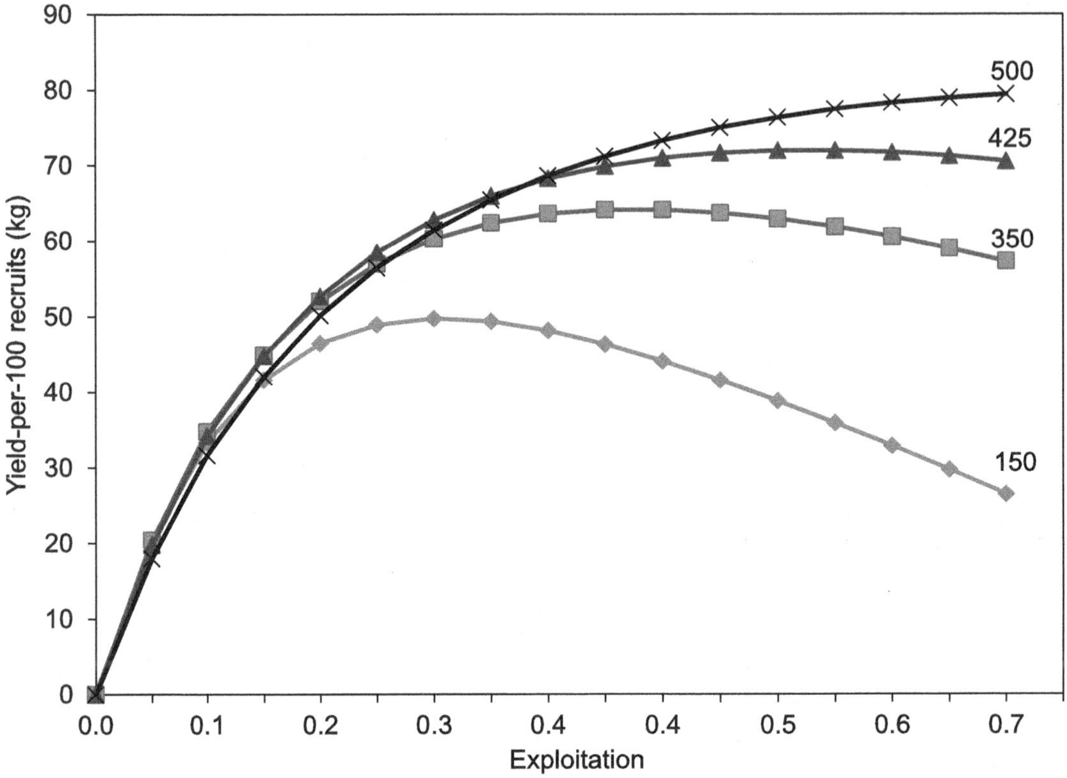

FIGURE 5. Yield-per-100 recruits modeled over a range of exploitation, a natural mortality of 0.25, and size limits from 150 mm total length ("no size limit") to 500 mm total length for palmetto bass in Monroe Lake, Indiana.

of 12.8 km. In summer, the fish moved down into the lower one-third of the lake and were never located further up lake than the Ramp Creek embayment after the first sampling event in summer (Figure 7). In general, hybrids traveled in circular patterns between the dam and Ramp Creek in the summer and were concentrated towards the dam at a mean distance of 4.1 km. In fall, especially in October when the surface temperature decreased from 25°C to 20°C, fish began to disperse throughout the lake (Figure 8). The fish were at a mean distance of 8.4 km from the dam during fall. By winter, nearly all remaining hybrids were located in the same shallow embayment in the upper basin of the lake (Figure 9), with a mean distance from the dam of 23.5 km. During spring 2007, fish dispersed again throughout the lake, many finding their way back to flooded brush in shallow water. Spring movements (501 m/d) differed significantly from other seasons (analysis of variance [ANOVA]; F = 15.28; df = 3; P < 0.0001). No significant difference in movement was detected among summer (207 m/d), fall (224 m/d), and winter (109 m/d).

Diel movements differed between May and August. Fish moved similarly during day and night periods in spring but differently during day (503 m/h) and night (65 m/h) periods during the summer 24-h sampling event. In late August, surface temperatures were approximately 26°C and fish moved more during the day than during dusk. Generally, fish moved in a horizontal circular pattern through the day, whereas at night they were all located near the dam where they were originally captured and tagged.

Areas utilized (utilization distribution) re-

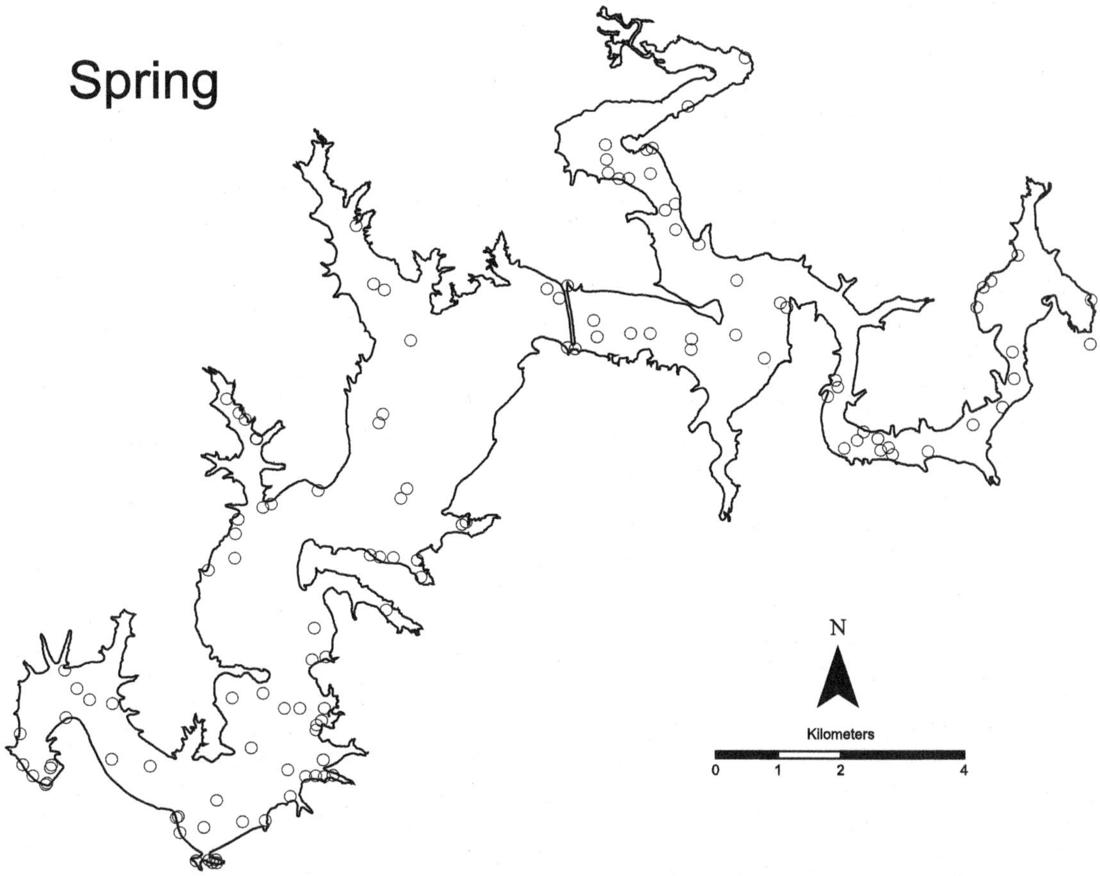

FIGURE 6. Spring (March–May) locations of palmetto bass, Monroe Lake, Indiana, 2006 and 2007. Each circle represents one location.

flected movement patterns. In general, fish used smaller areas of the lake during summer and winter. UD95 ranged from 752.5 ha in summer to 1248.7 ha in spring; however, there was no significant difference among seasons (ANOVA; F = 1.24; df = 3; P = 0.3028). Mean UD50 ranged from 476.6 ha in summer to 1,676.6 ha in spring. Spring UD50 was significantly greater than all other seasons (ANOVA; F = 12.86; df = 3; P < 0.0001).

TABLE 2. Seasonal movements and utilization distributions (UDs) for hybrid striped bass in Monroe Lake, 2006–2007.

				Utilization distribution (ha)					
	Minimum daily distance (km/d)			UD50			UD95		
Season	*N*	Mean	SE	*N*	Mean	SE	*N*	Mean	SE
Spring	107	0.5	0.053	17	1,676.6	170.0	17	1,248.7	282.4
Summer	119	0.207	0.019	23	476.6	81.5	23	752.5	104.7
Fall	66	0.224	0.028	17	788.9	94.8	17	932.3	155.5
Winter	47	0.109	0.016	15	704.0	208.3	15	893.3	231.7

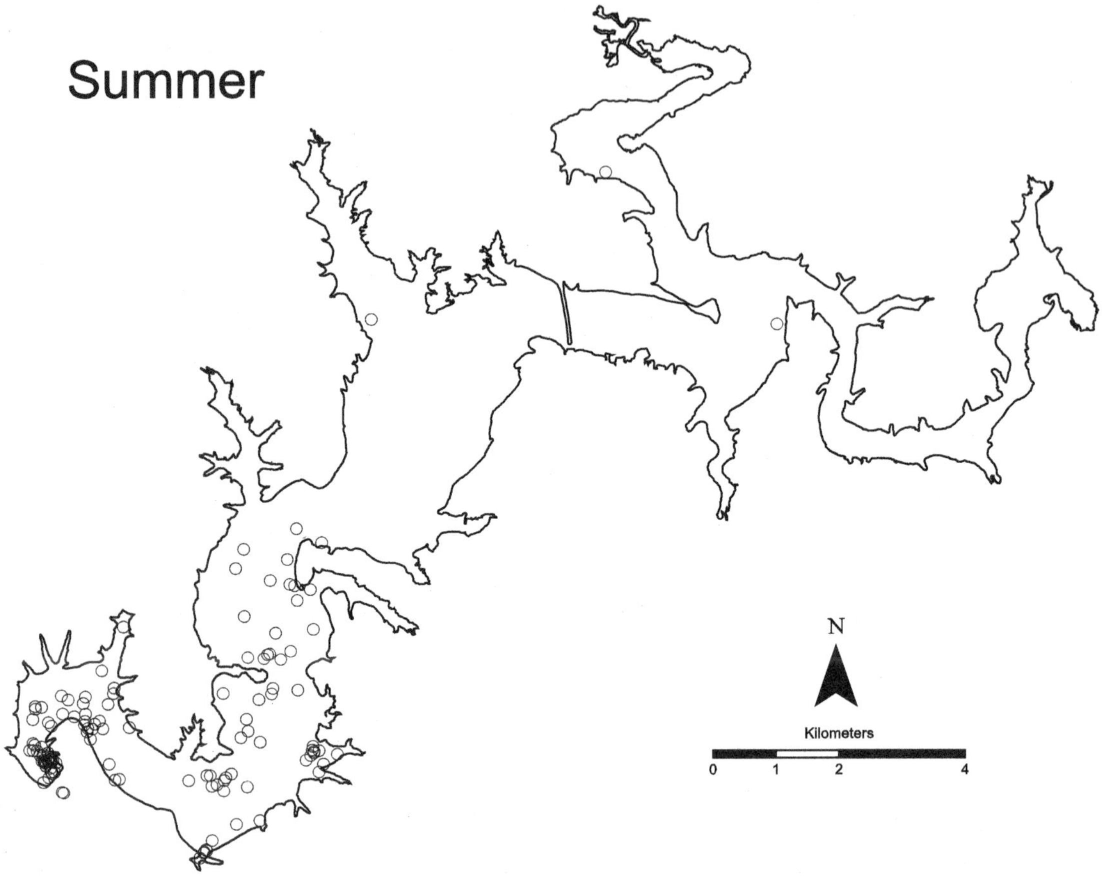

FIGURE 7. Summer (June–August) locations of palmetto bass, Monroe Lake, Indiana, 2006. Each circle represents one location.

Habitat

The volume of suitable habitat available to hybrids changed throughout the year. Temperature was the main limiting factor for the defined optimal habitat conditions throughout much of the year. During most of spring, the first part of summer, and winter, temperature was too cold (<21.5°C). In spring, dissolved oxygen levels were considered optimal (>4.5 mg/L) throughout the entire lake; however, temperature was suboptimal (<21.5°C) throughout the entire lake. Dissolved oxygen was a limiting factor only during the warmest parts of the summer when 37% of the lake volume was considered poor habitat because of low oxygen. On average, approximately 22% of the lake volume was optimal habitat for both oxygen and temperature during summer. Optimal conditions appeared during the first part of summer, disappeared during the warmest part of summer, then reappeared towards the end of summer and continued into fall. In fall, 47% of the lake was considered optimal and 52% was suboptimal. In winter, only temperature was a limiting factor and the entire lake was considered suboptimum. We were not able to determine the depth of individual fish, yet we recorded the maximum depth of fish locations as a surrogate for actual depth. Fish were located in the water column over deeper water (mean = 7.6 m) on average during summer (ANOVA; F = 1.60; df = 3; P = 0.0015) than other seasons. During winter, palmetto bass were located in shallower depths (4.7 m), but these depths were not significantly different

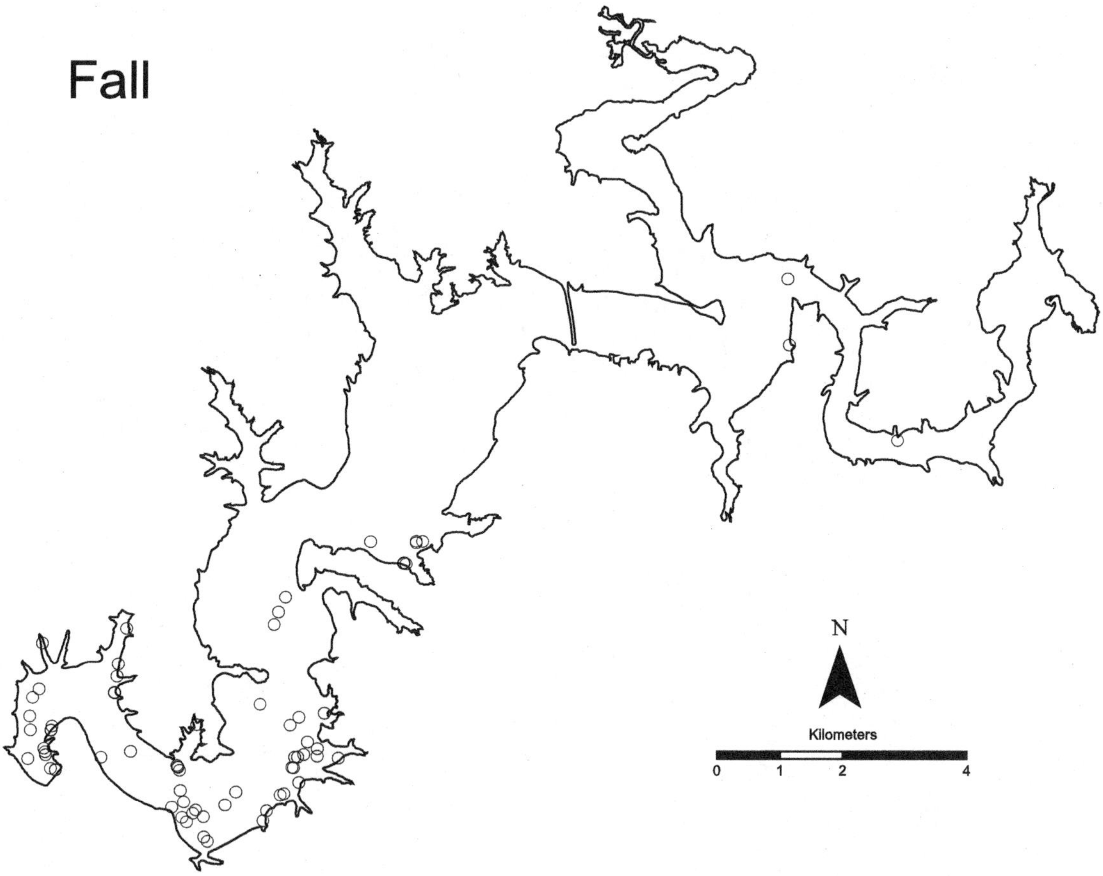

FIGURE 8. Fall (September–November) locations of palmetto bass, Monroe Lake, Indiana, 2006. Each circle represents one location.

than in other seasons, likely due to the shortage of sampling events in winter.

Creel survey

An estimated 38,019 anglers fished 223,978 h (51.5 h/ha) at Monroe Lake from 2 April to 31 October 2007 (Kittaka 2008). Anglers harvested an estimated 51,053 fish of all species (11 fish/ha) weighing a total of 14,650.7 kg (3.37 kg/ha). Hybrids accounted for 3.9% of the total harvest by number and 13.8% (ranked second) by weight. Anglers harvested an estimated 1,971 hybrids weighing a total of 2,015.3 kg. The mean length of harvested hybrids was 424 mm TL (range 152–673 mm TL). Anglers caught and released an additional 5,453 hybrids. Mean catch rate for hybrids was 0.03 fish/h, including all anglers, even those not targeting hybrids.

Discussion

Hybrid striped bass were initially stocked at Monroe Lake in 1983, with the primary goals of providing additional angling opportunities and establishing a quality fishery by taking advantage of the abundant gizzard shad forage base. Hybrids have become a major component of the Monroe Lake fishery, and their popularity among anglers seems to have increased in recent years. The IDNR previously evaluated the hybrid population in Monroe Lake annually with fall electrofishing catch rates, but the evaluations were not as extensive as in this study. Because of the lack of data published on hybrids in the primary litera-

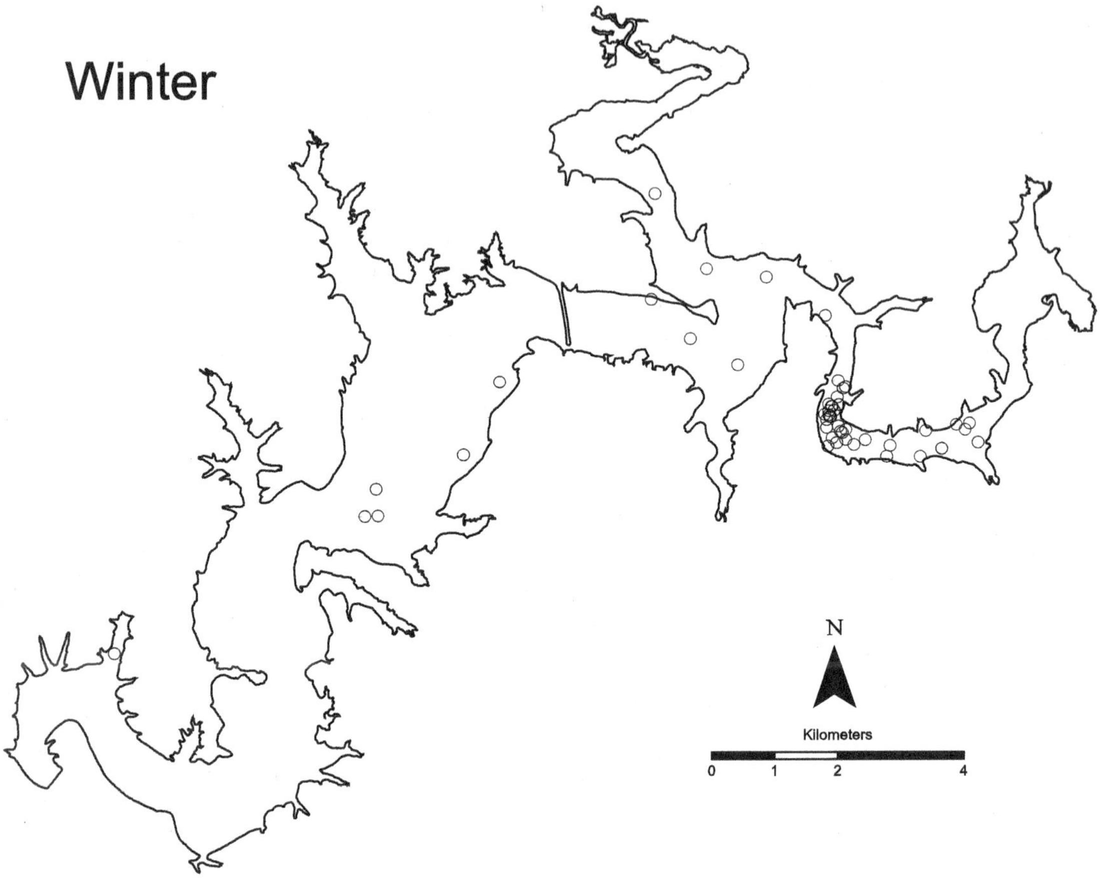

FIGURE 9. Winter (December–February) locations of palmetto bass, Monroe Lake, Indiana, 2006–2007. Each circle represents one location.

ture, we initiated this project to assess the hybrid fishery in Monroe Lake by analyzing population dynamics, seasonal and diel movement patterns, available habitat, and stocking efficacy.

We believe useful information was obtained from 1 year of population sampling data, although our population-dynamics assessment would have been strengthened with more years of data. Although sample sizes relative to other studies may have been small, we believe sample data adequately described the status of the hybrid population and fishery in Monroe Lake at the time. For example, population dynamics analysis relied on a sample of 229 fish, with a subsample of 100 fish aged by extracting otoliths. These fish were captured using a combination of experimental gill nets with multiple mesh sizes (which we assumed would minimize selectivity; but see McRae et al. 2013, this volume), and we caught a wide range of sizes, from 104 to 650 mm TL.

Hybrids in Monroe Lake grew similar to or better than other populations in North America. Growth data for hybrids are not abundant in the primary literature, but some comparisons can be made. Because hybrid growth data from Carlander (1997) were back-calculated and our data were not, direct comparisons of mean total length at age are not feasible. However, we can compare growth parameters (K) from von Bertalanffy growth functions, which is analogous to the slope of a regression line. Mean calculated total lengths at age from Carlander (1997) were fitted to a von Bertalanffy growth function and compared to mean total length at age for hybrids in this study. Our hybrids grew faster (K = 0.687) than the North American average (K = 0.497). Hybrids in Monroe Lake also grew faster than those in a 320-ha oxbow in Iowa (K = 0.21; Schultz and Dodd 2008). Monroe Lake hybrids grew fast, reaching 432 mm TL by fall of their third year. The oldest age-class observed in this study was 13 years old, which likely means the Monroe Lake population is underutilized by anglers. Once the fish reached age 5 in our study, length held steady around 600 mm TL for subsequent year-classes.

Our growth estimates may have been influenced by our capturing the fish in the fall and then assigning integer years of age. Because our fish were captured in the fall, they probably had substantial growth since their birth date (spring). A comparison using additional half years for our fish would have reduced the apparent growth rate and size at age of our fish. Nevertheless, our growth rates were quite good.

Estimates of annual mortality are an essential component needed by managers to evaluate a fishery. We estimated annual mortality using three methods, and all three produced similar estimates, yet the Robson and Chapman (1961) method provided the smallest error and the most reliable estimate compared to Heinke or catch-curve regression. Based on our 1-year sampling, annual mortality of hybrids in Monroe Lake was 32% (±5%), which is similar to annual mortality of hybrids in a 320-ha oxbow in Iowa (31%; Schultz and Dodd 2008). Using our estimates of annual mortality and natural mortality (25%), we can make some assumptions regarding the level of exploitation, even though exploitation was not directly measured in this study. Assuming an additive model (i.e., total mortality = natural mortality + fishing mortality), exploitation was fairly low (approximately 7%). If exploitation was compensatory (i.e., replacing natural mortality), exploitation could have been closer to 30%. Because we do not have an estimate of natural mortality in the presence of fishing mortality, our estimate of exploitation is probably a low estimate, considering that we do not know how natural mortality changed in the presence of fishing. Since we do have an estimate of total annual mortality, we can make the safe deduction that exploitation was no higher than 32%. Although there are methods of estimating exploitation from radio-tagging studies (e.g., Hightower et al. 2001), we chose not to estimate exploitation during the movement study to maximize the number of radio-tagged fish available and asked anglers to release a radio-tagged fish if caught.

We used our estimates of annual and natural mortality to simulate changes in yield by varying minimum size limits in FAST over a range of exploitation rates. Anglers began harvesting hybrids as small as 152 mm TL, so we used 150 mm TL to represent a value of "no minimum size limit." At rates of exploitation of 0 to approximately 15%, there was no substantial difference in yield among minimum size limits. At approximately 20% exploitation, hybrids would begin experiencing growth overfishing with essentially no minimum size limit. Yield per recruit may increase by more than 20% if a minimum size limit of at least 350 mm TL was established. Currently, hybrids in Indiana are managed with a 12-fish daily bag limit and only two can be above 432 mm TL. We believe the restricted bag limit of two fish longer than 432 mm TL is sufficient at low levels of exploitation; however, we will continue to evaluate the fishery and we suggest implementing a study to directly estimate exploitation to further refine our models.

In addition to population dynamics, we wanted to analyze the efficacy of the hybrid stocking program at Monroe Lake. Although hybrids are not 100% sterile (Forshage et al. 1988), annual stocking is required to maintain the fishery at Monroe Lake. Typically in Indiana, hybrid fry are obtained through interstate trade for another species (e.g., muskellunge *Esox masquinongy*) with another state agency (e.g., Tennessee Wildlife Resources Agency). The IDNR stocks the fingerlings (25–50 mm TL) in Monroe Lake in early June of each year. Due to a variety of factors, including fry availability from other states and low production returns of ponds, the number of hybrids stocked in Monroe Lake from 1997 to 2008 differed by over an order of magnitude (5,732–161,250). Because there was a large difference in stocking rates, we were able to evaluate stocking rates based on year-class strength using residuals from a catch-curve regression to investigate fluctuations in year-class strength (Maceina 1997). The relationship between year-class strength and the number of hybrid fingerlings initially stocked is a domed-shaped curve (Figure 4). However, the decline in year-class strength at high stocking densities is driven by one point, which represents the greatest number of fish stocked during the evaluation (n = 161,520). Whether this represents the true relationship (i.e., a downward curve beginning around 110,000 fingerlings) is irrelevant when determining optimal stocking rate. Another fit of the data could have been an asymptotic relationship beginning at 60,000 fish. Even with an asymptotic relationship, cost efficiency of the stocking program is maximized by staying to the left of the apex or just before the relationship becomes asymptotic. Therefore, we have determined an optimal stocking rate of approximately 12–23 fish/ha to achieve the maximum potential of this program and remain cost-efficient. The current stocking rate of 12 fish/ha falls within this range of producing an average to above average year-class.

We acknowledge that our population assessment is soft because we used only 1 year of data to produce a catch curve. There is often significant annual variability in recruitment in hybrids as environmental conditions as well as stocking rate can affect the population. The report by Miranda and Bettoli (2007) had larger sample sizes and combined data from four reservoirs to look at regional trends. Thus, we cannot say with assurance whether the catch curve is driven by annual mortality or annual variability in recruitment. Nonetheless, the fact that we found such a strong relationship between stocking rate and year-class strength suggests that stocking rate may be the dominant factor, at least in this time frame.

Another important component of this study was to analyze angler utilization of the hybrid fishery. We conducted a creel survey to assess effort and catch of hybrids by recreational anglers at Monroe Lake from April to the end of October 2007. Anglers made an estimated 38,019 trips during the creel survey, and 8% of anglers targeted palmetto bass; therefore, we estimate that anglers targeting hybrids made 3,042 fishing trips. The U.S. Department of the Interior, U.S. Fish and Wildlife Service, and U.S. Department of Commerce, U.S. Census Bureau (2006) determined that anglers spent an average of US$62.60 for each fishing trip in Indiana during 2006, which includes expenditures for food, lodging, transportation, equipment, and license fees. Using the estimated number of trips specifically targeting hybrids and the estimated cost per trip, anglers spent approximately $190,429 (3,042 trips × $62.60/trip) in 2007. The current annual stocking rate for Monroe Lake is 12 hybrid fingerlings per ha. The cost for raising these fish in 2005 was $0.10 per fingerling for a total cost of $5,375, which includes all costs related to receiving fry, extensively culturing fry in earthen ponds, and stocking fingerlings (Indiana Department of Natural Resources, unpublished data; although there may also be hidden costs for maintaining a hatchery program that were not included). If the stocking costs are compared to the economic value of the hybrid fishery, the cost–benefit ratio for maintaining a quality hybrid fishery at Monroe Lake is 1:35 annually.

As part of this study, we developed news releases and Web pages to inform and update the public during the study. This generated a lot of interest in the fishery, and it seems that interest among anglers increased from 2001, which was the last creel survey conducted at Monroe Lake previous to this study. It is likely that the popularity of the hybrid fishery at Monroe Lake will continue to increase as anglers learn to locate and become more efficient at catching these fish. IDNR biologists often received comments from the public that hybrids were difficult to locate in Monroe Lake on any particular day because of its large size (4,350 ha). This and the lack of data published on hybrid movement rates led us to radio-track palmetto bass in Monroe Lake to determine seasonal and diel movements.

Our hybrids used different areas of Monroe Lake during different seasons. In the spring, fish used the entire lake, typically in shallow water in close proximity to the shoreline and often in flooded brush. This most likely corresponds with an attempt to spawn, reflecting their white bass heritage. As water warms during summer, Monroe Lake stratifies thermally and dissolved oxygen is absent below the thermocline, consequently restricting fish to the volume of water above that depth (typically 5 m in Monroe Lake). During summer, tagged hybrids concentrated in the lower third of the lake, closest to the dam, where depths were greatest, similar to behavior seen for striped bass in many reservoir systems (Coutant 2013, this volume). They were not located above the Ramp Creek embayment, even though there was suitable habitat available above that location. We do not know why. Optimal habitat, as determined by the habitat suitability index, varied seasonally. Optimal habitat began appearing early in summer but disappeared as the water temperature increased in midsummer and returned in late summer and most of fall. Hybrids began fall migration back to the upper end of the lake as optimal habitat returned and water temperatures decreased. No difference was detected in movement rates among fall, winter, and summer. However, there was suggestion of a difference in winter movement, but we were unable to detect a difference at an alpha of 0.05 with the amount of our data. Winter conditions prevented us from tracking as often as the other seasons, consequently resulting in greater error in movement estimates. Unexpectedly, our tagged fish spent much of the winter in a small area in the upper end of the lake, which happened to be a waterfowl resting area established by IDNR managers. The waterfowl resting area is closed to any boat traffic during much of late fall and winter, and this area may be acting as a refuge from anglers for hybrids. Mean water temperature in that area of the lake was slightly warmer on average during winter than other areas of the upper basin, which may have been an attractant.

Only a few studies have tracked hybrid striped bass movements. Our diel movement results (0.065–0.503 km/h during two diel tracking events with similar movement between day and night periods in spring but more during day than night in summer) can be compared to two other studies. Jones and Rogers (1998) tracked diel movements of hybrids in a fluctuating Colorado irrigation lake. Hourly movement rates ranged from 0 to 0.30 km/h, and averaged from 0.027 to 0.061 km/h. In a similar study, Kilpatrick (2003) tracked seasonal movements in 1,810 ha Claytor Lake in southwest Virginia, USA. Hourly movement rates averaged between 0.313 and 0.425 km/h, and there was no difference between day and night hourly movement. Even fewer data are available to compare seasonal movements (our fish ranged from 0.109 to 0.501 km/d). Kilpatrick (2003) found that hybrids in Claytor Lake displayed spring and fall migrations and spent the warmer months in the lower part of the lake near the dam, similar to hybrids in Monroe Lake. Seasonal movements in our study represented displacement between tracking events and thus underrepresent true, total movement, which is typical for seasonal fish tracking studies. Utilization distributions are calculated to determine areas of concentration (Worton 1989) for radio-tagged animals. Pooled, mean utilization distributions of hybrids in Claytor Lake ranged from 505.7 to 1,039.1 ha (Kilpatrick 2003) and 752.5–1248.7 ha in Monroe Lake. Utilization distributions were slightly greater in our study compared to those for Claytor Lake; however, Monroe Lake is nearly 2.5 times the size of Claytor Lake (1,810 ha), which may account for larger utilization distribution values observed in our study. Seasonal patterns of pooled utilization distributions mimicked seasonal movement patterns, as described above.

Our goal in this study was to thoroughly evaluate the hybrid fishery in Monroe Lake by analyzing population dynamics, seasonal movements, available habitat, and stocking efficiency. We determined that current stocking rates (12 fish/ha) were adequate to maintain a quality fishery. If excess fish are available, stocking rates can be increased up to approximately 23 fish/ha, which should translate into slightly stronger year-class strength. However, rates greater than this may not provide further increase in year-class strength. We believe that the hybrid fishery had been underutilized by anglers prior to this study, not because of popularity but because the fish were difficult for the anglers to locate in a large lake on any particular day. Keeping anglers informed of the tracking project and posting maps of fish locations on the Web site seemed to spark interest among anglers, and we received positive feedback concerning the program.

Monroe Lake has developed a good fishery for hybrid striped bass using the current stocking and regulatory program. There is no plan to alter management strategies for hybrids at Monroe Lake at this time based on our study. Data from this study can be used to establish or maintain hybrid fisheries in new locations in Indiana, and throughout the Midwest. Future studies at Monroe Lake may include using hydroacoustics to estimate forage biomass (gizzard shad) and relate it to fish movements and condition (see Thompson and Rice 2013, this volume) and better determining exploitation rates of hybrids by recreational anglers.

Acknowledgments

The Federal Aid in Sportfish Restoration Program and the Indiana Department of Natural Resources provided funding for this study. We thank the many IDNR employees that assisted with this study, and special thanks to M. Bartley for assisting with radio-tracking data collection. We also thank anonymous reviewers for their comments that greatly improved this manuscript.

References

Avise, J. C., and M. J. Van Den Avyle. 1984. Genetic analysis of reproduction of hybrid white bass × striped bass in the Savannah River. Transactions of the American Fisheries Society 113:563–570.

Busacker, G. P., I. R. Adelman, and E. M. Goolish. 1990. Growth. Pages 363–387 *in* C. B. Schreck and P. B. Moyle, editors. Methods for fish biology. American Fisheries Society, Bethesda, Maryland.

Carlander, K. D. 1997. Handbook of freshwater fishery biology, volume 3. Life history data on ichthyopercid and percid fishes of the United States and Canada. Iowa State University Press, Ames.

Collier, W. R., P. W. Bettoli, G. D. Scholten, and T. N. Churchill. 2013. Regulation of striped bass and hybrid striped bass fisheries in the United States. Pages 449–460 *in* J. S. Bulak, C. C. Coutant, and J. A. Rice, editors. Biology and management of inland striped bass and hybrid striped bass. American Fisheries Society, Symposium 80, Bethesda, Maryland.

Coutant, C. C. 2013. When is habitat limiting for striped bass? Three decades of testing the temperature–oxygen squeeze hypothesis. Pages 65–91 *in* J. S. Bulak, C. C. Coutant, and J. A. Rice, editors. Biology and management of inland striped bass and hybrid striped bass. American Fisheries Society, Symposium 80, Bethesda, Maryland.

DeVries, D. R., and R. V. Frie. 1996. Determination of age and growth. Pages 483–512 *in* B. R. Murphy and D. W. Willis, editors. Fisheries techniques, 2nd edition. American Fisheries Society, Bethesda, Maryland.

Forshage, A. A., W. D. Harvey, K. E. Kulzer, and L. T. Fries. 1988. Natural reproduction of white bass × striped bass hybrids in a Texas reservoir. Proceedings of the Annual Conference Southeastern Association of Fish and Wildlife Agencies 40(1986):9–14.

Hightower, J. E., J. R. Jackson, and K. H. Pollock. 2001. Using telemetry methods to estimate natural and fishing mortality of striped bass in Lake Gaston, North Carolina. Transactions of the American Fisheries Society 130:557–567.

Jennings, C. A., and G. L. Looney. 1998. Evaluation of two types of anesthesia for performing surgical procedures on striped bass, Morone saxatilis. North American Journal of Fisheries Management 18:187–190.

Jones, M. S., and K. B. Rogers. 1998. Palmetto bass movements and habitat use in a fluctuating Colorado irrigation reservoir. North

American Journal of Fisheries Management 18:640–648.

Kilpatrick, J. M. 2003. Habitat use, movements, and exploitation of striped bass and hybrid striped bass in Claytor Lake, Virginia. Master's thesis. Virginia Polytechnic Institute and State University, Blacksburg.

Kilpatrick, J. M., and J. J. Ney. 2013. Temperature and dissolved oxygen habitat use by striped bass and hybrid striped bass in Claytor Lake, Virginia. Pages 147–159 *in* J. S. Bulak, C. C. Coutant, and J. A. Rice, editors. Biology and management of inland striped bass and hybrid striped bass. American Fisheries Society, Symposium 80, Bethesda, Maryland.

Kittaka, D. S. 2008. Monroe Lake, Monroe and Brown counties: fishing pressure and fish harvest at Monroe Lake, 2007. Indiana Department of Natural Resources, Indianapolis.

Maceina, M. J. 1997. Simple application of using residuals from catch-curve regression to assess year-class strength in fish. Fisheries Research 32:115–121.

Maceina, M. J., and P. W. Bettoli. 1998. Variation in largemouth bass recruitment in four mainstream impoundments of the Tennessee River. North American Journal of Fisheries Management 18:998–1003.

McRae, B. J., J. S. Bulak, B. E. Taylor, and C. T. Waters. 2013. Evaluation of the use of gill nets for monitoring reservoir striped bass fisheries. Pages 263–278 *in* J. S. Bulak, C. C. Coutant, and J. A. Rice, editors. Biology and management of inland striped bass and hybrid striped bass. American Fisheries Society, Symposium 80, Bethesda, Maryland.

Miranda, L. E., and P. W. Bettoli. 2007. Mortality. Pages 229–278 *in* C. Guy and M. Brown, editors. Analysis and interpretation of freshwater fisheries statistics. American Fisheries Society, Bethesda, Maryland.

Pollock, K. H., C. M. Jones, and T. L. Brown. 1994. Angler survey methods and their applications in fisheries management. Pages 251–253 *in* K. H. Pollock, C. M. Jones, and T. L. Brown, editors. American Fisheries Society, Special Publication 25, Bethesda, Maryland.

Quinn, T. J., II, and R. B. Deriso. 1999. Quantitative fish dynamics. Oxford University Press, New York.

Rash, J. M., and J. J. Ney. 2013. Comparative ecology of juvenile striped bass and juvenile hybrid striped bass in Claytor Lake, Virginia. Pages 229–247 *in* J. S. Bulak, C. C. Coutant, and J. A. Rice, editors. Biology and management of inland striped bass and hybrid striped bass. American Fisheries Society, Symposium 80, Bethesda, Maryland.

Robson, D. S., and D. G. Chapman. 1961. Catch curves and mortality rates. Transactions of the American Fisheries Society 90:181–189.

Rodgers, A.R., A.P. Carr, H.L. Beyer, L. Smith, and J.G. Kie. 2007. HRT: home range tools for ArcGIS. Ontario Ministry of Natural Resources, Centre for Northern Forest Ecosystem Research, Thunder Bay, Ontario.

Schultz, R. D., and B. J. Dodd. 2008. Growth, mortality, and harvest of walleye and hybrid striped bass in an Iowa urban lake: simulated effects of minimum size limits. Pages 413–424 *in* R. T. Eades, J. W. Neal, T. J. Lang, K. M. Hunt, and P. Pajak, editors. American Fisheries Society, Symposium 67, Bethesda, Maryland.

Seaman, D. E., J. J. Misspaugh, B. J. Kernoohan, G. C. Brundige, K. J. Raedeke, and R. A. Gitzen. 1999. Effects of sample size on kernel home range estimates. Journal of Wildlife Management 63:739–747.

Slipke, J. W., and M. J. Maceina. 2000. Fishery analysis and simulation tools. Auburn University, Auburn, Alabama.

Thompson, J. S., and J. A. Rice. 2013. The relative influence of temperature and forage availability on growth of age 1–5 striped bass in two southeastern reservoirs. Pages 93–120 *in* J. S. Bulak, C. C. Coutant, and J. A. Rice, editors. Biology and management of inland striped bass and hybrid striped bass. American Fisheries Society, Symposium 80, Bethesda, Maryland.

Trimble Navigation Limited. 2004. Why postprocess GPS data? Trimble Navigation Limited, Westminster, Colorado.

U.S. Department of the Interior, U.S. Fish and Wildlife Service, and U.S. Department of Commerce, U.S. Census Bureau. 2006 National survey of fishing, hunting, and wildlife-associated recreation. U.S. Fish and Wildlife Service, Washington, D.C.

Worton, B. J. 1989. Kernel methods for estimating the utilization distribution in home-range studies. Ecology 70:164–168.

American Fisheries Society Symposium 80:333–349, 2013

Utilizing Anglers to Improve Striped Bass Management

Daniel M. Wilson*
Virginia Department of Game and Inland Fisheries
1132 Thomas Jefferson Road Forest, Virginia 24551, USA

Abstract.—Anglers have been an integral part of striped bass *Morone saxatilis* management at Smith Mountain Lake, Virginia. Since the mid-1990s, volunteer anglers assisted with agency data collection, maintained diaries, tagged fish for angler exploitation studies, and collected striped bass for growth analysis. Data from existing angler programs such as tournaments and trophy fish awards were also incorporated to track the striped bass population. Anglers were utilized to specifically fill data voids. Established gill-net and creel surveys were not sufficient to track trends in the striped bass population. However, by integrating anglers, substantial changes were documented in the Smith Mountain Lake striped bass population that were not detectable with established assessment tools and helped to make management decisions more compatible with angler interests. Incorporating volunteers into striped bass management required substantial interaction but filled data gaps and improved public relations. In addition, many anglers developed realistic expectations of the fishery and became more invested in responsible management of the striped bass population.

Introduction

Fisheries managers' ability to collect adequate data for sound striped bass *Morone saxatilis* management decisions is often compromised due to manpower and financial limitations. As a result, many sport fish management decisions are based primarily on stock assessment data and limited information about angler catches, effort, and catch rates (Isbell and Rawson 1989). Too often, available stock assessment data are insufficient to adequately justify appropriate stocking rates and regulations. To expand traditional agency data, some state and federal agencies have long histories of utilizing anglers to gather fishery information, which they do not have the ability to collect (Leslie et al. 2004).

Smith Mountain Lake developed a reputation for its striped bass fishery, including the establishment of an exceptional trophy fishery (Sutton et al. 2000). With declining numbers of preferred length (>761 mm) striped bass, anglers began lobbying Virginia Department of Game and Inland Fisheries (VDGIF) in the early 1990s to increase stocking rates. By the mid-1990s, disagreements on stocking rates began polarizing the agency and anglers. Anglers believed that substantially increasing stocking rates would improve overall catch rates in the short term and increase the catch of larger fish in the long term. Virginia Department of Game and Inland Fisheries was concerned that overstocking could reduce fish health resulting in fewer larger fish in the future and also negatively affect other species. A compromise was eventually reached where VDGIF would increase stocking rates while the Smith Mountain Striper Club (SMSC), a local angling organization, would provide data collection assistance. Tennessee Wildlife Resources Agency also incorporated various stakeholder groups for conflict resolution and the development of a management plan for a high profile striped

* Corresponding author: dan.wilson@dgif.virginia.gov

bass fishery at Norris Reservoir, Tennessee (Churchill et al. 2002).

The ability of VDGIF to measure changes in the Smith Mountain Lake striped bass fishery with traditional stock assessment tools, which included standardized gill netting and creel surveys, was limited. Annual angler-catch rates, fishing mortality, and growth rates for fish over age 4 were identified as data gaps for tracking the striped bass fishery. Beginning in 1998, volunteers from the SMSC were incorporated into various aspects of striped bass assessment at Smith Mountain Lake to fill data gaps. Striped bass anglers were initially recruited to establish an angler diary program, expand gill-net sampling, and obtain access to additional stocking locations. After establishing positive relationships with numerous volunteers, additional volunteers were recruited for otolith collection from larger striped bass, fish tagging, supplying tournament data, developing regulation revisions, and formulating new management objectives.

In April 2003, a substantial striped bass fish kill occurred at Smith Mountain Lake. Events leading to this occurrence included decreased fish health as evident by declining growth, a near complete winterkill of threadfin shad *Dorosoma petenense*, which dominated the forage base, and an outbreak of striped bass parasitic copepods *Achtheres* sp. How much any one of these factors contributed to the striped bass kill is still uncertain. Dead striped bass of all sizes were found throughout the reservoir, but the kill was dominated by large adult fish. This kill was observed for approximately 2 months before subsiding, and no major mortality of other species was observed. The extent of the fish kill was difficult to ascertain due to the longevity of the kill, the expansiveness of the reservoir, and the fact that striped bass were distributed throughout the reservoir during that time. The wide distribution of fish in April and May indicates that thermal and oxygen limitations were not a factor as seen in the summer and fall months in many reservoirs (Coutant 2013, this volume). The magnitude of the mortality event suggested a major impact to the striped bass fishery. Unfortunate as this event was, it provided an opportunity to track changes in the fishery with the monitoring that had been established.

Due to the loss of larger striped bass from the fish kill and to reduce fishing mortality, new regulations were enacted in the summer of 2006. Regulations changed from a 508-mm minimum length limit and two fish per day creel to a 660–914-mm protective slot limit from October to May with no length limit from June to September; the creel limit of two fish per day was unchanged. Stricter striped bass regulations were suggested by Thompson et al. (2007) to benefit the striped bass fishery by reducing high annual fishing mortality of 46–51% if managers and anglers utilizing the lake desired goals other than the existing fishery dominated by harvest and small fish. In this paper, I discuss integrating angler collected data with traditional assessment tools to make informative management decisions and the opportunities that may be available but not frequently utilized by fisheries biologists.

Study Area

Smith Mountain Lake is an 8,337-ha hydroelectric reservoir in south-central Virginia. The reservoir has mean and maximum depths of 16.8 and 63.7 m, respectively, and a retention time of 1.35 years. A substantial nutrient gradient is produced from eutrophic headwaters to oligotrophic conditions near the dam (Yurk 1989). Largemouth bass *Micropterus salmoides*, smallmouth bass *M. dolomieu*, striped bass, and black crappie *Pomoxis nigromaculatus* are the primary sport fishes. Gizzard shad *Dorosoma cepedianum*, alewife *Alosa pseudoharengus*, and threadfin shad are the principal pelagic prey species. Virginia Department of Game and Inland Fisheries has stocked 25–50-mm fingerling striped bass in Smith Mountain Lake annually since 1964; annual stockings averaged 300,000 prior to 1998, 450,000 in 1998–2002, and 345,000 in 2003–2008. Striped bass stock assessment, prior to 1998, consisted of gill net and angler creel surveys.

Methods

Traditional data collection

Monofilament gill nets were the primary method to collect and track Smith Mountain Lake striped bass populations prior to 1998 and were used in conjunction with angler data through the study period. Two single panel gill nets (30.5 m long × 2.4 m deep) with bar-mesh sizes of 51 and 64 mm were set two to four nights per month in October–December. These mesh sizes were only effective at consistently sampling striped bass approximately 300–550 mm in total length. However, the objective of gill-net sampling was to annually monitor the abundance and the characteristics of this segment of the population. Gill nets were deployed at fixed sites each month and across years. Length and weight were recorded, and otoliths were removed for each striped bass captured. Catch per unit effort was reported as catch per net night.

Roving creel surveys (Schlechte et al. 2013, this volume) during the study period were conducted 20 d each month in 1998 (January–December), 2003 (March–September), and 2006 (March–December). Survey times (morning versus evening), sample days, and sample areas were selected by a stratified-random method. Creel survey data provided estimates for striped bass angling effort and catch rates.

Management plan

A committee comprised of anglers from the SMSC and VDGIF personnel was organized to direct future management for striped bass at Smith Mountain Lake. The committee was an avenue for both parties to document their goals and expectations for the fishery, which resulted in the development of the Smith Mountain Lake Striped Bass Management Plan. This document attempted to establish a roadmap for future direction by establishing goals, measurable objectives, and methods of evaluation. Both parties agreed that this document would not cover all facets of long-term management, and modifications would be needed over time. Although neither party knew if all the goals in this plan could be achieved, the foundations were established for both parties to work toward common ground and improvement of the fishery.

The development of the management plan identified data gaps and the inability to quantify and monitor stated goals with existing data collection. As a result, volunteers were recruited to acquire more consistent angler catch rate and length-frequency data than periodic VDGIF creel surveys provided, age and growth from striped bass exceeding age 4, and tag-return estimates. The management plan was available to all interested anglers.

Angler diaries

To supplement limited creel survey data, SMSC anglers volunteered to participate in an angler diary program from 1998 to 2008 (average participants per year = 62; range = 25–116). Only striped bass data and angler trips targeting striped bass were utilized. Diaries were sent to anglers in January or February of each year with instructions printed on the inside cover of each diary. Diaries measured 15 × 10.5 cm and were printed on water-resistant paper. Each diary contained adequate space to record 96 trips and contained carbon-copy paper between pages for anglers to keep for their records after returning the diary.

Anglers were asked to record the lake they were fishing, date, number of anglers, hours fished, day or night fishing, bait type, number, and size of fish harvested and released. Anglers were instructed to record information from each fishing trip, even if no fish were caught. As a broad measure of experience, anglers were also asked how many years they have fished for striped bass at Smith Mountain Lake. Return mailing information was included in each diary. If needed, anglers were provided with a 106-cm vinyl adhesive measuring tape to assist with recording lengths. Anglers were requested to indicate lengths in approximate categories by inches. A letter including any other pertinent information and deadlines for the diary returns was included in the diary mailings. Diaries were provided to all anglers upon request.

Fish tagging

A fish-tagging study was initiated to estimate fishing mortality. Volunteer anglers were recruited from the SMSC to tag striped bass based on recommendations from the club president, their ability to catch fish, fishing experience, and their positive reputation. Only six anglers were originally recruited for this project to reduce confusion and simplify data collection and updating. Other anglers were recruited to replace original participants that dropped out of the program. The biologist met with all the prospective taggers and provided an overview of the project goals and objectives. Each tagger was shown how to tag fish and required to practice on a striped bass. Tagging was conducted from October to May, and all taggers reported their information (date, length, water temperature, tag number, and location) to VDGIF at regular intervals.

Anchor tags used in the study contained a narrow 75-mm monofilament strand that passes through a piece of vinyl tubing at one end and terminates with a 25-mm-long anchor of the same material. Tags were attached to the abdominal area of each fish. The address for anglers to return the tag, an identification number, and the phrase "Reward $5–$50" were printed on each tag. Anglers were not required to harvest the fish to collect the reward. Informational posters advising anglers of the project stated that tags could be "clipped" from fish not harvested. Tag return rates were calculated for 365 d post-tagging, were unadjusted for tag loss or tagging mortality, and were estimated as (number of returned tags/reporting rate)/number of fish tagged.

Additional summer tagging was conducted by the same taggers and procedures as in the cooler months to address controversies concerning summer catch-and-release mortality at Smith Mountain Lake. Summer tagging data were not included in the annual tagging data but were only utilized to resolve these disputes. I used chi-square (χ^2) analysis to compare summer and winter tag returns.

A postcard program to determine nonreporting of tags by anglers was conducted in 2001–2002 (Maceina et al. 1998). Seventy-five postage-paid postcards were distributed by boat to active striped bass anglers both years. Postcards contained four multiple-choice questions, a space for their return address, and the same reward offered for the return of fish tags.

Information packets were distributed to marinas and local sporting goods stores to make anglers aware of the tagging study. Information packets included posters, study information, and postage-paid envelopes for tag returns. The back of each tag-return envelope requested information on date of capture, fish length, location of catch, and if the fish was harvested or released.

A reward system was important for encouraging long-term angler cooperation but problematic because issuing a large number of reward checks through VDGIF was not desirable. However, the SMSC volunteered to set up a special reward account where VDGIF would deposit one or two checks annually and SMSC would issue reward checks as needed. Anglers received their reward check after the appropriate information was sent from the biologist to the SMSC, who in turn issued the appropriate checks. As an additional incentive for tag returns, reward checks were accompanied with information on date and location of tagging.

Tournament data

Smith Mountain Lake striped bass tournaments were conducted in April and September of each year. Information requested from tournaments included number of boats and anglers participating, start and stop times, total number and weight of all fish entered, length and weight of largest fish and the specific number, and weights of the top five place winners. Only tournaments conducted during similar time frames each year were used for comparisons. Average weight of fish caught in each tournament was calculated.

Trophy fish program

Virginia Department of Game and Inland Fisheries has maintained a list of trophy fish awards (citations) issued for specific species and waters for many years through the Freshwater Trophy Fish Award Program. This program is similar to that of many other states where anglers can submit information on large fish they caught and in return receive a certificate documenting their accomplishment. Each fish species has a minimum size to qualify and, for striped bass, an angler can qualify with a striped bass that is at least 9.07 kg or 940 mm. Even though this program is strictly on a volunteer basis, if the program receives substantial participation, general trends in the number of trophy fish caught in a given body of water can reflect major changes in the fishery over time.

Otolith collection

Because VDGIF did not have the ability to collect adequate numbers of older striped bass for age and growth determinations with existing gill-net configuration, anglers were utilized to collect otoliths from harvested fish. Beginning in 2001, anglers collected and stored striped bass heads for otolith extraction and analysis by a biologist at a later date. The program was expanded to include drop-off points at a local sporting goods store and marina. Virginia Department of Game and Inland Fisheries provided a freezer at the marina, which contained a fish-cleaning station frequented by striped bass guides. Anglers collecting striped bass heads were required to attach, at a minimum, the fish length and date caught. As an incentive for angler participation, information on the age and year stocked of each fish submitted was sent to anglers who included the appropriate return address information with their fish head.

Results

Traditional stock assessment

Gill netting provided an annual index of striped bass recruitment to the fishery. Mean gill-net catch rates increased from 7.9 fish/net in 1998 to 20.3 fish/net in 2002–2008 and indicated that higher numbers of striped bass were recruited into the fishery after increased stocking began in 1998 (Figure 1). Gill-net data did show a moderate decline in catch rates in 2003, but it was not evident in 2004 samples. Striped bass were stocked at a much smaller size in 2005, resulting in very poor survival and reduced catch rates the following year. Age 1–2 striped bass were the only age-groups collected consistently across years to calculate meaningful growth analysis (Figure 2). Reduced growth was observed for age 1–2 fish after stocking was increased in 1998 but improved after the fish kill and stocking was reduced in 2003 (Figure 3).

Creel survey catch rates for striped bass increased from 0.10 fish per angler-hour in 1998 to 0.18 and 0.17 fish per angler-hour in 2003 and 2006, respectively. However, overall effort decreased from 25.1 h/ha in 1998 to 16.1 h/ha in 2003 and 6.6 h/ha in 2006.

Management plan

The initial Smith Mountain Lake Striped Bass Management Plan was finalized in 1999. This plan provided a brief history of the Smith Mountain Lake striped bass fishery, outlined future direction, and recorded angler and VDGIF commitments. Two primary objectives were established in the first management plan. The first objective was to improve skilled (i.e., fished at least 5 years for striped bass at Smith Mountain Lake) angler catch rates of fish greater than 507 mm to 0.25 fish/h within 3–6 years and to 0.50 fish/h in 7–10 years. The second objective was to enhance the number of VDGIF citations issued for striped bass at Smith Mountain Lake to 235 annually in the near term and enhance to 300 annually in the long term. The management plan also outlined specific methods to achieve these objectives.

In 2003, the Smith Mountain Lake Striped Bass Management Plan was revised using data collected since 1999 and from all striped bass anglers instead of only skilled anglers. The

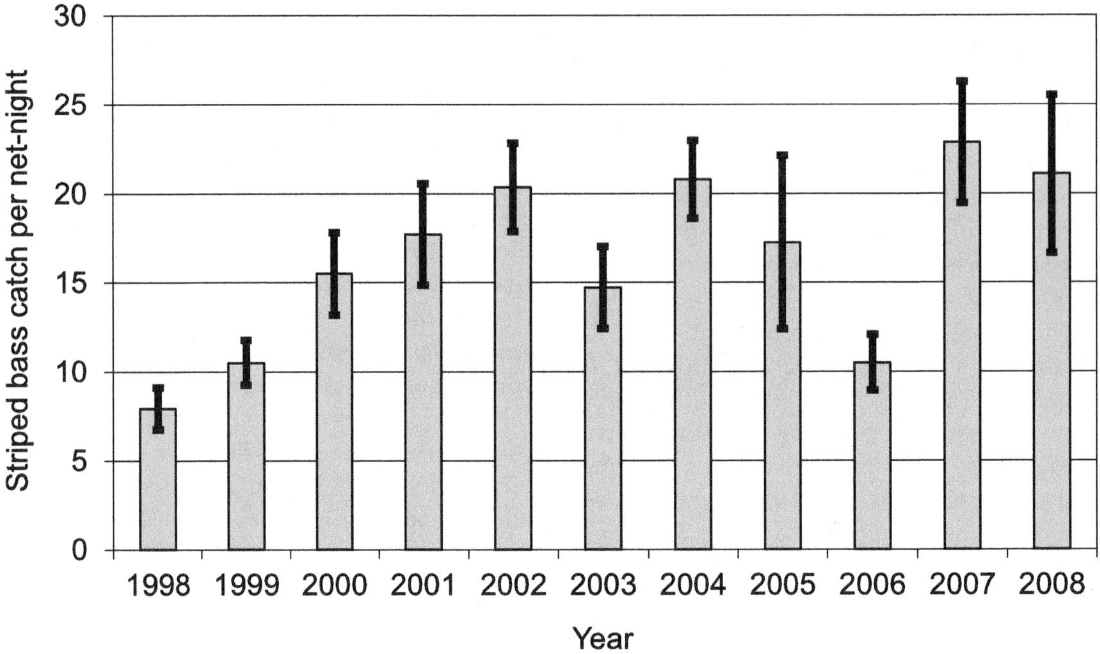

FIGURE 1. Gill-net catch-per-unit-effort data (mean catch per net-night) for striped bass at Smith Mountain Lake, Virginia (1998–2008). Error bars denote ±1 SE from the mean.

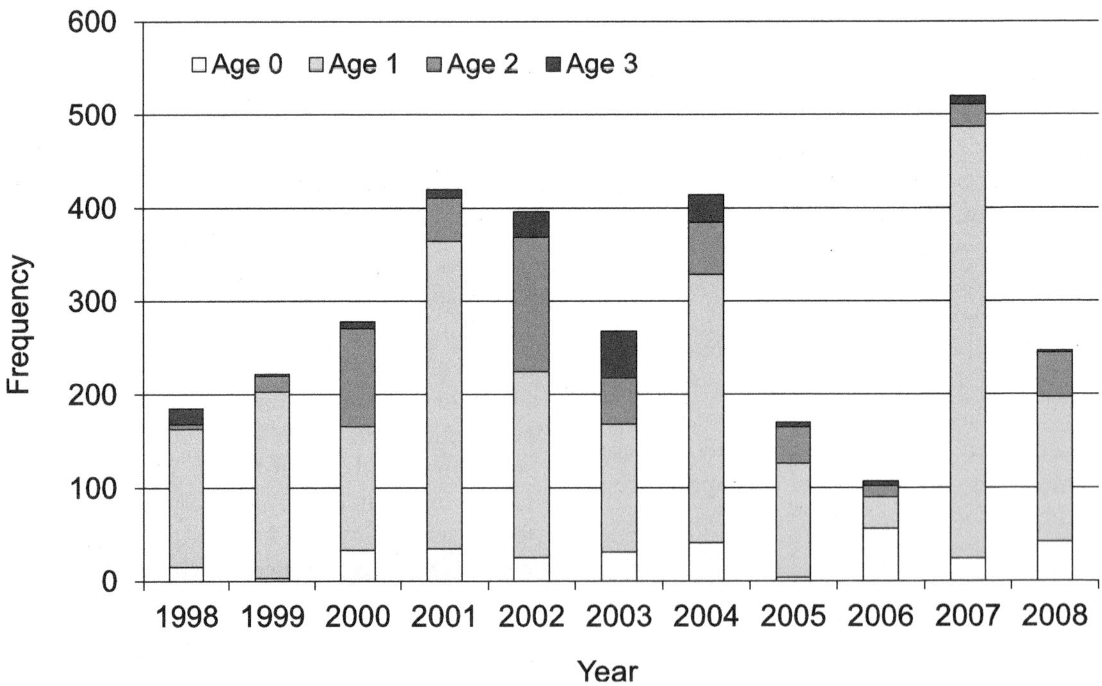

FIGURE 2. Gill-net catch frequency by age-group for striped bass at Smith Mountain Lake, Virginia (1998–2008).

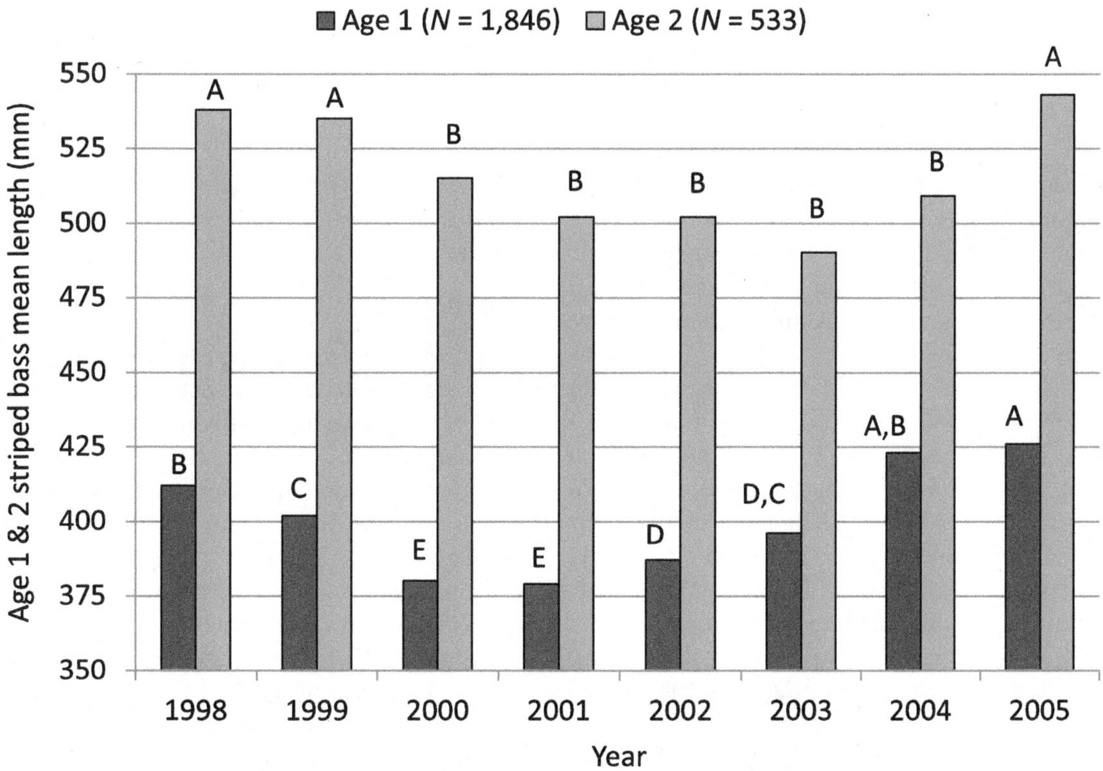

FIGURE 3. Duncan's multiple range test for mean lengths (mm) of age-1 and age-2 striped bass collected in fall gill nets from Smith Mountain Lake, Virginia (1998–2005). Means were significantly different among years for age 1 and age 2 (analysis of variance: F = 13.77, 53.68; df = 7, P < 0.001). Means with different letters were significantly different at alpha = 0.05.

2003 revision included maintaining an angler catch rate for striped bass exceeding 507 mm in length of 0.37 fish/h, increasing percentage of preferred length striped bass (>761 mm) caught to 30%, and issuing 150 trophy fish citations annually. An annual stocking rate of 350,000 fingerlings was established with specific guidelines for modification based on changes in striped bass and black bass growth rates, forage base dynamics, and VDGIF electrofishing catch rates for largemouth bass. Both parties were in agreement that substantial stocking reductions or additions were possible and should be based on these guidelines. The 2003 plan also included management actions and accomplishments during the previous 5 years.

Due to the fish kill in 2003, the Smith Mountain Lake Striped Bass Management Plan was revised a third time in 2005 to account for the substantial changes in the striped bass population and redirect a number of the management strategies. However, goals established by the management plan in 2003 remained the same.

Angler diaries

Angler diaries provided annual length frequencies, angler catch rates for a broad range of sizes, harvest-and-release habits, and seasonal variations in data that were not previously available. Angler diaries recorded 108,372 angler-hours, 28,841 angler trips, and 31,089 total lengths for striped bass by volunteer striped bass anglers from 1998 to 2008. In contrast, creel surveys recorded 692 striped bass fishing trips, and only 244 striped bass were measured in the three creel surveys conducted during the same time period. These robust angler data also provided repeatable information on seasonal harvest-and-release habits that were utilized to determine additional restrictive regulations. Additionally, the diary program had economic advantages providing annual data when annual creel surveys are limited due to financial constraints. Costs associated with these two programs were estimated at $100 annually for the angler diary program and $40,000 annually for the creel surveys, not including administrative and data-analysis costs.

Increased stocking rates beginning in 1998 were linked with increasing numbers of fish in gill nets. However, overall angler catch rates showed little change, and the percent of striped bass caught exceeding 761 mm initially remained stable but started declining in 2003 and subsided after 2004 (Table 1). With no additional fish kills observed after the spring of 2003 and additional restrictive regulations enacted in 2006, angler diary data indicated an increase in the percentage of striped bass exceeding 761 mm. These data documented positive changes in striped bass length frequencies and provided information on the percentage of fish that were released or harvested (Figure 4).

Participation in the angler diary program did decrease over time from a high of 116 in 2001 to a low of 25 in 2006. An incentive program was initiated in 2007 to prevent further declines in participation by offering a US$20–50 reward for all returned diaries. Participation began to improve in 2008 with 31 diaries returned.

Fish tagging

A total of 1,390 striped bass exceeding 507 mm were tagged from 2001 to 2005. Striped bass above the minimum length limit of 508 mm were targeted for tagging. Including adjustments for a 48% nonreporting rate derived from postcard returns, the tag return rate for striped bass tagged in 2001 was 43% with a 91% harvest rate (Figure 5). Tag return rates declined from 2001 to a low of 7% in 2004 and increased to 20% in 2005. Harvest rates for tagged fish declined each year of the study.

Tag return rates from summer-tagged striped bass were significantly lower ($\chi^2 = 9.92$, df = 1, $P = 0.0016$) than those tagged in October through May. A total of 377 striped bass were tagged from 2002 to 2005 during the

TABLE 1. Striped bass data from 1998 to 2008 for angler diaries collected from Smith Mountain Lake, Virginia. Catch per unit of effort (CPUE) is the mean number of striped bass caught per hour of fishing, *N* is the number of angler trips, and percent of catch greater than 761 mm is the percentage of captured striped bass greater than 761 mm. CI = confidence interval.

		CPUE (*n*/h; all sizes)		CPUE (*n*/h; >507 mm)		Percent of catch
	N	Mean	90% CI	Mean	90% CI	> 761 mm
1998	2,535	0.33	0.32–0.35	0.27	0.25–0.28	17%
1999	2,440	0.46	0.43–0.49	0.31	0.28–0.33	17%
2000	2,354	0.27	0.25–0.28	0.17	0.16–0.18	17%
2001	2,593	0.35	0.33–0.37	0.23	0.22–0.25	16%
2002	2,028	0.44	0.42–0.47	0.32	0.30–0.34	17%
2003	871	0.43	0.39–0.47	0.33	0.30–0.36	10%
2004	667	0.52	0.49–0.56	0.34	0.31–0.37	2%
2005	720	0.47	0.43–0.50	0.39	0.36–0.42	8%
2006	696	0.34	0.31–0.36	0.29	0.26–0.31	10%
2007	1,025	0.35	0.33–0.38	0.24	0.22–0.26	13%
2008	1,102	0.40	0.37–0.43	0.31	0.28–0.34	19%

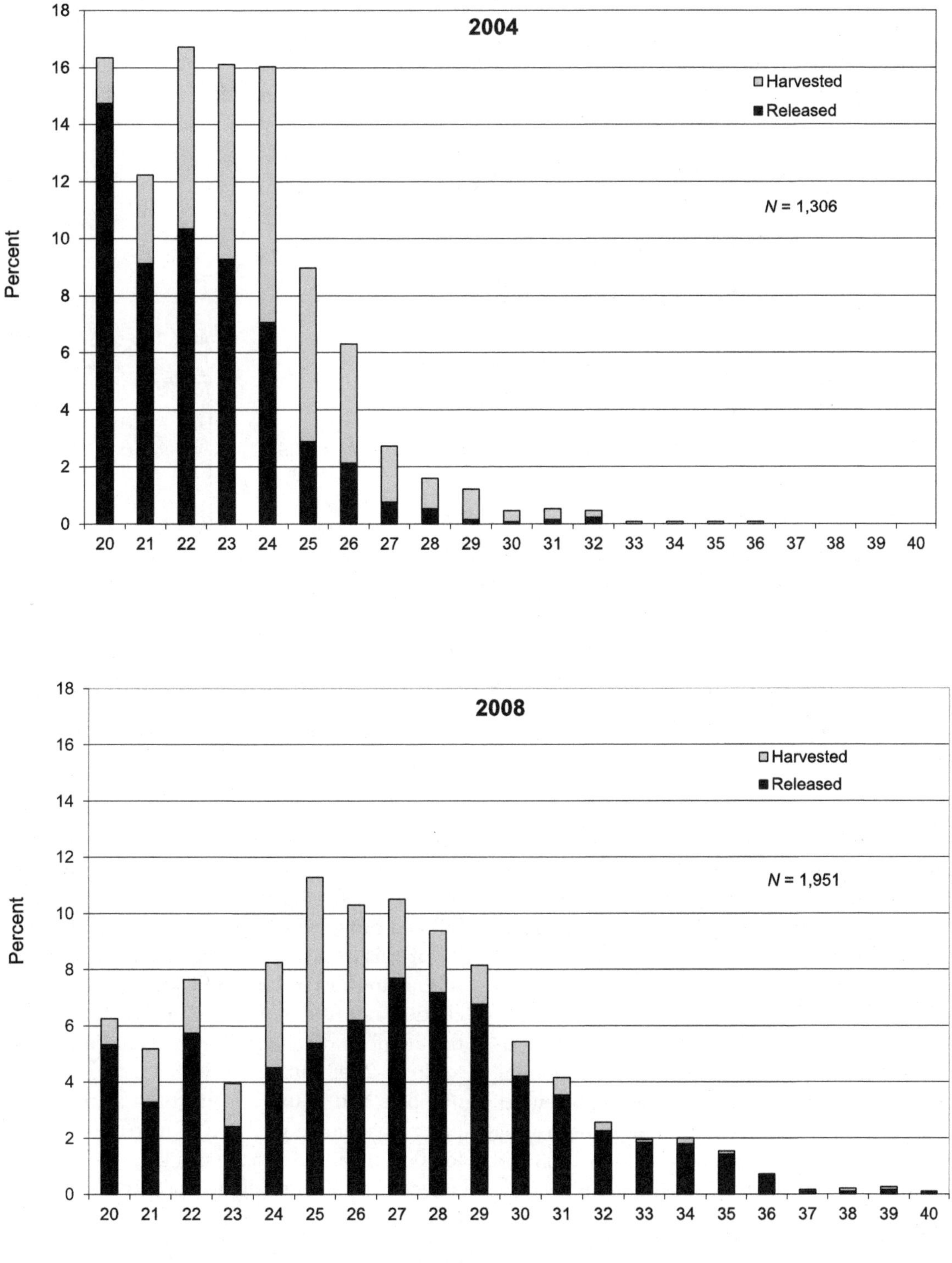

FIGURE 4. Striped bass data collected from Smith Mountain Lake angler diaries in 2004 and 2008. Each bar represents the percent of captured striped bass that was between 20 and 40 in in total length.

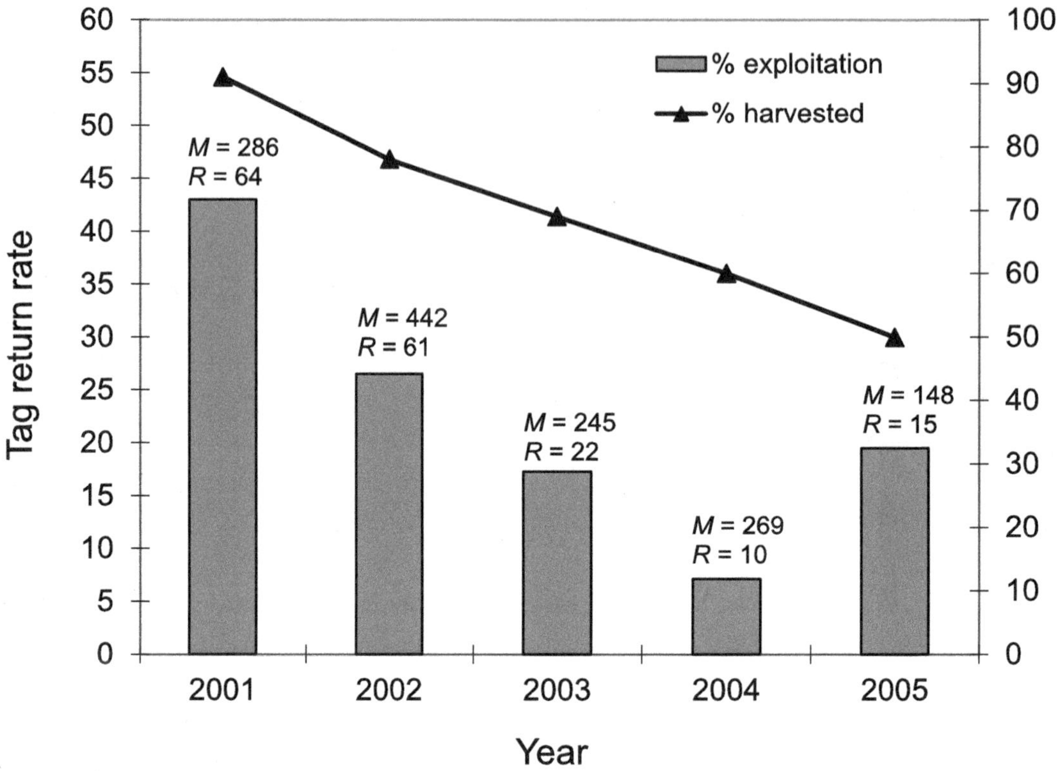

FIGURE 5. Adjusted tag return rate and percent harvest of tagged striped bass caught by anglers in 2001 to 2005 at Smith Mountain Lake, Virginia. Tag return rate was adjusted by a tag reporting rate of 52%. *M* is the number of fish tagged each year, and *R* is the number of tags returned within 1 year after tagging.

months of June–September. First-year tag returns for all summer-tagged striped bass were 8% (accounting for nonreporting), as compared to 24% (accounting for nonreporting) for the remaining months.

Tournament data

Mean sizes of striped bass entered in the tournaments varied depending on season. Combining the two seasons was less informative, but when analyzed seasonally, the data showed decreasing mean weights over time from 2000 to 2003 and increasing weights after 2004 (Figure 6).

Trophy fish program

From 1995 to 2002, citations issued to striped bass anglers at Smith Mountain Lake remained relatively consistent at approximately 150 citations per year. There were only seven citations issued in 2004 after the fish kill (Figure 7). This enormous drop in citations was very insightful when assessing the effects of the 2003 fish kill and provided confirmation as to the severity of this occurrence. Continued low citation submissions after 2004 indicated that this low citation rate was not an anomaly. Since 2004, the number of citations issued has increased to 46 in 2008 but is still lower than historical numbers and management goals. Initially, the first management goal was to increase citations to 300 annually. After the fish kill, this goal was reduced to 150 annually to make the goal more realistic.

Otolith collections

Anglers supplied more than 200 striped bass heads and 74% of available age data for fish exceeding age 4 from 2001 to 2009 (Figure 8).

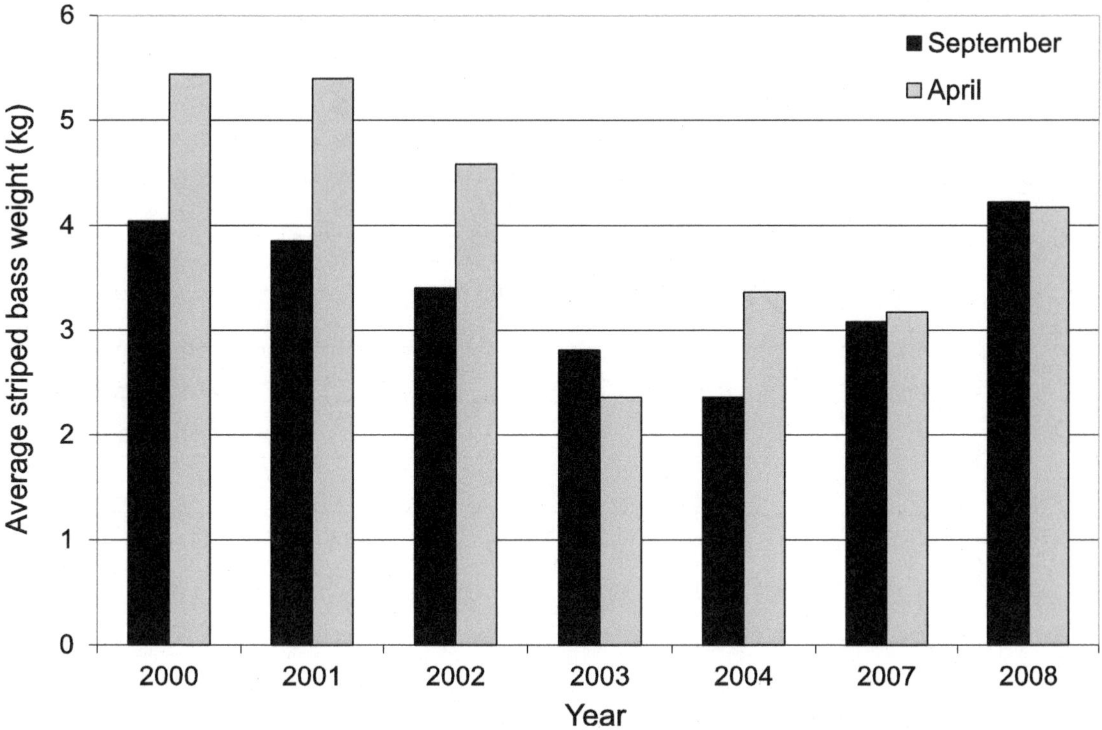

FIGURE 6. Average weights for striped bass entered in tournaments from 2000 to 2008 at Smith Mountain Lake, Virginia.

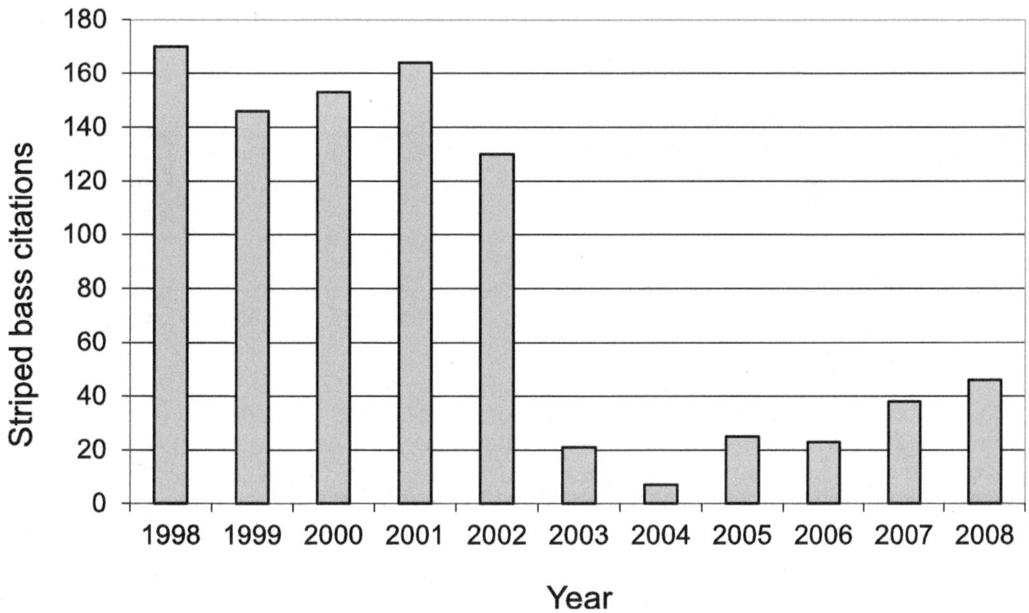

FIGURE 7. Striped bass citations issued through the Virginia Freshwater Fish Reward Program from 1998 to 2008 at Smith Mountain Lake, Virginia. Striped bass must be at least 9.07 kg or 940 mm to qualify.

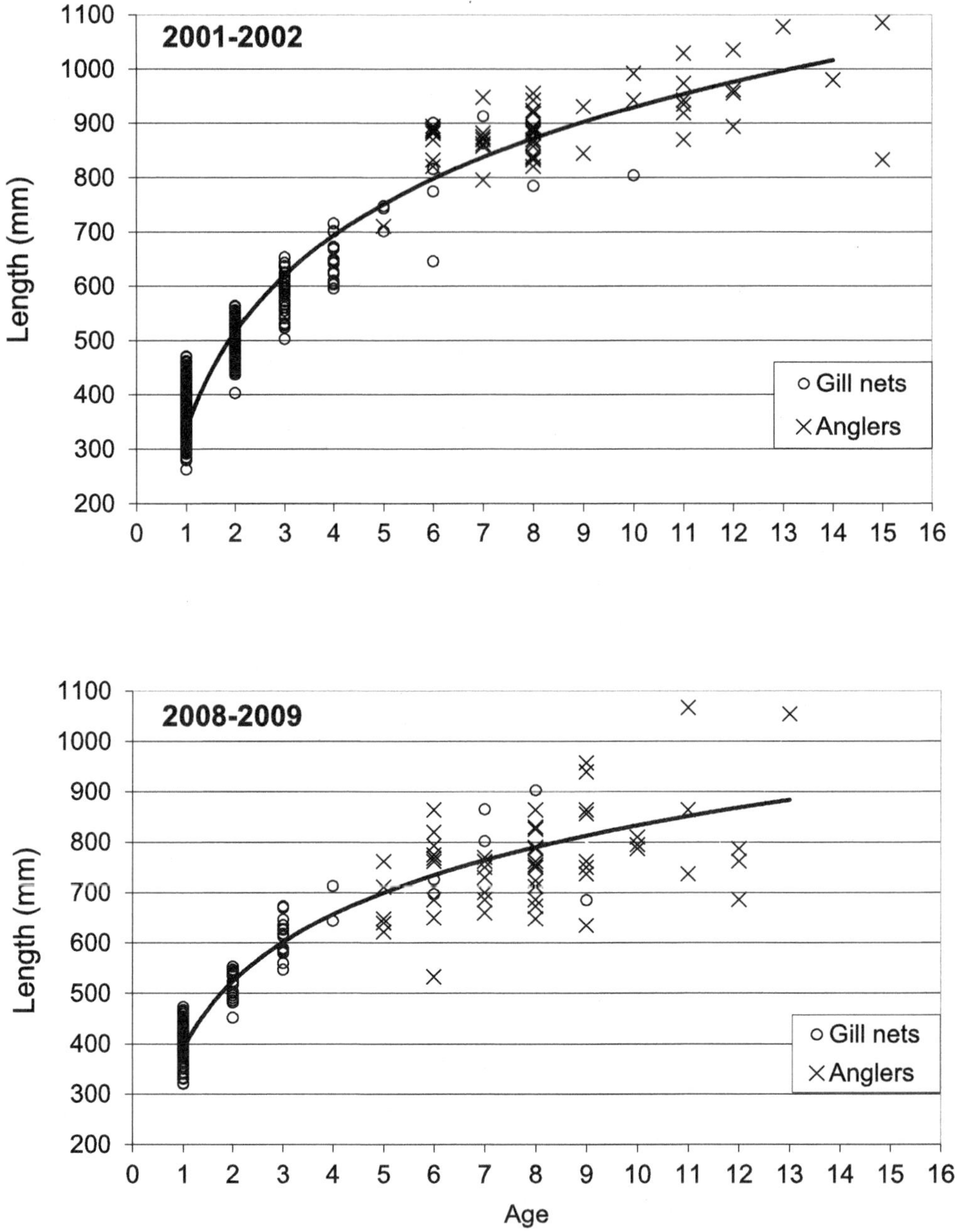

FIGURE 8. Relationship between length and age of striped bass collected with gill nets and by anglers in 2001–2002 and 2008–2009 at Smith Mountain Lake, Virginia. Regression lines were used to describe striped bass length at age for the two time periods with the following equation y = slope($\log_e x$) + intercept; y is fish length (total length, mm) and x is fish age (years).

Fifty-two fish were collected in 2001–2002 and considered baseline growth for future comparisons. Growth rates from striped bass ages 1–2 collected in gill nets showed a growth reduction from 1998 to 2001 with a rebound to historical rates by 2005 (Figure 3). Angler collected striped bass did not show a similar recovery and were still below targeted growth rates in

2008–2009 (Figure 8). Growth rates were more variable among individuals in 2008–2009 than in 2001–2002.

Discussion

Angler contributions to data collection substantially improved biological understanding of striped bass in Smith Mountain Lake. These contributions ranged from management planning to fish sampling and tagging. Participation by anglers allowed the collection and analysis of information beyond that which could have been carried out by VDGIF's staff. In addition, angler participation provided enhanced understanding and public acceptance of the information.

Increasing striped bass stocking rates was the first step outlined in the Smith Mountain Lake Striped Bass Management Plan to improve the fishery. Gill-net data were outlined as the primary tool for monitoring year-class strength at age 1 while angler diary data were used to monitor changes in angler catch rates. Gill nets documented an increasing number of striped bass after stocking increased in 1998 (Wilson 2004). Gill netting has been noted as a good indicator of angler catch rates in subsequent years for walleye *Sander vitreus* (Isbell and Rawson 1989). However, unlike many walleye populations, striped bass populations contain a broader size range making gill-net sampling often impractical to sample fish sizes preferred by anglers. Despite an increasing population of 300–550-mm striped bass documented with gill nets, angler diary catch rates remained stable from 1998 to 2002. While gill nets were useful to document an increasing population of young fish, angler diaries indicated a moderate increase in angler catch rates, but desired improvements in angler catch rates for preferred length fish were not realized.

Angler-related journals and programs have been useful in other areas. Volunteer anglers provided usable data sets for 124 Texas reservoirs, which was determined to be an acceptable method for supplementing traditional surveys and could be useful as a warning system to alert managers to problems (Prentice et al. 1995). For Lake Erie, angler diaries monitored fisheries where conventional creel surveys were limited (Sztramko et al. 1991), and angler catch statistics were considered essential to effective recreational fisheries management throughout Mississippi (Bray and Schramm 2001). Trophy-fish programs provided documentation of the development of a trophy largemouth bass fishery at Briery Creek Lake, Virginia (Wilson and DiCenzo 2002), and Quinn (1987) advocated angler recognition data as an indicator of fishing quality. At Smith Mountain Lake, a striped bass fish kill occurred in 2003 and was evident in angler-collected data. However, striped bass gill-net and creel survey data were not as useful for detecting kill-induced striped bass population changes because gill-net configurations only targeted young fish and creel surveys collected limited data on larger fish. In addition to angler diaries, citations submitted for Smith Mountain Lake striped bass dropped from 130 in 2002 to 7 in 2004, and average weights of striped bass submitted in fall tournaments dropped from 3.4 to 2.4 kg during the same time period. Creel survey data showed an increase in angler catch rates from 1998 to 2003, but angler diary catch rates declined moderately after the fish kill in 2003 for several years. However, creel survey data did document that overall angler effort substantially decreased. Because catch rates were not severely reduced, the loss of large striped bass was assumed the primary reason for the decline in fishing pressure. Numerous anglers complained about the lack of large fish at Smith Mountain Lake after the fish kill and reported that they either stopped or reduced their fishing effort for striped bass.

Harvest rates of tagged fish in 2001 may have been higher than those of nontagged fish. Initially, many anglers were not equipped to clip off tags or thought harvest was necessary for reward collection. During the first year of tagging, some anglers commented that they were uncertain if the fish had to be harvested to receive a reward. Beginning in 2003, anglers became more hesitant to harvest fish infested with parasites and were interested in returning more fish to help restoration efforts after the fish kill. The decline in tag returns for 2002 was unexpected but was expected starting in 2003 due to the loss of many tagged fish during the fish kill and a reduction in angler effort. Tag returns increased in 2005 as anglers started to return to the fishery after the fish kill in 2003.

Tag loss and tagging mortality for cooler months were not estimated but assumed to be low based on a study conducted by Dunning et al. (1987). Generated tag return rates were considered to be substantial, warranting more restrictive regulations to achieve management goals. Accounting for reductions of available tagged fish for capture by anglers due to tag loss or tagging mortality would only have increased tag return rates. The postcard survey nonreporting rate was lower than the 73% nonreporting rate observed in a sauger *Sander canadensis* study by Maceina et al. (1998). After years of informational programs aimed to reduce summer release rates, angler diary data indicated that anglers had not changed their behavior. Reducing high catch-and-release rates during the summer months was considered important based on other studies indicating that striped bass catch and release mortality increases during summer months (Wilde et al. 2000; Bettinger and Wilde 2013, this volume). Many members of the SMSC maintained that studies documenting the discrepancies in survival across seasons were not applicable to Smith Mountain Lake. Consequently, prior to additional regulations, release rates were similar for all months. Angler diary data highlighted that without additional regulations, most anglers refused to adjust catch-and-release practices based on seasonality. Based on findings by Bettoli and Osborne (1998), we attributed the low summer tag returns to poor catch-and-release survival after tagging due to high water temperatures. Low summer tag returns were beneficial in convincing some anglers to alter their summer angling practices and provided support for developing new regulations to address the problem.

Management changes enacted after 2003 were directed at the restoration of striped bass in the preferred and citation length-groups. Declining growth rates observed with otoliths collected from gill nets and anglers indicated that a reduction in the stocking rate was needed to restore growth rates to pre-1998 levels. Fishing mortality estimates indicated that sustained high fishing mortality, especially for preferred lengths, would further limit increased catch rates for preferred size fish as outlined in the management plan. Beginning in 2003, stocking rates were reduced to improve growth rates, and discussions were initiated to introduce more restrictive regulations for preferred length striped bass. In 2006, a new restrictive slot limit was initiated to replace the 508-mm minimum length limit. This regulation provided additional protection of larger fish when catch-and-release survival was most favorable and allowed for additional harvest of smaller fish. Reducing high fishing mortality with more restrictive striped bass size limits was also suggested to benefit the fishery in Badin Lake, North Carolina (Thompson et al. 2007). Improvements in the striped bass fishery by 2008 were evident with increased angler diary catch rates for preferred lengths, citations, and average weights of striped bass entered in tournaments. However, traditional gill-net configuration and creel survey data again showed little change in the fishery due to limitations in the amount of data available.

One objection to new seasonal restrictive slot limit regulations was that additional protection of larger striped bass during the cooler months would be negated by additional harvest of those fish during the warmer months, when they were not protected. Angler diary data documented that annual release rates of striped bass exceeding 660 mm increased from 27% before new regulations (2003) to 72% after the new regulations (2008), and length frequencies showed that a more desirable size structure was realized after the new regulations. These data documented the success of the new regulations and provided additional support for the program.

Smith Mountain Lake experienced numerous changes in the striped bass population from 1998 to 2008. Initially, stocking rates were increased in 1998, resulting in a growing striped bass population. Higher striped bass densities negatively impacted growth rates and did not produce desired angler catch rates. Declining fish health and forage combined with an infestation of parasitic copepods culminated in a severe striped bass fish kill, primarily for striped bass exceeding 660 mm, in 2003. A similar fish kill occurred for Chinook salmon *Oncorhynchus tshawytscha* in Lake Michigan. Holey et al. (1998) hypothesized that slower growth and declining fitness due to reduced food availability, increased salmon densities, and heavy parasite infections were stressors that triggered epizootics and a reduction by 50% or more of adult Chinook salmon.

After the 2003 fish kill, management changes were directed at the restoration of preferred and citation size-groups, producing marginal improvements to the fishery as evidenced by angler-collected data. Traditional sampling with gill nets and creel surveys would not have allowed managers to track all of the population changes that occurred (especially at larger sizes), make sound management changes, and measure success of these changes. Combining various angler data sets with VDGIF data and utilizing volunteers to expand data collection for a variety of fish species, managers were provided a unique opportunity to document population changes in a timely manner and make appropriate management decisions to improve angler success and satisfaction. Alexander and Holbrook (1979), Gabelhouse and Willis (1986), Ebbers (1987), and Dolman (1991) also found that anglers can provide usable data for fisheries assessment.

Developing a striped bass management plan for Smith Mountain Lake was the initial step toward resolving a host of conflicts (McMullin 2013, this volume). Prior to this study period, public relations between VDGIF and striped bass anglers were tumultuous. Striped bass anglers at Smith Mountain Lake held views opposing those of managers, which prevented progress toward a resolution. The jointly prepared management plan was successful in bringing anglers and managers together to identify specific goals and fill data gaps. Having documented goals and objectives prevented the agency and anglers from changing their standards as time progressed without agreement from both sides. Additionally, opposing viewpoints and mistrust in agency data were moderated when a significant amount of the data was acquired through angler programs; acceptance of agency data by anglers is often met with skepticism (Churchill et al. 2002). Data collected by anglers on Smith Mountain Lake were considered credible and were generally well accepted by the public.

Volunteer programs do not come without a cost to an agency. To successfully institute and sustain angler programs, positive relations must be built, which can entail a considerable investment of time and effort. Without a willingness to compromise and develop angler relationships, partnerships of this nature have the potential to foster deeper feeling of mistrust and further separate the two parties. Including anglers in the management process allows them to become part of the solution to agency limitations and improve fisheries management.

Acknowledgments

I thank the numerous volunteers from the Smith Mountain Striped Bass Club for their assistance with field work, data collection, and lively debate on many management issues. Manuscript review and comments were provided by Vic Dicenzo and Scott Smith.

References

Alexander, C. M., and J. A. Holbrook. 1979. Comparison of bass club records with creel censuses. Proceedings of the Annual Conference Southeastern Association of Fish and Wildlife Agencies 32(1978):446–448.

Bray, G. S., and H. L. Schramm Jr. 2001. Evaluation of a statewide volunteer angler diary program for use as a fishery assessment tool. North American Journal of Fisheries Management 21:606–615.

Bettinger, J. M., and G. R. Wilde. 2013. Catch-and-release mortality of inland striped bass and hybrid striped bass. Pages 473–499 *in* J. S. Bulak, C. C. Coutant, and J. A. Rice, editors. Biology and management of inland striped bass and hybrid striped bass. American Fisheries Society, Symposium 80, Bethesda, Maryland.

Bettoli, P. W., and R. S. Osborne. 1998. Hooking mortality and behavior of striped bass following catch and release angling. North American Journal of Fisheries Management 18:609–615.

Churchill, T. N., P. W. Bettoli, D. C. Peterson, W. C. Reeves, and B. Hodge. 2002. Angler conflicts in fisheries management. Fisheries 27(2):10–19.

Coutant, C. C. 2013. When is habitat limiting for striped bass? Three decades of testing the temperature–oxygen squeeze hypothesis. Pages 65–91 *in* J. S. Bulak, C. C. Coutant, and J. A. Rice, editors. Biology and management of inland striped bass and hybrid striped bass. American Fisheries Society, Symposium 80, Bethesda, Maryland.

Dolman, W. B. 1991. Comparison of bass-club tournament reports and creel survey data from Texas reservoirs. North American Journal of Fisheries Management 11:177–184.

Dunning, D. J., J. R. Waldman, and M. T. Mattson. 1987. Tag retention by, and tagging mortality of, Hudson River striped bass. North American Journal of Fisheries Management 7:535–538.

Ebbers, M. A. 1987. Vital statistics of a largemouth bass population in Minnesota from electrofishing and angler-supplied data. North American Journal of Fisheries Management 7:252–259.

Gabelhouse, D. W. Jr., and D. W. Willis. 1986. Biases and utility of angler catch data for assessing size structure and density of largemouth bass. North American Journal of Fisheries Management 6:481–489.

Holey, M. E., R. F. Elliott, S. V. Marcquenski, J. G. Hnath, and K. D. Smith. 1998. Chinook salmon epizootics in Lake Michigan: possible contributing factors and management implications. Journal of Aquatic Health 10:202–210.

Isbell, G. L., and M. R. Rawson. 1989. Relations of gill-net catches of walleyes and angler catch rates in Ohio waters of western Lake Erie. North American Journal of Fisheries Management 9:41–46.

Leslie, L. L., C. Velez, and S. Bonar. 2004. Utilizing volunteers on fisheries projects: benefits, challenges, and management techniques. Fisheries 29(10):10–14.

Maceina, M. J., P. W. Bettoli, S. D. Finely, and V. J. DiCenzo. 1998. Analyses of the sauger fishery with simulated effects of a minimum size limit in the Tennessee River of Alabama. North American Journal of Fisheries Management 18:66–75.

McMullin. S. L. 2013. Managing user conflicts on reservoirs supporting striped bass or hybrid striped bass fisheries. Pages 521–530 *in* J. S. Bulak, C. C. Coutant, and J. A. Rice, editors. Biology and management of inland striped bass and hybrid striped bass. American Fisheries Society, Symposium 80, Bethesda, Maryland.

Prentice, J. A., B. W. Farquhar, and W. E. Whitworth. 1995. Comparison of volunteer angler-supplied fisheries catch and population structure data with traditional data. Proceedings of the Annual Conference Southeastern Association of Fish and Wildlife Agencies 47(1993):666–678.

Quinn, S. P. 1987. The status and usefulness of angler recognition programs in the United States. Fisheries 12(2):10–16.

Schlechte, J. W., N. G. Smith, and J. B. Taylor. 2013. Options for estimating striped bass catch and harvest: effectiveness of creel surveys. Pages 291–312 *in* J. S. Bulak, C. C. Coutant, and J. A. Rice, editors. Biology and management of inland striped bass and hybrid striped bass. American Fisheries Society, Symposium 80, Bethesda, Maryland.

Sutton, T. M., K. A. Rose, and J. J. Ney. 2000. A model analysis of strategies for enhancing stocking success of landlocked striped bass populations. North American Journal of Fisheries Management 20:841–859.

Sztramko, L. K., W. I. Dunlop, S. W. Powell,

and R. G. Sutherland. 1991. Applications and benefits of an angler diary program on Lake Erie. Pages 520–528 *in* D. Guthrie, J. M. Hoenig, M. Holliday, C. M. Jones, M. J. Mills, S. A. Moberly, K. H. Pollock, and D. R. Talhelm, editors. Creel and angler surveys in fisheries management. American Fisheries Society, Symposium 12, Bethesda, Maryland.

Thompson, J. S., D. S. Waters, J. A. Rice, and J. E. Hightower. 2007. Seasonal natural fishing mortality of striped bass in a southeastern reservoir. North American Journal of Fisheries Management 27:681–694.

Wilde, G. R., M. I. Muoneke, P. W. Bettoli, K. L. Nelson, and B. T. Hysmith. 2000. Bait and temperature effects on striped bass hooking mortality in freshwater. North American Journal of Fisheries Management 20:810–815.

Wilson, D. M. 2004. Effects of expanded stocking locations on striped bass survival. Proceedings of the Annual Conference Southeastern Association of Fish and Wildlife Agencies 56(2002):79–85.

Wilson, D. M., and V. J. DiCenzo. 2002. Profile of a trophy largemouth bass fishery in Briery Creek Lake, Virginia. Pages 583–592 *in* D. P. Philipp and M. S. Ridgway, editors. Black bass: ecology, conservation, and management. American Fisheries Society, Symposium 31, Bethesda, Maryland.

Yurk, J. J. 1989. Assessment of total phosphorus concentration as a predictor and determinant of fishery productivity in southern Appalachian reservoirs: application to Smith Mountain Lake, Virginia. Master's thesis. Virginia Polytechnic Institute and State University, Blacksburg.

American Fisheries Society Symposium 80:351–364, 2013

Population Models for Assessment and Management of Inland Striped Bass Fisheries

Micheal S. Allen* and Daniel Gwinn
School of Forest Resources and Conservation, The University of Florida
7922 NW 71st Street, Gainesville, Florida 32653, USA

Abstract.—Population models are frequently used to assess stocks and inform management, but use of models for inland striped bass *Morone saxatilis* fisheries is rare in the literature. We summarize three common types of population models and describe how they could aid management of striped bass and hybrid striped bass fisheries. Yield-per-recruit models are ideal for identifying growth overfishing and optimum size at harvest to maximize yield. Time-dynamic, age-structured models are useful for evaluating effects of variable recruitment on angler catch and effects of temporary changes in model parameters (e.g., fish kills, changes in growth) on model output. Catch-at-age models can be used to measure recruitment to age 1 for inland striped bass fisheries and for quantifying fishing mortality rates. Catch-at-age models can also improve the utility of creel survey data and help evaluate the efficacy of stocking programs by estimating age-1 recruits (from model) per stocked fish (from hatchery). Population models force investigators to be explicit about their hypotheses regarding fisheries systems, identify data gaps, and allow assessment of potential impacts of management actions on the fishery. The examples shown here can be used to improve striped bass and hybrid striped bass monitoring and management.

Introduction

Fish population models are valuable tools to assess fish stocks, and results from models are typically used to set harvest regulations for commercial and recreational fisheries. Population models range from very simple change-in-biomass (e.g., surplus-production) models to complex age- and length-structured dynamic models. Each model type has advantages and disadvantages, and choosing the appropriate model depends on the management questions at hand and the type of data that are available for the assessment. We reviewed several common fish population model types and here describe how they could be used to improve management of striped bass *Morone saxatilis* and hybrid striped bass fisheries.

A population model is a simplistic depiction of a fishery, and some fisheries managers are skeptical of using modeling results to develop management strategies. However, most fisheries managers responsible for a major fishery already have a model; it's just located in their minds! For example, an astute biologist may know (1) the conditions (e.g., years and sizes of fish) for most effective stocking of striped bass or hybrid striped bass, (2) the rainfall and temperature conditions when summer habitat is most limited, and (3) the factors that drive levels of harvest (fishing mortality) each year. Each of these mental summaries of experience represents a simplistic depiction of the natural system, and thus even if the model is not in a computer, the biologist typically has a working "model" of the system. Each component of

* Corresponding author: msal@ufl.edu

the biologist's beliefs about the system can be thought of as a working hypothesis of the factors influencing the quality of the fishery.

Likewise, a population model should be considered as a working hypothesis about the set of factors (e.g., growth, mortality, recruitment) that interact to produce observed results (e.g., harvest, total catch). It is not realistic, however, to consider the model a tool to predict exactly what will happen in the future. Instead, use of a population model forces the investigator to specify assumptions about how the fishery operates and the parameters that are acting on the fishery. By specifying a model structure and associated parameters, fishery managers formalize their hypotheses about the system. They can then compare predictions based on those hypotheses with existing or future data or use the predictions based on the stated assumptions as a basis for management. This process of compiling all available data into a population model often results in new insights and serves as a guide for future field collections, helping to eliminate data gaps and reduce uncertainty. Thus, the advantages of building population models go way beyond (and should not always include) predictions about what will happen in the future. Models can help us conceptualize a fishery, identify key needs for future work, and hopefully improve management strategies in the future by reducing those uncertainties.

Although models for stock assessment of coastal striped bass fisheries are common (ASMFC 2003; Shepherd 2006), there are surprisingly few published examples of population modeling for striped bass fisheries in reservoirs and other inland waters. Bulak et al. (1995) built an age structured model to evaluate optimal harvest regulations and the effects of stocking on a reproducing striped bass population in the Santee–Cooper Reservoir system, South Carolina. Thompson et al. (2007) used a yield-per-recruit model to assess a North Carolina reservoir striped bass fishery. Most other quantitative analysis of striped bass stocks has occurred in coastal systems. Rose and Cowan (1993) constructed an individual-based model to identify recruitment bottlenecks during early life. Richards and Rago (1999) reviewed the recovery of striped bass in Chesapeake Bay based on a range of stock assessment evaluations. Although the use of population models for inland striped bass fisheries is not common, there is a variety of potential modeling approaches that could aid management of these striped bass or hybrid striped bass fisheries. We reviewed several types of common population models and explored how they could be used to improve management of inland striped bass and hybrid striped bass fisheries.

Types of Models

Equilibrium yield-per-recruit models

An equilibrium yield-per-recruit (YPR) model is probably the most common type of population model used in freshwater and marine fisheries. The primary use of a YPR model is to diagnose growth overfishing. Growth overfishing is the condition where fish are harvested at a higher fishing mortality rate and smaller size than that which produces the maximum yield per recruit. Basic YPR models can identify the combination of minimum size limit and fishing mortality that maximizes yield (Table 1). This analysis is very useful for recreational fisheries where the harvest of fish is an important motivation for anglers, such as striped bass fisheries. Thus, development of a simple YPR model is commonly the first step in quantitative evaluations of fish stocks.

Data requirements.—A YPR model uses information about weight-at-age population structure, fishing mortality, and natural mortality rates. Growth parameters describing the length-at-age structure, typically L_∞, K, and t_0 from the von Bertalanffy growth equation combined with weight-length equations are often used to estimate the weights at different ages (Table 1). At least an approximate estimate of natural mortality (M) and annual exploitation (U) are needed. Mortality parameters are best estimated with passive and

TABLE 1. Population model type, data requirements, and utility for striped bass fisheries in reservoirs. The parameters are instantaneous natural mortality (*M*), annual exploitation (*U*), asymptotic length (L_∞), the Brody growth coefficient (*K*), age at length zero (t_o), and weight-length parameters *a* and *b*.

Model type	Data requirements	Outputs/uses	Advantages	Disadvantages
Yield per recruit (YPR)	M, U, L_∞, K, t_o, W-L relationship	Diagnose growth overfishing, use of size limits, maximum	Simple, not data intensive yield	Equilibrium model, not predictive
Age structured/time dynamic	Same as YPR, recruitment indices, year-specific M, U	Same as YPR, changes in harvest via recruitment variation, time-specific predictions yield and catch	Predictive, more realism, incorporate variable U or M	More data needed
Catch at age	M, creel survey harvest estimates, age composition of the	U, fish vulnerability recruitment to age 1, reconstruct stock history	Utilizes creel survey data for assessment, estimates recruitment	Long time series needed

active tagging methods (Hightower and Pollock 2013, this volume). If it is not feasible to estimate mortality values, estimates of natural mortality (M) could be borrowed from other locations for striped bass (e.g., Hightower et al. 2001; Thompson et al. 2007). Surrogate values based on growth parameters, such as M = 1.5K (Jensen 1996), or temperature and growth (Pauly 1980) will also provide approximations of M that are adequate for an initial YPR investigation. Estimates of annual exploitation (U) are required to identify the specific condition of the fishery, but YPR models are insightful without estimates of U because they show the yield responses for a wide range of potential fishing mortality rates. The weight-length (W-L) parameters a and b (W = aL^b) can be obtained from field measurements of fish weight and length or borrowed from Brown and Murphy (1991) for striped bass if fish weight measurements are not available. Uncertainty in parameters such as M or the W-L relationship will influence the YPR model output but usually will not strongly influence the relative value of a set of harvest policy options (e.g., minimum length limits).

Formulation and construction.—Existing YPR model formulations range from what appear as complex equations (Beverton and Holt 1957) to more streamlined methods that are easy to build using common spreadsheet software (Walters and Martell 2004). The FAST software (Slipke and Maceina 2005) is probably the most user-friendly application for YPR modeling. The software also includes routines for estimating growth parameters, W-L relationships, and natural mortality surrogates mentioned above. This software is very easy to learn and includes a helpful user's manual. We encourage readers to explore its capabilities. However, like any software, the FAST code is static, which makes the software inflexible to handle specific issues that may be important in certain situations. Examples include modeling specific effects of discard mortality, which can be particularly important during summer for striped bass fisheries (described below), or modeling effects of changes in fish growth rates (L_∞ and K) on model outputs. As an alternative, building YPR and other models in spreadsheets allows the investigator to form the model to the specific case under evaluation. Spreadsheet models make it easy to visualize and identify how changes in model parameters influence the output.

A YPR model uses fish size and mortality schedules to predict the size and level of fishing mortality that will maximize yield. Walters and Martell's (2004) Box 3.1 describes a simple method for building YPR models using an approach initiated by Botsford and Wickham (1979). The method constructs the survivorship schedules on a per-recruit (i.e., single recruit to age 1) basis. We have found their model formulation to be extremely useful, transparent, and easy to explain to students. Survivorship is the proportion of fish from some initial recruits that survive to each age. Starting with a single recruit at age 1 (lx_1 = 1), the survivorship to each age is

$$\text{lx}_a = \text{lx}_{a-1} S_o (1 - U v_{a-1}), \qquad (1)$$

where lx_a is the survivorship to age a, S_o = e^{-M} (i.e., the annual survival from natural mortality, M), U is the annual exploitation rate (i.e., the fraction of the vulnerable fish population harvested in a year), and v is the vulnerability to fishing. Values of S_o could be held constant for all ages, as indicated above, or set to vary with age, if information is available. Values of v range from 0, if fish are not vulnerable to fishing (e.g., below a minimum length limit), to 1, for those fully vulnerable to harvest. In a fishery with a minimum length limit, values of v_a are typically modeled with a logistic function based on fish length as

$$v_a = \frac{1}{\left[1 + e^{-\frac{(\text{TL} - L_{\text{ref}})}{sigma}}\right]}, \qquad (2)$$

where TL is the mean total length (mm) of age a fish, L_{ref} is the reference length, set at the minimum length limit, and the parameter *sigma* represents a measure of the variance in length at a particular age and influences the rate of ascent of the curve from zero to one. Values of *sigma* are often set at some percentage of L_{ref} (e.g., 10%) to simulate a recruitment of fish to fishing over 1–3 ages. Figure 1 shows an example of how this equation can vary vulnerability with fish length and thus age.

Equation (1) implies that values of lx_2 and lx_3 represent the theoretical proportion of a single recruit ($lx_a = 1$) that would be alive at ages 2 and 3, based on survival from natural mortality and fishing. For example, if S_o were 0.85, U were 0.5, and v were 0.3, survivors to age 2, starting with 100 age-1 recruits, would be 100(0.85)(1 – 0.3*0.5) = 72.3 fish. The survivorship function in equation (1) makes predicting the number of fish at any age (N_a) as simply R recruits*lx_a, where R could be any arbitrary number of recruits (e.g., 1,000, 10,000, etc.).

Equation (1) and the others used in this chapter assume that fishing and natural mortalities occur separately within a series of discrete time intervals (years). There are yield-per-recruit model formulations that account for simultaneous fishing and natural mortality rates (Quinn and Deriso 1999). However, in most fisheries applications, including striped bass fisheries, these methods are not required. For example, Hightower et al. (2001) found relatively constant natural mortality but seasonally varying fishing mortality for striped bass in Lake Gaston, North Carolina. Further, differences between the mortality expressions used here versus those for continuous fishing and natural mortalities are minor and not likely to influence management recommendations unless fishing and natural mortality are very high (Walters and Martell 2004).

We can extend equation (1) to predict total biomass and yield. Vulnerable biomass per recruit is found as

$$\varphi_{VB} = w_a v_a + lx_{a+1} w_{a+1} v_{a+1} + lx_{a+2} w_{a+2} v_{a+2} + \ldots$$

where w_a is the average fish weight at age a. Calculations such as φ_{VB} are called Botsford incidence functions, which in this case indicates the vulnerable biomass per recruit across all ages. Yield per recruit is then the fraction of the vulnerable biomass per recruit that is harvested: $Y = \varphi_{VB} U$. Yield per recruit can then be evaluated for a range of size limits (different vulnerability schedules) and exploitation rates (U).

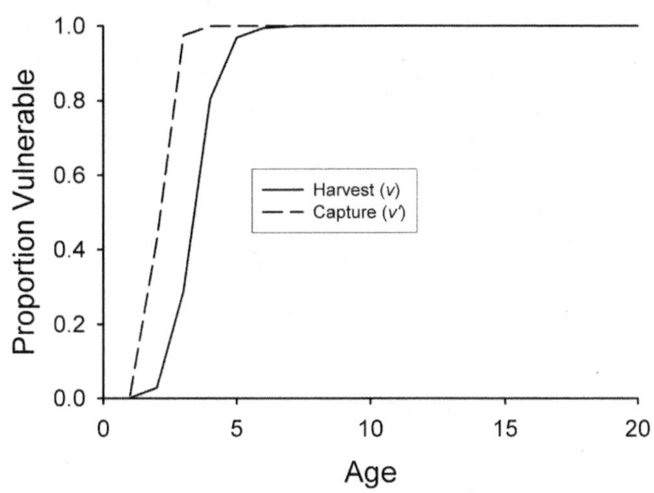

FIGURE 1. Harvest (v) and capture (v') vulnerability (y-axis) for ages of striped bass (x-axis) used in the dynamic model. The harvest vulnerability represents a 450-mm minimum length limit. Vulnerabilities range from 0 (not vulnerable to fishing) to 1 (fully vulnerable).

The Botsford-incidence-function approach provides an equilibrium solution to the population-age structure and predicted yield for any given value of parameters (*M*, *U*, *K*, etc.). These equilibrium estimates are found instantly, rather than having the population model run through many years to stabilize following any change in *M* or *U*. This saves much computer time and makes the model easy to construct in a spreadsheet. Functions other than yield, such as number of fish caught, can be estimated as $C = \varphi_N U$, where the vulnerable abundance per recruit is

$$\varphi_N = v_a + lx_{a+1}v_{a+1} + lx_{a+2}v_{a+2} + \ldots$$

Thus, the Botsford calculations can be adapted to a range of model output metrics that may be of interest to the investigator.

Example.—We constructed a YPR model as described above based on parameters from a striped bass population in the Santee–Cooper Reservoir system, South Carolina (Bulak et al. 1995; Table 2). The yield isopleths (Figure 2) show that maximum yield per recruit is expected at a minimum length limit of about 550 mm TL and a corresponding *U* of about 0.25. At *U* values higher than 0.25 and minimum length limits below 550 mm, yield declines owing to growth overfishing. At *U* values less than 0.25, the yield declines because fish are not harvested at a high enough rate to maximize yield. Notice that very large minimum length limits also reduce yield, owing to biomass dying from natural mortality before it can be harvested. This example demonstrates the utility of YPR models for evaluating length limits.

Dynamic age-structured models

Time-dynamic models allow parameters such as recruitment or mortality to vary across years, whereas the YPR models described above represent equilibrium conditions (i.e., constant model parameters across years). Time-dynamic models with all parameters held constant give identical results to an equilibrium model. Thus, dynamic models can have the same format and expressions as YPR models, but they allow parameters (e.g., mortality rates, growth) to vary through time. Time-dynamic models are typically used in two ways: (1) as a simulation, to evaluate "what if" scenarios of how the fish population could respond to changes in management (e.g., stocking rates or length limits); and (2) as a formal stock assessment model that can be used to estimate parameters (e.g., the fishing mortality rate) based on data collected from the fishery. In this paper, we focus primarily on the simulation aspects of population models; extensive overviews of stock assessment methods are already available (Hilborn and Walters 1992; Quinn and Deriso 1999). Simulation modeling is useful

TABLE 2. Parameters for the striped bass population from the Santee–Cooper system, South Carolina used for yield per recruit and dynamic model examples. Data were obtained from Bulak et al. (1995).

Parameter	Value
L_∞	1,245 mm total length (TL)
K	0.133
t_0	−1.04
M	0.20/year (M = 1.5K, Jensen 1996)
S_0	0.82/year
U	Ranged from 0 to 0.7
v_a	Equation 2, L_{ref} ranged from 200 to 800 mm TL
a	3.63×10^{-5}
b	2.83

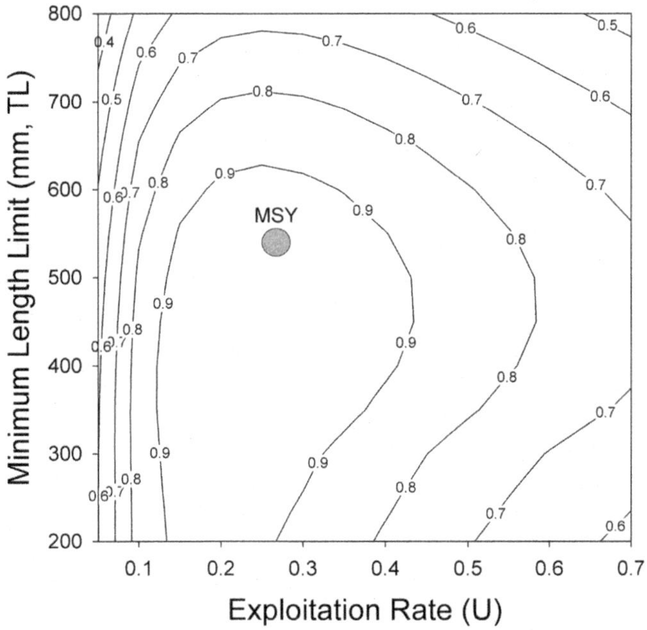

FIGURE 2. Yield isopleths for minimum length limit (*y*-axis; TL = total length) and exploitation rate (*x*-axis) in a hypothetical example based on data for striped bass in the Santee–Cooper system (Table 2). Yield values are scaled to a maximum value of 1.0.

for exploring questions such as (1) how does variable recruitment influence angler harvest through time?, (2) what would be the recovery time for a striped bass population following severe natural mortality events due to summer temperature and oxygen limitations?, or (3) how would a change in the minimum length limit influence angler harvest in the short term (1–5 years)? Here, we explore how a time-dynamic model simulation can be used to address questions of this type.

Data requirements.—Time-dynamic models use the same parameters as described for the YPR models but usually incorporate additional data sources to address time-specific questions. For simulation purposes, it is not required to have annual estimates of all model parameters, but information about how variable parameters are in time (e.g., variation in recruitment) will help inform the simulations. A wide range of data from the fish population can be used directly in simulation models. For example, to identify the effects of variable stocking numbers on angler catch, a time series of age-1 fish catch per effort from sampling programs could be used to represent age-1 recruits in the model such that the modeled recruits mimic the relative abundance of age-1 fish. High natural mortality in a given year due to temperature-oxygen issues can be included, particularly if the magnitude of the mortality is measured (Hightower and Pollock 2013).

Formulation and construction.—Time-dynamic models can use the Botsford formulation described above for YPR models, but the model is set up to follow cohorts diagonally down a table with years as rows and ages as columns. For example, the model begins with age-1 fish in the year 2000, and this cohort is followed diagonally down the table through its life. This is done for all cohorts in a table containing multiple year-classes followed for some number of years (e.g., 50 or 100). Output metrics such as yield are calculated for each year, exactly as described in the YPR model section. Setting up the model this way allows the user to see each year of the simulation and make predictions of how model outputs (e.g., harvest) are expected to vary through time.

Recruitment in the model can be handled in several ways depending on the application. Recruitment to age 1 can be set to mimic some field-estimated measure of the relative abundance of age-1 fish through time, as described above. This would be useful for cases where a long time series of recruitment abundance is available and the goal is to reconstruct the history of the fishery. Alternately, recruitment can be set to vary randomly around some arbitrary (e.g., 1,000) average value using a random number generator. This allows inspection of how the model output (e.g., angler catch) is expected to vary owing to variable recruitment, particularly if the coefficient of variation around age-1 fish abundance is known or approximated from sampling data.

Discard mortality of fish caught and released is an important consideration in striped bass fisheries because catch-and-release mortality rates can be high (Bettoli and Osborne 1998; Wilde et al. 2000), especially in summer, and can have substantial impacts on striped bass abundance and size structure (Coggins et al. 2007). Thus, catch-and-release mortality should be considered in many modeling exercises, both for fish below the minimum length limit and those that are legal to harvest but voluntarily released. Equation (1) can be modified to include discard mortality as

$$lx_a = lx_{a-1}S_o(1 - Uv_{a-1})[1 - (U_o v'_{a-1})d], \quad (3)$$

where U_o is the fraction of the fish population vulnerable to angling that are caught, U is the harvest rate (proportion of vulnerable fish harvested), v' is the vulnerability to being captured by angling, and d is the discard mortality of fish caught and released. Other terms are defined above. Values of v' should be chosen based on angler catch data to reflect the sizes of fish that are caught by anglers, which is typically at smaller sizes than those where fish recruit to harvest, depending on the minimum length limit in place (Figure 1). As an example for equation (3), if 30% of the age-4 fish are caught by anglers but only 20% of the fish in this age-class are actually harvested, that leaves 10% that were released and subjected to discard mortality (d). The use of U_o in equation (3) allows the model to account for fish that are legal to harvest but voluntarily released by anglers. In cases where voluntary release of fish legal to harvest is nil, U_o can be set equal to U. The last term $(U_o v'_{a-1} - Uv_{a-1})d$ models deaths due to discard mortality for fish caught below the minimum length limit and those that are legal to harvest but are voluntarily released by anglers. Note that under conditions of no minimum length limit ($v' = v$) and no voluntary release ($U_o = U$), this term goes to zero resulting in equation (1).

Example.—We extended the spreadsheet YPR model above to be a time-dynamic model based on parameters in Table 2. For values of v and v', we assumed that fish would recruit to capture at L_{ref} of 300 mm TL and recruit to harvest based on a 450-mm-TL minimum size limit (L_{ref} = 450) using equation (2), with *sigma* values set at 10% of the L_{ref} values. This resulted in fish recruiting to capture by anglers between ages 2 and 3 and recruiting to harvest via the minimum length limit between ages 3 and 5 (Figure 1). We assumed a U of 0.2 and that $U_o = U$ (i.e., no voluntary release of fish legal to harvest).

For this example, we simulated a case of high natural mortality (M) in 1 year, as would occur via a large fish kill during one summer. The objective was to evaluate the effect of this event on angler harvest and explore the recovery time for the fishery. We ran the model from 1990 to 2030 and implemented a 10-fold increase in M (M = 2.0, S_o = 0.13) after the fishing season in 1999 to simulate the fish kill. We assumed an arbitrary average value of 1,000 age-1 fish entering the population each year and a 40% coefficient of variation around this average across years. This would simulate a stocking program with variable stocking numbers (or survival) across years. The variation in recruits was simulated using a lognormal distribution from the PopTools add-in for Excel (Hood 2009).

The time-dynamic model predicted that the fish kill would immediately reduce harvest from an average of about 300 fish down to about 60 (Figure 3). Five model runs showed that recovery times ranged from 6 to 9 years (Figure 3), assuming no other changes in mortality or average recruitment. A Monte Carlo analysis of 1,000 model iterations using PopTools (Hood 2009) similarly showed that recovery times would average 6 years. Fishery managers in this situation might consider increasing stocking rates to speed up recovery of the fishery. We increased the average age-1 fish abundance by 50% (i.e., to 1,500) for 4 years to simulate higher stocking rates in response to the fish kill (Figure 3). This reduced average recovery times slightly to 5 years. Thus, our example showed how a time dynamic model can be used to address issues that occur for striped bass fisheries.

Catch-at-age models

Most marine fish stock assessments in the United States rely on some type of catch-at-age model. The purpose of a catch-at-age model

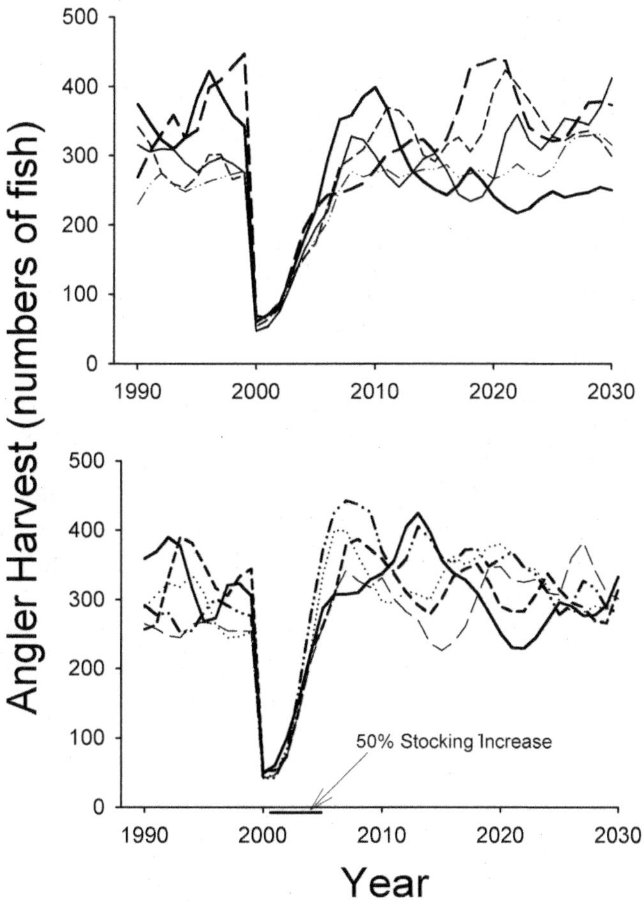

FIGURE 3. Dynamic model simulations of angler harvest (*y*-axis) by year (*x*-axis). The lines represent five model runs, each with a new set of random recruitment values. Natural mortality (*M*) was increased 10-fold in 1999 to simulate a fish kill.

is to estimate fishing mortality, recruitment to the adult population, and the status of the population. The approach reconstructs cohort abundances based on age composition and total catch information. This method has much potential for use in inland striped bass and hybrid striped bass fisheries but has some challenges as well. Here, we present a brief overview of a virtual population analysis (VPA), a common type of catch-at-age method, and discuss how it could be used to inform management in inland striped bass or hybrid striped bass fisheries. More detail on this method can be found in Hilborn and Walters (1992) and Walters and Martell (2004).

Data requirements.—The data requirements for a VPA include estimates of total annual harvest (numbers of fish; in reservoirs, typically from creel surveys), age composition of the annual harvest, and an approximate value for natural mortality (M), all of which are usually treated as known. These data types allow for modeling the yearly abundances of cohorts for which harvest data exist for their entire lifespan. However, for incomplete cohorts that have not reached their maximum age, additional information is necessary to model their annual abundances. For these cases, estimates of the terminal year fishing mortality rates, or an additional index of annual total population abundance such as catch per unit effort, are required to "tune" the model and inform the abundance of incomplete cohorts relative to complete cohorts. We do not cover these types of model tuning procedures, thus limiting our discussion to cohorts with complete information (Hilborn and Walters 1992).

Formulation and construction.—A VPA reconstructs each cohort based on expected natural survival ($S_o = e^{-M}$) and the estimates of removals (i.e., catches) from that cohort throughout its life. The number of fish for each age and time is found by

$$N_{a,t} = \frac{N_{a+1,t+1}}{S_o} + C_{a,t}, \tag{4}$$

where $C_{a,t}$ is the catch of age a fish in year t. As an example, equation (4) says that the cohort abundance at the start of age 5 is the population abundance at age 6, adjusted upwards for a year of natural mortality (divided by S_o) plus the catch of age-5 fish during that year. The abundance estimate for each cohort begins at the oldest age and works backwards through time. Annual exploitation for each cohort and year can be found as $U = C_{a,t}/N_{a,t}$. With multiple years of catch data for a range of cohorts, it is possible to reconstruct cohort abundances to age 1 for each year ($N_{1,t}$). Thus, the VPA method can provide estimates of recruitment, which could be compared to striped bass stocking records to assess stocking efficacy for striped bass fisheries. Equation (4) assumes that fishing and natural mortality occur separately within the year (similar to the models described above); these model formulations become more complicated if fishing and natural mortality are modeled concurrently (Hilborn and Walters 1992).

Example.—We illustrate this method with a simple VPA for a single 1989 cohort that completed its life in a hypothetical fishery (Table 3). By assuming the same M (S_o = 0.82) as the previous models, the VPA predicted that the catches from this cohort were produced by about an initial 12,400 age-1 fish in 1990. Estimates of U for the last 4–5 ages are not reliable due to low catches, and they should be ignored (Hilborn and Walters 1992). Fish recruited to the fishery at ages 2–3, and annual exploitation rates were 0.5–0.7 for fish ages 3–6. By following the catches of this cohort throughout its life, we were able to obtain estimates of recruitment to age 1, exploitation, and information about the size and age selectivity of the fishery.

Use of VPA analyses requires estimates of the total number of fish harvested and age composition. Schlechte et al. (2013, this volume) described creel survey data for reservoir striped bass fisheries in the southern United

TABLE 3. Example of a virtual population analysis (VPA) for a hypothetical striped bass cohort. The $C_{a,t}$ values represent catch of fish at each age and year. The natural mortality rate M was 0.2. Values of $N_{a,t}$ are the VPA predicted population size for this cohort at each age. The $U_{a,t}$ values are annual exploitation for each age and year, but they are not reliable for the last 4–5 years of the cohort's life due to low catches.

Year	Age	$C_{a,t}$	$N_{a,t}$	$U_{a,t}$
1990	1	5	12,446	<0.001
1991	2	2,400	10,202	0.23
1992	3	3,450	6,397	0.54
1993	4	1,521	2,417	0.63
1994	5	502	735	0.68
1995	6	121	191	0.63
1996	7	21	57	0.37
1997	8	6	30	0.20
1998	9	4	19	0.21
1999	10	8	13	0.63
2000	11	2	4	NA
2001	12	0	1	NA
2002	13	0	1	NA
2003	14	1	1	NA
2004	15	0	0	NA

States, and many states conducted creel surveys and obtained estimates of total harvest. However, we found no data sets where age composition of the harvest estimates was obtained for consecutive years, which would be required for VPA analysis. Doing so may be difficult except for the most popular striped bass fisheries, as striped bass often represent a relatively low fraction of the total angler interviews in creels (Schlechte et al. 2013). Thus, for some fisheries, it could require new creel surveys to target striped bass anglers, as recommended by Schlechte et al. (2013). Age composition of the catch can be obtained with fishery independent methods as long as a sample of fish lengths is obtained from the fishery. Despite these challenges, we believe that a VPA approach has much promise for assessment of fishing mortality in inland systems. Further, results of a VPA could be used to assess stocking efficacy (i.e., estimates of age-1 fish abundance from VPA compared to stocking records) for striped bass or hybrid striped bass fisheries.

Incorporating Uncertainty and Sensitivity Analyses

Population models are a simplified representation of the biological processes driving population dynamics, and as such, uncertainty is inherent and expected. Two important types of uncertainty are structural and parameter uncertainty. Structural uncertainty is defined as the uncertainty associated with the processes that drive the model outputs. Examples of structural uncertainty could include (1) uncertainty in whether growth is best represented by a linear growth model or a nonlinear growth model, (2) uncertainty in the possible sources of recruitment (e.g., immigration), or (3) uncertainty in whether natural mortality (M) is best represented as fixed or should vary across ages. All of these cases represent uncertainty because the model may not consider an important factor influencing the population (e.g., time dynamics, shape of relationships). To account for this type of uncertainty, it is good practice to consider multiple model structures as different hypotheses about the processes that govern the population. Reviewing the literature can inform the best choice of model structures, while evaluating the output from different model structures can reveal the trade-offs associated with using one model versus another. Considering alternate model structures also reveals data gaps and can guide future research.

Parameter uncertainty originates because many input parameters such as mortality parameters (e.g., M, U, d), growth parameters (e.g., K, L_∞, t_0) or weight-length parameters (e.g., a, b) are estimated from field data, which contain noise often associated with error in the observation process. For example, natural mortality (M) and associated uncertainty (e.g., confidence intervals) has been estimated for southeastern reservoir striped bass fisheries (Hightower et al. 2001; Thompson et al. 2007). The uncertainty in the applied estimate of M could be accounted for in the models presented here using Monte Carlo simulations. For example, the user could specify an average M and standard deviation, then use deviates from a specified distribution (e.g., normal distribution) to account for the possible influence of uncertainty in M on the model output (e.g., yield). The model would then produce a distribution of possible yield outcomes that consider uncertainty in M. Further, this can be done for multiple parameters simultaneously (e.g., L_∞, M) to account for all known sources of parameter uncertainty.

Estimating the uncertainty in model outputs due to parameter uncertainty is not always possible, however. Sometimes estimates of the input parameters and associated uncertainty are unavailable. For these circumstances, sensitivity analysis can be a powerful tool to evaluate the influence of these parameters on the model outputs. A simple sensitivity analysis is performed by applying a percent change in a given input parameter while holding all other input parameters constant and evaluating the change in the model output. This is repeated for each input parameter, and the percent change in output metrics is recorded. These results reveal the relative influence of each input parameter on the output of the model. This is an effective way to identify key parameters to estimate in field surveys (high sensitivity and uncertainty = high priority for field work). Caution must be practiced, however, when parameters evaluated are correlated, such as the growth parameters L_∞ and K and the length-weight parameters a and b. For example, when fitting von Bertalanffy growth models to length-at-age data, underestimating the K parameter is often a corollary to overestimating the L_∞ parameter. More complex sensitivity analyses can be adopted to include correlation among parameters (e.g., Bartell et al. 1986).

We constructed a sensitivity analysis for our yield-per-recruit model. We increased and decreased values of L_∞, K, t_0, M, a, b, and d by 10% and evaluated the percent change in yield per recruit. This showed that effects of some parameters were not linear on the model outputs. Increasing L_∞ caused a 27% increase in yield, whereas decreasing this parameter caused yield to decline by only 23%. This occurs because of nonlinear effects of some parameters on model outputs. Further, the sensitivity analysis revealed that parameters such as the von Bertalanffy growth parameter t_0 and discard mortality rate d had very little influence on the results of the yield-per-recruit model (Table 4). Alternately, parameters such as the von Bertalanffy growth parameter L_∞ and the weight-length parameter b had large influences over the results of the yield-per-recruit analysis (Table 4). This information could guide future field work, but the investigator should consider what is known about each parameter. For example, we may have good information about the length-weight parameters a and b from other studies but very poor information about the M or the growth parameters K and L_∞ parameter. Thus, estimating M or the von Bertalanffy growth parameters could become a priority for future field investigations.

Management Implications

Our goal was to demonstrate a simple framework for building population models and show

TABLE 4. Sensitivity analysis of the yield per recruit model. A 10% increase and decrease in each input parameter L_∞, K, t_0, M, a, b, and d was invoked to investigate the percent change in the yield per recruit (YPR) of striped bass.

Parameter	10% parameter increase percent change in YPR	10% parameter decrease percent change in YPR
L_∞	+27	−23
K	+14	−13
t_0	+2	−2
M	−12	+11
a	+12	−10
b	+650	−85
d	+1	−1

how they could be used to address questions of interest when managing inland striped bass or hybrid striped bass fisheries. The effects of fishing mortality, stocking, and size limits on yield and angler catch are important factors to consider when managing harvest-oriented fisheries such as striped bass. Models can be easily adapted to assess unanticipated issues, such as how major natural mortality events will influence fisheries. Stocking can be included in the model to evaluate its efficacy to improve angler catches and shorten recovery times. Catch-at-age methods have promise to improve the value of creel survey data and serve as a useful tool for monitoring fishing mortality and stocking efficacy.

However, regardless of the predictions and insights from the model, the model building process itself is valuable. Constructing a model causes investigators to compile their data into one framework and specify their "hypotheses" about the fishery. This process nearly always illuminates key limitations of the data, reveals missing information on parameters, and identifies other uncertainties that warrant field work for investigation. This iterative feedback between having the data inform the model and the model process identify key data needs for fieldwork can allow information to accrue in a manner that reduces overall uncertainty in a meaningful and systematic way. All models used in this paper are available at http://fishweb.ifas.ufl.edu/allenlab/courses.html, and we encourage readers to try these methods as they evaluate striped bass fisheries.

Acknowledgments

We thank Jim Bulak, Jim Estes, Joe Hightower, Jim Rice, and Warren Schlechte for conversations that improved this paper. This work was supported by The University of Florida.

References

ASMFC (Atlantic States Marine Fisheries Commission). 2003. Amendment 6 to the Interstate Fishery Management Plan for Striped Bass. Atlantic States Marine Fisheries Commission, Fisheries Management Report #41, Washington, D.C.

Bartell, S. M., R. H. Breck, R. H. Gardner, and A. L. Brenkert. 1986. Individual parameter perturbation and error analysis of fish bioenergetics models. Canadian Journal of Fisheries and Aquatic Sciences 43:160–168.

Bettoli, P. W., and R. S. Osborne. 1998. Hooking mortality and behavior of striped bass following catch and release angling. North American Journal of Fisheries Management 18:609–615.

Beverton, R. J. H., and S. J. Holt. 1957. On the dynamics of exploited fish populations. Ministry of Agriculture and Fisheries, Fisheries Investigation Series 2, Report Number 19, Lowestoft, UK.

Botsford, L. W., and D. E. Wickham. 1979. Population cycles caused by inter-age, density-dependent mortality in young fish and crustaceans. Pages 73–82 *in* E. Naybr and R. Hartnoll, editors. Cyclic phenomena in marine plants and animals. Proceedings of the 13th European marine biology symposium. Permagon, New York.

Brown, M. L., and B. R. Murphy. 1991. Standard weight (W_s) development for striped bass, white bass, and hybrid striped bass. North American Journal of Fisheries Management 11:451–467.

Bulak, J. S., D. S. Wethey, and M. G. White, III. 1995. Evaluation of management options for a reproducing striped bass population in the Santee–Cooper system, South Carolina. North American Journal of Fisheries Management 15:84–94.

Coggins, L. C., Jr., M. J. Catalano, M. S. Allen, W. E. Pine, III, and C. J. Walters. 2007. Effects of cryptic mortality and the hidden costs of length limits in fishery management. Fish and Fisheries 8:196–210.

Hightower, J. E., and K. H. Pollock. 2013. Tagging methods for estimating population size and mortality rates of inland striped bass populations. Pages 249–262 *in* J. S. Bulak, C. C. Coutant, and J. A. Rice, editors. Biology and management of inland striped bass and hybrid striped bass. American Fisheries Society, Symposium 80, Bethesda, Maryland.

Hightower, J. E., J. R. Jackson, and K H. Pollock. 2001. Use of telemetry methods to estimate natural and fishing mortality of striped bass in Lake Gaston, North Carolina. Transactions of the American Fisheries Society 130:557–567.

Hilborn, R., and C. J. Walters. 1992. Quantitative fisheries stock assessment: choice, dynamics and uncertainty. Chapman and Hall, New York.

Hood, G. 2009. PopTools: software for ecological models. Version 3.1. Available: www.cse.csiro.au/poptools. (February 2012)

Jensen, A. L. 1996. Beverton and Holt life history invariants result from optimal trade-off of reproduction and survival. Canadian Journal of Fisheries and Aquatic Sciences 53:820–822.

Pauly, D. 1980. On the interrelationship between natural mortality, growth parameters, and mean environmental temperature in 175 fish stocks. Journal du Conseil International pour l'Exploration de la Mer 39:175–192.

Quinn, T. J., II, and R. B. Deriso. 1999. Quantitative fish dynamics. Oxford, New York.

Richards, R. A., and P. J. Rago. 1999. A case history of effective fishery management: Chesapeake Bay striped bass. North American Journal of Fisheries Management 19:356–375.

Rose, K. A., and J. H. Cowan, Jr. 1993. Individual-based model of young-of-year striped bass population dynamics. I. Model description and baseline simulations. Transactions of the American Fisheries Society 122:415–438.

Schlechte, J. W., N. G. Smith, and J. B. Taylor. 2013. Options for estimating striped bass catch and harvest: effectiveness of creel surveys. Pages 291–312 *in* J. S. Bulak, C. C. Coutant, and J. A. Rice, editors. Biology and management of inland striped bass and hybrid striped bass. American Fisheries Society, Symposium 80, Bethesda, Maryland.

Shepherd, G. 2006. Status of fishery resources off the northeastern US: Atlantic striped bass. National Oceanic and Atmospheric Association, Resource Evaluation and Assessment Division, Washington D.C.

Slipke J. W., and M. J. Maceina. 2005. Fisheries analysis and simulation tools (FAST), user's guide. Auburn University, Auburn, Alabama.

Thompson, J. S., D. S. Waters, J. A. Rice, and J. E. Hightower. 2007. Seasonal natural and fishing mortality of striped bass in a southeastern reservoir. North American Journal of Fisheries Management 27:681–694.

Walters, C. J., and S. J. D. Martell. 2004. Fisheries ecology and management. Princeton, New Jersey.

Wilde, G., Muoneke, M. I., P. W. Bettoli, K. L. Nelson, and B. T. Hysmith. 2000. Bait and temperature effects on striped bass hooking mortality in freshwater. North American Journal of Fisheries Management 20:810–815.

American Fisheries Society Symposium 80:365–382, 2013

Biotic and Abiotic Determinants of Stocking Success for Striped Bass in Inland Waters

TRENT M. SUTTON*
University of Alaska Fairbanks, School of Fisheries and Ocean Sciences
905 Koyukuk Drive, Fairbanks, Alaska 99775, USA

DANIEL M. WILSON
Virginia Department of Game and Inland Fisheries
1132 Thomas Jefferson Road, Forest, Virginia 24551, USA

JOHN J. NEY
Virginia Polytechnic Institute and State University
Department of Fisheries and Wildlife Sciences
Blacksburg, Virginia 24061, USA

Abstract.—The stocking of fingerling striped bass *Morone saxatilis* in freshwater impoundments has led to the development of successful put-grow-take fisheries throughout the southern United States. However, first-year survival of stocked fingerlings is often low. To enhance stocking success of striped bass, a better understanding is needed on the impacts of different stocking strategies on early life-history dynamics. In this review paper, we first examined the existing literature on the role of abiotic and biotic factors on recruitment dynamics of stocked piscivores in inland freshwater systems. Second, we compiled the results of a progressive series of studies that were completed over a 25-year period in Smith Mountain Lake, Virginia, that focused on understanding the relationship between striped bass stocking success and biotic interactions, forage-fish prey availability and dynamics, and first-year recruitment. This case study demonstrated that differential intra-cohort growth and poor first-year winter survival are the primary factors limiting stocking success and that stocking fingerlings at a greater number of sites throughout the lake at lower densities improved recruitment to age 1. With this information, we provide stocking size, time, density, and location strategy recommendations that should yield increased survival and stocking success of striped bass in freshwater impoundments.

Introduction

The discovery in the mid-1950s that the striped bass *Morone saxatilis* could successfully complete its life cycle in freshwater (Scruggs 1957) resulted in the development of a nationally recognized sport fishery in Santee–Cooper Reservoir, South Carolina (Stevens 1958). The popularity of this fishery led other state fisheries management agencies to examine the utility of this anadromous sport fish for stocking in the large, multipurpose impoundments that had been created throughout the southern United States (Axon and Whitehurst 1985; Farquhar and Gutreuter 1989; Schorr et al. 1995; Van Horn 2013, this volume). Because most of these systems contained populations of gizzard shad

* Corresponding author: tmsutton@alaska.edu

Dorosoma cepedianum and threadfin shad *D. petenense*, ostensibly underutilized by indigenous, shore-oriented predators such as largemouth bass *Micropterus salmoides*, fisheries managers were hopeful that striped bass would not only control overabundant populations of these forage fishes, but also fill the vacant ecological niche (i.e., the pelagic zone) created during the formation of these impoundments (Ney and Orth 1986).

Although introductions of newly hatched striped bass fry were initiated by a number of state agencies throughout the southern United States, several years of evaluation demonstrated that stocking this life stage was ineffective. As a result, artificial propagation and rearing techniques were developed that permitted rearing fry to fingerling size prior to stocking (Stevens 1967; Bailey 1975). Because striped bass typically require long stretches of free-flowing river for successful reproduction (Crance 1984; but see Gustaveson et al. 2013, this volume), a characteristic lacking from most impoundments, self-sustaining populations have rarely been established (Mensinger 1971; Schaffler and Winkelman 2008; Baker et al. 2009). Consequently, most landlocked striped bass fisheries have been maintained on a put-grow-take basis through annual stockings of fingerlings. These stockings have largely been a trial-and-error process with regard to stocking densities, stocking locations, and the size and time at stocking. As a result, large numbers of small (<50 mm total length [TL]) fingerlings are often stocked at one or two locations to offset low (<20%) early life stage survival rates often observed for this species (Bailey 1975; Moore et al. 1991).

Although striped bass fisheries have been established in many reservoirs throughout the southern United States, a paucity of information exists on appropriate stocking strategies. However, the influence of stocking density, size at stocking, and time of year at stocking on first-year growth and survival has been well documented for other piscivores, such as walleye *Sander vitreus* (Fielder 1992; Santucci and Wahl 1993; Olson et al. 2000; Brooks et al. 2002; Hoxmeier et al. 2006; Zweifel et al. 2009), largemouth bass (Olson 1996; Hoffman and Bettoli 2005; Mesing et al. 2008), muskellunge *Esox masquinongy* (McKeown et al. 1999), and tiger muskellunge (muskellunge × northern pike *E. lucius*) (Carline et al. 1986). As a result, there is a critical need to better understand the impacts of stocking strategies on the early life history dynamics of striped bass stocked into freshwater impoundments. Not only should this type of evaluation make stocking efforts more efficient, but management will be less uncertain. The objectives of this paper were to (1) assess the importance and impacts of key abiotic and biotic variables in freshwater impoundments on recruitment of stocked striped bass; (2) illustrate how understanding and alleviating the sources of a recruitment bottleneck led to enhanced stocking success for striped bass in Smith Mountain Lake, Virginia; and (3) identify and discuss strategies for improving stocking success of striped bass in inland reservoirs.

Recruitment Dynamics of Stocked Juvenile Piscivores

Management of inland fish populations can be improved by gaining knowledge of the factors controlling growth, survival, and recruitment of early life stages. Identification of the mechanisms affecting early growth and survival of fishes should not only allow improved prediction of future recruitment, but may also permit fisheries managers to take appropriate remedial actions that will increase recruitment, thereby dampening population fluctuations or reversing long-term population declines. For fisheries biologists managing populations of stocked piscivores, this is particularly true because a considerable degree of control exists in managing populations through the regulation of size, time, location, and density at stocking.

Variability in recruitment has been identified as a central problem in fisheries research (Sissenwine et al. 1988). Although it has long been understood that year-class strength of most fish populations is set early in life, knowledge of the mechanisms affecting recruitment of age-0 fishes is generally poor. While fish recruitment can be affected by many biotic and abiotic factors, size-dependent mechanisms such as growth, cannibalism, predation, and starvation-induced winter mortality have been hypothesized to explain survival and recruitment variability of age-0 fish populations (Gutreuter and Anderson 1985; Miller et al. 1988; Cushing 1995). Because recruitment is affected by such a large suite of factors, their impacts will vary on a system-by-system and temporal basis for many species.

For juvenile piscivores in lakes and reservoirs, an important factor limiting first-year growth is the availability of forage-fish prey (Noble 1981). The problem of forage-fish availability is immediate and acute for fingerlings that become piscivorous at or soon after stocking but are limited to small forage fish, usually of the same year-class (Ney and Orth 1986). For these fingerlings, a lack of small, ingestible fish prey can have a tremendous impact on the size distribution of juveniles by the end of the growing season (Shelton et al. 1979; Timmons et al. 1980; Keast and Eadie 1985; Garvey and Stein 1998) because many individuals are relegated to prey types, such as invertebrates, that have a lower caloric content per unit body mass than fish prey (Cummins and Wuycheck 1971). As a result, the development of a bimodal or multimodal length distribution is often observed in these populations (Forney 1976; Shelton et al. 1979; Timmons et al. 1980; Keast and Eadie 1985; Ludsin and DeVries 1997). Thus, size-dependent differences in fish prey availability to juvenile piscivores can significantly influence first-year growth, physiological condition, and size attained by the end of the growing season.

For piscivorous fishes in temperate climates, rapid growth during the first months of life is critical for survival. Age-0 fishes that grow slowly are vulnerable to predation for longer periods of time, and as a result, recruitment is typically poorer than for populations of faster-growing individuals (Ricker and Foerster 1948). This premise has been observed for walleye, in which cannibalism by older fish on slow-growing juveniles was one of the most important factors governing recruitment (Chevalier 1973; Forney 1976). Size-dependent predation has also been shown to have a significant influence on recruitment variability in other freshwater fishes (Nielsen 1980; Garvey et al. 1998; Buckmeier et al. 2005; Hoxmeier et al. 2006).

Size-dependent predation may be particularly acute in systems that rely on stocked fingerlings, not only because these fish may be "naive," lacking the necessary predator avoidance behaviors that wild fish exhibit (Stein et al. 1981; Santucci and Wahl 1993), but because they may remain concentrated at high densities in nearshore regions, further exacerbating the problem by attracting predators. For example, Carline et al. (1986) found that tiger muskellunge stocked in Ohio ponds and impoundments suffered losses as high as 100% within 50 d poststocking from predation by largemouth bass. The authors hypothesized that mortality events were a result of two factors: (1) tiger muskellunge <250 mm TL were within the gape limits of most adult largemouth bass; and (2) thermal stratification, coupled with an anoxic hypolimnion, forced largemouth bass and tiger muskellunge to concentrate in nearshore areas. Comparable results on poststocking predation by nearshore piscivores (primarily largemouth bass) have also been reported for other stocked esocids (Stein et al. 1981; Wahl and Stein 1989; Szendrey and Wahl 1996), centrarchids (Buckmeier et al. 2005), and percids (Forney 1976; Johnson et al. 1988; Santucci and Wahl 1993; Hoxmeier et al. 2006). As a result, size-dependent predation on juvenile fishes, particularly smaller, slower-growing individuals, can be a significant factor regulating first-year recruitment.

Size-dependent, first-year survival may also result from differential starvation. Body size of age-0 fishes at the end of the first growing season is a critical factor influencing winter survival (Oliver et al. 1979; Shuter et al. 1980; Adams and DeAngelis 1987; Garvey et al. 1998; Hurst and Conover 1998; Schultz and Conover

1999). Smaller individuals within a cohort have higher maintenance requirements per unit body mass than larger fish. During winter at temperate latitudes, smaller individuals may exhaust stored energy earlier to meet basal metabolic requirements (Shuter et al. 1980). As a consequence, smaller individuals are more vulnerable to winter mortality due to starvation than larger fish of the same cohort (Oliver et al. 1979; Garvey et al. 1998; Hurst and Conover 1998; Schultz and Conover 1999).

A Case Study: Striped Bass Stocking in a Virginia Reservoir

Smith Mountain Lake, an 8,343-ha hydroelectric impoundment in south-central Virginia, is in many respects typical of other freshwater systems in which striped bass populations have been supported on a put-grow-take basis. The reservoir has suitable habitat for all age-classes of striped bass but does not contain adequate spawning habitat for natural reproduction. As a result, annual stockings are required to maintain the fishery. Since 1964, stocking striped bass (between 180,000 and 800,000 fingerlings annually) has led to the development and maintenance of a successful sport fishery. Initial stockings of striped bass throughout the 1970s established an exceptional trophy fishery, with growth and harvest rates among the highest reported for freshwater systems throughout the United States (Hart 1978). Despite the success of this fishery, fingerling stocking density and survival to age 1 have varied inversely from 1973 to 1984; based on marginal-yield analysis, it was determined that an annual stocking density between 250,000 and 300,000 fingerlings would optimize the number of year-class recruits in this system (Moore et al. 1991). In response, the number of striped bass stocked annually in Smith Mountain Lake by the Virginia Department of Game and Inland Fisheries (VDGIF) fluctuated between 200,000 and 400,000 fingerlings through 1995. Although stocking practices were changed in 1996, the inverse relationship between stocking density and first-year survival continues to the present.

Abiotic factors have been shown to limit recruitment of juvenile piscivores in freshwater ecosystems (Maceina and Bettoli 1998; Slipke et al. 1998; Sammons et al. 1999; Hoxmeier et al. 2006; Ward et al. 2007) and juvenile striped bass in estuarine environments (Kimmerer et al. 2001; Martino 2008). In particular, pumpback-storage reservoirs, such as Smith Mountain Lake, have the potential to limit recruitment through fluctuations in water levels, which may also alter temperature and dissolved oxygen regimes. Cultural oligotrophication of reservoirs can also impact recruitment success through declines in primary production and fish biomass following reductions in nutrient availability (Ney 1996). In Smith Mountain Lake, daily and seasonal variability in reservoir hydrology due to pumpback operation has had minimal impact on water temperature and dissolved oxygen regimes (Ney et al. 1990). Although mean annual total phosphorus concentrations and total fish standing stock declined 79% and 76%, respectively, from 1974 to 1984 before stabilizing following completion of an advanced wastewater treatment facility, there was only a weak relationship between total phosphorus concentration and piscivore standing crop (Yurk and Ney 1989; Ney et al. 1990). However, oligitrophication has doubled the available summer habitat for striped bass in Smith Mountain Lake, which may have mitigated any negative impacts associated with nutrient abatement (Ney 1996). Based on these results, abiotic factors may not limit first-year survival of stocked juvenile striped bass.

In Smith Mountain Lake, the recruitment bottleneck appears to be trophic in nature, with size- and density-dependent biotic interactions, such as predation, cannibalism, and competition, more likely contributing to poor year-class strength. In Smith Mountain Lake, predation and cannibalism on juvenile striped bass by piscivorous sport fishes was minimal. Striped bass and walleye are virtual trophic equivalents, with alewives *Alosa pseudoharengus* comprising 84% and 97%, respectively, of their diets by weight; in contrast, largemouth bass and smallmouth bass *M. dolomieu* maintained more diverse diets, feeding primarily on gizzard shad, crayfish, and other prey, such as sunfishes (Figure 1; Moore 1988). However, juvenile striped bass were not found in the stomach contents of any of these piscivores. Michaelson et al. (2001) found that predation on juvenile striped bass by largemouth bass immediately poststocking until dispersal from nearshore areas was negligible (1994: 0.1%; 1995: 1.2%). Moore (1988) and Sutton (1997) examined the stomach contents of 708 adult striped bass (1983–1984) and 324 adult and 2,299 juvenile striped bass (1994–1996), respectively, and found no evidence of cannibalism. In addition, diet overlap between juvenile striped bass and largemouth bass in nearshore areas was insufficient to suggest trophic competition (Sutton and Ney 2002). As a result, biotic interactions for trophic resources involving juvenile striped bass do not appear to adversely impact recruitment success in Smith Mountain Lake.

Alewives and gizzard shad are the principal forage-fish species utilized by piscivorous sport fishes in Smith Mountain Lake, and these two species are compatible and complementary prey sources in this system (Tisa and Ney 1991). Gizzard shad spawn earlier in the growing season and their larvae grow faster than alewives, providing a high degree of temporal isolation that limits trophic competition between these species. These fishes also exhibit strong spatial segregation; alewives are located primarily in the lower lake and are pelagic while gizzard shad are predominantly located in the upper lake in littoral areas. Although the temporal and spatial segregation of alewives and gizzard shad has expanded feeding opportunities for sport fishes in Smith Mountain Lake, the available prey supply for age-0 juvenile piscivores may limit recruitment (Ney 1990). Based on growth-projection analysis, Ney and Orth (1986) and Tisa (1988) determined that striped bass could not utilize alewives and giz-

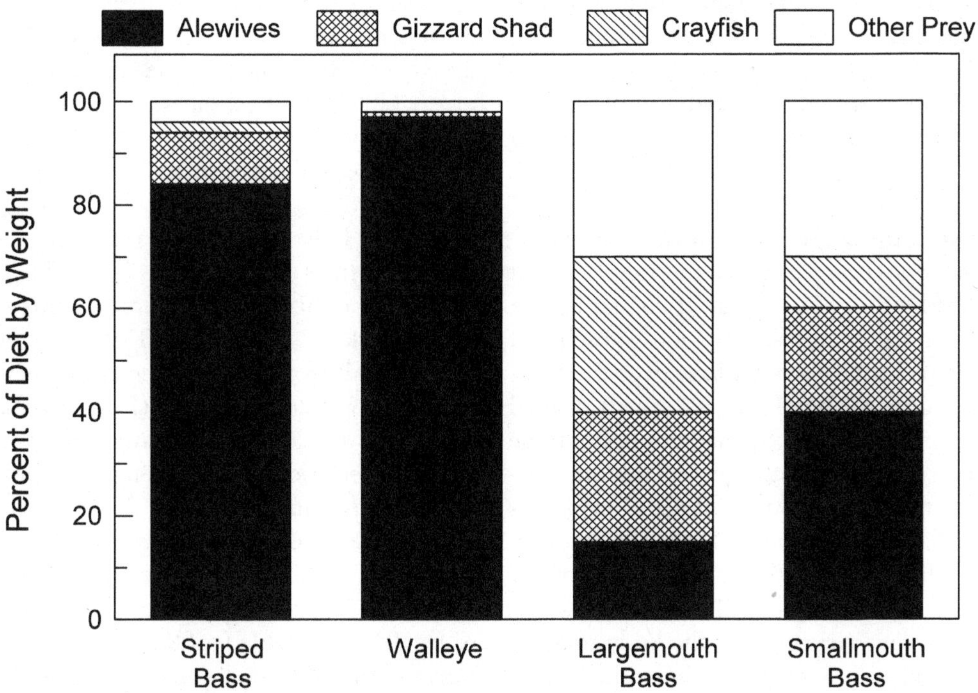

FIGURE 1. Percent diet composition by weight of alewife, gizzard shad, crayfish, and other prey for adult striped bass, walleye, largemouth bass, and smallmouth bass in Smith Mountain Lake from April through September 1983 and 1984 (from Moore 1988 and Ney et al. 1990).

zard shad in the same cohort until years 2 and 4, respectively. In a follow-up evaluation, Cyterski (1999) determined that 94% of the available clupeid supply in Smith Mountain Lake was utilized to meet predator demand. For striped bass, prey supply exceeded predator demand by 30% for gizzard shad but only by 4% for alewives. As a result, the forage-fish prey base for striped bass in Smith Mountain Lake during their first year poststocking appears to be insufficient and may contribute directly to the limited stocking success.

Limitations in the forage-fish prey base in Smith Mountain Lake appear to be directly linked to first-year growth of striped bass. Sutton and Ney (2001) found that the length range of fingerling striped bass was narrow at stocking in 1994 and 1995 with a unimodal distribution (Figure 2). The length distribution diverged over the growing season, and by fall, a bimodal distribution was evident both years. The size distribution continued to diverge and, by the end of the growing season, large-mode striped bass (median = 225 mm, range = 182–266 mm) were significantly larger than small-mode fish (median = 94 mm, range = 80–140). Only a single size mode, which consisted almost entirely of fish greater than 150 mm, was present during spring sampling in 1995 and 1996 (Figure 3; Sutton and Ney 2001).

Given the observed growth disparities, what factors contributed to the development of bimodality within the age-0 cohort by the end of the growing season and subsequent loss of small mode striped bass over the winter in Smith Mountain Lake? Small striped bass maintained a mixed diet throughout the growing season, consuming zooplankton, benthic aquatic insects, and small age-0 cyprinids (Figure 4; Sutton and Ney 2001). In contrast, larger individuals (>140 mm TL) were strictly piscivorous, eating primarily age-0 alewives. For piscivorous striped bass, the body mass of alewives consumed by larger individuals was, on average, three times greater than the body mass of cyprinids consumed by smaller striped bass (Sutton and Ney 2001). Because striped bass that ate larger, more energetically profitable alewife prey grew at an accelerated rate, these fish reached a larger size and accumulated greater lipid stores by winter than individuals that maintained a mixed diet (Figure 5; Sutton and Ney 2001). The lower nutritional quality diet maintained by striped bass in the smaller mode limited their energy reserves and likely reduced their growth, thereby facilitating the formation of bimodality. Consequently, the significantly lower lipid stores and visible signs of emaciation observed for the few small-mode survivors collected during spring sampling periods suggests that the loss of these smaller fish overwinter most likely resulted from starvation due to depleted energy reserves. Although other factors, such as predation, osmotic stress, and compromised immune system response, may have contributed to size-selective winter mortality of stocked, juvenile striped bass in Smith Mountain Lake, differential patterns of growth and forage-fish prey availability have synergistically acted to limit first-year recruitment.

Using the aforementioned information, Sutton et al. (2000) developed an individual-based model to evaluate the influence of size at stocking, day of year at stocking, and stocking density on first-year growth, survival, and the number of age-1 recruits. Model results indicated that stocking striped bass in Smith Mountain Lake at a median total length of 52 mm on 8 June (rather than 42 mm TL on 15 June, as was historically done) and maintaining a stocking density at 300,000 fingerlings would result in the largest percent increases in the three output parameters (Figure 6). Cyterski (1999) developed a simulation model to examine the relationship between prey supply and predator demand in Smith Mountain Lake. This model predicted that an annual stocking rate of 300,000 fingerlings had the highest probability of producing the largest numbers of legally harvestable (>508 mm in TL) striped bass and that an annual stocking rate of 150,000 fish produced the greatest number of citation (>940 mm in TL or 9.07 kg in mass) striped bass. However, increasing stockings to 450,000

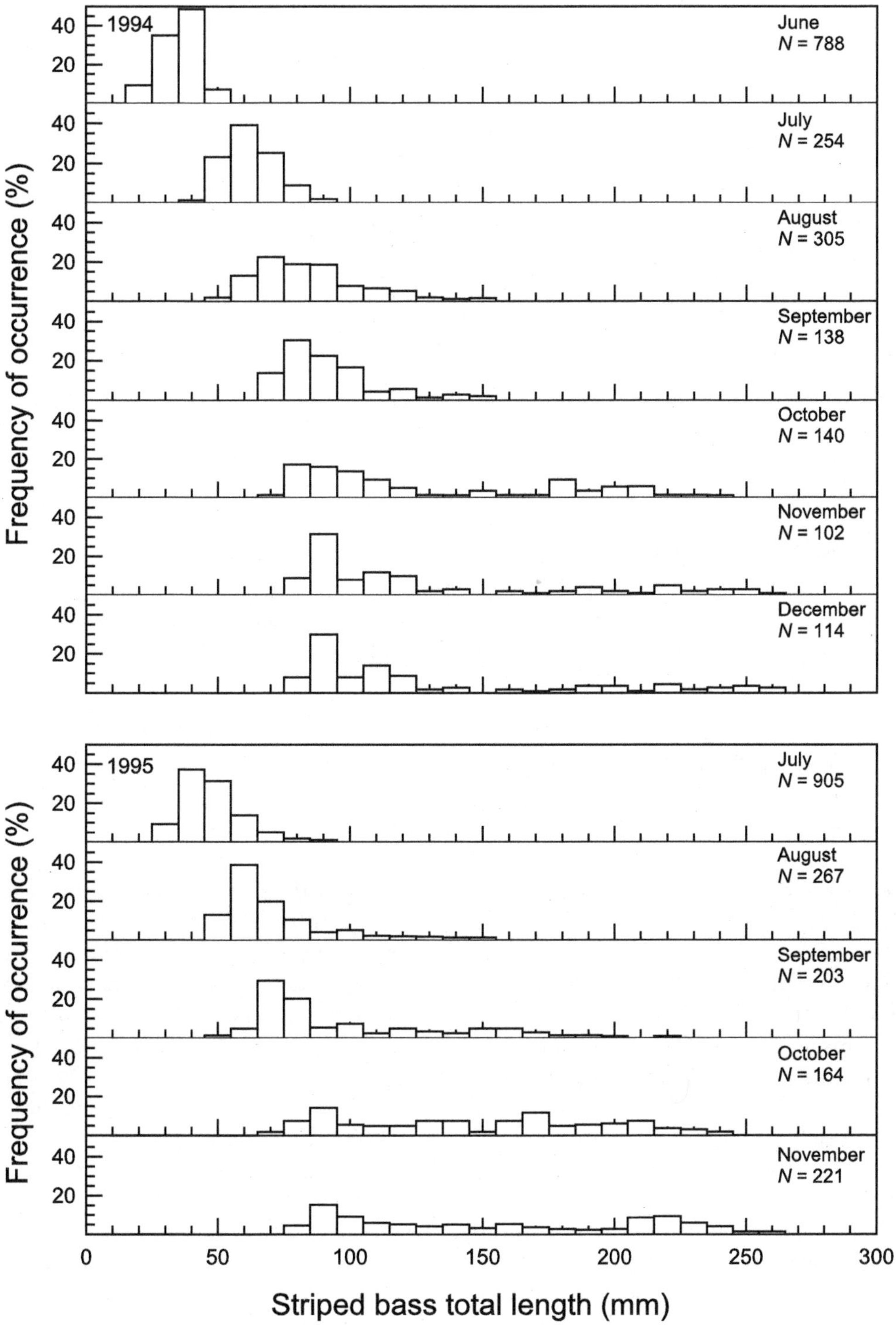

FIGURE 2. Monthly length-frequency distributions of juvenile striped bass in Smith Mountain Lake over the 1994 (top panel series) and 1995 (bottom panel series) growing seasons (from Sutton and Ney 2001).

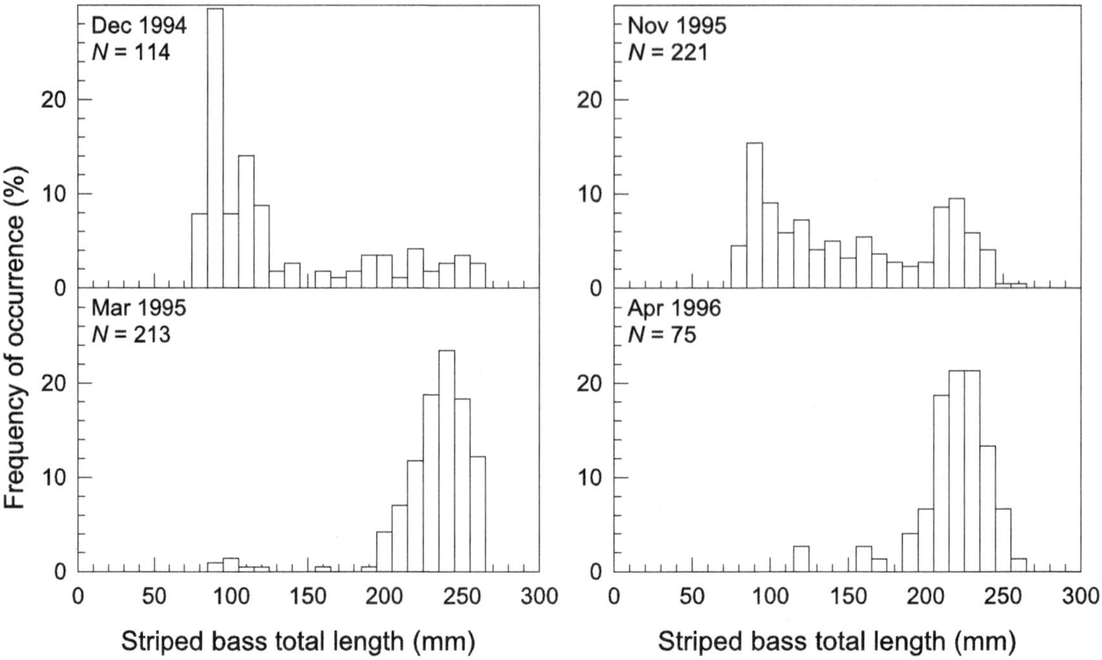

FIGURE 3. Comparison of juvenile striped bass length-frequency distributions before and after the winters of 1994–1995 (left panels; December 1994 versus March 1995) and 1995–1996 (right panels; November 1995 versus April 1996; from Sutton and Ney 2001).

fingerlings annually would lead to declines in the number of legal and citation striped bass. As a result, these modeling efforts suggested maintaining a stocking density of 300,000 fingerlings and stocking fish 10 mm larger and 1 week earlier in the growing season to maximize recruitment.

Despite this recommendation to the VDGIF, anglers felt that increasing the number of stocked striped bass would negate increased angler pressure by sustaining catch rates and increasing the number of large fish they were accustomed to catching. In Smith Mountain Lake, striped bass anglers accounted for 37% of the total angler pressure during a 1998 creel survey (Virginia Department of Game and Inland Fisheries, unpublished data), and according to previous creel surveys, angling pressure for striped bass had steadily increased since the 1980s. The popularity of this fishery is an important economic component of the local communities, with estimates of trip expenditures in 1998 by Smith Mountain Lake anglers totaling \$3.1 million (Wilson 2004). The concerns from striped bass anglers about the declining fishery spread to the local business community, and both groups began to pressure the VDGIF to stock additional striped bass to compensate for the perceived decline in the fishery. Consequently, the VDGIF worked in concert with the Smith Mountain Lake Striper Club (SMSC) to develop a striped bass management plan to identify specific goals and objectives to address angler concerns. A major component of the plan was to increase survival of stocked fish and to increase stocking rates from 300,000 to 450,000 fish annually.

To achieve recruitment success at a higher fingerling striped bass stocking density in Smith Mountain Lake, a decision was made by the VDGIF to stock fish at a lower density at each stocking site at multiple locations throughout the reservoir. As a result, the VDGIF and the SMSC collaborated to identify and gain access to numerous private access areas to distribute stocked fish more evenly throughout the reser-

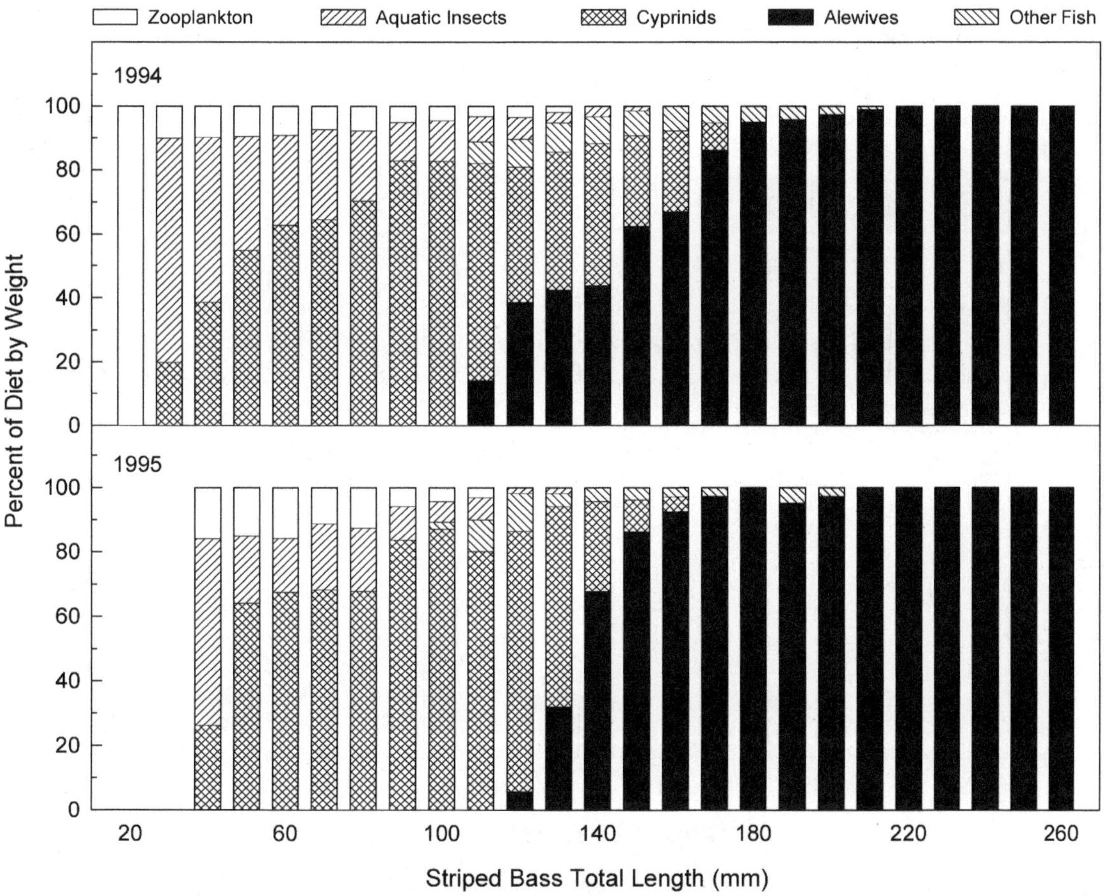

FIGURE 4. Length-dependent patterns in age-0 striped bass food habits during the 1994 (top panel) and 1995 (bottom panel) growing seasons in Smith Mountain Lake (from Sutton and Ney 2001).

voir. Prior to 1996, juvenile striped bass were always stocked at two sites at a density ranging from 125,000 to 175,000 fingerlings per site (Table 1). During 1996 and 1997, fingerlings were stocked at four sites, with the density per site ranging from 17,000 to 78,000 fish. Since 1997, age-0 striped bass (mean TL range = 12.7–34.3 mm) have been stocked at anywhere from 9 to 16 sites in the lake at a density of 27,000–45,000 fish per site. The reduction in striped bass stocking density at each site, despite maintaining a larger total stocking density, allowed for greater first-year survival and recruitment to age 1 than when fish were stocked at higher densities at fewer locations (Figure 7). Wilson (2004) attributed the observed increase in stocking success to lower striped bass density at each stocking site. In addition, higher prey availability led to rapid first-year growth and larger fall body size; this outcome was achieved by selecting stocking locations with high nutrient levels and sufficient forage-fish prey density. As a result of this change in stocking strategy, high first-year survival of stocked fingerlings was achieved, which allowed for the accommodation of a greater total annual stocking density and an improvement in the striped bass fishery in Smith Mountain Lake.

Stocking Recommendations for Inland Striped Bass

There are many factors that ultimately determine the success or failure of a stocking event,

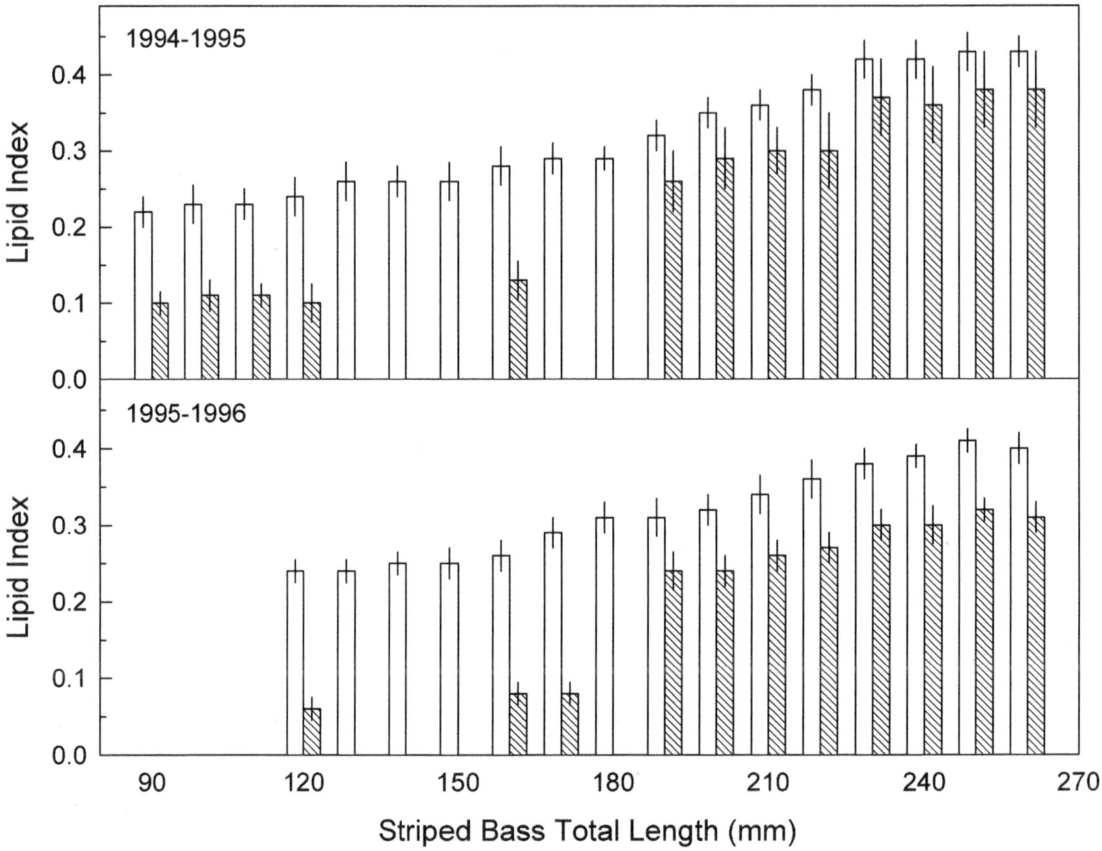

FIGURE 5. Comparison of juvenile striped bass lipid index by 10-mm length intervals before (open bars) and after (shaded bars) wintering in Smith Mountain Lake during 1994–1995 (top panel) and 1995–1996 (bottom panel). Values for each 10-mm length class are the medians and error bars represent the quartile ranges (from Sutton and Ney 2001).

many of which are related to the specific characteristics of individual water bodies. Understanding the relationship among first-year growth, forage-fish prey availability, energy stores, and winter survival lends particular insights into those factors that limit recruitment. Because improving survival of stocked fish is the primary focus of any put-grow-take fishery, identifying those stocking strategies that maximize year-class strength of stocked fishes is critically important to enhance stocking program efficiency. Potential options to increase stocking success include the following strategies, which may be used singly or in concert with each other and must be considered on a system-by-system basis.

Size at stocking

Numerous studies have been conducted to determine which size of juvenile fish at the time of stocking is most effective for maintaining put-grow-take fisheries (Carline et al. 1986; Santucci and Wahl 1993; Olson et al. 2000; Hoffman and Bettoli 2005). A fundamental assumption of most stocking programs is that fish stocked at larger sizes will exhibit greater survival than smaller fish. While it is generally accepted that survival of fish stocked as fry or small fingerlings is low, stocking larger fingerlings does not always lead to greater recruitment success (Miller et al. 1988; Ludsin and DeVries 1997; Olson et al. 2000). For stocked juvenile

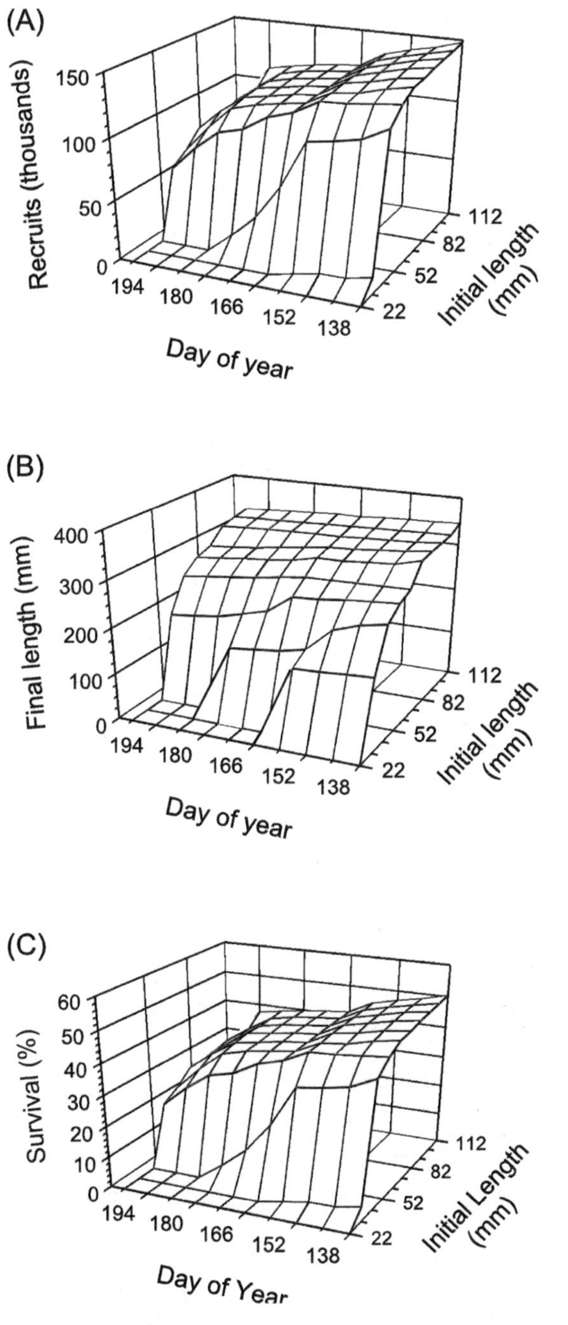

FIGURE 6. Simulated response surfaces for (A) number of age-1 recruits, (B) median total length, and (C) first-year survival of juvenile striped bass on 1 May to changes in median total length of fingerlings (initial length) and day of year at stocking (from Sutton et al. 2000).

piscivores, perhaps the most important determinant of stocking success is the relationship between piscivore (predator) and forage-fish body size (Shelton et al. 1979; Timmons et al. 1980; Ludsin and DeVries 1997; Garvey and Stein 1998). Although there have been no evaluations to date that have directly examined the relationship between size at stocking and recruitment success for striped bass in inland reservoirs, it has been documented that an abundance of small, age-0 clupeid prey is a prerequisite for rapid first-year growth and subsequent recruitment to age 1 (Van Den Avyle and Higginbotham 1980; Ney and Orth 1986; Sutton and Ney 2001). However, size at stocking may be difficult to regulate when rearing fish extensively in grow-out ponds, and hatchery production constraints may limit the ability to stock a particular size of fish. Because of the strong interplay between body size and time of year at stocking, the relationship between these factors must be considered in any stocking program.

Timing of stocking

To facilitate rapid growth of stocked age-0 piscivores, fingerling stockings should occur before age-0 forage fishes have exceeded their gape constraints. Survival of stocked fingerlings can be further enhanced if stocking is timed to coincide with the peak of forage-fish density, which typically occurs 2 to 3 weeks postspawning (Tisa and Ney 1991; Stahl et al. 1996; Zweifel et al. 2009). Timing the stocking of age-0 striped bass to occur when clupeids are in nearshore areas to spawn may also minimize the potential for predation on newly stocked fish. For example, Michaelson et al. (2001) found that the presence of spawning alewives in Smith Mountain Lake most likely buffered predation on striped bass. In order to determine the most appropriate time to stock juvenile striped bass into reservoirs, information on the timing and duration of clupeid spawning periods and temperature relationships must be understood by fisheries managers. For example, alewife spawning in Smith Mountain Lake ranges

TABLE 1. Number of stocked fingerling striped bass, number of stocking sites, gill-net catch per unit effort (CPUE) of age-1 striped bass, survival index of age-1 striped bass (CPUE adjusted for stocking rate), and the mean number of fingerling striped bass stocked per site in Smith Mountain Lake, Virginia, 1990–2007. All age-1 striped bass were collected during fall gill-net surveys the year following the stocking event.

Stocking year	Number of stocked fingerling striped bass	Number of stocking sites	Gill-net CPUE	Survival index	Stocking density per site
1990	300,131	2	1.56	0.52	150,066
1991	301,346	2	1.59	0.53	150,673
1992	282,281	2	1.72	0.61	141,141
1993	287,054	2	1.39	0.48	143,527
1994	339,337	2	1.29	0.38	169,669
1995	236,734	2	2.36	1.00	118,367
1996	66,857	4	0.69	1.03	16,714
1997	311,300	4	4.86	1.56	77,825
1998	453,063	11	6.22	1.37	41,188
1999	450,769	10	6.61	1.47	45,077
2000	459,996	14	10.06	2.32	32,857
2001	461,023	15	7.17	1.56	30,735
2002	500,815	16	5.53	1.20	31,301
2003	252,550	9	10.14	4.02	28,061
2004	451,374	14	8.33	1.85	32,241
2005	354,455	11	1.89	0.53	32,223
2006	349,963	13	12.92	3.69	26,920
2007	410,423	11	8.61	2.10	37,311

from mid-May through early to late August at 18–29°C while gizzard shad spawn from early May to early July at 17–28°C (Tisa and Ney 1991). Based on these data and annual variability in larval clupeid availability and surface temperatures, Sutton et al. (2000) determined that striped bass should be stocked 1 week earlier than the historical average date of stocking (8 June rather than 15 June). Although this stocking date recommendation was found to be robust to annual variations in alewife population dynamics and water temperature regimes, it should be noted that it is a site-specific stocking strategy for Smith Mountain Lake. As a result, it is imperative that fishery biologists communicate with hatchery managers regarding the timing of clupeid spawning events and subsequent larval fish availability in order to identify the most appropriate time at stocking for their specific system.

Stocking density

Stocking density is an important factor that can be directly linked to stocking program success. High stocking density can be detrimental to stocked fingerlings by depressing first-year growth and subsequent fall body size through intraspecific competition for limited prey resources (Laarman and Schneider 1986; Fielder 1992). Further, predation may be high because stocked fish frequently remain concentrated at high densities in nearshore areas for a period of time poststocking and generally lack the necessary predator avoidance behaviors exhibited by wild fish (Stein et al. 1981; Santucci and Wahl 1993; Hoxmeier et al. 2006). However, stocking

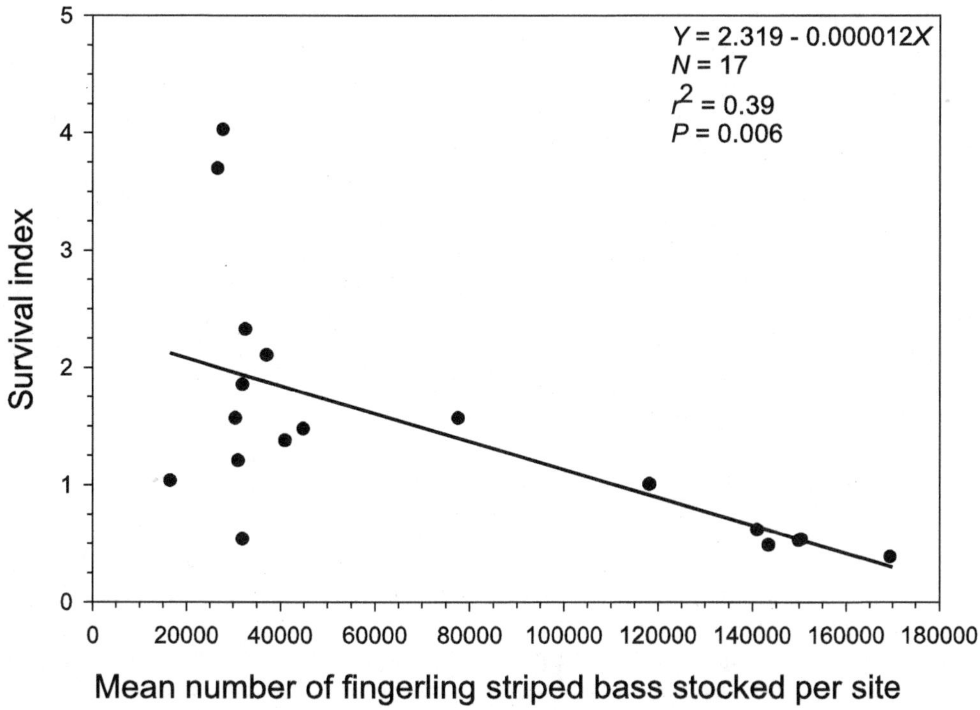

FIGURE 7. Relationship between the mean number of stocked fingerling striped bass stocked per site and survival index in Smith Mountain Lake, Virginia, 1990–2007. The solid line through the data points is the best-fit linear regression for that relationship.

densities that are too low may not allow for the development of the optimal capacity of a body of water, thereby resulting in a fishery with unrealized potential (Jacobson and Anderson 2007). Few studies have compared stocking densities of age-0 piscivores, and to date, no studies have evaluated the effect of different stocking densities on recruitment success of striped bass in impounded systems. However, more broadly distributing fingerlings throughout a system reduces the potential for overstocking fish at release sites and lessens the impact of highly localized stocking densities. Consequently, fisheries managers must couple decisions on stocking density with the choice of appropriate stocking locations to maximize the potential for rapid first-year growth and survival to age 1.

Stocking locations

If the goal is to stock juvenile striped bass at lower densities at each site while maintaining a high overall stocking rate, then fish must be stocked at multiple release locations throughout the lake or reservoir. Not only will stocking fish at multiple locations minimize the potential for overstocking each release site, but it will also facilitate fish dispersion throughout the lake. While the number of stocking locations chosen will vary depending on total stocking density, target stocking density at each release location, forage-fish prey composition and dynamics, and system-specific characteristics, in all cases stocking locations should be chosen in areas that have high productivity (i.e., high nutrient concentration and primary productivity) and support an abundant forage-fish prey base (Wilson 2004). In reservoirs, these areas typically occur in upstream areas as system fertility tends to decrease in a downstream direction (Yurk and Ney 1989; Ney et al. 1990; Ney 1996). However, fisheries managers must accurately identify these locations prior to the stocking of striped bass during early summer. In addition, fisheries managers must also be able to predict confidently the abundance and productivity of these age-0 forage-fish cohorts and be willing to adjust stocking locations and/or densities to ensure that predator demand does not exceed the available prey supply (Ney and Orth 1986). If these determinations are not possible or if there is the potential for a numerical response of nearshore predators to these high productivity areas in the upper reservoir where juvenile striped bass are stocked, then an alternative approach would be to adjust the overall stocking density based on expected system productivity and broadly distribute fingerlings at lower densities at each location throughout the entire reservoir.

Logistical constraints associated with stocking fish must also be considered when determining stocking locations. For example, multiple small loads from the hatchery or offloading fish for individual stocking sites from a central location at the lake to be transported and stocked elsewhere can lead to high handling stress and subsequent mortality during transport or post-stocking. As a consequence, potential gains in faster first-year growth and higher survival that result from stocking lower densities of fish at each location may be offset by losses associated with handling mortality, thereby leading to no net gain in stocking success.

Conclusions

Developing a better understanding of the ecological processes that regulate fish population structure and dynamics is a critical determinant for identifying and selecting the most appropriate stocking strategies for maintaining successful put-grow-take fisheries. Juvenile piscivores, particularly those stocked as small fingerlings, often encounter deficiencies in fish prey during their first year of life due to inherent differences in life-history characteristics. Whether prey is available to predators has morphological, distributional, and behavioral components, each of which can reduce the usable supply from the potential quantity generated by prey production. If the availability of ingestible-sized forage-fish prey to juvenile piscivores is inadequate, then growth of stocked fingerlings during the first summer of life will not be rapid and these fish will not grow large enough and will not accumulate sufficient lipid reserves to survive over the winter. Because small differences in body size between stocked juvenile piscivores and their forage-fish base can lead to large differences in prey availability, choosing a strategy that will allow striped bass to maintain a higher degree of piscivory, grow at a fast rate, and reach a large body size by winter should be the highest priority of any stocking program.

Enhancing first-year piscivory and growth of striped bass in reservoirs should improve winter survival and recruitment to age 1. The Smith Mountain Lake case study provides an example that demonstrates how stocking fingerlings at a size and time of the year that allowed them to maintain a size advantage over age-0 clupeid prey, coupled with a new emphasis on stocking fish at an increased number of locations at a lower density per site, enhanced first-year growth, survival, and recruitment. Because the specifics associated with the aforementioned stocking strategies are unique to Smith Mountain Lake, they might not yield the same results in a different system. As a result, understanding the vagaries associated with first-year dynamics of stocked striped bass within the context of system-specific attributes, in particular forage-fish prey dynamics, is a necessary prerequisite to identifying those stocking strategies that will most likely lead to recruitment success. While this may require some trial-and-error stocking evaluations until the most appropriate strategy or combination of strategies has been identified for a particular system, the ultimate outcome is the production of a striped bass fishery that fully realizes its potential and provides the necessary benefits to associated stakeholder groups.

References

Adams, S. M., and D. L. DeAngelis. 1987. Indirect effects of early bass-shad interactions on predator population structure and food web dynamics. Pages 103–117 *in* C. W. Kerfoot and A. H. Sih, editors. Predation: direct and indirect effects on aquatic communities. University of New England Press, Hanover, New Hampshire.

Axon, J. R., and D. K. Whitehurst. 1985. Striped bass management in lakes with emphasis on management problems. Transactions of the American Fisheries Society 114:8–11.

Bailey, W. M. 1975. An evaluation of striped bass introductions in the southeastern United States. Proceedings of the Annual Conference Southeastern Association of Game and Fish Commissioners 28(1974):54–68.

Baker, W. P., J. Boxrucker, and K. E. Kuklinski. 2009. Determination of striped bass spawning locations in the two major tributaries of Lake Texoma. North American Journal of Fisheries Management 29:1006–1014.

Brooks, R. C., R. C. Heidinger, R. J. H. Hoxmeier, and D. H. Wahl. 2002. Relative survival of three sizes of walleyes stocked into Illinois lakes. North American Journal of Fisheries Management 22:995–1006.

Buckmeier, D. L., R. K. Betsill, and J. W. Schlechte. 2005. Initial predation of stocked fingerling largemouth bass in a Texas reservoir and implications for improving stocking efficiency. North American Journal of Fisheries Management 25:652–659.

Carline, R. F., R. A. Stein, and L. M. Riley. 1986. Effects of size at stocking, season, largemouth bass predation, and forage abundance on survival of tiger muskellunge. Pages 151–167 *in* G. E. Hall, editor. Managing muskies: a treatise on the biology and propagation of muskellunge in North America. American Fisheries Society, Special Publication 15, Bethesda, Maryland.

Chevalier, J. R. 1973. Cannibalism as a factor in first year survival of walleye in Oneida Lake. Transactions of the American Fisheries Society 102:739–744.

Crance, J. H. 1984. Habitat suitability index models and instream flow suitability curves: inland stocks of striped bass. U.S. Fish and Wildlife Service, FWS/OBS-82/10.85, Washington, D.C.

Cummins, K. W., and J. C. Wuycheck. 1971. Calorific equivalents for investigations in ecological energetics. Mitteilungen Internationale Vereinigung fuer Theoreische und Angewandte Limnologie 18:1–158.

Cushing, D. 1995. Population production and regulation in the sea: a fisheries perspective. Cambridge University Press, Cambridge, UK.

Cyterski, M. J. 1999. Analysis of the trophic support capacity of Smith Mountain Lake, Virginia, for piscivorous fish. Doctoral dissertation. Virginia Polytechnic Institute and State University, Blacksburg.

Farquhar, B. W., and S. Gutreuter. 1989. Distribution and migration of adult striped bass in Lake Whitney, Texas. Transactions of the American Fisheries Society 118:523–532.

Fielder, D. G. 1992. Relationship between walleye fingerling stocking density and recruitment in lower Lake Oahe, South Dakota. North American Journal of Fisheries Management 12:346–352.

Forney, J. L. 1976. Year-class formation in the walleye (*Stizostedion vitreum*) population of Oneida Lake, New York, 1966–73. Journal of the Fisheries Research Board of Canada 33:783–792.

Garvey, J. E., and R. A. Stein. 1998. Linking bluegill and gizzard shad prey assemblages to growth of age-0 largemouth bass in reservoirs. Transactions of the American Fisheries Society 127:70–83.

Garvey, J. E., R. A. Wright, and R. A. Stein. 1998. Overwinter growth and survival of age-0 largemouth bass (*Micropterus salmoides*): revisiting the role of body size. Canadian Journal of Fisheries and Aquatic Sciences 55:2414–2424.

Gustaveson, W., and G. Blommer. 2013. History of striped bass management in the Colorado River. Pages 15–23 *in* J. S. Bulak, C. C. Coutant, and J. A. Rice, editors. Biology and management of inland striped bass and hybrid striped bass. American Fisheries Society, Symposium 80, Bethesda, Maryland.

Gutreuter, S. J., and R. O. Anderson. 1985. Importance of body size to the recruitment process in largemouth bass. Transactions of the American Fisheries Society 114:317–327.

Hart, L. G. 1978. Smith Mountain Lake research study. Virginia Commission of Game and Inland Fisheries, Virginia Dingell-Johnson Project F-30-R, Richmond.

Hoffman, K. P., and P. W. Bettoli. 2005. Growth,

dispersal, mortality, and contribution of largemouth bass stocked into Chickamauga Lake, Tennessee. North American Journal of Fisheries Management 25:1518–1527.

Hoxmeier, R. J. H., D. H. Wahl, R. C. Brooks, and R. C. Heidinger. 2006. Growth and survival of age-0 walleye (*Sander vitreus*): interactions among walleye size, prey availability, predation, and abiotic factors. Canadian Journal of Fisheries and Aquatic Sciences 63:2173–2182.

Hurst, T. P., and D. O. Conover. 1998. Winter mortality of young-of-the-year Hudson River striped bass (*Morone saxatilis*): size-dependent patterns and effects on recruitment. Canadian Journal of Fisheries and Aquatic Sciences 55:1122–1130.

Jacobson, P. C., and C. S. Anderson. 2007. Optimal stocking densities for walleye fingerlings in Minnesota lakes. North American Journal of Fisheries Management 27:650–658.

Johnson, B. M., D. L. Smith, and R. F. Carline. 1988. Habitat preferences, survival, growth, foods, and harvest of walleyes and walleye × saugeye hybrids. North American Journal of Fisheries Management 8:292–304.

Keast, A., and J. M. Eadie. 1985. Growth depensation in year-0 largemouth bass: the influence of diet. Transactions of the American Fisheries Society 114:204–213.

Kimmerer, W. J., J. H. Cowan, Jr., L. W. Miller, and K. A. Rose. 2001. Analysis of an estuarine striped bass population: effects of environmental conditions during early life. Estuaries 24:557–575.

Laarman, P. W., and J. C. Schneider. 1986. Walleye stocking experiments and fish population studies at Manistee Lake 1972–84. Michigan Department of Natural Resources, Fisheries Research Report 1938, Lansing.

Ludsin, S. A., and D. R. DeVries. 1997. First-year recruitment of largemouth bass: the interdependency of early life stages. Ecological Applications 7:1024–1038.

Maceina, M. J., and P. W. Bettoli. 1998. Variation in largemouth bass recruitment in four mainstream impoundments of the Tennessee River. North American Journal of Fisheries Management 18:998–1003.

Martino, E. J. 2008. Environmental controls and biological constraints on recruitment of striped bass *Morone saxatilis* in Chesapeake Bay. Doctoral dissertation. University of Maryland, College Park.

McKeown, P. E., J. L. Forney, and S. R. Mooradian. 1999. Effects of stocking size and rearing method on muskellunge survival in Chautaqua Lake, New York. North American Journal of Fisheries Management 19:249–257.

Mensinger, G. C. 1971. Observations on the striped bass, *Morone saxatilis*, in Keystone Reservoir, Oklahoma. Proceedings of the Annual Conference Southeastern Association of Game and Fish Commissioners 24(1970):447–463.

Mesing, C. L., R. L. Cailteux, P. A. Strickland, E. A. Long, and M. W. Rogers. 2008. Stocking of advanced-fingerling largemouth bass to supplement year-classes in Lake Talquin, Florida. North American Journal of Fisheries Management 28:1762–1774.

Michaelson, D. P., J. J. Ney, and T. M. Sutton. 2001. Largemouth bass predation on stocked striped bass in Smith Mountain Lake, Virginia. North American Journal of Fisheries Management 21:326–332.

Miller, T. J., L. B. Crowder, J. A. Rice, and E. A. Marschall. 1988. Larval size and recruitment mechanisms in fishes: toward a conceptual framework. Canadian Journal of Fisheries and Aquatic Sciences 45:1657–1670.

Moore, C. M. 1988. Food habits, population dynamics, and bioenergetics of four predatory fish species in Smith Mountain Lake, Virginia. Doctoral dissertation. Virginia Polytechnic Institute and State University, Blacksburg.

Moore, C. M., R. J. Neves, and J. J. Ney. 1991. Survival and abundance of stocked striped bass in Smith Mountain Lake, Virginia. North American Journal of Fisheries Management 11:393–399.

Ney, J. J. 1990. Trophic economics in fisheries: assessment of demand–supply relationships between predators and prey. Reviews in Aquatic Sciences 2:55–81.

Ney, J. J. 1996. Oligotrophication and its discontents: effects of reduced nutrient loading on reservoir fisheries. Pages 285–295 *in* L. E. Miranda and D. R. DeVries, editors. Multidimensional approaches to reservoir fisheries management. American Fisheries Society, Symposium 16, Bethesda, Maryland.

Ney, J. J., C. M. Moore, M. S. Tisa, J. J. Yurk, and R. J. Neves. 1990. Factors affecting the sport fishery in a multiple-use Virginia reservoir. Lake and Reservoir Management 6:21–32.

Ney, J. J., and D. J. Orth. 1986. Coping with future shock: matching predator stocking programs to prey abundance. Pages 81–92 *in* R. H. Stroud, editor. Fish culture in fisheries management. American Fisheries Society, Bethesda, Maryland.

Nielsen, L. A. 1980. Effect of walleye (*Stizostedion vitreum vitreum*) predation on juvenile mortality and recruitment of yellow perch (*Perca flavescens*) in Oneida Lake, New York. Canadian Journal of Fisheries and Aquatic Sciences 37:11–19.

Noble, R. L. 1981. Management of forage fishes in impoundments of the southern United States. Transactions of the American Fisheries Society 110:738–750.

Oliver, J. D., G. F. Holeton, and K. E. Chua. 1979. Overwinter mortality of fingerling smallmouth bass in relation to size, relative energy stores, and environmental temperature. Transactions of the American Fisheries Society 108:130–136.

Olson, M. H. 1996. Ontogenetic niche shifts in largemouth bass: variability and consequence for first year growth. Ecology 77:179–190.

Olson, M. H., T. E. Brooking, D. M. Green, A. J. VanDeValk, and L. G. Rudstam. 2000. Survival and growth of intensively reared large walleye fingerlings and extensively reared small fingerlings stocked concurrently in small lakes. North American Journal of Fisheries Management 20:337–348.

Ricker, W. E., and R. E. Foerster. 1948. Computation of fish production. Bulletin of the Bingham Oceanography Collection, Yale University 11(4):173–211.

Sammons, S. M., L. G. Dorsey, P. W. Bettoli, and F. C. Fiss. 1999. Effects of reservoir hydrology on reproduction by largemouth bass and spotted bass in Normandy Reservoir, Tennessee. North American Journal of Fisheries Management 19:78–88.

Santucci, V. J., and D. H. Wahl. 1993. Factors influencing survival and growth of stocked walleye (*Stizostedion vitreum*) in a centrarchid-dominated impoundment. Canadian Journal of Fisheries and Aquatic Sciences 50:1548–1558.

Schaffler, J. J., and D. L. Winkelman. 2008. Temporal and spatial variability in otolith trace-element signatures of juvenile striped bass from spawning locations in Lake Texoma, Oklahoma–Texas. Transactions of the American Fisheries Society 137:818–829.

Schorr, M. S., J. Sah, D. F. Schreiner, M. F. Meador, and L. G. Hill. 1995. Regional economic impact of the Lake Texoma (Oklahoma–Texas) striped bass fishery. Fisheries 20(5):14–18.

Schultz, E. T., and D. O. Conover. 1999. The allometry of energy reserve depletion: test of a mechanism for size-dependent winter mortality. Oecologia 119:474–483.

Scruggs, G. D. 1957. Reproduction of resident striped bass in Santee–Cooper Reservoir, South Carolina. Transactions of the American Fisheries Society 85:144–159.

Shelton, W. L., W. D. Davies, T. A. King, and T. J. Timmons. 1979. Variation in the growth of the initial year class of largemouth bass in West Point Reservoir, Alabama and Georgia. Transactions of the American Fisheries Society 108:142–149.

Shuter, B. J., J. A. MacLean, F. E. J. Fry, and H. A. Regier. 1980. Stochastic simulation of temperature effects on first-year survival of smallmouth bass. Transactions of the American Fisheries Society 109:1–34.

Sissenwine, M. P., M. J. Fogarty, and W. J. Overholtz. 1988. Some fisheries management implications of recruitment variability. Pages 129–152 *in* J. A. Gulland, editor. Fish population dynamics. Wiley, Chichester, UK.

Slipke, J. W., M. J. Maceina, D. R. DeVries, and F. J. Snow. 1998. Effects of shad density and reservoir hydrology on the abundance and growth of young-of-year crappie in Alabama reservoirs. Journal of Freshwater Ecology 13:87–95.

Stahl, T. P., G. P. Theide, R. A. Stein, E. M. Lewis, M. R. Austin, and D. A. Culver. 1996. Factors affecting survival of age-0 saugeye *Stizostedion vitreum* × *S. canadense* stocked in Ohio reservoirs. North American Journal of Fisheries Management 16:378–387.

Stein, R. A., R. F. Carline, and R. S. Hayward. 1981. Largemouth bass predation on tiger muskellunge. Transactions of the American Fisheries Society 110:604–612.

Stevens, R. E. 1958. The striped bass of the Santee–Cooper Reservoir. Proceedings of the Annual Conference Southeastern Association of Game and Fish Commissioners 11(1957):253–264.

Stevens, R. E. 1967. A final report on the use of hormones to ovulate striped bass. Proceedings of the Annual Conference Southeastern Association of Game and Fish Commissioners 18(1964):525–538.

Sutton, T. M. 1997. Early life history dynamics of a stocked striped bass (*Morone saxatilis*) population and assessment of strategies for improving stocking success in Smith Mountain Lake, Virginia. Doctoral dissertation. Virginia Polytechnic Institute and State University, Blacksburg.

Sutton, T. M., K. A. Rose, and J. J. Ney. 2000. A model analysis for enhancing stocking success of landlocked striped bass populations. North American Journal of Fisheries Management 20:841–859.

Sutton, T. M., and J. J. Ney. 2001. Size-dependent mechanisms influencing first-year growth and winter survival of stocked striped bass in a Virginia mainstream reservoir. Transactions of the American Fisheries Society 130:1–17.

Sutton, T. M., and J. J. Ney. 2002. Influence of trophic competition on first-year survival of stocked striped bass in Smith Mountain Lake, Virginia. North American Journal of Fisheries Management 22:550–559.

Szendrey, T. A., and D. H. Wahl. 1996. Size-specific survival and growth of stocked muskellunge: effects of predation and prey availability. North American Journal of Fisheries Management 16:395–402.

Timmons, T. J., W. L. Shelton, and W. D. Davies. 1980. Differential growth of largemouth bass in West Point Reservoir, Alabama–Georgia. Transactions of the American Fisheries Society 109:176–186.

Tisa, M. S. 1988. Compatibility and complementarity of alewife and gizzard shad in Smith Mountain Lake, Virginia. Doctoral dissertation. Virginia Polytechnic Institute and State University, Blacksburg.

Tisa, M. S., and J. J. Ney. 1991. Compatibility of alewives and gizzard shad as reservoir forage fish. Transactions of the American Fisheries Society 120:157–165.

Van Den Avyle, M. J., and B. J. Higginbotham. 1980. Growth, survival, and distribution of striped bass stocked into Watts Bar Reservoir, Tennessee. Proceedings of the Annual Conference Southeastern Association of Fish and Wildlife Agencies 33(1979):361–370.

Van Horn., S. L. 2013. A brief history of inland striped bass management. Pages 1–13 *in* J. S. Bulak, C. C. Coutant, and J. A. Rice, editors. Biology and management of inland striped bass and hybrid striped bass. American Fisheries Society, Symposium 80, Bethesda, Maryland.

Wahl, D. H., and R. A. Stein. 1989. Comparative vulnerability of three esocids to largemouth bass (*Micropterus salmoides*) predation. Canadian Journal of Fisheries and Aquatic Sciences 46:2095–2103.

Ward, M. T., S. J. Fisher, and D. W. Willis. 2007. Environmental influences on walleye fingerling production in southwestern Minnesota shallow lakes. North American Journal of Fisheries Management 69:297–304.

Wilson, D. M. 2004. Effects of expanded stocking locations on striped bass survival. Proceedings of the Annual Conference Southeastern Association of Fish and Wildlife Agencies 56(2002):79–85.

Yurk, J. J., and J. J. Ney. 1989. Phosphorus-fish community biomass relationships in southern Appalachian reservoirs: can lakes be too clean for fish? Lake and Reservoir Management 5:83–90.

Zweifel, R. D., R. S. Hale, D. B. Bunnell, and M. T. Bremigan. 2009. Hatch timing variations among reservoir gizzard shad population: implications for stocked *Sander* spp. fingerlings. North American Journal of Fisheries Management 29:488–494.

American Fisheries Society Symposium 80:383–393, 2013

Survey of Transportation Practices for Striped Bass and Striped Bass Hybrids

David L. Yow*
North Carolina Wildlife Resources Commission
57 Hilltop Road, Asheville, North Carolina 28803, USA

Brian J. McRae
North Carolina Wildlife Resources Commission
2312 Summit Drive, Hillsborough, North Carolina 27278, USA

Douglas A. Besler
North Carolina Wildlife Resources Commission
645 Fish Hatchery Road, Marion, North Carolina 28752, USA

Winthrop E. Taylor
Equinox Environmental Consultation & Design, Inc.
37 Haywood Street, Asheville, North Carolina 28801, USA

Abstract.—We surveyed production managers of state natural resource agencies with striped bass *Morone saxatilis* and white bass *M. chrysops* × striped bass hybrid stocking programs to identify information needs and opportunities for shared innovations. Among 18 states responding, 17 reported ongoing fish transportation and stocking activities involving striped bass or striped bass hybrids. Oxygen diffusion was the most commonly reported treatment of fingerlings during transport, and the majority of respondents reported using agitators or aerators in hauling units. Sodium chloride was the most common additive used to condition hauling unit water, and five states also reported use of calcium chloride. Water quality monitoring procedures were less consistent among respondents, especially for receiving waters at stocking sites; only water temperature measurement was reported by all respondents. Although the majority of responding agencies reported monitoring fingerling survival during hauls, postrelease monitoring of fingerling survival was uncommon, in spite of the potential impact of release site water quality conditions on fingerling survival. Survey responses generally reflected improvements in consistency, with established guidelines since earlier surveys, but revealed an ongoing focus on prerelease condition of fingerlings as the primary measure of success of striped bass and striped bass hybrid stocking programs. We recommend that postrelease fingerling survival should be more widely monitored and used to determine success of stocking programs.

* Corresponding author: david.yow@ncwildlife.org

Introduction

Transportation and stocking of hatchery-raised fingerlings is a vital link between fish production and the management of sport fisheries. Among the factors affecting the rate of return of stocked fingerlings to the recreational angler, fishery managers can directly control the influences of handling, loading, transporting, and stocking procedures on short-term survival of fingerlings. In the southeastern and midwestern United States, the majority of state fishery management agencies culture, transport, and stock fingerling striped bass *Morone saxatilis* or their hybrids with white bass *M. chrysops* to support sport fisheries in public waters; however, current information is lacking on handling, transportation, and stocking technologies employed. Long-standing warmwater fish culture guidelines (Bonn et al. 1976; McCraren and Millard 1978) have stressed the importance of pretransport handling procedures and water quality conditions in transport units to the ultimate success of striped bass stocking efforts. Routine handling procedures directly affect the amount of stress on fingerlings (Parker et al. 1990) and can have significant effects on postrelease survival (Pitman and Gutreuter 1993).

Established standards for transportation of warmwater fish species were founded largely upon compilation of the experiences of production managers (McCraren and Millard 1978), but Carmichael and Tomasso (1988) subsequently found that specific transportation techniques continued to vary among striped bass culture facilities and were often inconsistent with published standards. Geiger and Parker (1985) identified further research into transportation issues as a top priority among hatchery managers in the Southeast. Current information was needed on handling and transportation techniques employed by production managers in states culturing striped bass and hybrids. Additionally, a compilation of agency-specific innovations in technologies and procedures would hold potential benefit for culturists of striped bass and other species. Therefore, our objective was to survey production managers involved in transporting striped bass or striped bass hybrid fingerlings to determine the status of hauling methods and technologies and identify areas in need of refinement.

Methods

Fish production managers from natural resource agencies in 23 southeastern, midwestern, and southwestern states were surveyed by e-mail to compare handling, transport, and release procedures related to striped bass and striped bass hybrid culture. The survey consisted of 20 questions and was designed to take less than 30 min to complete. Production managers were given 30 d to complete the survey.

The number and size of fingerlings stocked, mean and maximum hauling times, and mean and maximum hauling densities were queried for striped bass and striped bass hybrid stocking trips. Other requested information included special protocols used during longer hauls, amount of oxygen and type of aeration employed, water conditioners or additives used, release methods and acclimation procedures, and frequency of water quality monitoring during fingerling transport. Each respondent was asked what water quality variables, if any, were measured at the stocking site and if percent survival of fingerlings was recorded upon arrival at stocking points. Respondents were also asked if poststocking survival had ever been monitored and to provide the duration of any survival monitoring efforts. Each respondent was asked to note any special stocking procedures or innovations used and to list any published protocols relating to striped bass or striped bass hybrid culture. Additionally, each survey respondent was asked about specific water quality variables monitored, including water temperature, dissolved oxygen, pH, salinity, conductivity, hardness, ammonia, and nitrite/nitrate. For each variable, respondents were asked to rate the importance (high, moderate, low, or not measured) of monitoring; separate variable ratings were obtained for water quality monitoring in hauling units and in receiving waters at stocking points. To facilitate communication among production managers, all survey responses were compiled and tabulated by responding agency.

Results and Discussion

Production managers from 17 states reported ongoing fish transportation and stocking activities involving striped bass or striped bass hybrids and provided survey responses. Two states (Iowa and Missouri) transported only striped bass hybrids, whereas Pennsylvania transported only striped bass; the 14 remaining states transported both striped bass and hybrids (Table 1). One additional state (Utah) reported no striped bass or hybrid transporting activity, as all of their culture and stocking of these species was contracted to private vendors. Numbers of fingerlings transported annually by each state ranged from 10,000 to 3.5 million among states culturing striped bass and from 52,000 to 2.5 million for those culturing hybrids. Size of fingerlings transported ranged from 25 to 76 mm, with most agencies transporting 25- to 51-mm fingerlings; size at transport did not differ within states between striped bass and hybrids except in Virginia and West Virginia, where hybrid fingerlings were reported to be slightly larger than striped bass at time of transport. Among all states responding to our survey, only Texas reported publishing standard agency guidelines related to transportation and stocking of striped bass and hybrids (Lyon et al. 2006). Although our survey did not require respondents to specify whether striped bass hybrids were palmetto bass (male white bass × female striped bass) or sunshine bass (female white bass × male striped bass), palmetto bass culture was assumed; one respondent specified palmetto bass, and none specified sunshine bass or "reciprocal" hybrids.

Mean hauling times ranged from 1 to 6 h for striped bass fingerlings and from 1 to 12 h for hybrids; maximum hauling times ranged from 3 to 10 h and from 2 to 18 h, respectively, reflecting the dispersed nature of managed waters, especially in larger states (Table 1). The majority of states reported mean hauling times of 4 h or less as in past surveys (Carmichael and Tomasso 1988).

Hauling densities (Table 1) were reported in pounds of fingerlings per gallon of hauling unit water (lbs/gal or 0.12 kg/L) and in most states reflected rates recommended in published guidelines (Parker et al. 1990); 0.25 lbs/gal (0.030 kg/L) was the modal value given by survey respondents. Reported mean fingerling densities ranged from 0.06 to 0.75 lbs/gal (0.007–0.090 kg/L), and 11 of 17 states indicated that densities were modified based on length of stocking runs. In most cases, hauling densities for striped bass hybrid fingerlings were similar to those of striped bass, although several states reported higher hauling densities for hybrids. Reported maximum hauling densities ranged from 0.15 to 1.5 lbs/gal (0.018–0.180 kg/L) for striped bass and 0.2–1.75 lbs/gal (0.024–0.210 kg/L) for hybrids; Texas reported using variable hauling density guidelines ranging from 0.25 lbs/gal (0.030 kg/L) for transport durations exceeding 8 h to 0.5 lbs/gal (0.060 kg/L) for transport durations of 4 h or less (Lyon et al. 2006) and further reducing hauling densities at higher water temperatures or in situations where fingerlings are otherwise stressed. In addition to protocols for reducing fingerling densities on longer hauls, two states reported additional precautions for extended hauls; Arkansas reported using additional oxygen for high-density hauls and Texas reported using ice in hauling tanks to alleviate fingerling stress. Overall, reported mean hauling densities were lower than those reported in earlier surveys (Geiger and Parker 1985; Carmichael and Tomasso 1988). Reductions in maximum fingerling hauling densities in recent years likely have resulted from practical experiences of production managers and improved understanding of density-related stress effects on fingerling health in striped bass (Davis and Parker 1990; Parker et al. 1990) and sunshine bass (Weirich et al. 1992).

Management of dissolved gases in hauling tank water during transport of striped bass and

Table 1. Striped bass (SB) and striped bass hybrid (SBH) annual production, sizes of fingerlings produced, mean and maximum hauling times and densities, and special protocols for long hauls reported by state agency survey respondents. Hauling densities are given in lbs/gal (kg/L × 8.3454) as measured and reported by agency culturists. Protocols are listed as D = fingerling densities reduced for long hauls, or number of trips increased to avoid high densities; O = additional oxygenation used when higher densities of fingerlings are hauled; I = ice added to reduce stress from high hauling densities; N = no special protocols for high-density hauls.

State	Number stocked annually (million)		Length at stocking (mean or range, mm)		Mean/maximum hauling time (h)		Mean/maximum hauling density (kg/L × 8.3454)	Protocol for long hauls
	SB	SBH	SB	SBH	SB	SBH		
Alabama	0.5–0.75	0.3–0.4	25–51	25–51	2.5/5	2.5/5	0.25/0.4	N
Arkansas	0.45	0.35	38	38	2.5/6	2.5/6	0.25/0.5	D, O
Florida[a]	0.25–0.5	0.25–0.5	35–50	35–50	2/7	2/4	0.25/0.5	D
Georgia	1.5	1.5	25	25	5/9	5/9	0.3/0.5	D
Iowa	0	0.075	NA	51	NA	2/3.5	0.25/0.5	D
Illinois	0.015–0.05	0.25–1.0	32–51	32–51	4/7.5	4/7.5	0.25/0.5	N
Kentucky	0.9	0.8	38	38	4/6	3.5/6	0.5/1.5	D
Missouri	0	0.3	NA	51	NA	3/3	0.25/0.3	N
North Carolina	1.2	0.06	25–51	25–51	4/6	4/6	0.15/0.25	D
Ohio	0.1–0.2	0.5–0.75	30–40	30–40	3/3	3/6	0.33/0.33	N
Oklahoma	0.025–0.03	0.6–0.7	38–51	38–51	2/4	3/8	0.2/0.4[b]	D
Pennsylvania	0.025–0.3	0	25-51	NA	6/8	NA	0.06/0.15	N
South Carolina	3.5	0.6	25–51	25–51	1/6	5/6	0.2/0.25	N
Tennessee[a]	1.0–1.2	0.2–0.24	51–76	51–76	1.3/3	1/2	0.37/0.5	N, D[e]
Texas	2.5	2.5	38	38	5/10	5/10	Variable[c]	D, I
Virginia	0.6–1.0	0.052	25–45	51–64	1/6	12/18	0.07/0.2	D
West Virginia	0.01	0.125	25–51	51	3/4	2/4	0.75/1.5[d]	D

[a] Florida and Tennessee responses incorporate feedback from multiple hatchery managers; where applicable, overall total, mean, or maximum values are shown.

[b] Oklahoma hauling densities in table are for striped bass; hauling densities for hybrid bass averaged 0.3 lb/gal (0.036 kg/L) and ranged as high as 0.6 lb/gal (0.072 kg/L).

[c] Texas guidelines adjust hauling densities based on a number a variables, including size and condition of fish, water temperature on stocking truck, and hauling distance.

[d] West Virginia hauling densities in table are for striped bass; hauling densities for hybrid bass averaged 1 lb/gal (0.120 kg/L) and ranged as high as 1.75 lbs/gal (0.210 kg/L).

[e] Two Tennessee hatcheries reported no protocols for longer hauls; one hatchery reported reducing densities on longer hauls.

hybrid fingerlings varied among state agencies but typically included oxygenation and various methods of water circulation (Table 2). Use of oxygen was reported by all states except Florida and was measured either by a set rate of flow (L/min) or a targeted concentration in tank water (ppm). Reported flow rates varied from 0.3 to 4.0 L/min, whereas oxygen concentrations ranged from 7 to 20 ppm. Oxygen diffusion was the most common treatment employed among all survey respondents; similarly, Carmichael and Tomasso (1988) found that oxygen was used in fish transport by 91% of surveyed production facilities. Aeration was employed by 11 of 17 states, most frequently using vertical-pump aerators or paddle agitators; Alabama and South Carolina reported using compressed air and air stones in at least some fingerling hauls. As with oxygen, frequency of use of aeration or agitation among our survey respondents was similar to that reported previously (Carmichael and Tomasso 1988).

The use of hauling additives, particularly sodium and calcium chlorides, was reported by a majority of responding states (Table 2). Sodium chloride was the most common salt treatment, used during hauls by 14 of 17 states and as a prehaul treatment by Georgia. Five states also reported using calcium chloride during hauls. Two states reported using defoamers, Missouri reported using treatments to prevent spread of zebra mussels *Dreissena polymorpha*, and Pennsylvania reported using a multipurpose additive incorporating defoamer, electrolyte supplement, and ammonia and nitrite suppression. The use of sodium and calcium salts for hauling fingerling striped bass and hybrids is a long-standing recommendation of researchers (McCraren and Millard 1978; Geiger and Parker 1985; Grizzle et al. 1985; Weirich et al. 1992; Pitman and Gutreuter 1993) and established professional (Parker et al. 1990) and extension (Jensen 1990) guidelines. However, benefits of salts for alleviating physiological stress during transport may not extend to improved postrelease survival of fingerlings stocked into freshwater environments, possibly due to metabolic imbalances from handling stress (Mazik et al. 1991), interactions of multiple physiological and environmental stressors (Pitman and Gutreuter 1993; Wallin and Van Den Avyle 1995), or unknown synergistic effects of multiple variables on postrelease physiology, behavior, and vulnerability to predation (Port et al. 2010).

Acclimation and release procedures reported by survey respondents varied by conditions encountered at release sites (Table 2). Quick-release stocking (flushing of hauling unit water and fingerings into receiving waters following acclimation) was the primary method reported by 15 of 17 states, whereas two states reported that they frequently stocked fingerlings by boat. Standard acclimation procedures were reported by 15 of 17 states and typically followed established guidelines for acclimation times and temperature ranges (Davies 1973; Parker et al. 1990), although rates of acclimation differed among states, ranging from 5°C/h to 20°C/h. Several states reported using a standard acclimation time (0.5–1.0 h) or maximum temperature difference between hauling unit and receiving waters (0.1–2.0°C) rather than a standard acclimation rate.

Reported monitoring of water quality in hauling units and at stocking sites varied substantially among state agency respondents (Table 3). Two states reported constant monitoring during hauls, and three states reported no water quality monitoring after loading of fingerlings and departure from culture stations; the remainder of states reported monitoring at least once during hauls, usually at some established interval ranging from 0.5 to 2.0 h. All responding agencies reported measuring at least one water quality variable, typically temperature, at stocking sites. Recording of percent survival of fingerlings during hauls was reported by 11 of 17 states; conversely, only five states reported monitoring of fingering survival after release, and only Florida reported ever having monitored fingerling survival more than 24 h after release.

The perceived importance of monitoring specific water quality variables also varied

TABLE 2. Hauling treatments, release method, and acclimation procedures reported by state agency survey respondents. Additives included sodium chloride (NaCl), calcium chloride (CaCl2), and listed commercial products. Release methods included quick-release (Q) and boat release (B).

State	Oxygen flow (L/min) or concentration (ppm)	Type of aeration	Other additives	Release method	Acclimation procedure
Alabama	1.0–4.0 L/min	Compressor/air stones or none	NaCl, $CaCl_2$, exchange water	Q	Fully acclimate; defoamer
Arkansas	1.5–2.0 L/min	None	NaCl	Q	Exchange water
Florida	None	12-V agitator	NaCl, $CaCl_2$	Q	Fully acclimate; 30–45 min
Georgia	8 ppm	None	NaCl[b], $CaCl_2$	Q	Minimum 30 min; <0.1°C; equalize salinity, pH
Iowa	0.3 L/min	None	NaCl	Q	Fully acclimate
Illinois	8–12 ppm	None	NaCl	Q	Acclimate 30–45 min if temperature difference >10°F
Kentucky	20 ppm	None	None	Q	None
Missouri	0.25 L/min	12-V agitator	Zebra mussel treatment	B	Acclimate to within 2°F
North Carolina	0.5–2.0 L/min	Vertical pump aerator	NaCl	Q	Acclimate 1 h/10°F of temperature difference
Ohio	0.5–1.0 L/min	Vertical pump aerator	NaCl	Q	None
Oklahoma	3.0 L/min	12-V agitator	NaCl, $CaCl_2$	Q	Acclimate 1 h/5°F of temperature difference
Pennsylvania	2.0–3.0 L/min	Vertical pump aerator	NaCl, defoamer/ NH_3 reducer	Q	Fully acclimate
South Carolina	3.0 L/min	Compressor/air stones	NaCl	Q	Acclimate 1 h
Tennessee[a]	0.5–2.0 L/min	Paddle agitator, aerator, or none	NaCl, defoamer	Q	Minimum 20 min; within 1°C, equal DO
Texas	8–12 ppm	12-V agitator	NaCl	B	Acclimate 20 min/4°C
Virginia	1.5–4.0 L/min	Vertical pump aerator	NaCl, $CaCl_2$	Q	Acclimate 15–20 min/5°C
West Virginia	7–10 ppm	None	NaCl	Q	If >5°F difference, acclimate 1 h

[a] Tennessee oxygen flow based on range reported by three hatcheries; other Tennessee responses based on two of three hatcheries that used acclimation procedures.

[b] Georgia reported overnight pre-treatment of fingerlings with 4 parts per thousand sodium chloride; calcium chloride used during haul.

TABLE 3. Frequency of water quality monitoring (if any) during fingerling hauls, water quality monitoring at stocking sites, fingerling survival monitoring during hauls, and fingerling short-term survival monitoring reported by state agency survey respondents.

State	Frequency of water quality monitoring during haul	Water quality monitoring at stocking site	Percent survival of fingerlings during haul recorded	Short-term fingerling survival monitored	Duration of short-term survival monitoring (h)
Alabama	None	Y	Y	N	NA
Arkansas	1–2 h	Y	N	N	NA
Florida	Constant	Y	N	Y	48–72
Georgia	1 h	Y	Y	Y	<1
Iowa	Prior to departure	Y	N	N	NA
Illinois	None	Y	Y	N	NA
Kentucky	3 times per trip	Y	Y	Y	24
Missouri	1 h	Y	N	N	NA
North Carolina	1 h	Y	Y	Y	24
Ohio	Prior to departure, 30 min	Y	Y	N	NA
Oklahoma	Monitored; no standard interval	Y	Y	N	NA
Pennsylvania	1 h	Y	Y	N	NA
South Carolina	Constant	Y	N	N	NA
Tennessee[a]	Once per trip	Y	N	N	NA
Texas	1 h	Y	Y	N	NA
Virginia	Once, 1 h after departure	Y	Y	Y	24
West Virginia	2 h or prior to stocking	Y	Y	N	NA

[a] Tennessee responses based on two of three hatcheries that reported monitoring conditions.

among respondents; however, all respondents placed greater importance on water quality conditions in hauling tanks than those in receiving waters (Table 4). All states reported monitoring temperature in hauling units and at stocking sites, and 15 of 17 states considered temperature to be a top priority among measurable water quality variables. Dissolved oxygen monitoring on hauling units was considered a high priority by 15 of 17 states, but only 9 of 17 states prioritized dissolved oxygen monitoring at stocking sites. Among the remaining water quality variables included in our survey, the majority of monitoring priorities reported by states concerned ionic concentrations in hauling units. Salinity was considered of moderate to high importance by 7 of 17 states, hardness was considered of moderate importance by three states, and one state considered conductivity monitoring of moderate importance; no

TABLE 4. Perceived importance of monitoring of water quality parameters reported by state agency survey respondents. Parameters listed are: Temp = Water temperature; DO = dissolved oxygen; pH; Sal = Salinity; Cond = conductivity; Hard = hardness; Amm = ammonia; Nitr = nitrite/nitrate. Separate perception values are given for hauling unit followed by receiving waters. Perception values are: H = high importance; M = moderate importance; L = low importance; N = not important and not monitored.

State	Temp	DO	pH	Sal	Cond	Hard	Amm	Nitr
Alabama	H, M	N, N	N, N	N, N	N, N	N, N	N, N	N, N
Arkansas	H, H	H, N	N, N	M, N	N, N	N, N	N, N	N, N
Florida	M, M	H, H	L, L	M, M	N, N	N, N	N, N	N, N
Georgia	H, H	H, H	N, N	M, M	M, M	M, M	N, N	N, N
Iowa	H, H	H, L	N, N	H, N	N, N	N, N	N, N	N, N
Illinois	H, H	H, N	N, N	N, N	N, N	N, N	N, N	N, N
Kentucky	H, H	H, H	N, N	N, N	N, N	N, N	N, N	N, N
Missouri	H, H	H, N	N, N	N, N	N, N	N, N	N, N	N, N
North Carolina	H, H	H, H	M, N	H, N	N, N	M, N	N, N	N, N
Ohio	H, H	H, N	N, N	N, N	N, N	N, N	N, N	N, N
Oklahoma	H, H	H, M	N, N	L, N	N, N	M, N	N, N	N, N
Pennsylvania	H, H	N, N	N, N	M, N	N, N	N, N	N, N	N, N
South Carolina	M, M	H, H	N, N	H, N	N, N	N, N	N, N	N, N
Tennessee[a]	H, H	H, H	N, N	N, N	N, N	N, N	N, N	N, N
Texas	H, H	H, H	H, H	M, M	N, N	N, N	N, N	N, N
Virginia	H, H	H, M	M, M	N, L	N, N	N, N	M, N	N, N
West Virginia	H, H	H, H	M, L	L, L	L, L	L, L	L, L	L, L

[a] Tennessee responses based on highest rating from two of three hatcheries that reported perceptions.

respondents considered these variables to be of high importance for monitoring at stocking sites, although three states considered salinity of moderate importance and one state considered conductivity and hardness to be of moderate importance. Monitoring of pH in hauling units was given moderate priority by three states and high priority by only one state; and one state each gave moderate and high priority to pH monitoring at stocking sites. No respondents gave priority to monitoring ammonia or nitrite/nitrate at stocking sites, and only one state gave moderate priority to ammonia monitoring in hauling units.

The importance of controlling and monitoring temperature and oxygen levels during transport was recognized by all respondents in our survey. Both variables must be managed to prevent immediate prerelease fingerling stress and mortality during hauls; recommendations for their management were established in early production guidance (Bonn et al. 1976; McCraren and Millard 1978) and remain prominent among striped bass transport recommendations (Parker et al. 1990). Temperature and oxygen can both be measured remotely and continually with existing transport technologies, although continuous monitoring was still uncommon among surveyed agencies. Maintaining relatively cool temperatures is particularly crucial to lower fingerling metabolism, reduce stress, and avoid synergistic effects on ammonia accumulation (Emerson et al. 1975; Wedemeyer 1996), and consistent management of these two variables likely has reduced incidence of observed fingerling survival issues associated with other water quality variables.

The low perception of importance of ammonia and pH monitoring among survey respondents may result from the assumption, based on experience of production managers, that ammonia toxicity may be avoided through management of fingerling densities, haul times, and water temperatures. Additionally, the accumulation of carbon dioxide during transport tends to reduce pH in hauling units (Amend et al. 1982; Treasurer 2010) and suppress the accumulation of un-ionized ammonia (Thurston et al. 1981), thereby reducing the risk of ammonia toxicity. However, interaction of pH and accumulated ammonia may produce lethal conditions for striped bass (Oppenborn and Goudie 1993), white bass (Ashe et al. 1996), palmetto bass (Oppenborn and Goudie 1993; Bergerhouse 1994), and sunshine bass (Barkoh et al. 2004) and may warrant increased monitoring of one or both variables, especially during long hauls. Also, elevated pH levels may occur in receiving waters as a result of algal or macrophyte growth in nutrient-rich impoundments (Halstead and Tash 1982; Murray and Ziebell 1984) and are potentially most severe late on summer afternoons (Halstead and Tash 1982) when striped bass fingerlings are typically stocked. Consequently, late-afternoon summer stocking of fingerlings should be avoided if possible, and pH of receiving water should be checked at stocking points before adding it to hauling units to acclimate fingerlings.

In our investigations at three North Carolina reservoirs (Taylor 2005), elevated pH in receiving waters was associated with substantial mortality of striped bass fingerlings. In 2002, we monitored percent survival of fingerlings 24 h after stocking in three replicate holding cages at release locations on three reservoirs and concurrently measured water quality variables at release locations and in hauling units before and after temperature acclimation. At one reservoir, pH at the release point approached 9.5 on the afternoon of stocking, and survival of fingerlings averaged 71.2%, significantly lower than the 98.3% survival observed among fingerlings held at the originating hatchery or the survival rates of 92.6–94.3% observed at other reservoirs stocked earlier in the day. Although the elevated mortality could have resulted directly from the high pH of receiving waters, we suspected an interaction of pH with accumulated ammonia in the hauling unit when reservoir water was added during acclimation. Unfortunately, levels of un-ionized ammonia present during the acclimation procedure could not be precisely estimated, and an attempt in 2003 to repeat the procedure with remediation of pH in acclimation water was inconclusive because elevated pH levels were not encountered in any of the reservoirs stocked. However, our results demonstrated the potential of water quality conditions in receiving waters to adversely affect the success of fingerling stocking.

In light of our field observations, the current measure of success for striped bass and hybrid fingerling stocking trips, as reported by respondents to our survey, may require adjustment. Among survey respondents, prerelease fingerling survival monitoring was relatively common compared to postrelease studies, indicating a focus of trip success criteria on prerelease condition of fingerlings. Our postrelease study revealed a significant impairment of short-term fingerling survival in a stocking situation where no impairment was evident from prerelease observations, indicating a need for expanded postrelease investigations. Additionally, hauling additives that improve conditions for fingerlings during transport have produced neutral or negative responses in postrelease survival studies of fingerling striped bass (Mazik et al. 1991; Wallin and Van Den Avyle 1995) and hybrids (Sink and Neal 2009; Port et al. 2010), further indicating that pre- and postrelease assessments of stocking success, and their implications on management of associated sport fisheries, may differ substantially.

Conclusions

Survey responses of state agency production managers reflected improvements in consistency with established guidelines since earlier surveys but reveal an ongoing focus on prerelease condition of fingerlings as the primary measure of success of striped bass and hybrid striped bass stocking programs. Because observed fingerling survival during transport may not reveal issues associated with postrelease mortality, managers should monitor short-term survival, particularly at stocking locations involving extended haul times or where adverse conditions in receiving waters are anticipated. Furthermore, perceptions of stocking trip success should be reexamined to fully consider the fate of stocked fingerlings following release. Future investigations should focus on postrelease responses of fingerlings to established hauling practices and treatments across a broad range of water quality conditions in receiving waters.

Acknowledgments

This project was funded under the Federal Aid in Sport Fish Restoration Program administered cooperatively by the North Carolina Wildlife Resources Commission (NCWRC) and U.S. Fish and Wildlife Service. The authors thank the following state agency fish production managers for their participation in the hauling practices survey: T. Beisser, Ohio Department of Natural Resources; R. Cook and R. Terrell, Missouri Department of Conservation; J. Evans and K. Briggs, NCWRC; J. Fleming, Georgia Department of Natural Resources; E. Gwathmey, Virginia Department of Game and Inland Fisheries; S. Krueger, Illinois Department of Natural Resources; G. Kurten, Texas Parks and Wildlife Department; R. Middleton, Kentucky Department of Fish and Wildlife Resources; J. Mitchell and K. Winningham, Arkansas Game and Fish Commission; W. Nichols, Alabama Department of Conservation and Natural Resources; C. O'Bara, West Virginia Division of Natural Resources; T. Overly, Pennsylvania Fish and Boat Commission; F. Sessions, South Carolina Department of Natural Resources; M. Smith, D. Roddy, and L. Mason, Tennessee Wildlife Resources Agency; G. Sobotka, Iowa Department of Natural Resources; S. Spade, Oklahoma Department of Wildlife Conservation; C. Stence, Maryland Department of Natural Resources; and R. Stout and D. Yeager, Florida Fish and Wildlife Conservation Commission. G. Vaughan and the Duke Power Company analytical laboratory staff processed all water quality samples collected during the North Carolina fingerling survival study.

References

Amend, D. A., T. R. Croy, B. A. Goven, K. A. Johnson, and D. H. McCarthy. 1982. Transportation of fish in closed systems: methods to control ammonia, carbon dioxide, pH, and bacterial growth. Transactions of the American Fisheries Society 111:603–611.

Ashe, D., J. R. Tomasso, A. G. Eversole, and T. I. J. Smith. 1996. Toxicity of ammonia to fingerling white bass: effect of selected environments on uptake dynamics. Progressive Fish-Culturist 58:277–280.

Barkoh, A., D. G. Smith, J. W. Schlechte, and J. M. Paret. 2004. Ammonia tolerance by sunshine bass fry: implication for use of ammonium sulfate to control *Prymnesium parvum*. North American Journal of Aquaculture 66:305–311.

Bergerhouse, D. L. 1994. Lethal effects of elevated pH and ammonia on early life stages of hybrid striped bass. Journal of Applied Aquaculture 2(3–4):81–100.

Bonn, E. W., W. M. Bailey, J. D. Bayless, K. E. Erickson, and R. E. Stevens, editors. 1976. Guidelines for striped bass culture. American Fisheries Society, Southern Division, Striped Bass Committee, Bethesda, Maryland.

Carmichael, G. J., and J. R. Tomasso. 1988. Survey of fish transportation equipment and techniques. Progressive Fish-Culturist 50:155–159.

Davies, W. D. 1973. Rates of temperature acclimation for hatchery reared striped bass fry and fingerlings. Progressive Fish-Culturist 35:214–217.

Davis, K. B., and N. C. Parker. 1990. Physiological stress in striped bass: effect of acclimation temperature. Aquaculture 91:349–358.

Emerson, K., R. C. Russo, R. E. Lund, and R. V. Thurston. 1975. Aqueous ammonia equilibrium calculations: effect of pH and temperature. Journal of the Fisheries Research Board of Canada 32:2379–2383.

Geiger, J. G., and N. C. Parker. 1985. Survey of striped bass hatchery management in the southeastern United States. The Progressive Fish-Culturist 47:1–13.

Grizzle, J. M., A. C. Mauldin, I. I., D. Young, and E. Henderson. 1985. Survival of juvenile striped bass (*Morone saxatilis*) and *Morone hybrid bass* (*Morone chrysops* x *Morone saxatilis*) increased by addition of calcium to soft water. Aquaculture 46:167–171.

Halstead, B. G., and J. C. Tash. 1982. Unusual diel pHs in water as related to aquatic vegetation. Hydrobiologia 96:217–224.

Jensen, G. L. 1990. Transportation of warmwater fish: procedures and loading rates. Louisiana State University, Southern Region Aquaculture Center Publication 392, Baton Rouge.

Lyon, D. D., A. Barkoh, D. Smith, and D. Begley. 2006. Guidelines for the culture of striped bass and palmetto bass. Texas Parks and Wildlife Department, Management Data Series No. 243, Austin.

Mazik, P. M., B. A. Simco, and N. C. Parker. 1991. Influence of water hardness and salts on survival and physiological characteristics of striped bass during and after transport. Transactions of the American Fisheries Society 120:121–126.

McCraren, J. P., and J. L. Millard. 1978. Transportation of warmwater fishes. Pages 43–88 *in* Manual of fish culture, section G. U.S. Fish and Wildlife Service, Washington, D.C.

Murray, C. A., and C. D. Ziebell. 1984. Acclimation of rainbow trout to high pH to prevent stocking mortality in summer. The Progressive Fish-Culturist 46:176–179.

Oppenborn, J. B., and C. A. Goudie. 1993. Acute sublethal effects of ammonia on striped bass and hybrid striped bass. Journal of the World Aquaculture Society 24:90–101.

Parker, N. C., G. T. Klar, T. I. J. Smith, and J. H. Kerby. 1990. Special considerations in the culture of striped bass and striped bass hybrids. Pages 191–215 *in* R. M. Harrell, J. H. Kerby, and R. V. Minton, editors. 1990. Culture and propagation of striped bass and its hybrids. American Fisheries Society, Southern Division, Striped Bass Committee, Bethesda, Maryland.

Pitman, V. M., and S. Gutreuter. 1993. Initial poststocking survival of hatchery-reared fishes. North American Journal of Fisheries Management 13:151–159.

Port, P. R., S. E. Lochmann, and J. W. Neal. 2010. Poststocking survival of hybrid bass after changes in water hardness and temperature. North American Journal of Aquaculture 72:118–123.

Sink, T. D., and J. W. Neal. 2009. Stress response and posttransport survival of hybrid striped bass transported with or without clove oil. North American Journal of Aquaculture 71:267–275.

Thurston, R. V., R. C. Russo, and G. A. Vinogradov. 1981. Effect of pH on the toxicity of the un-ionized ammonia species. Environmental Science and Technology 15:837–840.

Taylor, W. E. 2005. Fingerling striped bass poststocking survival evaluation. North Carolina Wildlife Resources Commission, Division of Inland Fisheries, Project F-24, Final Report, Raleigh.

Treasurer, J. W. 2010. Remediation of ammonia accumulation during live transport of juvenile cod, *Gadus morhua* L., and the effects of fast period on ammonia levels and water quality Aquaculture 308:190–195.

Wallin, J. E., and M. J. Van Den Avyle. 1995. Interactive effects of stocking site salinity and handling stress on survival of striped bass fingerlings. Transactions of the American Fisheries Society 124:736–745.

Wedemeyer, G. A. 1996. Transportation and handling. Developments in Aquaculture and Fisheries Science 29:727–758.

Weirich, C. R., J. R. Tomasso, and T. I. J. Smith. 1992. Confinement and transport-induced stress in white bass *Morone chrysops* × striped bass *M. saxatilis* hybrids: effect of calcium and salinity. Journal of the World Aquaculture Society 23:49–57.

American Fisheries Society Symposium 80:395–411, 2013

Use of Genetic Microsatellite Markers to Identify Factors Affecting Stocking Success in Striped Bass

TANYA L. DARDEN*
South Carolina Department of Natural Resources
Marine Resources Research Institute
215 Fort Johnson Road, Charleston, South Carolina 29412, USA

FORREST SESSIONS
South Carolina Department of Natural Resources
Dennis Wildlife Center
305 Black Oak Road, Bonneau, South Carolina 29431, USA

MICHAEL R. DENSON
South Carolina Department of Natural Resources
Marine Resources Research Institute
215 Fort Johnson Road, Charleston, South Carolina 29412, USA

Abstract.—The emerging use of genetic markers for stock identification is advantageous as this technique utilizes an inherent permanent mark, it eliminates stress associated with conventional tagging methods, and mark recovery is nonlethal. In controlled experiments, increased accuracy and reliability have been demonstrated with genetic tags, as compared to other commonly used markers. While a variety of genetic markers exist for fish identification, microsatellites are preferred because of their polymorphic nature and versatile use, including identification of population structure, measures of genetic diversity, and parentage analysis. Microsatellites provide a cost-effective tool for a variety of applications, including their use in complex experimental designs, as well as responsible genetic population management. We provide examples of the use of microsatellites as genetic tags in South Carolina striped bass *Morone saxatilis*, describing an experimental restoration effort in the Ashley River and a large-scale stock enhancement effort for the Santee-Cooper reservoirs.

Introduction to Genetic Tagging

The use of definitive marks to identify released larvae and juveniles is necessary in properly assessing the effects of stocking. The use of stocking as a management tool requires implementation of quantitative measures of success to evaluate effects (Blankenship and Leber 1995). However, when stocking small juveniles and larvae, the use of physical tags can be problematic. External tags are usually too large for early life stages, and handling small fish individually can add significant stress (Van Den Avyle and Wallin 2001). To address this issue, several otolith marking approaches have been used. Temperature manipulations and out-of-season stocking have been used to produce unique morphological patterns on otoliths of larvae that can be used to identify stocked fish when they are adults (Volk et al. 1990; Hendricks et al. 1994; Schroder et al. 1995; Jenkins et al. 1997). Im-

* Corresponding author: dardent@dnr.sc.gov

mersion-marking chemicals, such as one of the tetracycline congeners (Secor et al. 1991; Dabrowski and Tsukamoto 1986; Denson and Smith 2008), alizarin complexone (Secor and Houde 1995; Thomas et al. 1995), and calcein (Wilson et al. 1987; Thomas et al. 1995), have also been used to identify stocked fish. Secor and Houde (1995) were able to identify alizarin-marked stocked striped bass *Morone saxatilis* larvae that were recaptured 10 d after release. Reinert et al. (1998) marked larval striped bass with oxytetracycline (OTC) and, 3 years later, detected marks from 74.2% of known marked adults. In South Carolina, Denson and Smith (2008) marked 1.5-mm larval red drum *Sciaenops ocellatus* by immersion in OTC; monthly visual evaluation over a 2-year period showed that the marks were only accurately readable for 3 months. The main disadvantage of using otolith characteristics for fish identification is that animals have to be sacrificed to evaluate mark presence.

Recent investigations demonstrate that microsatellites provide nonlethal genetic tags for all life stages of fish (Chapman et al. 2002; Renshaw et al. 2006; Robbins et al. 2008; Fountain et al. 2009; Saillant et al. 2009). Microsatellites are short stretches of DNA containing repeated nucleotide sequences that are assumed to be neutral and inherited in Mendelian fashion. The population level variation resulting from high mutation rates within sites provides sufficient resolution to differentiate hatchery individuals from wild fish (Blouin et al. 1996; Liu and Cordes 2004). Reviews of genetic techniques and their application to fisheries research consistently indicate that microsatellite markers are the primary choice for parentage assignment (Liu and Cordes 2004; Jones et al. 2010). Although RFLP (restriction fragment length polymorphism analysis) and multi-locus nuclear probes have been successfully applied toward population-level stock identification (Wirgin and Maceda 1991; Wirgin et al. 1991; Bosworth et al. 1994; Waldman et al. 1997), these technologies do not harbor sufficient variation for individual-level identification as is needed for genetic tag applications. Several researchers have reported using nuclear microsatellite markers to discriminate released hatchery fish from their wild cohorts by matching hatchery offspring to broodstock with a low probability of mismatch (Ferguson and Danzmann 1998; Perez-Enriquez and Taniguchi 1999; McDonald et al. 2004; Gold et al. 2008; Karlsson et al. 2008; Robbins et al. 2008; Saillant et al. 2009). An advantage of using nonlethal genetic tools such as microsatellites is that the animals do not have to be sacrificed, allowing repeated evaluation of the sample population (i.e., sampling with replacement). The utilization of genetic markers for stock identification also provides multiple benefits to research and management programs, as it alleviates the stress associated with conventional tagging methods and the mark is never lost. In addition, fisheries scientists can evaluate stock structure, hybridization, inbreeding, and outbreeding depression using microsatellites. Therefore, the incorporation of genetic markers into a stocking program allows for a single data collection effort (i.e., genotyping) that provides conclusive hatchery identification along with a wealth of population information that is necessary for effective management.

Due to the economic importance of striped bass, as well as a need to incorporate genetic information into responsible management plans, substantial research effort has led to the development of a large number of genetic markers. Although striped bass have shown very low levels of variation in terms of both allozymes (Bulak et al. 2004) and mitochondrial DNA (Chapman 1990; Stellwag and Rulifson 1994; Waldman et al. 1998; Brown et al. 2005), they exhibit relatively high levels of polymorphism at microsatellite loci (Laughlin and Turner 1996; Brown et al. 2005). Over the past 10 years, more than 500 microsatellite loci have been isolated and characterized for striped bass from geographically diverse sample sources. A summary of the potentially most useful loci based on heterozygosity estimates includes 72 markers of a variety of repeat motifs, allele sizes ranging from 108 to 404 base pairs, and polymorphism range of 3–28 alleles (Table 1). The availability of markers provides the opportunity

TABLE 1. Summary of available polymorphic microsatellite markers that have been isolated and characterized for striped bass. Criteria for locus inclusion was expected heterozygosity (H_E) > 0.70; * indicates observed heterozygosity (H_O) with inclusion criteria >0.65.

Source	Locus	Repeat motif	Size range (base pairs)	Alleles (#)	H_E
Han et al. 2000	SB83	(GT)	163–209	12	0.75*
	SB6	(GT)	183–247	9	0.73*
	SB11	(GT)	115–195	11	0.78*
Roy et al. 2000	SB113	(CT)(GT)	183–261	28	0.96
	SB117	(CT)(N)	184–232	16	0.76
	SB91	(GT)(N)(CT)	132–154	9	0.87
Toombs 2000	Msa2	TTTA	~142	6	0.70
Brown et al. 2003	AG25-1#1	CTTT	154–308	26	0.89
	AC25-6#12	(TG)	161–209	18	0.70
	AT150-2#21	(GT)	168–196	13	0.80
Ross et al. 2004	Hsb1B	(GT)	228–248	7	0.85
	Hsb7C	(AC)	263–313	6	0.75
Couch et al. 2006	MSM1067	(CA)(GT)	190–210	5	0.80
	MSM1085	(CA)	144–189	13	0.90
	MSM1092	(CA)	183–216	7	0.79
	MSM1094	(CA)	125–157	7	0.83
	MSM1095	(TG)	155–188	8	0.81
	MSM1096	(CA)	182–200	7	0.80
	MSM1107	(GT)	144–225	8	0.76
	MSM1137	(CA)	153–240	16	0.92
	MSM1138	(TG)	184–192	5	0.75
	MSM1139	(AC)	171–201	8	0.82
	MSM1140	(AC)	179–209	8	0.78
	MSM1144	(AC)	115–150	10	0.88
	MSM1145	(CA)	155–293	17	1.00
	MSM1152	(CA)	197–249	10	0.87
	MSM1155	(CA)	154–181	8	0.83
	MSM1157	(CA)	165–200	12	0.90
	MSM1161	(CT)(CA)	154–188	8	0.88
	MSM1165	(AC)	213–231	6	0.80
	MSM1166	(CA)	176–207	7	0.84
	MSM1168	(CA)	132–155	6	0.82
	MSM1186	(CA)	192–212	6	0.75
	MSM1193	(CA)	122–145	7	0.79
	MSM1194	(GT)	223–258	7	0.78
	MSM1208	(TA)	171–195	7	0.81
	MSM1229	(GT)(AT)	122–140	8	0.86
	MSM1230	(GT)	108–170	12	0.83
	MSM1239	(CA)	224–250	8	0.81
	MSM1243	(CA)	222–244	6	0.81
	MSM1246	(CA)	214–238	8	0.83

TABLE 1. Continued.

Source	Locus	Repeat motif	Size range (base pairs)	Alleles (#)	H_E
Rexroad et al. 2006	MSM1334	(ATAG)	301–365	10	0.69*
	MSM1355	(CA)	304–353	10	0.91*
	MSM1357	(GATA)	242–282	11	0.91*
	MSM1359	(ATCT)	361–381	6	0.81*
	MSM1362	(ATCT)	343–396	11	0.75*
	MSM1375	(CT)	206–267	7	0.69*
	MSM1402	(CTT)	194–200	10	0.69*
	MSM1429	(GAA)	247–359	17	0.82*
	MSM1451	(GAT)(GT)	182–184	2	1.00*
	MSM1453	(GAT)	183–202	5	0.75*
	MSM1491	(TGA)	196–208	5	0.66*
	MSM1520	(CA)	172–222	15	1.00*
	MSM1524	(TACA)	337–345	3	0.75*
	MSM1526	(GT)	159–176	8	0.91*
	MSM1536	(GT)	254–296	10	0.91*
	MSM1547	(TATG)	280–353	15	0.78*
	MSM1556	(AGAT)	211–293	18	0.73*
	MSM1557	(AGAT)	356–404	10	0.75*
	MSM1558	(TAGA)	325–373	12	0.75*
	MSM1559	(AGAT)	320–376	11	0.91*
	MSM1560	(AGAT)	333–364	8	0.73*
	MSM1577	(CTAT)	335–380	9	0.83*
	MSM1579	(ATCT)	269–286	5	0.87*
	MSM1584	(ATCT)	251–303	11	0.95*
	MSM1589	(TAGA)	236–309	10	0.73*
	MSM1591	(TAGA)	293–361	11	0.78*
	MSM1592	(TAGA)	172–237	12	0.81*
	MSM1598	(AGAT)	356–388	9	0.93*
	MSM1602	(CTAT)	214–267	11	0.72*
	MSM1604	(ATCT)	188–246	10	0.94*
	MSM1626	(TAGA)	280–353	15	0.78*
	MSM1628	(TAGA)	172–221	11	0.83*

for relatively rapid selection and optimization of a suite of microsatellite markers with multiplexing potential for a wide variety of program applications. Surprisingly, this wealth of information is just beginning to be utilized in research and management programs for striped bass (Liu and Ely 2009; Fountain 2010).

Our paper describes research conducted in South Carolina using genetic markers for stocked striped bass. In the late 1970s and 1980s, striped bass populations throughout their range experienced declines and have been supplemented by stocking programs designed to augment natural reproduction. Researchers at the South Carolina Department of Natural Resources (SCDNR) currently utilize microsatellite markers to monitor and evaluate enhancement success of striped bass. For this effort, we selected a suite of three multiplexed microsatellite panels containing a total of 12 loci that have been optimized and tested on two populations of striped bass in South Carolina (Fountain et al. 2009). Microsatellite loci were chosen based on their level of polymorphism, allowing versatile use in applications with the capability to achieve not only discrimination of stocked fish from their wild counterparts and determine their parentage, but also obtain information on the genetic diversity and structure of the population (i.e., "responsible approach" fisheries management, Blankenship and Leber 1995).

Methods of Hatchery Identification

A wide variety of methods are available for genetic analysis of parentage to assist in the identification of hatchery individuals that have been released into the wild. All methods assume conformity to Mendelian inheritance expectations in co-dominant, autosomal, diploid systems. Most available programs employ exclusion-based methods, which allow for assessment of parentage, with probabilities determined when mismatches occur. PROBMAX3 (Danzmann 1997) is an example of one of the basic computer programs that determines the probability of progeny assignment to a pool of potential parents of known genotypes (i.e., broodstock). Exclusion-based methods are typically capable of identifying a single parental pair for each offspring, but in some cases, multiple candidate parents may result due to common hatchery practices of spawning several males with a single female (which would still be sufficient for determining "wild" versus "hatchery" distinction). However, if it is of interest to determine which parental pair of the multiple candidates is the true parent, likelihood-based methods provide an advantage through their use of allele frequency information and parental heterozygosity/homozygosity states. These methods are used to calculate the likelihood ratio of each parental pair being the true parent to not being the true parent, with larger ratios indicating a higher probability that the pair is the offspring's true parent. The determination of a single parental pair provides information suitable for evaluating individual reproductive success, pedigree analysis and first generation migration (e.g., FAMOZ, Gerber et al. 2003; GENECLASS2, Piry et al. 2004; CERVUS, Kalinowski et al. 2007). Likelihood-based methods usually allow for a substantial degree of flexibility in terms of the basis for identification, whether it is limited to paternal or maternal contribution or inclusive of both parents, with many providing the option of identifications with incomplete parental information. The more well defined the broodstock group, both in terms of individuals included and sex, the higher the confidence in the ultimate parental assignments. However, data with missing genotypes and incomplete sex assignments can still be useful to extract some level of information. More recently, Bayesian and Monte Carlo resampling approaches have been developed for parentage analysis (Cornuet et al. 1999; Baudouin and Lebrun 2001; Tringali 2006) and have been incorporated as options in a number of analysis programs (e.g., GENECLASS2). A Bayesian approach to parentage analysis can be advantageous because of its insensitivity to both the total numbers of breeders in the population (both hatchery and wild) and unequal family sizes among hatchery and wild breeders, which is expected in stock enhancement. For a more complete description and comparison of parentage analysis techniques, see Jones and Ardren (2003) and Jones et al. (2010).

The evaluation of stocking contribution can also greatly benefit from the use of genetic markers in identity analyses, in terms of the capability to identify single individuals and siblings. Similar to parentage analysis, exclusion methods are capable of identifying individual samples with matching genotypes (e.g., MS Excel Microsatellite Toolkit Add-In) and likelihood methods are an improvement as they provide probabilities of identity even in the event of nonprecisely matching genotypes (e.g., CERVUS, FAMOZ). The nonlethal nature of sampling implemented with genetic marking makes this type of identification important to avoid pseudoreplication when evaluating within-year contributions (i.e., when the same fish is collected multiple times throughout the sample period), but also highly informative when evaluating among-year contributions in terms of movement, growth, and so forth of individuals.

As with the application of any analytical tool in science, an assessment of the statistical confidence of its results is necessary. We provide a brief overview of general techniques here; please see Jones and Ardren (2003) for a thorough, critical review of specific applications. Bravington and Ward (2004) provide a method to determine the probability of the wild population producing an offspring whose genotype is consistent with one potentially produced in the hatchery (i.e., probability of compatibility by chance [PCC]). Probability calculations are based on allele frequencies and genotypes of broodstock. An alternative exhaustive exclusion-based approach in the software FAP (Taggart 2007) allows for predictive resolving capabilities of known broodstock groupings at the individual group level, in contrast to a project or marker level. One disadvantage to both of these methods is that they require confidence calculations with each group (i.e., year) of broodstock and, due to hatchery logistics (require actual broodstock genotypes), are often limited to post hoc evaluation. Simulation-based assessments represent an important improvement by allowing a priori evaluation of genetic marker detection capabilities. Both CERVUS and FAMOZ can provide estimates of nonexclusion probabilities for a set of genetic markers based on allele frequencies and estimated numbers of broodstock. Additionally, these applications determine critical test statistic values for varying levels of confidence to be used during the actual application of the markers for likelihood-based parentage analysis. Although Bayesian approaches to confidence estimates have been developed (PATRI, Signorovitch and Nielsen 2002; PARENTE, Cercueil et al. 2002) to improve the simulation-based approaches, the interpretation of the posterior probabilities is difficult without a mechanism of calculating critical values.

Case Studies

Stocking of freshwater systems with striped bass juveniles from the Santee–Cooper River system has been conducted in South Carolina since the early 1960s (Stevens 1975). Today, striped bass remains an important and popular recreational fish species in South Carolina, and stocking efforts continue as a key component of SCDNR's management program. Researchers utilize "responsible approach" protocols (Blankenship and Leber 1995) to implement a methodical scientific approach for broodstock selection, animal health, and verification of marked animals to document stocking success. We present two case studies using microsatellite genetic tags that are ongoing efforts in South Carolina; the first has an experimental restoration emphasis while the second has direct application to an ongoing management program. Both projects utilize 12 multiplexed microsatellite loci (Fountain et al. 2009), which show high locus heterozygosities and polymorphism (Table 2).

Our quality assurance/quality control protocol for genotyping samples included two independent readers scoring each sample, and discrepancies were reconciled in conference. Samples were reanalyzed if necessary and discarded if there was no consensus. Additionally, all broodstock samples were independently genotyped at least twice to ensure that the correct genotype had been assigned. We have directly verified the Mendelian inheritance, Hardy Weinberg equilibrium, and lack of linkage for these markers. The confidence estimates for this suite of markers based on 2006–2008 adult collections (n = 436) include the postbroodstock case PCC of 99.999%, average parent pair nonexclusion probability of 2.7×10^{-7}, and average identity nonexclusion probability of 2.5×10^{-12}. Nonexclusion probabilities refer to the

TABLE 2. Multiplexed microsatellite locus sets used for SCDNR striped bass research programs. Included are the size ranges, polymorphisms and observed heterozygosities (H_O) for each locus. Summary statistics are based on 436 adult striped bass from the Santee-Cooper system in South Carolina sampled during 2006–2008.

Multiplexed group	Locus	Allele size range	No. alleles	H_O
1	MSM1144	118–164	15	0.81
	MSM1095	168–198	12	0.76
	MSM1096	179–205	10	0.75
	MSM1243	229–247	8	0.63
2	MSM1094	127–161	7	0.79
	MSM1526	139–161	10	0.62
	MSM1208	184–198	7	0.80
	MSM1067	193–211	5	0.46
3	MSM1168	140–156	4	0.60
	MSM1139	155–213	11	0.66
	MSM1592	155–211	13	0.82
	MSM1357	217–273	13	0.84

probability of not excluding an unrelated parent pair during parentage analysis or the probability that the genotypes are the same between two randomly chosen individuals in the case of identity analysis. Due to the low numbers of recaptures typical in stocking research, all statistical comparisons among treatment groups were conducted using a Monte Carlo approach in the statistical program R (R Development Core Team 2009) to test the null hypothesis that the percentage of fish returned from the treatment groups were no different from the percentage of fish that were stocked from those treatments. For each analysis, we created a histogram by resampling the data (10,000 permutations with replacement) to simulate a null distribution and performed a one-tailed probability test (p = 0.05) to compare the observed results to the simulated distribution.

Ashley River, South Carolina

The small, tidally influenced Ashley River drains into the Charleston Harbor estuary in South Carolina. Historically, this estuary supported a striped bass population (South Carolina State Board of Fisheries 1924), but very few young-of-year and adult fish have been observed since monthly sampling began in the early 1990s (J. Archambault, South Carolina Department of Natural Resources, personal communication). The SCDNR has set a goal of re-establishing historical biodiversity levels, and restoring a key indicator species like striped bass is an important step toward meeting this conservation goal.

Initial work included 2 years of stocking to determine whether stocked striped bass juveniles were able to survive within the system (Fountain 2010). The research design included stocking two size-classes during two seasons (phase I: 25–50 mm total length [TL], 30–45 d old, early summer and phase II: 100–150 mm TL, 180–210 d old, late fall) at the freshwater–saltwater interface. Phase-II fish were reared at both fresh and brackish water production facilities to determine which rearing salinity produced the highest postrelease survival. A standardized stratified random sampling survey was conducted monthly to monitor abundance in the river and obtain tissue samples for genetic analysis (see Fountain 2010 for survey details).

Approximately 99,000 phase-I striped bass were stocked in 2007 and 22,000 phase-II fish were released during two stocking years (2006 and 2007), with each produced from unique genetic groups. Sampling from November 2006 through May 2010 resulted in the collection of 127 striped bass from the 2006 and 2007 year-classes, with genetic evaluation indicating a 100% hatchery contribution for both year-classes. The age-0 results indicate that phase-II fish reared at the brackish water production facility had better postrelease survival than those raised in freshwater (Table 3). However, by age 2, the contribution from each production facility was similar, suggesting that it was likely that the fish stocked from each facility utilized different areas of the river which were not sampled. Interestingly, no phase-I fish were recaptured during electrofishing in the brackish reaches of the river; however, a single phase-I fish was recaptured during a directed sampling effort upstream in the freshwater portion of the river. The initial results provided good evidence for the survival potential of young stocked striped bass in the Ashley River system and have simultaneously raised two new hypotheses. First, these preliminary results suggest that stocking phase-II fish from a brackish water production facility produces the highest contribution levels. Second, the lack of phase-I fish in the brackish water reach of the river and its presence in the freshwater section suggest that habitat partitioning occurs among different size-classes.

The first 2 years of this project provided evidence that stocking increased the abundance of juvenile striped bass in the Ashley River and optimistic preliminary indications for the potential for successful re-establishment of the striped bass population. However, as female striped bass do not reach reproductive maturity until 4–5 years of age, the ultimate goal of the project to determine the self-sustaining potential for a population of striped bass will not be fully realized for 2–3 more years as the stocked fish recruit into the adult reproductive pool. As studies continue, genetic markers will continue to play a vital role in assessing the design and success of the restoration effort.

Lakes Marion and Moultrie, South Carolina

A large effort is aimed at managing the striped bass population in the lower Santee–Cooper River system, which includes lakes Marion and Moultrie. Striped bass are collected annually using a standardized winter gill-net sampling regime to evaluate wild and hatchery-released striped bass abundance (Lamprecht et al. 2013, this volume). Although the goal of SCDNR's program is to manage striped bass populations in terms of optimal sport fishing harvest while maximizing the possibility of conserving genetically diverse populations, the Freshwater Fisheries Section (FFS) also incorporates ex-

TABLE 3. Proportion of phase-II striped bass released and recaptured in the Ashley River, South Carolina from brackish (BW) and freshwater (FW) hatcheries. No. indicates number of fish recaptured from each treatment; % indicates the proportion of recaptured fish from each treatment. Asterisks indicate significant differences ($p < 0.05$) in proportions recaptured in relation to proportions stocked based on Monte Carlo random simulation analyses.

			Age 0		Age 1		Age 2	
	No. stocked	% stocked	No.	%	No.	%	No.	%
2006 year-class								
BW	5,788	52	19	79*	27	53	12	63
FW	5,302	48	5	21*	24	47	7	37
Total	11,090	100	24	100	61	100	19	100
2007 year-class								
BW	5,681	51	8	80*	10	91*	1	50
FW	5,395	49	2	20*	1	9*	1	50
Total	11,076	100	10	100	11	100	2	100

perimental research into its program to improve the understanding of habitat utilization, recruitment limitations, and environmental influences on striped bass survival. In 2009, SCDNR began using genetic markers for tracking all striped bass hatchery releases within the state. However, the FFS has been collecting and archiving genetic samples of broodstock since 2006, and as a result, staff have been able to retroactively determine the genetic characteristics of the 2006–2008 stocking cohorts that were resampled by gill netting (Darden et al. 2009). From these data, a number of questions have been examined regarding striped bass stocking success within the lakes, as well as hatchery performance and movement of hatchery produced striped bass.

Overall contribution.—In both the 2006 and 2007 year-classes, contributions of stocked fish were high (70–75%, Table 4) at age 0 and 1 and increased to 100% at age 2 for the 2006 year-class. The sampling of the 2008 year-class (age 0) indicated higher contribution, ~85%, than the previous two year-classes. As sampling continues in succeeding years, data will become available to determine the most successful release strategies (size, season, location, etc.) so future stockings can be optimized to improve survival and eventual contribution to the fishery. Data obtained from these efforts can also be used to evaluate trends in contribution that can be put into the context of year-class strength and annual environmental variability.

Influence of stocking habitat.—In recent years, researchers have been evaluating the influence of stocking habitat on the effectiveness of striped bass stocking within the system (Lamprecht and colleagues, South Carolina Department of Natural Resources, unpublished data). Predation can play a significant role in initial post release mortality of stocked hatchery fish (Buckmeier et al. 2005; Mesing et al. 2008), and striped bass fingerling predators may be more abundant in shallow nearshore habitats. A study was designed to determine if mid-lake stocking would produce better survival than traditional shoreline stocking through a reduction of initial postrelease predation. Phase-I striped bass were stocked in either shallow shoreline or mid-lake waters. Based on two year-classes (2006, 2007) evaluated during three sampling seasons, the results indicated that stocked fish were being recaptured in the same proportion as they were initially stocked, with similar stocking survival among the mid-lake and shoreline stocking treatments (Table 5). These results imply that there is no advantage to stocking phase-I fish mid-lake versus nearshore and suggest that there is no difference in initial predation mortality among these habitats and/or that phase-I striped bass are capable of quickly dispersing into appropriate habitat following stocking.

Phase I versus phase II.—In 2007, a study was initiated to compare the stocking effectiveness of phase-I and phase-II striped bass within the lower Santee–Cooper system (Lamprecht,

TABLE 4. Stocking contribution summary for striped bass within Lakes Marion and Moultrie, South Carolina over multiple stocking years and ages at recapture. Stocking contribution refers to the percentage contribution stocked fish made to the whole striped bass population (hatchery + wild). Sample sizes for each sample year are shown in parentheses.

	Percent stocking contribution (*n*)		
Year-class	Age 0	Age 1	Age 2
2006	72.0 (25)	74.0 (53)	100.0 (16)
2007	69.6 (23)	74.4 (126)	73.3 (15)
2008	84.5 (191)	81.9 (127)	83.3 (78)

TABLE 5. Stocking impact summary of the 2006 and 2007 year-classes, comparing shore and mid-lake stocked phase-I striped bass by age at recapture in Lakes Marion and Moultrie, South Carolina. No. indicates number of fish recaptured from each treatment; % indicates the percentage of recaptured fish from each treatment. There were no significant differences detected among treatments in proportion recaptured relative to proportion stocked based on Monte Carlo random simulation analyses.

			Age 0		Age 1		Age 2	
	No. stocked	% stocked	No.	%	No.	%	No.	%
2006 year-class								
Shore	1,538,829	74.7	14	87.5	30	88.2	10	83.3
Mid-lake	519,897	25.3	2	12.5	4	11.8	2	16.7
Total	2,058,726	100.0	16	100.0	34	100.0	12	100.0
2007 year-class								
Shore	965,196	85.6	4	100.0	26	83.9	5	100.0
Mid-lake	162,636	14.4	0	0.0	5	16.1	0	0.0
Total	1,127,832	100.0	4	100.0	31	100.0	5	100.0

personal communication). Evaluations of contributions by the 2007 year-class over two sampling seasons provided conflicting results. In the age-0 collections, phase-II fish were recaptured in significantly higher proportions than they were stocked (Table 6), suggesting that phase-II stocking was a more effective management strategy. However, in age-1 collections, phase-II fish were recaptured in the same proportions relative to their initial stocking, suggesting no difference in the effectiveness of phase-I and phase-II stockings. Although the initial results are variable, both suggest that phase-II fish are recruiting at least as effectively as phase-I fish, which is contrary to results of Hayes and Penny (1993), who concluded that phase-I stocking was the more effective stocking strategy in Lake Murray, South Carolina. Perhaps, as Wallin and Van Den Avyle (1995) suggest, the early phase-II results were not due to stocking strategy, but rather handling methods used during the initial experiment since fish were harvested, tagged, transported, and released within a 2-d period. To further evaluate these stocking strategies, phase-II fish were also stocked in 2008 and were to be released again in fall 2009. Contributions based on the 2008 and 2009 replicate stocking years will be evaluated to elucidate a clearer picture of relative contribution based on size and season of release. However, the final

TABLE 6. Stocking impact summary of the 2007 year-class fish comparing phase-I and phase-II striped bass stocked within Lakes Marion and Moultrie by age at recapture. No. indicates number of fish recaptured from each treatment; % indicates the percentage of recaptured fish from each treatment. Asterisks indicate significant differences ($p < 0.05$) in proportions recaptured in relation to proportions stocked based on Monte Carlo random simulation analyses (Age 1 comparison: $p = 0.067$).

			Age 0		Age 1	
	No. stocked	% stocked	No.	%	No.	%
Phase I	2,490,139	99.0	14	87.5*	68	97.1
Phase II	26,152	1.0	2	12.5*	2	2.9
Totals	2,516,291	100.0	16	100.0	70	100.0

determination of the optimum strategy will need to take into account financial aspects of rearing, as phase-I production may remain the more cost-effective strategy within freshwater systems due to the increased production and transportation costs associated with stocking phase-II fish.

Hatchery production.—The first 2 years (2006 and 2007 year-classes) of genetic-based data are also providing some insight into hatchery production of striped bass. For instance, success at the level of genetic groups (spawning of a single female with three males) can be determined as well as success of individual males and females (Table 7). The 2006 evaluation indicated that within a genetic group, there often appeared to be a disproportionately high contribution from one male among recaptured offspring (Table 7). Hatchery protocol during 2006 included strip-spawning females into a single container and then simultaneously adding milt from three males prior to water activation. In response to the genetic results, 2009 hatchery protocols were changed so that the eggs collected from each female were divided into three containers and milt was added from a single male to each, effectively eliminating any sperm competition and potentially increasing the hatchery's effective population size, which is important for this naturally reproducing population. We also verified the genetic distinctiveness of the striped bass broodstock that were used from 2006 to 2008, with each adult only represented in 1 year during these 3 years.

Movement.—Results from the genetic identification of offspring have shown that striped bass within the Santee–Cooper River system are capable of long-distance movement, even within their first year. Age-0 fish recaptured in Lakes Marion and Moultrie were genetically identified as having been stocked as far upstream as Lake Wateree and the Catawba River (~100 river kilometers). Therefore, these initial evaluations using multiple distinct family groups as proxies for treatment tags allowed for finer-scale evaluation of the effects of stocking groups and provided more information than conventional tags.

Implications for Study Design and Hatchery Operations

Our examples have shown there are multiple benefits of incorporating genetic markers into both experimental and management stocking applications for striped bass. From an experimental design standpoint, their use allows for the development of complex designs with potential for both within- and among-year treatment evaluation. As typical with any successful experimental design, a priori determination of treatments and genetic assignments is necessary and early hatchery coordination is vital due to the stocking component. Likely the most important implication for both study design and hatchery operations is the critical nature of organized and accurate record keeping—both in terms of genetic samples and hatchery production.

Within the hatchery, broodstock need to be carefully identified and tracked. Genetic samples need to be taken from all broodstock and accurate spawning design records are a necessity. These requirements may produce a need for a short-term, visual tag on hatchery broodstock, depending on how long fish are maintained and how often they are used. One of the most important aspects of using genetic markers is the tracking of genetic groups of offspring continuously from spawning through hatching, pond production, harvesting, and release. Development of a good program begins with broodstock management, and it is advisable, if possible, to evaluate broodstock combinations for genetic uniqueness in advance. If broodstock are rotated and used for multiple production years, it is highly beneficial to verify genetic distinction between groups both within and among years. In the case that new broodstock are acquired each year, initial analyses are warranted to verify that sufficient levels of genetic diversity exist in the population to yield distinct genetic groups from random sampling. The ul-

TABLE 7. Parental contribution of 2006 year-class hatchery-produced striped bass in recapture efforts across multiple years in the Santee-Cooper system, South Carolina.

	Age 0				Age 1				Age 2		
Mother	Fathers (# offspring)			Total	Fathers (# offspring)			Total	Fathers (# offspring)		Total
BSF0602	0614(1)			1	0609(1)			1	0614(1)		1
BSF0603	0605(3)			3	0605(3)	0609(1)	0614(5)	9	0605(1)	0614(1)	2
BSF0604	0605(3)			3	0605(4)			4			
BSF0614	0646(1)			1					0646(1)		1
BSF0620	0616(1)	0637(1)		2							
BSF0623									0607 (1)		1
BSF0624					0639(1)			1			
BSF0626	0611(2)	0614(11)	0647(1)	14	0614(2)			2			
BSF0627	0617(1)			1							
BSF0628	0651(1)	0653(1)	0656(2)	4							
BSF0630	0616(4)	0628(4)	0638(30)	38					0616(1)	0638(2)	3
BSF0631	0621(1)	0631(4)	0640(6)	11	0640(1)			1			
BSF0633					0628(1)	0647(1)		2			
BSF0636	0661(1)			1							
BSF0637	0645(1)			1							
BSF0640	0673(1)	0677(2)		3	0677(4)			4	0673(2)		2
BSF0642					0670(1)	0673(1)		2			
BSF0643	0677(1)			1	0673(1)			1			
BSF0644	0678(5)	0682(1)		6	0678(3)			3	0678(1)	0682(1)	2
BSF0645	0692(1)			1	0679(1)	0692(1)		2			
BSF0648	0694(16)			16	0693(1)	0694(2)		3	0694(2)	0693(1)	3
BSF0649					0692(1)			1			
BSF0656	0690(2)			2							
BSF0657	0688(1)			1	0695(1)			1			
BSF0658	0690(1)			1							
			Total	110			Total	37		Total	15

timate goal of a project should guide the overall hatchery production design. For instance, with an experimental focus of small scale over a finite time period where treatments are composed of discrete genetic groups, fewer parents per treatment would be preferable while still maintaining high statistical power through replication. However, in the case of large-scale, long-term population management, ideally more parents are preferable to increase the effective population size of the hatchery and thereby minimize negative impacts on the wild effective population size. Therefore, study design and hatchery operations are intimately tied together in the use of genetic markers and there needs to be a balance between the management of genetic diversity in a stocking or research program, detection power of the marker suite, and the logistics that affect hatchery production.

Further Management Applications

There are two main components of comprehensive management plans for striped bass that can benefit from the application of genetic markers. The first component includes the improvements to stocking programs provided by the advantages of genetic markers. As they are an inherent property of the fish, genetic tags can be utilized with any size fish, no tag recovery bias or mortality is introduced due to tagging effects either in the hatchery or once stocked in the wild, and there is absolutely no tag loss. The nonlethal retrieval of tag information, such as minute tissue samples from a fin, provides a situation where no population reduction is required for evaluation, and therefore long-term, repeated detection and assessment is possible. The availability of previously identified genetic markers for striped bass provides unlimited potential for experimental stocking research, whether investigating the effects of stocking size, hatchery production variables or environmental variation associated with season or recruitment to better understand and manage the species. The second component of a comprehensive management program that can be improved by utilizing genetic markers is the cost-effective use of the same genetic data generated from evaluation of a stocking program to genetically characterize and monitor the managed population. Population genetic assessments can provide information on structure and gene flow, effective population sizes, ancestral population sizes, bottlenecks, and overall genetic diversity. Therefore, there are substantial benefits to using a single genetic marker to integrate the various components of comprehensive management plans for striped bass.

Acknowledgments

Our work was funded by the National Fish and Wildlife Foundation and the state of South Carolina and conducted in collaboration with the SCDNR Waddell Mariculture Center, SCDNR Dennis Wildlife Center, U.S. Fish and Wildlife Service Orangeburg National Fish Hatchery and Hollings Marine Laboratory. We thank Wally Jenkins, Scott Lamprecht, and Jim Bulak for their extensive contribution to and support of this striped bass research. Leo Rose was invaluable in recreating the genetic hatchery and release records necessary for the Lakes Marion and Moultrie studies. Karl Brenkert, Doug Cooke, Beth Cushman, Jen Fountain, Jim Glenn, Chad Holbrook, Kim Hughes, Mary Martin, Matt Perkinson, Matt Walker, and Justin Yost contributed significantly to various aspects as well. We would like to thank the dedicated staff at all of our collaborating institutions for their striped bass efforts that made this project successful. This is publication number 672 from the Marine Resources Division, South Carolina Department of Natural Resources.

References

Baudouin, L., and P. Lebrun. 2001. An operational Bayesian approach for the identification of sexually reproduced cross-fertilized populations using molecular markers. Acta Horticulturae (ISHS) 546:81–93.

Blankenship, H., and K. Leber. 1995. A responsible approach to marine stock enhancement. Pages 167–175 *in* H. Schramm, Jr.

and R. Piper, editors. Uses and effects of cultured fishes in aquatic ecosystems. American Fisheries Society, Symposium 15, Bethesda, Maryland.

Blouin, M., M. Parsons, V. Lacaille, and S. Lotz. 1996. Use of microsatellite loci to classify individuals by relatedness. Molecular Ecology 5:393–401.

Bosworth, B., E. Dunnington, G. Libey, and L. Stallard. 1994. Restriction enzyme/multilocus probe combinations useful for DNA fingerprinting of the striped bass and their F1 hybrid. Aquaculture 123:205–215.

Bravington, M., and R. Ward. 2004. Microsatellite markers: evaluating their potential for estimating the proportion of hatchery-reared offspring in a stock enhancement programme. Molecular Ecology 13:1287–1297.

Brown, K., G. Baltazar, and M. Hamilton. 2005. Reconciling nuclear microsatellite and mitochondrial marker estimates of population structure: breeding population structure of Chesapeake Bay striped bass (*Morone saxatilis*). Heredity 94:606–615.

Brown, K., G. Baltazar, B. Weinstein, and M. Hamilton. 2003. Isolation and characterization of nuclear microsatellite loci in the anadromous marine fish *Morone saxatilis*. Molecular Ecology Notes 3:414–416.

Buckmeier, D., R. Betsill, and J. Schlechte. 2005. Initial predation of stocked fingerling largemouth bass in a Texas reservoir and implications for improving stocking efficiency. North American Journal of Fisheries Management 25:652–659.

Bulak, J., C. Thomason, K. Han, and B. Ely. 2004. Genetic variation and management of striped bass populations in the coastal rivers of South Carolina. North American Journal of Fisheries Management 24:1322–1329.

Cercueil, A., E. Bellemain, and S. Manel. 2002. PARENTE: computer program for parentage Analysis. Journal of Heredity 93:458–459.

Chapman, R. 1990. Mitochondrial DNA analysis of striped bass populations in Chesapeake Bay. Copeia 2:355–366.

Chapman, R., A. Ball, and L. Mash. 2002. Spatial homogeneity and temporal heterogeneity of red drum, *Sciaenops ocellatus*, microsatellites: effective populations sizes and management implications. Marine Biotechnology 4:589–603.

Cornuet, J., S. Piry, G. Luikart, A. Estoup, and M. Solignac. 1999. New methods employing multilocus genotypes to select or exclude populations as origins of individuals. Genetics 153:1989–2000.

Couch, C., A. Garber, C. Rexroad, III, J. Abrams, J. Stannard, M. Westerman, and C. Sullivan. 2006. Isolation and characterization of 149 novel microsatellite markers for striped bass, *Morone saxatilis*, and cross-species amplification in white bass, *Morone chrysops*, and their hybrid. Molecular Ecology Notes 6:667–669.

Dabrowski, K., and K. Tsukamoto. 1986. Tetracycline tagging in coregonid embryos and larvae. Journal of Fish Biology 29:691–698.

Danzmann, R. 1997. PROBMAX: A computer program for assigning unknown parentage in pedigree analysis from known genotypic pools of parents and progeny. Journal of Heredity 88:333–334.

Darden, T., F. Sessions, S. Lamprecht, and L. Rose. 2009. Retrospective application of genetic tags for evaluation of striped bass stocking in the Santee–Cooper system from 2006–2008. South Carolina Department of Natural Resources, Internal Report, Columbia.

Denson, M., and T. Smith. 2008. Use of oxytetracycline to mark larval red drum: survival and mark persistence. North American Journal of Fisheries Management 28:1779–1789.

Ferguson, M., and R. Danzmann. 1998. Role of genetic markers in fisheries and aquaculture: useful tools or stamp collecting? Canadian Journal of Fisheries and Aquatic Sciences 55:1553–1563.

Fountain, J. 2010. Assessment of multiple stocking strategies of striped bass in the Ashley River, South Carolina using multiplexed microsatellite panels. Master's thesis. College of Charleston, Charleston, South Carolina.

Fountain, J., T. Darden, W. Jenkins, and M. Denson. 2009. Three multiplexed microsatellite panels for striped bass, *Morone saxatilis*. Southeastern Naturalist 8:671–676.

Gerber, S., P. Chabrier, and A. Kremer. 2003. FAMOZ: a software for parentage analysis using dominant, codominant and uniparentally inherited markers. Molecular Ecology Notes 3:479–481.

Gold, J., L. Ma, E. Saillant, P. Silva, and R. Vega. 2008. Genetic effective size in populations of hatchery-raised red drum, *Sciaenops ocellatus*, released for stock enhancement. Transactions of the American Fisheries Society 137:1327–1334.

Han, K., L. Li, G. Leclerc, A. Hays, and B. Ely. 2000. Isolation and characterization of microsatellite loci for striped bass (*Morone saxatilis*). Marine Biotechnology 2:405–408.

Hayes, W., and G. Penny. 1993. Evaluation of phase I versus phase II striped bass introduced into Lake Murray. Pages 78–85 *in* Fisheries investigations in lakes and streams – District III. South Carolina Wildlife and Marine Resources Department, Internal Annual Progress Report F-18-19, Charleston.

Hendricks, M. L., D. L. Tarsello, and T. W. H. Backman. 1994. Use of otolith microstructure to distinguish wild from hatchery-reared American shad in the Susquehanna River. North American Journal of Fisheries Management 14:151–161.

Jenkins, W., M. Denson, M. Collins, and T. Smith. 1997. Differentiating between hatchery and wild red drum by the position of the first annulus on sagitta. North American Journal of Fisheries Management 17:1001–1004.

Jones, A., and W. Ardren. 2003. Methods of parentage analysis in natural populations. Molecular Ecology 12:2511–2523.

Jones, A. G., C. M. Small, K. A. Paczolt, and N. L. Ratterman. 2010. A practical guide to methods of parentage analysis. Molecular Ecology Resources 10:6–30.

Kalinowski, S., M. Taper, and T. Marshall. 2007. Revising how the computer program CERVUS accommodates genotyping error increases success in paternity assignment. Molecular Ecology 16:1099–1006.

Karlsson, S., E. Saillant, B. Bumguardner, R. Vega, and J. Gold. 2008. Genetic identification of hatchery-released red drum (*Sciaenops ocellatus*) in Texas bays and estuaries. North American Journal of Fisheries Management 28:1294–1304.

Lamprecht, S. D., M. D. Mauck, and V. J. DiCenzo. 2013. Natural reproduction: how it has affected striped bass management of John H. Kerr and Santee–Cooper reservoirs and Lake Texoma. Pages 413–430 *in* J. S. Bulak, C. C. Coutant, and J. A. Rice, editors. Biology and management of inland striped bass and hybrid striped bass. American Fisheries Society, Symposium 80, Bethesda, Maryland.

Laughlin, T., and B. Turner. 1996. Hypervariable DNA markers reveal high genetic variability within striped bass populations of the lower Chesapeake Bay. Transactions of the American Fisheries Society 125:49–55.

Liu, Z., and J. Cordes. 2004. DNA marker technologies and their applications in aquaculture genetics. Aquaculture 238:1–37.

Liu, J. X., and B. Ely. 2009. Sibship reconstruction demonstrates extremely low effective population size of striped bass *Morone saxatilis* in the Santee–Cooper system, South Carolina, USA. Molecular Ecology 18:4112–4120.

McDonald, G., R. Danzmann, and M. Ferguson. 2004. Relatedness determination in the absence of pedigree information in three cultured strains of rainbow trout (*Oncorhynchus mykiss*). Aquaculture 233:65–78.

Mesing, C., R. Cailteux, P. Strickland, E. Long, and M. Rogers. 2008. Stocking of advanced-fingerling largemouth bass to supplement year-classes in Lake Talquin, Florida. North American Journal of Fisheries Management 28:1762–1774.

Perez-Enriquez, R., and N. Taniguchi. 1999. Genetic structure of red sea bream (*Pagrus major*) population off Japan and the southwest Pacific, using microsatellite DNA markers. Fisheries Science 65:23–30.

Piry, S., A. Alapetite, J. Cornuet, D. Paetkau, L. Baudouin, and A. Estoup. 2004. GENECLASS2: a software for genetic assignment and first-generation migrant detection. Journal of Heredity 95:536–539.

R Development Core Team. 2009. R: a language and environment for statistical computing. R Foundation for Statistical Computing, Vienna, Austria.

Reinert, T., J. Wallin, M. Griffin, M. Conroy, and M. Van Den Avyle. 1998. Long-term retention and detection of oxytetracycline marks applied to hatchery-reared larval striped bass,

Morone saxatilis. Canadian Journal of Fisheries and Aquatic Sciences 55:539–543.

Renshaw, M., E. Saillant, R. Broughton, and J. Gold. 2006. Application of hypervariable genetic markers to forensic identification of 'wild' from hatchery-raised red drum, *Sciaenops ocellatus*. Forensic Science International 156:9–15.

Rexroad, C., R. Vallejo, I. Coulibaly, A. Garber, M. Westerman, and C. Sullivan. 2006. Identification and characterization of microsatellites for striped bass from repeat-enriched libraries. Conservation Genetics 7:971–982.

Robbins, S., T. Darden, A. Ball, and M. Denson. 2008. Genetic parentage analysis of red drum in South Carolina and the use of genotypic data from archived otoliths. North American Journal of Fisheries Management 28:1796–1801.

Ross, K., X. Wang, K. O'Malley, D. Gatlin, III, and J. Gold. 2004. Microsatellite DNA markers for parental assignment in hybrid striped bass (*Morone saxatilis* × *Morone chrysops*). Molecular Ecology Notes 4:156–159.

Roy, N., L. Maceda, and I. Wirgin. 2000. Isolation of microsatellites in striped bass *Morone saxatilis* (Teleostei) and their preliminary use in population identification. Molecular Ecology 9:827–829.

Saillant, E., M. Renshaw, D. Gatlin, III, W. Neill, R. Vega, and J. Gold. 2009. An experimental assessment of genetic tagging and founder representation in hatchery-reared red drum (*Sciaenops ocellatus*) used in stock enhancement. Journal of Applied Ichthyology 25(1):108–113.

Schroder, S., C. Knudsen, and E. Volk. 1995. Marking salmon fry with strontium chloride solutions. Canadian Journal of Fisheries and Aquatic Sciences 52:1141–1149.

Secor, D., and E. Houde. 1995. Temperature effects on the timing of striped bass egg production, larval viability, and recruitment potential in the Patuxent River (Chesapeake Bay). Estuaries 18:527–544.

Secor, D., M. White, and J. Dean. 1991. Immersion marking of larval and juvenile hatchery-produced striped bass with oxytetracycline. Transactions of the American Fisheries Society 120:261–266.

Signorovitch, J., and R. Nielsen. 2002. PATRI - paternity inference using genetic data. Bioinformatics 18:341–342.

South Carolina State Board of Fisheries. 1924. Annual Report of the Board of Fisheries to the Governor and General Assembly. South Carolina State Library, Columbia.

Stellwag, E., and R. Rulifson. 1994. Mitochondrial DNA diversity of Roanoke River striped bass. Transactions of the American Fisheries Society 123:321–334.

Stevens, R. 1975. Current and future considerations concerning striped bass culture and management. Proceedings of the Annual Conference Southeastern Association of Game and Fish Commissioners 28(1974):69–73.

Taggart, J. 2007. FAP: an exclusion-based parental assignment program with enhanced predictive functions. Molecular Ecology Notes 7:412–415.

Thomas, L., S. Holt, and C. Arnold. 1995. Chemical marking techniques for larval and juvenile red drum (*Sciaenops ocellatus*) otoliths using different fluorescent markers. Pages 703–711 *in* D. Secor, J. Dean and S. Campana, editors. Recent developments in fish otolith research. University of South Carolina Press, Columbia, South Carolina.

Toombs, A. 2000. Development of microsatellite loci for examination of hatchery-sustained and wild assemblages of striped bass, *Morone saxatilis*. Master's thesis, Virginia Commonwealth University, Richmond.

Tringali, M. 2006. A Bayesian approach for the genetic tracking of cultured and released individuals. Fisheries Research 77:159–172.

Van Den Avyle, M., and J. Wallin. 2001. Retention of internal anchor tags by juvenile striped bass. North American Journal of Fisheries Management 21:656–665.

Volk, E., S. Schroder, and K. Fresh. 1990. Inducement of unique otolith banding patterns as a practical means to mass-mark juvenile Pacific salmon. Pages 203–215 *in* N. C. Parker, A. E. Giorgi, R. C. Heidinger, D. B. Jester, Jr., E. D. Prince, and G. A. Winans, editors. American Fisheries Society, Symposium 7, Bethesda, Maryland.

Waldman, J., R. Bender, and I. Wirgin. 1998. Multiple population bottlenecks and DNA diversity in populations of wild striped bass, *Morone saxatilis*. Fishery Bulletin 96:614–620.

Waldman, J., R. Richards, W. Schill, I. Wirgin, and M. Fabrizio. 1997. An empirical comparison of stock identification techniques applied to striped bass. Transactions of the American Fisheries Society 126:369–385.

Wallin, J., and M. Van Den Avyle. 1995. Interactive effects of stocking site salinity and handling stress on survival of striped bass fingerlings. Transactions of the American Fisheries Society 124:736–745.

Wilson, C., D. Beckman, and J. Dean. 1987. Calcein as a fluorescent marker of otoliths of larval and juvenile fish. Transactions of the American Fisheries Society 116:668–670.

Wirgin, I., and L. Maceda. 1991. Development and use of striped bass-specific RFLP probes. Journal of Fish Biology 39(Supplement A):159–167.

American Fisheries Society Symposium 80:413–430, 2013

Natural Reproduction: How It Has Affected Striped Bass Management of John H. Kerr and Santee–Cooper Reservoirs and Lake Texoma

Scott D. Lamprecht*
South Carolina Department of Natural Resources
305 Black Oak Road, Bonneau, South Carolina 29431, USA

Matt D. Mauck
Oklahoma Department of Wildlife Conservation
2021 Caddo Highway, Caddo, Oklahoma 74729, USA

Victor J. DiCenzo
Virginia Department of Game and Inland Fisheries
107 Foxwood Drive, Farmville, Virginia 23901, USA

Abstract.—Striped bass *Morone saxatilis* have been introduced into hundreds of inland impoundments, and in several instances, natural reproduction and recruitment has developed. Three of these impoundments are Lake Texoma, John H. Kerr Reservoir, and Santee Cooper Reservoir, whose natural recruitment is dependent on their tributary rivers. Historic management of these populations has required addressing issues of excess reproduction, excess harvest, and spawning flow alteration. Initial high abundance fostered a harvest oriented mind set among striped bass anglers in these reservoirs, which today can only be sustained on Lake Texoma. Supplemental stocking of both Kerr and Santee-Cooper reservoirs has been used to replace declines in natural recruitment. Mortality estimates suggest that reproductive overfishing may be the underlying cause of natural recruitment declines. Recent adjustments to harvest regulations at both Kerr and Santee-Cooper were required to increase the number of striped bass reaching reproductive age. Both management approaches reduce fishing mortality through reduced daily bag and higher size limits, but also recognize that high catch-and-release mortality makes length limits counterproductive during the summer.

Introduction

When striped bass *Morone saxatilis* were introduced into freshwater reservoirs from their native coastal rivers (Van Horn 2013, this volume), it was assumed that population size could be controlled through stocking rate adjustment. Unexpectedly, several reservoir systems proved capable of sustaining natural reproduction. The loss of the ability to exert control over recruitment, solely through stocking, led to a number of management challenges, which are somewhat unique to each reservoir system. In their native range, anadromous striped bass migrate far enough up coastal rivers to allow current to suspend their semibuoyant eggs throughout their 2-d incubation period. The generally recognized minimum flow rate need to keep striped bass eggs in suspension is 0.3 m/s (Albrecht 1964), but this is dependent on both the density of the water and the density of the

* Corresponding author: lamprechts@dnr.sc.gov

water hardened eggs. Upstream migrations in excess of 200 river kilometers are not uncommon. Although natural reproduction has occurred within deep, oligotrophic western reservoirs (Gustaveson et al. 1984; Gustaveson and Blommer 2013, this volume), we concentrate on three southeastern U.S. reservoirs where riverine-based reproduction has historically occurred. We examine the evolution of the striped bass fisheries, historical management actions, and future direction of each system.

Reservoir Descriptions

Our three focal reservoirs, Santee–Cooper, John H. Kerr, and Texoma, are relatively old, having been built in the 1940s and 1950s. Santee–Cooper Reservoir is located in central South Carolina while John H. Kerr Reservoir is located on Virginia's southern border and is shared with North Carolina. Also shared by two states, Lake Texoma straddles the border of Oklahoma and Texas. Each reservoir has two tributary rivers that provide sufficient unimpeded length to successfully incubate striped bass eggs.

The Santee–Cooper system was impounded in 1941 for the purpose of power generation and to providc a public works project (Figure 1). Constructed in the coastal plain of the Atlantic slope, it is composed of two reservoirs, Lakes Marion and Moultrie, that span the lower Santee and Cooper River watersheds, covering an area of 68,000 ha. It has relatively soft water with a specific conductance range of 100–120 µS/cm. The confluence of the Congaree and Wateree rivers form the headwaters of the reservoir system. Flow in both rivers is influenced by upstream peaking hydroelectric projects, the Wateree flowing 122 km from the first upstream dam and the Congaree flowing 85 km from the first of two upstream tributary dams. The system has fish passage facilities on the Cooper and lower Santee rivers, which annually provide upstream access to anadromous American shad *Alosa sapidissima* and blueback herring *A. aestivalis*. The reservoir is classified as mesotrophic. Its shape and depth promote wind mixing that prevents strong stratification, though weak temperature and oxygen gradients are intermittently observed throughout a typical year. Minimum summer water temperatures typically exceed 27°C for more than 3 months, while minimum oxygen levels rarely fall below 3 mg/L in the deeper areas of the reservoir.

John H. Kerr is also an Atlantic slope reservoir that was constructed in the Roanoke River drainage by the U.S. Army Corps of Engineers for flood control and the production of hydroelectricity (Figure 2). It was completed in 1952 and totals nearly 20,000 ha. Kerr's two tributaries, the Dan and Roanoke rivers, join at the upper reaches of the reservoir. The Dan River flows 68 km from Brantley Steam Plant Dam before entering the reservoir, while the Roanoke River (also known as the Staunton River) flows 103 km from the outfall of upstream Leesville Dam. Like Santee–Cooper, Kerr Reservoir's water is relatively soft with specific conductance ranging from 105 to 120 µS/cm, and it has been classified as mesotrophic to slightly eutrophic. It becomes strongly stratified during the summer months, with epilimnetic temperature exceeding 27°C and portions of the hypolimnion experiencing dissolved oxygen levels below 2 mg/L.

Lake Texoma is a 36,000-ha impoundment of the Red River (Figure 3). It was constructed by the U.S. Army Corps of Engineers in 1944 and is authorized for flood control, hydroelectric power, water supply, and recreation. Its two highly mineralized tributaries, the Red (2,380–5,460 µS/cm) and the Washita (449–1,911 µS/cm) rivers, each flow unimpeded for more than 500 km before entering the reservoir in widely separated arms. Lake Texoma is more heavily mineralized than its Atlantic slope counterparts, with specific conductance ranging from 1,005 to 3,781 µS/cm and is classified as eutrophic. It becomes strongly stratified during the summer months with epilimnetic temperatures exceeding 27°C and the late summer hypolimnetic oxygen level falling below 2 mg/L.

Initial Fisheries Establishment

A striped bass fishery developed without augmentation in the Santee–Cooper system after

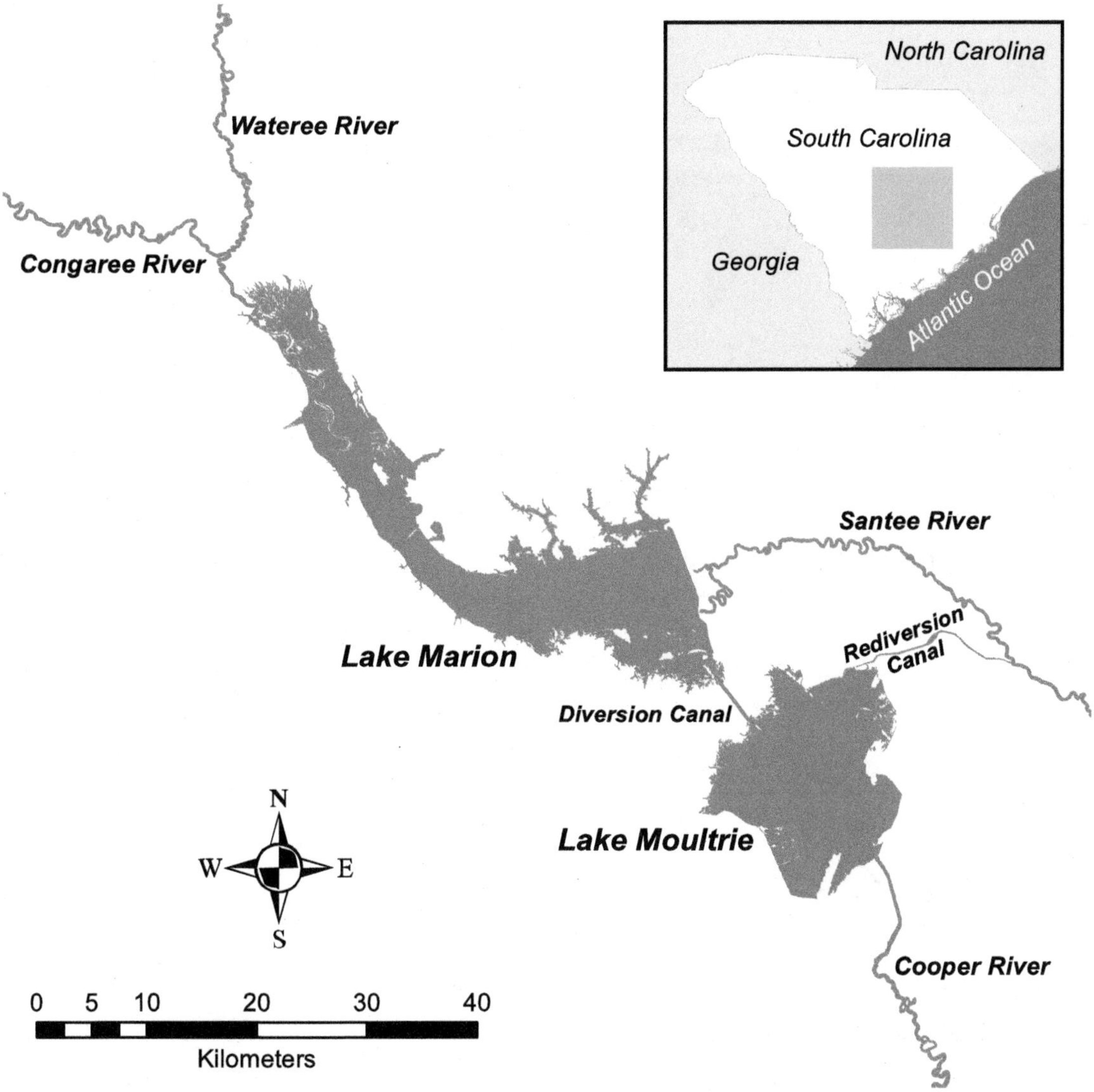

FIGURE 1. Map of Santee–Cooper Reservoir, South Carolina. The inset shows the location in the state of South Carolina, USA.

impoundment. Investigations of this population first demonstrated that striped bass could complete its life cycle entirely in freshwater (Scruggs 1957; Van Horn 2013). There is some debate about how the Kerr Reservoir population was established, as striped bass were known to occur in the river system prior to impoundment and fry stockings were conducted after impoundment. Regardless of the source, natural reproduction was documented in the late 1950s (Neal 1969). Lake Texoma's striped bass population was established through Oklahoma Department of Wildlife Conservation (ODWC) stockings made from 1965 to 1974, with natural reproduction first detected in 1973 (Mauck 1991).

Santee–Cooper's striped bass population quickly became established in the new reservoir, and by 1950, Stevens (1958) reported that anglers were targeting schooling fish and experimenting with various baits and techniques for catching them. Like its estuarine striped bass

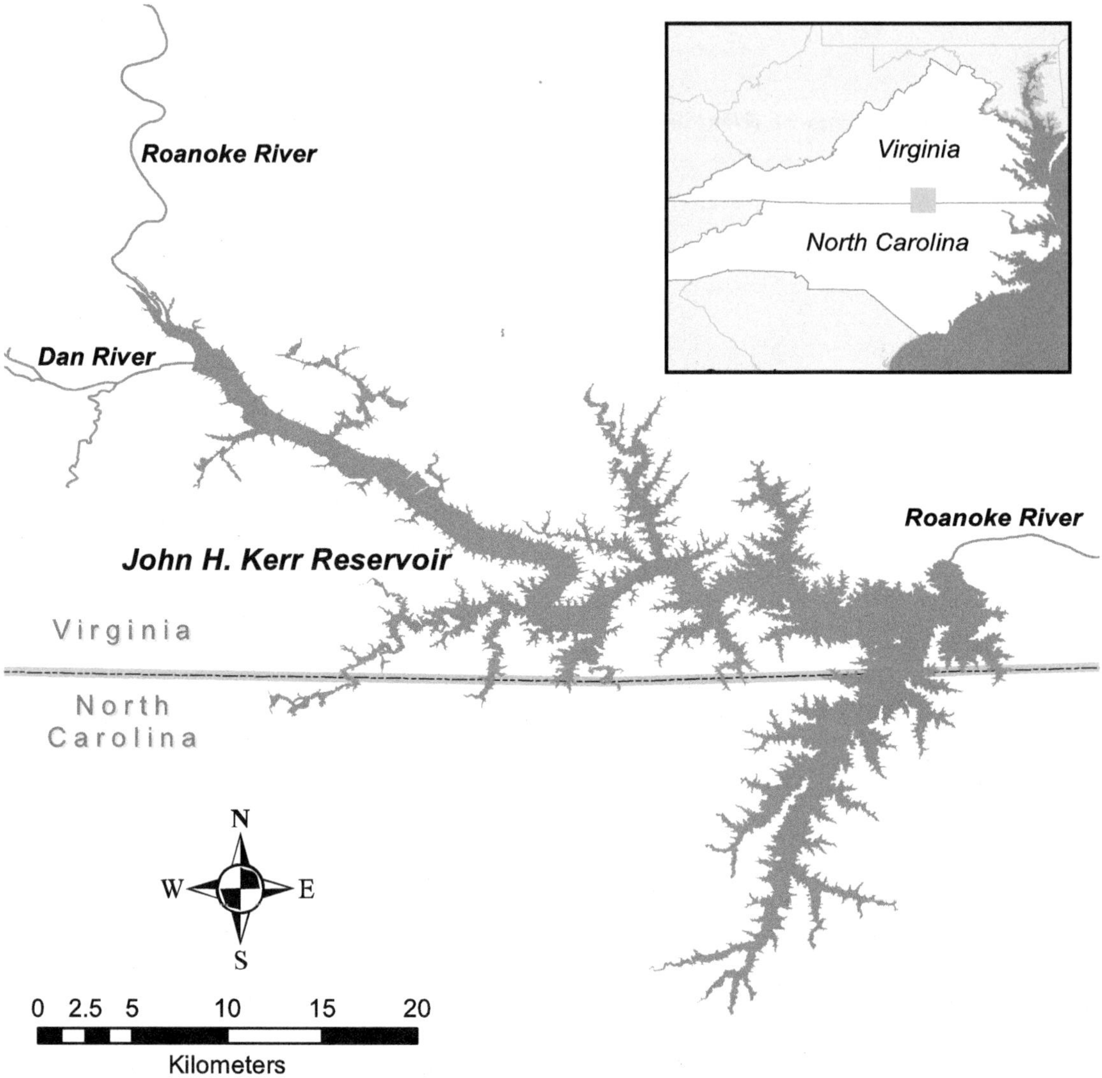

FIGURE 2. Map of John H. Kerr Reservoir, North Carolina/Virginia. The inset shows the location relative to the state of Virginia, USA.

counterparts, Santee–Cooper's landlocked population experienced variation in year-to-year recruitment. As gear and techniques were relatively unsophisticated, there was little attention given to harvest management. Ten years passed after impoundment before the Kerr striped bass fishery began to be noticed and an eight fish per day, 305 mm total length (TL) minimum length limit was established. The development of Texoma's striped bass fishery was quite different from either of the preceding systems. The population was established through multi-year stocking beginning in 1965, and it developed under higher catch efficiencies produced by a better equipped and informed angling public. Harvest management began with a one fish/d bag limit in 1967 and continued for 10 years when it was raised to three fish/d in 1977. During this time period, natural reproduction began and stocking was discontinued (Moczygemba and Hysmith 1996). Following initial establishment of self-sustaining, land-locked

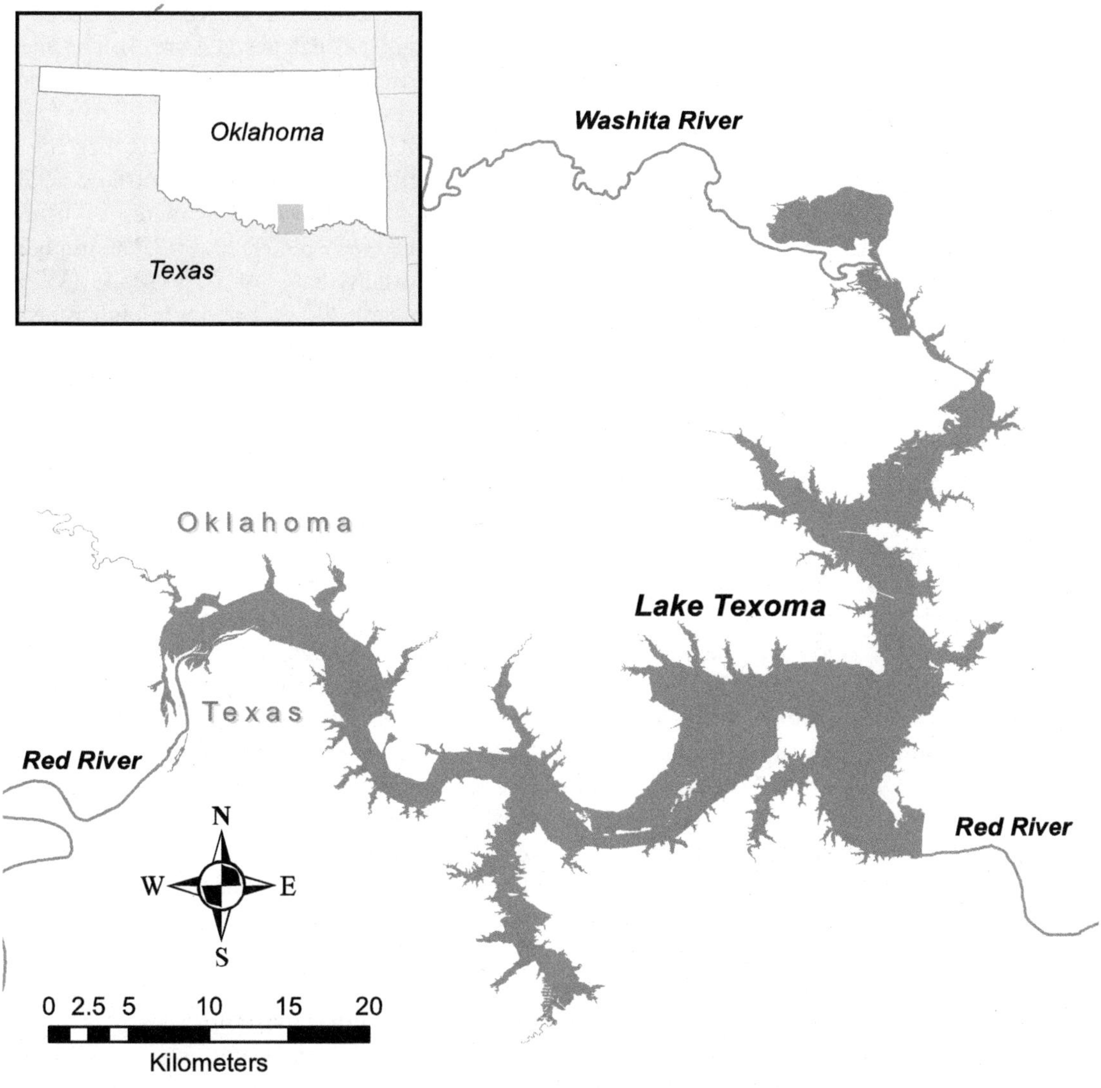

FIGURE 3. Map of Lake Texoma, Oklahoma/Texas. The inset shows the location relative to the state of Oklahoma, USA.

populations, the management opportunities and challenges at each reservoir took different trajectories.

Santee–Cooper

During the first half of its 68-year existence, the Santee–Cooper striped bass population was self-sustaining. Management first began in the early 1960s when the Pinopolis Dam springtime boat lock operations were increased in an effort to elevate anadromous fish passage as a means of providing additional forage for an underweight and overabundant striped bass population (Stevens 1960). A threefold increase in fishing pressure from 1972 to 1982 (White 1974; Bulak et al. 1983) and lowered levels of recruitment over the same time span led to a decline in the quality of the fishery (White 1989). To address this decline, a minimum size limit of 456 mm was established in 1984, with no change to the 10 fish/d limit. Supplemental stocking was also initiated in 1984 at a rate of 12 fingerlings/ha. In 1989, the daily harvest limit was reduced to five fish/d and stocking was increased to 37 fingerlings/ha. Five years later, 1994, the minimum size limit was raised to 533 mm.

Numerous studies of Santee–Cooper striped bass spawning have been conducted since Scruggs (1957) first reported landlocked reproduction. May and Fuller (1965) reported that most striped bass spawned in the Congaree River, and this tendency was reported again 28 years later (Bulak et al. 1993). Finger and Bulak (1988) documented an episode of acute and chronic toxicity to striped bass larvae in both tributaries. Bulak (1994) also suggested that operation of the two hydroelectric dams could disrupt the timing of spawning and Bulak et al. (1997) showed that most recruitment resulted from flows that delivered eggs at the right time to high-quality nursery habitat in upper Lake Marion. This biological information helped to construct a voluntary flow agreement with the hydroelectric project on the Wateree River and has proven very useful in the current Federal Energy Regulatory Commission (FERC) relicensing process on both the Wateree and Congaree River drainages.

The supplemental stocking that began in 1984 with 25–45-mm-long fingerlings was initiated to decrease the annual variability in striped bass abundance that was attributed to fluctuations in natural reproduction. Oxytetracycline marking of hatchery produced striped bass was initiated in 1990, and during years when hatchery contribution estimates were possible, the contribution has ranged between a high of 89% to a low of 26%. Without significant changes in methodology, stocked fingerling recruitment has varied as much as sevenfold among years, suggesting that in some years, significant environmental limitations occur at juvenile life stages as well as the earlier life stages affecting natural recruits.

The combined use of a minimum size limit, lowered bag limit, and supplemental stocking did not lead to an increase in harvestable fish, though the number of younger fish increased in winter gill-net catches (Figure 4). Despite a high rate of supplemental stocking, annual recruitment remained highly variable. Recent estimates of total mortality for fish older than age 1+ (~20 months) have ranged from 60% to 69.3% (Bulak et al. 1995) with estimates of fishing mortality between 39.3% and 43.2%, implying natural mortality rates of 26.3–30.2% (White and Lamprecht 2002). This high adult mortality rate coupled with several consecutive years of poor natural recruitment and poor recruitment of stocked fingerling resulted in a severe decline in the fishery, reaching its low in 2007.

Biologists' and anglers' concern over the recent decline provided the momentum to initiate a stakeholders' advisory process. Presented with the biological information, the stakeholders group proposed and analyzed potential harvest management actions to rebuild the fishery. The group presented their proposed harvest restrictions to the state legislature, which passed legislation restricting Santee–Cooper striped bass harvest to three fish/d with a 610-mm-TL minimum size limit and a closed harvest season from June 1 to September 30 (fishing regulations are set by legislation in South Carolina). This law became effective July 1, 2008. The objective was to allow more females to reach reproductive age and to address the high catch-and-release mortality that occurs in warm water when minimum-length limits are used. Because nearly 50% of the Santee–Cooper harvest occurs in its tributary and connected downstream rivers (White and Lamprecht 2002), this regulation has been applied to the entire system.

Striped bass anglers have recently accounted for approximately 11% of the annual reservoir fishing-effort estimates (White and Lamprecht 2002). This undervalues the total fishery because a significant portion of the effort and harvest occurs in the reservoir's unsurveyed tributary rivers. Average striped bass harvest size has varied as strong year-classes dominate the catch over several years and historical comparisons are obfuscated by the implementation

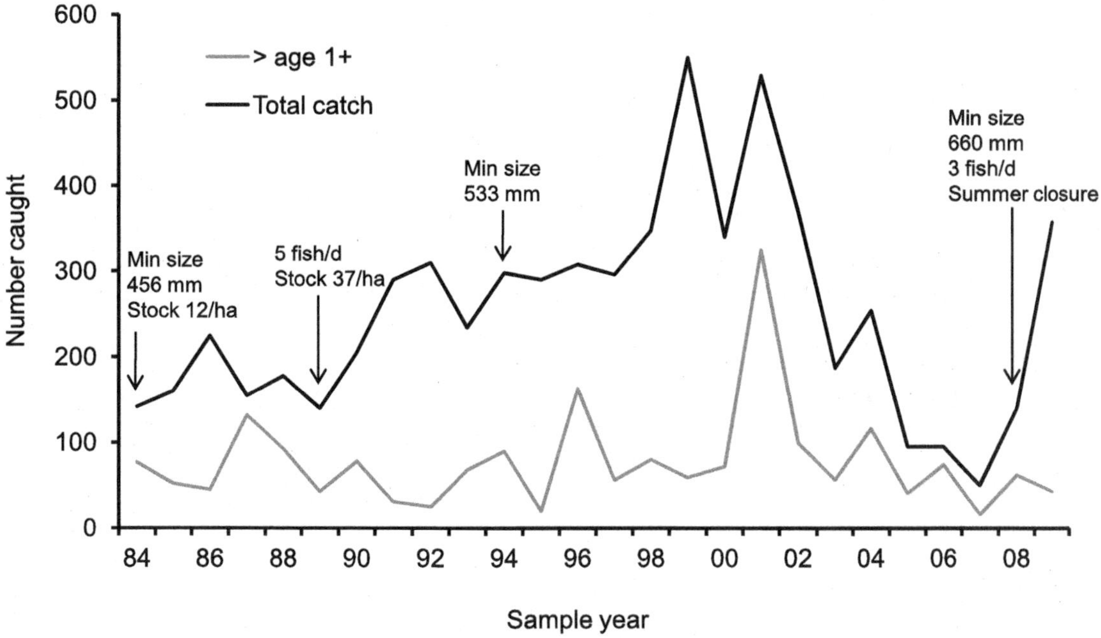

FIGURE 4. The catch of striped bass in the annual Santee–Cooper winter gill-net survey with changes in minimum size, bag limits, and stocking indicated.

of minimum size limits. During a 1999 creel survey, when fishing quality was satisfactory, 86% of interviewed striped bass anglers indicated that they would not support a 610-mm limit that would increase the average size of the catch, indicating a preference for numbers rather than quality.

Santee–Cooper striped bass grow relatively rapidly (Table 1). Since standardized annual sampling began in 1984, no significant growth suppression at age 2 has been observed during years of high natural and stocked fingerling recruitment. Adult winter condition has remained high throughout the current 25-year data set, indicating that adult forage has not been limiting during the fall and winter months. Conversely, during years of poor natural and stocked recruitment, the average size of young of year caught in winter gill nets is lower than average, suggesting to managers that the absence of appropriate juvenile forage affects recruitment in those years.

TABLE 1. Striped bass length at age data from winter reservoir gill-net surveys conducted in the Santee–Cooper, Texoma, and John H. Kerr systems.

	Average length (mm)		
Age	Santee–Cooper	Texoma	John H. Kerr
0+	284	196	240
1+	434	348	431
2+	553	472	537
3+	613	541	609
4+	658	589	668
5+	691	665	708

Santee–Cooper's summer water temperatures exceed 24°C, the upper limit of optimal striped bass habitat as defined by Young and Isely (2002) for another South Carolina reservoir, for more than 3 months (for thermal habitat limitations, also see Coutant 2013; Rice et al. 2013; Thompson and Rice 2013; all this volume). The system's main thermal refuge is created by a cool hypolimnetic release from an upstream tributary dam where a portion of the adult population has been shown to oversummer in this 16-km tailwater river segment. While the system has not experienced acute summer mortality events, varying degrees of chronic mortality have been observed. Adult striped bass that are resident in the reservoir in summer experience their lowest annual condition level during August and September. Chronic summer striped bass mortality has been attributed to the stresses of high water temperature combined with either disease or catch-and-release fishing. In the short time period since summer catch-and-release angling was banned, observations of summer mortalities have appeared to decline.

John H. Kerr Reservoir

The establishment of the John H. Kerr Reservoir striped bass population occurred through the intentional stocking of 2.2 million fry from 1953 to 1955 or from unintentional trapping of resident striped bass when Kerr Dam was closed on the Roanoke River in 1952. The first report of anglers catching striped bass in Kerr Reservoir occurred 2 years after impoundment. The fishery developed slowly, and when the Virginia Game Commission began their fish citation program in 1963, no citations recognizing a striped bass 4.5 kg or larger were issued for Kerr Reservoir angled fish. However, 1 year later, 19 striped bass citations were issued. The first creel survey was conducted in 1968; striped bass made up 5.1% of the harvest by number and were comprised mostly of fish greater than 600 mm TL. The absence of small fish was presumed to have resulted from year-class failures of natural reproduction (1963–1967) that corresponded to Roanoke River flow alterations associated with the completion of the Leesville-Smith Mountain Lake hydroelectric facility upstream of Kerr Reservoir (Neal 1969). The Leesville-Smith Mountain Reservoir pumpback storage project, completed in 1963, lies approximately 103 km upstream of the headwaters of Kerr Reservoir. In fact, one conclusion reported at the end of the 1968 striped bass monitoring was that "Kerr Reservoir striped bass population remains dangerously near extinction" (Neal 1969).

From 1968 to 1981, striped bass studies were conducted to estimate location, time and duration of natural spawning, egg production, egg transport rates, and survival (Neal 1969, 1976; Whitehurst 1982). Results from these studies continue to guide flow management in this reach of the Roanoke River. Neal (1976) suggested that upstream spawning migration was stimulated by simultaneously rising discharge and water temperature, while erratic stream flows, with or without rising water temperature, appeared to deter upstream migration. From 1968 to 2008, minimum flows at Brookneal, Virginia (60 km downstream) were set at 1,150 ft^3/s (32.6 m^3/s) during striped bass spawning season (April 15–June 1). A recent change to the FERC license has increased the minimum flow to 1,500 ft^3/s (42.5 m^3/s) during the same time period. Minimum flows are achieved by operating a Leesville Dam 4,500 ft^3/s (127.4 m^3/s) capacity turbine for a portion of each hour in order to eliminate fluctuating flow at downstream spawning sites.

Striped bass harvest was first regulated in the early 1960s with a 304-mm minimum length limit and an eight fish/d bag limit. In July 1980, this regulation was changed to a 508-mm minimum length limit with a four fish/d bag. This regulation remained in effect until July 1, 2006 when the Virginia Department of Game and Inland Fisheries (VDGIF) passed new striped bass harvest restrictions that account for seasonal differences in striped bass catch-and-release survival. From June 1 to September 30, there is no minimum length limit and the daily creel limit is four striped bass. The minimum length limit from October 1 through May 31 is 660 mm with a daily creel limit of two fish. With this approach, the VDGIF is encouraging anglers to be more harvest-oriented during the summer months when mortality of released striped bass is high and to be more conservative in the cooler months when angled fish can be successfully released.

The Kerr Reservoir striped bass population has predominantly been sustained by natural reproduction. In 1963, the Brookneal Hatchery was constructed on the Roanoke River to take advantage of the striped bass run upstream of the reservoir. From 1975 to 2002, surplus fry and fingerlings were intermittently returned to the system. Year-class strength of striped bass, measured initially by summer seining and later by fall gill-net sampling, has been highly variable. During the 1990s, consistently poor recruitment reduced angler catch rates and anglers began requesting additional stocking. Beginning in 2002, approximately 350,000 striped bass fingerlings (18/ha) have been stocked annually to supplement natural reproduction.

Creel surveys conducted in 2001 and 2007 indicated that striped bass were the fourth most sought after Kerr Reservoir species behind largemouth bass *Micropterus salmoides*, catfishes *Ictalurus* spp., and crappies *Pomoxis* spp. Estimates of directed striped bass fishing effort were 75,057 angler-hours in 2001 and 51,914 angler-hours in 2007, which made up less than 7% of the total fishing effort in each year. Less than 1% of these anglers fished with guides and striped bass angling activity accounted for about 6% of total trip expenditures. Striped bass catch rate estimates were 0.10 fish/h in 2001 and 0.25 fish/h in 2007, with a harvest rate of 0.03 fish/h in each year. Total number of striped bass harvested was 10,490 (2001) and 2,416 (2007). Changes in catch rates and number of striped bass harvested can be attributed to the supplemental stocking that began in 2002 and the harvest restrictions implemented in 2006.

Kerr Reservoir striped bass anglers can be characterized as more harvest-oriented than largemouth bass anglers. A 2001 angler opinion mail survey indicated that Kerr Reservoir striped bass anglers were highly supportive of stocking and moderately supportive of restrictive regulations. They opposed regulations restricting number of rods used, bait type restrictions, and seasonal closures. Though striped bass angling is fourth in the number of hours expended, Kerr Reservoir striped bass anglers remain the most active in fisheries management activities. When striped bass fishing was perceived to be in decline in 1999, anglers formed a local club with the stated intention of becoming actively involved in improving the striped bass fishery. Their contributions include assisting with sampling, facilitating stocking, educating anglers about summertime catch-and-release mortality, and teaching others how to fish for striped bass.

Kerr Reservoir typically stratifies from July through September. During years of above average temperatures and below average rainfall, optimum striped bass habitat, defined as less than 25°C and greater than 5 mg/L dissolved oxygen (Coutant 1985), is eliminated during the summer. While suboptimal summer habitat has not resulted in acute mortality events, it is believed to contribute to reduced growth and increased mortality and ultimately a population that does not produce striped bass weighing more than 8 kg.

Lake Texoma

The development of the Lake Texoma striped bass fishery required intensive management and regulation adaptation to account for high rates of recruitment, flood-related emigration, winter-kill-associated prey deficiencies, and increasing angler efficiencies. The ODWC introduced approximately 1 million striped bass into the reservoir from 1965 to 1974 to create an additional sport fishery and to provide a biological control on shad populations (Harper and Namminga 1986). Natural reproduction was verified in 1973, 1975, and each year thereafter (Mauck 1986). While sharing some limnological characteristics with western reservoirs that experience striped bass reproduction within the reservoir (Gustaveson et al. 1984; Gustaveson and Blommer 2013), Lake Texoma recruitment has repeatedly been shown to result from riverine spawning migrations (Baker et al. 2009). During the population-development years, spawning peaks were noted in 4- to 5-year cycles as strong year-classes reached sexual maturity. To protect the developing population, a one fish/d bag limit was imposed in 1967. Ten years later, this bag limit was raised to three fish/d and later to five fish/d in 1980. Populations of striped bass and forage fish were in excellent condition and yielded numerous trophy fish in the early 1980s (Mauck 1986). By 1982, liberal bag limits were instituted allowing anglers to harvest 15 fish/d of which only five could exceed 508 mm.

Concurrent with the liberalized bag limits, the population experienced an adjustment period marked by environmental events and new fishing techniques. Mauck (1986) reported that the discharge of flood waters from the dam and winter kill of threadfin shad *Dorosoma petenense* negatively impacted the striped bass population, particularly larger-sized individuals. Flood water releases spilled large numbers of striped bass into the tailwater in 1982, 1987, and 1990, and the harsh winters of 1981–1982, 1987–1988, and 2000–2001 led to threadfin shad die-offs that reduced the abundance of exceptionally strong striped bass year-classes the following growing seasons (Mauck 1991; Oklahoma Department of Wildlife Conservation, unpublished observations).

By 1980, Lake Texoma had become nationally recognized for its excellent striped bass fishery and experienced increased fishing pressure. The current lake size record, 15.93 kg, was caught in 1984. Artificial lures were the primary method of fishing prior to 1985 when bait fishing made its debut (Mauck 1986). This technique allowed anglers and guides to more effectively target large fish and regularly harvest the daily bag limit. By the late 1980s, it was apparent that a combination of environmental factors and fishing pressure was having a negative impact on larger fish in the population, and in 1989, the regulation was adjusted, once again, to15 fish/d, only one greater than 508 mm TL.

A joint angler creel survey, conducted by the Texas Parks and Wildlife Department (TPWD) and the ODWC from 1987 to 1999, was initiated to provide estimates of catch rates, fishing pressure, and harvest (Hysmith et al. 2000). During this 13-year survey, striped bass anglers accounted for 63.6% of the total angling effort and harvested an estimated 854,032 striped bass annually (range 474,459–1,233,066) supported entirely by natural reproduction. Weight of harvested striped bass averaged 774,154 kg annually but saw yearly estimates as high as 1,307,858 kg. Unlike Santee–Cooper and J. H. Kerr reservoirs, minimal spring angling effort is directed at striped bass on their spawning runs in the Red and Washita rivers.

Despite efforts to reduce harvest of the population's larger fish, striped bass were not reaching their historical sizes and regulations were ineffective in reviving numbers of striped bass greater than 508 mm (Moczygemba and Hysmith 1996). While cognizant of the system's underlying summer temperature and oxygen limitations, managers focused on the mortality factors they believed they could affect. The opinion that live bait fishing was a major contributing factor to the lack of large striped bass prompted a fishing mortality survey. Hysmith et al. (1994) demonstrated that postrelease mortality of striped bass caught on live bait (57.6%) far exceeded mortality of fish caught with treble-hooked artificial baits (32.3%) and single-hooked artificial baits (15.7%). These results further showed that this discrepancy was higher for fish greater than 508 mm with mortality estimates as high as 71% when caught with live bait during spring and summer months. To reduce this apparent waste but still protect the fishery as a whole, the ODWC and the TPWD adopted a regulation decreasing the bag limit to 10 fish/d while increasing the number greater than 508 mm to two fish/d. This regulation was adopted in 1996 and remains in effect today.

Subsequent to the 1996 regulation change, an angler opinion study concluded that 84% of striped bass anglers were moderately to extremely satisfied with striped bass fishing at Lake Texoma (Hunt and Ditton 1998). This study also demonstrated that 62% of independent anglers and 92% of guided anglers used live bait on trips. Regulation scenarios that prohibited live-bait fishing were not well supported (21%). Less than 25% of anglers listed "to obtain a trophy fish" as a very or extremely important reason for fishing the reservoir. Although the biological information indicates that a reduction in live bait use would increase large fish numbers, the majority of anglers do not appear willing to change their behavior.

Striped bass spawning investigations have documented spawning migrations in both the Washita and Red rivers. Egg collections indicate that spawning takes place between 70 and 160 km upstream of the reservoir (Baker et al. 2009). These areas were associated with boulder and bedrock substrates in the rivers otherwise dominated by sand substrate. Relative annual reproductive contribution of each river is largely influenced by the magnitude and timing of discharge events within and between the two systems. High concentration of dissolved solids in the Red River, and to a lesser extent in the Washita River, arrests the swelling of egg chorions prior to water hardening, resulting in smaller diameter eggs (1.2–1.8 mm) than reported from the East Coast and Gulf Coast (2.0–3.0 mm) (Baker et al. 2009). Though the smaller diameter eggs have a higher specific gravity than their East Coast counter parts, Lake Texoma's tributaries have provided the necessary flow velocities and river lengths required to produce consistent striped bass recruitment since the mid-1970s (Mauck 1987). Managers are concerned that any alteration of the flow or water chemistry in these rivers could have negative effects on the striped bass population (Baker et al. 2009).

Winter gill-net data collected since 1993 indicates that the striped bass population has stabilized (Figure 5). The percentage of fish ≥508 mm in the population has averaged 21.7% since the regulation change in 1996. Slower growth rates in comparison to the Santee–Cooper and Kerr populations, especially of young ages, are attributable to the high abundance of striped bass within the lake produced by natural reproduction (Table 1). First-year winter survival has been described as a recruitment bottleneck in other striped bass populations where size-selective mortality is common (Hurst and Conover 1998; Sutton and Ney 2001). In contrast, striped bass less than 100 mm have been shown to overwinter within Lake Texoma, and analysis indicates that size-selective overwinter mortality is not likely limiting recruitment (Mauck 2004). Total mortality estimates from annual population samples are approximately 50% for striped bass age 1+ (22 months) and older (Oklahoma Department of Wildlife Conservation, unpublished data). Natural mortality apparently increases in years following a winter threadfin shad kill and was evident in the decline in numbers of fish collected in gill nets during 2002 following the harsh 2000–2001 winter.

Lake Texoma develops strong stratification from July through September that affects habitat suitability for striped bass. Habitat containing optimal oxygen concentration (>5 mg/L) and water temperature (<25°C) (Coutant 1985) is severely limited or ceases to exist by midsummer, and fish seeking dissolved oxygen concentration of 2 mg/L or better will be subject to water temperatures greater than 27°C (Matthews et al. 1985). Late-summer chronic mortality is observed in some years prior to lake turnover and has been attributed to suboptimal habitat stress, catch-and-release-related stress (as noted above), and a combination of both. Higher than normal summer mortality has been observed following threadfin shad winter kill events that led to striped bass entering the summer in lowered physical condition and having lowered prey abundance. The combination of high metabolic demands, caused by high temperatures, coupled with reduced prey avail-

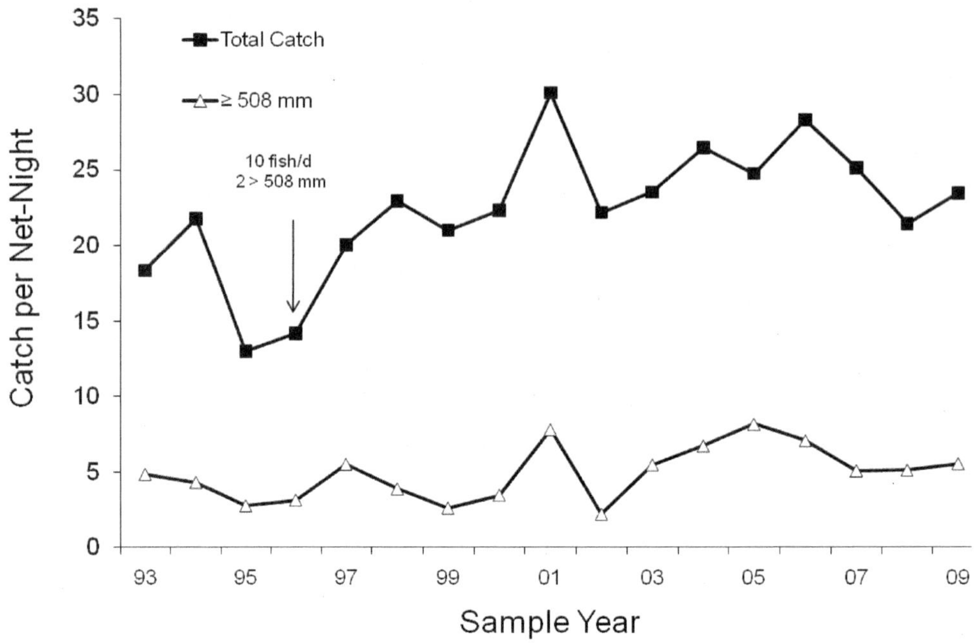

FIGURE 5. The winter gill-net catch rates of Lake Texoma striped bass with the last harvest restriction change indicated.

ability can result in an energetic deficit resulting in weight loss and late-summer mortality (Thompson and Rice 2013). This type of lowered-condition-related mortality involved all size ranges and not just larger fish that are more vulnerable to temperature stress. In combination with high population density, flood-related emigration of large striped bass, and winter-kill-associated prey deficiencies, the quality of summer habitat is believed to limit Lake Texoma's potential to produce large, trophy-sized striped bass.

The high level of striped bass natural reproduction and recruitment in Lake Texoma has permitted exceptionally liberal harvest regulations because of the need to reduce numbers of small fish and maintain acceptable growth rates. Approximately 200 striped bass guides operate on the lake, resulting in a level of harvest that few fisheries can support. In a 1997 to 2000 survey, guided trips accounted for 60% of the directed effort for striped bass and comprised 77% of the total harvest (Moczygemba et al. 2005). Since the establishment of the current regulations, the striped bass population remains stable and the majority of anglers are satisfied with their Lake Texoma fishing experience—one of the few examples of the desires of the angling public aligning with the production capabilities of the resource. Lake Texoma's striped bass fishery is estimated to provide in excess of US$25 million annually to the local economy (Schorr et al. 1995) making it arguably the single-most valuable fishery resource in the state of Oklahoma.

Discussion

Natural striped bass reproduction has affected the development of striped bass populations and fisheries management in these three reservoirs in several ways. Initial high abundance, provided by natural reproduction, promoted harvest-oriented angling. The size structure of each population shifted towards younger and smaller fish as directed effort and angler efficiencies increased. High mortality rates have been a concern in all three, with catch-and-release mortality, particularly in warm water, shaping present harvest regulations.

Lake Texoma is the only one of these reservoirs in which natural recruitment has remained high enough to satisfy a large fishery despite concerns about the low abundance of fish exceeding 508 mm. Striped bass reproductive potential can be adversely affected by suboptimum summer habitat when it leads to poor body condition early in the cycle of ovary development (Coutant 1987). However, suboptimum summer habitat and periodic prey deficiencies has not deterred a high level of reproductive success in Lake Texoma. Both Kerr and Santee–Cooper reservoirs provide broodstock for successful striped bass production programs, which suggests that sub-optimum summer habitat is not a primary factor in limiting natural recruitment in either system. In the Santee–Cooper Reservoir system, recruitment overfishing has been a recognized issue since the early 1990s (Bulak 1994). During this same time frame, Kerr Reservoir also experienced a decline in the ability of natural recruitment to satisfy the fishery, suggesting similar increases in angling efficiencies and interest in striped bass fishing.

Declining natural recruitment in both Santee–Cooper and Kerr reservoirs has led managers to initiate annual supplemental stocking. Only resident broodstocks are used, and no striped bass hybrids are stocked in either system. The South Carolina Department of Natural Resources began supplemental fingerling stocking in Santee–Cooper at a rate of 12/ha in 1984, increasing the rate to 37/ha in 1989 and maintaining this rate since. Results of this effort suggest that in Santee–Cooper, year-class strength is determined not only at early life history stages, but in some years at the juvenile life stage, too. Genetic diversity concerns led to the development of hatchery protocols that maximized the number of pairings used in each year's stocking to more closely simulate natural reproduction. Investigations into the number of parents contributing to a naturally recruited year-class (Diaz et al. 2000) led to the understanding that stocking could increase the effective population size in years of relatively poor natural recruitment. Prior to 2002, the Kerr system received irregular stocking of surplus fry and fingerlings, but beginning in 2002, deliberate stocking at a rate of 18 fingerlings/ha was begun. Assessment of year-class strength demonstrated improvement 4 years after this stocking effort began and new harvest regulations were implemented (Figure 6).

Each of these three reservoirs and their striped bass populations are distinctive, yet they share several characteristics. Lake Texoma's high natural recruitment rate and associated slower growth rates separate it from the other two. Its highly mineralized and euthrophic water conditions appear to greatly aid the system's natural recruitment, as does its free-flowing tributary rivers suitable for spawning. Kerr and Santee–Cooper reservoirs are Atlantic slope reservoirs whose relatively soft and infertile tributary river flows are all affected by upstream hydroelectric projects. Santee–Cooper is unique in receiving anadromous shad and herring inputs that, because of their timing and magnitude, may lead to increased competition at larval and juvenile life stages, negatively affecting striped bass recruitment in some years (Lamprecht et al. 2010). All three reservoirs have two major tributary rivers that would appear to increase the likelihood of successful riverine reproduction. Populations in all three reservoirs experience high total mortality rates, where few individuals survive long enough to achieve sizes exceeding 9 kg. These populations are similar to those constrained by suboptimal summer habitat, which were described by Thompson et al. (2010). Higher than optimal summer water temperatures also elevate catch-and-release mortality (Wilde et al. 2000; Bettinger et al. 2005; Bettinger and Wilde 2013, this volume). Regardless of the cause, high mortality rates create concern over selection for younger age and size at maturity, which threatens the long-term stability and value of the resource (Trippel 1995; Swain 2010).

The importance of natural striped bass reproduction is most clearly illustrated at Lake

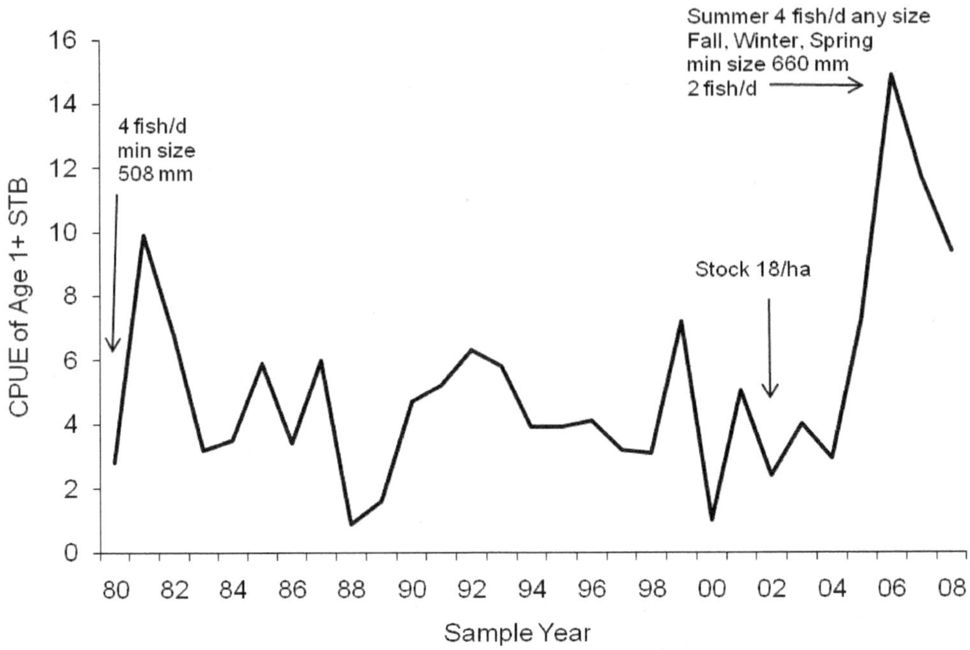

FIGURE 6. John H. Kerr Reservoir index of year-class strength as indicated by the catch rate (#/net-night) of age 1+ striped bass with changes in size and bag limits and stocking indicated.

Texoma where it has created the single most economically significant fishery in the state of Oklahoma with little input from either Oklahoma or Texas fishery management agencies. Natural reproduction in Kerr Reservoir, while not sufficient to sustain the popular fishery, reduces the stocking demand and provides a genetically diverse parental stock for the Brookneal hatchery. Santee–Cooper system broodstock progeny are used to maintain three other popular piedmont reservoir fisheries, emphasizing the importance of maintain the genetic diversity provided by natural reproduction.

Natural striped bass reproduction in all three reservoirs faces potential physical and biological threats. Increased demands on water resources have the potential to alter reproductive success by diverting water flow and altering water chemistry. Examples include a proposed water diversion and chloride reduction project on Lake Texoma's Red River and an ongoing interstate conflict over interbasin water transfers out of a Santee–Cooper tributary. Oligotrophication has become a concern at Santee–Cooper not only because of upstream operation of advanced waste treatment facilities, but also because drought has lengthened retention periods in upstream reservoirs and has elevated sequestration of the watershed's nutrient load. Invasive species of fish, invertebrates, and plants all have the potential to negatively affect natural reproduction. On the Atlantic slope, invasive white perch *Morone americana* populations compete with young striped bass and consume striped bass eggs (Harris 2006). The potential introduction of bighead carp *Hypophthalmichthys nobilis* and silver carp *H. molitrix* is an obvious threat, but even endemic juvenile American shad that fail to out-migrate normally pose a potential competitive threat to recruitment in Santee–Cooper Reservoir. Zebra mussels *Dreissena polymorpha* and corbicula clams *Corbicula fluminalis* have the potential to compete for and tie up a significant portion of a system's productivity, particularly in the less productive Atlantic slope reservoirs. Toxic algae blooms have become more prevalent in western systems, and today's mobile anglers have the potential to spread the algae species responsible for these blooms (Sager et al. 2007).

The naturally reproducing striped bass populations in these three reservoirs have been challenged by both environmental changes and changes in angling efficiencies over the past 30 years. Lake Texoma managers have addressed angler attitudes and matched management to the productivity of the system to achieve an equilibrium. Harvest regulation adjustments at both Kerr and Santee–Cooper hold promise for increasing the number of striped bass reaching reproductive age, something that minimum length limits and supplemental stocking have not been able to achieve over the past 20 years. A management theme common to all three systems has been active involvement in protection and enhancement of spawning season water flows, which water resource development pressures make a continual necessity. Experience suggests that environmental change and the evolving preferences of the angling public can be expected to provide future challenges to managing naturally reproducing inland striped bass populations. However, the experience we have gained over the past 40 years provides a substantial foundation on which to meet these challenges.

Acknowledgments

The authors thank Jeff Boxrucker, Jim Bulak, Bruce Hysmith, Paul Mauck, and Miller White for their assistance over the years in developing our present-day understanding of the systems discussed. This manuscript was improved through comments provided by reviewers Wayne Gustaveson and John Ney as well as the book's editors.

References

Albrecht, A. G. 1964. Some observations on factors associated with survival of striped bass eggs and larvae. California Fish and Game 50:100–113.

Baker, W. P., J. Boxrucker, and K. E. Kuklinski. 2009. Determination of striped bass spawning locations in two major tributaries of Lake Texoma. North American Journal of Fisheries Management 29:1006–1014.

Bettinger, J. M., J. R. Tomasso, Jr., and J. J. Isely. 2005. Hooking mortality and physiological responses of striped bass angled in freshwater and held in live-release tubes. North American Journal of Fisheries Management 25:1273–1280.

Bettinger, J. M., and G. R. Wilde. 2013. Catch-and-release mortality of inland striped bass and hybrid striped bass. Pages 473–499 *in* J. S. Bulak, C. C. Coutant, and J. A. Rice, editors. Biology and management of inland striped bass and hybrid striped bass. American Fisheries Society, Symposium 80, Bethesda, Maryland.

Bulak, J. S. 1994. Factors affecting recruitment of striped bass *Morone saxiatilis* in the Santee–Cooper system, South Carolina. Doctoral dissertation. University of South Carolina, Columbia.

Bulak, J. S., J. S. Crane, and D. W. Gates. 1983. Creel survey—Congaree and Wateree rivers and Lakes Marion and Moultrie. South Carolina Wildlife and Marine Resources Department, Completion Report, Federal Aid Project WC-2, Columbia.

Bulak, J. S., N. M. Hurley, Jr., and J. S. Crane. 1993. Production, mortality, and transport of striped bass eggs in Congaree and Wateree rivers, South Carolina. Pages 29–37 *in* L. A. Fuiman, editor. Water quality and the early life stages of fishes. American Fisheries Society, Symposium 14, Bethesda, Maryland.

Bulak, J. S., D. S. Wethey, and M. G. White, III. 1995. Evaluation of management options for a reproducing striped bass population in the Santee–Cooper system, South Carolina. North American Journal of Fisheries Management 15:84–94.

Bulak, J. S., J. S. Crane, D. H. Secor, and J. M. Dean. 1997. Recruitment dynamics of striped bass in the Santee–Cooper system, South Carolina. Transactions of the American Fisheries Society 126:133–143.

Coutant, C. C. 1985. Striped bass, temperature, and dissolved oxygen: a speculative hypothesis for environmental risk. Transactions of the American Fisheries Society 114:31–61.

Coutant, C. C. 1987. Poor reproductive success

of striped bass from a reservoir with reduced summer habitat. Transactions of the American Fisheries Society 116:154–160.

Coutant, C. C. 2013. When is habitat limiting for striped bass? Three decades of testing the temperature–oxygen squeeze hypothesis. Pages 65–91 *in* J. S. Bulak, C. C. Coutant, and J. A. Rice, editors. Biology and management of inland striped bass and hybrid striped bass. American Fisheries Society, Symposium 80, Bethesda, Maryland.

Diaz, M., D. Wethey, J. Bulak, and B. Ely. 2000. Effect of harvest and effective population size on genetic diversity in a striped bass population. Transactions of the American Fisheries Society 129:1367–1372.

Finger, S. E., and J. S. Bulak. 1988. Toxicity of water from three South Carolina rivers to larval striped bass. Transactions of the American Fisheries Society 117:521–528.

Gustaveson, A. W., T. D. Pettengill, J. E. Johnson, and J. R. Wahl. 1984. Evidence of in-reservoir spawning of striped bass in Lake Powell, Utah–Arizona. North American Journal of Fisheries Management 4:540–546.

Gustaveson, W., and G. Blommer. 2013. History of striped bass management in the Colorado River. Pages 15–23 *in* J. S. Bulak, C. C. Coutant, and J. A. Rice, editors. Biology and management of inland striped bass and hybrid striped bass. American Fisheries Society, Symposium 80, Bethesda, Maryland.

Harris, J. L. 2006. Impacts of the invasive white perch on the fish assemblage of Kerr Reservoir, Virginia. Master's thesis. Virginia Polytechnic Institute and State University, Blacksburg.

Harper, J. L. and H. E. Namminga. 1986. Fish population trends in Lake Texoma Reservoir following establishment of striped bass. Pages 156–165 *in* G. E. Hall and M J. Van Den Avyle, editors. Reservoir fisheries management: strategies for the 80's. American Fisheries Society, Southern Division, Reservoir Committee, Bethesda, Maryland.

Hunt, K. M. and R. B. Ditton. 1998. Characteristics of anglers and guides at Lake Texoma, their fishing participation patterns, and attitudes towards management of the recreational fishery. Texas A&M University, Human Dimensions of Fisheries Research Laboratory, College Station.

Hurst, T. P., and D. O. Conover. 1998. Winter mortality of young-of-the-year Hudson River striped bass (*Morone saxatilis*): size-dependent patterns and effects on recruitment. Canadian Journal of Fisheries and Aquatic Sciences 55:1122–1130.

Hysmith, B. T., J. H. Moczygemba, and G. R. Wilde. 1994. Hooking morality of striped bass in Lake Texoma, Texas–Oklahoma. Proceedings of the Annual Conference Southeastern Association of Fish and Wildlife Agencies 46(1992):413–420.

Hysmith, B. T., J. H. Moczygemba, and T. R. Robinson. 2000. Statewide freshwater fisheries monitoring and management program: survey report for Lake Texoma, 1999. Texas Parks and Wildlife Department, Performance Report, Federal Aid Project F-30-R, Austin.

Lamprecht, S. D., S.C. Holbrook, and M. C. Martin. 2010. Fisheries investigation in lakes and streams, region IV. South Carolina Department of Natural Resources, Study Completion Report, Federal Aid Project F63–15-4, Columbia.

Matthews, W. J., L. G. Hill, and S. M. Schellhaass. 1985. Depth distribution of striped bass and other fish in Lake Texoma (Oklahoma–Texas) during summer stratification. Transactions of the American Fisheries Society 114:84–91.

Mauck, M.D. 2004. Factors influencing first year recruitment of striped bass in Lake Texoma. Master's thesis. Oklahoma State University, Stillwater.

Mauck, P. E. 1986. Striped bass research study: population trends (Texoma Reservoir) March 1, 1981 to February 29, 1986. Oklahoma Department of Wildlife Conservation, Federal Aid Project F-29-R, Job No. 3. Oklahoma City.

Mauck, P. E. 1987. Fish management and survey recommendations for Texoma Reservoir, 1986. Oklahoma Department of Wildlife Conservation, Federal Aid Project F-44-D-1, Norman.

Mauck, P. E. 1991. Fish management surveys and recommendations for Texoma Reservoir. Oklahoma Department of Wildlife Conservation, Job Performance Report, Federal Aid Project F-44-D-5, Oklahoma City.

May, O. D., Jr., and J. C. Fuller, Jr. 1965. A study on striped bass egg production in the Congaree and Wateree rivers. Proceedings of the Annual Conference Southeastern Association of Game and Fish Commissioners 16(1962):285–301.

Moczygemba, J. H., and B. T. Hysmith. 1996. Evaluation of reduced striped bass bag limit, Lake Texoma, Texas and Oklahoma. Proceedings of the Annual Conference Southeastern Association of Fish and Wildlife Agencies 48(1994):433–440.

Moczygemba, J. H., B. T. Hysmith, and J. W. Schlechte. 2005. Comparison of guided and unguided fishing trips from two Texas reservoirs. Pages 17–30 *in* C. C. Bonds, editor. Annual Proceedings of the Texas Chapter of the American Fisheries Society, volume 27. American Fisheries Society, Texas Chapter, Bethesda, Maryland.

Neal, W. E. 1976. Landlocked striped bass survey. Virginia Commission of Game and Inland Fisheries, Richmond.

Neal, W. E. 1969. Landlocked striped bass of John H. Kerr Reservoir, a ten year summary of research and management activities. Virginia Commission of Game and Inland Fisheries, Richmond.

Rice, J. A., J. S. Thompson, J. A. Sykes, and C. T. Waters. 2013. The role of metalimnetic hyposia in striped bass summer kills: consequences and management implications. Pages 121–145 *in* J. S. Bulak, C. C. Coutant, and J. A. Rice, editors. Biology and management of inland striped bass and hybrid striped bass. American Fisheries Society, Symposium 80, Bethesda, Maryland.

Sager, D. S., L. Fries, L. Singhurst, and G. Southard. 2007. Guidelines for golden alga *Prymnesium parvum* management options for ponds and small reservoirs (public waters) in Texas. Inland Fisheries Texas Parks and Wildlife Department. Available: www.tpwd.state.tx.us/publications/pwdpubs/media/pwd_rp_t3200_1404.pdf. (September 2010).

Schorr, M. S., J. Sah, D. F. Schreiner, M. R. Meador, and L. G. Hill. 1995. Regional economic impact of the Lake Texoma (Oklahoma–Texas) striped bass fishery. Fisheries 20(5):14–18.

Scruggs, G. D. 1957. Reproduction of resident striped bass in Santee–Cooper Reservoir, South Carolina. Transactions of the American Fisheries Society 85:144–159.

Stevens, R. E. 1958. Striped bass of the Santee–Cooper reservoirs. Proceeding of the Annual Conference Southeastern Association of Game and Fish Commissioners 11(1957):253–264.

Stevens, R. E. 1960. Combined Quarterly-Annual Progress Report, F-1-R-8. South Carolina Wildlife and Marine Resources Department, Columbia.

Sutton, T. M., and J. J. Ney. 2001. Size-dependent mechanisms influencing first-year growth and winter survival of stocked striped bass in a Virginia mainstream reservoir. Transactions of the American Fisheries Society 130:1–17.

Swain, D. P. 2010. Life-history evolution and elevated natural mortality in a population of Atlantic cod (*Gadus morhua*). Evolutionary Applications 4:18–29.

Thompson, J. S., J. A. Rice, and D. S. Waters. 2010. Striped bass habitat selection rules in reservoirs without suitable summer habitat offer insight into consequences for growth. Transactions of the American Fisheries Society 139:1450–1464.

Thompson, J. S., and J. A. Rice. 2013. The relative influence of thermal experience and forage availability on growth of age 1–5 striped bass in two southeastern reservoirs. Pages 93–120 *in* J. S. Bulak, C. C. Coutant, and J. A. Rice, editors. Biology and management of inland striped bass and hybrid striped bass. American Fisheries Society, Symposium 80, Bethesda, Maryland.

Trippel, E. A. 1995. Age at maturity as a stress indicator in fisheries. BioScience 45:759–771.

Van Horn, S. L. 2013. A brief history of inland striped bass management. Pages 1–13 *in* J. S. Bulak, C. C. Coutant, and J. A. Rice, editors. Biology and management of inland striped bass and hybrid striped bass. American Fisheries Society, Symposium 80, Bethesda, Maryland.

White, M. G., III. 1974. Fisheries studies in district five. South Carolina Wildlife and Marine Resources Department, Progress Report, Federal Aid Project F-16–17, Columbia.

White, M. G., III. 1989. Fingerling densities produced by a spawning population of striped bass in the Santee–Cooper Reservoir, South Carolina. Proceedings of the Annual Conference Southeastern Association of Game and Fish Commissioners 42(1988):44–50.

White, M. G., III, and S. D. Lamprecht. 2002. Fisheries investigations in lakes and streams—District V. South Carolina Department of Natural Resources, Study completion report, Federal Aid Project F63–7-5, Columbia.

Whitehurst, D. K. 1982. Kerr Reservoir striped bass population study. Virginia Commission of Game and Inland Fisheries, Federal Aid Project F-34, Richmond.

Wilde, G. R., M. I. Muoneke, P. W. Bettoli, K. L. Nelson, and B. T. Hysmith. 2000. Bait and temperature effects on striped bass hooking mortality in freshwater. North American Journal of Fisheries Management 20:810–815.

Young, S. P., and J. J. Isely. 2002. Striped bass annual site fidelity and habitat utilization in J. Strom Thurmond Reservoir, South Carolina–Georgia. Transactions of the American Fisheries Society 119:72–76.

American Fisheries Society Symposium 80:431–448, 2013

Striped Bass Dispersion and Effects on Fisheries Management in Lakes Mohave and Pleasant, Colorado River Basin

William T. Stewart*
Arizona Game and Fish Department
5000 West Carefree Highway Phoenix, Arizona 85086, USA

Mike Burrell
Nevada Department of Wildlife
1557 Foothill # A-110 Boulder City, Nevada 89005, USA

Abstract.—The striped bass *Morone saxatilis* was introduced into the lower Colorado River in the late 1950s and into Lake Mead, Nevada and Arizona, in the late 1960s. The unintended immigration of striped bass into Lake Mohave, Nevada and Arizona, on the main stem, and Lake Pleasant, a tributary reservoir in central Arizona, has resulted in changing management practices. Striped bass entered Lake Mohave via downstream emigration from Lake Mead through Hoover Dam at various life stages, and the newly established population quickly became the primary sport fish in the reservoir. Predation from the striped bass population in Lake Mohave coincided with elimination of threadfin shad *Dorosoma petenense* and a rapid decline in the survival of stocked rainbow trout *Oncorhynchus mykiss*. Striped bass are also believed to be hindering ongoing efforts to reestablish the native endangered species razorback sucker *Xyrauchen texanus* and bonytail chub *Gila elegans*. Striped bass gained access to Lake Pleasant via Lake Havasu, California and Arizona, by way of the Central Arizona Project (CAP) Canal. Operation of the CAP Canal began in 1985 and the canal was fully connected to Lake Pleasant in 1992. In 1986 and 1989, striped bass population densities in the CAP Canal were estimated at 70 ± 37 fish/ha and 3 ± 1 fish/ha, respectively. Striped bass were first captured in Lake Pleasant in 1998 during a gill-netting survey. Catch per unit effort increased almost yearly from 0.13 fish/net-night in 1998 to 6.74 fish/net-night in 2005. Since their unintended introduction into both reservoirs, striped bass have established viable reproducing populations. Management efforts have emphasized promoting harvest and minimizing the impacts of striped bass on existing fisheries. These experiences provide guidance for evaluating unintended dispersion of striped bass elsewhere.

Introduction

The discovery of landlocked populations of striped bass *Morone saxatilis* in impounded waters in the Santee and Cooper rivers in the 1950s significantly changed the way reservoirs are managed (Van Horn 2013, this volume). Due to success in the eastern United States, striped bass were introduced into the Colorado River in the late 1950s (Gustaveson and Blommer 2013, this volume). From 1959 to 1972,

* Corresponding author: bstewart@azgfd.gov

an estimated 93,000 striped bass were planted into the lower Colorado River (Grabowski et al. 1984). Stocking locations included two of the three major reservoirs along the lower Colorado River: Lake Mead and Lake Havasu. At the time, natural reproduction was considered unlikely because literature stated that striped bass eggs and larvae must be suspended by current to prevent settling and death by suffocation (Gustaveson and Blommer 2013). However, self-sustaining populations soon became established creating one of the most popular sport fisheries in the southwest. Lake Mohave, located between Mead and Havasu, was never intentionally stocked with striped bass. Nonetheless, striped bass were soon found in Lake Mohave, having likely passed through Hoover Dam (Lake Mead). Striped bass then found their way to Lake Pleasant, likely through a water storage pumping project on Lake Havasu serving Phoenix and southern Arizona.

Inadvertent dispersion of striped bass through dam and pumping plant intake structures is not uncommon. For example, the A.D. Edmonstom Pumping Plant carries water from Northern California along the western side of the San Joaquin Valley over the Telachapi Mountains to Southern California. A 2-year study on the movement and survival of fish passing through the California State Water Project's A.D. Edmonstom Pumping Plant in 1972 and 1973 documented several species of fish including striped bass surviving the 587-m lift and 13.5-km-long journey. An estimated 20,500 striped bass ranging from 250 to 410 mm total length (TL) were pumped into Southern California during the 2-year study (Aasen et al. 1982). Additionally, striped bass have immigrated into Gene Wash, Copper Basin, Lake Mathews, Lake Skinner, San Vincente, and El Capitan reservoirs (all in California) through the Colorado River Aqueduct, a diversion canal that provides water from Lake Havasu to Southern California (M. Giusti, California Fish and Game Department, personal communication). This paper addresses the magnitude, impacts, and implications of the inadvertent dispersion of striped bass on two popular sport fisheries in the southwestern United States: Lake Mohave and Lake Pleasant. It concludes with points to be considered for evaluating the likelihood of striped bass emigration from the intended water body and dispersion to unintended locations where fisheries management may be affected.

Lake Mohave

Striped bass source reservoir (Lake Mead)

Lake Mead was created in 1935 by the construction of Hoover Dam. Hoover Dam was built by the U.S. Bureau of Reclamation (USBR) in the Black Canyon of the Colorado River near Las Vegas, Nevada (Figure 1). At full pool, Lake Mead is 63,900 ha with a maximum depth of 177 m and maximum length of approximately 184 km. The Colorado River is the primary water source for Lake Mead; additionally there are two secondary inflows: the Virgin-Muddy River system and the Las Vegas Wash. The water resources of Lakes Mead and Mohave are managed by the USBR. Lake Mead and Lake Mohave are part of the Lake Mead National Recreation Area administered by the National Park Service (NPS). The Nevada Department of Wildlife (NDOW), the Arizona Game and Fish Department (AGFD), the NPS, the U.S. Fish and Wildlife Service (USFWS), and the USBR are all involved in managing the aquatic wildlife resources of Lakes Mead and Mohave.

The fish assemblage in Lake Mead is comprised mostly of nonnative fish, including largemouth bass *Micropterus salmoides*, smallmouth bass *M. dolomieu*, striped bass, channel catfish *Ictalurus punctatus*, yellow bullhead *I. natalis*, black bullhead *I. melas*, bluegill *Lepomis macrochirus*, green sunfish *L. cyanellus*, black crappie *Pomoxis nigromaculatus*, gizzard shad *Dorosoma cepedianum*, threadfin shad *D. petenense*, red shiner *Cyprinella lutrensis*, blue tilapia *Oreochromis aureus*, and common carp *Cyprinus carpio* (Allen and Roden 1978; Nevada Department of Wildlife, unpublished report, 2008).

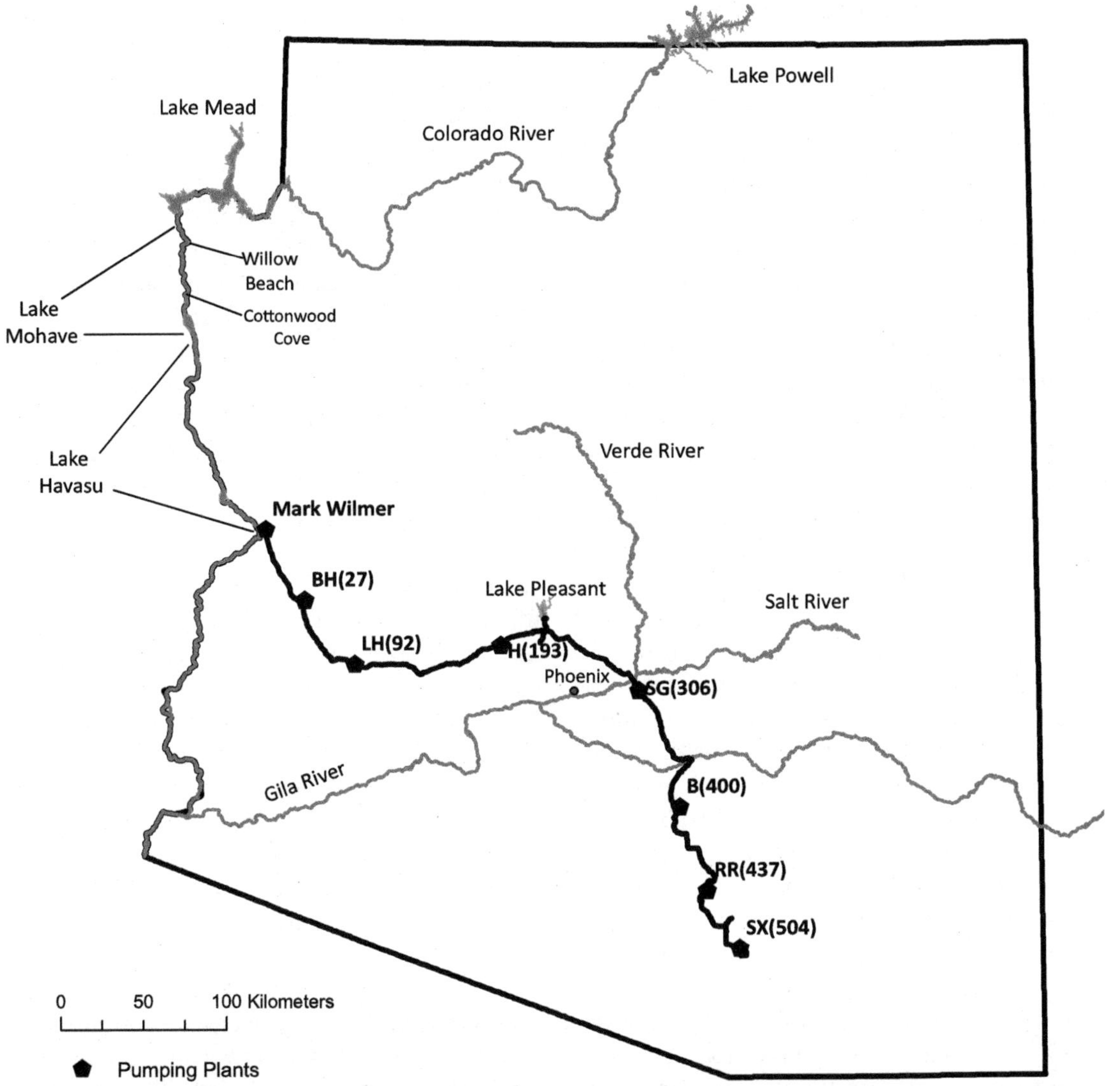

FIGURE 1. Map of lower Colorado River reservoirs, Lake Pleasant, and location of Bouse Hills (BH), Little Harquahala (LH), Hassayampa (H), Salt-Gila (SG), Brady (B), Red Rock (RR), and San Xavier (SX) pumping plants along the Central Arizona Project Canal. Numbers in parentheses represent distance from Lake Havasu in kilometers.

A remnant population of the native razorback sucker *Xyrauchen texanus* can also be found in Lake Mead (Allen and Roden 1978).

Given the success of striped bass in the eastern United States and the belief that they would not reproduce, stocking striped bass in Lake Mead seemed appropriate. Between 1969 and 1972, approximately 64,000 striped bass, ranging from 50 to 150 mm TL, were stocked into Lake Mead (Allen and Roden 1978). Successful reproduction of striped bass in Lake Mead was first documented in 1973 (Allen and Roden 1978), and by 1976, striped bass comprised 1.4% of the angler catch. Seven years later, in 1983, striped bass harvest increased to 50% of the total angler catch (Allen and Roden 1978; Nevada Department of Wildlife, unpublished report, 1983).

Striped bass receiving reservoir (Lake Mohave)

Located southeast of Las Vegas, Nevada, Lake Mohave was created in 1951 by the USBR. The reservoir was impounded by Davis Dam approximately 103 river km south of Hoover Dam on the Colorado River (Figure 1). At full pool, the north end of Lake Mohave backs into the tailrace of Hoover Dam. Located in the Mohave Desert, the lake forms the common boundary between Clark County, Nevada and Mojave County, Arizona. The 11,396-ha reservoir is 103 km long, with a maximum width of 6.4 km and 322 km of irregular shoreline. The maximum depth is 42 m, with a mean depth of 19 m. The only significant inflow into Lake Mohave is the cold, clear Colorado River passing through Hoover Dam. The upper 23 km of the reservoir is confined within the narrow walls of the Black Canyon. The remainder of the reservoir is comprised of two open basins interspersed with the flooded canyons of the Colorado River channel. Water elevations generally fluctuate between 193 and 196 m above sea level. Lake Mohave water levels are generally on the rise through the winter and peak in the late spring to early summer. Elevations remain high through the summer and are declining by late summer to early fall. Through the winter months, Lake Mohave is isothermal, with water temperatures ranging between 11°C and 13°C. As summer approaches, the temperature regime becomes more complex. The Hoover Dam discharge is taken from the hypolimnion of Lake Mead, which is 11–13°C. Because of this cold inflow, the upper 24–32 km of the reservoir remains cold, 11–13°C, all year. The remainder of the reservoir becomes seasonally stratified with surface temperatures above 26°C (Jonez and Sumner 1954).

Prior to the passage of striped bass into Lake Mohave, the reservoir provided both a warm- and coldwater sport fishery. The coldwater fishery was supported by stocked salmonids, primarily rainbow trout *Oncorhynchus mykiss*. The warmwater fishery was supported by largemouth bass and channel catfish. There was and remains an abundant population of bluegill and green sunfish, and threadfin shad were an abundant forage fish. Common carp were present and abundant both before and after the establishment of striped bass (Allen and Roden 1978; Nevada Department of Wildlife, unpublished report, 1979–1981). Two native cyprinids, the bonytail chub and the razorback sucker, are found in Lake Mohave (Minckley 1973) and are federally endangered species.

Lake Mohave is a popular sport fishery. In 1980, the estimated lake-wide total angler use was 340,000 angler-days. Those anglers were estimated to have caught 380,000 rainbow trout, 130,000 largemouth bass, and 24,000 channel catfish. A creel census in 1980 showed that at Willow Beach, 98% of the anglers were fishing for trout while at Cottonwood Cove, 45% of the anglers were fishing for trout, 47% for largemouth bass, and 4% for channel catfish (Figure 1; Nevada Department of Wildlife, unpublished report, 1980).

Striped bass were never intentionally introduced into Lake Mohave but were first documented there in 1981 (Liles 1991). They likely entered Lake Mohave through Hoover Dam as eggs or larvae. Larger life stages may have passed through the dam, and a few may have survived the pressure change and turbines. Additional introductions probably occurred in 1983–1984, when Lake Mead filled to capacity and flowed over the Hoover Dam spillways. Striped bass eggs were first documented in the reservoir through ichthyoplankton sampling by AGFD in 1986 (Liles 1991). Additional ichthyoplankton sampling by NDOW through the late 1980s found striped bass eggs and larvae from April through September. Spatially, striped bass eggs and larvae were found throughout the entire seasonally warm part of the reservoir (Nevada Department of Wildlife, unpublished reports, 1984–1992). While some of the eggs and larvae may have passed through Hoover Dam, the numbers collected, the widely dispersed samples (64 km of reservoir), and the rapid increase in the striped bass population indicated that they were spawned in Lake Mohave.

As the striped bass population grew, so did its popularity with anglers. By 1990, striped bass had become the largest sport fishery on Lake Mohave when the harvest estimate exceeded 100,000 fish. The harvest estimate peaked in 1994 at more than 400,000 fish (Nevada Department of Wildlife, unpublished reports, 1990–2008). Population growth as measured by annual gill-net sampling accelerated in 1989 and continued upward to peak in 1994. After the 1994 peak, the population began to decline until 2006 when it leveled off at a density similar to the late 1980s (Figure 2). The cessation of a strong year-class entering the population every few years was not anticipated. This change in the population dynamic coincided with the invasion of quagga mussels *Dreissena bugensis* into Lake Mohave (Moore et al. 2009). The direct effect that mussels are having on the striped bass population is unknown at this time.

Until recently, Lake Mohave exhibited excellent recruitment of striped bass. The proportional stock densities (PSD) show that the population was dominated by small fish, at least through 2007 (Figure 3). Following the sharp decline in gill-net catches (Figure 2), a major increase in PSD appeared in 2008 when it more than doubled from 2007. A further increase was noted in 2009 when PSD exceeded 40 for the first time since striped bass arrived in the lake. The reasons for the population decline and subsequent PSD changes are not yet known.

Lake Mohave lacks an abundant, easily available, small to mid-size forage fish (Nevada Department of Wildlife, unpublished reports, 1990–2009). The forage deficit likely inhibits the recruitment of fish into larger size-classes. While the number of fish that recruit past 650 mm TL is small, the fish that do recruit have the potential to grow to exceptional size by foraging on stocked trout, carp, and stocked native fish (the razorback sucker and bonytail chub, both listed as endangered are stocked for

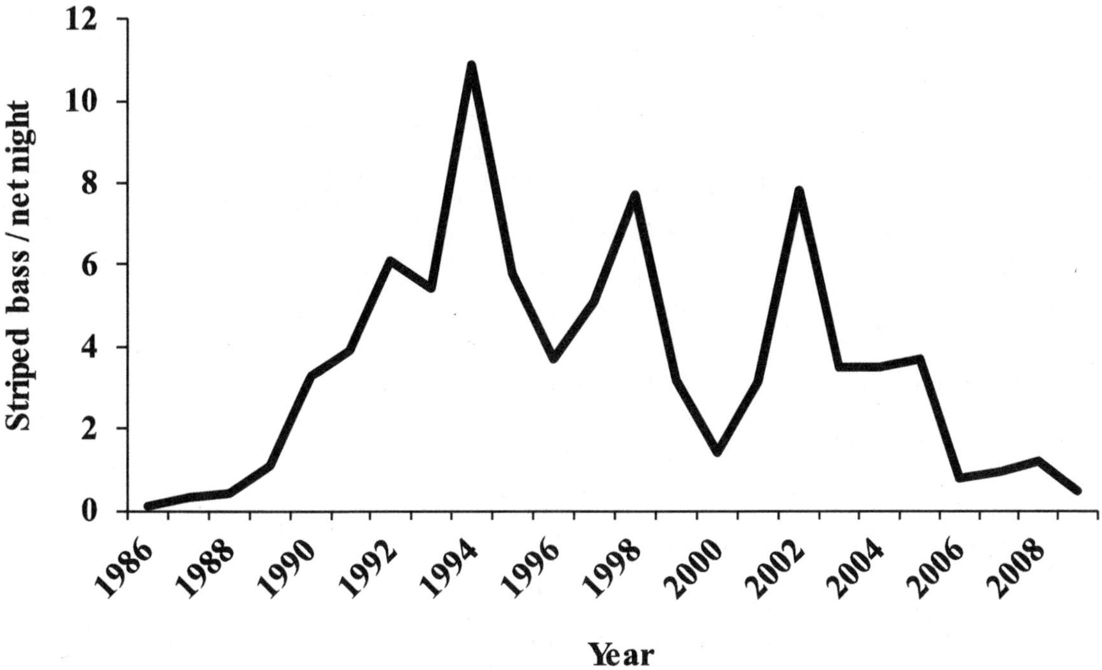

FIGURE 2. Catch per unit effort (CPUE; fish per net-night) of striped bass in gill-net surveys of Lake Mohave, 1986–2009 (Nevada Department of Wildlife, unpublished reports, 1986–2009).

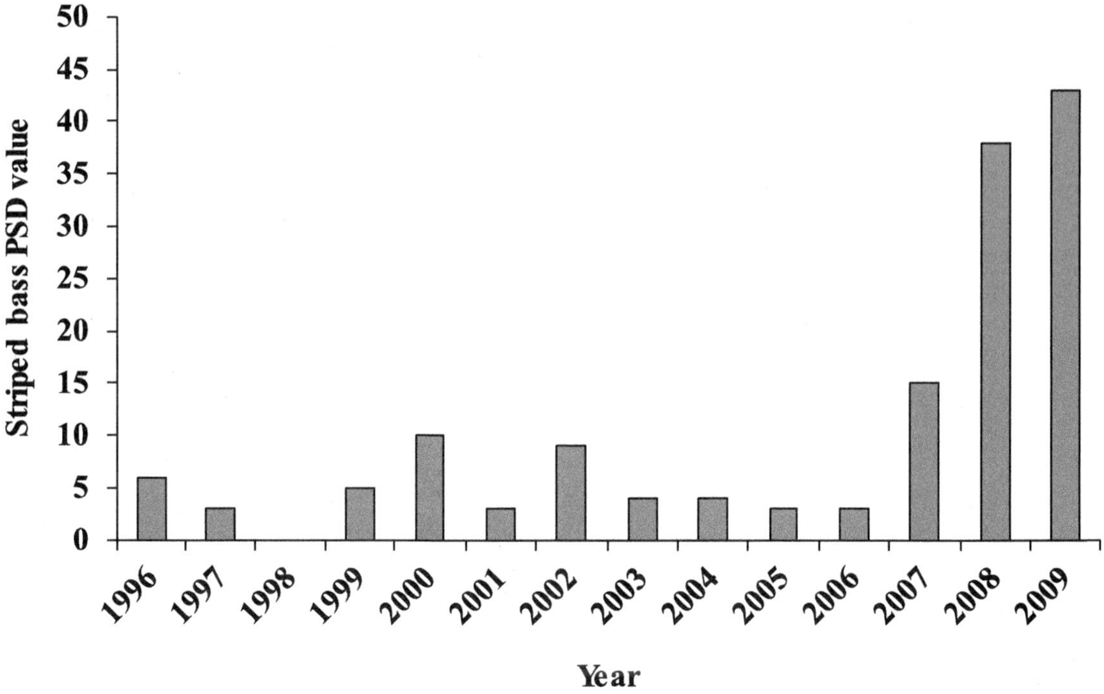

FIGURE 3. Lake Mohave striped bass proportional stock densities (PSD) from gill-netting surveys, 1996–2009. No data from 1998. (Nevada Department of Wildlife, unpublished reports, 1986–2009).

population restoration purposes). Lake Mohave has become known as a reservoir where large striped bass can be found. Individual fish up to 30 kg have been caught by anglers (Nevada Department of Wildlife, unpublished reports, 1990–2009).

Impacts of striped bass on Lake Mohave

Establishment of striped bass likely contributed, to some extent, to a decline in the rainbow trout population and fishery. However, rainbow trout harvest was in decline prior to the striped bass population expansion (Figure 4). The decline was attributed to several other factors, including a decrease in angler effort for trout, temporary trout hatchery shut downs, and reduction in stocking. When the striped bass harvest began to accelerate in 1990, the estimated trout harvest had begun to level off between 50,000 and 100,000 fish, which was down significantly from an estimated annual harvest of 500,000 in the early 1980s (Figure 4). In response to declining trout harvest, the AGFD initiated a study in the early 1990s to investigate the trout fishery decline in Lake Mohave's Hoover Dam tailwater (Walters et al. 1996). In that study, Walters et al. concluded that stocked trout were the main prey item of striped bass in Lake Mohave and that striped bass predation could quickly deplete the stocked trout population.

As the striped bass population continued to increase rapidly, there was a major shift in angler interest. In 1987, at Cottonwood Cove, 72% of anglers were fishing for trout and 2.5% were fishing for striped bass. By 1991, the ratio had reversed to 3% of anglers fishing for trout and 76% angling for striped bass (Nevada Department of Wildlife, unpublished reports, 1987–1991). At Willow Beach changes also occurred but to a lesser degree. In 1987, 96% of anglers were fishing for trout and only 1.5% for striped bass. By 1991, that ratio had changed to 27% trout anglers and 30% striped bass anglers (Figure 5).

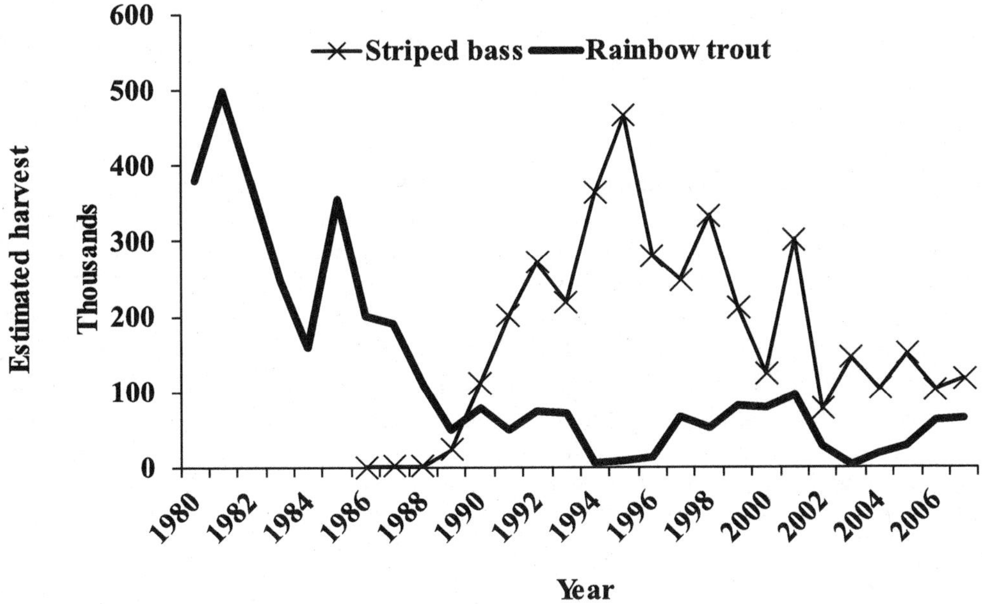

FIGURE 4. Estimated rainbow trout and striped bass harvest in Lake Mohave from 1980 to 2007 (Nevada Department of Wildlife, unpublished reports, 1986–2009). Trout hatcheries were shut down in 1984, 1994, and 2003.

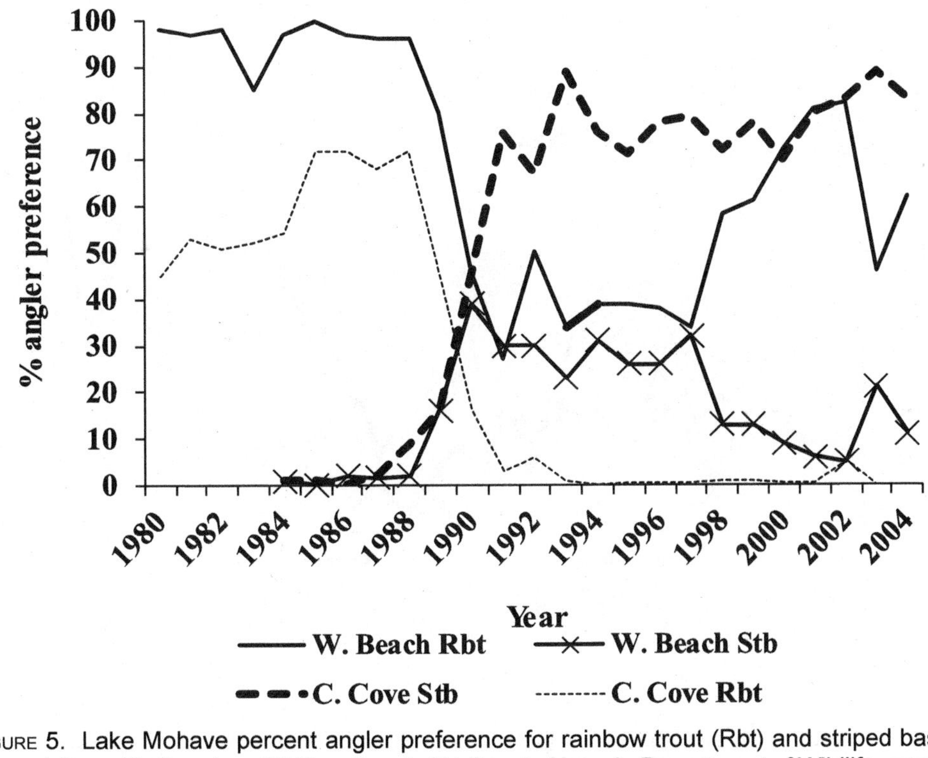

FIGURE 5. Lake Mohave percent angler preference for rainbow trout (Rbt) and striped bass (Stb), Cottonwood Cove (C. Cove) and Willow Beach (W. Beach; Nevada Department of Wildlife, unpublished reports, 2006–2009).

Because of constant, coldwater temperatures and continued trout stocking, Willow Beach remained the center of trout angling on Lake Mohave and retained a significant amount of trout angler interest. The trout fishery, however, has become primarily a put-and-take fishery. Very few striped bass less than or equal to 500 mm TL are present in the upper reaches of Lake Mohave, presumably because of the cold water temperatures. The continued stocking of trout, however, does attract striped bass greater than 500 mm TL. The Willow Beach area is known by anglers as an area where large striped bass can be caught year-round (Nevada Department of Wildlife, unpublished reports, 1980–2009). Between 2006 and 2009, the mean length of striped bass measured by creel census at Willow Beach was 635 mm TL (n = 116). By contrast, between 2007 and 2008 (no data available for 2006 and 2009), striped bass caught at Cottonwood Cove, located in the seasonally warm part of the lake, had a mean length of 353 mm TL (n = 916) (Nevada Department of Wildlife, unpublished reports, 2006–2009).

Angler catch rates on Lake Mohave were also significantly affected by striped bass population growth. At Cottonwood Cove, the combined catch rate for all species more than doubled from a pre-striped-bass average of 0.23 fish per angler-hour (f/h) (1970–1990) to a post-striped-bass average of 0.57 f/h (1991–2004) (Figure 6). The increase in catch rate was largely the result of increasing numbers of striped bass in the catch (Nevada Department of Wildlife, unpublished reports, 1990–2004). At Willow Beach, the angler catch rates actually declined. The combined catch rate for all species dropped from a pre-striped-bass average of 0.37 f/h (1970–1990) to a post-striped-bass average of 0.26 f/h (1991–2008) (Figure 6; Nevada Department of Wildlife, unpublished reports, 1970–2004). This drop can be attributed to decreased trout stocking and, as described by Walters et al. (1996), striped

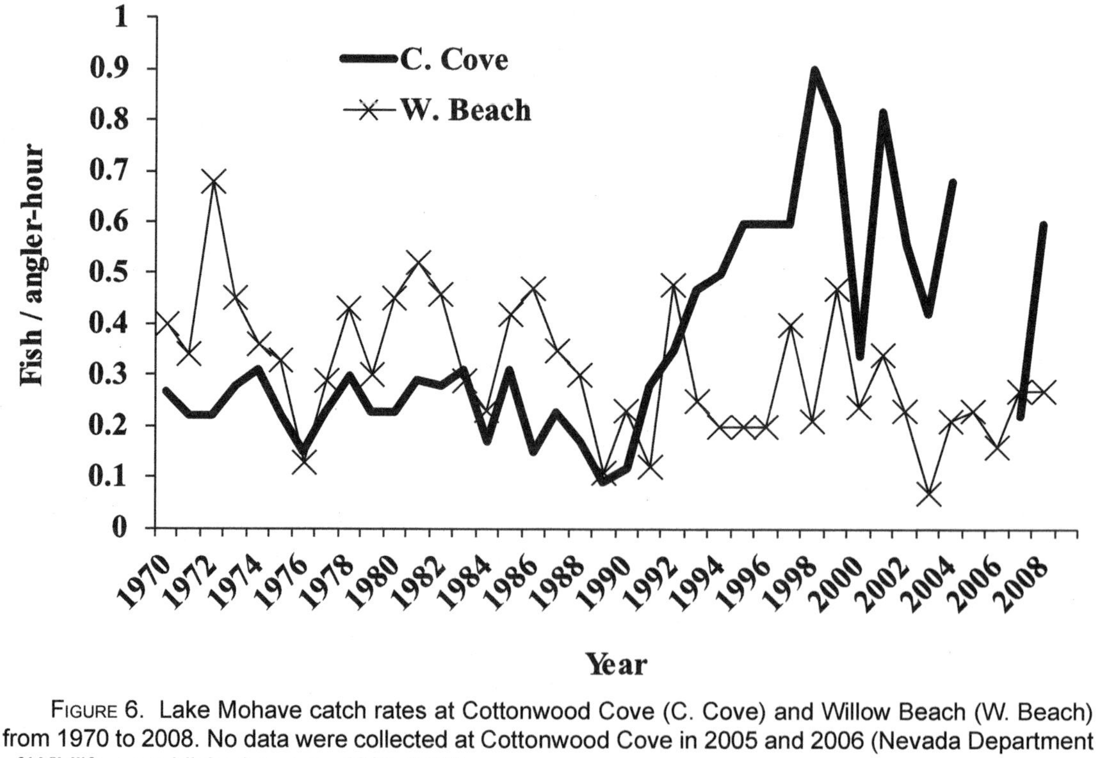

FIGURE 6. Lake Mohave catch rates at Cottonwood Cove (C. Cove) and Willow Beach (W. Beach) from 1970 to 2008. No data were collected at Cottonwood Cove in 2005 and 2006 (Nevada Department of Wildlife, unpublished reports, 1986–2009).

bass predation on stocked trout in the Hoover Dam tailwater.

In Lake Mohave, threadfin shad abundance declined as the striped bass population expanded. Threadfin shad production in Lake Mohave was monitored by open-water trawls, which began in 1987 when striped bass numbers were still fairly low. By 1991, striped bass numbers had increased significantly (Figure 2) and shad production became undetectable (Figure 7). In 1995, after several years of no production, larval shad were again noted in the reservoir and shad production continued through 1997. After 1997, no threadfin shad were observed in Lake Mohave by any method until 2007 (Nevada Department of Wildlife, unpublished reports, 1997–2007). In that year, a number of separate shad schools were noted visually in dispersed locations by agency personnel (NDOW, AGFD, USBR) and by the general public. However, in 2008 and 2009, threadfin shad were again undetectable, with no visual observations or collections in annual gill netting or electrofishing surveys. We hypothesize that because of low water levels in Lake Mead, a pulse or pulses of shad larvae passed through the upper intakes of Hoover Dam and into Lake Mohave in 2007 without continuing survival and recruitment.

Predation by nonnative species is negatively affecting the recruitment of native fish in Lake Mohave, particularly the razorback sucker. There were an estimated 44,000 razorback suckers in Lake Mohave in 1991; by 2001, the estimate was fewer than 3,000 (Marsh et al. 2003). Spawning and the production of razorback sucker larvae continue. Predation by nonnative species has eliminated recruitment (Marsh and Langhorst 1988; Minckley and Deacon 1991; Marsh and Pacey 2005). The Native Fish Work Group (NFWG) was formed in the early 1990s to address the razorback recruitment problem. The NFWG consists of state, federal, academic, and private biologists actively involved in the Lake Mohave fisheries

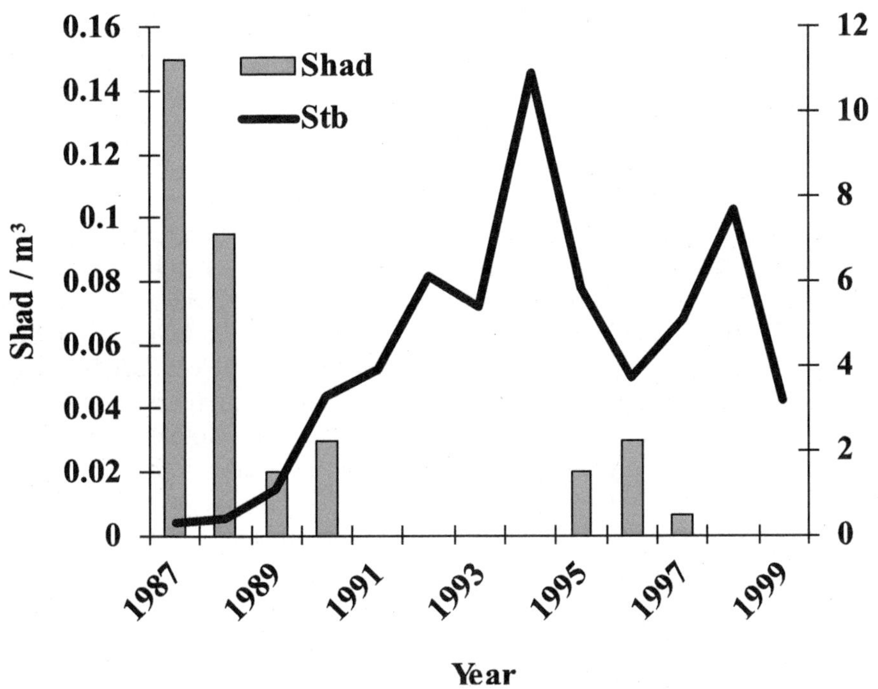

FIGURE 7. Lake Mohave threadfin shad trawl summary from 1987 to 1999. In 1993 and 1994, a light was used to determine presence or absence; no shad were found and no trawls attempted in those years (Nevada Department of Wildlife, unpublished reports, 1986–2009). Stb = striped bass.

program. The goals of the program are to maintain a population of 50,000 razorback suckers in the reservoir and to maintain the genetic diversity of the original wild population. A program of collecting wild razorback sucker larvae, rearing them in a predator-free environment, and repatriating them back into the lake began in the early 1990s and continues with modifications today. The survival rate of stocked fish has been disappointing, less than 3%. While the program has been unable to maintain the desired population size of 50,000, it has been successful in maintaining the genetic diversity of the original population (Dowling et al. 2005).

In summary, the dispersion of striped bass into Lake Mohave has had unintended consequences for fisheries management. Striped bass were never stocked into Lake Mohave because managers and the public preferred to maintain the stocked trout fishery. There was less concern for native fish at the time striped bass dispersed from Lake Mead, but concern has increased over the past 20 years. To compensate, management efforts have been directed at promoting striped bass harvest to minimize the predation impact of striped bass on the existing fishes. From the beginning, striped bass daily harvest limits have been liberal, 20 fish, and by 2001 the limit was totally removed for fish less than or equal to 500 mm TL (NDOW 2001). The ability of Lake Mohave to produce young fish and the desire to control those numbers precipitated the limit removal. The 20-fish limit on fish greater than 500 mm TL was kept.

The trout stocking program has been modified to optimize returns to anglers. The number of trout stocked has been cut by 70%. Trout stocking has been limited to only areas with concentrated angler use. Stocking trout in high-use areas also serves to increase the harvest of striped bass, as large striped bass also concentrate in these areas because of the trout stocking. As recommended by Walters et al. (1996), the sizes of the stocked trout were increased to 300 mm TL. It was thought by Walters et al. that the larger trout would be more adept at escaping predation. The native fish stocking size has also gradually increased over time, up to 500 mm TL. When possible, native fish are stocked in areas of the lake where trout are not stocked to avoid concentrations of striped bass and other predators and in areas likely to provide escape cover for those fish.

Lake Pleasant

Striped bass source reservoir (Lake Havasu)

Lake Havasu is located downstream from Lake Mohave between Arizona and California on the lower Colorado River (Figure 1). Formed behind Parker Dam in 1938, the 8,470-ha lake is a storage reservoir from which water is pumped into two major aqueducts, one feeding California and the other feeding Arizona. Lake Havasu extends 40 km north from Parker Dam where it receives water from the Colorado River. The river extends another 81 km to the base of Davis Dam, which forms Lake Mohave.

Striped bass were first stocked into Lake Havasu in 1962. At the time of the introduction, it was generally accepted that striped bass would not be able to reproduce sufficiently to sustain a self-maintaining population (Gustaveson and Blommer 2013). Evidence of reproduction in Lake Havasu was found in the late 1960s and a self-maintaining population was established (Edwards 1974) that continues to persist.

During the period of initial striped bass stocking, plans were being developed for the Central Arizona Project (CAP) Canal to divert water to central and southern Arizona. Authorized in 1968 and completed in 1993, the CAP Canal diverts water from the Colorado River at Lake Havasu to central Arizona (Mueller 1990). The concrete-lined canal system is fairly extensive, extending 521 km across the state (Figure 1). Water from the Colorado River at the south end of Lake Havasu is lifted 251 m by the Mark Wilmer Pumping Plant and runs through 11.3 km of tunnel under the Buckskin Mountains before entering the CAP Canal. The pumping plant has six pumps producing a maximum discharge of 85 m^3/s. The canal has three sequential aqueducts: Hayden-Rhodes, Fannin-McFarland, and Tucson. Water in the Hayden-Rose Aqueduct is pumped 306 km from the outlet portal at the Buckskin Mountains through three pumping stations (Bouse Hills, Little Harquahala, and Hassayampa) to the city of Phoenix. At kilometer 239, the Waddell Canal diverts water 7.9 km from the Hayden-Rhodes Aqueduct and passes through the Waddell Pumping Plant before entering Lake Pleasant.

The development of the canal system provided striped bass the opportunity to disperse downstream to Lake Pleasant and into central and southern Arizona. The potential effects of the CAP Canal project on Lake Pleasant fisheries were assessed during the construction period. Despite the challenges to fish passage presented by the series of aqueducts and pumping plants, concern over the impacts of striped bass from Lake Havasu on the existing ichthyofauna in Lake Pleasant prompted Grabowski et al. (1984) to assess potential impacts before the canal was filled. They concluded that despite predation, pump intake stresses, and temperatures exceeding 30°C, juvenile striped bass would likely move down the canal and enter Lake Pleasant. They also predicted that if striped bass did survive to maturity in Lake Pleasant, reproductive success would likely be low (Grabowski et al. 1984).

Striped bass were found in the CAP Canal almost immediately after it began to fill. Initial water diversions into the CAP Canal occurred in 1985. Fish sampling from 1986 to 1989 indicated that striped bass were widely distributed throughout the canal, but the majority of them remained in the first 27 km of the canal to the Bouse Hills pumping station, the first main impediment to further passage. As the canal system reached operational capacity, abundance of all fish species declined. Mark–recapture population density estimates of striped bass in 1986 were 70 ± 37 fish/ha and in 1989 were 3 ± 1 fish/ha (Mueller 1990). The decline in striped bass and other species in the canal was attributed to higher operational velocities and associated scour. From 1986 to 1989, striped bass represented up to 14% of the fish community. However, no striped bass larvae and only five eggs were collected in the canal, suggesting that striped bass populations within the canal are dependent upon entrainment rather than in-canal reproduction (Mueller 1996).

Further analysis and observations showed larger fish were capable of dispersing through the canal and that they did so. Grabowski et al. (1984) showed that striped bass less than or equal to 25 mm TL would not be able to avoid intake velocities at the Mark Wilmer Pumping Plant and likely would be entrained. They also calculated that a striped bass up to 305 mm TL could be pumped and transported down the canal based on minimum clearance of water passage through impellers. However, in 1986, 18 months after the canal began filling, a 500-mm-TL striped bass was collected in the first reach, suggesting that larger fish were entering the canal. From 1986 to 1989, sizes of striped bass ranged from 115 to 613 mm TL. Fish greater than 305 mm TL may have swum through the turbines when pumps were inactive and flushed downstream once operations resumed (Mueller 1990).

In 1995, the USBR developed and implemented a long-term monitoring program to determine the presence and distribution of nonnative fish in the CAP Canal (Clarkson 1996). Seven sampling stations located at the Bouse Hills, Little Harquahala, Hassayampa, Salt-Gila, Brady, Red Rock, and San Xavier pumping plants were selected throughout the canal (Figure 1). With the exception of 2004, fewer than 50 striped bass were collected in any one annual survey. In 2004, more than 1,200 striped bass were observed at the Bouse Hills station during a dewatering event in the first reach of the canal. Nearly 80% of the captured fish that year were young of year. Striped bass have been captured at all of the sampling stations, with the exception of the farthest downstream station, San Xavier (Table 1). Most of the striped bass captured during the annual surveys are at

TABLE 1. Presence (+) or absence (–) of striped bass from 1995 to 2008 at each of seven sampling stations along the Central Arizona Project canal (USBR 1995–2008), in order from Lake Havasu downstream: Bouse Hills (BH), Little Harquahala (LH), Hassayampa (H), Salt-Gila (SG), Brady (B), Red Rock (RR), and San Xavier (SX). Blank cells indicate no sampling conducted.

	Location						
Year	BH	LH	H	SG	B	RR	SX
1995	+	+	–	+	+	–	–
1996	–	–	–	–	–	–	–
1997	+	+	+	+	+	+	–
1998	+	+	+	+	+	+	–
1999	+	–	–	+	+	–	–
2000	–	+	–	–	–	+	–
2001	+		–	–	–	–	–
2002	+		–	+	+	+	–
2003	+	+	+		+	+	–
2004	+	+	+	+	+		–
2005				+	–	+	–
2006	+	+	+				
2007	+	–	–	+	–	+	–
2008				+	–	–	–

the first station, Bourse Hill Pumping Plant (USBR 1995–2008). This was also true during Mueller's (1990) study, where only 2% of the striped bass taken from the canal were collected downstream of the first reach. Nonetheless, striped bass dispersal has been extensive in the CAP Canal system.

Striped bass receiving reservoir (Lake Pleasant)

Originally formed by the Waddell Dam in 1927, Lake Pleasant was a 1,497-ha irrigation water storage facility located 48 km northwest of the city of Phoenix (Figure 1). At the time, the lake was primarily fed by the Agua Fria River, an intermittent stream, and several smaller ephemeral streams on the north end of the reservoir. Over the years, Arizona Game and Fish stocked largemouth bass, bluegill, channel catfish, white crappie *Pomoxis annularis*, black crappie, white bass *Morone chrysops*, and threadfin shad. Threadfin shad has been the primary forage base that supports the only self-sustaining white bass fishery in Arizona (Morgensen 1990).

In 1992, the New Waddell Dam was constructed 800 m downstream from the Waddell Dam to increase water storage at Lake Pleasant. Following the completion of the New Waddell Dam, water from the CAP Canal was pumped into Lake Pleasant and nearly tripled its size to 4,168 ha. Lake Pleasant is now filled primarily by CAP water at the south end of the reservoir. The pattern of water level fluctuation changed drastically. Prior to CAP water, the lake fluctuated less than 4 m throughout the year and was considered a mesotrophic reservoir. Since CAP water has been entering the lake, water levels fluctuate seasonally up to 23 m, based on water needs, and the lake has become meso-oligotrophoic (Bryan 2005). As demand for water increases during the summer, water is released through the pump-generation plant back into the canal. In the winter, when water demand is low, Colorado River water is pumped from the CAP Canal into the lake. The Waddell Pumping Station lifts water 58 m into the lake and has a capacity to pump water at 85 m^3/s.

In December 1992, the old dam was breached and, for the first time, water from the CAP Canal was pumped into the original reservoir. No striped bass were collected during the first gill-netting survey conducted 2 years after CAP water began entering the reservoir (Arizona Game and Fish Department, unpublished data). The first striped bass were detected during a 1998 gill-netting survey and ranged from 395 to 451 mm TL (Arizona Game and Fish Department, unpublished data). Striped bass density gradually increased until a substantial population expansion in 2005 (Table 2; Bryan 2005; Stewart et al. 2007). During early spring, striped bass use the intermittent Agua Fria River located on the north end of the reservoir for reproduction, but successful spawning appears to occur only during years of high inflow (Stewart et al. 2007). Discharge during spring 2005 in the Agua Fria River was nearly three and a half times the average discharge since striped bass were introduced into the lake, and 2005 was the first year that substantial recruitment was detected in the reservoir. Stewart et al. (2007) showed that striped bass reproductive success is highly dependent upon the infrequent flows in the Agua Fria River. The upper reaches of the Agua Fria River run dry most of the year, and the lower reach becomes inundated when the reservoir is at full pool. Hydroacoustic surveys along with gill-netting surveys were conducted in 2005 and 2006 to obtain density estimates. The estimated striped bass population was 40,535 ± 2,980 and 27,229 ± 2,112 in 2005 and 2006, respectively (Stewart et al. 2007).

Impacts of striped bass on Lake Pleasant

Striped bass introduction has impacted Lake Pleasant fish populations and angler attitudes towards available fish. According to a 2001 statewide survey, Lake Pleasant was the highest-use lake in Arizona with more than 520,000 user days per year (Pringle 2004). It was a premier largemouth bass fishery holding approximately 150 tournaments per year. A creel survey conducted from 2000 to 2004 showed that less than 1% of anglers were targeting striped bass (Bryan 2005). Since the population expansion in 2005, however, that number has likely increased.

Anglers' concern over competition with largemouth bass, and to a lesser degree white bass, for food (primarily threadfin shad) made striped bass an unpopular sport fish. Most studies, however, have reported little to no predation on other game fish (Miranda and Raborn 2013, this volume). Studies at Lake Pleasant found relatively low diet overlap between largemouth bass and striped bass (Stewart et al. 2007). For white bass, anglers' concerns about competition

TABLE 2. Catch per unit effort (CPUE) with SE, number caught (*N*), and mean total length (TL) and range (mm) of striped bass during gill-netting surveys at Lake Pleasant from 1994 to 2006. Catch per unit effort was calculated as the number of fish caught per net per 12 h (Bryan 2005; Stewart et al. 2007; Arizona Game and Fish Department, unpublished data)

Year	CPUE	SE	*N*	Mean TL	Range
1994	0.00	0	0		
1998	0.13	0.06	5	419	395–451
2000	0.48	0.22	14	455	220–622
2001	0.47	0.17	27	414	215–722
2002	0.72	0.17	36	443	238–686
2003	1.25	0.17	68	314	210–735
2004	1.14	0.37	53	492	214–762
2005	6.74	2.03	230	323	215–678
2006	3.85	1.01	192	366	236–597

have been supported. From the time striped bass immigrated into Lake Pleasant, white bass relative abundance decreased (Bryan 2005; Stewart et al. 2007). This was especially evident in 2006, the year following the striped bass population expansion, when relative abundance of white bass decreased from 7.62 to 0.52 fish/net-night and relative weights decreased from 93 to 80 (Stewart et al. 2007). Unlike Lake Mohave, threadfin shad do not appear to be declining in Lake Pleasant as indicated by gill-netting surveys (Stewart et al. 2007). Lake Pleasant is composed of numerous embayments and coves that provide refuge for threadfin shad. Lake Pleasant provides the only self-sustaining white bass fishery in the state.

Management of striped bass in Lake Pleasant and CAP Canal has focused on minimizing unwanted impacts. In 1994, the AGFD implemented a special regulation to allow unlimited take of striped bass to control the striped bass population, to reduce competition with the existing fish populations, and to preserve the only white bass fishery in Arizona. However, the impacts of angling pressure on the striped bass population are unknown.

CAP Canal Downstream of Lake Pleasant

Downstream from Lake Pleasant, the Hayden-Rose Aqueduct continues to carry water though Phoenix before reaching the Fannin-McFarland Aqueduct. Prior to reaching the Fannin-McFarland Aqueduct, water must pass through the Granite Reef Diversion Dam, which is an interconnect between the CAP Canal and the Salt River Project (SRP) canals, which connects water in the Salt and Verde rivers. Striped bass escapement out of the CAP into these river drainages could adversely affect native fish. To control the spread of striped bass and other nonnative fishes from the CAP Canal to other river systems, electrical barriers were constructed as a control strategy to prevent nonnative species from moving upstream once outside the CAP. The electrical barriers were not entirely effective (Clarkson 2004), but striped bass have not been detected above the barriers.

The Fannin-McFarland Aqueduct has only one pumping station, Salt-Gila Pumping Plant, and extends 93 km before reaching the Tucson Aqueduct. The Tucson Aqueduct has nine pumping plants and extends 140 km to the end of the system. In total, the CAP Canal extends 521 km from Lake Havasu to Tucson, Arizona. Striped bass have dispersed almost entirely throughout the canal system and have been detected as far downstream as the Red Rock Pumping Plant (km 437).

Discussion

Unintended dispersal of striped bass from Colorado River reservoirs where they were stocked has had major impacts on fishery management in Lake Mohave and bodies of water connected to the CAP Canal, including Lake Pleasant. When initial stocking of striped bass took place in the 1960s, managers viewed the introduction of nonindigenous species much differently than they do today. At that time, striped bass, along with other limnetic predators, were stocked in Lake Mead to utilize the large population of threadfin shad. It was assumed that striped bass would not successfully reproduce in the reservoirs in the lower Colorado River and that populations and fisheries could be controlled by stocking. The thought at the time was that despite potential spawning runs into tributaries, the reach of river that runs from Davis Dam down to Lake Havasu would not allow adequate time for eggs to hatch before reaching the lake, and they would die. This turned out not to be the case. In addition, spill of striped bass from Lake Mead into Lake Mohave was not anticipated. Thus, the dispersal began.

Striped bass populations in Lake Mohave and Lake Pleasant have been able to succeed largely because of their ability to reproduce locally. Due to the low numbers of striped bass successfully passing Hoover Dam and found in the CAP Canal, numbers entering from the source reservoirs are inadequate to support striped bass populations large enough to impact other sport fish species by direct predation and competition for common prey items. Striped bass were never intended for these reservoirs, and management strategy has been to reduce the numbers with liberal catch limits to lessen the impact on other species. It is difficult to determine how effective the liberal harvest has been at controlling striped bass numbers. Certainly, many striped bass have been removed by anglers, but the ability of these reservoirs to produce striped bass may outstrip the ability of anglers to remove them.

In Lake Mohave, a new and popular fishery developed for striped bass. Angler effort and harvest estimates for striped bass have declined over time, but catch per unit effort has remained high. The population density as measured by the annual gill-net surveys has declined. How much of this decline is the result of angler harvest as opposed to other factors is unclear and may warrant further studies. There has been no recovery of the threadfin shad population, which might be expected with declining striped bass densities. The result has been an excellent fishery for small striped bass and development of a limited trophy fishery, which has become extremely popular with the angling public.

Despite the lack of in-canal reproduction, dispersion of striped bass throughout the CAP Canal system was extensive and rapid, likely because fish are exposed to higher degrees of downstream drift compared with naturally flowing system. The potential impacts of striped bass dispersing in the CAP Canal and into Lake Pleasant were anticipated and evaluated. Predictions were made that striped bass would enter the CAP Canal and inevitably make their way to Lake Pleasant. It was unknown at the time whether they would establish in the lake, but they certainly did. As a proactive strategy, managers set unlimited bag limits at Lake Pleasant even before the first striped bass was captured. Not knowing whether they would become established in the lake, managers took a wait-and-see approach. Gradually striped bass numbers increased until a very successful recruitment occurred in 2005. As striped bass popularity increased managers continued to promote and encourage harvest even amending regulations to allow for spear fishing (a rather unique liberalization of harvest for a game fish). Natural hybridization of striped bass and white bass has been recorded in several reservoirs across the United States (Harrell 2013, this volume). As the only self-sustaining white bass fishery in Arizona, the introduction of striped bass has provided opportunity for natural hybridization to occur but is of little concern to managers as white bass are not a highly prized sport fish in Arizona.

The well-being of native fishes is of much greater concern today than it was when striped bass were introduced to the Southwest in the 1960s. After recognition that introduced striped bass disperse well beyond stocked waters and impact existing fish populations, efforts were made to protect and restore native fishes. Native fish have been propagated, but survival of stocked fish has been poor (Mueller and Burke 2005). The degree to which poor survival can be attributed to striped bass is under evaluation. The USFWS in the 1990s anticipated striped bass escaping through the CAP Canal and initiated installation of electrical barriers to prevent them from entering the Salt and Gila rivers. With dispersion of striped bass through the CAP Canal system recognized as inevitable, a "recovery in lieu of threat removal" concept was developed that encompassed several aspects: fish barriers on high-value native fish streams, native fish recovery where striped bass have invaded, and direct control of nonnatives such as striped bass (R. Clarkson, U.S. Bureau of Reclamation, personal communication).

The future of the striped bass and other fisheries in Lakes Mohave and Pleasant after the inadvertent dispersal of striped bass to these lakes is unclear. Recent dispersal of quagga mussels into Lake Mohave and Lake Pleasant will likely be a significant agent of ecological change, possibly affecting striped bass and other fish species being managed. Additionally, the recent and undesired establishment of gizzard shad and blue tilapia in Lake Mead (Nevada Department of Wildlife, unpublished report, 2008) will further complicate management of Colorado River fisheries, as these species may disperse through the system like striped bass did.

Management recommendations

The lower Colorado River was initially stocked with striped bass more than 50 years ago. Prior to their introduction in the lower Colorado River and during the construction of the CAP Canal, varying degrees of risk analysis were conducted that evaluated the impacts of striped bass and their potential to disperse. While some predictions came true (dispersal to Lake Pleasant), others did not (reproduction would not occur in Lake Mead). Managers wanting to stock striped bass or develop water diversion projects must consider the ability of striped bass to reproduce in both intended and unintended water bodies. Dispersion of striped bass from Lakes Mead and Havasu can largely be attributed to successful reproduction in those reservoirs and a bountiful source for downstream dispersal of striped bass at all life stages. Furthermore establishment of striped bass in unintended waters would not be possible in the absence of reproduction and an adequate prey source such as threadfin shad. It is also evident that dams and pumping plants do not act as barriers to dispersal and managers wanting to prevent dispersal should consider additional measures, which likely depend on the status and importance of fish resources in waters that might be invaded.

Acknowledgments

We thank the many agencies and personnel that provided data and support, including Arizona Game and Fish Department, Arizona State University, Marsh and Associates, National Park Service, Lake Mead National Recreation Area, Nevada Department of Wildlife, U.S. Bureau of Reclamation, U.S. Fish and Wildlife Service, University of Nevada Las Vegas, Willow Beach National Fish Hatchery, Rob Clarkson of the USBR, and Mike Guisti of the California Department of Fish and Game.

References

Aasen, K. D., G. E. Reiner, and A. W. Baracco. 1982. The movement and survival of fish through the A.D. Edmunston Pumping Plant. California Department Fish and Game, Inland Fisheries Administrative Report No. 82–2, Sacramento.

Allen, R. C., and D. L. Roden. 1978. Fishes of Lake Mead and Lake Mohave. Nevada Department of Wildlife, Biological Bulletin No. 7, Reno.

Bryan, S. D. 2005. Phase II: Limnological and fisheries investigation of Lake Pleasant. Final Report to the U.S. Bureau of Reclamation, Phoenix Area Office, Cooperative Agreement No. 00-FC-32–0160. Arizona Game and Fish Department, Research Branch, Phoenix.

Clarkson, R. W. 1996. Long-term monitoring plan for fish populations in selected waters of the Gila River basin, Arizona. U.S. Bureau of Reclamation, Revision No. 2, Phoenix, Arizona.

Clarkson, R. W. 2004. Effectiveness of electrical fish barriers associated with the Central Arizona Project. North American Journal of Fisheries Management 24:94–105.

Dowling, T. E., P. C. Marsh, A. Kelsen, and C. A. Tibbets. 2005. Genetic monitoring of wild and repatriated populations of endangered razorback sucker (*Xyrauchen texanus*, Catostomidae, Teleostei) in Lake Mohave, Arizona–Nevada. Molecular Ecology 14:123–135.

Edwards, G. B. 1974. Biology of the striped bass Morone saxatilis in the lower Colorado River (Arizona-California-Nevada). Master's thesis. Arizona State University, Tempe.

Grabowski, S. J., S. D. Hiebert, and D. M. Lieberman. 1984. Potential for introduction of three species of nonnative fishes into central Arizona via the Central Arizona Project. U.S. Bureau of Reclamation Report, Denver.

Gustaveson, W., and G. Blommer. 2013. History of striped bass management in the Colorado River. Pages 15–23 *in* J. S. Bulak, C. C. Coutant, and J. A. Rice, editors. Biology and management of inland striped bass and hybrid striped bass. American Fisheries Society, Symposium 80, Bethesda, Maryland.

Harrell, R. M. 2013. Releasing hybrid *Morone* in natural waters with congeneric species: implications and ethics. Pages 531–549 *in* J. S. Bulak, C. C. Coutant, and J. A. Rice, editors. Biology and management of inland striped bass and hybrid striped bass. American Fisheries Society, Symposium 80, Bethesda, Maryland.

Jonez, A., and R. C. Sumner. 1954. Lake Mead and Mohave investigations. Nevada Fish and Game Commission, Final Report, D-J Project F-1-R, Las Vegas.

Liles, T. A. 1991. The status of striped bass in Lake Mohave. Arizona Game and Fish Department, Phoenix.

Marsh, P. C., and D. R. Langhorst. 1988. Feeding and fate of wild larval razorback sucker. Environmental Biology of Fishes 21:59–67.

Marsh, P. C., C. A. Pacey, and B. R. Kesner. 2003. Decline of the razorback sucker in Lake Mohave, Colorado River, Arizona and Nevada. Transactions of the American Fisheries Society 132:1251–1256.

Marsh, P. C., and C. A. Pacey. 2005. Immiscibility of native and nonnative fishes. Pages 59–63 *in* M. J. Brouder, C. L. Springer, and S. C. Leon, editors. Proceedings of two symposia: Restoring native fish to the lower Colorado River: interactions of native and nonnative fishes, and Restoring natural function within a modified riverine environment—the lower Colorado River. U.S. Fish and Wildlife Service, Albuquerque, New Mexico.

Minckley, W. L. 1973. Fishes of Arizona. Arizona Game and Fish Department, Phoenix.

Minckley, W. L., and J. E. Deacon, editors. 1991. Battle against extinction: native fish management in the American West. University of Arizona Press, Tucson.

Miranda, L. E., and S. W. Raborn. 2013. Interactions between striped bass and other game fish in reservoirs. Pages 501–519 *in* J. S. Bulak, C. C. Coutant, and J. A. Rice, editors. Biology and management of inland striped bass and hybrid striped bass. American Fisheries Society, Symposium 80, Bethesda, Maryland

Moore, B., S. Gerstenberger, and D. Wong. 2009. Quagga mussel invasion into Lakes Mead and Mohave in 2007: abundance, distribution and size frequency. Lake Mead Science Symposium. University of Nevada, Las Vegas.

Morgensen, S. A. 1990. Phase I: Baseline limnological and fisheries investigation of Lake Pleasant. Final Report to the U.S. Bureau of Reclamation. Arizona Game and Fish Department, Research Branch, Phoenix.

Mueller, G. 1990. Fisheries investigations in the Central Arizona Project canal system: final report 1986–1989. U.S. Bureau of Reclamation Report, Boulder City, Nevada.

Mueller, G. 1996. Establishment of a fish community in the Hayden-Rhodes and Salt Gila aqueducts, Arizona. North American Journal of Fisheries Management 16:795–804.

Mueller, G., and T. Burke. 2005. Survival of young razorback sucker in relation to stocking rates (fish/ha) and in the presence or absence of predator communities in Lake Mohave, Arizona–Nevada. Pages 155–163 *in* M. J. Brouder, C. L. Spring, and S. C. Leon, editors. Proceedings from two symposia: Restoring native fish to the lower Colorado River: interactions of native and non-native fish, and Restoring natural function within a modified riverine environment—the lower Colorado River. Albuquerque, New Mexico: U.S. Fish and Wildlife Service, Southwest Region.

NDOW (Nevada Department of Wildlife). 2001. 2001 Nevada fishing seasons and regulations. Nevada Department of Wildlife, Fisheries Division, Reno.

Pringle, T. 2004. Statewide survey of 2001 Arizona anglers. Arizona Game and Fish Department, Fisheries Technical Report 03–01, Phoenix.

Stewart W. T., M. M. Meding, and D. R. Rogers. 2007. Lake Pleasant striped bass. Arizona Game and Fish Department, Research Branch, Technical Guidance Bulletin No. 11, Phoenix.

USBR (U.S. Bureau of Reclamation). 1995–2008. Annual sampling summary of fish surveys in behalf of a long-term monitoring plan for fish populations in selected waters of the Gila River Basin, Arizona. Submitted in partial fulfillment of the U.S. Bureau of Reclamation Agreement Number 04-CS-32–0170, Denver.

Van Horn, S. L. 2013. A brief history of inland striped bass management. Pages 1–13 *in* J. S. Bulak, C. C. Coutant, and J. A. Rice, editors. Biology and management of inland striped bass and hybrid striped bass. American Fisheries Society, Symposium 80, Bethesda, Maryland.

Walters, J. P., T. D. Fresques, S. D. Bryan and B. R. Vlach. 1996. Factors affecting the rainbow trout fishery in the Hoover Dam tailwater, Colorado River. Arizona Game and Fish Department Technical Report 22, Phoenix.

American Fisheries Society Symposium 80:449–460, 2013

Regulation of Striped Bass and Hybrid Striped Bass Fisheries in the United States

William R. Collier*
Tennessee Cooperative Fishery Research Unit[1]
205 Pennebaker Hall, Tennessee Technological University
Cookeville, Tennessee 38505, USA

Phillip W. Bettoli
U.S. Geological Survey
Tennessee Cooperative Fishery Research Unit[1]
205 Pennebaker Hall, Tennessee Technological University
Cookeville, Tennessee 38505, USA

George D. Scholten[2] and Timothy N. Churchill
Tennessee Wildlife Resources Agency
Post Office Box 40747, Nashville, Tennessee 37204, USA

Abstract.—Inland populations of striped bass *Morone saxatilis* and hybrid striped bass (white bass *M. chrysops* × striped bass) are intensively managed across the United States for their recreational value using a variety of regulations; however, the full extent of different regulatory strategies is unknown. This paper describes regulations used by states to manage their inland striped bass and hybrid striped bass fisheries. An Internet-based survey seeking information on issues such as stocking dependency, trophy potential, catch rates, and statewide and specialized regulations for both striped bass and hybrid striped bass fisheries was created and distributed to all state agencies. The survey found that 32 states have inland populations of striped bass, hybrid striped bass, or both. Daily creel limits for striped bass and hybrid striped bass ranged from 2 to 30; the most liberal creel limits were found in southern states. Half of the states reported that their statewide length limit for both taxa has few or no exceptions for conditions in particular water bodies, and 85% of states with striped bass and 92% of states with hybrid striped bass enforce a minimum length limit as their statewide length limit. Although both taxa were generally managed with similar regulations, striped bass were usually considered a more important sport fish species and were managed more intensively than hybrid striped bass. A literature review confirmed that little research has been published on the response (or lack thereof) of striped bass and their hybrids to any of the various management schemes reported herein.

* Corresponding author: willrcollier@gmail.com
[1] The unit is jointly sponsored by the Tennessee Wildlife Resources Agency, the U.S. Geological Survey, the U.S. Fish and Wildlife Service, and Tennessee Technological University.
[2] Present address: Iowa Department of Natural Resources, 502 East 9th Street, Des Moines, Iowa 50319, USA

Introduction

The introduction of striped bass *Morone saxatilis* and hybrid striped bass (white bass *M. chrysops* × striped bass) into inland reservoirs, lakes, and rivers has allowed both taxa to become important game fish throughout the United States (Etnier and Starnes 1993). Populations of both taxa have increased dramatically since culture techniques were refined in the 1960s. As of 1970, only seven inland striped bass fisheries existed in the United States; by 1978, that number had increased to 53 striped bass fisheries and 26 hybrid striped bass fisheries in the southeastern United States, alone (Axon and Whitehurst 1985). This expansion was partly due to a change in stocking regimes and an increase in angling popularity. The expansion of striped bass and hybrid striped bass fisheries in the United States has continued, and angling for moronids has remained popular over the decades, with approximately 19% of U.S. anglers seeking them and a relatively high rate of angler avidity, especially for striped bass (U.S. Department of the Interior, Fish and Wildlife Service, and U.S. Department of Commerce, U.S. Census Bureau 2006; Van Horn 2013, this volume).

Axon and Whitehurst (1985) developed a striped bass and hybrid striped bass questionnaire in conjunction with the Striped Bass Committee (Southern Division of the American Fisheries Society) and the U.S. Fish and Wildlife Service to evaluate accomplishments and problems associated with management of striped bass and hybrid striped bass. Their survey examined stocking programs, harvest rates, limiting factors, and management problems but did not consider harvest regulations. The survey was distributed to all 50 states, and additional information was solicited via telephone interviews. As of 1981, 34 states had stocked inland striped bass or hybrid striped bass, and the largest stocking programs (and fisheries) were found in the southern United States. Their study recommended additional research on predator–prey relationships and the habitat requirements of both taxa, and a large body of literature has been created since then on those and related topics (e.g., Cheek et al. 1985; Coutant 1985; Bjorgo et al. 2000; Coutant this volume, Rice et al. this volume; Thompson and Rice this volume).

Large surveys of inland fisheries regulations are not common. Bettoli et al. (2009) surveyed fishery agencies in the Mississippi River basin to document the various management schemes in place to manage and conserve paddlefish *Polyodon spathula*. Paukert et al. (2001) performed a similar survey of fisheries agencies in North America to reveal the breadth of regulations in place to manage northern pike *Esox lucius*. A review of striped bass and hybrid striped bass harvest regulations throughout the United States was lacking, which prompted this study.

The objectives of this study were to (1) quantify the number of states that regulate and stock striped bass and hybrid striped bass, (2) quantify the various state regulations for both taxa, (3) determine the rationales justifying these regulations, (4) quantify the various special regulations for both taxa, and (5) gain insight into how managers view their fisheries.

Methods

An online-based survey concerning the regulations of striped bass and hybrid striped bass fisheries was emailed to the fisheries chiefs of all 50 states in early April 2009. The e-mail linked respondents to the Web site surveymonkey.com, where they were asked 58 questions. These 58 questions allowed the authors to determine how many states regulated and stocked landlocked striped bass and hybrid striped bass, when regulations were first established, the types of statewide creel and size limits and the rationales behind them, the abundance and types of special regulations that differed from the statewide regulation, and the importance and time spent on management of both taxa. Respondents were asked to answer questions relevant only to significant landlocked fisheries, and the term *significant fisheries* implied fisheries in which significant (or directed) striped bass or hybrid striped bass angling took place, but was left for each respondent to define. Questions concerning rationales for establishing length limits and creel limits asked respondents to respond to a list of possible rationales on a five-point Likert scale, where 1 was "strongly disagree," 2 was "disagree," 3 was "neutral," 4 was "agree," and 5 was "strongly agree." The popularity of a particular rationale was based on the percentage of respondents who "agreed" or "strongly agreed" with it. A second e-mail was sent to nonrespondents 7 weeks after the original e-mail requesting their responses, and follow-up emails were sent to any nonrespondents after 2 weeks. A total of 47 completed surveys were received, and after visiting states' Web sites, we confirmed that the remaining three states did not possess any inland striped bass or hybrid striped bass fisheries.

Results

Statewide regulations for striped bass

Twenty-four state agencies responded that they managed at least one of the 140 significant striped bass fisheries reported nationally (Table 1). Of those 24 states, 4 had striped bass fisheries only while the other 20 states reported both striped bass and hybrid striped bass fisheries (Figure 1). Only three states with significant striped bass fisheries (Arizona, Nevada, and Utah) claimed that they had never stocked striped bass in their state, and two other states (California and Maryland) reported that they had stocked striped bass in the past but not within the past decade. Most state agencies (68%) reported that they began regulating inland striped bass populations between the 1960s and 1980s, but a few coastal states reported dates as early as the 1930s. These coastal states stocked inland striped bass at a later date, but first managed them under the existing regulations for anadromous populations.

Statewide daily creel limits for striped bass varied between 2 and 30 (Figure 2), with the most liberal daily creel limits found in southern states. Only five states (Kansas, Louisiana, Mississippi, New Jersey, and Pennsylvania) managed all of their inland striped bass fisheries with the same statewide creel limit; the rest had fisheries with exceptions to the statewide daily creel limit or no statewide daily creel limit. The top four rationales behind statewide daily creel limits were as follows, in descending order of popularity: (1 and 2 [tied]) strive for uniformity of regulations among water bodies whenever possible (75%) and equitable distribution of harvest (81%), (3) maintain quality fisheries (76%), and (4) prevent overfishing (75%). When exceptions to statewide daily creel limits were enacted, the most popular reason was because system potential differed among water bodies. Seventy-one percent of the statewide striped bass creel limits were in combination with hybrid striped bass only or other moronids.

Nearly all states (92%) with a statewide length limit used a minimum length limit. Statewide minimum length limits ranged from 381 to 711 mm (Figure 3). Half of states reported that the statewide length limit applied to all or most of their fisheries; the other half reported that each major fishery had its own length limit or that there were no length limits at all. The top rationales behind statewide length limits were (1) to maintain desired age and size structure of the population (85%), (2) to maintain desirable yield (69%), and (3) to maintain large fish in the population (54%). The top rationales behind specific length limits based on mean rankings were (1) to maintain desired yield (100%), (2) to maintain desired age/size structure (75%), and (3) system potential differed from other systems (75%). Sixty-two percent of states with striped bass fisheries reported that there has been some change in regulations in the past decade, and 41% of states reported that their regulations have become more restrictive.

Statewide regulations for hybrid striped bass

The 203 significant hybrid striped bass fisheries reported in Table 1 were found in 28 states.

TABLE 1. Striped bass and hybrid striped bass regulations in the United States. Special regulations codes are (1) fishery-specific creel limit, (2) fishery-specific minimum length limit, (3) "some over" length limit, (4) "some under" length limit, (5) maximum length limit, and (6) protected slot length limit.

	Number of significant fisheries		Striped bass regulations			Hybrid striped bass regulations		
State	Striped bass	Hybrid striped bass	Creel limit	Minimum length limit (mm)	Special regulations	Creel limit	Minimum length limit (mm)	Special regulations
Alabama	7	12	30[a]	none	1,3	30[a]	none	1,3
Arizona	5	0	10	none	1	–	–	–
Arkansas	3	2	6[b]	none	1	6[c]	none	none
California	7	1	2[b]	457	1,2,5	2[c]	457	none
Colorado	0	13	10[a]	none	none	10[a]	none	1,2
Florida	2	5	none	none	1,2,3	20[a]	none	2,3
Georgia	25	11	15[a]	none	1,3	15[a]	none	1,3
Illinois	6	20	3[a]	432	1,2	3[a]	432	1,2
Indiana	3	5	2	none	1,3	12[a]	none	1,3
Iowa	0	4	–	–	–	none	none	none
Kansas	3	15	2	none	2	2	none	1,2
Kentucky	2	8	5	381	1,2	15	none	1,3
Louisiana	1	4	5[b]	none	3	5[c]	none	3
Maryland	6	0	2[b]	457	1,3	2[c]	457	1,3
Mississippi	1	2	6[b]	381	none	6[c]	381	none
Missouri	0	4	15[a]	none	1,2,3	15[a]	none	1,2,3
Nebraska	0	27	15[a]	none	1,3	15[a]	none	1,3
Nevada	2	4	none	none	1,3	none	none	1,2,3
New Jersey	3	3	2[b]	711	1	2[c]	711	none
New Mexico	1	0	3	none	1	25[a]	none	none
North Carolina	13	5	8[b]	406	1,2,4,5	8[c]	406	4
Ohio	0	5	none	none	1,3	none	none	1,2,3
Oklahoma	6	9	15	none	1,3	20[a]	none	3
Oregon	0	1	–	–	–	none	none	1,2

Table 1. Continued.

State	Number of significant fisheries		Striped bass regulations			Hybrid striped bass regulations		
	Striped bass	Hybrid striped bass	Creel limit	Minimum length limit (mm)	Special regulations	Creel limit	Minimum length limit (mm)	Special regulations
Pennsylvania	4	0	2[b]	508	none	2[c]	508	none
South Carolina	10	3	10[b]	none	1,2	10[c]	none	1,2
Tennessee	14	4	2[b]	381	1,2,6	2[c]	381	none
Texas	10	27	5[b]	457	1,2	5[c]	457	1
Utah	1	2	none	none	none	6	none	none
Virginia	5	1	4[b]	508	1,2,5,6	4[c]	508	none
West Virginia	0	5	4[a]	none	1,3	4	381	1,2,3
Wisconsin	0	1	–	–	–	none	none	1,2

[a] In combination with other moronids.
[b] In combination with hybrid striped bass only.
[c] In combination with striped bass only.

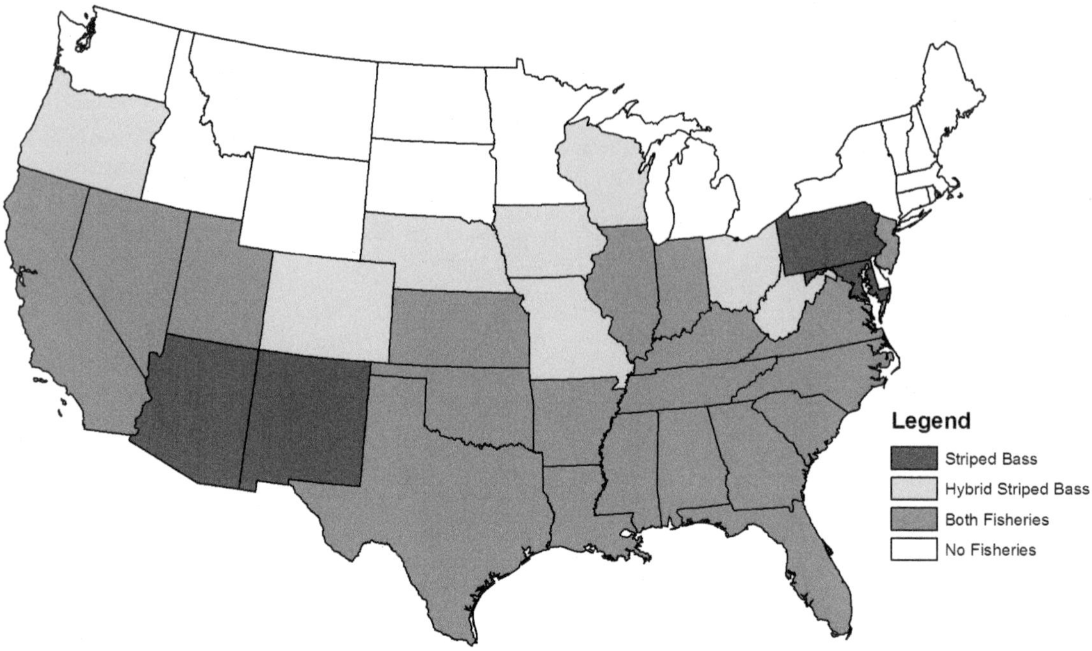

FIGURE 1. States that possess inland fisheries for striped bass and hybrid striped bass, as revealed by a survey of states.

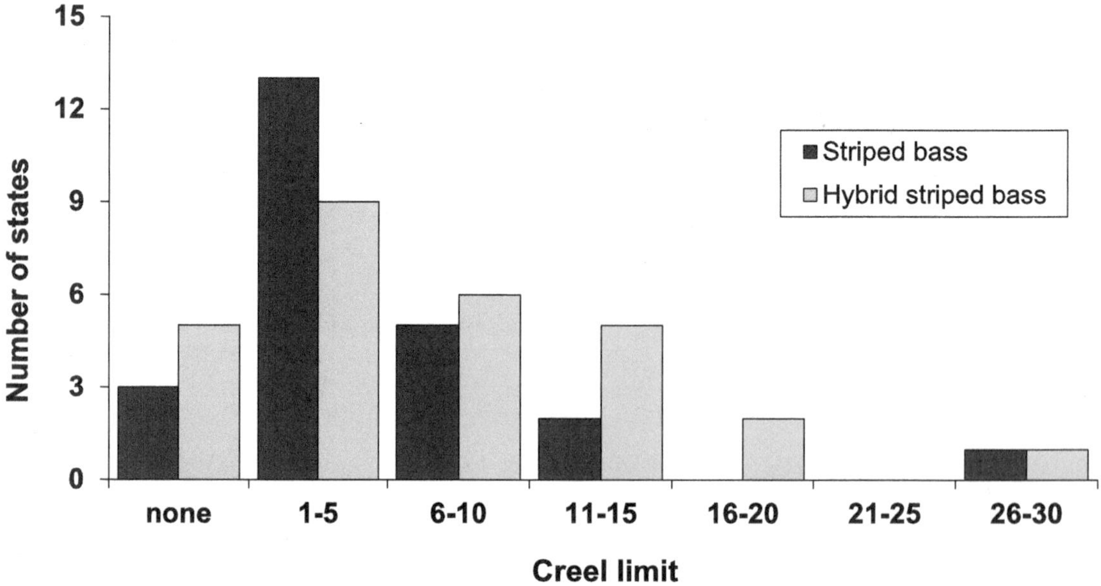

FIGURE 2. Daily creel limits for striped bass and hybrid striped bass within the United States, as revealed by a survey of states.

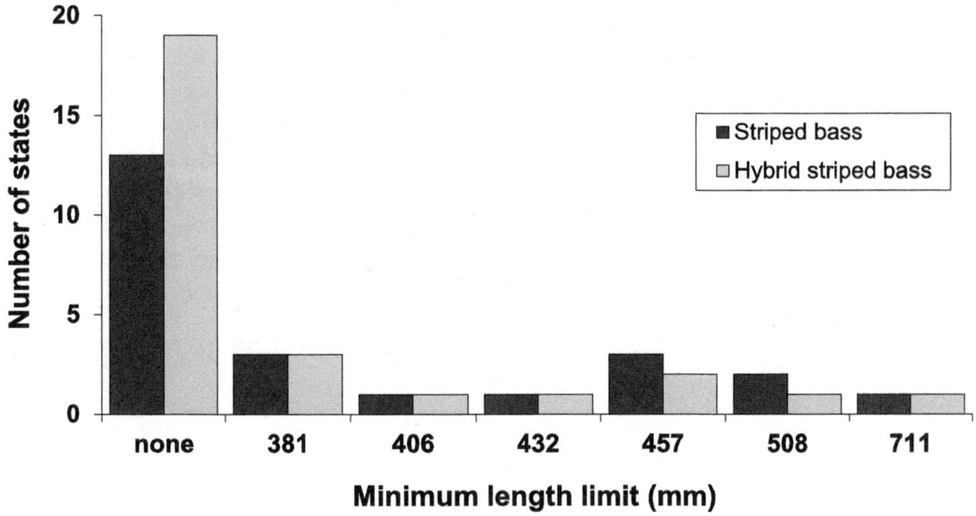

FIGURE 3. Minimum length limits for striped bass and hybrid striped bass within the United States, as revealed by a survey of states.

Eight of the 28 states reported only hybrid striped bass fisheries, and the other 20 states reported both striped bass and hybrid striped bass fisheries (Figure 1). States began regulating hybrid striped bass in the 1960s, with the majority (67%) establishing regulations in the 1980s and 1990s.

As with striped bass fisheries, statewide daily creel limits for hybrid striped bass ranged from 2 to 30, with the most liberal daily creel limits found in southern states (Figure 2). However, unlike striped bass fisheries, more states (n = 11; Arkansas, California, Florida, Louisiana, Mississippi, New Jersey, North Carolina, Oklahoma, Tennessee, Utah, and Virginia) managed all of their hybrid fisheries with the same statewide daily creel limit. The top four rationales based on mean ranks for statewide daily creel limits for hybrid striped bass were similar to those for striped bass: (1) equitable distribution of harvest, (2) striving for uniformity of regulations among water bodies whenever possible, (3) to maintain quality fisheries, and (4) to prevent overfishing. These same four rationales were ranked similarly when viewed as the percentage of respondents who "agreed" or "strongly agreed" with each rationale (83% to 65%, respectively). When exceptions to the statewide creel limits for hybrid striped bass were enacted, the most popular reason was not because system potential differed in some water bodies (as for striped bass); rather, it was to avoid confusion with striped bass. Seventeen state agencies manage both striped bass and hybrid striped bass or all other moronids with the same statewide creel limit (Table 1).

Similar to striped bass fisheries, half of state agencies reported that the statewide length limit for hybrid striped bass applied to all or most of their fisheries, and the majority of states (85%) with a statewide length limit utilized a minimum length limit. The statewide minimum length limits, as for striped bass, ranged from 381 to 711 mm (Figure 3). The top three rationales behind the statewide length limits were (1) to avoid confusion with striped bass or other temperate basses (85%), (2) to maintain desired yield (62%), and (3) to maintain desired size/age structure (57%). The top rationale behind specific length limits for hybrid striped bass was to "maintain large fish in the population"; this rationale ranked fourth in the striped bass portion of the survey. Compared to their management of striped bass, somewhat fewer (39%) states with hybrid striped bass fisheries reported that there had been some change in regulations in the past decade, and only 28% of states reported that regulations had become more restrictive.

Special regulations for striped bass

Almost all states with striped bass fisheries used at least one type of special regulation or had at least one exception to their statewide regulations (Figure 4). In one fishery in New Jersey, anglers were restricted to the use of circle hooks, and in four states (California, Illinois, Tennessee, and Utah), anglers were prohibited from using umbrella rigs (i.e., an array of multiple lures with an additional hooked lure or lures trailing behind) in all fisheries. Five states (Maryland, North Carolina, South Carolina, Tennessee, and Virginia) changed their regulations (size, creel limits, or both) during summer months on certain fisheries to adjust for high discard (postrelease) mortality. Seventy-nine percent of states utilized a fishery-specific daily creel limit, and 42% of states used a fishery-specific minimum length limit. Thirty-three percent of states had a "some over" length limit (e.g., no more than two over 18 in), and only one state (North Carolina) had a "some under" length limit (e.g., no more than two under 18 in). Three states (California, North Carolina, and Virginia) used maximum length limits, and two states (Tennessee and Virginia) used protected slot length limits. Only three states (Mississippi, Pennsylvania, and Utah) reported no exceptions to their statewide striped bass creel or length limits.

Special regulations for hybrid striped bass

New Jersey, in one fishery, was also the only state that restricted anglers to the use of circle hooks for hybrid striped bass, and six states (California, Illinois, Iowa, Ohio, Tennessee, and Utah) prohibited the use of umbrella rigs in all hybrid striped bass fisheries. In only one fishery in Tennessee did the regulations (size or creel limits) change in summer months to adjust for high discard mortality of hybrid striped bass, but that change was targeted at striped bass, which were managed with the same harvest limits. Fishery-specific daily creel limits were used in 57% of the states, and

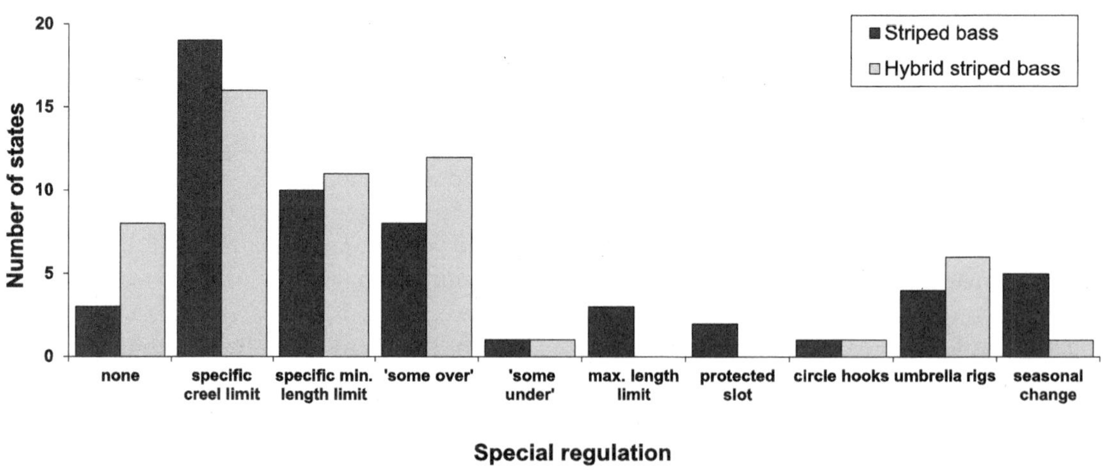

FIGURE 4. Special regulations or exceptions to statewide creel or length limits of striped bass and hybrid striped bass within the United States, as revealed by a survey of states. In "some over" regulations, anglers are allowed to harvest at least one fish over a certain length, and "some under" regulations allow anglers to harvest at least one fish under a certain length. Circle hook regulations require anglers to using only circle hooks, umbrella rig regulations prohibit anglers from using umbrella rigs, and "seasonal change" regulations refer to agencies changing their regulations (size limits, creel limits, or both) during summer months to account for high discard (postrelease) mortality.

39% used fishery-specific length limits. Forty-three percent of states had a "some over" length limit, and "some under" length limits existed in only one state (North Carolina). Maximum and protected slot length limits were not reported for any states, and 29% of states had no exceptions to their statewide hybrid striped bass creel or length limits.

Management perspectives

When asked "How important are striped bass in your state relative to other species in terms of recreational angling?", 58% of states responded either "important" or "very important"; in contrast, only 21% of states ranked hybrid striped bass as "important" or "very important" (Figure 5). In fact, 42% of states considered their hybrid striped bass fisheries to be of little or no importance.

The amounts of time that agencies devoted to managing their respective moronid fisheries mirrored their opinions on the importance of their fisheries. When asked, "How much time is spent addressing issues related to your striped bass program relative to other sport fish programs?", 33% of states responded either "a great deal" or "much." Only 7% of states claimed they spent "much" time addressing issues related to hybrid striped bass, and no states responded that "a great deal" of time was spent managing hybrid striped bass (Figure 6).

Discussion

In summary, 32 state agencies in the United States regulated 343 fisheries for striped bass or hybrid striped bass. Four states regulated striped bass only, eight states regulated hybrid striped bass only, and 20 states regulated both taxa. Only three of these states (Arizona, Nevada, and Utah) claimed to have never stocked striped bass in their fisheries (see Gustaveson et al. 1984; Stewart et al. 2007; Gustaveson and Blommer 2013; Stewart and Burrell 2013; both this volume), while all hybrid striped bass fisheries have been stocked. Statewide creel limits for striped bass and hybrid striped bass ranged from 2 to 30, and the majority of striped bass (86%) and hybrid striped bass (65%) creel limits were between 2 and 10 fish per day. State agencies regulated both taxa primarily with a statewide minimum length limit ranging from 381 to 711 mm, with the majority of striped bass (67%) and hybrid striped bass (75%) statewide minimum length limits being between 381 and 457 mm. Agency personnel gave similar rationales for creel limits and length limits among striped bass and hybrid striped bass fisheries, and both taxa were managed with comparable special regulations.

It appears that the vast majority of state agencies regulated striped bass and hybrid striped bass under similar management plans;

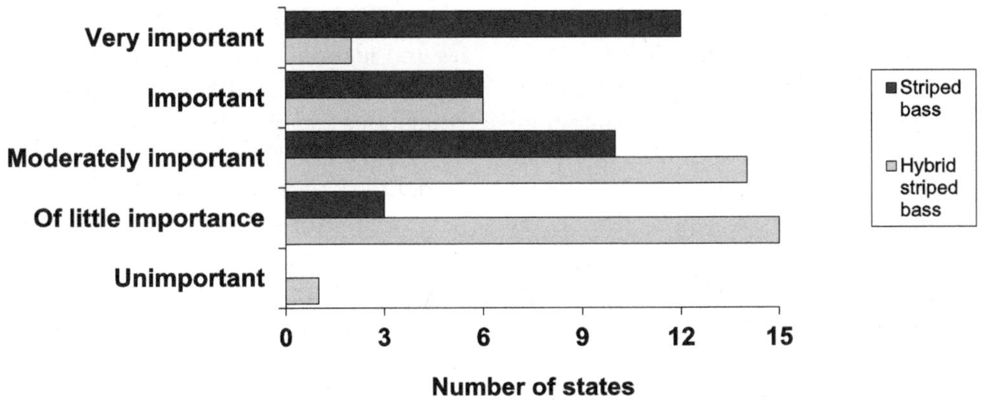

FIGURE 5. The importance of striped bass and hybrids relative to other species, as revealed by a survey of states.

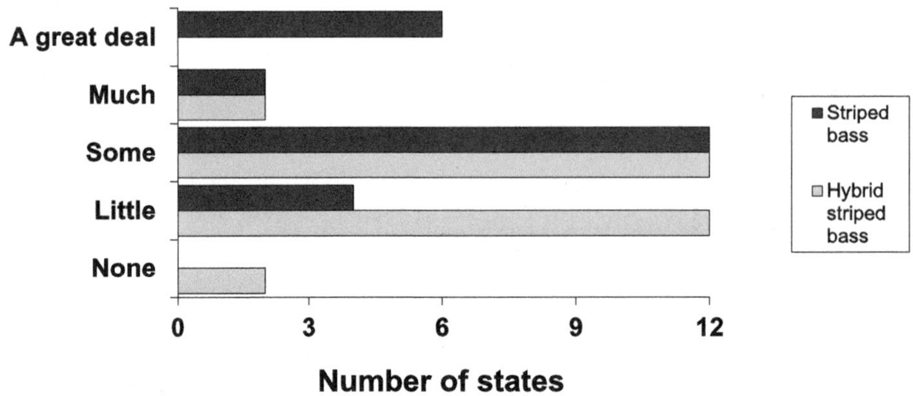

FIGURE 6. The amount of time spent managing striped bass and hybrid striped bass relative to other sport fish programs as revealed by a survey of states.

however, their inherent value to agencies was greatly skewed. Striped bass were ranked highly important and intensively managed while hybrid striped bass fisheries were viewed as less important and managed less intensively. An explanation for such bias is unknown, but it may be attributed to the potential size difference between taxa. The maximum size of a striped bass exceeds three times that of a hybrid striped bass; however, more states managed hybrid striped bass, due in part to their rapid growth, infrequent natural reproduction, and adaptability to freshwater (Etnier and Starnes 1993).

The ambiguity of why anglers view certain fisheries differently is a prime example of survey limitation. Lack of responses, inability to ask all necessary questions, and attaining quantitative rationales were some of the limitations of this survey. Perhaps the key deficiency of this survey is that we did not receive justification for survey responses, specifically an explanation of regulations in terms of actual studies of efficacy.

Implicit in the rationale for regulations is the assumption that they actually control population structure and yield the desired fishery. The responses we received did not, however, fully justify the regulations. A paucity of literature exists on the response of populations of striped bass or their hybrids to different regulations. Bulak et al. (1995) discussed how a higher minimum size limit would protect a naturally reproducing stock of inland striped bass from recruitment overfishing, but those findings were based on mathematical simulations. Moczygemba and Hysmith (1996) concluded that the abundance and size structure of striped bass in Lake Texoma, Texas–Oklahoma did not change after the maximum number of fish longer than 508 mm that could be kept was lowered from five to one fish. In recent decades, research on the efficacy of Morone spp. regulations was much more likely to deal with white bass (e.g., Muoneke 1994; Lovell and Maceina 2002; Schultz and Robinson 2002) than striped bass or hybrid striped bass fisheries. We reviewed all articles published since 2000 in the North American Journal of Fisheries Management, the Proceedings of the Southeastern Association of Fish and Wildlife Agencies, and Fisheries and identified 37 papers that discussed the response of inland fish populations to harvest regulations and only one dealt with striped bass or hybrid striped bass. In that paper, Van Horn et al. (2001) did not detect changes in growth or condition of striped bass in Lake Norman, North Carolina, after the minimum size limit was raised, due to several confounding factors (i.e., higher stocking rate; expanding blue catfish Ictalurus furcatus and flathead catfish *Pylodictis olivaris* populations).

After several decades of published research on predator–prey relationships, culture techniques, catch-and-release mortality, and habitat requirements, there is a need for information on how striped bass and their hybrids respond to harvest regulations. Virtually all inland sport fisheries are managed with a daily creel limit, and as we observed in our survey, striped bass and hybrid bass are no exception. However, definitive assessments of the efficacy of daily creel limits to protect or enhance striped bass and hybrid striped bass stocks are lacking and needed. Likewise, size limits are imposed universally and have been the subject of numerous investigations with other species over the decades, but reports on their efficacy with striped bass and their hybrids (if they exist) are buried in gray literature. It is our hope that this survey of how agencies across the country are managing inland striped bass and hybrid striped bass will stimulate formal assessments of when and how regulations function to achieve explicit management goals.

Acknowledgments

We greatly appreciate the timely, thorough, and thoughtful responses to our survey questions by all of the agency personnel we contacted.

References

Axon, J. R., and D. K. Whitehurst. 1985. Striped bass management in lakes with emphasis on management problems. Transactions of the American Fisheries Society 114:8–11.

Bettoli, P. W., J. A. Kerns, and G. D. Scholten. 2009. Status of paddlefish in the United States. Pages 23–37 *in* C. P. Paukert and G. D. Scholten, editors. Paddlefish management, propagation, and conservation in the 21st century: building from 20 years of research and management. American Fisheries Society, Symposium 66, Bethesda, Maryland.

Bjorgo, K. A., J. J. Isely, and C. S. Thomason. 2000. Seasonal movement and habitat use by striped bass in the Combahee River, South Carolina., Transactions of the American Fisheries Society 129:1281–1287.

Bulak, J. S., D. S. Wethey, and M. G. White, III. 1995. Evaluation of management options for a reproducing striped bass population in the Santee–Cooper system, South Carolina. North American Journal of Fisheries Management 15:84–94.

Cheek, T. E., M. J. Van Den Avyle, and C. C. Coutant. 1985. Influences of water quality on distribution of striped bass in a Tennessee River impoundment. Transactions of the American Fisheries Society 114:67–76.

Coutant, C. C. 1985. Striped bass, temperature, and dissolved oxygen: a speculative hypothesis for environmental risk. Transactions of the American Fisheries Society 114:31–61.

Etnier, D. A., and W. C. Starnes. 1993. The fishes of Tennessee. The University of Tennessee Press, Knoxville.

Gustaveson, W., and G. Blommer. 2013. History of striped bass management in the Colorado River. Pages 15–23 *in* J. S. Bulak, C. C. Coutant, and J. A. Rice, editors. Biology and management of inland striped bass and hybrid striped bass. American Fisheries Society, Symposium 80, Bethesda, Maryland.

Gustaveson, W. A., T. D. Pettengill, J. E. Johnson, and J. R. Wahl. 1984. Evidence of in-reservoir spawning of striped bass in Lake Powell, Utah–Arizona. North American Journal of Fisheries Management 4:540–546.

Lovell, G. R., and M. J. Maceina. 2002. Population assessment and minimum length limit evaluations for white bass in four Alabama reservoirs. North American Journal of Fisheries Management 22:609–619.

Moczygemba, J. H., and B. T. Hysmith. 1996. Evaluation of reduced striped bass bag limit, Lake Texoma, Texas and Oklahoma. Proceedings of the Annual Conference Southeastern Association of Fish and Wildlife Agencies 48(1994):433–440.

Muoneke, M. I. 1994. Dynamics of a heavily exploited Texas white bass population. North American Journal of Fisheries Management 14: 415–422.

Paukert, C. P., J. A. Klammer, R. B. Pierce, and T. D. Simonson. 2001. An overview of northern pike regulations in North America. Fisheries 26:6–13.

Schultz, R. D., and D. A. Robinson Jr. 2002. Exploitation and mortality rates of white bass in

Kansas reservoirs. North American Journal of Fisheries Management 22:652. -658.

Stewart, W. T., and M. Burrell. 2013. Striped bass dispersion and effects on fisheries management in Lakes Mohave and Pleasant, Colorado River basin. Pages 431–448 *in* J. S. Bulak, C. C. Coutant, and J. A. Rice, editors. Biology and management of inland striped bass and hybrid striped bass American fisheries Society, Symposium 80, Bethesda, Maryland.

Stewart, W. T., M. M. Meding, and D. R. Rogers. 2007. Lake Pleasant striped bass. Arizona Game and Fish Department, Research Branch, Technical Guidance Bulletin No. 11, Phoenix.

U.S. Department of the Interior, Fish and Wildlife Service, and U.S. Department of Commerce, U.S. Census Bureau. 2006 national survey of fishing, hunting, and wildlife-associated recreation. Available: www.census.gov/prod/2008pubs/fhw06-nat.pdf (April 2012).

Van Horn, S. L. 2013. A brief history of inland striped bass management. Pages 1–13 *in* J. S. Bulak, C. C. Coutant, and J. A. Rice, editors. Biology and management of inland striped bass and hybrid striped bass. American Fisheries Society, Symposium 80, Bethesda, Maryland.

Van Horn, S. L., B. K. Baker, and M. Rash. 2001. Growth and condition response of Lake Norman striped bass to increased stocking rates and more restrictive harvest regulations. Proceedings of the Annual Conference Southeastern Association of Fish and Wildlife Agencies 53(2001):193–219.

American Fisheries Society Symposium 80:461–472, 2013

Size-Based Mortality Caps as Thresholds for Managing Hybrid Striped Bass in Kansas Reservoirs

Randall D. Schultz*
Iowa Department of Natural Resources
24570 U.S. Highway 34, Chariton, Iowa 50049, USA

Jason M. Goeckler
Kansas Department of Wildlife, Parks and Tourism
1820 Merchant Street, Emporia, Kansas 66801, USA

Michael C. Quist
U.S. Geological Survey – Idaho Cooperative Fish and Wildlife Research Unit
Department of Fish and Wildlife Sciences, University of Idaho
Moscow, Idaho 83844, USA

Abstract.— Total annual mortality and mortality caps (maximum mortality thresholds) were estimated for hybrid striped bass (striped bass *Morone saxatilis* × white bass *M. chrysops*) populations sampled in six large Kansas reservoirs spanning 1995–2001; all reservoirs lacked a minimum length limit but had a 2/d creel limit. Total annual mortality of age-1 and older hybrid striped bass was estimated using a weighted catch curve. Mortality caps were modeled by varying growth rates and management objectives. Mortality rates that approach the cap signal the need for increased monitoring to determine what additional management action should be taken (e.g., harvest regulations to reduce mortality) or to reevaluate the size objective (targeted mean length of harvested fish) for the fishery. Estimated mortality caps were then compared with observed growth and mortality rates. Total annual mortality rates for all reservoirs varied from 22% to 52% (mean = 38%; SE = 4.9). All but two reservoirs had total annual mortality rates greater than 40%. Analysis of mortality caps indicated that when the objective was to maintain the mean length of harvested hybrid striped bass at current levels, total annual mortality was lower than the mortality cap for most reservoirs if a 381 mm total length (TL) minimum length limit (i.e., preferred length) were to be imposed. A management objective of 500 mm TL for the mean length of harvested hybrid striped bass was realistic for all reservoirs modeled with a 457-mm-TL length limit (special regulation available to managers) but only for two reservoirs with a 381-mm-TL length limit. Likewise, only two of the study reservoirs could support a management objective of 550 mm TL (herein defined as trophy length) for mean length of harvested hybrid striped bass when modeled with a 457-mm-TL length limit. Our analysis illustrates the potential of mortality caps for monitoring and establishing realistic management goals for hybrid striped bass fisheries.

* Corresponding author: randy.schultz@dnr.iowa.gov

Introduction

Hybrid striped bass is a cross between striped bass *Morone saxatilis* and white bass *M. chrysops*. The "original" cross (male white bass × female striped bass), also referred to as palmetto bass, was developed in South Carolina in the mid-1960s (Bayless 1968; Bishop 1968). The reciprocal cross (female white bass × male striped bass), or sunshine bass, is also produced and stocked by some state agencies. Hybrid striped bass have been stocked as an alternative to striped bass, which in some reservoirs experience a temperature–dissolved oxygen squeeze that can constrain growth or result in mortalities (Van Horn 2013, this volume). Hybrid striped bass have been successfully stocked into ponds, lakes, streams, and reservoirs for their sport fish potential. Additionally, these fish have been stocked in an attempt to restructure panfish communities (Layzer and Clady 1984; Jahn et al. 1987; Neal et al. 1999; Hutt et al. 2008) and control invasive species (Iowa Department of Natural Resources, unpublished data). Hybrid striped bass have also been introduced to help control abundant prey species such as gizzard shad *Dorosoma cepedianum* (Dettmers et al. 1996). Dettmers et al. (1998) indicated control of gizzard shad could only occur when hybrid striped bass are stocked at high densities in impoundments with low natural mortality and intermediate densities of gizzard shad.

The three primary factors regulating fish population dynamics are growth, recruitment, and mortality (Ricker 1975) and are particularly important with regard to the management of sport fishes. Growth is important because it integrates the effects of environmental conditions (e.g., temperature and water quality), prey availability, and genetic factors, making it a useful metric to evaluate habitat suitability or the effect of management strategies targeting sport fishes and their prey (Avise and Van Den Avyle 1984; Miranda and Durocher 1986; McInerny and Cross 1999; Leitner et al. 2002; Quist et al. 2003; Schultz et al. 2008; Thompson and Rice 2013, this volume). In addition, growth assessments are commonly used to assess potential problems (e.g., overfishing) and management manipulations, such as regulations, lake renovations, and prey stockings (Heman et al. 1969; Noble 1981; Maceina et al. 2007; Paukert et al. 2007; Schultz and Dodd 2008). Thus, knowledge of growth is important for management of sport fish populations and their prey. Recruitment of many fish populations is highly variable, depending greatly on environmental vagaries and density and size structure of the parental stock. These recruitment fluxes interact with both growth and mortality. In hybrid striped bass fisheries, the variability of recruitment is generally constrained as juvenile fish are stocked, often at a constant density.

Information on mortality is important for the management of fish populations. Changes in either natural or fishing mortality may be related to environmental factors, predator–prey interactions, or exploitation. However, mortality limits are not typically used by management agencies to manage freshwater fisheries. Miranda (2002) suggested that resource managers develop size-based mortality caps, that is, mortality rates above which mean length objectives for the fishery will not be attained. Although size objectives could vary by different management goals, here we describe size objectives as the targeted mean length of harvested hybrid striped bass.

Mortality rates that approach the cap signal the need for increased monitoring to determine what additional management action should be taken to allow further growth of the fish (e.g., harvest regulations to reduce mortality) or reevaluate the mean length objective for the fishery (Miranda 2002). Since this call to action, the use of mortality caps in freshwater systems is increasing. Quist et al. (2004) determined that a 500 mm mean length objective for eight Kansas walleye *Sander vitreus* populations was only realistic if a 457 mm minimum length limit were to be imposed. Scholten and Bettoli (2005) combined the Beverton–Holt equilibrium yield equation and spawning potential ratios with mortality caps to substantiate paddlefish *Polyodon spathula* commercial fishing regulation changes. Makinster and Paukert (2008) used mortality caps to determine levels of total annual mortality (*A*) for flathead catfish that would sustain size structure in separate reaches of the Kansas River.

Although the popularity of hybrid striped bass to recreational anglers and resource managers continues after more than four decades, most studies have focused on recruitment (i.e., stocking rates) but only limited effort has been placed on measuring growth and mortality. Our objectives were to evaluate the utility of mortality caps for managing hybrid striped bass in large Kansas reservoirs and provide guidance for resource managers on effective use of mortality caps in fisheries management.

Methods

Hybrid striped bass were sampled from Cedar Bluff, Cheney, Lovewell, Milford, Norton, and Webster reservoirs in Kansas during fall, 1995–2001. Typical of most large federal reservoirs in Kansas, the study reservoirs (mean size 2,553 ha, SE 897 ha) are relatively shallow (mean depth 6.3 m, SE 0.6 m), turbid, and seldom thermally stratified due to strong, persistent winds. Most of the reservoirs have rocky shorelines with gradual slopes and little aquatic vegetation. The watersheds (mean size 5,397 km^2, SE 2,209 km^2) are dominated by grassland habitat. Gizzard shad are the primary prey species in all reservoirs, with dominant sport fish populations including channel catfish *Ictalurus punctatus*, hybrid striped bass, largemouth bass *Micropterus salmoides*, white bass, white crappie *Pomoxis annularis*, and walleye.

Sampling consisted of a gill-net complement comprising four separate monofilament gill nets, each measuring 30.5 × 2.4 m with mesh sizes of 25, 38, 64, and 102 mm bar measure (Mosher and Willis 1997). Scale samples were taken from all captured hybrid striped bass at the point where the tip of the pectoral fin meets the body. Scales were selected for aging due to their ease of collection, avoidance of sacrificing large hybrid striped bass, and because no special setup was required prior to aging, compared to using otoliths (Isely and Grabowski 2007). Scales are the most common structure used to age moronids, although it is widely recognized that scales fail to accurately age fish to their maximum age (Maceina et al. 2007). Although otoliths have been shown to provide more precise age estimates than scales for striped bass (Welch et al. 1993), precision is similar between otoliths and scales for white bass (Soupir et al. 1997). Scales were read independently by two readers. When disagreements occurred (less than 5% of the samples), scales were analyzed by both readers until consensus was reached.

Mortality estimates were calculated for fish captured in each reservoir. Based on catch curve analyses, age-1 hybrid striped bass were considered fully recruited to the sampling gear and were used to index recruitment. Catch curves were limited to age-1 and older hybrid striped bass because age-0 hybrid striped bass were not fully recruited to the sampling gear. We assumed that the standard gill-net complement put equal pressure on the age cohorts of interest. Total annual mortality (*A*) was estimated from the descending limb of catch curves using weighted regression (Ricker 1975; Slipke and Maceina 2000). A von Bertalanffy growth function was used to estimate the theoretical maximum length (L_∞), growth coefficient (*K*), and age at which total length is zero (t_0; Ricker 1975). Mortality caps for hybrid striped bass were estimated following model 1 as described by Miranda (2002). This model allows for the incorporation of different management objectives and harvest regulations, based on growth parameter estimates from the von Bertalanffy growth model:

$$\text{Model 1: } Z = K \times [(L_\infty - L_{mean}) \times (L_{mean} - L_x)^{-1}],$$

where *Z* is instantaneous total annual mortality, *K* is the growth coefficient and L_∞ is the theoretical maximum length from the von

Bertalanffy growth model, L_{mean} is the management objective for the average length of harvested fish, and L_x is the minimum length at which fish become vulnerable to capture or harvest (e.g., length limit). Estimates of Z were transformed into A by the equation $A = 1 - e^{-Z}$ (Ricker 1975). Mortality caps were estimated for each reservoir with annual and overall estimates of L_∞ and K. Because L_{mean} (length objectives) have not been established for Kansas hybrid striped bass, we chose length objectives of 500 and 550 mm (all lengths hereafter are millimeters total length). These lengths correspond roughly to memorable and trophy lengths for hybrid striped bass, respectively (Anderson and Neumann 1996). We modeled these lengths because many agencies cite the fast growth and trophy capability of hybrid striped bass as reasons for popularity among anglers and continued stocking. Additionally, we also modeled mortality caps with L_{mean} equal to the mean length of harvested hybrid striped bass, which equates to default current length objectives. The mean length of harvested hybrid striped bass was obtained from creel data concurrent with fish collections (1994–2002). Five years of creel survey data were available for Cedar Bluff and Norton reservoirs; four for Cheney, Lovewell, and Webster reservoirs; and three for Milford Reservoir.

We used 381 and 457 mm as the minimum lengths available for harvest (L_x). Surveys indicated that 305 mm was about the minimum length at which anglers were willing to harvest hybrid striped bass (Kansas Department of Wildlife and Parks, unpublished data). Additionally, Kansas does not have a statewide length limit for hybrid striped bass, but managers have the option of implementing a 457 mm minimum length limit as a special regulation, which had not been implemented on the study impoundments at the time of this study. As such, empirical total annual mortality rates were plotted against mortality cap estimates to determine whether length objectives were realistic given current growth and mortality rates.

Results and Discussion

A total of 1,770 hybrid striped bass was sampled and aged from the study reservoirs (Table 1), and von Bertalanffy growth parameter estimates were determined (Slipke and Maceina 2000; Table 2). Maximum age of hybrid striped bass in this study was 9 years, but 69% of the fish were less than 4 years old.

Total annual mortality of age-1 and older hybrid striped bass was relatively high and variable among reservoirs (mean 38%, SE 4.9%, range 22–52%; Figure 1). Based on our specified size objectives, all of the reservoirs would likely maintain current harvest rates if a 381 mm minimum length limit were imposed (Figure 2). For Milford, Norton and Webster reservoirs, the current mean length of harvested hybrid striped bass was above 457 mm (Figure 2) and thus able to support current harvest lengths at this minimum length limit.

Managing hybrid striped bass with an objective of memorable length (500 mm) as the mean length of harvested fish utilizing a 381 mm minimum length limit was only realistic in Cedar Bluff and Norton reservoirs. All other reservoirs had the capability to reach this management objective only when minimum length for harvest was 457 mm. At current total annual mortality, growth rates, and mean lengths available for harvest, again, only Cedar Bluff and Norton reservoirs are likely to realize a targeted mean length of harvested hybrid striped bass equal to trophy length (defined herein as 550 mm) but only if a 457 mm length limit were imposed. This is not to imply that some fish of this size do not exist in the study reservoirs because 15% of sampled hybrid striped bass were 550 mm or larger.

State agencies have been lax in evaluating population dynamics of hybrid striped bass (Collier et al. 2013, this volume). Currently, Kansas lacks a statewide minimum length limit for hybrid striped bass, but managers do have the option of a 457 mm minimum length limit as a special regulation. Mortality cap analysis illustrated several trends in

TABLE 1. Mean total length (mm) at age of hybrid striped bass collected from six Kansas reservoirs during 1995–2001. Parenthetic values are standard errors; N is number per age-class.

	Age							
Reservoir	1	2	3	4	5	6	7	8
Cedar Bluff	336 (3.0) N = 108	428 (6.3) N = 50	505 (3.9) N = 71	543 (4.3) N = 53	587 (4.6) N = 38	602 (9.2) N = 16		
Cheney	350 (4.3) N = 61	434 (5.1) N = 49	483 (5.5) N = 33	495 (14.3) N = 12	559 (16.0) N = 2			572 – N = 1
Lovewell	345 (3.3) N = 63	414 (4.4) N = 62	456 (6.9) N = 49	516 (5.7) N = 14	529 (5.0) N = 16	624 (47.5) N = 2		
Milford	274 (6.0) N = 60	361 (8.6) N = 70	458 (9.0) N = 62	501 (5.7) N = 56	550 (8.1) N = 23	599 (7.9) N = 16	560 – N = 1	687 (23) N = 2
Norton	368 (3.1) N = 63	452 (2.3) N = 67	500 (2.4) N = 62	558 (4.4) N = 29	599 (4.1) N = 26	623 (6.4) N = 26	675 (22.8) N = 28	675 (8.4) N = 11
Webster	333 (11.8) N = 42	438 (5.2) N = 57	484 (5.3) N = 62	526 (3.4) N = 110	616 (7.2) N = 53	627 (9.7) N = 25	715 (6.5) N = 6	

Kansas hybrid striped bass populations, suggesting that it may be a useful assessment tool for management of these and similar fisheries. This study revealed that although a minimum length limit of 381 mm will sustain a status quo management objective (maintain current mean length of harvested hybrid striped bass), a minimum length limit of 457 mm has the potential to allow biologists to manage for memorable-length (500 mm) hybrid striped bass. The possibility of maintaining harvest with a mean length of 550 mm (i.e., trophy length) is not likely without reducing fishing mortality through highly restrictive harvest regulations (>457 mm), decreasing natural mortality, or increasing growth rates.

A major concern with minimum length limits, particularly in Kansas' shallow, turbid reservoirs, is the effect of delayed mortality (e.g., hooking mortality) from catch-and-release angling in summer (Bettinger and Wilde 2013; Coutant 2013; both this volume). Muoneke and Childress (1994) report that at temperatures of 25–30°C, palmetto bass had a 29% hooking mortality rate. Carlson et al. (1995) reported that hybrid striped bass held at 29°C showed signs of thermal stress; this temperature is commonly exceeded in Kansas reservoirs from July to September. Therefore, in the event of imposition of length limits in Kansas' hybrid striped bass populations, monitoring hooking mortality will be essential in determining the overall effect of increased regulations on this fishery.

Miranda (2002) and Quist et al. (2003) recommended the use of mortality caps to assist with management of important sport fish species, recognizing that caution should be

TABLE 2. Von Bertalanffy growth parameter estimates (theoretical maximum length [L_∞], growth coefficient [K], and theoretical age when length equals zero [t_0]) from hybrid striped bass sampled in six Kansas reservoirs. The coefficient of determination (R^2) and P-values signify model fit.

Reservoir Year	L_∞ (mm)	K	t_0	R^2	P
Cedar Bluff					
1996	810	0.19	−1.54	0.96	0.0001
1997	740	0.23	−1.45	0.96	0.0005
1998	725	0.25	−1.33	0.98	0.0001
1999	791	0.15	−3.89	0.68	0.0001
2000	679	0.33	−1.01	0.98	0.0001
2001	679	0.3	−1.35	0.94	0.0001
All years	665	0.34	−1.06	0.99	0.0001
Cheney					
1995	633	0.25	−3.37	0.93	0.002
1996	587	0.44	−0.91	0.99	0.0001
1998	601	0.31	−1.69	0.93	0.002
1999	692	0.28	−0.98	0.95	0.0001
2000	619	0.42	−0.92	0.91	0.0001
2001	582	0.67	0.27	0.72	0.0001
All years	583	0.69	−1.05	0.99	0.0001
Lovewell					
1995	773	0.19	−1.89	0.93	0.002
1996	637	0.38	−0.84	0.93	0.002
1997	772	0.17	−2.52	0.99	0.006
1998	621	0.27	−1.97	0.99	0.0001
1999	615	0.31	−1.79	0.96	0.003
2000	888	0.11	−3.81	0.98	0.0002
2001	953	0.16	−1.09	0.93	0.002
All years	712	0.19	−2.53	0.99	0.0009
Milford					
1996	741	0.2	−1.47	0.98	0.0002
1998	718	0.19	−1.37	0.98	0.0001
1999	650	0.33	−0.99	0.98	0.0002
2000	677	0.42	−0.56	0.94	0.0001
2001	688	0.36	−0.11	0.92	0.0002
All years	875	0.15	−1.56	0.97	0.0001
Norton					
1995	935	0.16	−1.83	0.96	0.0001
1996	688	0.35	−1.2	0.98	0.0002
1998	681	0.34	−1.24	0.97	0.0001
1999	713	0.3	−1.27	0.96	0.0001
2001	792	0.16	−2.98	0.93	0.0001
All years	731	0.27	−1.39	0.99	0.0001

TABLE 2. Continued.

Reservoir Year	L_∞ (mm)	K	t_0	R^2	P
Webster					
1995	640	0.52	–0.15	0.92	0.003
1996	891	0.18	–1.31	0.99	0.0001
1998	711	0.3	–1.03	0.97	0.0003
1999	948	0.19	–1.29	0.96	0.0001
2000	913	0.18	–1.37	0.99	0.0001
2001	1,025	0.12	–1.86	0.96	0.0001
All years	843	0.2	–1.29	0.98	0.0001

used in interpreting results for populations where observed mortality approaches calculated mortality caps (Miranda 2002). In such situations, it is appropriate to assess a population's total mortality accurately prior to instituting more stringent harvest regulations. Churchill and Black (2007) recommended that more frequent estimation of total mortality in Tennessee's black bass populations be conducted when observed mortality is ≤5% of the mortality cap at a desired mean harvest length goal. This is exemplified in the current study, whereas yearly variability in total annual mortality (Figure 2) depicted that a 1-year snapshot of many of the Kansas hybrid striped bass populations may have differed greatly from the mean of several yearly estimates. Sammons and Maceina (2008) noted a lack in agreement between mortality caps and yield models in predicting the effectiveness of length limits to improve size structure of redbreast sunfish *Lepomis auritus* populations. Explanation provided by the authors surrounded size-selective predation on redbreast sunfish by flathead catfish *Pylodictis olivaris*, leading to high natural mortality unaccounted for in the yield models and uncertainty in the ability of minimum length limits to restructure redbreast sunfish populations where heavy flathead catfish predation on redbreast sunfish may be occurring. Again, results from Sammons and Maceina (2008) depict the necessity of accurate mortality assessments before imposing length limit changes. We concur with Churchill and Black (2007) that regulation changes should not be undertaken without numerous years of population dynamics data.

Growth of hybrid striped bass is sufficient in Kansas reservoirs to create an exceptional angling opportunity. Similar to flathead catfish (Quinn 1993; Arterburn et al. 2002), the trophy potential of hybrid striped bass may lend angler support for harvest regulations to protect and encourage this potential. Continued monitoring of growth, mortality, and lengths of harvested fish should accompany regulation changes to determine if population dynamics change due to regulation changes. The current study can be useful to managers for assessing current conditions, establishing realistic management goals, and providing guidance on effective use of mortality caps in management of hybrid striped bass fisheries. Similar studies should be conducted in other areas where hybrid striped bass are stocked to provide further insight into the application of mortality caps to manage this popular fishery.

Acknowledgments

This study was funded through the Federal Aid in Sport Fish Restoration Program. Appreciation is extended to Gordon Schneider, Kyle Austin, Tommie Berger, Lynn Davignon, John Reinke, and Steve Price for assistance with data collection.

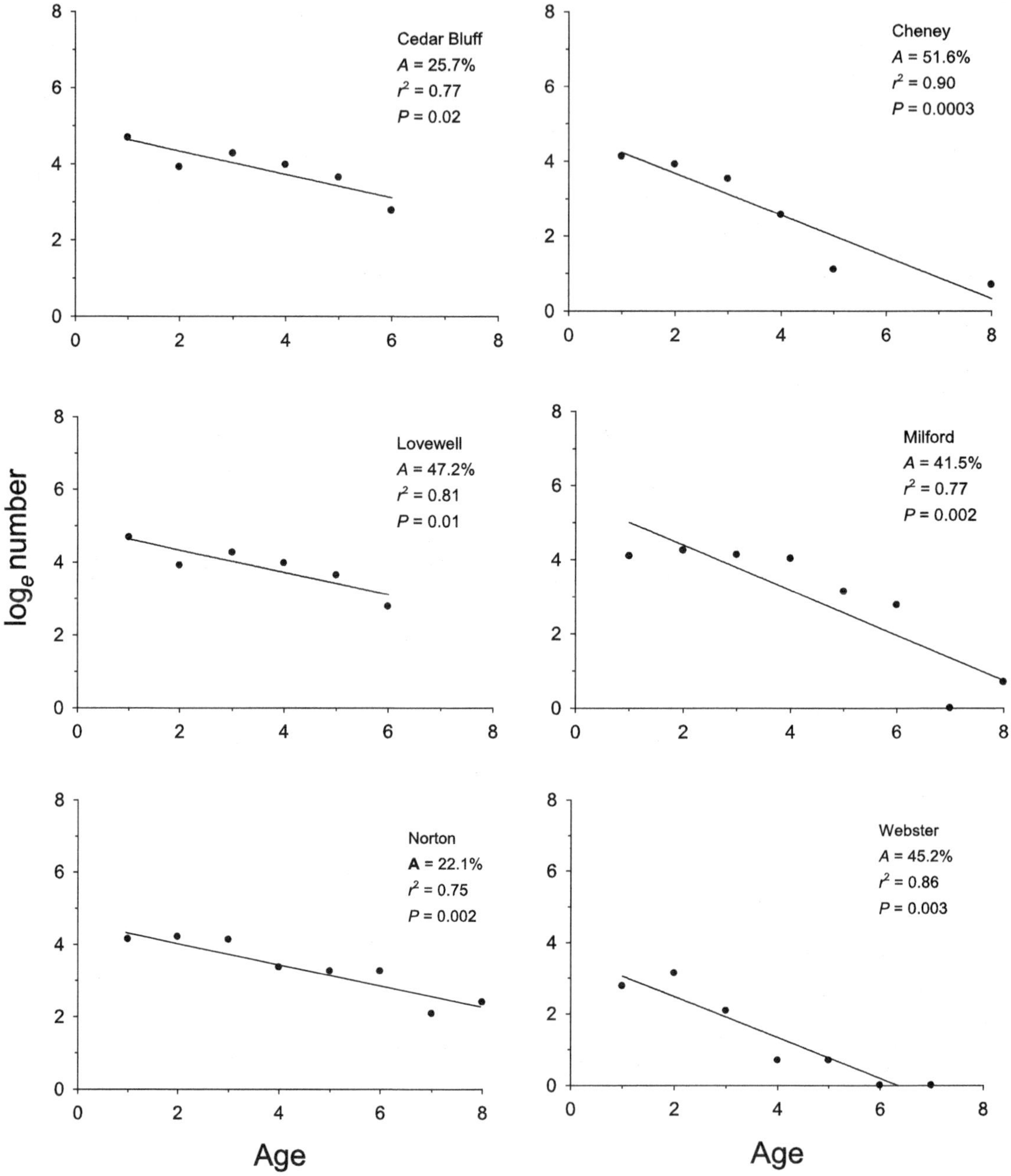

FIGURE 1. Weighted catch-curve regressions and total annual mortality (*A*) for age-1 and older hybrid striped bass sampled from six Kansas reservoirs between 1995 and 2001. The coefficient of determination (R^2) and *P*-values signify model fit.

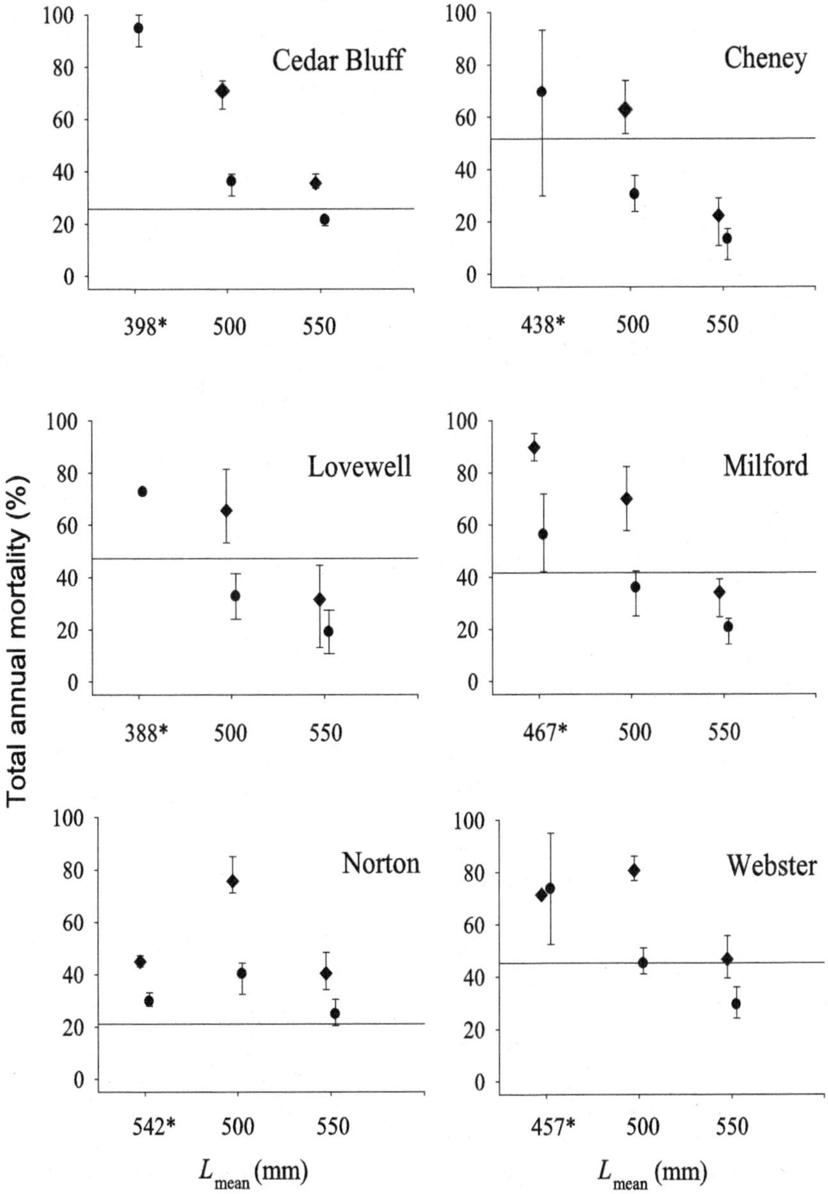

FIGURE 2. Estimated mortality caps for hybrid striped bass in six Kansas reservoirs. Overall mortality cap estimates (symbols) and year-specific mortality cap estimates (bars are maximum and minimum estimates) were calculated based on theoretical maximum length (L_∞) and the growth coefficient (K). Mortality caps were estimated for a mean length objective (L_{mean}) equal to the current mean length of harvested hybrid striped bass (asterisks) and for values of 500 mm total length (TL) and 550 mm TL. The minimum length available for harvest (L_x) was set at 381 mm TL (circles), and 457 mm TL (diamonds), the special regulation length available to Kansas managers for regulating hybrid striped bass. In Cedar Bluff, Cheney, and Lovewell reservoirs, the mortality cap for the current mean length of harvested hybrid striped bass (i.e., status quo) was not plotted when L_x = 457 mm because the current mean was below 457 mm for harvested hybrid striped bass in these reservoirs. The horizontal lines represent mean total annual mortality for each reservoir.

References

Anderson, R. O., and R. M. Neumann. 1996. Length, weight, and associated structural indices. Pages 447–482 *in* B. R. Murphy and D. W. Willis, editors. Fisheries techniques, 2nd edition. American Fisheries Society, Bethesda, Maryland.

Arterburn, J. E., D. J. Kirby, and C. R. Berry. 2002. A survey of angler attitudes and biologist opinions regarding trophy catfish and their management. Fisheries 27(5):10–21.

Avise, J. C., and M. J. Van Den Avyle. 1984. Genetic analysis of reproduction of hybrid white bass × striped bass in the Savannah River. Transactions of the American Fisheries Society 113:563–570.

Bayless, J. D. 1968. Striped bass hatching and hybridization experiments. Proceedings of the Annual Conference Southeastern Association of Game and Fish Commissioners 21(1967):233–244.

Bettinger, J. M., and G. R. Wilde. 2013. Catch-and-release mortality of inland striped bass and hybrid striped bass. Pages 473–499 *in* J. S. Bulak, C. C. Coutant, and J. A. Rice editors. Biology and management of inland striped bass and hybrid striped bass. American Fisheries Society, Symposium 80, Bethesda, Maryland.

Bishop, R. D. 1968. Evaluation of the striped bass and white bass hybrids after two years. Proceedings of the Annual Conference Southeastern Association of Game and Fish Commissioners 21(1967):245–253.

Carlson, R. E., E. P. Baker, and R. E. Fuller. 1995. Immunological assessment of hybrid striped bass at three culture temperatures. Fish and Shellfish Immunology 5:359–373.

Churchill, T. N., and W. P. Black. 2007. Use of mortality caps to determine effectiveness of black bass harvest restrictions in Tennessee reservoirs. Proceedings of the Annual Conference Southeastern Association of Fish and Wildlife Agencies 59(2005):217–226.

Collier, W. R., P. W. Bettoli, G. D. Scholten, and T. N. Churchill. 2013. Regulation of striped bass and hybrid striped bass fisheries in the United States. Pages 449–460 *in* J. S. Bulak, C. C. Coutant, and J. A. Rice editors. Biology and management of inland striped bass and striped bass hybrids. American Fisheries Society, Symposium 80, Bethesda, Maryland.

Coutant, C. C. 2013. When is habitat limiting for striped bass? Three decades of testing the temperature–oxygen squeeze hypothesis. Pages 65–91 *in* J. S. Bulak, C. C. Coutant, and J. A. Rice, editors. Biology and management of inland striped bass and hybrid striped bass. American Fisheries Society, Symposium 80, Bethesda, Maryland.

Dettmers, J. M., D. R. DeVries, and R. A. Stein. 1996. Quantifying responses to hybrid striped bass predation across multiple trophic levels: implications for reservoir biomanipulation. Transactions of the American Fisheries Society 125:491–504.

Dettmers, J. M., R. A. Stein, and E. M. Lewis. 1998. Potential regulation of age-0 gizzard shad by hybrid striped bass in Ohio reservoirs. Transactions of the American Fisheries Society 127:84–94.

Heman, M. L., L. C. Redmond, and R. S. Campbell. 1969. Manipulation of fish populations through reservoir drawdown. Transactions of the American Fisheries Society 98:293–304.

Hutt, C. P., J. W. Neal, and T. J. Lang. 2008. Stocking harvestable hybrid striped bass in an urban fishing program: angling success, angler satisfaction, and influence on bluegill size structure. Pages 403–412 *in* R. T. Eades, J. W. Neal, T. J. Lang, K. M. Hunt, and P. Pajak, editors. Urban and community fisheries programs: development, management, and evaluation. American Fisheries Society, Symposium 67, Bethesda, Maryland.

Isely, J. J., and T. B. Grabowski. 2007. Age and growth. Pages 187–228 *in* C. S. Guy and M. L. Brown, editors. Analysis and interpretation of freshwater fisheries data. American Fisheries Society, Bethesda, Maryland.

Jahn, L. A., D. R. Douglas, M. J. Terhaar, and G. W. Kruse. 1987. Effects of stocking hybrid striped bass in Spring Lake, Illinois. North American Journal of Fisheries Management 7:522–530.

Layzer, J. B., and M. D. Clady. 1984. Evaluation of the striped bass × white hybrid for controlling stunted bluegills. Proceedings of the Annual

Conference Southeastern Association of Fish and Wildlife Agencies 35(1981):297–310.

Leitner, J., J. Bulak, and R. Dunham. 2002. A comparison of first and third year growth of two strains of largemouth bass in South Carolina. Pages 365–370 *in* D. P. Philipp and M. S. Ridgway, editors. Black bass: ecology, conservation, and management. American Fisheries Society, Symposium 31, Bethesda, Maryland.

Maceina, M. J., J. Boxrucker, D. L. Buckmeier, R. Scott Gangl, D. O. Lucchesi, D. A. Isermann, J. R. Jackson, and P. J. Martinez. 2007. Current status and review of freshwater fish aging procedures used by state and provincial fisheries agencies with recommendations for future directions. Fisheries 32:329–340.

Makinster, A. S., and C. P. Paukert. 2008. Effects and utility of minimum length limits and mortality caps for flathead catfish in discrete reaches of a large prairie river. North American Journal of Fisheries Management 28:97–108.

McInerny, M. C., and T. K. Cross. 1999. Effects of lake productivity, climate warming, and intraspecific density on growth and growth patterns of black crappie in southern Minnesota lakes. Journal of Freshwater Ecology 14:255–264.

Miranda, L. E. 2002. Establishing size-based mortality caps. North American Journal of Fisheries Management 22:433–440.

Miranda, L. E., and P. P. Durocher. 1986. Effects of environmental factors on growth of largemouth bass in Texas reservoirs. Pages 115–121 *in* G. E. Hall and M. J. Van Den Avyle, editors. Reservoir fisheries management: strategies for the 80's. American Fisheries Society, Southern Division, Reservoir Committee, Bethesda, Maryland.

Mosher, T. D., and D. W. Willis. 1997. Fish survey techniques for small lakes and reservoirs, 3rd edition. Kansas Department of Wildlife and Parks, Emporia.

Muoneke, M. I., and W. M. Childress. 1994. Hooking mortality: a review for recreational fisheries. Reviews in Fisheries Science 2:123–156.

Neal, J. W., R. L. Noble, and J. A. Rice. 1999. Fish community response to hybrid striped bass introduction in small warmwater impoundments. North American Journal of Fisheries Management 19:1044–1053.

Noble, R. L. 1981. Management of forage fishes in impoundments of the southern United States. Transactions of the American Fisheries Society 110:738–750.

Paukert, C. P., M. A. McInerny, and R. D. Schultz. 2007. Historical trends in creel limits, length-based limits, and season restrictions for black basses in the United States and Canada. Fisheries 32:62–72.

Quinn, S. P. 1993. Description of a multiuse fishery for flathead catfish. North American Journal of Fisheries Management 13:594–599.

Quist, M. C., C. S. Guy, R. D. Schultz, and J. L. Stephen. 2003. Latitudinal comparisons of walleye growth in North America and factors influencing growth of walleyes in Kansas reservoirs. North American Journal of Fisheries Management 23:677–692.

Quist, M. C., J. L. Stephen, C. S. Guy, and R. D. Schultz. 2004. Age structure and mortality of walleyes in Kansas reservoirs: use of mortality caps to establish realistic management objectives. North American Journal of Fisheries Management 24:990–1002.

Ricker, W. E. 1975. Computation and interpretation of biological statistics of fish populations. Fisheries Research Board of Canada, Bulletin 191.

Sammons, S. M., and M. J. Maceina. 2008. Evaluating the potential effectiveness of harvest restrictions on riverine sunfish populations in Georgia, USA. Fisheries Management and Ecology 15:167–178.

Scholten, G. D., and P. W. Bettoli. 2005. Population characteristics and assessment of overfishing for an exploited paddlefish population in the lower Tennessee River. Transactions of the American Fisheries Society 134:1285–1298.

Schultz, R. D., and B.J. Dodd. 2008. Growth, mortality, and harvest of walleye and hybrid striped bass in an Iowa urban lake: simulated effects of minimum size limits. Pages 413–424 *in* R. T. Eades, J. W. Neal, T. J. Lang, K. M. Hunt, and P. Pajak, editors. Urban and community fisheries programs: development, management, and evaluation. American Fisheries Society, Symposium 67, Bethesda, Maryland.

Schultz, R. D., Z. J. Jackson, and M. C. Quist. 2008. Relating impoundment morphometry and water quality to black crappie, bluegill, and largemouth bass populations in Iowa. Pages 479–491 *in* M. S. Allen, S. M. Sammons, and M. J. Maceina, editors. Balancing fisheries management and water uses for impounded river systems. American Fisheries Society, Symposium 62, Bethesda, Maryland.

Slipke, J. W., and M. J. Maceina. 2000. Fisheries analyses and simulation tools (FAST). Auburn University, Department of Fisheries and Allied Aquacultures, Agricultural Experiment Station, Auburn, Alabama.

Soupir, C. A., B. B. Blackwell Scientific Publications, and M. L. Brown. 1997. Relative precision among calcified structures for white bass age and growth assessment. Journal of Freshwater Ecology 12:531–534.

Thompson, J. S., and J. A. Rice. 2013. The relative influence of thermal experience and forage availability on growth of age 1–5 striped bass in two southeastern reservoirs. Pages 93–120 *in* J. S. Bulak, C. C. Coutant, and J. A. Rice, editors. Biology and management of inland striped bass and hybrid striped bass. American Fisheries Society, Symposium 80, Bethesda, Maryland.

Van Horn, S. L. 2013. A brief history of inland striped bass management. Pages 1–13 *in* J. S. Bulak, C. C. Coutant, and J. A. Rice, editors. Biology and management of inland striped bass and hybrid striped bass. American Fisheries Society, Symposium 80, Bethesda, Maryland.

Welch, T. J., M. J. Van Denm Avyle, R. K. Betsill, and E. M. Driebe. 1993. Precision and relative accuracy of striped bass age estimates from otoliths, scales, and anal fin rays and spines. North American Journal of Fisheries Management 13:616–620.

American Fisheries Society Symposium 80:473–499, 2013

Catch-and-Release Mortality of Inland Striped Bass and Hybrid Striped Bass

Jason M. Bettinger*
South Carolina Department of Natural Resources
1921 Vanboklen Road, Eastover, South Carolina 29044, USA

Gene R. Wilde
Department of Biological Sciences, Texas Tech University
Lubbock, Texas 79409, USA

Abstract.— We conducted a meta-analysis of existing studies of striped bass *Morone saxatilis* and hybrid striped bass (striped bass × white bass *M. chrysops*) catch-and-release mortality in freshwaters. Water temperature or season (P < 0.0001) and anatomical site of hooking (P < 0.0001) were the most consistently significant predictors of striped bass catch-and-release mortality. Bait type (P = 0.0007), presence of bleeding in hooked fish (P = 0.0017), and handling time (P = 0.0363) also affected catch-and-release mortality. Only two studies examined catch-and-release mortality of hybrid striped bass. We conducted surveys of fishery management agencies and striped bass fishing clubs to assess their opinions about the importance of catch-and-release mortality in striped bass and hybrid striped bass fisheries. Compared with fishery management agencies, a greater proportion of fishing clubs believed that catch-and-release mortality impacted striped bass fisheries. In general, fishery management agencies believed that striped bass catch-and-release mortality was low, although most had little information, and viewed catch-and-release mortality as a social, rather than biological, issue. To assess the potential impacts of catch-and-release mortality on striped bass fisheries, we constructed a simple model that, based on average summer release rates and water temperatures, estimated the proportion of striped bass total summer catch lost to catch-and-release mortality. Estimates ranged from 16% to 60% (mean = 36%) if all fish were captured on natural baits and 11% to 31% (mean = 19%) if all fish were captured on artificial baits. Although most fishery management agencies view catch-and-release mortality of striped bass and hybrid striped bass as mostly a social issue, it appears that they currently underestimate the magnitude of the biological problems presented by catch-and-release fishing for striped bass and hybrid striped bass.

Introduction

Striped bass *Morone saxatilis* and hybrid striped bass (striped bass × white bass *M. chrysops*) have been widely introduced into reservoirs throughout the United States to create recreational and trophy fisheries and to control populations of pelagic prey fishes (Axon and Whitehurst 1985; Dettmers et al. 1996; Vatland et al. 2008; Van Horn 2013, this volume). Striped bass and hybrid striped bass fisheries commonly are man-

* Corresponding author: bettingerj@dnr.sc.gov

aged with minimum length and possession limits that are intended to protect and enhance fish stocks and promote growth of large individuals. These regulations may require anglers to release under- or oversized fish or those in excess of daily possession limits. In addition to fish that are released due to regulatory requirements, there is a growing prevalence of striped bass fishing tournaments, many of which promote live release of captured fish (e.g., Bettinger et al. 2005; Young and Isely 2006). The increase in catch-and-release angling for striped bass underscores the importance of understanding catch-and-release mortality in striped bass.

Recreational angling has the potential to substantially impact freshwater fish stocks (McPhee et al. 2002; Post et al. 2002; Cooke and Cowx 2006). These impacts may be exacerbated by catch-and-release fishing. For example, Bartholomew and Bohnsack (2005) reported that catch-and-release mortality in some species can be as great as 95%, which exceeds the 20% that Muoneke and Childress (1994) suggested as a threshold for preventing fishery impacts. More recently, Coggins et al. (2007) estimated that release mortality rates as low as 5%, depending on the species life history, can have negative impacts on fisheries. Consequently, Aalbers et al. (2004) and Millard et al. (2005), among others, have suggested that this source of mortality should be included in fishery assessments. Given the popularity and economic importance of striped bass and hybrid striped bass fisheries (e.g., Schorr et al. 1995; U.S. Fish and Wildlife Service and U.S. Bureau of the Census 2006), surprisingly few studies (13, by our count) have examined some aspect of catch-and-release mortality of striped bass in freshwaters, and only two studies (Childress 1989; Walters 1997) have examined survival of hooked and released hybrid striped bass. In this paper, we synthesize previous studies of striped bass catch-and-release mortality in freshwaters, present results of a survey of fishery management agencies' and striped bass fishing clubs' opinions about striped bass and hybrid striped bass hooking mortality, and model the potential magnitude of catch-and-release mortality in several southeast U.S. reservoirs. Our goals are to evaluate the current state of knowledge about catch-and-release mortality in freshwater striped bass and hybrid striped bass stocks, assess the potential fishery effects of this source of mortality, and provide a general review of the factors that impact catch-and-release mortality.

Methods

Meta-analysis of previous studies

Previous studies have examined the effects of a number of environmental conditions, biological variables, and handling- and gear-related variables on striped bass catch-and-release mortality in freshwater. To provide a quantitative overall assessment of which factors consistently influenced catch-and-release mortality, we extracted results of statistical tests from these studies (Table 1) and used the observed significance (P) value as a measure of the reliability of a variable in predicting catch-and-release mortality. Because the specific null hypotheses varied in each study, and several studies presented P-values rather than test statistics needed to conduct a more formal meta-analysis, we used Fisher's method of combining probabilities from independent tests (Sokal and Rohlf 1981) to combine probabilities for each catch-and-release mortality predictor:

$$\chi^2 = -2\sum_{i=1}^{k} \log_e(p_i),$$

where χ^2 is a chi-square variate with $2k$ degrees of freedom, k is the number of independent tests, and p_i is the observed probability for the ith test. For P-values presented as indeterminate quantities (e.g., <0.05, <0.0001) we used the nominal value of that quantity.

Several studies examined variation in catch-and-release mortality as a function of water temperature or season (without also presenting information on water temperature). We grouped studies that assessed, separately, the effects of water temperature and season because

TABLE 1. Summary of potential predictors of striped bass survival after being caught and released in freshwaters. Observed levels of significance are based on paired *t*-tests (Bettoli and Osborne 1998) or χ^2 or goodness of fit tests (all others). All tests are based on the null hypothesis that the predictor does not result in increased mortality.

Predictor	*P*	Habitat	Source	Summary
Bait type	0.0001	Reservoir	Hysmith et al. (1994)	Mortality was greatest for striped bass caught on live bait (58%) followed by those captured on artificial lures with treble hooks (32%) and artificial lures with a single hook (16%).
Bait type	>0.0500	Pond	Harrell (1988)	There were no differences in mortality attributable to method of capture.
Bait (hook) type	0.3323	River	Millard et al. (2005)	Mortality averaged 16% for fish captured with traditional J-hooks and 5% for fish captured on circle hooks.
Bait type	0.1601	River	Nelson (1998)	Mortality did not differ between fish captured on artificial lures (7.3%) and live bait (5.3%).
Bait type	0.9203	River	Millard et al. (2005)	Mortality did not differ between fish captured on cut bait (17.0%) and live bait (9.0%).
Bleeding	0.0008	River	Millard et al. (2005)	The incidence of bleeding was a significant predictor of mortality based on logistic regression. However, there was no difference in bleeding between fish captured on circle (5.4%) or J-hooks (7.4%) or between fish captured on cut (7.5%) or live bait (0.0%).
Bleeding	0.2261	River	Millard et al. (2003)	Bleeding was not a significant predictor of mortality based on logistic regression.
Capture depth	0.8517	Reservoir	Bettinger et al. (2005)	Capture depth was not a significant predictor of mortality based on logistic regression.
Handling time	0.0640	Reservoir	Bettoli and Osborne (1998)	Mortality was not related to handling time.
Handling time	0.0924	River	Nelson (1998)	Handling time was not a significant predictor of mortality based on logistic regression.
Anatomical site of hooking	0.0002	River	Nelson (1998)	Mortality was much greater among fish hooked in the pharynx, esophagus, or gills than among fish hooked externally or in the mouth.
Anatomical site of hooking	0.0005	River	Millard et al. (2005)	There was a direct increase in mortality with increased depth of hook location (lip, mouth, gill, and gut).

TABLE 1. Continued.

Predictor	*P*	Habitat	Source	Summary
Anatomical site of hooking	0.0118	River	Millard et al. (2003)	Mortality was greater in deep-hooked fish (69%) than in shallow-hooked fish (27%).
Playing time	<0.0500	River	Thompson et al. (2002)	Mortality of fish captured at 26°C increased with playing time, from about 20% among fish played for 1 min to greater than 50% for fish played more than 3 min.
Landing time	0.9622	River	Nelson (1998)	Landing time was not a significant predictor of mortality based on logistic regression.
Landing time	0.1615	River	Millard et al. (2005)	Mortality was not related to landing time between fish played for <5 min compared with those played for longer periods.
Landing time	0.5620	Reservoir	Bettoli and Osborne (1998)	Mortality was not related to landing time.
Landing time	0.7916	River	Millard et al. (2003)	Landing time was not a significant predictor of mortality based on logistic regression.
Length	0.0001	Reservoir	Hysmith et al. (1994)	Predicted mortality increased directly with total length (TL), from 20% in fish 250 mm TL to >70% in fish larger than 700 mm TL.
Length	0.0578	River	Millard et al. (2005)	Total length was not a significant predictor of mortality based on logistic regression.
Length	0.6960	Reservoir	Childress (1989)	Mortality did not differ between fish <457.2 mm TL and those >457.2 mm TL
Length	0.2120	River	Nelson (1998)	Total length was not a significant predictor of mortality based on logistic regression.
Length	<0.2800	Reservoir	Bettinger et al. (2005)	Total length was not a significant predictor of mortality based on logistic regression.
Length	0.5520	Reservoir	Bettoli and Osborne (1998)	Mortality was not related to fish length.
Length	0.6121	River	Millard et al. (2003)	Total length was not a significant predictor of mortality based on logistic regression.
Sex	0.1552	River	Millard et al. (2005)	Sex was not a significant predictor of mortality based on logistic regression.

TABLE 1. Continued.

Predictor	P	Habitat	Source	Summary
Sex	0.4488	River	Nelson (1998)	Sex was not a significant predictor of mortality based on logistic regression.
Season	<0.0001	Reservoir	Hysmith et al. (1994)	Mortality was greater in spring (69%) and summer (47%) than in fall (8%) and winter (13%).
Season	<0.0001	Reservoir	Childress (1989)	Mortality was greater among fish captured in summer (69%) than in those captured in winter (11%).
Season	0.0010	Reservoir	Bettinger et al. (2005)	Mortality was greater during summer (83%) than spring (0%).
Water temperature	0.0005	River	Nelson (1998)	Logistic regression showed a positive relationship between mortality and water temperature.
Season	<0.0500	Pond	Harrell (1988)	Mortality was greater for fish caught during August on natural bait (40%) and artificial lures (36%) than for fish caught during June, October, or February on either bait type (<21%).
Air temperature	0.0070	Reservoir	Bettoli and Osborne (1998)	Mortality was directly, linearly related to air temperature.
Water temperature	0.0810	Reservoir	Bettoli and Osborne (1998)	Mortality was not related to water temperature.
Water temperature	0.2774	Reservoir	Bettinger et al. (2005)	No mortality was observed among fish captured in the spring (surface water temperatures 8°C to 16°C); however, during summer, overall mortality was 83% (surface water temperatures 28°C to 29°C).

we assumed season was a surrogate measure of temperature. We omitted results from Wilde et al. (2000) from Table 1 and from the meta-analysis because they are based on the results of Childress (1989), Hysmith et al. (1994), Bettoli and Osborne (1998), and Nelson (1998). Similarly, our meta-analysis omitted the results for air temperature reported by Bettoli and Osborne (1998) because we assumed that they were not independent from the water temperatures presented in that study.

Agency survey

We conducted a survey of agencies in 33 states with striped bass or hybrid striped bass fisheries in reservoirs (Figure 1) to assess their views on the importance of catch-and-release mortality in striped bass and hybrid striped bass and whether these views guided management actions. Prior to conducting the survey, we contacted state agencies responsible for management of these fisheries, explained the purpose of the survey, and asked an agency representative to complete an online survey. Our initial contacts with state agencies were through their Southern Division American Fisheries Society representative to the Striped Bass Technical Committee, a biologist within the agency who was known to work with striped bass, or the agency's fishery chief. The 18-question survey instrument was designed and administered using Survey Monkey, an online program, and included both open-ended and closed-ended questions. The survey was conducted from April 2009 to May 2009.

Club survey

We conducted a survey of striped bass fishing clubs to assess the prevalence of catch-and-release angling in striped bass fishing tournaments and to determine whether these clubs actively encouraged catch-and-release angling. We were able to locate contact information for 24 striped bass fishing clubs, identified through an Internet search, and one national coalition of striped bass clubs. All of these organizations were asked to participate in an online survey. The president or vice president of each organization was sent an e-mail message that explained the purpose of the survey and asked them to complete a 21-question survey instrument at surveymonkey.com. The survey was conducted from May 2009 to June 2009.

Mortality model

To assess potential effects of catch-and-release mortality on striped bass fisheries, we developed a predictive model. We combined the original data from four studies of striped bass catch-and-release mortality conducted in freshwater reservoirs in which water temperature at capture and bait type were known for each fish (Table 2). We chose to study these variables based on the results of Wilde et al. (2000) and our meta-analysis. Three of those studies (Childress 1989; Hysmith et al. 1994; Bettoli and Osborne 1998) were previously reviewed by Wilde et al. (2000), and the fourth study was conducted by Bettinger et al. (2005). We used logistic regression to model the effects of water temperature at capture and bait type on striped bass catch-and-release mortality. In our logistic regression model, bait type was treated as a classification variable coded as +1 for natural baits and −1 for artificial baits. The model was developed using Proc Logistic in SAS (SAS Institute, Cary, North Carolina).

We obtained creel data from roving daytime surveys that included catch, release, and harvest rates for striped bass in seven southeastern reservoirs (Table 3). Hourly rates, based on targeted striped bass effort, for catch, harvest, and release were reported from J. Strom Thurmond Reservoir, South Carolina and Georgia, and during 2000 and 2001 from Lake Murray, South Carolina. All other surveys reported estimates of total catch and harvest based on total angler effort. Creel data were either summarized by season or month. Because of the relationship between striped bass catch-and-release mortality and temperature (Wilde et al. 2000), we considered only results from the summer, which was defined as 1 June through 31 August, except for Lake Murray where sum-

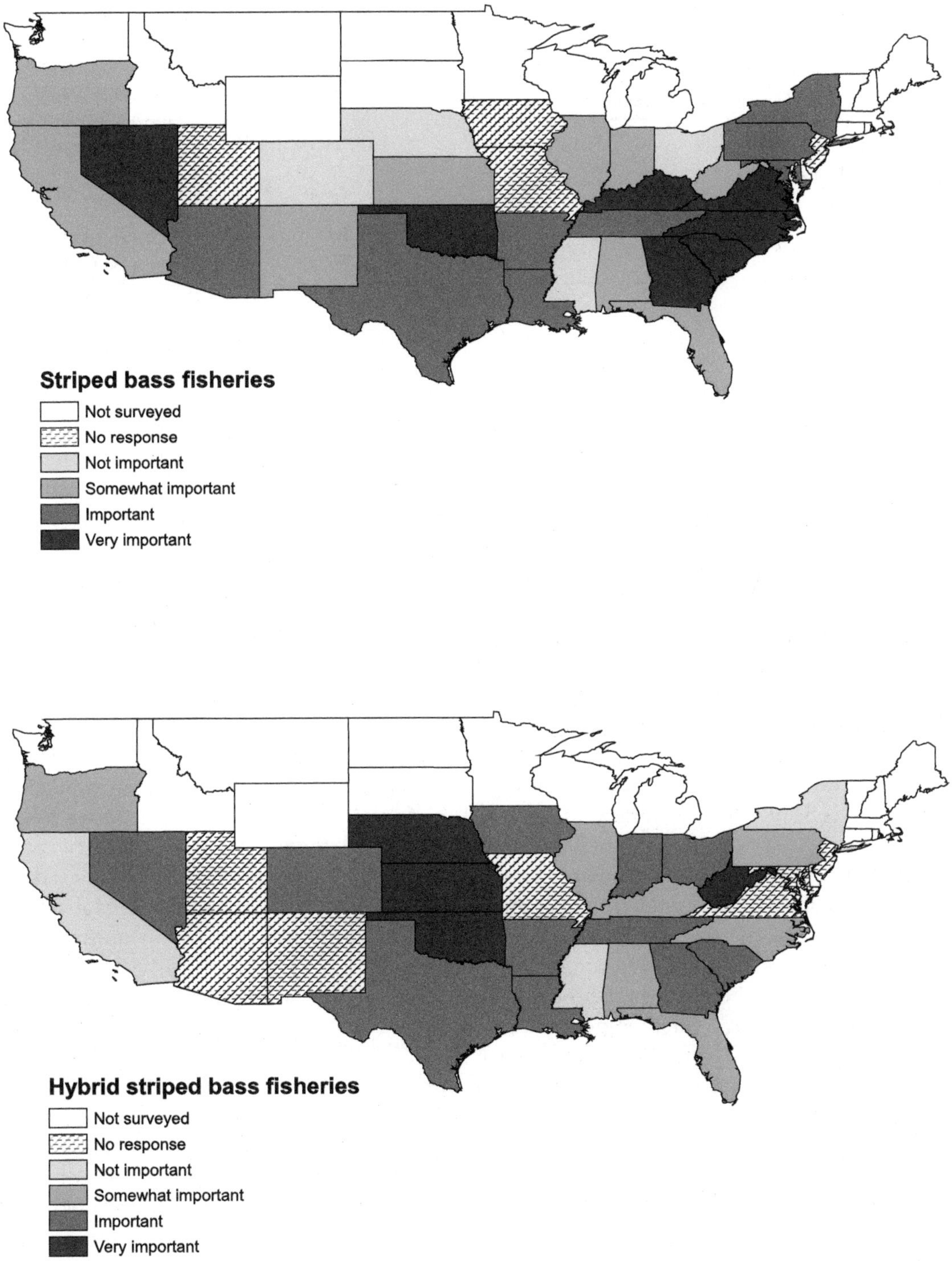

FIGURE 1. States surveyed in 2009 for catch-and-release mortality (all shadings), with their reported importance of striped bass and hybrid striped bass fisheries.

TABLE 2. Summary of striped bass hooking mortality studies used to develop logistic regression model for reservoir striped bass populations, number of observations for each bait type, and range of water temperatures at capture.

Study site	Bait type		Water temperature (°C)	Reference
	Artificial	Natural		
Tims Ford Reservoir, TN	55	34	11.0–31.0	Bettoli and Osborne (1998)
Lake Buchanon, TX	7	0	22.5	Childress (1989)
Lake Texoma, OK–TX	0	106	8.3–25.8	Childress (1989)
Lake Texoma, OK–TX	182	125	6.0–30.0	Hysmith et al. (1994)
Lake Murray, SC	0	53	8.0–29.0	Bettinger et al. (2005)

mer results also included those for the month of September. When data were reported by month, total catch and harvest were summed across June, July, and August. Striped bass release rates were calculated as

$$\text{Proportion released} = (N_{\text{caught}} - N_{\text{harvested}})/N_{\text{caught}},$$

where N_{caught} is the number (or hourly rate) of striped bass captured by anglers, whether released or harvested, and $N_{\text{harvested}}$ is the number (or hourly rate) of striped bass harvested by anglers.

We obtained surface (<1 m) water temperature for each reservoir from which striped bass creel data were available. Water temperature data were collected from a variety of sources. Most of these data were unpublished and were collected by state agencies and power companies. We calculated mean summer temperatures for the period 1 June through 31 August, except for Lake Murray for which summer temperature was determined for the period 1 June through 30 September. Water temperature data collected from Lake Norman, North Carolina; Smith Mountain Lake, Virginia; and J. Strom Thurmond Reservoir were single observations collected at least monthly from one or more sites in the lower section of the reservoir. In Lake Cumberland, Kentucky, mean water temperature data were collected several times each month at four areas in the lower lake. At Lake Gaston, North Carolina; Lake Murray; and Lake Moultrie, South Carolina, surface water temperature data were collected every 15 min from a single location and mean daily temperatures were used to calculate mean summer surface temperatures. Water temperature data were collected in the same years as were creel data except in Smith Mountain Lake and Lake Moultrie in which water temperature data were not available for the 1998 and 2000 creel years. Consequently, we combined all available data to estimate mean annual summer water temperatures for those reservoirs.

To estimate the proportion of fish lost as a result of catch-and-release angling during summer, we used mean summer water temperatures in our predictive regression equation to estimate reservoir-specific mortality rates. We then multiplied these estimated mortality rates by the summer release-rate observed in each reservoir to obtain the proportion of total catch lost as a result of catch-and-release mortality.

Results

Meta-analysis of previous studies

Catch-and-release mortality of striped bass in freshwaters is affected by a number of factors (Table 1). Across all available studies, water temperature or season (χ^2 = 79.44, df = 14, P < 0.0001) and anatomical site of hooking (χ^2 = 41.12, df = 6, P < 0.0001) were the most consistent significant predictors of striped bass catch-and-release mortality. Based on five and two studies, respectively, bait type (χ^2 = 30.44,

TABLE 3. Summary of expanded striped bass creel data collected from roving daytime surveys and used to calculate release rates for seven southeastern reservoirs.

Reservoir	Year	Summer creel period	Blocking	Estimates expanded from	Estimate type	Source
Lake Norman, NC	2007	June–August	Seasonal	Total effort	Catch and harvest	North Carolina Wildlife Resources Commission (unpublished data)
Lake Gaston, NC	2008	June–August	Seasonal	Total effort	Catch and harvest	North Carolina Wildlife Resources Commission (unpublished data)
Smith Mountain Lake, VA	1998, 2003, 2006	June–August	Monthly	Total effort	Catch and harvest	Virginia Department of Game and Inland Fisheries (unpublished data)
Lake Cumberland, KY	1998, 2001, 2004, 2008	June–August	Monthly	Total effort	Catch and harvest	KDFWR (1999, 2002, 2005, 2009)
J. Strom Thurmond Reservoir, SC	2005–2008	June–August	Seasonal	Directed effort	Hourly rate	Brown (2006), Houck (2007, 2008, 2009)
Lake Moultrie, SC	2000	June–August	Monthly	Total effort	Catch and harvest	White and Lamprecht (2002)
Lake Murray, SC	1999	June–September	Seasonal	Total effort	Catch and harvest	Hayes and Penny (2000)
Lake Murray, SC	2000–2001	June–August	Seasonal	Directed effort	Hourly rate	Hayes and Penny (2001, 2002)

df = 10, P = 0.0007) and presence of bleeding in hooked fish (χ^2 = 17.23, df = 4, P = 0.0017) were significant predictors of mortality. Handling time was marginally significant (χ^2 = 10.26, df = 4, P = 0.0363), based on only two studies, but landing time was not a significant predictor of mortality (χ^2 = 16.81, df = 10, P = 0.0786). Across seven studies, fish length was a significant predictor of mortality (χ^2 = 32.66, df = 14, P = 0.0032), but this is due to results of a single study (Hysmith et al. 1994), and if these results are excluded, length is not a significant predictor of mortality (χ^2 = 14.24, df = 12, P = 0.2854). Sex was not a significant predictor of mortality (χ^2 = 5.33, df = 4, P = 0.2552) based on results of two studies.

Agency survey

Twenty-nine (88%) of the 33 state agencies invited to participate in our survey completed the online instrument (Figure 1), though not all respondents answered every question. In 28 states, either striped bass (24 states) or hybrid striped bass (22 states) fisheries were considered to be somewhat important, important, or very important. Those responding states that indicated striped bass or hybrid striped bass fisheries were not important were not included in the survey results for striped bass and hybrid striped bass, respectively.

Catch-and-release angling is known to cause mortality of striped bass in 17 of the responding states (71%) and mortality of hybrid striped bass in 12 states (55%) (Table 4). Sixty-four percent of responding states reported observing some level of striped bass catch-and-release mortality in their reservoirs, whereas the remaining eight (36%) states did not know the proportion of reservoirs in which catch-and-release mortality occurred. Hybrid striped bass catch-and-release mortality in reservoirs was reported by 12 (55%) responding states, although 10 (45%) states were unsure of the proportion of reservoirs in which catch-and-release mortality occurred. Catch-and-release mortality in rivers also was common, with 58% and 33% of states reporting that mortality for striped bass and hybrid striped bass, respectively, occurred in at least some rivers. In 17 responding states, noticeable catch-and-release mortality of striped bass was observed most frequently during summer (16 states), but also occurred during spring (4 states) and fall (3 states). Similarly, 10 of 11 responding states reported noticeable catch-and-release mortality of hybrid striped bass during summer, 3 states reported catch-and-release mortality in fall, and 1 state reported hybrid striped bass catch-and-release mortality during each of winter and spring.

Many responding states did not believe that catch-and-release mortality negatively impacted reservoir striped bass (59%) or hybrid striped bass (41%) fisheries in their state (Table 4). Only two states, Virginia and South Carolina, reported that catch-and-release mortality had a negative impact on reservoir striped bass fisheries, and two states, Kansas and Nebraska, reported that catch-and-release mortality had a negative impact on reservoir hybrid striped bass fisheries. A substantial proportion of states did not know whether catch-and-release mortality had a negative impact on reservoir striped bass (32%) or hybrid striped bass (50%) fisheries. Among river fisheries, only three states (New York, North Carolina, and South Carolina) reported that catch-and-release mortality had a negative impact on striped bass fisheries, and none reported a negative impact on hybrid striped bass fisheries.

Most (≥50%) states did not know the extent to which catch-and-release mortality contributed to total annual mortality. A majority of the remaining states believed that catch-and-release mortality was a minor portion of the total annual mortality experienced by striped bass and hybrid striped bass in both reservoirs and rivers. In only four states, Georgia, New York, South Carolina, and Virginia, was catch-and-release mortality believed to be an important component of total annual striped bass mortality (Table 4).

Catch-and-release mortality of striped bass was viewed as being more of a social issue than a biological issue (Table 5). Social and biologi-

TABLE 4. Percentage of agency responses to statements and questions in a survey related to the magnitude of catch-and-release mortality and its impact on striped bass and hybrid striped bass fisheries. (Number of respondents in parentheses.)

Response	Striped bass	Hybrid striped bass
"Mortality from catch-and-release angling of striped bass or hybrid striped bass occurs in what proportion of your reservoirs in which they occur?"		
None	0% (0)	0% (0)
Fewer than half	18% (4)	23% (5)
Half	5% (1)	0% (0)
More than half	5% (1)	5% (1)
All	36% (8)	27% (6)
Don't know	36% (8)	45% (10)
"Mortality from catch-and-release angling of striped bass or hybrid striped bass occurs in what proportion of your rivers in which they occur?"		
None	5% (1)	22% (4)
Fewer than half	26% (5)	11% (2)
Half	0% (0)	0% (0)
More than half	0% (0)	0% (0)
All	32% (6)	22% (4)
Don't know	37% (7)	44% (8)
"Catch-and-release mortality has a negative impact on striped bass and/or hybrid striped bass fisheries in one or more reservoirs in your state."		
Strongly agree	5% (1)	0% (0)
Agree	5% (1)	9% (2)
Disagree	50% (11)	27% (6)
Strongly disagree	9% (2)	14% (3)
Don't know	32% (7)	50% (11)
"Catch-and-release mortality has a negative impact on striped bass and/or hybrid striped bass fisheries in one or more rivers in your state."		
Strongly agree	0% (0)	0% (0)
Agree	13% (3)	0% (0)
Disagree	52% (12)	40% (8)
Strongly disagree	9% (2)	15% (3)
Don't know	26% (6)	45% (9)
"What proportion of total annual mortality is related to catch-and-release mortality in your striped bass and hybrid striped bass fisheries?"		
Minor	33% (8)	33% (7)
Moderate	13% (3)	5% (1)
Significant	4% (1)	0% (0)
Don't know	50% (12)	62% (13)

TABLE 5. Percentage of agency responses to the survey questions dealing with the influence of catch and release mortality on management decisions. (Number of respondents in parentheses.)

Response	Striped bass	Hybrid striped bass
"Catch-and-release mortality of striped bass is more of a social management issue than a biological management issue in your state"		
Strongly agree	12% (3)	
Agree	67% (16)	
Disagree	21% (5)	
Strongly disagree	0% (0)	
"How important of a consideration is catch-and-release mortality in the management of striped bass in your state?"		
Not important	50% (12)	
Somewhat important	33% (8)	
Important	8% (2)	
Very important	8% (2)	
"Has catch-and-release mortality influenced management of striped bass or hybrid striped bass?"		
Yes	33% (8)	18% (4)
No	67% (16)	82% (18)

cal issues were not defined for survey respondents; as a result, there may have been some confusion as to what constituted a social or biological issue. Seventy-nine percent of 24 responding states agreed or strongly agreed with the statement "Catch-and-release mortality of striped bass is more of a social management issue than a biological management issue in your state." Only five states (California, Florida, New York, South Carolina, and Virginia) reported that catch-and-release mortality of striped bass was more of a biological issue than a social issue. There was no consensus as to whether catch-and-release mortality is an important management consideration: 50% of responding states considered catch-and-release mortality to be of some importance in striped bass management (Table 5). Among these states, four considered catch-and-release mortality of striped bass to be very important (New York and Virginia) or important (Maryland and South Carolina).

Management strategies for striped bass and hybrid striped bass fisheries have been influenced by catch-and-release mortality in eight and four states, respectively (Table 5). Among the management strategies used by the states to account for catch-and-release mortality of striped bass, four states (North Carolina, Tennessee, South Carolina, and Virginia) seasonally relaxed regulations to allow increased harvest of fish that would likely die if released, three (Maryland, North Carolina, and South Carolina) attempted to reduce or eliminate fishing pressure without prohibiting angling, and one (Maryland) reported that it prohibited striped bass angling during periods of high mortality. Although only one state reported that it prohibited striped bass angling during periods of high mortality as a general management strategy, other states (e.g., Tennessee) use seasonal and spatial closures to protect striped bass from catch-and-release-related mortality in certain reservoirs. (Editor's note added in proof: Before publication, South Carolina had instituted a prohibition of catch-and-release fishing for striped bass, June–September, in all coastal rivers including the Santee–Cooper lakes [Marion and Moultrie] but excluding the lower Saluda River.) When asked whether these management efforts resulted in any perceived or measured benefit, five states indicated that the public perceived that there was a benefit, four states indicated that there was a perceived benefit by the agency, and two states responded that there was a measured benefit. No state believed that these strategies were without benefit. Most (64%) states did not make a publicized effort for anglers to limit catch-and-release fishing during periods of high mortality, but half of the states do suggest handling practices for fish anglers intend to release.

Club survey

Completed surveys were received from 15 of the 24 striped bass clubs contacted. Responses were obtained from clubs in a total of eight states; most (46%) respondents were from Georgia and South Carolina. Clubs had an average of 120 members (range = 23–340) and held an average of eight tournaments per year (range = 1–12) that averaged 26 boats (range = 7–100) per tournament. Most (67%) striped bass clubs conduct tournaments during the summer. Most (71%) striped bass clubs reported that anglers were allowed to cull fish without limit, 21% of clubs indicated their tournament rules prohibit culling during all seasons, and one (8%) club indicated that culling was seasonally prohibited.

Live release of striped bass after weigh-in was common during tournaments. The majority of clubs encouraged the live release of striped bass, but few clubs required live release of tournament-caught fish. Seventy-three percent of striped bass clubs reported some live release in their tournaments, and 53% of clubs reported an increase in the prevalence of live release in their tournaments during the past decade (Table 6). Sixty percent of striped bass clubs encouraged live release of striped bass captured in tournaments. Three clubs not only encouraged live release, but required fish to be released immediately after being photographed for length verification. When asked, "How does your club encourage the live release of striped bass entered in tournaments?" 83% of clubs that required striped bass to be weighed-in selected "special recognition for live release, but does not impact tournament standings." Only the national striped bass coalition awarded points for live release of striped bass captured in tournaments it sponsored.

There was no consensus among striped bass fishing clubs as to whether catch-and-release mortality during summer posed a threat to striped bass fisheries. In response to the question, "Does your club consider catch-and-release mortality during summer a significant threat to striped bass fisheries?" 50% responded "yes" and 50% responded "no"; the national striped bass conservation coalition responded "yes." All three

TABLE 6. Percentage of striped bass clubs' responses to questions dealing with the amount and trends of catch-and-release angling. (Number of respondents in parentheses.)

Question	None	<25%	25–50%	50–75%	>75%
What proportion of fish weighed in at your tournaments are released alive?	27% (4)	27% (4)	20% (3)	0% (0)	27% (4)
How much of an increase in catch-and-release angling have you noticed at your tournaments over the past decade?	47% (7)	20% (3)	13% (2)	7% (1)	13% (2)

clubs in South Carolina, two of three from Georgia and the single Texas and Nevada clubs, did not consider catch and release during summer to be a significant threat to striped bass fisheries. Both clubs in Tennessee and Virginia, and one club each in North Carolina, Pennsylvania, and Georgia did consider summer catch-and-release mortality to be a significant threat to striped bass fisheries.

Mortality model

Water temperature and bait type were significant predictors of catch-and-release mortality in striped bass captured from reservoirs (Table 7). Striped bass hooking mortality was positively related to water temperature and was greater among fish captured with natural baits than among those captured with artificial baits (Figure 2). Striped bass catch-and-release mortality at 30°C was estimated to be 76% for fish captured on natural baits and 45% for fish captured on artificial baits. Combining the intercept and bait-type coefficients allows prediction of striped bass catch-and-release mortality, for each bait type, as a function of water temperature:

$$P_{\text{natural}} = e^{-3.31 + 0.15 \cdot \text{water temp (°C)}}/[1 + e^{-3.31 + 0.15 \cdot \text{water temp (°C)}}], \text{ and}$$

$$P_{\text{artificial}} = e^{-4.69 + 0.15 \cdot \text{water temp (°C)}}/[1 + e^{-4.69 + 0.15 \cdot \text{water temp (°C)}}],$$

where P_{natural} is the predicted catch-and-release mortality of striped bass captured on natural baits and $P_{\text{artificial}}$ is the predicted catch-and-release mortality of striped bass captured on artificial baits. In general, this model predicts a higher catch-and-release mortality rate for striped bass caught on natural baits and a slightly lower catch-and-release mortality for striped bass caught on artificial baits than that presented by Wilde et al. (2000).

Average striped bass release rate among reservoirs was 51% and ranged from 21% to 95% (Table 8). Mean release rates were relatively high at Lake Murray and Lake Gaston, exceeding 70%. Lake Cumberland and Lake Norman had relatively low release rates of around 30%. Moderate rates of release were observed at the other reservoirs. Mean summer water temperature among reservoirs was 28°C, and ranged from 26.3°C to 29.6°C (Table 8).

Mean catch-and-release mortality rates, based on the average summer temperature in each lake, were 70% for fish caught on natural baits and 38% for fish caught on artificial baits (Table 8). Overall proportion of total catch lost to catch-and-release fishing during summer ranged from 16% to 60% for fish caught on natural bait and 11% to 31% for fish caught on artificial bait. In Lake Murray and Lake Gaston the predicted number of fish lost to catch-and-release fishing with natural bait exceeded the number of fish harvested. Temperature and release rate interact nonlinearly to affect the number of fish lost to catch-and-release fishing (Figure 3). When release rates exceed 50% and water temperatures exceed 28°C, more than one-third of total striped bass caught on natural baits would be lost to catch-and-release fishing.

Discussion

Factors affecting striped bass and hybrid striped bass catch-and-release mortality

An extensive literature relates catch-and-release mortality in a wide variety of fishes to the use of various angling techniques, baits, and gears (Siewert and Cave 1990; Taylor and White 1992; Diodati and Richards 1996; Lukacovic

TABLE 7. Analysis of deviance table for the logistic regression of striped bass hooking mortality on bait type and water temperature for striped bass caught and released in reservoirs.

Factor	χ^2	df	P	Coefficient	SE
Intercept				–4.00	0.479
Water temperature	63.1	1	0.0001	0.15	0.018
Bait type	47.8	1	0.0001	0.69	0.099

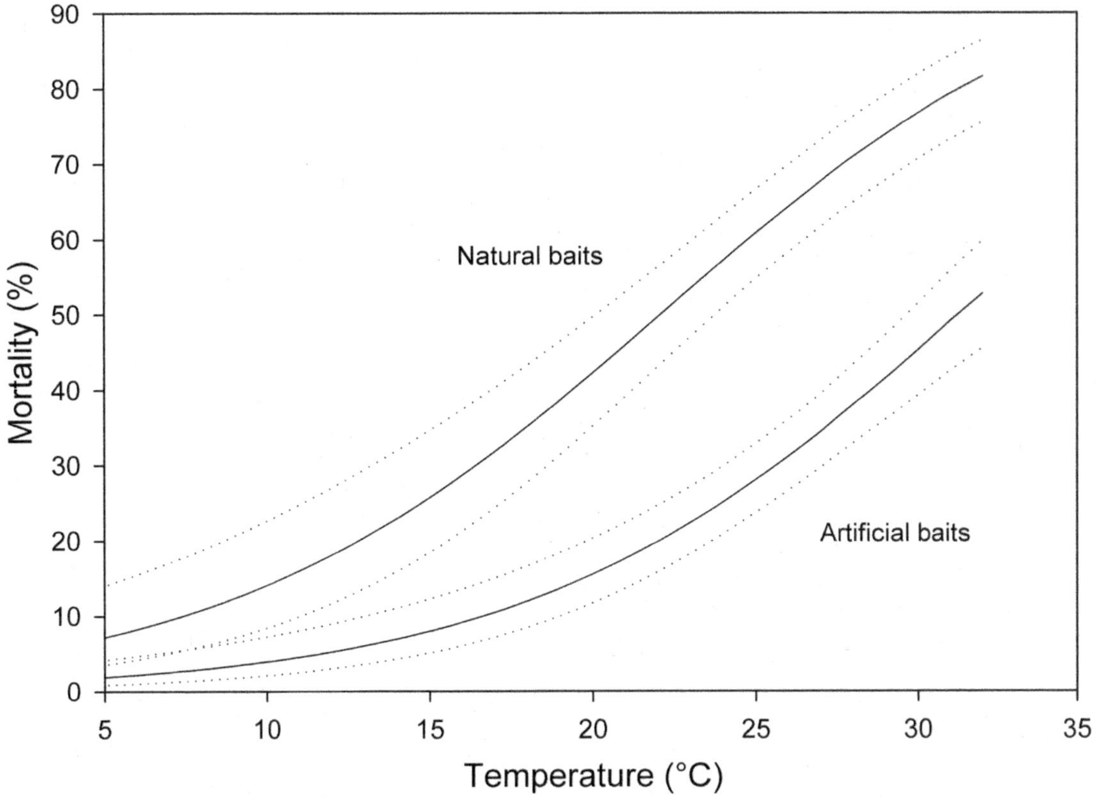

FIGURE 2. Striped bass hooking mortality as a function of bait type (natural and artificial) and water temperature with associated 95% confidence intervals (dotted lines). Predictive equations derived from logistic regression are presented in the text.

and Florence 1998; Cooke and Suski 2004), water temperature (Marnell and Hunsaker 1970; Nelson 1998; Wilde 1998), handling (Cooke et al. 2001; Millard et al. 2003), and depth of capture (Feathers and Knable 1983; Gravel and Cooke 2008; Jarvis and Lowe 2008), among other factors. These studies have consistently shown that severity of hooking wounds, determined chiefly by where fish are hooked, and water temperature are the most important determinants of mortality (Muoneke and Childress 1994; Bartholomew and Bohnsack 2005; Arlinghaus et al. 2007). Our meta-analysis shows that the most consistently significant factors affecting catch-and-release mortality in freshwater striped bass are anatomical site of hooking (whether fish are hooked in the stomach and esophagus or elsewhere) and water temperature. This is consistent with results of our mortality simulations, which showed that catch-and-release mortality of striped bass captured at high water temperatures on bait may exceed the mortality due to harvest. Increased water temperatures result in higher metabolic rates in striped bass (Hartman and Brandt 1995) and increase the probability that striped bass will exceed their maximum metabolic rate and be unable to recover from the physiological stress of capture and handling. Although bait type also was a significant predictor of striped bass catch-and-release mortality, this appears to be due to the tendencies of various baits and gears to differentially hook fish in locations that are associated with greater or lesser mortality. The significant relationships between mortality and bleeding and handling time also appear to be a function of site of hooking: fish

TABLE 8. Total catch estimated from summer creel surveys, release rate for caught striped bass, average summer water temperatures, mortality rates based on temperature and bait type, and proportion of total catch lost to catch and release (C&R) angling for seven southeastern impoundments.

					C&R mortality rates		Proportion of total catch lost to C&R mortality	
Year	System	Total catch	Release rate	Temperature (°C)	Natural	Artificial	Natural	Artificial
1998	Smith Mt Lake, VA	15,715	0.60	26.3	0.65	0.32	0.39	0.19
2003	Smith Mt Lake, VA	20,957	0.58	25.8	0.63	0.31	0.37	0.18
2006	Smith Mt Lake, VA	6,167	0.56	27.1	0.68	0.35	0.38	0.20
1998	Lake Cumberland, KY	6,203	0.28	28.3	0.72	0.39	0.20	0.11
2001	Lake Cumberland, KY	12,857	0.31	26.5	0.66	0.33	0.20	0.10
2004	Lake Cumberland, KY	11,383	0.21	27.7	0.70	0.37	0.18	0.08
2008	Lake Cumberland, KY	14,549	0.43	27.6	0.69	0.37	0.30	0.16
2008	Lake Gaston, NC	5,201	0.95	26.3	0.65	0.32	0.60	0.31
2007	Lake Norman, NC	744	0.30	29.9	0.76	0.45	0.23	0.13
2000	Lake Moultrie, SC	6,143	0.57	29.5	0.75	0.43	0.43	0.25
1999	Lake Murray, SC	45,793	0.70	27.7	0.70	0.37	0.49	0.26
2000	Lake Murray, SC	99,007	0.82	27.8	0.70	0.37	0.57	0.31
2001	Lake Murray, SC	37,760	0.68	28.0	0.71	0.38	0.48	0.26
2005	Lake Thurmond, SC	9,155	0.49	29.3	0.75	0.48	0.37	0.21
2006	Lake Thurmond, SC	15,585	0.47	29.6	0.75	0.43	0.35	0.21
2007	Lake Thurmond, SC	11,468	0.21	29.4	0.75	0.44	0.16	0.09
2008	Lake Thurmond, SC	17,152	0.48	28.9	0.73	0.41	0.35	0.20
	Mean	19,755	0.51	28.0	0.70	0.38	0.36	0.19

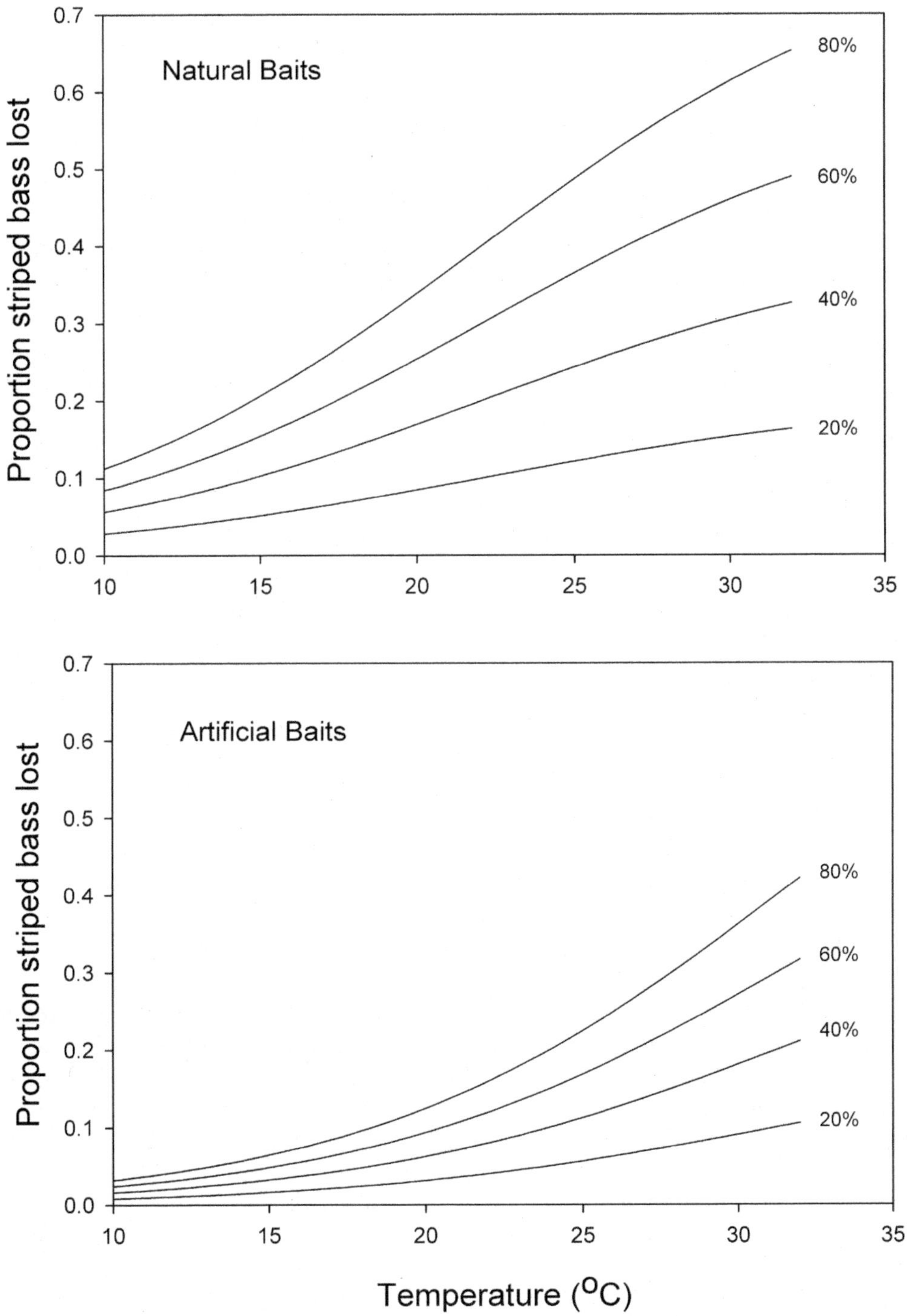

FIGURE 3. Proportion of striped bass total catch that is predicted to die due to catch-and-release mortality based on temperature, and four release rates (20, 40, 60, and 80%) for striped bass caught on natural baits (top panel) and artificial baits (bottom panel). Proportions were estimated by multiplying release rate by the mortality rate estimated from the bait-specific logistic regression equations presented in the text.

hooked in the stomach and esophagus require more time for dehooking and are more likely to bleed than fish hooked in superficial locations (Dextrase and Ball 1991; Wilde and Pope 2008), and fish hooked in the stomach and esophagus are more likely to suffer wounds to the heart and other internal organs (Muoneke and Childress 1994; Caruso 2000; Aalbers et al. 2004). Thus, the simple model of Wilde et al. (2000), which predicts mortality of striped bass as a function of bait type and water temperature and guided construction of our predictive model, captures the two most important predictors of striped bass catch-and-release mortality. The model developed herein is specific to freshwater impoundments, and therefore we recommend it over that of Wilde et al. (2000) in these habitats.

Limiting landing and handling time may reduce striped bass release mortality. Playing time is positively related to physiological disturbance in striped bass (Tomasso et al. 1996; Thompson et al. 2002), as well as other fishes (Gustaveson et al. 1991; Thorstad et al. 2003). Thompson et al. (2002) noted a positive relationship between angling duration and mortality of striped bass. Similarly, air exposure increases physiological disturbances (Cooke et al. 2001; Suski et al. 2004) and recovery time (Cooke et al. 2001) and can increase mortality (Ferguson and Tufts 1992; Schisler and Bergersen 1996) of released fish. Barbless single hooks can be removed more quickly than barbed hooks or treble hooks, reducing handling time and tissue damage (Muoneke and Childress 1994; Cooke et al. 2001) and can make it easier to release fish without removing them from the water. Natural bait is typically associated with more deeply hooked fish that require longer hook removal times (Nelson 1998; Siewert and Cave 1990; Cooke et al. 2001). Not surprisingly, deeply hooked striped bass have higher mortality rates than striped bass hooked in the mouth (Diodati and Richards 1996; Nelson 1998; Millard et al. 2003, 2005). Circle hooks are well suited for use with natural bait as they typically hook fish in the mouth, causing less tissue damage and allowing for quicker release (Cooke and Suski 2004). A review of circle hooks studies found that mortality rates were about 50% less, overall, for fish captured on circle hooks than on J-style hooks (Cooke and Suski 2004).

Several studies have demonstrated that survival rates are higher when deeply embedded hooks are left in the fish by cutting the line as opposed to removing them (Mason and Hunt 1967; Schisler and Bergersen 1996; Fobert et al. 2009). Wilde and Sawynok (2009) assessed the potential effects of leaving hooks embedded in 27 species of angler-caught fish and determined that recapture rates were 18% higher for fish in which the hook was not removed compared to those in which the hook was removed. The coating of the hook left embedded in the fish could influence mortality of striped bass. A laboratory experiment evaluating hook rejection in striped bass found that hook plating material significantly affected striped bass mortality. After 120 d, 20% of striped bass embedded with hooks coated with tin/cadmium died, 3% of fish embedded with bronze or nickel coated hooks died, and no mortality was observed when stainless steel hooks were used (May 1992). Hooks can be shed relatively quickly in some species. Tsuboi et al. (2006) estimated that the average time for hook evacuation was 53 d for whitespotted char *Salvelinus leucomaenis*, and Fobert et al. (2009) found that 71% of bluegill *Lepomis macrochirus* expelled hooks embedded in their esophagus within 10 d.

Although striped bass mortality has not been directly related to depth of capture, it likely contributes to the high mortality experienced by striped bass angled from warm water. Depth of capture has been positively related to mortality in several freshwater fish species (Schreer et al. 2009) and marine species (see review by Bartholomew and Bohnsack 2005). During the summer, striped bass are restricted to deeper portions of stratified reservoirs, especially in southern impoundments where suitable temperature and dissolved oxygen may be confined to a thin layer of the metalimnion or localized refuges (Coutant 1985; Coutant 2013, this volume; Rice et al. 2013, this volume), or absent altogether, forcing fish into warm water in the lower epilimnion (Thompson and Rice 2013, this volume). Striped bass restricted to these habitats are often above thermal preferences; are in poor condition due to food limitation, cessation of feeding, or separation from prey resources; and, after enduring the physiological stress of capture, are released into surface waters with even higher temperatures. Additionally, striped bass angled from these deeper waters almost certainly experience the effects of depressurization, even if the effects are not apparent. Physoclistous fish species caught from depths greater than 5 m exhibit signs of barotrauma (Feathers and Knable 1983; Schreer et al. 2009) and physiological disturbance (Morrissey et al. 2005). Injuries suffered during decompression can be severe enough to directly lead to mortality or may lead to mortality indirectly (e.g., not allowing fish to return to capture depths). Many anglers attempt to mitigate the effects of depressurization by puncturing the swim bladder (i.e., "fizzing") to allow the fish to return to capture depths. Forty percent of the clubs we surveyed that had suggested handling practices advised fizzing striped bass with distended swim bladders. Although fizzing has increased survival of fish in some studies (e.g., Feathers and Knable 1983), a recent meta-analysis of 17 studies concluded that there was little evidence that fizzing increased fish survival and that fizzing should be discouraged (Wilde 2009). At this time, it is unknown whether fizzing would increase survival of striped bass angled from deep water; therefore, fisheries management agencies should discourage catch-and-release angling of striped bass caught from deep water (>10 m).

There have been few studies of catch-and-release mortality of hybrid striped bass. Walters (1997) found that hooking mortality of hybrid striped bass in experimental ponds was greater for fish hooked at high (24°C or 26°C) water temperatures (40%) versus low (14°C) water temperatures (7%) and greater for esophagus-hooked fish (55%) as opposed to jaw-hooked fish (19%). Childress (1989) reported that catch-and-release mortality in hybrid striped bass was greater among fish captured in summer (29%), compared to winter (1%), and during summer was greater among fish captured with natural baits (75%) than with artificial baits (10%). However, his summer sample size was small (combined N for both bait types = 14). Based on temperatures reported by Childress (1989), we used our predictive equation to estimate catch-and-release mortality in hybrid striped bass. Our model predicted winter catch-and-release mortality to be 12% (assuming all fish were captured with artificial baits). In summer, catch-and-release mortality was predicted to be 79% for fish captured with live baits and 32% for fish captured with artificial baits. Our model is in qualitative agreement with Childress's (1989) results, but it appears to substantially overestimate mortality in hybrid striped bass captured with artificial baits. Additional studies of catch-and-release mortality in hybrid striped bass would be necessary either to establish a predictive relationship or establish the utility of our model.

Most studies of catch-and-release mortality, including those for striped bass and hybrid striped bass reviewed herein, have been conducted in the field. Such assessments have the benefit of allowing researchers to study the responses of fish in natural settings; however, these studies generally suffer interactions between variables and confounding of effects of variables endogenous and exogenous to the study. As an example, we used S-Plus (MathSoft 1999) to develop a classification tree (Breiman et al. 1984) for striped bass hooking mortality based on the data of Hysmith et al. (1994). This tree is analogous to a dichotomous key and, based on season, bait type, and fish length, allows one to predict the proportion of striped bass mortality due to each factor (Figure 4). Based on this analysis, the greatest determinant of mortality was the season (i.e., water temperature) in which a fish was captured. In spring and summer, fish length was more important than bait type, but the opposite result occurred in fall and winter. Thus, there was evidence of an interaction between these variables that precluded any simple interpreta-

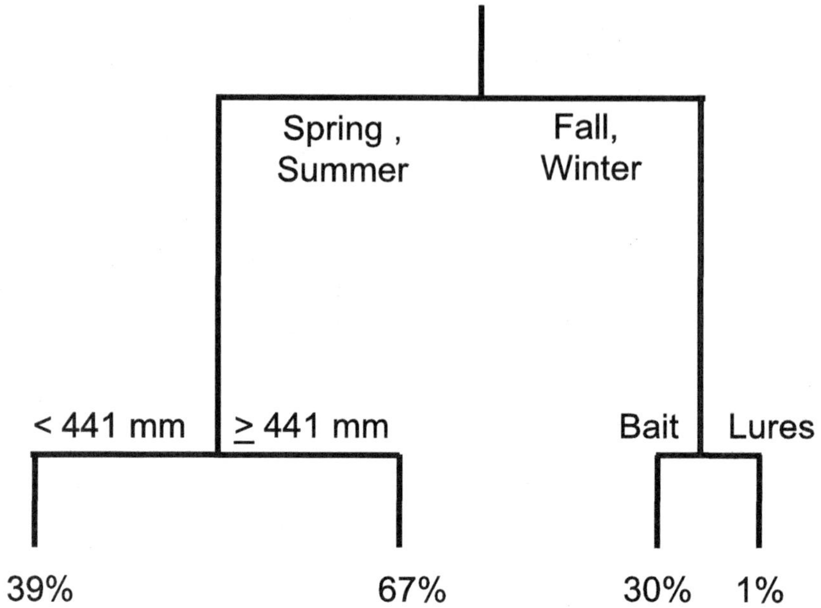

FIGURE 4. Regression tree showing probability of mortality among caught-and-released striped bass based on season of capture, length, and bait type. This figure is based on the data of Hysmith et al. (1994).

tion. Further, because larger fish inhabit deeper waters, bait type, fish length, temperature, and barotrauma, all are confounded. Season of capture (water temperature), per se, may not be the most important determinant.

Agency views on striped bass and hybrid striped bass catch-and-release mortality

Despite the widespread occurrence of catch-and-release mortality in striped bass and hybrid striped bass fisheries and occasional acknowledgment of its importance, most state agencies viewed this mortality as a social issue. This may be a consequence of the view held by many states that catch-and-release mortality is a small proportion of total annual mortality. However, most agencies have little idea of what proportion of annual mortality is attributable to catch-and-release fishing. We suggest that potential biological problems arising from excess catch-and-release mortality are not fully appreciated.

It has been known for more than a decade that angler gear choice and behavior can affect survival of caught and released striped bass, yet not all agencies provide handling guidelines for striped bass anglers or publicize the high mortality of fish angled from warm water. Forty-three percent of states that responded to our survey do not publicize efforts to limit catch and release during periods of high mortality or provide anglers with suggested handling and release practices; however, most states do have general guidelines for catch-and-release practices (Pelletier et al. 2007). Because anglers likely rely on state agencies to provide them with information about their fisheries, it is important such information is available and clearly conveyed.

Striped bass catch-and-release guidelines should advise anglers of responsible catch-and-release practices, recommend terminal tackle choices that can reduce mortality, and advise anglers of the high mortality rates associated with releasing striped bass angled from warm and deep water. Catch-and-release guidelines for striped bass should include recommendations to limit landing and handling time. Use of the proper weight rods, line, and terminal tackle will allow anglers to land and release fish quickly. Use of barbless single hooks on artificial lures can reduce handling time, and tissue damage, and employing circle hooks with natural bait will reduce the number of deeply hooked fish. Anglers should be advised to cut the line rather than attempt to remove deeply embedded hooks. Attempting to remove deeply embedded hooks can lead to increased tissue damage and handling time. Cadmium and tin plated hooks should be avoided, especially when fishing natural bait that is more likely to be deeply ingested requiring the line to be cut and the hook left embedded in the fish. Reducing or eliminating air exposure by leaving the fish in the water will allow for the best chance of survival. Anglers should be discouraged from catch-and-release angling of striped bass caught from deep (>10 m) or warm water. Pelletier et al. (2007) provides an excellent review of catch-and-release guidelines for those interested in developing or updating their suggested handling practices for catch-and-release striped bass anglers.

Striped bass fishing club survey

Our survey of striped bass fishing clubs revealed certain practices that can result in substantial catch-and-release mortality. Fifty percent of striped bass clubs believed that catch-and-release mortality during summer posed a threat to striped bass fisheries, a belief shared by the national striped bass conservation coalition. Nevertheless, most of the responding clubs conducted tournaments during the summer, nearly three-fourths of these clubs allowed anglers to cull fish, and about three-fourths reported that fish were released by at least some anglers, a practice that has increased in prevalence during tournaments in the past decade.

Striped bass captured and held during tournaments can be released with excellent survival rates (Bettinger et al. 2005; Young and Isely 2006). However, survival is considerably reduced in summer (13%), compared to spring (100%) and fall (87%), which is consistent with our predictive model. Consequently, potential fishery effects of tournaments, including culling and capture and release of fish that are not retained, cannot be dismissed. Whether tournaments impact striped bass populations would depend, in large part, on the proportion of the total angling effort they represent and the season in which the tournaments are conducted.

Mortality model

Our mortality model included catch-and-release mortality of sublegal fish sizes, which are not often included in estimates of natural mortality. Low telemetry-based estimates of natural mortality in freshwater striped bass populations, ranging from 0.09 to 0.16 (Hightower et al. 2001; Young and Isely 2004; Thompson et al. 2007), may have created the perception that catch-and-release mortality is low. However, natural mortality estimates in these studies were based on legal-sized (>473 mm total length) striped bass (Thompson et al. 2007; Hightower et al. 2001) or large striped bass (>3.4 kg) in a lake with no minimum length limit (Young and Isely 2004). In some reservoirs, the vast majority of released fish can be below minimum length limits, and catch-and-release mortality of those individuals would not be accounted for in natural mortality estimates based on legal-sized fish. For example, creel data that included number of released fish greater than the minimum length limits were available for three reservoirs (Lake Norman, Lake Gaston, and Lake Cumberland) used in our mortality model; in each of those reservoirs, 90% or more of the fish released during summer were below the minimum length limit and would not have been accounted for in natural mortality estimates based on fish greater than the minimum length limit.

Evidence gathered in this study demonstrates that catch-and-release mortality is an important mortality component in some reservoirs. Based on creel surveys conducted on several southeast U.S. reservoirs, we estimate that approximately 21–95% (mean = 51%) of striped bass captured by anglers during the summer are released, either due to regulatory requirements or for conservation purposes.

Given typical summer temperatures observed in these reservoirs, and using our predictive equation, we estimate that striped bass catch-and-release mortality could range from 63% to 76% (mean = 70%) if all fish were captured on natural baits and 31–48% (mean = 38%) if all fish were captured on artificial baits. When release rates are considered with mortality rates, the proportion of total catch lost due to catch-and-release fishing could range from 16% to 60% (mean = 36%) if all fish were captured on natural baits and 8–31% (mean = 19%) if all fish were captured on artificial baits. These predicted losses exceed losses due to harvest on two reservoirs.

Catch-and-release practices appear to be increasing in prominence. Among the striped bass fishing clubs that responded to our survey, most indicated that they promoted catch-and-release fishing for striped bass and observed an increased prevalence of catch and release in the past decade. It would be reasonable to expect a similar increase in catch-and-release angling among the general fishing population, as has been observed in other fisheries, such as those for largemouth bass *Micropterus salmoides* (e.g., Allen et al. 2008). Indeed, a review of striped bass entered into angler recognition programs in Virginia and Tennessee (the only states with angler recognition programs we found that collected information on whether or not striped bass were harvested) shows such an increase in proportion of fish released. Between 1995 and 2002, only 8% of striped bass entered into the Virginia angler recognition program were released alive, whereas between 2003 and 2009, 30% of fish entered in to that program have been released alive. Similarly, during 2003–2008, 58% of all striped bass entered into the Tennessee angler recognition program have been released alive. Therefore, catch-and-release mortality, which may already be substantial in some waters, is likely to become a more important source of mortality in the future.

Poor nutritional status likely influences high catch-and-release mortalities in summer. Summer mortality of striped bass has been observed in a number of reservoirs (e.g., Matthews 1985; Zale et al. 1990; Young and Isely 2004; Thompson et al. 2007), presumably as result of high temperatures and low dissolved oxygen concentration (Coutant 1985). Such mortality may represent a substantial portion of annual natural mortality in freshwater striped bass populations (e.g., Young and Isely 2004) and appears, ultimately, to be due to a cessation of feeding or separation from prey resources, which results in death by starvation at a time of high metabolic rate due to warm water temperatures (Lochmiller et al. 1989). Environmental conditions that result in reduced feeding may exacerbate catch-and-release mortality. Ridley et al. (1997) examined several physiological responses to catch and release in fed and unfed striped bass and suggested that recovery from catch and release may be related to the nutritional condition of the fish. Because fishing pressure is greatest in summer (e.g., Young and Isely 2004; Thompson et al. 2007), the likelihood that a fish will be captured and released is greatest during the very season in which it may be least likely to recover from capture effects.

Conclusions

Catch-and-release mortality of striped bass and hybrid striped bass is primarily related to water temperature and bait type (i.e., location of hooking). Given information on water temperature and bait type, catch-and-release mortality of striped bass can be fairly accurately predicted (Wilde et al. 2000; this study), but it is as yet unclear whether similar predictions for hybrid striped bass can be made. Because studies of striped bass and hybrid striped bass catch-and-release mortality have been conducted in the field, the exact relationship between mortality and various predictor variables, especially temperature, is unknown. Seasonal differences in water temperature affect fish habitat use, particularly depth, by different ages of fish, which may result in seasonal differences in angler behavior and gears used, which may affect mortality. Laboratory studies (e.g., Wilde and Pope 2008) would allow the effects of potential interactions and confounding variables to be minimized or eliminated.

Currently, most fishery management agencies do not believe that catch-and-release mortality has a negative impact on striped bass and/or hybrid striped bass fisheries. Catch-and-release mortality likely has little biological impact in fisheries where few fish are released or total catch is small relative to the population size. However, our modeling exercise shows that high catch-and-release rates for striped bass, particularly in the summer when release mortality is greatest, has the potential to negatively impact fisheries. Given the growing prevalence of catch-and-release fishing for striped bass and hybrid striped bass, the relationship between catch-and-release mortality and water temperature, and the possibility that fish captured in summer may, based on nutritional condition, be less capable of surviving capture and release, it appears that some management agencies may be underestimating the magnitude of the biological problems presented by catch-and-release fishing for striped bass and hybrid striped bass.

Acknowledgments

Marcy Anderson, Weston Houck, Scott Lamprecht, Brian McRae, Kirk Rundle, and Dan Wilson provided striped bass creel data from their respective states. We thank the state representatives that participated in our online survey on the importance of striped bass catch-and-release mortality. We thank the striped bass fishing clubs that responded to our online survey, including, Claytor Lake Striper Club, Lake Gaston Striper Club, Lanier Striper Club, Middle Tennessee Striper Association, Midlands Striper Club, North Georgia Striper Club, Oakwood Striper Club, Clark Hill Striper Club, Raystown Striper Club, Smith Mountain Striped Bass Club, Striper Kings, Tennessee Striped Bass Association, Texas Striper Association, and the Striped Bass Conservation Coalition. The manuscript benefited from the suggestions and comments of two anonymous reviewers.

References

Aalbers, S. A., G. M. Stutzer, and M. A. Drawbridge. 2004. The effects of catch-and-release angling on the growth and survival of juvenile white seabass captured on offset circle and J-type hooks. North American Journal of Fisheries Management 24:793–800.

Allen, M. S., C. J. Walters, and R. Myers. 2008. Temporal trends in largemouth bass mortality, with fishery implications. North American Journal of Fisheries Management 28:418–427.

Arlinghaus, R., S. J. Cooke, J. Lyman, D. Policansky, A. Schwab, C. Suski, S. G. Sutton, and E. B. Thorstad. 2007. Understanding the complexity of catch-and-release in recreational fishing: an integrative synthesis of global knowledge from historical, ethical, social, and biological perspectives. Reviews in Fisheries Science 15:75–167.

Axon, J. R., and D. K. Whitehurst. 1985. Striped bass management in lakes with emphasis on management problems. Transactions of the American Fisheries Society 114:8–11.

Bartholomew, A., and J. A. Bohnsack. 2005. A review of catch-and-release angling mortality with implications for no-take reserves. Reviews in Fish Biology and Fisheries 15:129–154.

Bettoli, P. W., and R. S. Osborne. 1998. Hooking mortality and behavior of striped bass following catch and release angling. North American Journal of Fisheries Management 18:609–615.

Bettinger, J. M., J. R. Tomasso, Jr., and J. J. Isely. 2005. Hooking mortality and physiological responses of striped bass angled in freshwater and held in live-release tubes. North American Journal of Fisheries Management 25:1273–1280.

Breiman, L., J. H. Friedman, R. A. Olshen, and C. J. Stone. 1984. Classification and regression trees. Chapman and Hall, New York.

Brown, R. L. 2006. Lake Richard B. Russell and Lake J. Strom Thurmond creel survey report 2005. South Carolina Department of Natural Resources, Columbia.

Caruso, P. G. 2000. A comparison of catch and release mortality and wounding for striped bass (*Morone saxatilis*), captured with two baited hook types. Massachusetts Division of Marine Fisheries, Sportfisheries Research Project (F-57-R), Completion Report for Job 12, Boston.

Childress, W. M. 1989. Hooking mortality of striped bass, white bass, and hybrid striped bass. Texas Parks and Wildlife Department, Final Report F-31-R-15, Austin.

Coggins, L. G., Jr., M. J. Catalano, M. S. Allen, W. E. Pine, and C. J. Walters. 2007. Effects of cryptic mortality and the hidden costs of using length limits in fishery management. Fish and Fisheries 8:196–210.

Cooke, S. J., and I. G. Cowx. 2006. Contrasting recreational and commercial fishing: searching for common issues to promote unified conservation of fisheries resources and aquatic environments. Biological Conservation 128:93–108.

Cooke, S. J., and C. D. Suski. 2004. Are circle hooks an effective tool for conserving marine and freshwater recreational catch-and-release fisheries? Aquatic Conservation: Marine and Freshwater Ecosystems 14:299–326.

Cooke, S. J., D. P. Philipp, K. M. Dunmall, and J. F. Schreer. 2001. The influence of terminal tackle on injury, handling time, and cardiac disturbance of rock bass. North American Journal of Fisheries Management 21:333–342.

Coutant, C. C. 1985. Striped bass, temperature, and dissolved oxygen: a speculative hypothesis for environmental risk. Transactions of the American Fisheries Society 114:31–61.

Coutant, C. C. 2013. When is habitat limiting for striped bass? Three decades of testing the temperature–oxygen squeeze hypothesis. Pages 65–91 *in* J. S. Bulak, C. C. Coutant, and J. A. Rice, editors. Biology and management of inland striped bass and hybrid striped bass. American Fisheries Society, Symposium 80, Bethesda, Maryland.

Dettmers, J. M., D. R. DeVries, and R. A. Stein. 1996. Quantifying responses to hybrid striped bass predation across multiple trophic levels: implications for reservoir biomanipulation. Transactions of the American Fisheries Society 125:491–504.

Dextrase, A. J., and H. E. Ball. 1991. Hooking mortality of lake trout angled through the ice. North American Journal of Fisheries Management 11:477–479.

Diodati, P. J., and R. A. Richards. 1996. Mortality of striped bass hooked and released in salt water. Transactions of the America Fisheries Society 125:300–307.

Feathers, M. G., and A. E. Knable. 1983. Effects of depressurization upon largemouth bass. North American Journal of Fisheries Management 3:86–90.

Ferguson, R. A., and B. L. Tufts. 1992. Physiological effects of brief air exposure in exhaustively exercised rainbow trout (*Oncorhynchus mykiss*): implications for "catch and release" fisheries. Canadian Journal of Fisheries and Aquatic Sciences 49:1157–1162.

Fobert, E., P. Meining, A. Colotelo, C. O'Connor, and S. J. Cooke. 2009. Cut the line or remove the hook? An evaluation of sublethal and lethal endpoints for deeply hooked bluegill. Fisheries Research 99:38–46.

Gravel, M., and S. J. Cooke. 2008. Severity of barotrauma influences the physiological status, postrelease behavior, and fate of tournament-caught smallmouth bass. North American Journal of Fisheries Management 28:607–617.

Gustaveson, A. W., R. S. Wydoski and G. A. Wedemeyer. 1991. Physiological response of largemouth bass to angling stress. Transactions of the American Fisheries Society 120:629–636.

Harrell, R. M. 1988. Catch and release mortality of striped bass caught with artificial lures and baits. Proceedings of the Annual Conference Southeastern Association of Fish and Wildlife Agencies 41(1987):70–75.

Hartman, K. J., and S. B. Brandt. 1995. Comparative energetics and the development of bioenergetics models for sympatric estuarine piscivores. Canadian Journal of Fisheries and Aquatic Sciences 52:1647–1666.

Hayes, W. E., and G. M. Penny. 2000. Fisheries investigations in lakes and streams – District III. South Carolina Department of Natural Resources, Annual Progress Report, F-63-5-3, Columbia.

Hayes, W. E., and G. M. Penny. 2001. Fisheries investigations in lakes and streams – District III. South Carolina Department of Natural

Resources, Annual Progress Report, F-63-6-3, Columbia.

Hayes, W. E., and G. M. Penny. 2002. Fisheries investigations in lakes and streams – District III. South Carolina Department of Natural Resources, Annual Progress Report, F-63-7-3, Columbia.

Hightower, J. E., J. R. Jackson, and K. H. Pollock. 2001. Use of telemetry methods to estimate natural and fishing mortality of striped bass in Lake Gaston, North Carolina. Transactions of the American Fisheries Society 130:557–567.

Houck, W. E. 2007. Lake Richard B. Russell and Lake J. Strom Thurmond creel survey report 2006. South Carolina Department of Natural Resources, Columbia.

Houck, W. E. 2008. Lake Richard B. Russell and Lake J. Strom Thurmond creel survey report 2007. South Carolina Department of Natural Resources, Columbia.

Houck, W. E. 2009. Lake Richard B. Russell and Lake J. Strom Thurmond creel survey report 2008. South Carolina Department of Natural Resources, Columbia.

Hysmith, B. T., J. H. Moczygemba, and G. R. Wilde. 1994. Hooking mortality of striped bass in Lake Texoma, Texas–Oklahoma. Proceedings of the Annual Conference Southeastern Association of Fish and Wildlife Agencies 46(1992):413–420.

Jarvis, E. T., and C. G. Lowe. 2008. The effects of barotrauma on the catch-and-release survival of southern California nearshore and shelf rockfish (Scorpaenidae, *Sebastes* spp.). Canadian Journal of Fisheries and Aquatic Sciences 65:286–1296.

KDFWR (Kentucky Department of Fish and Wildlife Resources). 1999. Annual Performance Report. KDFWR, Sport Fish Restoration, Grant F-50–21, Frankfort.

KDFWR (Kentucky Department of Fish and Wildlife Resources). 2002. Annual Performance Report. KDFWR, Sport Fish Restoration, Grant F-50–24, Frankfort.

KDFWR (Kentucky Department of Fish and Wildlife Resources). 2005. Annual Performance Report. KDFWR, Sport Fish Restoration, Grant F-50–27, Frankfort.

KDFWR (Kentucky Department of Fish and Wildlife Resources). 2009. Annual Performance Report. KDFWR, Sport Fish Restoration, Grant F-50–31, Frankfort.

Lochmiller, R. L., J. D. Weichman, and A. V. Zale. 1989. Hematological assessment of temperature and oxygen stress in a reservoir population of striped bass (*Morone saxatilis*). Comparative Biochemistry and Physiology 93:535–541.

Lukacovic, R., and B. Florence. 1998. Mortality rates of striped bass caught and released with artificial lures during spring on the Susquehanna Flats. Maryland Department of Natural Resources, Annapolis.

Marnell, L. F., and D. Hunsaker, I. I. 1970. Hooking mortality of lure-caught cutthroat trout (*Salmo clarki*) in relation to water temperature, fatigue, and reproductive maturity of released fish. Transactions of the American Fisheries Society 99:684–688.

Mason, J. W., and R. L. Hunt. 1967. Mortality rates of deeply hooked rainbow trout. Progressive Fish-Culturist 29:87–91.

MathSoft. 1999. S-Plus 2000 users guide. MathSoft, Seattle.

Matthews, W. J. 1985. Summer mortality of striped bass in reservoirs of the United States. Transactions of the American Fisheries Society 114:62–66.

May, E. B. 1992. An analysis of the effect of hooks on striped bass (*Morone saxatilis*) when left embedded in the pharyngeal and gastric region. Maryland Department of Natural Resources, Tidewater Administration, Annapolis.

McPhee, D. P., D. Leadbitter, and G. A. Skilleter. 2002. Swallowing the bait: is recreational fishing ecologically sustainable? Pacific Conservation Biology 8:40–51.

Millard, M. J., S. A. Welsh, J. W. Fletcher, J. Mohler, A. Kahnle, and K. Hattala. 2003. Mortality associated with catch and release of striped bass in the Hudson River. Fisheries Management and Ecology 10:295–300.

Millard, M. J., J. W. Mohler, A. Kahnle, and A. Cosman. 2005. Mortality associated with catch-and-release angling of striped bass in the Hudson River. North American Journal of Fisheries Management 25:1533–1541.

Morrissey, M. B., C. D. Suski, K. R. Esseltine, and B. L. Tufts. 2005. Incidence and physiological consequences of decompression in smallmouth bass after live-release angling tournaments. Transactions of the American Fisheries Society 134:1038–1047.

Muoneke, M. I., and W. M. Childress. 1994. Hooking mortality: a review for recreational fisheries. Reviews in Fisheries Science 2:123–156.

Nelson, K. L. 1998. Catch-and-release mortality of striped bass in the Roanoke River, North Carolina. North American Journal of Fisheries Management 18:25–30.

Pelletier, C., K. C. Hanson, and S. J. Cooke. 2007. Do catch-and-release guidelines from state and provincial fisheries agencies in North America conform to scientifically based best practices? Environmental Management 39:760–773.

Post, J. R., M. Sullivan, S. Cox, N. P. Lester, C. J. Walters, E. A. Parkinson, A. J. Paul, L. Jackson, and B. J. Shuter. 2002. Canada's recreational fisheries: the invisible collapse? Fisheries 27(1):6–17.

Rice, J. A., J. S. Thompson, J. A. Sykes, and C. T. Waters. 2013. The role of metalimnetic hypoxia in striped bass summer kills: consequences and management implications. Pages 121–145 *in* J. S. Bulak, C. C. Coutant, and J. A. Rice, editors. Biology and management of inland striped bass and hybrid striped bass. American Fisheries Society, Symposium 80, Bethesda, Maryland.

Ridley, C. A., J. J. Isley, and J. R. Tomasso, Jr. 1997. Effects of supplemental feeding on physical characteristics and physiological responses to angling of striped bass. North American Journal of Fisheries Management 17:791–794.

Schisler, G. J. and E. P. Bergersen. 1996. Post-release hooking mortality of rainbow trout caught on scented artificial baits. North American Journal of Fisheries Management 16: 570–578.

Schorr, M. S., J. Sah, D. F. Schreiner, M. R. Meador, and L. G. Hill. 1995. Regional economic impact of the Lake Texoma (Oklahoma–Texas) striped bass fishery. Fisheries 20(5):14–18.

Schreer, J. F., J. Gokey, and V. J. DeGhett. 2009. The incidence and consequences of barotrauma in fish in the St. Lawrence River. North American Journal of Fisheries Management 29:1707–1713.

Siewert, H. F., and J. B. Cave. 1990. Survival of released bluegill, *Lepomis macrochirus*, caught on artificial flies, worms, and spinner lures. Journal of Freshwater Ecology 5:407–411.

Sokal, R. R., and F. J. Rohlf. 1981. Biometry, 2nd edition. Freeman, New York.

Suski, C. D., S. S. Killen, S. J. Cooke, J. D. Kieffer, D. P. Philipp and B. L. Tufts. 2004. Physiological significance of the weigh-in during live-release angling tournaments for largemouth bass. Transactions of the American Fisheries Society 133:1291–1303.

Taylor, M. J., and K. R. White. 1992. A meta-analysis of hooking mortality of nonanadromous trout. North American Journal of Fisheries Management 12:760–767.

Thompson, J. A., S. G. Hughes, E. R. May, and R. M. Harrell. 2002. Effects of catch and release on physiological responses and acute mortality of striped bass. Pages 139–143 *in* J. A. Lucy and A. L. Studholme, editors. Catch and release in marine recreational fisheries. American Fisheries Society, Symposium 30, Bethesda, Maryland.

Thompson, J. S., and J. A. Rice. 2013. The relative influence of thermal experience and forage availability on growth of age 1–5 striped bass in two southeastern reservoirs. Pages 93–120 *in* J. S. Bulak, C. C. Coutant, and J. A. Rice, editors. Biology and management of inland striped bass and hybrid striped bass. American Fisheries Society, Symposium 80, Bethesda, Maryland.

Thompson, J. S., D. S. Waters, and J. A. Rice, and J. E. Hightower. 2007. Seasonal natural and fishing mortality of striped bass in a southeastern reservoir. North American Journal of Fisheries Management 27:681–694.

Thorstad E. B., T. F. Nesje, P. Fiske, and B Finstad. 2003. Effects of hook and release on Atlantic salmon in the River Alta, northern Norway. Fisheries Research 60:293–307.

Tomasso, A. O., J. J. Isely, and J. R. Tomasso, Jr. 1996. Physiological responses and mortality of striped bass angled in freshwater. Transactions of the America Fisheries Society 125:321–325.

Tsuboi, J., K. Morita, and H. Ikeda. 2006. Fate of

deep-hooked white-spotted charr after cutting the line in a catch-and-release fishery. Fisheries Research 79:226–230.

U.S. Fish and Wildlife Service and U.S. Bureau of the Census. 2006. 2006 National survey of fishing, hunting, and wildlife-associated recreation. U.S. Government Printing Office, Washington, D.C.

Van Horn, S. L. 2013. A brief history of inland striped bass management. Pages 1–13 *in* J. S. Bulak, C. C. Coutant, and J. A. Rice, editors. Biology and management of inland striped bass and hybrid striped bass. American Fisheries Society, Symposium 80, Bethesda, Maryland.

Vatland, S., P. Budy, and G. Thiede. 2008. A bioenergetics approach to modeling striped bass and threadfin shad predator-prey dynamics in Lake Powell, Utah–Arizona. Transactions of the American Fisheries Society 137:262–277.

Walters, D. A. 1997. Impact of hooking mortality on hybrid striped bass. Ohio Department of Natural Resources, Federal Aid in Sport Fish Restoration, Project F-69-P, Final Report, Xenia.

White, M. G., and S. D. Lamprecht. 2002. Fisheries investigations in lakes and streams – District V. South Carolina Department of Natural Resources, Study Completion Report, F-63-7-5, Columbia.

Wilde, G. R. 1998. Tournament-associated mortality in black bass. Fisheries 23(10):12–22.

Wilde, G. R., M. I. Muoneke, P. W. Bettoli, K. L. Nelson, and B. T. Hysmith. 2000. Bait and temperature effects on striped bass hooking mortality in freshwater. North American Journal of Fisheries Management 20:810–815.

Wilde, G. R., and K. L. Pope. 2008. A model for predicting survival of angler-caught and released largemouth bass. Transactions of the American Fisheries Society 137:834–840.

Wilde, G. R., and W. Sawynok. 2009. Effect of hook removal on recapture rates of 27 species of angler-caught fish in Australia. Transactions of the American Fisheries Society 138:692–697.

Wilde, G. R. 2009. Does venting promote survival of released fish? Fisheries 34:20–27.

Young, S. P., and J. J. Isely. 2004. Temporal and spatial estimates of adult striped bass mortality from telemetry and transmitter return data. North American Journal of Fisheries Management 24:1112–1119.

Young, S. P., and J. J. Isely. 2006. Posttournament survival and dispersal of adult striped bass. North American Journal of Fisheries Management 26:1030–1033.

Zale, A. V., J. D. Weichman, R. L. Lochmiller, and J. Burroughs. 1990. Limnological conditions associated with summer mortality of striped bass in Keystone Reservoir, Oklahoma. Transactions of the American Fisheries Society 119:72–76.

American Fisheries Society Symposium 80:501–519, 2013

Interactions between Striped Bass and Other Game Fish in Reservoirs

Leandro E. Miranda*
U.S. Geological Survey, Mississippi Cooperative Fish and Wildlife Research Unit Post Office Box 9691, Mississippi State, Mississippi 39762, USA

Scott W. Raborn
*LGL Alaska Research Associates, Inc.
207 Pearce Road, Pineville, Louisiana 71360, USA*

Abstract.—Competitive interactions among reservoir fishes may be pronounced because fish assemblages in these artificial environments have had little time to develop niche-partitioning strategies that alleviate negative interspecific interactions. Such interactions may at times have been intensified by introductions of predators such as striped bass *Morone saxatilis*, introduced to create additional fisheries and control pelagic clupeids. Possible interactions between existing fish assemblages and striped bass include predation and competition. While there is a perception among angler groups that predation by striped bass on coexisting game fish is significant, most studies have reported little or no predation on game fish by striped bass and have considered predation rare and inconsequential. Moreover, predation that occurs will likely be compensatory and fail to reduce overall game fish survival. Any indirect effect of striped bass predation by restricting prey-sized game fish to limited refuge sites remains unknown. Exploitative competition may be more common. Although infrequently, introduced striped bass have depleted prey resources shared with other piscivores, particularly when stocking rates have been high, when there is a high rate of natural reproduction, or when prey supply has plunged in response to environmental fluxes. Fluctuation in prey supply, associated with ordinary environmental variability, and associated time lags in prey supply and predator demand, preclude adjusting predator densities to exactly balance demand with supply. The frequency of low supply–demand ratios varies across systems and exhibits seasonal trends. Nevertheless, chronic supply–demand imbalances are manageable where the predator assemblage is at least partially controlled through stocking, harvest regulations, or both. Because of the poor state of knowledge concerning the parameters defining balance and because uncontrollable annual fluctuations preclude exact management of alternating prey levels, we suggest adjusting stocking to manage demand so that it equals the median historical prey supply. Simulating the removal of striped bass and predicting the aftermath may be the most cost-effective way to provide decision support for stakeholders involved in determining if a striped bass stocking program is beneficial to most users.

* Corresponding author: smiranda@usgs.gov

Introduction

Stocking of striped bass *Morone saxatilis* has generated controversy among anglers that target other game fish (henceforth, we use the term "game fish" to mean all species other than striped bass that are targeted by anglers). Common among many anglers is the perception that fishing was better a long time ago. This premise, coupled with the knowledge that striped bass are now present in many of their preferred lakes and that striped bass are large and voracious predators, leads to the presumption that striped bass are hindering their harvest of other game species. Typically, anglers express two primary concerns (Churchill et al. 2002). Principally, they protest that temperate basses *Morone* spp. prey on game fish, including black bass *Micropterus* spp. and crappies *Pomoxis* spp., and thus predatory interactions reduce availability of target species. Second, some anglers recognize that piscivorous game fish vie for limited prey with introduced striped bass and contend that this exploitative feeding competition reduces body condition, growth, and overall standing crop of other game fish.

The artificial and recent origins of reservoir environments and their fish assemblages potentially produce atypical spatial and temporal overlaps that may alter predator diets to include each other and/or overlap so as to exacerbate exploitative competition. Given enough time, evolutionary changes may facilitate long-term responses such as niche partitioning by means of character displacement, thereby allowing species to coexist (Noble 1986). Some niche partitioning is evident because the function of each species, in the spectrum of ad hoc assemblages, is circumscribed by their evolutionary baggage. The stability of the assemblages that result remains even less understood than that of endemic assemblages.

Negative interactions among reservoir fish species may be pronounced and diverse given the variety of fish assemblages in reservoirs. Headwater reservoirs tend to include a few ubiquitous taxa with greater potential for niche overlap and may be more susceptible to negative interactions than downstream reservoirs that include progressively more speciose assemblages that add more riverine species (Miranda et al. 2008). Also, reservoirs continuously undergo anthropogenic influences that alter species densities (Miranda and Hunt 2010). Disturbances such as periodic water level drawdowns and disconnection from river systems coupled with rapid habitat degradation could favor some species in ways that offset more gradual population changes that would normally result from competitive exclusion and predator control.

Introductions and maintenance stocking of selected piscivorous species, such as the temperate basses, including striped bass and their hybrids, are used to create additional fisheries and control pelagic clupeids. These hatchery-supported populations tend to work against normal density-dependent mechanisms within and among populations in fish assemblages (Stewart et al. 1981). If these species cohabitated naturally without stocking programs, competition (both intra- and interspecific) may limit fecundity and increase mortality to reduce population densities and their demand for limited resources. The resulting adult fish community would experience fewer negative interactions as a result of more resources per individual. Stocking may intensify these interactions by circumventing these negative feedbacks that reduce recruitment.

Likewise, dominant predators may outcompete and reduce the population densities of weaker predators resulting in fewer negative interactions. Fisheries that harvest dominant predators would tend to counter these interactions and may allow weaker competitors to persist even when prey are limited.

Furthermore, in these assemblages, one or two prey species, usually clupeids, often account for most of the forage base. Simple predator–prey systems may be unstable because the entire predator assemblage can be affected by random fluctuations in a single forage species. If the collective biomass of the predator assemblage is dependent on clupeids, it is unlikely that other prey species with far less representation could compensate in the event clupeid production failed. Conversely, in systems where predators rely on multiple forage species, one prey may fill the deficit left by another and buffer the food web against a collapse in any one prey species (Stewart et al. 1981).

Predatory and competitive interactions are routinely accepted as mechanisms that influence fish populations and communities (Matthews 1998) but have been difficult to experimentally verify. While predation and competition represent part of the central dogma upon which many management and conservation decisions are based, this dogma remains among the hardest to test because of biases in field surveys, inability to document all confounding influences, and lack of realism in controlled laboratory and field experiments. Most evidence for interactions among piscivorous fishes has been indirect, such as diet analyses, correlations between prey density and predator growth, and analyses of condition factors (Ploskey and Jenkins 1982; Muth and Wolfert 1986; Marwitz and Hubert 1997; Cavalli et al. 1998). More definitive evidence of competition among fish species has been disclosed by field experiments with amenable species (e.g., Mittelbach 1988), and by lake manipulations (e.g., Hodson et al. 1991). Nevertheless, logistical limitations often prevent system-wide control and the mobility of some species prevents using small-scale experiments to represent the natural environment. Thus, proper experimentation is lacking and progress is unlikely to occur as long as the search for competition among predators in reservoirs continues to rely on surveys that provide indirect clues.

We review available information about predatory and competitive interactions between striped bass and cohabiting game fish in reservoirs. We first examine evidence for predation by striped bass on other game fish, and if direct losses to predation may limit game fish representation in fish assemblages. We propose an alternate predation-related hypothesis that striped bass may indirectly reduce game fish production even if game fish presence in the diets of striped bass is limited. Second, we examine evidence for the presence and severity of exploitative competition. For predators to compete, they must overlap in their use of the prey supply, and this supply must be limited and in high demand. To this end, we review diet overlap among predator species, as well as studies that attempt to quantify the ratio of prey supply and predator demand. Third, we offer a simpler and inexpensive approach to assess striped bass introductions; this approach simulates the removal of striped bass from a particular reservoir and models how much the biomass of other game fish might increase if striped bass stocking was discontinued. Last, we discuss the prognosis for investigating negative interactions between striped bass and game fish in reservoirs and the management actions that could alleviate detrimental interactions.

Predation by Striped Bass on Other Game Fish

Various studies have examined the food habits of striped bass in reservoirs to, in part, identify predation on other game fish. Juvenile striped bass depend generally on a mixed diet of invertebrates, larval fish, and small cyprinids (Van den Avyle and Higginbotham 1983; Sutton and Ney 2002), whereas adults depend largely on clupeids (Table 1). When present, species such as gizzard shad *Dorosoma cepedianum*, threadfin shad *D. petenense*, and alewife *Alosa pseudoharengus* typically represent well over 90% of the diet of adult striped bass by number. Representation of game fish in the diet is minor, mostly less than 5% (Kohler and Ney 1982; Gilliland and Clady 1984; Filipek and Tommey 1987; Matthews et al. 1988; Moore 1988; Thompson et al. 2005). Game fish species reported in the diet include sunfishes *Lepomis* spp., crappies, black bass, white perch *Morone americana*, yellow perch *Perca flavescens*, and rainbow trout *Oncorhynchus mykiss*. Miranda et al. (1998) combined a diet analysis, consump-

TABLE 1. Percentage composition (by number) of clupeids, other nongame fish, and game fish in stomachs of striped bass in various U.S. reservoirs.

Reservoir	Clupeids	Other nongame fish	Game fish	Source
Santee Cooper, South Carolina	95	3	<2	Stevens (1958)
Keystone, Oklahoma	78	18[a]	4	Combs (1978)
Watts Bar, Tennessee	96	3	<1	Higginbotham (1979)
Claytor, Virginia	92	0	8	Kohler and Ney (1980)
Powell, Utah	94	6	0	Persons and Bulkley (1982)
Hamilton, Arkansas	93	5	<2	Filipek and Tommey (1984)
Texoma, Oklahoma–Texas	94	3	3	Matthews et al. (1988)
Smith Mountain, Virginia	89	9	<2	Moore (1988)
Norris, Tennessee	91	5	4	Peterson et al. (1993)
Norris, Tennessee[b]	93	2	5	Miranda et al. (1998)
Watts Bar, Tennessee[b]	100	0	0	Miranda et al. (1998)
Weiss, Alabama	93	6	<1	Slipke et al. (2002)
Badin, North Carolina	92	4	4	Thompson et al. (2005)
Norman, North Carolina	97	2	1	Thompson et al. (2005)
Lewis Smith, Alabama	98	1	<1	Shepherd and Maceina (2009)

[a] Primarily silversides.
[b] Percentages reported on a weight basis.

tion estimates from a bioenergetics model, estimates of striped bass population density, and estimates of population densities to estimate the fractions of game fish populations consumed by striped bass. Their results suggested that a 4% consumption rate of sunfishes represented about 6% of the sunfish biomass in Norris Reservoir, Tennessee. No other game fish species was observed during the diet analysis despite the large sample size ($N = 590$ striped bass with food and 1,135 total). Using similar methodology and sample size ($N = 503$ with food and 718 total), Shepherd and Maceina (2009) estimated that black basses, sunfishes, and crappies accounted for 11% by weight of the annual adult striped bass diet in Lewis Smith Lake, Alabama, which is the highest estimate found in the peer-reviewed literature.

Striped bass adults for the most part consume clupeids 50–175 mm total length (TL), despite spatial and morphological (mouth gape) access to smaller and larger prey, including game fish (Kohler and Ney 1982; Filipek and Tommey 1987; Cyterski and Ney 2005; Vatland et al. 2008). Also, larger striped bass may feed on larger clupeids and a wider range of clupeid sizes (Combs 1979; Slipke et al. 2002). Nevertheless, they are also capable of preying on adult game fish. In Norris Reservoir, striped bass consumed sunfishes 53–203 mm TL (mean = 118; Miranda et al. 1998). In Lewis Smith Lake, striped bass consumed black bass ranging 25–305 mm TL (mean = 178 mm) and sunfishes and crappies ranging 71–290 mm TL (mean = 127 mm; Shepherd and Maceina 2009). Thus, striped bass are capable of consuming game fish within the lower range of sizes targeted by recreational fisheries, even though they rarely do so.

Although food habit studies have concluded that striped bass predation does not affect game fish populations because percent occurrence of game fish in the diet is low, modest representation in the diet may correspond to appreciable total consumption if the striped bass population is large. The metric of concern is the abundance of game fish removed through predation relative to game fish density.

Predation as a regulator of population density cannot be fully evaluated without assessing compensatory processes (McFadden 1977; Goodyear 1980). The potential for a source of mortality to be compensatory (i.e., causing no or less than a proportionate increase in overall mortality) as opposed to additive (i.e., causing a proportionate increase in overall mortality) is a function of the survival-density relationship for the population. If survival is density-dependent and decreases with increasing density, other sources of mortality (e.g., starvation, disease, harvest) may be reduced in response to predation, potentially resulting in no net change or even an increase in population size. Raborn et al. (2003) concluded that in Norris Reservoir, survival of black basses decreased with increasing density. As a result, predation was projected to be a compensatory source of mortality that failed to reduce overall survival, and thus the loss of individuals of these species to predation was not inevitably a limiting factor.

The loss of game fish to direct predation by striped bass has never been shown to be a limiting factor despite numerous investigations. But no study has focused on the indirect effects of striped bass predation on game fish. For reef fish species, survival is often density-dependent in the presence of predators and density independent in their absence (Forrester 1995; Hixon and Cari 1997; Steele 1998; Forrester and Steele 2000). When predators are around, reef fish density is limited by the availability of refuge sites. Moreover, competition among prey may be controlled by the presence of predators when prey is restricted to limited refuges (Mittelbach and Chesson 1987). So the argument can be made that if smaller game fish species or juveniles of larger species are restricted to littoral refuges to avoid predation by the more pelagic striped bass, then their growth and survival, and therefore their carrying capacity, is partly controlled by the interaction of striped bass presence and the availability of refuge sites. This possibility would be greatest in upper tributary reservoirs that typically have steep and narrow littoral zones, such as Norris Reservoir. This impoundment has limited littoral habitat, little to no sheltering aquatic vegetation, deteriorating submerged terrestrial vegetation, and a striped bass stocking program (see Miranda et al. 1998 for a detailed description of this reservoir). The pronounced density-dependent survival observed for black basses in Norris Reservoir (Raborn et al. 2003) may have resulted from striped bass and older cannibalistic conspecifics limiting juvenile largemouth bass *M. salmoides* to the maximum densities supported by refuge sites. This phenomenon could occur even if few or no game fish were found in a diet survey of striped bass and would create density-dependent survival in the game fish species that would negate any direct losses to predation. Perhaps survival and growth of game fish would be less density-dependent if juveniles had access to more habitats in the absence of striped bass. While juvenile centrarchids tend to be primarily littoral-oriented, they periodically move away from the littoral as part of seasonal or ontogenetic cycles. Moreover, striped bass may restrict older centrarchids more toward the littoral zone creating greater spatial and temporal overlap with, and thereby more cannibalism on their juveniles, which in turn would promote the situation just described above. This level of ecological complexity surrounding the indirect effects of striped bass on other game fish is yet to be studied.

Predation by Striped Bass on Nongame Fish Species

There is limited information available about predation by striped bass on less abundant nongame species, as most studies have focused on predation on dominant prey species (clupeids) and game species. In the limited number of studies available, nongame species are often not identified but rather grouped into broad taxa (e.g., minnows or other fish), unless identification to the species level is clear-cut. This practice likely reflects low representation of several nongame species in the diet of striped bass. The limited number of nongame species reported in the diets of reservoir striped bass include various cyprinids, silversides *Labidesthes* spp., darters *Etheostoma* spp., freshwater drum *Aplodinotus grunniens*, and common carp *Cyprinus carpio*, but their representation in the diet is usually small (Table 1). This minor level of representation is possibly related to low habitat overlap between the primarily pelagic striped bass and primarily littoral species that colonize reservoirs (Fernando and Holčík 1982). Low levels of consumption of nongame species by striped bass could potentially have large effects on small nongame populations, but no such investigations have been pursued, perhaps because no existing or impending consequences have been evident. However, Stewart and Burrell (2013, this volume) suspected that unanticipated reproduction by introduced striped bass populations in reservoirs of the Colorado River system may be negatively impacting two endangered, nongame fish species—razorback sucker *Xyrauchen texanus* and bonytail *Gila elegans*.

Competition between Striped Bass and Other Game Fish

Monitoring programs and focused studies have been designed to find direct losses to predation and have been inadequate for capturing at least the potential for indirect effects. The opposite is true with respect to competitive interactions between striped bass and other game fish. Direct interference competition would involve aggressive behavior by striped bass that precluded access to some resource by other game fish, whereas indirect competition from striped bass depleting a common resource would be called exploitative. All evidence regarding striped bass competition in the literature pertains to the latter. A hypothetical scenario of interference competition would be if smallmouth bass *M. dolomieu* were excluded from offshore shoals by the presence of large striped bass. We encountered this argument from anglers several times while studying Norris Reservoir. Exploitative competition for a common food resource is more amenable to quantification by comparing the type and amount of prey consumed by striped bass and other game fish. Not surprisingly, all published studies investigating competition from striped bass focus on exploitative feeding competition, which is still difficult to demonstrate from field surveys. Exploitative competition among species requires the simultaneous finding of substantial diet overlap, limited supply, and high demand (Wiens 1977, 1993). A true measure of competition may never be obtained for highly mobile species such as striped bass and many other game fish, as the intensity of competitive interactions is likely to vary annually along with diet overlap, prey supply, and predator demand. In most reservoirs, the probability of exploitative competition (both inter- and intraspecific) is expected to vary seasonally as well (Raborn et al. 2004). Because of a dearth of studies on interference competition, we reviewed only the existing evidence concerning exploitative competition.

Diet overlap

Diet overlap indexes are used to assess trophic competition and resource partitioning among sympatric fish (Gotelli and Graves 1996) but are not infallible indicators. For species to compete, they must overlap in their use of resources. Low overlap could, however, result from past competition that has caused predators to switch to alternative prey (i.e., ghost of competition past; Connell 1980). Also, competition can still occur despite low diet overlap when a prey species represents a small portion of the diet of an abundant predator but a substantial diet portion of a rare predator. Conversely, significant overlap may occur with no competition if predators are sharing abundant prey resources. Thus, diet overlap alone is insufficient to test the presence or severity of exploitative competition (Colwell and Futuyma 1971), and accompanying information about how well predators are meeting their resource needs is required to interpret overlap indices. Moreover, the extent of diet overlap in adult fish generally varies seasonally as prey availability and prey and predator distributions change (Raborn et al. 2004). Thus, while information about diet overlap can be useful for interpreting species interactions, it must be interpreted with caution and at the appropriate temporal resolution.

Seasonal trends in diets and diet overlap were shown in Norris Reservoir, Tennessee. There, alewife and *Dorosoma* spp. (threadfin shad and gizzard shad) were the dominant prey items for most predators during most seasons, although largemouth bass also relied substantially on sunfishes and cannibalized black bass (Raborn et al. 2004). Striped bass and walleye *Sander vitreus* exploited *Dorosoma* spp. heavily in winter but increased their consumption of alewife in other seasons, especially summer. This finding was consistent with the thermal squeeze hypothesis (Coutant 2013, this volume) if striped bass, walleye, and alewife were restricted to common pelagic depths of suitable temperature and dissolved oxygen during summer reservoir stratification. Black crappie *Pomoxis nigromaculatus* diets consisted almost entirely of *Dorosoma* spp., except during spring when their predation on alewife increased. There was substantial diet overlap among striped bass, largemouth bass, black crappie, and walleye over all seasons, except between striped bass and largemouth bass in summer. When the results from the diet overlap analysis and consumption patterns were combined, summer exhibited the least potential for interspecific competition (Raborn et al. 2004). Moore (1988) found similar seasonal trends in consumption of alewife and gizzard shad by striped bass, largemouth bass, and walleye in Smith Mountain Lake, Virginia.

In Lewis Smith Lake, Alabama diet overlap between striped bass and black bass (largemouth bass and spotted bass *M. punctulatus*) was generally low, but varied seasonally (Shepherd and Maceina 2009). Highest diet overlap occurred in April and June and was significant only between striped bass and spotted bass. Striped bass demand for shad was high, whereas black bass demand was low as these predators consumed more crayfish, crappies, and sunfish. Thus, the potential for competition between striped bass and black bass was low. Consumptive demand was partly divided temporally among the species as striped bass prey consumption was higher in fall and early winter, while black bass consumption was higher between spring and early fall. Relative weights of black bass did not decline, and the frequency of black bass empty stomachs did not increase during or after periods of high overlap with striped bass, suggesting that sufficient prey was available.

At Harlan County Reservoir, Nebraska, Olson et al. (2007) compared diet overlap of white bass *Morone chrysops*, hybrid striped bass (white bass × striped bass), and walleye, as well as their vertical distribution in the water column. All three species consumed similar prey (i.e., gizzard shad and Chironomidae), and diet overlap was high during all months. White bass were consistently located within the upper 3 m of water, whereas the vertical distribution of hybrid striped bass and walleyes varied from the surface to 10 m deep. Spatial overlap was therefore not as high as dietary overlap and was variable among species and months. Although diet overlap was high, Olson et al. (2007) reasoned that resource partitioning (i.e., different feeding locations and different sizes of gizzard shad eaten) reduced the negative interactions among the three predators.

Matthews et al. (1992) studied food and habitat use by juvenile largemouth bass, spotted bass, striped bass, and white bass in Lake

Texoma, Oklahoma–Texas. Diet overlap was highest between the two species of *Morone* and between the two species of *Micropterus* and was also substantial between largemouth bass and striped bass and between largemouth bass and white bass. This overlap suggested potential trophic competition between juveniles, but the two genera occupied substantially different habitats and spatial segregation seemed sufficient to prevent competition. Juvenile striped bass overlapped more in habitat and diet with juvenile white bass than with largemouth bass, but competition between the two *Morone* species could not be inferred with confidence.

Sutton and Ney (2002) also examined the food habits of juvenile striped bass and largemouth bass. As these species increased in length in Smith Mountain Lake, food for both species shifted from zooplankton to insects and then to fish as the principal prey items. Striped bass 50–100 mm TL ate primarily aquatic insects and cyprinids, whereas largemouth bass in the same length range ate aquatic and terrestrial insects. Age-0 alewives became the predominant prey for larger striped bass, whereas darters and age-0 sunfishes constituted the primary fish prey of largemouth bass regardless of body length. Except for fish smaller than 50 mm during summer, diet overlap was low and insufficient to suggest extant trophic competition. Although temporal and length-related segregation by prey size was sufficient to minimize trophic overlap, limitations in these resources may cause interspecific competition in smaller individuals.

Diet overlap is likely associated with prey supply and seasonal habitat distributions of both prey and predators. Generally, but not exclusively, striped bass are largely pelagic in distribution, whereas centrarchids are more littoral (but not exclusively) and *Dorosoma* have a wide distribution, including pelagic and littoral areas. Alewives, however, tend to be pelagic. For the most part, neither prey nor predators are entirely limited to a narrow selection of habitats. An exception would be the thermal squeeze of striped bass during summer (Coutant 2013; but see Thompson and Rice 2013, this volume). Prey limitations are probably compensated by the ability of either prey or predators to move between littoral and pelagic areas. Such movements are facilitated in canyon-type reservoirs because they are generally narrow without extensive open-water areas. In shallow reservoirs with more open-water habitat, diet overlap might be expected to be lower. Nevertheless, the argument that striped bass and game fish do not compete for prey resources because they occur in separate but adjacent habitats is weak even if not all prey are equally available to all predators all of the time.

Certainly, interference competition may have caused the partitioning of habitats, while exploitative competition is ever-present due to predator and prey mobility. Alternately, it is clear that striped bass can be separated from other predators and some prey if suitable habitat for striped bass is reduced vertically and horizontally (Rice et al. 2013, this volume) during summer. However, Thompson and Rice (2013) report striped bass surviving and growing in reservoirs where suitable habitat is not available during summer. Apparently, striped bass will tolerate warmer water than they prefer if the forage base is sufficient to offset the added metabolic costs associated with higher temperatures. Again, exploitative competition from other users of clupeids would work in tandem with the thermal squeeze hypothesis to stress striped bass, even though they are separated spatially from those other users. The reliance of striped bass and game fish on clupeid species, together with the seasonality of clupeid availability, suggests that substantial decreases in clupeid production, such as that associated with annual environmental fluctuations, could intensify diet overlap and consequently competition.

Supply-demand

Several studies have attempted to capture the simultaneous requirement of resource overlap and prey insufficiency to gain perspective on competition as an ecological mechanism limiting game fish production. These studies move beyond the simple determination of diet overlap and attempt to quantify the availability of prey (termed supply), which is restricted by prey density, behavior, morphology, and spatial distribution (Ney 1990). In addition, this supply is compared with predator demand, which is usually assumed to be equal to total consumption of prey over a period of time. The difficulties in estimating supply, demand, and the ideal ratio were thoroughly reviewed by Ney (1990), but few improvements to the current state of knowledge have occurred since.

Diet and consumption studies have shown that the greatest demand among reservoir predators is for clupeids, which, as it turns out, also represent the greatest prey supply. Clupeids have greater biomass and production in reservoirs than other prey fishes (at least in reservoirs with striped bass stocking programs), and such predominance probably influences prey choice (Lyons 1987; Sih and Moore 1990). Einfalt and Wahl (1997) suggested that the escape behavior of clupeids might render them more susceptible to predation than sunfishes; they found bluegill *L. macrochirus* to rely on maneuverability for escape, whereas gizzard shad relied on schooling. Morphologically, alewife and threadfin shad are more available than sunfishes because they possess soft rays instead of spines (Einfalt and Wahl 1997) and generally do not grow beyond the gape-width limitations of most predators. While cyprinids are similar to alewife and threadfin shad morphologically, their biomass is usually a fraction of that for clupeids.

Spatial distributions of predators and prey also affect prey availability. At Norris Reservoir, predation on alewife by striped bass and walleye increased during summer, a time when both predators may have been spatially restricted (Raborn et al. 2007). During summer reservoir stratification, striped bass and walleye can be vertically confined between heated surface waters above and hypoxic waters below (i.e., the thermal squeeze hypothesis; Coutant 2013). Alewife, a pelagic coolwater species, may be more spatially available to these predators, which are hindered in their potential foraging movements.

As the forage base of many reservoirs is dominated by one to three clupeid species, these simple prey–predator systems may be unstable because the entire predator assemblage can be affected by random fluctuations in abundance of a single forage species (Stewart et al. 1981). Stability may increase with prey diversity (Petchey 2000), as multiple weak predator–prey interactions create a more stable community than fewer strong interactions (McCann 2000). Thus, when clupeid supply is low, alternative prey may not be numerically sufficient to support the predator biomass and therefore may not offset the adverse effects of exploitative competition among predators.

Shortages in prey supply have been inferred from short-term population characteristics of predators. If prey is insufficient to meet demand, expected predator responses include diet shifts (Hodson et al. 1991), reduced consumption (Rice and Cochran 1984), slower growth (Muth and Wolfert 1986), low condition (Marwitz and Hubert 1997; Porath and Peters 1997), and diminished predator biomass (Ploskey and Jenkins 1982; Gustaveson and Blommer 2013, this volume). Although these population characteristics do not confirm a response to prey shortages, inferences to that effect are strengthened when they are related to concurrent or prior changes in prey abundance. Morris and Follis (1979) attributed changes in shad population structure and subsequent declines in striped bass growth to excessive predation on shad leading to prey depletion. Such response could have been exacerbated by declines in prey supply associated with natural cycles in recruitment. They recommended reductions in striped bass stocking. Striped bass stocking and uncontrolled natural reproduction in Lake Powell (Arizona–Nevada) coincided with reductions in shad abundance and decreases in striped bass body condition (Gustaveson and Blommer 2013). A similar response was seen in lower Colorado River reservoirs (Stewart and Burrell 2013, this volume). In reservoirs of the

Santee–Cooper River system, where striped bass reproduced naturally, Stevens (1979) reported that striped bass apparently depleted shad abundance and many lost body condition because they did not switch to another prey source. Conceivably, such depletion affected coexisting game fish that depend at least partially on clupeids. However, various studies in other reservoirs have not detected these signs of insufficient prey supply (e.g., Combs 1979; Gilliland and Clady 1984; TPWD 1986; Matthews et al. 1988; Nash et al. 1988).

Instead of searching for indications of prey shortages, a limited number of studies have estimated supply, demand, and the resulting ratio. In these studies, the authors usually define prey biomass as "surplus" if it is above and beyond what is needed for recruitment to maintain current biomass levels. In Smith Mountain Lake, the mean annual surplus biomass of alewives and gizzard shad was barely adequate to meet predator demand. Total annual predator demand was 70 kg/ha for alewives and 80 kg/ha for gizzard shad (Cyterski et al. 2002). Mean annual surplus biomass was 74 kg/ha for alewives and 116 kg/ha for gizzard shad (Cyterski et al. 2003). Given annual variability in clupeid recruitment and biomass, Cyterski et al. (2003) concluded that at Smith Mountain Lake, there would be years in which the forage base was sufficient and years in which it was not.

In Lewis Smith Lake, total available supply of shad was about 47 kg/ha and shad consumption by black bass and striped bass averaged 3 and 14 kg/ha, respectively (Shepherd and Maceina 2009). Thus, shad consumption by striped bass and black bass relative to the total annual supply was more sustainable, suggesting that competition for prey was less likely. Moreover, consumptive demand was temporally segregated to some extent between black bass and striped bass as prey consumption by striped bass was higher in the fall and early winter, while prey consumption by black bass was higher between spring and early fall.

In Norris Reservoir, median annual demand by striped bass, crappies, black basses, and walleyes combined for clupeids and sunfishes combined was 271 and 182 kg/ha in two consecutive years (only prey sizes available to predators were considered supply; Raborn et al. 2007). Median supply during these years was 848 and 413 kg/ha, resulting in supply:demand ratios of 2.9:1 and 2.3:1 for the two periods. A 22-year data set for this reservoir suggested that the median annual supply was 167 kg/ha (90% confidence limit = 49–982 kg/ha), demand 109 kg/ha (51–239 kg/ha), and the supply:demand ratio 1.5:1 (0.4:1–10.4:1). Historical values of the supply:demand ratio were less than 1:1 (i.e., supply was insufficient to meet demand) 32% of the time (Figure 1).

This estimate of 32% was considered minimal for two reasons. First, demand was assumed to be equal to consumption, which is at best a minimum estimate of demand. Second, although supply:demand ratios were often greater than 1:1, they do not imply that forage conditions were adequate. Search and capture efficiency by predators is less than 100%, and supply is not balanced (i.e., supply is sufficient to meet demand with no excess) with demand at a 1:1 ratio because predators require a certain amount of available prey in excess of their demand (Ney 1990). How much in excess is a key question, and determining when this ratio is balanced is, at best, difficult. Given the annual variability in clupeid recruitment and biomass in reservoirs (Adams et al. 1982; Storck 1986; Adams and DeAngelis 1987; Hale 1996; Michaletz 1997, 1998), there will likely be years in which the forage base is sufficient and years in which it is not. Even in reservoirs where annual supply has historically been sufficient to meet annual demand, occasional prey shortages may occur seasonally (Raborn et al. 2004).

Simulating Striped Bass Removal

Modeling might assist in resolving the complex interactions between striped bass and other game fish. If striped bass were negatively impacting other game fish because of severe exploitative competition, then the prey

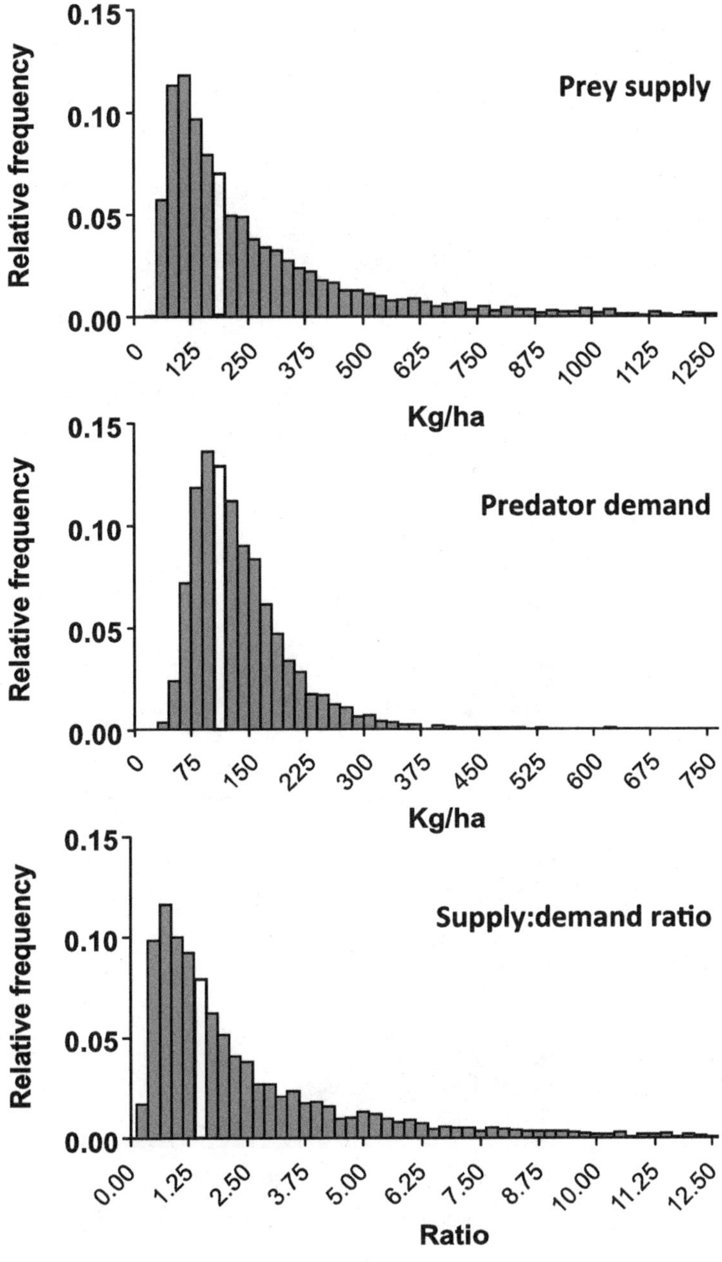

FIGURE 1. Distributions of prey supply, predator demand, and their ratio in Norris Reservoir from 1975 to 1997 (after Raborn et al. 2007). White bars indicate median values.

released from their predation if striped bass were removed would be available to other game fish to increase their collective biomass. How much would depend on how efficient game fish were in capturing the released prey (prey availability restrictions apply) and in converting the consumed prey into game fish biomass (physiological restrictions apply). Likewise, the amount of game fish eaten would be released to increase game fish recruitment or die from other causes with the probability of one outcome versus the other determined by the level

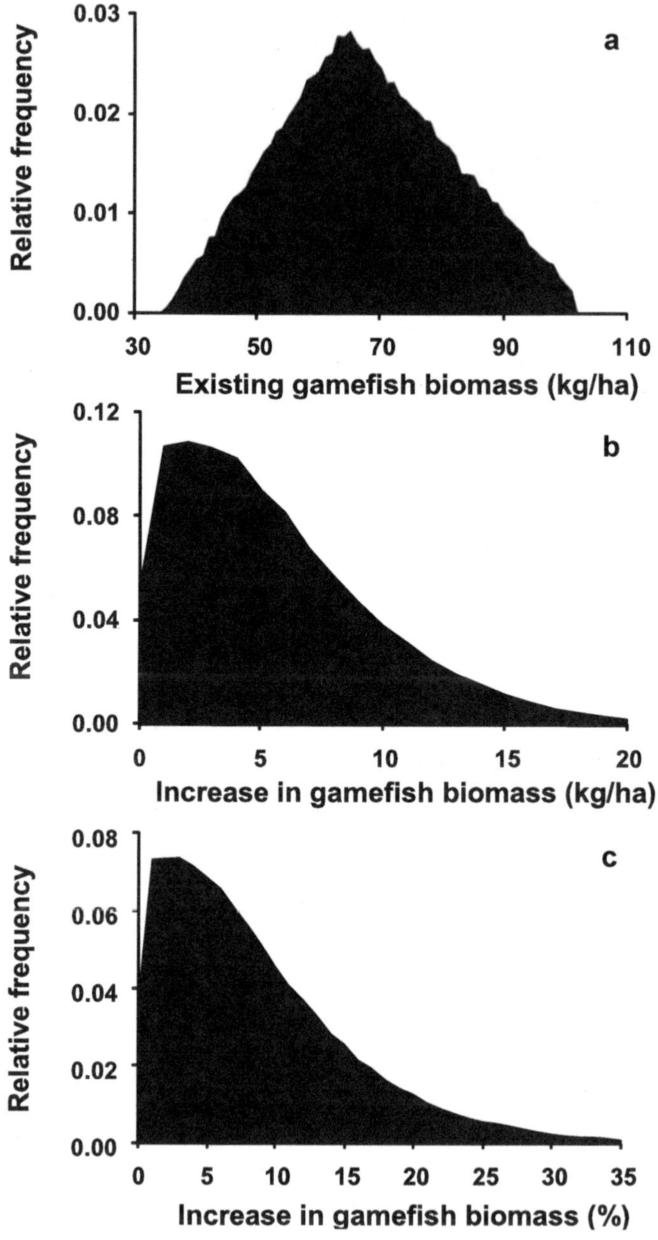

FIGURE 2. Frequency distributions of biomass and changes in biomass of game fish (black basses, black crappie, and percids) in Norris Reservoir, Tennessee estimated from model simulations: (a) estimated game fish biomass in 1996, (b) potential increase in game fish biomass if striped bass had been removed during this period, and (c) potential increase in game fish biomass given as a percentage. Distribution (c) = distribution (a) ÷ distribution (b) × 100 (after Raborn et al. 2002).

of density-dependence in game fish survival. To complicate matters, the resulting increase in game fish by removing both predation by and competition from striped bass may have a multiplicative effect. If the prey released increases carrying capacity and thereby decreases density dependence in game fish growth, fecundity, and survival, then more game fish released from predation would survive. Modeling this level of ecological complexity would be assuming too much resolution in model output given available data.

Disregarding the possible interaction of predation and competition effects, we can simply model separately how much the biomass of game fish might increase due to the ecological release of prey and how much game fish recruitment might increase from their juveniles not being eaten. A theoretical, maximum-case scenario would be to assume that all prey released by striped bass were consumed by other predators. Given that gross conversion efficiency was equal among striped bass and other game fish (an additional but not unreasonable assumption), this scenario would be tantamount to just assuming that game fish biomass would increase by the estimated biomass of the striped bass population being removed. In the Norris Reservoir study, this comes to a game fish increase of 21%. Raborn et al. (2002) accounted for differences in diets and gross conversion efficiencies among predator species with bioenergetic modeling to refine this estimate somewhat. They found that the most probable increase in the biomass of other game fish in Norris Reservoir, given the complete removal of striped bass, would be about 3%, with a 75% probability that it would be less than 12% (Figure 2). Still, these estimates were biased high because capture efficiency of the released prey by the remaining predators would be less than 100%. Regardless, this approach offers an upper bound for the expected benefits of abating a striped bass stocking program.

With respect to easement of striped bass predation on game fish, sunfishes were the only game fish consumed by striped bass in the Norris Reservoir study. Nevertheless, black bass survival was so compensatory that losses of black bass juveniles to predation by other black basses were actually beneficial to overall survival (Raborn et al. 2003). The density of black basses, as well as the losses to predation, were varied to help understand the chances that predation would be important from time to time. Modeling losses to striped bass predation would be no different for other systems if survival-density curves can be developed.

Prognosis for Research and Management

Population models are available for optimizing harvest of inland striped bass populations (Allen and Gwinn 2013, this volume), but a cost-effective multispecies model that accounts for food web interactions of striped bass with everything else in reservoir ecosystems does not exist. Studies to determine whether striped bass limit other game fish via predation need to expand their focus. The representation of game fish in the diets of striped bass is probably no longer worth quantifying by itself as this occurrence is well documented and only provides part of the answer. Without knowing the percentage of the game fish population that was removed and without accounting for density dependence in game fish survival, researchers cannot conclude the importance of specific amounts of predation. More research is needed on the potential for compensatory responses in game fish survival.

Further, an understanding of interference competition between striped bass and other game fish could have substantive management implications and fill a gap in the fisheries ecology literature. However, current sampling techniques and protocols offer little opportunity to test this mechanism. To do so would require a clear definition of fish distribution both vertically and horizontally and how fish relate to cover. These distributions would have to be defined both with and without striped bass to test for an effect. A linkage between behavioral changes in the fish community and presence of striped bass would require visual observation of species interactions. Innovative protocols are needed.

The weight of evidence is clearly on the side of there being low impact of striped bass on other game species, but if a comprehensive study is needed, there are ways to do it. Regardless of the mechanism, testing whether striped bass negatively impact other game fish is possible. A proper before-after-control-impact design would be needed and involve reservoir-wide estimates of game fish production before and after striped bass were introduced (or removed) from some reservoirs and not others. The information needed for estimating annual production (growth, mortality, and density) in multiple reservoirs is expensive to acquire and beyond budgets designed for routine monitoring and coarse-scale management. Estimating game fish relative weights would provide a cheaper alternative metric than estimating production. However, relative weights may be a short-term response to an increase in competition from striped bass as game fish species adjust their fecundity, growth, and survival to reduce densities, alleviate competition, and ultimately improve body condition of the remaining individuals. If monitoring such a system begins after these events unfold, there may be no apparent effect of striped bass on the relative weights of game fish, even though their overall production was reduced. Nevertheless, even if a striped bass effect was demonstrated, determining the causal mechanism of the effect is another expensive endeavor.

If a substantial effect is found, stakeholders and managers may opt to simply live with it and continue stocking striped bass because of the benefits of a striped bass fishery. Because many reservoir populations of striped bass must be maintained through stocking, the other extreme would be to simply discontinue the program and let exploitation and natural mortality remove the remaining striped bass. Presumably, other game fish would rebound. As a middle ground, game fish managers could try to tailor the harvest and stocking program of striped bass and game fish to alleviate the negative interactions, which would depend on the mechanism by which striped bass impede other game fish. Even if the mechanism is established (no easy task), this last option will prove arduous in and of itself. For instance, if direct predation was the problem, stocking rates of striped bass could be reduced (and increased for other game fish to augment their recruitment) to the point where losses to striped bass predation were acceptable to disconcerted anglers. Deciding stocking rates should be reduced is one thing, but determining the ideal amount would require data inputs that are not cost effective. Total consumption and diet composition for striped bass, density estimates of game fish, and the game fish survival-density curves must all be estimated. Variable annual survival of stocked striped bass would make the correct matching of stocking rates even more difficult.

If prey supply consistently falls short of predator demand, all game fish production will subside as a result of reduced fecundity, survival, and individual growth. Chronic supply:demand imbalance can be fixed in systems where striped bass account for enough demand that reducing their stocking rate will make a difference. Managers should not attempt to manipulate the fish assemblage towards the highest possible supply:demand ratio because in the absence of density-independent forces, sustained prey surpluses would likely elicit Holling's (1959) numerical response (i.e., predators increase in density rather than increase individual consumption levels). Occasional surplus prey is not easy to manage because there is a time lag of several years from when predators are stocked as fingerlings to when they are able to effectively consume the surplus (i.e., predator inertia, Stewart et al. 1981). The effectiveness of management actions to ameliorate acute supply:demand imbalance depends on future prey production (Ney and Orth 1986), which coupled with natural variability in prey and predator recruitment renders a bleak prognosis. Even in years of surplus prey, predators will probably have to endure seasonal food resource bottlenecks, as discussed above (e.g., striped bass separation from some prey during summer reservoir stratification). Managers are left with accepting random periods of poor prey production that are beyond their control.

Because the state of knowledge concerning the parameters defining balance remains minimal and because uncontrollable annual fluctuations preclude exact management of alternating prey levels, we suggest adjusting stocking to manage demand so it equals the median historical prey supply. Our rationale is that supply will exceed demand 50% of the time, thereby reducing chronic prey shortages. Conversely, the majority of surplus prey will be used the rest of the time, which precludes an overabundance of prey. Again, proper supply-demand studies are expensive and while they increase the state of ecological knowledge more than traditional diet overlap studies, the decision support they provide is too uncertain and untimely for objective management.

The collective evidence to date suggests that reservoir striped bass do not adversely affect other game fish through direct predation or exploitative competition. Expansive studies that capture all of the possible ecological complexities surrounding these interactions throughout the entire lake community is cost prohibitive. Simply quantifying the existing biomass of striped bass and assuming the theoretical, maximum-case scenario of this amount being added to the remaining game fish if striped bass were removed is far less expensive to ascertain and could help with decision support. Letting the stakeholders make informed value judgments and decide whether the majority of users want striped bass may be the best way to manage striped bass fisheries in reservoirs.

Acknowledgments

We thank Jonah Dagel, Dan Dembkowski, and Kyle Hartman for helpful reviews.

References

Adams, S. M., R. B. McLean, and J. A. Parrotta. 1982. Energy partitioning in largemouth bass under conditions of seasonally fluctuating prey availability. Transactions of the American Fisheries Society 111:549–558.

Adams, S. M., and D. L. DeAngelis. 1987. Indirect effects of early bass-shad interactions on predator population structure and food web dynamics. Pages 103–117 *in* C.W. Kerfoot and A. Sih, editors. Predation: direct and indirect impacts on aquatic communities. University Press of New England, Hanover, New Hampshire.

Allen, M. S., and D. Gwinn. 2013. Population models for assessment and management of inland striped bass fisheries. Pages 351–364 *in* J. S. Bulak, C. C. Coutant, and J. A. Rice, editors. Biology and management of inland striped bass and hybrid striped bass. American Fisheries Society, Symposium 80, Bethesda, Maryland.

Cavalli, L., R. Chappaz, and A. Gille. 1998. Diet of arctic charr (*Salvelinus alpinus* (L.)) and brown trout (*Salmo trutta* L.) in sympatry in two high altitude alpine lakes. Hydrobiologia 386:9–17.

Churchill, T. N., P. W. Bettoli, D. C. Peterson, W. C. Reeves, and B. Hodge. 2002. Angler conflicts in fisheries management: a case study of the striped bass controversy at Norris Reservoir, Tennessee. Fisheries 27(2):10–19.

Colwell, R. K., and D. J. Futuyma. 1971. On the measurement of niche breadth and overlap. Ecology 52:567–576.

Combs, D. L. 1979. Food habits of adult striped bass from Keystone Reservoir and its tailwaters. Proceedings of the Annual Conference Southeastern Association of Fish and Wildlife Agencies 32(1978):571–575.

Connell, J. H. 1980. Diversity and the coevolution of competitors, or the ghost of competition past. Oikos 35:131–138.

Coutant, C. C. 2013. When is habitat limiting for striped bass? Three decades of testing the temperature–oxygen squeeze hypothesis. Pages 65–91 *in* J. S. Bulak, C. C. Coutant, and J. A. Rice, editors. Biology and management of inland striped bass and hybrid striped bass. American Fisheries Society, Symposium 80, Bethesda, Maryland.

Cyterski, M., and J. Ney. 2005. Availability of clupeid prey to primary piscivores in Smith Mountain Lake, Virginia. Transactions of the American Fisheries Society 134:1410–1421.

Cyterski, M., J. Ney, and M. Duval. 2002. Predator demand for clupeid prey in Smith Mountain Lake, Virginia Fisheries Research 59:1–16.

Cyterski, M., J. Ney, and M. Duval. 2003. Estimation of surplus biomass of clupeids in Smith Mountain Lake, Virginia. Transactions of the American Fisheries Society 132:361–370.

Einfalt, L. M., and D. H. Wahl. 1997. Prey selection by juvenile walleye as influenced by prey morphology and behavior. Canadian Journal of Fisheries and Aquatic Sciences 54:2618–2626.

Fernando, C. H., and J. Holčík. 1982. The nature of fish communities: a factor influencing the fishery potential and yields of tropical lakes and reservoir. Hydrobiologia 97:127–40.

Filipek, S. P., and W. H. Tommey. 1987. The food habits of adult striped bass from Lake Hamilton, Arkansas, before and during an extreme drawdown. Proceedings of the Annual Conference Southeastern Association of Fish and Wildlife Agencies 38(1984):327–334.

Forrester, G. E. 1995. Strong density-dependent survival and recruitment regulate the abundance of a coral reef fish. Oecologia 103:275–282.

Forrester, G. E., and M. A. Steele. 2000. Variation in the presence and cause of density-dependent mortality in three species of reef fishes. Ecology 81:2416–2427.

Gilliland, E. R., and M. D. Clady. 1984. Diet overlap of striped bass × white bass hybrids and largemouth bass in Sooner Lake, Oklahoma. Proceedings of the Annual Conference Southeastern Association of Fish and Wildlife Agencies 35(1981):317–330.

Goodyear, P. C. 1980. Compensation in fish populations. Pages 253–280 *in* C. H. Hocutt and J. R. Strauffer, Jr., editors. Biological monitoring of fish. Lexington Books, Lexington, Massachusetts.

Gotelli, N. J., and G. R. Graves. 1996. Null models in ecology. Smithsonian Institution Press, Washington, D.C.

Gustaveson, W., and G. Blommer. 2013. History of striped bass management in the Colorado River. Pages 15–23 *in* J. S. Bulak, C. C. Coutant, and J. A. Rice, editors. Biology and management of inland striped bass and hybrid striped bass. American Fisheries Society, Symposium 80, Bethesda, Maryland.

Hale, R. S. 1996. Threadfin shad use as a supplemental prey in reservoir white crappie fisheries in Kentucky. North American Journal of Fisheries Management 16:619–632.

Higginbotham, B. J. 1979. Growth, food habits, maturation, and distribution of the striped bass (*Morone saxatilis*) in Watts Bar Reservoir, Tennessee. Master's thesis. Tennessee Technological University, Cookeville.

Hixon, M. A., and M. H. Cari. 1997. Synergistic predation, density dependence, and population regulation in marine fish. Science 277:946–949.

Hodson, J. R., C. J. Hodgson, and S. M. Brooks. 1991. Trophic interaction and competition between largemouth bass and rainbow trout in a manipulated lake. Canadian Journal of Fisheries and Aquatic Sciences 48:1704–1712.

Holling, C. S. 1959. Some characteristics of simple types of predation and parasitism. Canadian Entomologist 91:385–398.

Kohler, C. C., and J. J. Ney. 1982. Suitability of alewife as a forage fish for southeastern reservoirs. Proceedings of the Annual Conference Southeastern Association of Fish and Wildlife Agencies 34(1980):137–150.

Lyons, J. 1987. Prey choice among piscivorous juvenile walleyes (*Stizostedion vitreus*). Canadian Journal of Fisheries and Aquatic Sciences 44:758–764.

Marwitz, T. D., and W. A. Hubert. 1997. Trends in relative weight of walleye stocks in Wyoming reservoirs. North American Journal of Fisheries Management 17:44–53.

Matthews, W. J. 1998. Patterns in freshwater fish ecology. Chapman and Hall, New York.

Matthews, W. J., F. P. Gelwick, and J. J. Hoover. 1992. Food of and habitat use by juveniles of species of *Micropterus* and *Morone* in a southwestern reservoir. Transactions of the American Fisheries Society 121:54–66.

Matthews, W. J., L. G. Hill, D. R. Edds, J. J. Hoover, and T. G. Heger. 1988. Trophic ecology of striped bass, *Morone saxatilis*, in a freshwater reservoir (Lake Texoma, U.S.A.). Journal of Fish Biology 33:273–288.

McCann, K. S. 2000. The diversity-stability debate. Nature (London) 405:228–233.

McFadden, J. T. 1977. An argument supporting the reality of compensation in fish populations and a plea to let them exercise it. Pages 153–183 *in* W. Van Winkle, editor. Proceedings of the conference on assessing the effects of power-plant-induced mortality on fish populations. Pergamon New York.

Michaletz, P. H. 1997. Factors affecting abundance, growth, and survival of age-0 gizzard shad. Transactions of the American Fisheries Society 126:84–100.

Michaletz, P. H. 1998. Population characteristics of gizzard shad in Missouri reservoirs and their relation to reservoir productivity, mean depth, and sport fish growth. North American Journal of Fisheries Management 18:114–123.

Miranda, L. E., M. T. Driscoll, and S. W. Raborn. 1998. Competitive interactions between striped bass and other freshwater predators. Final Report submitted to Tennessee Wildlife Resources Agency, Nashville.

Miranda, L. E., M. Habrat, and S. Miyazono. 2008. Longitudinal gradients along a reservoir cascade. Transactions of the American Fishery Society 137:1851–1865.

Miranda, L. E., and K. M. Hunt. 2010. An index of habitat impairment for reservoirs. Environmental Monitoring and Assessment 172:225–234.

Mittelbach, G. G. 1988. Competition among refuging sunfishes and effects of fish density on littoral zone invertebrates. Ecology 69:614–623.

Mittelbach, G. G., and P. L. Chesson. 1987. Predation risk: indirect effects on fish populations. Pages 315–330 *in* C. W. Kerfoot and A. Sih, editors. Predation: direct and indirect impacts on aquatic communities. University Press of New England, Hanover, New Hampshire.

Moore, C. M. 1988. Food habits, population dynamics, and bioenergetics of four predatory fish species in Smith Mountain Lake, Virginia. Doctoral dissertation. Virginia Polytechnic Institute and State University, Blacksburg.

Morris, D. J., and B. J. Follis. 1979. Effects of striped bass predation upon shad in Lake E.V. Spence, Texas. Proceedings of the Annual Conference Southeastern Association of Fish and Wildlife Agencies 32(1978):697–702.

Muth, K. M., and D. R. Wolfert. 1986. Changes in growth and maturity of walleye associated with stock rehabilitation in western Lake Erie, 1964–1983. North American Journal of Fisheries Management 6:168–175.

Nash, V. S., W. E. Hayes, R. L. Self, and J. P. Kirk. 1988. Effect of striped bass introduction in Lake Wateree, South Carolina. Proceedings of the Annual Conference Southeastern Association of Fish and Wildlife Agencies 41(1987):48–54.

Ney, J. J. 1990. Trophic economics in fisheries: assessment of demand–supply relationships between predators and prey. Review of Aquatic Sciences 2:55–81.

Ney, J. J., and D. J. Orth. 1986. Coping with future shock: matching predator stocking programs to prey abundance. Pages 81–92 *in* R. H. Stroud, editor. Fish culture in fisheries management. American Fisheries Society, Bethesda, Maryland.

Noble, R. L. 1986. Predator–prey interactions in reservoir communities. Pages 137–143 *in* G. E. Hall and M. J. Van Den Avyle, editors. American Fisheries Society, Southern Division, Reservoir Committee, Bethesda, Maryland.

Olson, N. W., C. S. Guy, and K. D. Koupal. 2007. Interactions among three top-level predators in a polymictic Great Plains reservoir. North American Journal of Fisheries Management 27:268–278.

Persons, W. R., and R. V. Bulkley. 1982. Feeding activity and spawning time of striped bass in the Colorado River Inlet, Lake Powell, Utah. North American Journal of Fisheries Management 2:403–408.

Petchey, O. L. 2000. Prey diversity, prey composition, and predator population dynamics in experimental microcosms. Journal of Animal Ecology 69:874–882.

Peterson, D. C., D. A. Tomljanovich, C. J. O'Bara, J. W. Stair, D. G. Scott, J. M. Smith, I. Vanderpool, J. L. Wilson, and N. Woomer. 1993. Norris Reservoir Task Force progress report. Tennessee Wildlife Resources Agency and Tennessee Valley Authority, Nashville.

Ploskey, G. R., and R. M. Jenkins. 1982. Biomass model of reservoir fish and fish-food interactions, with implications for management. North American Journal of Fisheries Management 2:105–120.

Porath, M. T., and E. J. Peters. 1997. Walleye prey selection in Lake McConaughy, Nebraska: a comparison between stomach content analysis and feeding experiments. Journal of Freshwater Ecology 12:511–520.

Raborn, S. W., L. E. Miranda, and M. T. Driscoll. 2002. Effects of simulated removal of striped bass from a southeastern reservoir. North American Journal of Fisheries Management 22:406–417.

Raborn, S. W., L. E. Miranda, and M. T. Driscoll. 2003. Modeling predation as a source of mortality for piscivorous fishes in a southeastern U.S. reservoir. Transactions of the American Fisheries Society 132:560–575.

Raborn, S. W., L. E. Miranda, and M. T. Driscoll. 2004. Diet overlap and consumption patterns suggest seasonal flux in the likelihood for exploitative competition among piscivorous fishes. Ecology of Freshwater Fish 13:276–284.

Raborn, S. W., L. E. Miranda, and M. T. Driscoll. 2007. Prey supply and predator demand in a reservoir of the southeastern United States. Transactions of the American Fisheries Society 126:12–23.

Rice, J. A., and P. A. Cochran. 1984. Independent evaluation of a bioenergetics model for largemouth bass. Ecology 65:732–739.

Rice, J. A., J. S. Thompson, J. A. Sykes, and C. Waters. 2013. The role of metalimnetic hypoxia in striped bass summer kills: consequences and management implications. Pages 121–145 *in* J. S. Bulak, C. C. Coutant, and J. A. Rice, editors. Biology and management of inland striped bass and hybrid striped bass. American Fisheries Society, Symposium 80, Bethesda, Maryland.

Shepherd, M. D., and M. J. Maceina. 2009. Effects of striped bass stocking on largemouth bass and spotted bass in Lewis Smith Lake, Alabama. North American Journal of Fisheries Management 29:1232–1241.

Sih, A., and R. D. Moore. 1990. Interacting effects of predator and prey behavior in determining diets. Pages 771–796 *in* R. N. Hughes, editor. Behavioral mechanisms of food selection. North Atlantic Treaty Organization Advanced Studies Institute Series, Series G: Ecological Sciences Number 20. Springer-Verlag, Berlin.

Slipke, J. W., S. M. Smith, and M. J. Maceina. 2002. Food habits of striped bass and their influence in crappie in Weiss Lake, Alabama. Proceedings of the Annual Conference Southeastern Association of Fish and Wildlife Agencies 54(2000):88–96.

Steele, M. A. 1998. The relative importance of predation and competition in two reef fishes. Oecologia 115:222–232.

Stevens, R. E. 1958. The striped bass of the Santee–Cooper Reservoir. Proceedings of the Annual Conference Southeastern Association of Game and Fish Commissioners 11(1957):253–264.

Stevens, R. E. 1979. Discussion comments. Page 295 *in* R. H. Stroud and H. Clepper, editors. Predator–prey systems in fisheries management. Sport Fisheries Institute, Washington, D.C.

Stewart, D. J., J. F. Kitchell, and L. B. Crowder. 1981. Forage fishes and their salmonid predators in Lake Michigan. Transactions of the American Fisheries Society 110:751–763.

Stewart, W. T., and M. Burrell. 2013. Striped bass dispersion and effects on fisheries management in Lakes Mohave and Pleasant, Colorado River basin. Pages 431–448 *in* J. S. Bulak, C. C. Coutant, and J. A. Rice, editors. Biology and management of inland striped bass and hybrid striped bass. American Fisheries Society, Symposium 80, Bethesda, Maryland.

Storck, T. W. 1986. Importance of gizzard shad in the diet of largemouth bass in Lake Shelbyville, Illinois. Transactions of the American Fisheries Society 115:21–27.

Sutton, T. M., and J. J. Ney. 2002. Trophic resource overlap between age-0 striped bass and largemouth bass in Smith Mountain Lake, Virginia. North American Journal of Fisheries Management 22:1250–1259.

Thompson, J. S., and J. A. Rice. 2013. The relative influence of thermal experience and forage availability on growth of age 1–5 striped bass in two southeastern reservoirs. Pages 121–145 *in* J. S. Bulak, C. C. Coutant, and J. A. Rice, editors. Biology and management of inland striped bass and hybrid striped bass. American Fisheries Society, Symposium 80, Bethesda, Maryland.

Thompson J. S., J. A. Rice, D. S. Waters, J. C. Taylor, J. E. Hightower, L. A. Davias, and P. S. Rand. 2005. Energetics of reservoir striped bass populations: phase 1 final report. Federal

Aid in Sport Fish Restoration Project F-68–04, Final Report 499 submitted to North Carolina Wildlife Resources Commission, Division of Inland Fisheries, Raleigh.

TPWD (Texas Parks and Wildlife Department). 1986. A review of the striped bass and hybrid striped bass program. TPWD, Inland Fisheries Branch, Austin.

Van den Avyle, M. J., and B. J. Higginbotham. 1983. Habitat preferences and food habits of young-of-the-year striped bass, white bass, and yellow bass in Watts Bar Reservoir, Tennessee. North American Journal of Fisheries Management 3:163–170.

Vatland, S., P. Budy, and G. P. Thiede. 2008. A bioenergetics approach to modeling striped bass and threadfin shad predator–prey dynamics in Lake Powell, Utah–Arizona. Transactions of the American Fisheries Society 137:262–277.

Wiens, J. A. 1977. On competition and variable environments. American Scientist 65:590–597.

Wiens, J. A. 1993. Fat times, lean times and competition among predators. Trends in Ecology and Evolution 8:348–349.

American Fisheries Society Symposium 80:521–530, 2013

Managing User Conflicts on Reservoirs Supporting Striped Bass or Hybrid Striped Bass Fisheries

STEVE L. MCMULLIN*
Department of Fish and Wildlife Conservation, Virginia Tech
Blacksburg, Virginia 24061, USA

Abstract.—Managers of reservoirs that support striped bass *Morone saxatilis* or hybrid striped bass fisheries face conflicts among angler groups and between anglers and other reservoir users. Striped bass anglers frequently want more striped bass stocked in reservoirs while anglers who seek black bass *Micropterus* spp., crappies *Pomoxis* spp., and other species often see striped bass as predators or competitors with the species they desire to catch. Anglers also share these reservoirs with recreational boaters, including personal watercraft users, often resulting in conflicts.

I conducted a survey of reservoir fishery managers designed to assess the extent and severity of user conflicts on reservoirs that support striped bass or hybrid striped bass fisheries. Thirty-two fishery managers, from 18 of the 27 states contacted, responded to the survey (56% response rate). Twenty-two managers reported conflicts between striped bass anglers and other anglers, and 20 managers reported conflicts between anglers and recreational boaters or personal watercraft users on the reservoirs they managed. Nearly all conflicts among anglers had decreased in severity over the past 10 years, largely due to efforts by fishery managers to build more constructive relationships with angling stakeholders and to make the management process more transparent. Managers built more constructive relationships by involving anglers in fisheries management activities and by meeting with them frequently to discuss reservoir management. Despite attempts to involve stakeholders in decision-making processes, transparency of the management process remains an issue. Marine fisheries managers in the Southeastern United States collaborate with stakeholders to identify and review data and stock assessment procedures. The marine fisheries model could be adapted to increase the transparency of reservoir management.

Introduction

Fisheries managers sometimes refer to fisheries management as combat biology because management often involves resolution of conflicts among stakeholders as much as it does application of biology. Fishery managers frequently deal with stakeholder groups that have competing interests for preferred species management (Hunt and Ditton 2002), bag limits or regulations (Hutt and Bettoli 2007), and uses of the water itself (e.g., anglers versus boaters and personal watercraft users; Margenau and Petchenik 2004).

Anglers who direct their effort primarily toward striped bass *Morone saxatilis* can be especially difficult to satisfy. For example, O'Neill (2001) found that striped bass anglers were more likely to say that obtaining fish for

* Corresponding author: smcmulli@vt.edu

eating was a very important reason for fishing than anglers who directed their effort towards catfish Ictaluridae, largemouth bass *Micropterus salmoides*, smallmouth bass *M. dolomieu*, or trout Salmonidae. Only crappie *Pomoxis* spp. anglers placed more importance on keeping fish to eat. Striped bass anglers also placed greater importance on catching trophy fish than all other species specialists except for largemouth bass anglers. The inherent conflict between a strong desire to harvest fish and the desire to catch trophy fish makes the job of the fishery manager particularly challenging when the fish in question needs many years to reach trophy size, as is the case with striped bass.

In reservoirs containing striped bass, conflicts among angler groups have been common and widespread (Churchill et al. 2002; Miranda and Raborn 2013, this volume). In this paper, I describe the nature of conflicts between users in reservoirs containing striped bass, present results of a survey of managers of reservoirs containing striped bass (or *Morone* hybrids), and discuss approaches to successful management of conflicts in reservoir fisheries management.

History and nature of conflicts in reservoirs containing striped bass

Although managers of reservoir striped bass fisheries have experienced all of the types of conflicts previously described, the most common (and intense) conflicts have involved stakeholder groups interested in management of preferred species. In most cases, the conflicts have occurred between anglers desiring a management emphasis on striped bass (or *Morone* hybrids) and other anglers preferring a management emphasis on walleye *Sander vitreus*, black bass *Micropterus* spp., or crappies. Conflicts among angler groups favoring these species groups have occurred on reservoirs throughout the southeastern United States and, to a lesser extent, the Midwest.

The best documented case of conflict over management of preferred species is that of Norris Reservoir, Tennessee (Churchill et al. 2002). Norris Dam, the first dam built by the Tennessee Valley Authority, impounded the Clinch and Powell rivers (headwaters of the Tennessee River) in 1936. An initially productive fishery for sauger *Sander canadensis* declined within 15 years after impoundment, and by the 1960s, the Tennessee Game and Fish Commission (later to become the Tennessee Wildlife Resources Agency [TWRA]) initiated stocking programs designed to augment populations of native species and also to introduce new sport fish. The agency introduced striped bass to Norris Reservoir in 1966 to prey on abundant gizzard shad *Dorosoma cepedianum*. Managers of the reservoir increased the stocking rate in 1975 and again in 1991, eventually stocking striped bass at more than three times the initial rate (Peterson and Negus 1995).

The popular striped bass fishery in Norris Reservoir generated opposition from walleye, black bass, and crappie anglers, who blamed declining fisheries for those species on competition with and predation by striped bass. Angler groups quickly polarized over the issue of Norris Reservoir fisheries management, with one group demanding the end of striped bass stocking and the other defending it. The TWRA created the Norris Reservoir Task Force (Task Force), made up of anglers and agency professionals (including TWRA and Tennessee Valley Authority representatives), and charged it with developing a fisheries management plan for the reservoir. Although the Task Force succeeded in developing a management plan that the TWRA implemented, uneven representation of stakeholders on the Task Force and the strong influence of agency personnel in the Task Force's decisions heightened the suspicions of anti-striped bass stakeholders. The controversy continued to grow.

The Tennessee Sportsman's Association, a grassroots group of concerned citizens that formed specifically to influence fisheries management in the state, suggested that the TWRA did not respond adequately to stakeholder concerns. The Tennessee Sportsman's Association convinced some state legislators to introduce a bill that would have curtailed the ability of the TWRA to manage Norris and other reservoirs in Tennessee, as well as other bills that would have negatively affected the agency's funding base. Although the bills were defeated, the TWRA was forced to fund out-of-state researchers to conduct a study of striped bass interactions with other fish species in Norris Reservoir. The study, based on sound biological principles and data, concluded that striped bass had negligible impacts on other fish in the reservoir (Miranda and Raborn 2013), but by the time the study was completed, anti-striped bass stakeholders did not believe any information coming from the TWRA or the study (Raborn et al. 2002, 2003, 2007).

The issue was not resolved until the TWRA convened a new Norris Lake Fishery Advisory Committee (Advisory Committee), composed of equal numbers of representatives from the striped bass angler stakeholders and the anti-striped bass stakeholders. In addition, the Advisory Committee also had members representing nonaffiliated anglers, business interests, and governments of counties adjoining the reservoir. Neutral facilitators conducted meetings of the Advisory Committee and the TWRA ceded decision-making authority, serving only to inform and advise the Advisory Committee. A key factor in the ultimate dampening of controversy was the departure from the process of key individuals in the anti-striped bass community who had been unwilling to compromise, which allowed the remaining representatives to work more effectively with TWRA and other stakeholders. As this paper is written, more than a decade later, the Norris Lake Fishery Advisory Committee continues to meet annually and fisheries management on Norris Reservoir is a much more civil process.

Although Norris Reservoir provides one of the most extreme cases, reservoir managers in Virginia, Texas, Arkansas, and other states have experienced similar controversies between striped bass anglers and anglers targeting other species, particularly black bass. A biologist with primary responsibility for management of one reservoir containing striped bass grew so tired of dealing with the controversy there that he resigned and found more rewarding employment as a biologist in another state. In the remainder of this paper, I discuss the extent and intensity of conflicts between striped bass anglers and other reservoir users, and present strategies employed by fisheries managers in several areas of the country to address those conflicts. Finally, I present some ideas for addressing future conflicts.

Methods

I attempted to contact 67 fisheries managers or researchers in 27 states who have worked with striped bass or *Morone* hybrids via e-mail during summer 2009 (Table 1). The Striped Bass Technical Committee of the Southern Division of the American Fisheries Society provided the list of managers and researchers. Nine e-mail addresses returned undeliverable messages and one person declined to participate, yielding an effective sample size of 57 people.

The initial e-mail contact invited the striped bass managers and researchers to participate in the survey and contained a hyperlink to the online survey. One follow-up e-mail message was sent to nonrespondents 11 d after the initial contact. The survey asked the managers and researchers how many reservoirs they worked on, in their current positions, that contained striped bass. I then asked them to characterize the severity of conflicts between striped bass anglers and five other reservoir user groups: other anglers, recreational boaters, personal watercraft users, hunters, and other users. The respondents were asked to describe any conflicts they characterized as moderate or severe. I also asked the respondents if the conflicts had become more or less severe, or if they had stayed about the same over the past 10 years. If they indicated the severity of conflict had changed, I asked them to describe causes of the change. Next, I asked respondents to rate the relationship between fisheries managers or researchers and striped bass anglers on the reservoirs they managed or upon which they conducted research. I also asked if the relationship was bet-

TABLE 1. Number of fisheries professionals contacted and number of responses by state.

State	Number of professionals contacted	Number of responses
Alabama	2	2
Arizona	1	1
Arkansas	3	1
California	1	0
Florida	1	1
Georgia	2	2
Illinois	1	1
Indiana	2	1
Iowa	1	0
Kansas	1	1
Kentucky	2	1
Louisiana	1	0
Maryland	1	0
Mississippi	2	1
Missouri	1	0
Nebraska	1	0
Nevada	1	1
North Carolina	2	0
Ohio	1	1
Oklahoma	2	1
Pennsylvania	1	1
South Carolina	9	2
Tennessee	4	3
Texas	18	7
Utah	1	0
Virginia	3	2
West Virginia	1	0
Unknown	–	2[a]

[a] Two respondents gave no indication of where they worked. One unidentified respondent reported no conflicts between anglers and other groups while the other unidentified respondent reported only minor conflicts with recreational boaters and personal watercraft users.

ter or worse than the relationship with other anglers on the same reservoir. Finally, I asked if the respondent would be willing to participate in a follow-up telephone interview.

I conducted follow-up telephone interviews with six reservoir managers who indicated in the survey that they had experienced significant changes in the severity of conflicts. The telephone interviews were designed to identify strategies that were employed to manage conflicts and also the most significant factors that contributed to the changes in severity of conflicts.

Results and Discussion

Survey response rate

Thirty-two people responded to the survey, a response rate of 56%, from 18 of 27 states (Table 1). Since the objectives of the survey were to characterize the types of conflicts existing on reservoirs containing striped bass, as well as strategies being employed to manage those conflicts, I made no attempt to assess nonresponse bias.

Frequency and severity of user conflicts

Respondents worked on an average of four reservoirs that contained striped bass or their hybrids (range = 0–32). Respondents reported that conflicts occurred between striped bass anglers and all other user groups except for hunters (Table 2). Twenty respondents reported conflicts between striped bass anglers and other anglers while 17 respondents reported conflicts with both recreational boaters and personal watercraft users. Only three respondents reported conflicts between striped bass anglers and other reservoir users. One of those respondents indicated that conflicts with other users involved recreational boaters and personal watercraft users. The other two respondents described conflicts with other anglers (and were counted with other angler conflicts). Both of the respondents who identified severe conflicts between striped bass anglers and other anglers had worked on Norris Reservoir, Tennessee. Respondents from Texas (two), Oklahoma (one), and Tennessee (one) identified moderate conflicts between striped bass anglers and other anglers. The Tennessee respondent referred to Norris Reservoir but characterized the conflict as less severe than other Tennessee respondents. Texas and Oklahoma respondents identified moderately severe conflicts on six lakes with black bass anglers who believed that striped bass or hybrids competed with or preyed upon black bass. In all cases, respondents mentioned that the conflicts occurred because of lack of information or misinformation about interactions between striped bass and black bass. Misinformation often originated with fishing guides or fishing Web sites.

Respondents reported moderate or severe conflicts between striped bass anglers and recreational boaters or personal watercraft users on reservoirs in Virginia, Arizona, South Carolina, and Indiana. The most severe conflicts occurred with personal watercraft users on reservoirs in Virginia and South Carolina. The conflicts usually involved lack of courtesy towards anglers, especially lack of awareness of trolling lines and associated equipment (e.g., planer boards). Conflicts between anglers of all types and personal watercraft users became a major issue for fishery managers as sales of personal watercraft peaked in the mid-1990s. Although annual sales of personal watercraft have decreased since the mid-1990s, approximately 1.3 million were in use in the United States in 2009 (National Marine Manufacturers Association 2010).

Changes in severity of user conflicts

One Oklahoma fishery manager felt that conflicts between striped bass anglers and other anglers had become more severe over the past 10 years (Table 3). The manager attributed the increase in severity of conflict to anglers' participation in Internet chat rooms that promote negative feelings toward striped bass management. Fishery managers reported declining severity of conflicts between striped bass anglers and other anglers on reservoirs in Tennessee, Virginia, and Texas. Norris Reservoir managers

TABLE 2. Frequency and severity of conflicts between striped bass anglers and other reservoir users as reported by 32 managers and researchers of reservoirs containing striped bass or their hybrids.

User group	No conflicts	Minor conflicts	Moderate conflicts	Severe conflicts
Other anglers	10	16	4	2
Recreational boaters	15	13	4	0
Personal watercraft users	15	11	4	2
Hunters	32	0	0	0

TABLE 3. Changes in severity of conflicts between striped bass anglers and other reservoir users over the past 10 years as reported by 32 managers and researchers of reservoirs containing striped bass or their hybrids.

User group	More severe	Stayed about the same	Less severe	No response
Other anglers	1	24	6	1
Recreational boaters	3	26	0	3
Personal watercraft users	5	23	1	3
Hunters	0	29	0	3

attributed the lessened severity of conflicts to the collaborative approach to management previously described. In Virginia, the conflict has abated due to improved black bass fishing in recent years and due to collaboration between managers and anglers in research and management of the fishery. At one Texas reservoir, the conflicts diminished because the most vocal opponents of striped bass either passed away or simply quit complaining. Managers of several other Texas reservoirs attributed the decline in conflict to guides diversifying their efforts to include fishing for both striped bass hybrids and black bass. The guides, as opinion leaders among anglers, apparently had significant influence in quieting the controversies.

Conflicts between striped bass anglers and recreational boaters or personal watercraft users have either remained the same or worsened over the past 10 years in all but one case. Increased use of reservoirs by personal watercraft users led to most of the conflicts, as managers cited lack of courtesy or lack of awareness of anglers and their gear. In the only case in which conflicts appeared to become less severe (Lake Mead, Nevada–Arizona), the manager attributed the change to drastically lowered water levels. The lower water levels made some popular shore fishing areas, where conflicts had occurred previously, unfishable.

Relationships between anglers and fishery managers

Thirty of the 32 respondents indicated that the relationship between fishery managers and striped bass anglers was excellent (n = 6) or good (n = 24), and 30 respondents said the relationship with striped bass anglers was about the same (n = 23) or better (n = 7) than the relationship with other anglers.

A Texas fishery manager indicated that the relationship with striped bass anglers at Toledo Bend Reservoir (which borders on Louisiana) was fair and that the small group of striped bass anglers there had expressed frustration with Texas managers for not stocking striped bass. Louisiana managers stock the reservoir to maintain a broodstock source, but with only 1% of the angling effort directed at striped bass, Texas managers have not felt that stocking was justifiable.

On Santee–Cooper Reservoir, South Carolina, the relationship between fishery managers and striped bass anglers was characterized as fair and also as worse than the relationship with other anglers. The manager at Santee–Cooper Reservoir suggested that the less-than-desirable relationship resulted from significant declines in striped bass abundance due to declining productivity and high mortality of striped bass caught and released during the warm summer months. The relationship with some striped bass anglers has improved recently as they have participated in a stakeholder group created specifically to address issues related to management of striped bass on the reservoir. The stakeholder group played an important role in convincing legislators to adopt a summer moratorium on harvest of striped bass in the reservoir (in South Carolina, the legislature, not the management agency, adopts fishing regulations). The stakeholder group continued to wrestle with summer mortality issues because some anglers continued to target striped bass for catch-and-release fishing during the summer, causing significant mortality. As a result, a new regulation, established in 2010, prohibited "taking or intent to take" striped bass during June through September.

In summary, 22 out of 32 respondents reported experiencing at least minor conflicts between striped bass anglers and other anglers. Seventeen respondents reported experiencing at least minor conflicts between striped bass anglers and recreational boaters or personal watercraft users. Nearly all conflicts between striped bass anglers and other anglers resulted from the perceptions of black bass or crappie anglers that striped bass prey on or compete with their favorite species. Although most studies have shown that striped bass in reservoirs prey on black bass and crappie infrequently (Moore et al. 1991; Smollen 1999; Miranda and Raborn 2013), angler perceptions often were based on lack of information or misinformation that came from Internet sources or opinion leaders in the fishing community (such as fishing guides). In all but one case, the severity of conflicts between striped bass anglers and other anglers has lessened or remained the same over the past 10 years.

Reducing conflict by increasing transparency of the management process

All strategies fishery managers described for reducing conflicts among angler groups involved building constructive relationships with anglers and increasing the transparency of the management process. Fishery managers built constructive relationships with anglers by involving them in sampling, fish stocking, and other projects related to fisheries management. Angler involvement in management activities helped to fill some data gaps and also helped to increase acceptance of proposed management strategies. For example, fishery managers have involved anglers in collection of otoliths for age and growth analyses and collection of stomachs for food habits studies. Anglers often catch older, larger fish more effectively than fishery managers can collect through their sampling techniques.

Managers increased the transparency of the management process by meeting more frequently with anglers to share and evaluate biological data and by providing meaningful opportunities for anglers to be involved in making decisions about fisheries management. These actions also allowed fishery managers to separate fact from fiction and to counter some of the misinformation to which many anglers were exposed. Approaches to increasing the transparency of the management process included the efforts of Virginia fishery managers to involve members of a striped bass fishing club that had a history of animosity with the agency in evaluating the available data so that both managers and anglers agreed on its significance to management of the fishery. Anglers also worked jointly with fisheries managers to develop a fishery management plan.

One approach to increasing transparency of the management process that improved stakeholder–manager relationships for wildlife managers in Virginia focused stakeholder involvement on making value choices about wildlife management (setting broad goals for desired outcomes) while, at the same time, focusing wildlife manager involvement on making technical choices (developing and evaluating potential strategies to achieve desired outcomes). Lafon et al. (2004) found that stakeholders who were involved in making value choices during the development of a management plan for black bears in Virginia were more knowledgeable about bears, had a better image of wildlife managers, and generally became more tolerant of the points of view of other stakeholders as a result of their participation in the process. They also found that wildlife managers developed a more positive view of stakeholder involvement in the management process.

Stakeholders must trust fishery managers to separate value choices from technical choices for stakeholder involvement to succeed. Trust usually leads to stakeholder acceptance of the managers' judgments on the meaning of technical fisheries data. If stakeholders do not trust fishery managers, the probability decreases that stakeholders will be content with making value choices while leaving technical choices to the managers. This lack of trust often manifests itself in the form of challenges to fishery managers' methods of collecting data, their analyses of data, and their evaluations of potential management options.

Marine fisheries managers in the southeastern United States developed a process to increase transparency of the management process that could serve as a model for reservoir managers who want to build more trust with stakeholders. The Magnuson-Stevens Fishery Conservation and Management Act requires the National Oceanic and Atmospheric Administration (NOAA) to use the best available data in management of marine fisheries in U.S. waters. However, scientists and stakeholders frequently disagree on the definition of best available data. Scientists at NOAA's Southeast Fisheries Science Center and the South Atlantic Fishery Management Council developed the Southeast Data, Assessment, and Review (SEDAR) process to bring fisheries scientists and stakeholders together to increase transparency of the management process and to improve the quality and reliability of stock assessments (SAFMC, no date).

The SEDAR process involves stakeholders and scientists in a series of three collaborative workshops to identify the best available data and models for stock assessments. During the data workshop, participants review and compile fisheries, resource monitoring, and life history data. In the assessment workshop, participants develop assessment models to estimate population parameters, using data provided from the data workshop. Peer review of the data, assessment methods, and assessment products occurs in the review workshop. Commercial and recreational fishers throughout the southeastern United States who were interviewed during a marine fisheries population dynamics workshop for undergraduate students expressed favorable opinions of the SEDAR process despite being otherwise critical of fisheries managers.

Reservoir fishery managers who face skeptical or even hostile stakeholders may benefit from using a similar approach. Involving opposing groups of stakeholders (e.g., striped bass anglers and black bass anglers) in workshops designed to share data used by managers, explain analytic approaches, and develop agreement on the biological basis of fisheries management may help to dispel some of the misinformation that often contributes to controversies over reservoir fisheries management. Such workshops would have great educational value for stakeholders and would increase transparency of the management process. Furthermore, collaborative data workshops have the potential to improve relationships among anglers and fishery managers.

Conflict resolution and reservoir fisheries management

Among the principles of effective conflict resolution, two seem especially relevant to management of reservoirs that contain striped bass. First, conflict resolution specialists suggest "separating people from the problem" (Fisher et al. 1991). Fishery managers who focus on building constructive working relationships increase the trust between them and stakeholders. In especially contentious situations, the relationship may not be friendly, but it should enable the parties to proceed independent of trust. If the management process is transparent and participants trust each other enough to proceed, they can attack problems without attacking each other. Many of the strategies described by reservoir fishery managers who responded to the survey focused on building relationships that would allow stakeholders and managers to separate people from the problem. Fisheries managers generally have no authority to regulate recreational boaters and personal watercraft users in ways that will reduce conflicts with anglers. However, many states now require boating safety courses for operators of recreational watercraft. Fisheries managers should pursue opportunities to incorporate messages of etiquette for water-based recreation into those courses that may help to reduce conflicts.

The second key principle is recognition of the best alternative to a negotiated agreement (BATNA). The probability that a stakeholder's interests will be favorably addressed in resolving a conflict plays a major role in that stakeholder's willingness to participate in the process. If a stakeholder believes that he has a better chance of satisfying his interests by other means, such as going over the manager's head (e.g., to commissioners or legislators), then circumventing the management process appears to provide a better alternative to negotiating with the manager and other stakeholders. In contrast, if stakeholders view the management process as transparent and fair, then constructive participation and cooperation in the management process becomes the better alternative.

Fisheries managers (and other natural resource professionals) traditionally have been reluctant to share decision-making authority because their scientific training and the culture of the natural resource professions have promoted an attitude of "We are the experts, we should make the decisions" (Magill 1988). However, reluctance to engage stakeholders in decision making increases the probability that the stakeholders' BATNA will be something other than the manager's process. If managers of reservoirs containing striped bass employ the strategies described by respondents to the survey, and if they increase the transparency of the management process by engaging stakeholders in data analysis workshops to build more constructive relationships among themselves and their stakeholders, they will increase the probability of success in resolving conflicts over fisheries management.

Acknowledgments

Jim Bulak provided names and contact information for striped bass managers and researchers. Dave Miko, Ben Ricks, Dan Wilson, Randy Myers, Bill Wentroth, Tim Churchill, and Scott Lamprecht participated in telephone interviews that helped the author gain additional insight into the survey results.

References

Churchill, T. N., P. W. Bettoli, D. C. Peterson, W. C. Reeves, and B. Hodge. 2002. Angler conflicts in fisheries management: a case study of the striped bass controversy at Norris Reservoir, Tennessee. Fisheries 27(2):10–19.

Fisher, R., W. Ury, and B. Patton. 1991. Getting to yes: negotiating agreement without giving in. Penguin Books, New York.

Hunt, K. M. and R. B. Ditton. 2002. Freshwater fishing participation patterns of racial and ethnic groups in Texas. North American Journal of Fisheries Management 22:52–65.

Hutt, C. P. and P. W. Bettoli. 2007. Preferences, specialization, and management attitudes of trout anglers fishing in Tennessee tailwaters. North American Journal of Fisheries Management 27:1257–1267.

Lafon, N. W., S. L. McMullin, D. E. Steffen, and R. S. Schulman. 2004. Improving stakeholder knowledge and agency image through collaborative planning. Wildlife Society Bulletin 32:220–231.

Magill, A. W. 1988. Natural resource professionals: the reluctant public servants. The Environmental Professional 10:295–303.

Margenau, T. L. and J. B. Petchenik. 2004. Social aspects of muskellunge management in Wisconsin. North American Journal of Fisheries Management 24:82–93.

Miranda, L. E., and S. W. Raborn. 2013. Interactions between striped bass and other game fish in reservoirs. Pages 501–519 *in* J. S. Bulak, C. C. Coutant, and J. A. Rice, editors. Biology and management of inland striped bass and hybrid striped bass. American Fisheries Society, Symposium 80, Bethesda, Maryland.

Moore, C. M., R. J. Neves, and J. J. Ney. 1991. Survival and abundance of stocked striped bass in Smith Mountain Lake, Virginia. North American Journal of Fisheries Management 11:393–399.

National Marine Manufacturers Association. 2010. 2009 recreational boating statistical abstract

Available: http://nmma.net/assets/cabinets/Cabinet445/2010_abstract_preview.pdf. (February 2012)

O'Neill, B. M. 2001. Market segmentation, motivations, attitudes, and preferences of Virginia resident freshwater anglers. Master's thesis. Virginia Polytechnic Institute and State University, Blacksburg.

Peterson, D. C. and J. Negus. 1995. Norris Reservoir Annual Report 1994. Tennessee Wildlife Resources Agency, Nashville.

Raborn, S. W., L. E. Miranda, and M. T. Driscoll. 2002. Effects of simulated removal of striped bass from a southeastern reservoir. North American Journal of Fisheries Management 22:406–417.

Raborn, S. W., L. E. Miranda, and M. T. Driscoll. 2003. Modeling predation as a source of mortality for piscivorous fishes in a southeastern U.S. reservoir. Transactions of the American Fisheries Society 132:560–575.

Raborn, S. W., L. E. Miranda, and M. T. Driscoll. 2007. Prey supply and predator demand in a reservoir of the southeastern United States. Transactions of the American Fisheries Society 136:12–23.

SAFMC (South Atlantic Fishery Management Council). No date. SEDAR (Southeast Data, Assessment, and Review) stock assessment program. Available: www.safmc.net/Portals/6/Regulations/SEDAR_FactSheet012609.pdf. (August 2009)

Smollen, M. A. 1999. Food habits of adult predators in Norris Reservoir during winter drawdown. Master's thesis. University of Tennessee, Knoxville.

American Fisheries Society Symposium 80:531–549, 2013

Releasing Hybrid *Morone* in Natural Waters with Congeneric Species: Implications and Ethics

Reginal M. Harrell*
University of Maryland College Park
Department of Environmental Science and Technology
Queenstown, Maryland 21658, USA

Abstract.—Since the 1950s, sympatric, congeneric species of the *Morone* complex have been transplanted outside their native range. Also, since the late 1960s, conspecific, artificially produced *Morone* hybrids have been introduced in systems with native congeneric "pure" stocks. Following previous documentation of F_1 and F_2 hybrid *Morone* being artificially produced under hatchery conditions and introduced into natural and artificial systems (i.e., impoundments), these "introductions" have resulted in reports of introgression (outcrossing of conspecifics) or natural hybridization within the *Morone* complex. Even though natural hybridization has been documented, based on the number of years transplanted hybrids and pure *Morone* have coexisted in the same bodies of water, such events appear to be rare or at least undocumented. Unfortunately, because of lack of consistent empirical evidence to make definitive statements, the seriousness of the issue is unclear and points to a need for more intensive sampling effort for verification. A survey of the literature and various state agencies responsible for managing inland and coastal systems that contain sympatric populations of *Morone* and artificially produced hybrids indicate that such hybridization appears to be the exception and therefore, unless otherwise documented, appears to be of minor concern. However, given that artificially produced backcrosses (a hybrid crossed back to a pure parent) are difficult to visually (phenotypically) distinguish from the "pure" parents, it may be possible that management agencies are unaware of outcrossing or backcrossing occurring. Because the impact of gene introgression is difficult to track, introgressive hybridization could affect behavior, recruitment, and even predator–prey interactions. Thus, if one is concerned about the potential consequences of hybridization or outcrossing of *Morone* with its conspecifics or its congeneric hybrids either genetically, ecologically, or even ethically, then it may be prudent to follow the precautionary principle until verification that such stocking does not present sufficient risk.

Introduction

Striped bass *Morone saxatilis* and white perch *M. americana* are originally Atlantic coast drainage species found particularly in coastal areas (Lee et al. 1980). Striped bass are also indigenous to coastal tributaries of the Gulf of Mexico from western Florida to Louisiana (Raney 1952). The other two *Morone* species, white bass *M. chrysops* and yellow bass *M. mississippiensis*, are historically Mississippi River drainage species (Lee et al. 1980). Thus, species within the *Morone* complex are normally geographically

* Corresponding author: rharrell@umd.edu

or reproductively isolated, the latter including behavioral or temporal spawning differences. However, because of human intervention, all four congeneric members can be found today within certain watersheds in North America. In fact, currently at least one species of *Morone* or hybrid *Morone* can be found in most of the 48 contiguous states (Harrell and Webster 1997).

Striped bass became a very popular recreational freshwater fish in the 1950s after a sustainable land-locked population was discovered in the Santee–Cooper reservoir system in South Carolina (Scruggs and Fuller 1955; Van Horn 2013, this volume). From a reservoir management perspective, early in the management history it was found that striped bass proved to be an excellent tool for biological control of clupeid (*Alosa* spp. and *Dorosoma* spp.) populations.

Although not an absolute necessity, habitat requirements for successful establishment of inland striped bass populations appear to be relegated to large, mesotrophic impoundments with sufficient prey, suitable summer water temperatures (<25–27°C), and oxygen levels (>2 mg/L) (Coutant 1985, 2013, this volume; Lewis et al. 1990; Rice et al. 2013, this volume; Thompson and Rice 2013, this volume). For more detail on the history of inland stocking, management, and success of striped bass, see reviews by Bailey (1975), Axon and Whitehurst (1985), Whitehurst and Stevens (1990), Gustaveson and Blommer (2013, this volume), and Van Horn (2013).

Despite the management potential, there was inconsistent success in establishing striped bass populations in inland reservoir systems so many state agencies sought other options. One option was using interspecific, artificially produced *Morone* hybrids. Biologists hoped that the hybrid *Morone* would exhibit the heterosis (hybrid vigor) typically found in such genetic crosses and that they would be better adapted to shallow, eutrophic reservoirs. Specifically, the state agencies were looking for a fish that tended to be more tolerant of warmer temperatures, lower oxygen concentrations, and smaller reservoirs than those where striped bass fisheries were well established (Axon and Whitehurst 1985; Lewis et al. 1990; Van Horn 2013).

Bishop (1968) first reported that hybrids between striped bass and white bass (male white bass males × female striped bass) met all the objectives set forth when the cross was first made. In other words, they would be similar to striped bass but more adaptable to smaller reservoirs, attain larger size than white bass, have faster growth than striped bass in the early stages, exhibit high survival of stocked fry in reservoirs, and serve as a biological control of gizzard shad *Dorosoma cepedianum*.

Armed with the knowledge that hybrid *Morone* did well in freshwater systems and had potential as a management tool, biologists began to produce and stock hybrids in inland reservoirs. Ultimately, it was confirmed that hybrid *Morone* had excellent potential for management, and they provided similar exceptional recreational angling opportunities as those found with inland populations of striped bass. Not surprisingly, the demand for hybrids increased rapidly as the culture methodologies for producing and rearing them were perfected.

By 1981, more than 30 states reported having stocked striped bass, hybrid *Morone*, or both within their inland systems (Axon and Whitehurst 1985). By 1986, that number had increased to 19 states reporting stocking striped bass and 26 stocking hybrids (Smith and Reeves 1986). Today, according to U.S. Geological Survey Nonindigenous Aquatic Species maps and a survey conducted with the various state agencies responsible for the management of aquatic resources, *Morone*, in the form of endemic or transplanted species or congeneric hybrids, can now be found in 48 of the 50 states. Only Idaho and Alaska do not have any record of *Morone* ever being present or transplanted in their state (USGS 2009).

Of concern among these stockings and transplantations is the potential loss of genetic integrity (introgression) of pure stock progeny through the blending of DNA from two different stocks, subspecies, or species, or with one of its hybrids. This concern is that such introgression could have a genetic impact that could influence the long-term sustainability of the naturally reproducing populations. This chapter principally focuses on the issue of interspecific hybridization and examines introgression in *Morone* from backcrossing only (i.e., mating of striped bass hybrids with natural stocks of pure striped bass). For specific discussions of intraspecific introgression within *Morone* and a larger discussion of its potential impact on effective population size with respect to genetic diversity and population health, interested readers should review Diaz et al. (1997, 1998), Dunham et al. (1989), Harrell (2002), Bulak et al. (2004), Liu and Ely (2009), Wirgin et al. (1997, 2005), and Long et al. (2013, this volume).

Overall, the purpose of this synthesis is (1) to assess the evidence, or lack thereof, of how widespread *Morone* backcrossing may be in natural and man-made systems, (2) address why this may be occurring, (3) discuss why it may be an issue ecologically or genetically, and (4) interject ethical points management biologists should consider in managing natural systems with nonnative or hybrid stocks of *Morone* for utility as a management tool.

Artificial *Morone* hybridization

For purposes of clarity, throughout this chapter, I use the following convention of terminology: hybridization—crossing between two congeneric species of *Morone* (e.g., striped bass with white bass creating an F_1 fish); outcrossing—crossing between two conspecific stocks of *Morone* (e.g., an Atlantic coast stock striped bass crossed with a Gulf Coast striped bass (see Long et al. 2013) or between two F_1s or later F_xs); backcrossing—crossing between a hybridized *Morone* with its pure parental stock (e.g., an F_1 or F_x crossed back to a pure striped bass, white bass, white perch, or yellow bass depending on the parental cross that created the F_1 or F_x); and introgression—genes of a pure stock or hybrid are mixed with another pure stock of different origin (i.e., outcrossed or backcrossed fish result in introgression).

The most common hybrid crosses produced have been the palmetto bass (male white bass × female striped bass) and the sunshine bass (female white bass × male striped bass). Other artificially produced hybrids include Virginia bass (male white perch × female striped bass), Maryland bass (female white perch × male striped bass), and paradise bass (male yellow bass × female striped bass) (Harrell et al. 1990; Robins et al. 1991).

In nature, many interspecific hybrid crosses can be found in fishes as a group, and they are more common among freshwater than marine species (Hubbs 1955; Billington 2003). As mentioned, the original idea for hybrid *Morone* was that they would express similar traits to striped bass with respect to growth and management utility for controlling prey species, the white bass's ability to adapt to warm, shallow inland reservoirs, and their ability to reproduce in the wild would be limited (Bishop 1968). Unfortunately, however, it is not unusual for fish hybrids to be reproductively viable (Hubbs 1955).

As it turns out, *Morone* hybrids are functionally able to reproduce and have reproductively viable offspring, at least in the F_1 generation. Bayless (1968, 1972) and Bishop (1968) successfully produced outcrossed second-generation hybrids (F_2s—an $F_1 \times F_1$ cross) and backcrosses in the laboratory. Since that time, others have successfully spawned (artificially and naturally in tanks) various hybrids and reared them to maturity (Bonn et al. 1976; Harrell 1984; Smith and Jenkins 1984; Rees and Harrell 1990; Woods et al. 1995; Salek et al. 2001).

Natural Hybridization in *Morone*

To address the question of whether natural hybridization, outcrossing, or backcrossing was occurring in the wild, I conducted an extensive literature review and surveyed the 48 state fisheries agencies where documented records show that *Morone* are native or have been intentionally stocked. The questionnaire was simple and asked if there was any concern of natural hybridization or *Morone* outcrossing occurring in their jurisdictional waters; if there was a concern, did they believe the impact could be genetic or ecological (i.e., competition, predation, or displacement) or have management implications (i.e., reduction or increased targeted fishery); if there was concern, had they been actively trying to document evidence of it occurring, and if so, could such documentation be provided.

Literature findings

There are reported cases of *Morone* hybridization and outcrossing occurring, but there is not an overabundance of documented information from peer-reviewed sources or the gray literature. The first documented case of natural (nonhatchery produced) *Morone* hybridization was with yellow bass and white bass found in 1955 in Lake Wingra, Wisconsin, a part of the Yahara River system. This evidence was not confirmed until it was reported by Waldman and Bailey (1992), who examined the fish from a museum collection at the University of Michigan. Waldman and Bailey (1992) reported that historically, yellow bass were not native to the system and were introduced in the 1930s. Hybrids began showing up in the early to mid-1950s when yellow bass populations vastly outnumbered the native white bass (Waldman and Bailey 1992). Subsequent reports of hybrid *Morone* within this system were from Lake Mendota, Wisconsin (also part of the Yahara River system) in 1960 (Wright and Hasler 1967) and 1975 (Becker 1983).

Hybrid backcrossing or introgressive hybridization, was reported (Williams 1972) within 3 years after F_1 hybrid introductions were made in inland reservoirs, and biologists began to express concerns of the potential of hybrid reproduction and backcrossing. Williams (1972) first observed stocked F_1s (palmetto bass) participating in spawning activity in a South Carolina reservoir, although he did not confirm fertilization or hatch success. Beginning in the 1980s, it became more apparent that natural hybridization, outcrossing, and backcrossing in *Morone* had occurred, albeit infrequently (Avise and Van Den Avyle 1984; Todd 1986; Crawford et al. 1987; Forshage et al. 1988; Fries and Harvey 1989; Harrell et al. 1993; J. Epifanio, The Center for Aquatic Ecology, Illinois Natural History Survey, personal communication; Table 1).

In 1981, natural hybridization between introduced striped bass and native white bass was found in two reservoirs in Arkansas: Lake Maumelle and Beaver Lake (Crawford et al. 1987). No stocking evidence could account for the presence of those hybrids. Electrophoretic analysis could not determine which cross constituted the captured hybrids, but meristically they were determined to be sunshine bass (Crawford et al. 1987).

Todd (1986) collected 11 suspect hybrid *Morone* from Lake Erie in the summers of 1983 and 1984. Using starch gel electrophoresis, he was able to determine that nine of the suspect fish were F_1 hybrids between white bass and white perch and two were either F_2s or backcrosses to white perch, with the latter being the greater probability (Todd 1986). White perch and white bass are normally allopatric species in this system. He speculated that white perch immigration into Lake Erie was effected through various shipping canals. He also examined seven presumptive white bass juveniles from Maumee Bay, Ohio and found that five were F_1s and one was a backcross to white bass.

Avise and Van Den Avyle (1984) examined *Morone* (striped bass, white bass, hybrids, and suspect fish) from the Savannah River system on the Georgia–South Carolina border and found evidence that approximately 1% (6 of 642) of the fish sampled had recombinant genotypes, indicating introgressive hybridization or hybrid outcrossing. The Savannah River system is where Williams (1972) reported observing hybrid spawning activity in impounded upstream reservoirs. Avise and Van Den Avyle (1984) found that one fish was an F_2 resulting from hybrid parents successfully reproducing or an introgressed F_1 backcross resulting from an F_1 crossing back to a pure striped bass or white bass parent, with direction of backcross not being known. The five other fish might have been

Table 1. Confirmed natural *Morone* hybridization, outcrossings (F_x), or backcrossings (BC[a]) by state and cross (SB—striped bass, WB—white bass, WP—white perch, YB—yellow bass).

State	Year	Hybrid cross	F_x or backcross	Citation
WI	1955	WB × YB[b]		Waldman and Bailey (1992)
AR	1981	SB × WB[c]		Crawford et al. (1987)
Lake Erie, Maumee Bay, OH	1983–1984	WB × WP[d]	F_2 and BC	Todd (1986)
GA, SC, TN, TX/LA	1983		F_2, F_x, and BC[e]	Avise and Van Den Avyle (1984)
TX	1985		F_2 and BC[f]	Forshage et al. (1988)
TX	1986–1988	YB × WB[g]		Fries and Harvey (1989)
MD	1991	WP × SB	BC[h]	Harrell et al. (1993)
TN	2007–2008	WB × SB	BC[i]	J. Epifanio, The Center for Aquatic Ecology, Illinois Natural History Survey, personal communication

[a] With current technology, it is difficult to definitively tell which hybrid crossed back to a pure parent.
[b] Hybridization among introduced allopatric congeneric *Morone*—yellow bass were introduced into the system in the 1930s where white bass were the indigenous species.
[c] Hybridization among introduced allopatric congeneric *Morone*—striped bass were introduced into Lake Maumelle and Beaver Lake where white bass were the indigenous species.
[d] Hybridization among introduced allopatric congeneric *Morone*—where white perch moved through shipping channels into the Great Lakes and white bass were the indigenous species. Backcrosses are believed to be to white perch – maternal direction of the backcross was unknown. No information was given about the hybrids in Ohio.
[e] The F_x represents a later hybrid reproduction; the backcross direction was suspected to be to white bass (Cherokee Reservoir, TN). The TX/LA site was Toledo Bend Reservoir.
[f] F_2s are believed to have been the result of reproduction among stocked F_1 hybrids. The backcrosses are introduced white bass that are believed to have been introduced by anglers—maternal direction of backcross was unknown.
[g] Hybridization among sympatric congeneric white bass and yellow bass.
[h] The white perch × striped bass cross could have resulted from an earlier intentional stocking. The backcross was to a striped bass from stocked F_1 striped bass × white bass hybrids—maternal direction was unknown.
[i] The backcross was of a striped bass × white bass F_1 hybrid of undocumented origin x yellow bass—maternal direction was unknown.

the result of hybrid reproduction, but the authors were hesitant to make such a definitive claim as the fish were only heterozygous at one allozyme (Esterase) and homozygous for striped bass allozymes elsewhere (Avise and Van Den Avyle 1984). In addition, several suspect fish were sent to the authors for analysis from several locations throughout the South

(Alabama, Tennessee, and the Toledo Bend Reservoir on the Texas–Louisiana border). Their analysis revealed that the fish were later generation hybrids (i.e., F_x) or F_1 backcrosses to white bass (Avise and Van Den Avyle 1984).

In 1985, young presumptive hybrids were collected from Lake Palestine, Texas, where no hybrids had been stocked since 1982 (Forshage et al. 1988). Using electrophoresis, half of the (21 of 41) presumptive hybrids collected were identified as white bass, but 11 were classified as F_1s and 9 were determined to be nonF_1 hybrids (e.g., backcrosses). By means of isoelectric focusing, there were 21 white bass, 13 F_1 hybrids, and 7 nonF_1 hybrids. Technique agreement was incongruent with eight of the F_1s or nonF_1 hybrids, and therefore it was decided that if a fish was identified as a nonF_1 by either technique, it was classified as a nonF_1. With this caveat, the total number of nonF_1 hybrids was 12 or 29% (Forshage et al. 1988). It was determined that if backcrossing had occurred, the direction of the cross was with a female white bass. The white bass in the reservoir are reported to have been introduced by anglers (Forshage et al. 1988).

Similarly, in three other Texas reservoirs, Fries and Harvey (1989) found, between 1986 and 1988, that unusually large yellow bass caught by anglers were actually F_1 hybrids between naturally sympatric species of yellow bass and white bass. Hybrid determination was validated by isoelectric focusing (Fries and Harvey 1989). Because these are man-made reservoirs Fries and Harvey (1989) suggested that the habitat perturbation going from a lotic to a lentic system may have allowed normal reproductive isolation mechanisms to be disrupted.

In Chesapeake Bay, 2 of 82 field-identified F_1 hybrids collected in the winter of 1991 by Maryland Department of Natural Resources biologists could not be explained by previous stockings of palmetto and Maryland bass in the Chesapeake Bay's tributaries as a "marked population," F_1 hybrids stocked in the Susquehanna River drainage by the state of Pennsylvania, or F_1s stocked in small impoundments in Virginia—all part of the Chesapeake Bay watershed (Harrell et al. 1993). Seventy-nine were identified as F_1 striped bass × white bass hybrids, one was a striped bass × white perch, and the other two were backcrossed fish. Identification of all fish was via nDNA genomic probes at five different alleles (Harrell et al. 1993).

Recently (2007–2008), examination of presumptive white bass from Kentucky Lake, Tennessee, using allozyme markers, identified one individual as a striped bass × white bass backcross to a yellow bass (Epifanio, personal communication). A second individual fish was also misidentified as a yellow bass or a backcross. In Lake Barkley, Tennessee, five presumptive but unknown hybrids were assayed, and it was determined that three were F_1 striped bass × white bass hybrids and two were probable striped bass × white bass backcrosses to a yellow bass. In both Lakes Barkley and Kentucky, there were no records of recent state stockings that could account for the year-class of the F_1s collected. Mature F_1s would have also had to have been present to effect backcrossing to the yellow bass (Epifanio, personal communication). Not only is the presence of backcrosses a concern, but also the presence of F_1 hybrids that could not be explained by stocking records is an issue.

With two exceptions, all cases of natural hybridization appear to be the result of intentional stockings of formerly allopatric congeneric *Morone* or hybrids of the species. The only case of hybridization with naturally sympatric species was reported by Fries and Harvey (1989) in three Texas reservoirs, but there the habitat had been altered and the species coexisted in artificial reservoirs. In the Lake Erie case it is believed that white perch invaded the Great Lakes by gaining access through shipping channels that connect to Atlantic drainage system where white perch are indigenous (Todd 1986).

Questionnaire results

Of the 40 responses (83%) to the survey questionnaire, results reveal that outside of the above documented cases, only the states of Arizona, Kansas, and Nevada have unconfirmed reports of presumptive hybrid *Morone* being found in rivers or reservoirs where they were not previously stocked. These incidents do not exclude the possibility of escapees from aquaculture operations (a question not addressed in the survey) or potential unauthorized stockings by anglers, which have been known to occur (e.g., the stocking of white bass in Lake Palestine, Texas, as mentioned by Forshage et al. 1988).

Arizona reported that they have not stocked *Morone* hybrids in their watersheds (K. Young, Arizona Game and Fish Commission, personal communication). However, there are occasional reports by anglers of hybrid *Morone* being caught in their systems. None of these reports have been confirmed, nor have their state biologists reported any findings of *Morone* hybrids in Arizona waters.

Nevada reported three of their larger reservoirs have been stocked annually since 1993 with sunshine bass, and two reservoirs (Lahontan and Rye Patch) have reproducing introduced white bass populations (P. Sollbeger, Nevada Department of Wildlife, personal communication). In 2006, presumptive backcrosses appeared in Lahontan Reservoir, but no genetic testing was completed to confirm the results. The size and number did not correspond with stocked F_1 information. In 2008, plans were made to conduct genetic testing but a severe draw-down on the reservoir resulted in high angler catch rates, and many of the remaining fish died so the testing was not conducted (Sollberger, personal communication).

Kansas has stocked and managed *Morone* hybrids (sunshine and palmetto bass) since the 1970s, but the state has not evaluated introgression or outcrossing with parental stocks. It has, however, collected presumptive backcrosses between the stocked F_1s and white bass but has not genetically tested these fish (T. Mosher, Kansas Department of Wildlife and Parks, personal communication). They are currently using F_1s as a biological control for introduced white perch in several of their reservoirs.

While there was no effort of sampling for *Morone* outcrossing and backcrossing reported by the Nebraska Game and Parks Commission, there is limited evidence of *Morone* introgression in two reservoirs and a state hatchery (Lueckenhoff 2011; K. Pope, Nebraska Cooperative Research Unit, personal communication). It was not determined if the small number of fish (4 out of 967) from the reservoirs were unintentionally produced in a hatchery and stocked or if they were the result of natural hybrid outcrossing.

Outside of the documented literature and limited concerns in the state reports mentioned above, none of the state responses, except Kansas and Nevada, indicated that there was current reason to be concerned that hybridization or introgression among *Morone* was occurring, and therefore did not have any active effort ongoing to look for evidence. Also, none of the states, except Kansas, reported any significant concern about hybrid predation or competition with other species other than what was being specifically targeted (i.e., clupeids and white perch).

The reported case where sympatric populations have hybridized (Fries and Harvey 1989) demonstrates that the hybridization occurred after the habitat in which the fish were naturally found had been significantly altered (i.e., river systems impounded). Likewise, in the case of Lake Erie, artificial structures (shipping channels) afforded white perch access to the Great Lakes that resulted in a native white bass population being subject to hybridization (Todd 1986) and a reduction in recruitment (Madenjian et al. 2000). The other reported cases of allopatric congeneric *Morone* natural hybridization and further outcrossings are found only where the fish have been intentionally stocked by state agencies or by anglers.

Clearly, outcrossing, hybridization, and backcrossing have occurred with *Morone* in natural systems, with as many as 29% of suspected fish confirmed as nonF_1 hybrids (Forshage et al. 1988) and 13 different states reporting suspected or documented cases. Yet, the low overall incidence of these cases and lack of direct concerns by the different state agencies may indicate that it is not a widespread problem. Given the possibility of hybridization or backcrossing occurring, as evidenced from the few documented and suspected cases, it seems prudent to ask whether this is a case of "evidence of absence" or "absence of evidence." Based on the information at hand, one cannot know the answer due to insufficient evidence from which to draw a conclusion, which, in itself, begs the question about the need for a more proactive evaluation where *Morone* populations coexist either artificially or naturally.

Determining if outcrossing and backcrossing is occurring may be as simple as identification of the introgressed fish themselves. Intuitively one would think finding introgressed fish would provide the answer to whether it is a case of lack of evidence. Conversely, not finding hybridized or introgressed fish in a large sample may indicate that no introgression or outcrossing is occurring (evidence of absence). Yet, simply identifying the fish by gross observation is not as straightforward as one would expect because phenotypic identification of backcrossed or introgressed fish is difficult (Harrell and Dean 1988; Harrell 1997). Typically, meristically and morphometically F_1s are intermediate to the two congeneric fish used in the cross (Hubbs 1955), but backcrosses and other outcrossed fish are not a 50:50 allelelic combination. Therefore, depending on the direction of the cross, the resultant progeny may be more like one or the other original parents (Harrell and Dean 1988). Therefore, it is possible that occurrences may be more widespread than expected or reported. Conversely, difficulty in identification does not warrant concluding that introgressive hybridization is a more common occurrence than evidenced. Definitive identification requires some form of genotypic analysis, and few states are conducting such analyses.

While the answer to the question of occurrence may be as simple as definitive identification, the issue of potential impacts of even localized hybridization and introgressive hybridization in *Morone* is more complex. Protocols and sample size needed to be able to state with statistical confidence the biological and genetic significance of hybridization or outcrossing are contingent on many site-specific factors. These include whether the populations in question are naturally reproducing pure stocks, if it is a put-grow-and-and-take fishery, whether the concern is at the individual or population level, if the system is self-contained (i.e., a small municipal lake versus a mainstream reservoir versus the Great Lakes or an open coastal system), the local spawning behaviors, whether there is immigration or emigration into or out of a site, whether there is evidence of coadapted gene complexes, and other issues such as competition with other species. These are just some of the complexities that confound the issue, some of which are addressed below.

Why Does Hybridization or Introgression in *Morone* Occur?

Speculation exists as to why *Morone* may be outcrossing or backcrossing in nature, much of which appears to be related to anthropogenic factors similar to that described in an overview of hybridization by Scribner et al. (2001). Basically, when allopatric, congeneric species are mixed, natural isolating mechanisms are, or may be, eliminated and those congeneric species with similar reproductive habits have the ability to cross with one another creating hybrids (Hubbs 1959; Scribner et al. 2001). For example, when the *Morone* natural spawning habitat, such as a river, becomes impounded, the resultant new habitat creates opportunity for spatial spawning overlap of normally segregated congeneric or conspecific populations thereby creating hybrid zones, as is probably evidenced in the Fries and Harvey (1989) report.

Immigration and emigration from given watershed systems are common in *Morone*, at least by the Atlantic coast natives, striped bass and white perch. Striped bass management biologists have observed downstream movement from impounded reservoirs (Stewart and Burrell 2013, this volume), and white perch have a strong proclivity to invade new habitat (Todd

1986). This movement may be due, in part, to the innate diadromy behavior found in the species. There is anecdotal information that hybrid *Morone* with a normally diadromous striped bass parent exhibits a similar trait. In such cases, this downstream movement provides additional hybrid zones where there may be problems with competition for food, spawning habitat, and mate pairing (Patrick and Moser 2001).

In sympatric populations, the isolating mechanisms are usually related to sexual, behavioral, or physiological differences, including temporal-mating differences (Willis et al. 2009). Physiological trait expressions that lead to reproductive isolation, which are genetically mediated, can become suppressed or incompletely expressed due to increased heterozygosity in congeneric hybrids. They also can be disrupted by subjection to environmental or habitat stressors. Thus, any cues that may have previously kept sympatric species isolated become blurred and no longer serve as reproductive barriers (Wikelski and Cooke 2006). Such may have been the case with sympatric white bass and yellow bass spawning in Texas reservoirs (Fries and Harvey 1989).

Further, breakdown in isolating mechanisms could ultimately lead to hybrid zones where two species are more prone to interbreed, resulting in potential hybrid swarms where the population is dominated by hybridized or introgressed fish (Hubbs 1959). Hybrid swarms are common when populations invade new habitats (in this case introduced) and the ecological conditions conducive to hybridization are also conducive to adaptive radiation (Seehausen 2004). Thus, hybrid swarms become more possible where the hybrid zone is suitable for repetitive spawning over several spawning seasons.

A good example of hybrid swarms in fish is where subspecies of bluegill *Lepomis macrochirus* formed significant hybrid swarms in northeast Georgia and most of South Carolina and both subspecies lost genetic fidelity and former reproductive isolating mechanisms that were unique to each subspecies (Avise et al. 1984). Such could be the initial beginnings of a similar case in the three Texas reservoirs (Fries and Harvey 1989) where sympatric white bass and yellow bass are hybridizing: in Lake Palestine, Texas, where as high as 29% of the samples taken of presumptive F_1s are actually introgressed or further outcrossed hybrids with introduced white bass (Forshage et al. 1988); in Lake Erie, with the invading white perch interbreeding with native white bass (Todd 1986); the Savannah River system, where hybrids are outcrossing with striped bass (Avise and Van Den Avyle 1984); in Lakes Barkley and Kentucky, where yellow bass are mating with hybrid *Morone* (Epifanio, personal communication); or even within striped bass races themselves where Gulf Coast-race striped bass appear to be panmictic with stocked Atlantic coast-race striped bass (Long et al. 2013). Before conclusions are drawn regarding this speculative hypothesis, significant sampling with genomic techniques of the populations in question and correlating that information with year-class structure is warranted.

Why Is This Problematic?

Numerous reports have discussed the effects of hybridization and introgression among fish species; for more detailed overview, refer to Campton (1987, 1995), Hindar et al. (1991), Thorgaard and Allen (1992), Epifanio and Nielsen (2001), Epifanio and Philipp (2001), Scribner et al. (2001), Utter (2001, 2003), and Harrell (2002). Outside of the competition concerns mentioned by Patrick and Moser (2001), the primary issue with *Morone* hybridization and introgression is the loss of genetic purity. Thus, those states that have naturally reproducing *Morone* populations and want to maintain that purity need to consider if outcrossing, hybridization, or backcrossing is occurring and what impact this genetic mixing may have on the overall population stability of the native, pure stock.

The population stability question resulting from introgression ultimately relates to fitness or the ability of organisms to transfer their genes to the next generation through successful reproduction. Through introgression, genetic admixtures from two stocks or species may erode evolutionary flexibility and impact adaptation of the organism to its environment, ultimately resulting in the failure to reproduce (Meffe 1986). Therefore, fitness and variability are important for an organism to be able to respond to fluctuating environmental changes and continue to sustain a viable population.

The appeal of hybrids is the expression of hybrid vigor (i.e., heterosis) due to dominant allele interaction at given loci, but it may not last through successive generations. Enhanced vigor expression in *Morone* hybrids was one of the key reasons leading to the original efforts to hybridize the fish. In most hybrid crosses, including hybrid *Morone*, breakdown in this vigor begins with the second generation (F_2) through loss of dominance expression, epitasis, and other gene interactions (Harrell 1984; Smith and Jenkins 1984).

While it is evident that there is viability breakdown occurring in successive hybrid generations (F_x), it is not clear in *Morone* how a reduction in viability is expressed in backcrossing where introgression occurs. It is clear that any factor that leads to a reduction in fitness may lead to demise of the population.

Theoretically, gene combinations resulting from introgression can lead to lowered fitness in species where two previously reproductively isolated populations interbreed and produce progeny less adapted to their local environment (Campton 1995). This loss of fitness, commonly known as outbreeding depression (Templeton 1986), can impact recruitment success. The question then becomes does outcrossing impact the overall population genetic integrity or is it just affecting the individual fish that are outcrossed? If one follows the argument that outbreeding depression is important wherein nonnative conspecifics (or hybrid or introgressed fish) are introduced to a locally adapted populations and outcrossing results in breakdown of these coadapted gene complexes, then the impact could be serious because it is at the population level (see Campton 1987, 1995; Phillipp 1991; Waples 1991, 1999; Philipp and Claussen 1995; Harrell 2002 for further discussion).

Although this phenomenon is somewhat controversial in fisheries circles, one should at least consider the impact if outbreeding depression does occur. Here, the precautionary principle may apply (The Precautionary Principle Project 2011). This principle deals with response to uncertainty in the face of risks to health or the environment (Science and Environmental Health Network 2000). Basically, the principle states that if an action or policy results in environmental harm, in the absence of scientific consensus that the action is harmful, the burden of proof that the action is not harmful falls on those taking action.

Thus, a concern with conspecific stockings or backcrossing of hybrids to pure stocks is the potential for outbreeding depression in locally adapted populations leading to loss of that population. Even though outbreeding depression has yet to be demonstrated in striped bass, where backcrossing has been documented (Harrell 2002), lack of demonstration does not warrant concluding that it does not occur in the wild or that if it does, it is not problematic.

To summarize, genetic impacts are relevant only if one wants to maintain pure stocks that can reproduce, adapt to a changing environment, and recruit into a fishery. Ecologically, hybrid swarms and successful reproduction can result in direct competition, predation, and competitive displacement. If a management biologist is not concerned with natural recruitment, the issue becomes academic, unless there is the possibility of emigration downstream to a system with natural recruitment occurring. If the responsible state agency is concerned about natural *Morone* recruitment in their tributaries or reservoirs (or interjurisdictional systems), then consideration of the potential genetic and ecological damage that introgression and hybridization can do to long-term population success and adaptive flexibility is important. This concern must be coupled with appropriately designed rigorous evaluation to determine if biological, ecological, or genetic concerns are indeed warranted before significant management actions are undertaken.

A recurring theme expressed with the responses to the questionnaire was that in those states where a successful hatchery-stocked hybrid programs existed, they were pleased with the fish as a management tool for biological control of excess clupeid or other high density, undesirable stocks. Likewise, they were pleased with the value-added component of a targeted recreational sport fishery that hybrid *Morone* provided in their reservoirs.

Management Options and Ethical Considerations

A critical component of a response to the question of whether the presence of hybrid *Morone* in natural and man-made systems is of concern is a clear understanding of what fully constitutes defining the problem. In its broadest sense, the "problem" relates not only to biological (genetic) and ecological considerations, but also in today's "environmentally conscious" society, it has an ethical component. No longer do resource management agencies have the luxury of using only science-based criteria for assessment and management planning; today, ethical and philosophical assessments must be considered as well (Dallmeyer 2005; Helfman 2007).

As agency funding continues to morph from historical "hook and bullet" user and formula-based sources to more general state or federal appropriations and corporate or foundational support, so too does the delineation of who constitutes the agency's stakeholders. Often more important than funding drivers is the legal mandate for natural resource agencies to be the stewards of the state's indigenous natural resources. How stewardship is defined is often linked to stakeholder opinion, which is often subjective. As such, there is a groundswell of nonconsumptive stakeholders rightfully demanding input into natural resource management decisions. To that end, savvy biologists and agencies take into consideration such stakeholder concerns, including their ethical implications.

Ethical duty is germane to the tenets of fisheries management and therefore has relevance for management decisions, especially for the use of genetically altered fishes (i.e., intentionally and artificially hybridized) and their impacts on native populations. Clearly, with respect to the focused discussion of this chapter, ethical duty can be linked to management decisions that impact genetic purity (by introgression leading to loss of evolutionary flexibility and adaptability). Thus, how that protection and management is defined, implemented, and evaluated is variable, and the diversity of stakeholder inputs must be weighed with respect to fiscal, physical, practical, and scientific limitations. One duty ethic seen more and more frequently among stakeholders is the desire for protection of native species populations.

From a utilitarian ethic perspective, one must consider the greatest good for the greatest number of people. A management biologist or agency may conclude that it is the consumptive user that represents their greatest number of stakeholders. Thus, establishing a targeted fishery with hybrid or allopatric populations of *Morone* in a newly created reservoir may serve the greatest good. Conversely, the perception that consumptive-user groups represent the greatest number of stakeholders may not be supported politically or in reality. In making such decisions, biologists should look at the inputs objectively and empirically to determine how stakeholders are identified and how benefit is calculated and evaluated. For example, is the potential of introgression with native *Morone* going to impact the long-term survival of the extant indigenous populations, a question that consumptive and nonconsumptive stakeholders may both value practically and ethically? Using rigorous social-science methodologies, one could approach the answer to these questions from a statistically valid perspective. Regardless, it is also ethically responsible to conduct a rigorous assessment of the potential risks associated with such management activities.

Consideration of the national trend toward fewer consumptive users of our natural resources and more nonconsumptive individuals becoming vocal stakeholders indicates that resource managers need to pay closer attention to the ethical implications of their management decisions. The ecological issues of hybrid *Morone* being used to fill an available open-water niche for fishing and control of overpopulated prey populations may not be ethically important for society if it is perceived that such activities are harmful to desirable fish (e.g., genetics of native populations, competing predators and prey). Hence, management decisions may become second-order, publicly debated, ethical and biological questions that are often removed from the need for science-based management options. Without proper consideration and response to ethical dilemmas, the outcome of such responses will effectively remove a management biologist's tools from the toolbox, thereby limiting resource management effectiveness.

Thus, from a philosophical, ethical perspective, and as stewards of a public resource (e.g., Hardin's [1968] discussion on "the tragedy of the commons"), thoughtful consideration of ethical implications needs to be a part of the formula in making decisions. This consideration is important whether it be artificially moving nonindigenous stocks, creating and stocking interspecific hybrids, or recommending a new regulation for creel limits or legal size limits. By including ethical considerations into the decision matrix, natural resource agencies demonstrate that they are not only fulfilling their state-chartered responsibility, they are responsibly recognizing and addressing their social contract with all the citizens of the state to whom they answer. It is not only politically naïve to not consider ethical implications, it is irresponsible as it will only result in limiting options.

From the science perspective, the answers are more objective and empirical. When scientific rigor rejects a hypothesis about a problem, then it historically has been a nonissue. However, if rigorous work cannot reject the hypothesis and the agency has concerns about unforeseen impacts, then the agency might want to consider a variety of questions to help address the situation. For example, one may need to decide if the issue is a concern of introgression and at what level introgression is biologically significant. Has outbreeding depression occurred, or can it even be determined as having occurred? Are there other management options available? How will not stocking or taking steps to eliminate nonnative *Morone* impact the agency's management mandate? How would reduction of overpopulated prey species from the watershed be accomplished without the added predators? Will recreational anglers be impacted, and how will that affect license revenues and federal-formula funding? Will the public or special interest groups express concern that native stocks are being hybridized? What does the cost–benefit analysis show, and what is the marginal cost of the different considerations? These and other questions need to be considered when trying to determine if hybrid *Morone* present more of a risk than a benefit, and to address these issues some form of risk analyses needs to be undertaken.

Traditionally, risk assessment in fisheries has been ecologically oriented and applies to invasions of exotic or nonnative species or escape of aquacultured species; however, risk assessment principles apply to the issues addressed in this chapter. For an overview of ecological risk assessment and its issues, including economic assessment, review the works of Kolar and Lodge (2001, 2002), Keller et al. (2007), Gertzen et al. (2008), and Leung and Dugeon (2008). With respect to risk assessment with genetic impacts and associated gene flow, the literature outlining the impacts of transgeneic fish is invaluable, and interested readers should review Kapuscinski et al. (2007), Hayes et al. (2007), and Senanan et al. (2007). These documents should provide management biologists the questions that should be framed, the appropriate levels of statistical and biological significance, and how to approach adequately addressing the situation that is unique to their state or system.

If there is a problem of risk (real or perceived), then there are only a limited number of options and most follow the tenets of the precautionary principle (The Precautionary Principle Project 2011). If the concern is about maintaining purity of indigenous populations and the water body has not yet received nonindigenous *Morone*, the first option is to not stock any conspecific or congeneric *Morone* in watersheds that provide access to reproducing populations. Because many of the established self-sustaining populations of *Morone* in the United States exist in watersheds that are interjurisdictional and cross state lines, working with adjacent states toward a common purpose is essential.

If nonindigenous, conspecific *Morone* or congeneric hybrids already exist within systems with reproducing populations, the options are more limited, especially if outcrossing is occurring. The only viable options appear to be to stop stocking hybrids and hope no (or limited) introgression has occurred or will further occur. Over time, with natural and fishing mortalities, the chance of further outcrossing occurring should become progressively less due to Allee affects. Unfortunately, there are limited options with self-reproducing conspecifics where natural isolating mechanisms have been broken down.

In large systems such as mainstream reservoirs, coastal rivers, or the Great Lakes, the option of commercial fishing for the hybrids or nonnatives to speed their removal could be considered. This approach is highly problematic because of bycatch, and it may be impractical politically, economically, and biologically. In a smaller watershed, consideration could be given to a reclamation approach through use of fish toxicants or drawdown followed by increasing angler creels, netting, or toxicant use, if appropriate. However, such efforts are rarely 100% effective, are likely cost-prohibitive, and can carry with them a negative public perception.

The use of sterile fish could be considered in systems with self-sustaining *Morone* populations where there is a desire to also introduce hybrids for additional management purposes and recreational fishing. It has been demonstrated previously that triploid hybrid striped bass are sterile with vestigial gonad development (Harrell et al. 1998; Kerby and Harrell 1990; Kerby et al. 2002). However, the technology is not 100% guaranteed because the standard method for producing triploidy is to mechanically force retention of the second polar body after fertilization. Success is influenced by many factors such as timing, temperature, and treatment. Unless 100% validation of ploidy is required before release, only a percentage of the fish need to be evaluated to provide an idea of the risk potential. This decision should be based on known species mortality rates, estimated size of the spawning stock, and spawning success to establish confidence intervals of an acceptable range.

Because of the lack of demand, hatchery biologists have yet to pursue creating natural triploids on a large scale by crossing artificially created tetraploids with normal diploids. Creating natural triploids through the use of tetraploids is expensive in that a captive tetraploid broodstock would initially have to be produced (mechanically), verified as tetraploids, reared to maturity, induced to spawn, and then successfully crossed with the natural diploid fish. The effective use of this approach on a scale large enough to stock progeny for reservoir management and put-grow-and-take recreational fishing, while plausible, is undetermined.

The last option to consider is in those systems where there are no self-reproducing native *Morone* stocks and there is a need for some management control of prey species or a desire to provide recreational fishing opportunities. In this case, stocking diploid hybrids have proven to be a success from most every perspective. Alternatively, if the food supply is adequate although the thermal–dissolved oxygen habitat may be suboptimal, according to the "rules" laid out by Thompson and Rice (2013) one might consider stocking pure striped bass instead of hybrids. In this latter case, it may be possible to avoid the concern of introgressive hybridization within a given system since hybrids are not being used and striped bass rarely reproduce successfully in reservoirs. It is, however, important to consider what *Morone* populations exist downstream from your stockings if escapement or possible downstream migration (Stewart and Burrell 2013) could impact other stocks.

Summary and Conclusions

From the literature and the questionnaire, there is a modicum of evidence that hybridization among, and introgression within, *Morone* species has occurred in the wild. It appears from this evidence that the frequency of occurrence is low enough to be considered rare. It is not clear, however, whether the rarity of evidence is due to the difficulty of visually identifying outcrossed and introgressed fish or because there is no deliberate effort to genetically identify suspect fish. Alternatively, is available information an evidence of absence because hybridization and introgression are truly not widespread? Obviously, only proper sampling and assessment will provide that answer.

As mentioned, empirically and definitively determining if hybridization or introgression has occurred is not a simple task. Regardless, the potential for hybridization and outcrossing exists. It is prudent to consider this possibility before implementing management options that involve stocking hybrid or nonnative *Morone* into any natural body of water where extant populations of *Morone* occur.

Considering the potential genetic and ecological risks the management biologist faces in the use of hybrid *Morone* as a management tool, it would also be prudent to consider the ethical implications of such decisions before implementing action. Ethical considerations, and common sense, argue that we need a rigorous evaluation of the potential risks associated with *Morone* stocking, and until that is accomplished, we should follow the precautionary principle.

Acknowledgments

Funding support was provided by the University of Maryland Agricultural Experiment Station, College of Agriculture and Natural Resources in association with Multistate Project NC_temp118.

References

Avise, J. C., E. Bremingham, L. G. Kessler, and N. C. Saunders. 1984. Characterization of mitochondrial DNA variability in a hybrid swarm between subspecies of bluegill sunfish (*Lepomis macrochirus*). Evolution 38(5):931–941.

Axon, J. R., and D. K. Whitehurst. 1985. Striped bass management in lakes with emphasis on management problems. Transactions of the American Fisheries Society 114:8–11.

Avise, J. C., and M. J. Van Den Avyle. 1984. Genetic analysis of reproduction of hybrid white bass × striped bass in the Savannah River. Transactions of the American Fisheries Society 113:563–570.

Bailey, W. M. 1975. An evaluation of striped bass introductions in the southeastern United States. Proceedings of the Annual Conference Southeastern Association of Game and Fish Commissioners 28(1974):54–68.

Bayless, J. D. 1968. Striped bass hatching and hybridization experiments. Proceedings of the Annual Conference Southeastern Association of Game and Fish Commissioners 21(1967):233–244.

Bayless, J. D. 1972. Artificial propagation and hybridization of striped bass, *Morone saxatilis* (Walbaum). South Carolina Wildlife and Marine Resources Department, Columbia.

Becker, G.C. 1983. Fishes of Wisconsin. University of Wisconsin Press, Madison

Billington, N. 2003. Mitochondrial DNA. Pages 83–84 *in* E. M. Hallerman, editor. Population genetics: principles and applications for fisheries scientists. American Fisheries Society, Bethesda, Maryland.

Bishop, R. D. 1968. Evaluation of the striped bass (*Roccus saxatilis*) and white bass (*R. chrysops*) hybrids after two years. Proceedings of the Annual Conference of Game and Fish Commissioners 21:245–254.

Bonn, E. W., W. M. Bailey, J. D. Bayless, K. E. Erickson, and R. E. Stevens. 1976. Guidelines for striped bass culture. American Fisheries Society, Southern Division, Striped Bass Committee, Bethesda, Maryland.

Bulak, J. S., C. S. Thomason, K. Han, and B. Ely. 2004. Genetic variation and management of striped bass populations in the coastal rivers of South Carolina. North American Journal of Fisheries Management 24:1132–1329.

Campton, D. E. 1987. Natural hybridization and introgression in fishes: methods of detection and interpretation. Pages 333–334 *in* N. Ryman and F. Utter, editors. Population genetics and fishery management. University of Washington Press, Seattle.

Campton, D. E. 1995. Genetic effects of hatchery fish on wild populations of Pacific salmon and steelhead: what do we really know? Pages 337–353 *in* H. L. Schramm, Jr., and R. G. Piper, editors. Uses and effects of cultured fishes in aquatic ecosystems. American Fisheries Society, Symposium 15, Bethesda, Maryland.

Coutant, C. C. 1985. Striped bass, temperature, and dissolved oxygen: a speculative hypothesis for environmental risk. Transactions of the American Fisheries Society 114:31–61.

Coutant, C. C. 2013. When is habitat limiting for striped bass? Three decades of testing the temperature–oxygen squeeze hypothesis. Pages 65–91 *in* J. S. Bulak, C. C. Coutant, and J. A. Rice, editors. Biology and management of inland striped bass and hybrid striped bass. American Fisheries Society, Symposium 80, Bethesda, Maryland.

Crawford, T., M. Freeze, R. Fourt, S. Henderson, G. O'Bryan, and D. Philipp. 1987. Suspected natural hybridization of striped bass in two Arkansas reservoirs. Proceedings of the Annual Conference Southeastern Fish and Wildlife Agencies. 38:455–469.

Dallmeyer, D. G. 2005. Toward a sea ethic. Pages 410–421 *in* E. A. Norse and L. B. Crowder, editors. Marine conservation biology; the science of maintaining the sea's biodiversity. Marine Conservation Biology Institute, Island Press, Washington, D.C.

Diaz, M., G. LeClerc, and B. Ely. 1997. Nuclear DNA markers reveal low levels of genetic divergence among Atlantic and Gulf of Mexico populations of striped bass. Transactions of the American Fisheries Society 126:163–165.

Diaz, M., J. MacPherson, and B. Ely. 1998. Striped bass population subdivision within the Santee–Cooper system, South Carolina. Molecular Marine Biology and Biotechnology 7(3):191–196.

Dunham, R. A., R. O. Smitherman, R. W. Chapman, T. Hess, and C. Messing. 1989. Mitochondrial DNA polymorphism of striped bass in the Southeastern United States. Proceedings of the Annual Conference Southeastern Association of Fish and Wildlife Agencies 42(1988):21–26.

Epifanio, J., and D. Philipp. 2001. Simulating the extinction of parental lineages from introgressive hybridization: the effects of fitness, initial proportions of parental taxa, and mate choice. Reviews in Fish Biology and Fisheries 10:339–354.

Epifanio, J., and J. Nielsen. 2001. The role of hybridization in the distribution, conservation, and management of aquatic species. Reviews in Fish Biology and Fisheries 10:245–251.

Forshage, A. A., W. D. Harvey, K. E. Kulzer, and L. T. Fries. 1988. Natural reproduction of white bass × striped bass in a Texas reservoir. Proceedings of the Annual Conference Southeastern Fish and Wildlife Agencies 40:9–14.

Fries, L. T., and W. D. Harvey. 1989. Natural hybridization of white bass with yellow bass in Texas. Transactions of the American Fisheries Society 118:87–89.

Gertzen, E., O. Familiar, and B. Leung. 2008. Quantifying invasion pathways: fish introductions from the aquarium trade. Canadian Journal of Fisheries and Aquatic Sciences 65:1265–1273.

Gustaveson, W., and G. Blommer. 2013. History of striped bass management in the Colorado River. Pages 15–23 *in* J. S. Bulak, C. C. Coutant, and J. A. Rice, editors. Biology and management of inland striped bass and hybrid striped bass. American Fisheries Society, Symposium 80, Bethesda, Maryland.

Hardin, G. 1968. The tragedy of the commons. Science 162:1243–1248.

Harrell, R. M. 1984. Tank spawning of first generation striped bass × white bass hybrids. The Progressive Fish-Culturist 46(2):75–78.

Harrell, R. M. 1997. Hybridization and genetics. Pages 217–234 *in* R. M. Harrell, editor. Striped bass and other *Morone* culture. Developments in Aquaculture and Fisheries Science, volume 30. Elsevier, New York.

Harrell, R. M. 2002. Genetic implications of escaped and intentionally-stocked cultured fishes. Pages 167–196 *in* J. Tomasso, editor. Aquaculture and the environment in the United States. United States Aquaculture Society/World Aquaculture Society, Baton Rouge, Louisiana.

Harrell, R. M., and D. W. Webster. 1997. An overview of *Morone* culture. Pages 1–10 *in* R. M. Harrell, editor, Striped bass and other Morone culture. Developments in Aquaculture and Fisheries Science, volume 30. Elsevier, New York.

Harrell, R. M., J. H. Kerby, and R. V. Minton, editors. 1990. Culture and propagation of striped bass and its hybrids. American Fisheries Society, Southern Division, Striped Bass Committee, Bethesda, Maryland.

Harrell, R. M., and J. M. Dean. 1988. Identification of juvenile hybrids of *Morone* based on meristics and morphometrics. Transactions of the American Fisheries Society 117:529–535.

Harrell, R. M., W. Van Heukelem, and J. H. Kerby. 1998. A comparison of triploid induction validation techniques. The Progressive Fish-Culturist 60:221–226.

Harrell, R. M., X. L. Xu, and B. Ely. 1993. Evidence of introgressive hybridization in Chesapeake Bay *Morone*. Molecular Marine Biology and Biotechnology 2(5):291–299.

Hayes, K. R., H. M. Regan, and M. A. Burgman. 2007. Introduction to the concepts and methods of uncertainty analysis. Pages 188–208 *in* A. R. Kapuscinski, K. R. Hayes, S. Li, and G. Dana, editors. Environmental risk assessment in genetically modified organisms. Volume 3: Methodologies for transgenic fish. CABI, Cambridge, Massachusetts.

Helfman, G. S. 2007. The ethics of exploitation and intervention: do we have the right? Pages 441–454 *in* G. S. Helfman, editor. Fish conservation: a guide to understanding and restoring global aquatic biodiversity and fishery resources. Island Press, Washington, D.C.

Hindar, K., N. Ryman, and F. Utter. 1991. Genetic effects of culture fish on natural fish populations. Canadian Journal of Fisheries and Aquatic Sciences 48:945–957.

Hubbs, C. L. 1955. Hybridization between fish species in nature. Systematic Zoology 4:1–20.

Hubbs, C. 1959. Population analysis of a hybrid swarm between *Gambusia affinis* and *G. heterochir*. Evolution 13:236–246.

Kapuscinski, A. R., J. J Hard, K. M. Paulson, R. Neira, A. Ponniah, W. Komanrat, W. Mwanja, I. A. Fleming, J. Gallardo, R. H. Devlin, and J. Trisak. 2007. Approaches to assessing gene flow. Pages 112–150 *in* A. R. Kapuscinski, K. R. Hayes, S. Li, and G. Dana, editors. Environmental risk assessment in genetically modified organisms. Volume 3: Methodologies for transgenic fish. CABI, Cambridge, Massachusetts.

Keller, R. P., D. M. Lodge, and D. C. Finnoff. 2007. Risk assessment for invasive species produces net bioeconomic benefits. Proceedings of the National Academy of Sciences of the United States of America 104:203–207.

Kerby, J. H., J. M. Everson, R. M. Harrell, J. G. Geiger, C. C. Starling, and H. Revels. 2002. Performance comparisons between diploid and triploid sunshine bass in fresh water ponds. Aquaculture 211:91–108.

Kerby, J. H., and R. M. Harrell. 1990. Hybridization, genetic manipulation, and gene pool conservation of striped bass. Pages 159–190 *in* R. M. Harrell, J. H. Kerby, and R. V. Minton, editors. Culture and propagation of striped bass and its hybrids. American Fisheries Society, Southern Division, Striped Bass Committee, Bethesda, Maryland.

Kolar, C. S., and D. M. Lodge. 2001. Progress in invasion biology: predicting invaders. Trends in Ecology and Evolution 16:199–204.

Kolar, C. S., and D. M. Lodge. 2002. Ecological predictions and risk assessment for alien fishes in North America. Science 298:1233–1236.

Lee, D. S., C. R. Gilbert, C. H. Hocutt, R. E. Jenkins, D. E. McAllister, and J. R. Stauffer, Jr. 1980. Atlas of North American freshwater fishes. North Carolina Museum of Natural History, North Carolina Biological Survey, Publication No. 1980-12, Raleigh.

Leung, K. M. Y., and D. Dugeon. 2008. Ecological risk assessment and management of exotic organisms associated with aquaculture activities. Pages 67–100 *in* M. G. Bondad-Reantaso, J. R. Authur, and R. P. Suasinghe, editors. Understanding and applying risk analysis in aquaculture. Food and Agricultural Organization of the United Nations, Fisheries and Aquaculture Technical Paper No. 519, Rome.

Lewis, R. E., T. E. Cheek, F. D. Leckie, Jr., R. L. McCabe, J. L. Moss, and G. L. Summers. 1990. Overview and considerations for management of striped bass and striped bass hybrids. Pages 239–242 *in* R. M Harrell, J. H. Kerby, and R. V. Minton, editors. Culture and propagation of striped bass and its hybrids. American Fisheries Society, Southern Division, Striped Bass Committee, Bethesda, Maryland.

Liu, J. X., and B. Ely. 2009. Sibship reconstruction demonstrates the extremely low effective population size of striped bass *Morone saxatilis* in the Santee–Cooper system, South Carolina, USA. Molecular Ecology 18:4112–4120.

Long, E. A., C. L. Mesing, K. J. Herrington, R. R. Weller, and I. I. Wirgin. 2013. Restoration of Gulf striped bass: lessons and management implications. Pages 25–63 *in* J. S. Bulak, C. C. Coutant, and J. A. Rice, editors. Biology and management of inland striped bass and hybrid striped bass. American Fisheries Society, Symposium 80, Bethesda, Maryland.

Lueckenhoff, R. W. 2011. Morphological variation between juvenile white bass and juvenile hybrid striped bass. Master's thesis. University of Nebraska, Lincoln.

Madenjian, C. P., R. L. Knight, M. T. Bur, and J. L. Forney. 2000. Reduction in recruitment of white bass in Lake Erie after invasion of white perch. Transactions of the American Fisheries Society 129:1340–1353.

Meffe, G. K. 1986. Conservation genetics and the management of endangered fishes. Fisheries 11(1):14–23.

Patrick, W. S., and M. L. Moser. 2001. Potential competition between hybrid striped bass (*Morone saxatilis* × *M. americana*) and striped bass (*M. saxatilis*) in the Cape Fear River estuary, North Carolina. Estuaries 24:425–429.

Phillipp, D. P. 1991. Genetic implications of introducing Florida largemouth bass *Micropterus salmoides floridanus*. Canadian Journal of Fisheries and Aquatic Sciences 48(Supplement 1):58–65.

Philipp, D. P., and J. E. Claussen. 1995. Fitness and performance differences between two stocks of largemouth bass from different river drainages within Illinois. Pages 236–243 *in* H. L. Schramm, Jr. and R. G. Piper, editors. Uses and effects of cultured fishes in aquatic ecosystems. American Fisheries Society, Symposium 15, Bethesda, Maryland.

Raney, E. C., 1952. The life history of the striped bass. *Roccus saxatilis* (Walbaum). Bulletin of the Bingham Oceanographic Collection (Yale University) 14:5–97.

Rees, R. A., and R. M. Harrell. 1990. Artificial spawning and fry production of striped bass and hybrids. Pages 43–72 *in* R. M. Harrell, J. H. Kerby, and R. V. Minton, editors. Culture and propagation of striped bass and its hybrids. American Fisheries Society, Southern Division, Striped Bass Committee, Bethesda, Maryland.

Rice, J. A., J. S. Thompson, J. A. Sykes, and C. T. Waters. 2013. The role of metalimnetic hypoxia in striped bass summer kills: consequences and management implications. Pages 121–145 *in* J. S. Bulak, C. C. Coutant, and J. A. Rice, editors. Biology and management of inland striped bass and hybrid striped bass. American Fisheries Society, Symposium 80, Bethesda, Maryland.

Robins, C. R., R. M. Bailey, C. E. Bond, J. R. Brooker, E. A. Lachner, R. N. Lea, and W. B. Scott. 1991. Common and scientific names of fishes from the United States and Canada, 5th edition. American Fisheries Society, Special Publication 20, Bethesda, Maryland.

Salek, S. J., J. Godwin, C. V. Sullivan, and N. E. Stacey. 2001. Courtship and tank spawning behavior of temperate basses (Genus *Morone*). Transactions of the American Fisheries Society 130:833–847.

Science and Environmental Health Network. 2000. A common sense way to protect health and the environment. Science and Environmental Health Network. Available: www.mindfully.org/Precaution/Precautionary-Principle-Common-Sense.htm. (February 2011).

Scribner, K. T., K. S. Page, and M. L. Bartron. 2001. Hybridization in freshwater fishes: a review of case studies and cytonuclear methods of biological inference. Reviews in Fish Biology and Fisheries 10:293–323.

Scruggs, G. D., Jr., and J. C. Fuller, Jr. 1955. Indications of a freshwater population of striped bass, *Roccus saxatilis* (Walbaum), in Santee–Cooper reservoirs. Proceedings of the Southeastern Association of Game and Fish Commissioners 8(1954):64–69.

Seehausen, O. 2004. Hybridization and adaptive radiation. Trends in Ecology and Evolution 19:198–207.

Senanan, W., J. J. Hard, A. Alcivar-Warren, J. Trisak, M. Zakararia-Ismail, and M. Lorenzo Hernandez. 2007. Risk management: post approval monitoring and remediation. Pages 239–271 *in* A. R. Kapuscinski, K. R. Hayes, S. Li, and G. Dana, editors. Environmental risk assessment in genetically modified organisms. Volume 3: Methodologies for Transgenic Fish. CABI, Cambridge, Massachusetts.

Smith, B. W., and W. C. Reeves. 1986. Stocking warm-water species to restore or enhance fisheries. Pages 17–29 *in* R. H. Stroud, editor. Fish culture in fisheries management. American Fisheries Society, Fish Culture Section and Fisheries Management Section, Bethesda, Maryland.

Smith, T. I. J., and W. E. Jenkins. 1984. Controlled spawning of F_1 hybrid striped bass (*Morone saxatilis* × *M. chrysops*) and rearing of F_2 progeny. Journal of the World Mariculture Society 15:145–161.

Stewart, W. T., and M. Burrell. 2013. Striped bass dispersion and effects on fisheries management in Lakes Mohave and Pleasant, Colorado River basin. Pages 431–448 *in* J. S. Bulak, C. C. Coutant, and J. A. Rice, editors. Biology and management of inland striped bass and hybrid striped bass. American Fisheries Society, Symposium 80, Bethesda, Maryland.

Templeton, A. R. 1986. Coadaptation and outbreeding depression. Pages 105–116 *in* M. E. Soule, editor. Conservation biology: the science of scarcity and diversity. Sinauer Associates, Sunderland, Massachusetts.

The Precautionary Principle Project. 2011. Guidelines for applying the precautionary principle to biodiversity conservation and natural resource management. Available: www.pprinciple.net/PP_guidelines_brochure.pdf. (February 2011)

Thompson, J. S., and J. A. Rice. 2013. The relative influence of thermal experience and forage availability on growth of age 1–5 striped bass in two southeastern reservoirs. Pages 93–120 *in* J. S. Bulak, C. C. Coutant, and J. A. Rice, editors. Biology and management of inland striped bass and hybrid striped bass. American Fisheries Society, Symposium 80, Bethesda, Maryland.

Thorgaard, G. H., and S. K. Allen. 1992. Environmental impacts of inbred hybrid and polyploid aquatic species. Pages 281–288 *in* A. Rosenfield and R. Mann, editors. Dispersal of living organisms into aquatic ecosystems. University of Maryland, Maryland Sea Grant College, Publication UM-SG-TS-92–04, College Park.

Todd, T. T. 1986. Occurrence of white bass-white perch hybrids in Lake Erie. Copeia 1986:196–199.

USGS (United States Geological Survey). 2009. Nonindigenous aquatic species data base home page for *Morone* and subsequent hybrid crosses. Available: http://nas.er.usgs.gov/queries/SpeciesList.asp?Group=Fishes&Genus=Morone&Species=&ComName. (September 2009)

Utter, F. M. 2001. Patterns of subspecific anthropogenic introgression in two salmonids genera. Reviews in Fish Biology and Fisheries 10:265–279.

Utter, F. 2003. Genetic impacts of fish introductions. Pages 364–370 *in* E. M. Hallerman, editor. Population genetics: principles and applications for fisheries scientists. American Fisheries Society, Bethesda, Maryland.

Van Horn, S. L. 2013. A brief history of inland striped bass management. Pages 1–13 *in* J. S. Bulak, C. C. Coutant, and J. A. Rice, editors. Biology and management of inland striped bass and hybrid striped bass. American Fisheries Society, Symposium 80, Bethesda, Maryland.

Waldman, J. R., and R. M. Bailey. 1992. Early occurrence of natural hybridization within *Morone* (Perciformes). Copeia 1992:552–559.

Waples, R. S. 1991. Genetic interaction between hatchery and wild salmonids: lessons from the Pacific Northwest. Canadian Journal of Fisheries and Aquatic Sciences 48:124–133.

Waples, R. S. 1999. Dispelling some myths about hatcheries. Fisheries 24(2):12–21.

Whitehurst, D. K., and R. E. Stevens. 1990. History and overview of striped bass culture and management. Pages 1–5 *in* R. M Harrell, J. H. Kerby, and R. V. Minton, editors. Culture and propagation of striped bass and its hybrids. American Fisheries Society, Southern Division, Striped Bass Committee, Bethesda, Maryland.

Wikelski, M., and S. J. Cooke. 2006. Conservation physiology. TRENDS in Ecology and Evolution 21(2):38–46.

Williams, H. M. 1972. Preliminary fecundity studies of the hybrid (striped bass × white bass) in two South Carolina reservoirs. Proceedings of the Annual Conference Southeastern Association of Game and Fish Commissioners 25(1971):536–542.

Willis, D. W., C. G. Scalet, and L. D. Flake. 2009. Introduction to fisheries and wildlife: an integrated approach. Freeman, New York.

Wirgin, I., D. Currie, N. Roy, L. Maceda, and J. R. Waldman. 2005. Introgression of nuclear DNA (nDNA) alleles of stocked Atlantic coast striped bass with the last remaining native Gulf of Mexico population. North American Journal of Fisheries Management 25:464–474.

Wirgin, I., L. Maceda, J. Strabile, and C. Messing. 1997. An evaluation of introgression of Atlantic coast striped bass mitochondrial DNA in a Gulf of Mexico population using formalin-preserved museum collections. Molecular Ecology 6:907–916.

Wright, T. D. and A. D. Hasler. 1967. An electrophoretics analysis of the effects of isolation and homing behavior upon the serum proteins of the white bass (*Roccus chrysops*) in Wisconsin. American Naturalist 101:401–413.

Woods, L. C., III, C. C. Kohler, R. J. Sheehand, and C. V. Sullivan. 1995. Volitional tank spawning of female striped bass with male white bass produces hybrid offspring.

American Fisheries Society Symposium 80:551–558, 2013

Economics of Recreational Fisheries for Inland Striped Bass and Hybrid Striped Bass

JOHN C. WHITEHEAD*
Department of Economics, Appalachian State University
3102 Raley Hall, Boone, North Carolina 28608, USA

Abstract.—This paper considers the economics of recreational inland striped bass *Morone saxatilis* and hybrid striped bass fisheries using the 2001 National Survey of Fishing, Hunting, and Wildlife-Associated Recreation database. First, the economics of recreational fishing is described using the basic demand model from introductory economics. I describe the data with a demographic summary of striped bass anglers and their targeting behavior, expenditures, and other relevant data. An average striped bass angler takes 24 fishing trips and specifically targets striped bass on nine fishing days per year. The aggregate economic expenditure is about US$1.2 billion dollars (2010 dollars) on striped bass fishing trips. I estimate a demand model for striped bass fishing days and use the results to infer the value of striped bass fishing day trips. The aggregate consumer surplus is about $1.85 billion.

Introduction

Striped bass *Morone saxatilis* and hybrid striped bass are important recreational fish in the United States. Striped bass are anadromous fish but have been introduced in inland water bodies for the primary purpose of recreational fishing (Van Horn 2013, this volume). While there has been much research on recreational striped bass fisheries in coastal waters (e.g., Patrick et al. 2006), little attention has been paid to inland striped bass fisheries (see Schorr et al. 1995 for an exception). In this paper, I describe the economic analysis of recreational fisheries, using data from the 2001 National Survey of Fishing, Hunting, and Wildlife-Associated Recreation (U.S. Department of the Interior, Fish and Wildlife Service, and U.S. Department of Commerce, U.S. Census Bureau 2002) to demonstrate these concepts. To the extent that the data allow, I provide estimates of national economic expenditures and benefits for inland striped bass, hybrid striped bass, and white bass *M. chrysops*. These fish are grouped in the database as one category, hereafter referred to as "striped bass."

Since much of the striped bass stocking activity is government funded, it is appropriate to consider whether the activity is in the public interest. One approach that economists take to address this question is benefit–cost analysis (Zerbe and Bellas 2006). Benefit–cost analysis is an information tool that sums all of the benefits of an activity and compares these to the sum of the costs of the activity. If the benefits exceed the costs, then the activity is considered economically efficient. In other words, the activity is a good way to allocate public resources. If the resources are allocated such that the additional benefits of an additional unit of the activity are just equal to the additional costs of one unit, then the activity has maximized the net benefits to society.

Monetary estimates of benefits and costs are needed in order to conduct benefit–cost

* Corresponding author: whiteheadjc@appstate.edu

analysis. Program costs are the opportunity costs of the activity. Opportunity costs are defined as the value of the next best alternative use of the consumed resources. Most of the government expenditure used to fund stocking programs comes from license fees and the federal excise tax on fishing gear. The next best alternative use of these funds is the goods and services that anglers would purchase if they did not pay fees and taxes. Fortunately, cost estimates can be estimated easily with little bias as the government expenditure on stocking programs.

Benefit estimates are more elusive since these occur outside of observable markets for tradeable goods and services. Nonmarket valuation methods have been developed to estimate the benefits of nonmarket behavior, such as recreational fishing (Champ et al. 2003). There are two types of nonmarket valuation methods. Revealed preference methods rely on actual behavior to generate data that are used to estimate economic benefits. Stated preference methods rely on answers to hypothetical questions made by survey respondents.

The travel cost method is a revealed preference method that is used most often to estimate the benefits of outdoor recreation (Parsons 2003). The travel cost method begins with the insight that individuals reside at varying distances from the recreation site. The negative relationship between travel cost and the number of trips taken is used to estimate a demand curve for the recreation site. The demand curve is then used to derive the consumer surplus (i.e., economic benefits) associated with using the site. With data on appropriate demand-curve-shift variables (i.e., independent variables such as measures of fish abundance), the economic benefits associated with changes in the shift variables (e.g., changes in catch rates) can be estimated.

The contingent valuation method (Boyle 2003) is a stated preference approach that directly elicits statements on willingness (and ability) to pay from survey respondents. Respondents are asked directly about their willingness to pay for a change in resource allocation. The method involves the development of a hypothetical market via in-person, telephone, mail, and Internet surveys (Champ 2003). In the hypothetical market, respondents are informed about the current problem and the policy designed to mitigate the problem. The choice is often framed as a referendum vote in order to make the situation more realistic. Respondents can be presented with multiple scenarios and make multiple choices. Statistical analysis of these data leads to the development of estimates of willingness to pay and consumer surplus.

There is a vast literature on the economic value of recreational fishing, with most studies using the travel cost method or the contingent valuation method (Johnston et al. 2006). However, studies that examine the recreational and related benefits of stocked fish are few. Patrick et al. (2006) examined the cost effectiveness of a stocking program for coastal striped bass in North Carolina. Recreational benefit estimates were obtained from a recreation demand model of striped bass fishing using the travel cost method. Rosenberger et al. (2005) estimated the benefits of a private trout-stocking program in West Virginia using the contingent valuation method. They conducted an intercept survey of anglers and supplemented with a less representative sample of area anglers. They found that the willingness to pay for the stocking program was between US$4 and $6 per fishing day (2010 dollars).

The goals of this chapter are more limited. Instead of conducting a benefit–cost analysis of striped bass stocking programs, I illustrate how the benefits would be estimated with currently available data. I leave cost estimation and benefit–cost analysis for future research. In the next section, I differentiate between economic impacts and benefits. Using data from the 2001 National Survey of Fishing, Hunting, and Wildlife-Associated Recreation, I profile striped bass anglers. Then, I estimate the economic expenditures and benefits of recreational striped bass fishing.

Economic Expenditures and Benefits

Whenever a government project or policy is implemented, there are winners and losers. The economic efficiency criterion requires that the gains to the winners exceed the losses imposed on the losers. Otherwise, the criterion for efficiency is not met. Benefit–cost analysis calculates and compares these monetary gains and losses (Zerbe and Bellas 2006). When government pursues a fisheries policy such as stocking or bag limits, gains and losses are distributed to consumers and firms.

Economic expenditures and benefits can be defined theoretically by using the concept of demand from introductory economics. The demand curve (D_1) is a downward sloping line with market price (*P*) on the vertical axis and quantity consumed (*Q*) on the horizontal axis (Figure 1). The demand curve slopes downward due to the fact that lower prices are required to convince current consumers to purchase larger quantities of a particular good (in this case, striped bass fishing) and as prices fall new consumers enter the market. The rectangle below the price and to the left of quantity, Q_1, is the initial expenditure on the goods (area *A*). Expenditures are the primary data for economic input–output models that are used to estimate the employment, wages, and taxes supported by the direct expenditures, as well as the indirect and induced effects (i.e., the multiplier effects) resulting from the direct spending (Miller and Blair 1985). This is commonly known as economic impact analysis.

Economic benefits are known theoretically as consumer surplus. The consumer surplus is the triangular area (area B, Figure 1) above the price and below the demand curve up to the quantity consumed. The consumer surplus is the difference between what the consumer is willing to pay and the amount actually spent on the good or service. Consumer surplus is a monetary measure of the net benefit that the consumer gains from the transaction.

Expenditures and benefits change along with changes in the demand curve. Changes in consumer income, prices of related goods, consumer tastes, or the quality of the good can cause shifts in demand. For example, an improvement

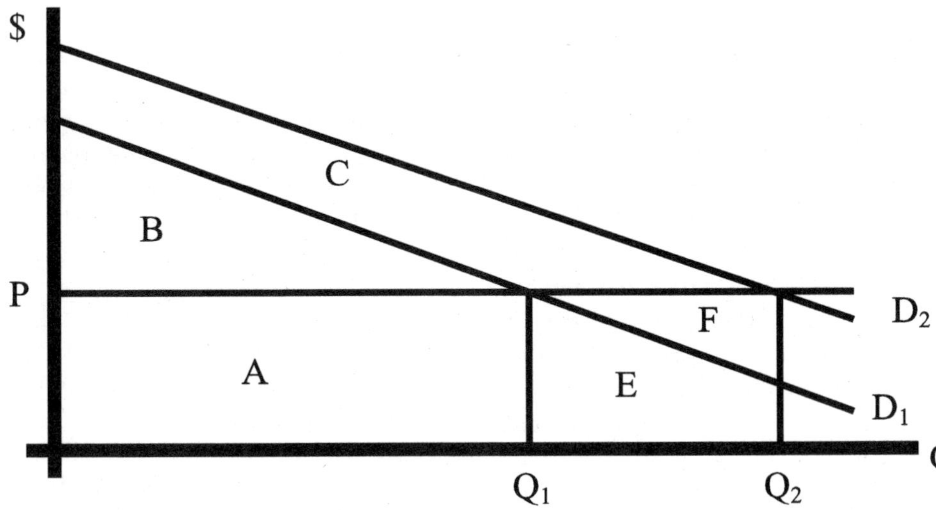

FIGURE 1. Initial (D_1) and increased (D_2) demand curves as a function of market price (P) and the quantity consumed (Q). Labeled areas under the curve indicate (*A*) initial expenditure on the good, (*B*) initial consumer surplus, (Q_1) initial quantity consumed, (*C*) increase in consumer surplus with increase in demand, (Q_2) increased quantity consumed with increase in demand, and (*E* + *F*) increase in expenditures with increase in demand.

in fishing quality would increase demand for fishing trips, shifting it to the right (from D_1 to D_2). When the demand curve shifts to the right, the associated expenditures and benefits increase. The change in expenditures is represented by areas $E + F$. The change in consumer surplus is represented by area C (Figure 1).

Recreational fishing trips are not bought and sold in organized markets but still provide consumer surplus. Suppose an angler is willing to pay $25 for each additional striped bass caught from a reservoir. In this case, the consumer's demand would be for the good (i.e., catch on a fishing trip) related to fish abundance. If a stocking program leads to greater abundance and the improvement enables the angler to catch more fish than usual, then the angler's demand for fishing trips would increase and consumer surplus would also increase. However, since the angler did not directly pay anything out of pocket for the improvement in fishing quality (i.e., there is no market price), the increase in consumer surplus is more difficult to measure.

Changes in expenditures are appropriate for use in economic impact analysis. For example, Schorr et al. (1995) estimated the economic impacts of the Lake Texoma striped bass sport fishery. They estimated that anglers contributed millions of dollars in fishing expenditures to the regional economy leading to higher incomes, tax revenues, and jobs. Economic impact analysis is useful for understanding the contributions of a government program to local economic activity, but it has limitations. Economic impacts are not appropriate for efficiency analysis (i.e., benefit–cost analysis) because they do not account for substitute activities. If all of the striped bass were taken from a reservoir, anglers would find substitute activities for striped bass fishing in that reservoir. They could fish the same reservoir for other target species, target striped bass at other locations, or pursue other alternative activities. The expenditures and resulting increases in incomes would potentially be reallocated, but the overall value of the impacts might not change.

Another aspect of economic impact analysis is that it estimates gross effects of a government policy. In an economic impact analysis, benefits and costs of a government policy are summed and all government policies appear to be good for society even if the costs exceed the benefits. In the case of recreational fishing, expenditures are positive economic impacts for the bait and tackle shop but negative economic impacts for the angler (since they are costs the angler would like to avoid). In a benefit–cost analysis, these impacts net out to zero. For these reasons, benefit–cost analysis is the preferred analytical tool for understanding the economic efficiency of government programs. Economic impact analysis, on the other hand, is appropriate for understanding the effects on a local economy.

Change in consumer surplus is the appropriate measure of benefits and costs for benefit–cost analysis (Chizinski et al. 2005). Demand curves take account of substitute activities, whereby a reduction in striped bass stocks would lead to a loss of consumer surplus that would not be fully replaced by other activities. An increase in consumer surplus that results from a striped bass stocking program would generate an increase in consumer surplus that would not take away from other activities. Since consumer surplus is the angler's gross willingness to pay minus expenditures spent on the trip, consumer surplus measures the benefits of recreational fishing and improvements in fishing with stocking programs.

Data

Extensive data on fish- and wildlife-related recreation are available from the 2001 National Survey of Fishing, Hunting, and Wildlife-Associated Recreation (FHWAR; U.S. Department of the Interior, Fish and Wildlife Service, and U.S. Department of Commerce, U.S. Census Bureau 2002). The U.S. Fish and Wildlife Service funds the FHWAR survey and the U.S. Bureau of the Census collects the data every 5 years. The FHWAR survey is conducted in two consecutive phases. First, screening interviews of a representative sample of the United States population are conducted. These interviews are by telephone or in person, according to preference of the respondent. A household representative 18 years old or older is asked to provide estimates of the wildlife-associated recreation activity of all household members 16 years old and older. The screening interview information is used to construct a representative sample of participants in wildlife-associated recreation for the detailed survey that follows. A sample of those household residents 16 years old and older who had participated recently or intended to participate in the upcoming year, or both, are selected for follow-up interviews.

The detailed interview phase of the survey asks respondents to recall their recreation activities and expenditures over a 4-month period. Questions included what kinds of wildlife-associated recreation respondents had participated in, how often they participated, and how much they spent for trip-related and equipment expenditures. In the last interview, direct willingness-to-pay questions were presented to bass or trout anglers; deer, moose, or elk hunters; and wildlife watchers. Unfortunately for this chapter, willingness-to-pay questions were not asked of striped bass anglers.

From the 2001 NHWAR survey, I identified the subset of data most relevant to the questions being addressed in this analysis. There are data on 11,064 anglers in the 2001 database. I deleted those who have missing data on key variables (e.g., trips, income, expenditures). In order to focus on inland striped bass and hybrid striped bass anglers, I considered only those who fished in "ponds or lakes or reservoirs" and who answered yes to the species target question "On your freshwater trip(s) in or to (state) from January 1, 2001 to December 31, 2001, was white bass, striped bass and striped bass hybrids one type of fish you primarily fished for?" I did not consider those anglers who took multiday fishing trips, to ensure that trips were for the primary purpose of fishing. Thirty-seven percent of the anglers (n = 4056) said they fished in ponds, lakes, or reservoirs. Fifteen percent (n = 606) of these day trip anglers targeted white bass, striped bass, and striped bass hybrids (i.e., the "striped bass" category) among other species.

Anglers were asked "In which state(s) did you freshwater fish from January 1, 2001 to December 31, 2001?" and "How many trips did you take in or to (state) to go freshwater fishing?", where (state) is the first state that the respondent reported, among up to seven states they could list. Striped bass anglers reported that they took 24 fishing trips in the past year to their primary state destination. The minimum number of trips was 1 and the maximum was 300. While I focused on the primary state destination, the number of trips to other destinations was minimal.

Striped bass anglers spent about $840 annually on all of their pond, lake, and reservoir fishing trips (this value and all following amounts are in 2001 dollars, unless indicated otherwise). This averages to $35 for each of their average of 24 annual trips. The largest components of total spending were for travel, $158; food and drink, $131; boating expenditures, $91; and bait, $53. Fixed capital costs, such as those made to purchase a boat, are not included. Using ordinary least squares regression analysis, I estimated the effects of income and trips on total spending. The simple model explains 30% of the variation in spending. The costs of fishing increased by $3.57 for each $1,000 in income. For example, an angler who earned $20,000 annually would spend $71.40 and an angler who earned $100,000 annually would spend $357 on all trips. Holding income constant, travel costs increased by $14 with each fishing trip. In other words, P = $14 in Figure 1.

Anglers were later asked "On how many different days did you fish for white bass, striped bass and striped bass hybrids in freshwater in (state)?" Since anglers were allowed to report targeting behavior for more than one species, there was a risk of double counting fishing days if anglers targeted more than one species on any given fishing day. Therefore, I adjusted the number of days spent fishing where "striped bass" was the specific target. For example, consider an angler who annually takes 25 fishing trips, spends 50 d targeting various fish species, and spends 10 d targeting the striped bass category. The adjustment equation is as follows:

$$\text{Days}' = \text{Days}\left(\frac{\text{Trips}}{\sum \text{Days}}\right),$$

where Days' is the number of days fished where striped bass is the only target category; Days is the number of days spent fishing for striped bass (along with other species), including partial days; Trips is the number of fishing day trips taken; and $\sum$ Days is the total number of fishing days summed over 14 species and species groups (which is greater than the actual number of days fished). Using the numbers for the hypothetical angler above, I would estimate that the number of days spent targeting striped bass is

$$\text{Days}' = 5 = 10\left(\frac{25}{50}\right).$$

Over the full sample (n = 606), the adjusted number of striped bass fishing days is 8.75 with a minimum of 1 and a maximum of 170.

Participation and intensity in striped bass fishing varied by region. Twenty-two percent of the striped bass anglers fished in the South Atlantic U.S. Census region (a breakdown of these regions can be found at www.census.gov/geo/www/us_regdiv.pdf), 19% in the East South Central region, 17% in the West North Central region, 11% in the West South Central region, 9% in the East North Central region, 8% in the Northeast region, 6% in the Mountain region, 3% in the Pacific region, and 3% in the Middle Atlantic region (the numbers do not sum to 100% due to rounding). The average number of days fished for striped bass was highest in the West South Central and East South Central regions at 10. South Atlantic striped bass anglers fished an average of 9 d, and West North Central anglers fished an average of 8 d for striped bass. The average number of days fished for striped bass ranged from 6 to 8 d in the rest of the United States.

Striped bass anglers did not differ from other pond, lake, and reservoir anglers in any socioeconomic variable in the FHWAR survey except age. Their average age was 41 years, 1.5 years younger than other pond, lake, and reservoir anglers (t = 1.66). Ninety-three percent of striped bass anglers were white, 80% were male, and 73% were married. Their average education level was 13 years, and their average household income was almost $50,000.

Analysis

The travel cost model is typically implemented with site-specific data (Parsons 2003). For example, a model such as $x_{ij} = f(p_{ij}, q_j, y_i, \theta_i)$ would be estimated, where x represents fishing trips, p represents the price of each of those trips, q represents trip quality or other site-specific variables, y represents income, and θ represents other relevant variables (e.g., demographics). The subscript i is for individuals and j is for recreation sites. The price of each trip is measured by the travel and time costs of the trips, $p = cd_{ij} + \gamma w_i (d_{ij} / \text{mph})$, where c is a constant cost per mile (e.g., $0.30), d is the round trip distance, $0 \leq \gamma \leq 1$ is the opportunity cost of time, w is the hourly wage rate, and mph is miles per hour.

In the FHWAR survey, site-specific and distance-traveled variables are not available, so I used a more general version of the travel cost demand model, $\text{Days}' = f(p_i, y_i, \theta_i)$, where travel cost is measured as the average transportation cost per fishing trip. Transportation cost is elicited with this question: "When you were fishing in (state) chiefly in freshwater from January 1, 2001 to December 31, 2001, how much was spent for your share of the round trip cost for transportation by private vehicle?" Average transportation cost is the total cost divided by the number of day trips. Transportation costs do not include depreciation and other transportation costs and may be conservative.

For illustration purposes, I estimated a simple ordinary least squares regression model with the natural log of striped bass fishing days as the dependent variable (see Haab and Mc-

Connell 2002 and Parsons 2003 for discussions of more complex econometric models, empirical specifications, and sources of potential bias). The estimated model is

$$\text{Log}_e \text{ Days}' = \underset{(9.91)}{2.70} - \underset{(-3.06)}{0.0092p} - \underset{(-0.59)}{0.0011\,y} - \underset{(-2.62)}{0.051 \text{ education}} - \underset{(-2.62)}{0.0084 \text{ age}}$$

$R^2 = 0.05$
$F = 7.85$

where the t-statistics are in parentheses. Striped bass fishing days decreased with increases in travel costs (p) and for respondents with more education and older respondents. Demand is highly inelastic with the percentage change in days due to a 1% change in price equal to –0.08. This is likely due to the fact that the model does not contain substitute activity prices, a well-known limitation.

With the semilog functional form, the consumer surplus is $CS = -\text{Days}'\ \beta_p$ where β_p is the coefficient on the travel cost variable (Bockstael and Strand 1987). With these data and simple model, the annual consumer surplus of striped bass fishing days per angler is estimated at $978. The consumer surplus of a single striped bass fishing day is $109. Similar to above, the lack of substitutes in the demand model likely biases the consumer surplus estimates upward.

The aggregate economic expenditure and benefit is the average expenditure and benefit scaled up to the population. Using the 2001 FHWAR survey, which had 11,162 freshwater anglers, the U.S. Fish and Wildlife Service (USFWS) estimated that there were 28.4 million freshwater anglers in the United States in 2001. Using the same scale factor, I estimated that there were about 1.54 million striped bass anglers in the United States in 2001. Considering that striped bass anglers spent about $840 annually, the aggregate economic expenditure was $1.3 billion dollars, about 3.6% of the $35.6 billion in total expenditures estimated by the USFWS. Input–output models can be used to estimate the economic impacts of these expenditures. Considering that the striped bass category angler consumer surplus was about $978 annually, the aggregate consumer surplus was $1.5 billion. Converting aggregate values to 2010 dollars using the consumer price index, the aggregate expenditure was $1.2 billion and aggregate consumer surplus was $1.85 billion.

Conclusions

In this chapter, I have discussed some economics of recreational fishing and illustrate these concepts with aggregated data for striped bass, striped bass hybrids, and white bass as presented in the 2001 National Survey of Fishing, Hunting, and Wildlife-Associated Recreation. The economics of recreational fishing can be used to estimate the importance of this striped bass group to local economies and to determine the economic efficiency of government stocking programs for striped bass and its hybrids that are designed to improve the quality of the fishing experience. With the FHWAR survey recreation data, I have provided a demonstration of the types of economic analysis that can be conducted. However, the consumer surplus results should be used with caution since they are likely biased. The FHWAR survey data are not designed for the travel cost method and the results should be considered illustrative, at best. Future research should be conducted that focuses on striped bass and hybrid striped bass fishing and collects the data necessary to estimate a more valid recreation demand model. Results from this future study could be compared to the costs of striped bass and hybrid stocking programs to determine their economic efficiency. The "striped bass" referred to in this chapter is actually the aggregate of striped bass, hybrid striped bass, and white bass. It is not known from the data what the actual species (or hybrid) breakdown in this category is. White bass might contribute significantly to the total values. If more accurate information on the economic impact of striped bass and hybrid striped bass fisheries is desired, more specific data will need to be collected in future surveys. Finally, after this analysis was conducted, the 2006 national survey data became available. Future research should update this analysis to determine if results are consistent over time.

References

Bockstael, N. E., and I. E. Strand. 1987. The effect of common sources of regression error on benefit estimates. Land Economics 6:11–20.

Boyle, K. J. 2003. Contingent valuation in practice. Pages 111–169 *in* P. A. Champ, K. J. Boyle and T. C. Brown, editors. A primer on nonmarket valuation. Kluwer, Boston.

Champ, P. A. 2003. Collecting survey data for nonmarket valuation. Pages 59–98 *in* P. A. Champ, K. J. Boyle and T. C. Brown, editors. A primer on nonmarket valuation. Kluwer, Boston.

Champ, P. A., K. J. Boyle and T. C. Brown. 2003. A primer on nonmarket valuation. Kluwer. Boston.

Chizinski, C. J., K. L. Pope, D. B. Willis, G. R. Wilde, and E. J. Rossman. 2005. Economic value of angling at a reservoir with low visitation. North American Journal of Fisheries Management 25:98–104.

Haab, T. C. and K. E. McConnell. 2002. Valuing environmental and natural resources: the econometrics of non-market valuation. Edward Elgar, Northampton, Massachusetts.

Johnston, R. J., M. H. Ranson, E. Y. B., and E. C. Helm. 2006. What determines willingness to pay per fish? A meta-analysis of recreational fishing values. Marine Resource Economics 21:1–32.

Miller, R. E., and P. D. Blair. 1985. Input-output analysis: foundations and extensions. Prentice Hall, Englewood Cliffs, New Jersey.

Parsons, G. R. 2003. The travel cost model. Pages 269–329 *in* P. A. Champ, K. J. Boyle and T. C. Brown, editors. A primer on nonmarket valuation. Kluwer, Boston.

Patrick, W. S., O. Bin, K. A. Schwabe, and P. W. Schuhmann. 2006. Hatchery programs, stock enhancement, and cost effectiveness: a case study of the Albemarle Sound/Roanoke River stocking program 1981–1996. Marine Policy 30:299–307.

Rosenberger, R. S., A. R. Collins, J. B. Svetlik. 2005. Private provision of a public good: willingness to pay for privately stocked trout. Society and Natural Resources 18:75–87.

Schorr, M. S., J. Sah, D. F. Schreiner, M. R. Meador, and L. G. Hill. 1995. Regional economic impact of the Lake Texoma (Oklahoma–Texas) striped bass fishery. Fisheries 20(5):14–18.

U.S. Department of the Interior, Fish and Wildlife Service, and U.S. Department of Commerce, U.S. Census Bureau. 2002. 2001 National survey of fishing, hunting, and wildlife-associated recreation. U.S. Fish and Wildlife, Washington, D.C.

Van Horn, S. L. 2013. A brief history of inland striped bass management. Pages 1–13 *in* J. S. Bulak, C. C. Coutant, and J. A. Rice, editors. Biology and management of inland striped bass and hybrid striped bass. American Fisheries Society, Symposium 80, Bethesda, Maryland.

Zerbe, R. O. and A. S. Bellas. 2006. A primer for benefit-cost analysis. Edward Elgar, Cheltenham, Massachusetts.

American Fisheries Society Symposium 80:559–574, 2013

Economic Analysis of Fish Consumption Advisories

Paul M. Jakus*
Department of Applied Economics, Utah State University
UMC 3530, Logan, Utah 84322, USA

Abstract.—As top predators, striped bass *Morone saxatilis* and their hybrids often accumulate pollutants and prompt recommendations for reduced consumption. Economists have estimated the economic effect of fish consumption advisories on anglers using nonmarket valuation techniques. Following a brief discussion of advisories policy in the United States and the theoretical underpinning for estimating economic losses they may cause, I contrast estimates of economic loss (consumer surplus) with measures of economic impact. In a comprehensive review of the economic costs of advisories in freshwater fisheries, I found that economists have focused almost exclusively on only the most extreme of angler's responses. These were changing the sites at which an angler fishes or changing the number of fishing trips during the season (ignoring losses associated with behavioral responses, such as changing target species, engaging in more careful food preparation, or strictly complying with recommended consumption limits). For decision makers needing to conduct benefit-cost analysis of an advisory policy but lacking the time or budget needed for an original study, I outline the method of benefit transfer. The paper concludes with a number of research topics not yet considered by economists: behavioral responses to advisories besides adjusting the site or number of trips, heterogeneity among anglers' preferences, and the effect of differences in perceived risk of consuming contaminated fish.

Introduction

Many species of freshwater and marine fish have been contaminated by pollutants and chemicals that occur both through natural processes and through human-caused releases. Beginning in the 1970s, various state authorities began to issue consumption advisories advising those who eat fish to either limit or cease their consumption of particular species. The evidence suggests that consumption advisories have altered behavior: for example, some anglers have chosen to change their consumption to follow the advice, some have engaged in more careful preparation of fish prior to consumption (such as removing fatty deposits to reduce exposure to polychlorinated biphenyls [PCBs]), some have changed the sites at which they fish, and perhaps a few anglers have decided to cease eating fish altogether. Any of these behavioral changes would imply that advisories have economic consequences. Indeed, there are also economic consequences to ignoring advisories, if the angler or others who consume contaminated fish develop illnesses associated with the contaminant.

In this chapter, I describe the nature and type of consumption advisories and outline in broad terms how economists have approached the advisory issue. Readers hoping to learn specifically about advisories for inland striped bass *Morone saxatilis* fisheries should note that most economic analyses, to date, have dealt with ad-

* Corresponding author: paul.jakus@usu.edu

visories in a general way, with little analysis of the effect of a single advisory for a single species (Jakus et al. [2002] is the only study to focus on striped bass). Therefore, this discussion of advisories will be at a very general level. To begin the discussion, it will be necessary to clarify the concept of economic value and how it is measured. A brief review of the two major methods of measurement, *revealed preference* models and *stated preference* models, will focus on how these methods have been applied within the context of fish consumption advisories (FCAs). Estimates of the economic losses associated with contamination are then presented. I conclude with a discussion of the economic loss estimates (what they do and do not measure), an introduction to a method by which decision makers may use the economic loss estimates reported in the literature for analysis of proposed policies elsewhere, and an outline of future research.

Consumption Advisories in the United States

The environmental contaminants of greatest concern to sport fisheries generally enter the environment through industrial pollution or, in the case of pesticides such as dioxin, chlordane, and dichlorodiphenyltrichloroethane (DDT), through agricultural runoff. While mercury is naturally present in many soils, mercury levels in the environment have increased as a result of industrial pollution sources such as coal-fired power plants and gold mining operations (Swain et al. 2007). Exposure to high enough concentrations of these contaminants in food can lead to a variety of human health effects, including impairment of nervous system development in infants and young children, reduced fertility, increased blood pressure, changes in behavior and psychology, and possibly cancer (Burger and Gochfeld 2006).

Contaminants in the aquatic environment enter fish through the process of bioaccumulation (Reinert et al. 1996). Fish readily absorb contaminants dissolved in the water column through their gill membranes, thus depositing contaminants directly into the blood to be distributed throughout the body. Some contaminants, like mercury, are deposited throughout the soft tissues of the fish's body and, in time, can be eliminated from the body if the contamination source is removed. PCBs are deposited only in fatty tissues, making them easier to remove from the fish during the cleaning process, but they cannot be eliminated from the body via excretion. Thus, PCB concentrations rise over the life of the animal. Biomagnification refers to the process through which organisms higher on the food chain accumulate higher concentrations of contaminants through the consumption of contaminated organisms lower on the food chain, so that larger predatory species, such as striped bass, are more prone to be the target of fish consumption advisories. Fish can gradually develop tissue concentrations of contaminants that are much greater than the levels found in the ambient environment, a phenomenon referred to as bioconcentration.

Since the 1970s, health advisories concerning the consumption of contaminated recreational and commercially caught wild fish have been issued in all 50 states (EPA 2009). As of 2008, there were 4,249 active advisories covering 43% of the nation's lake acreage and 39% of the nation's river miles. Ninety-seven percent of current fish consumption advisories are for fish contaminated with mercury, PCBs, dioxin, chlordane, or DDT; mercury and PCBs account for 80% of fish consumption advisories. An advisory generally refers to a particular species, a recommended consumption rate, and a target population. The most extreme advisories recommend no consumption of a given species by anyone, with less strict advisories recommending limited consumption of a species by sensitive subpopulations (generally pregnant women, women of child-bearing age, and children.)

Both the U.S. Environmental Protection Agency (EPA) and the U.S. Food and Drug Administration (FDA) are charged with assessing the risks associated with contaminated fish stocks and providing consumption advice. Both agencies use methods of risk assessment and risk management to set their respective fish consumption advisories, but differences in the human populations they are mandated to protect sometimes result in contradictory advice (Reinert et al. 1996). The FDA is responsible for protecting the general public from contaminated fish from commercial sources, whereas the EPA works with state officials to assess health risks to recreational anglers. The FDA is also mandated to consider the economic effects of consumption advisories on the commercial fishing industry, something the EPA need not consider. Recreational anglers tend to consume fish at much higher rates than the general public, leading the EPA to adopt more stringent guidelines for fish consumption advisories. For example, the EPA has set concentration limits for mercury and PCBs at 0.05 and 0.03 ppm, respectively, while the FDA set limits at 1.0 mg/L for mercury and 2.0 mg/L for PCBs. This can result in mixed signals being sent to the public on the safety of fish consumption. A recent example is the mercury advisory issued by the Maryland Department of the Environment for recreationally caught striped bass from Chesapeake Bay, while no such advisory exists for commercially caught striped bass from the same fish population. This perceived inconsistency can lead to confusion among anglers and the general public that may negatively influence the success of fish consumption advisories.

Numerous studies have been conducted to evaluate the public's awareness of and compliance with fish consumption advisories (see, for example, Velicer and Knuth 1994; MacDonald and Boyle 1997; Burger 2000; Campbell et al. 2002; Jakus and Shaw 2003; Knuth et al. 2003). These studies have identified several trends, but the key finding was that awareness of an advisory does not always translate into compliance with the advisory: some anglers comply, others may overreact by consuming much less than the recommended limit, whereas another group will continue to consume more fish than the advisory recommends. Some researchers have attributed this last trend to a deamplification of risk in the minds of anglers that view fishing and fish consumption as a familiar and enjoyable activity and, thus, view fish consumption as an activity that could not possibly be harmful to their health (Burger 2000). Burger (2000) has also noted that the consumption advisory approach to public health policy is complicated by efforts from state fish and game agencies to promote the sport of fishing in their efforts to increase license sales. I return to the issue of the perceived risks and benefits of fish consumption later in this chapter.

Methods: Measuring Economic Value

Economic value

Economic value is defined within the context of a trade-off, that is, the value of a good is determined by what one is willing to give up to obtain it. Although the most common metric is money, the trade-off could be "denominated" in any number of ways: for example, how many baseball caps one is willing to give up to obtain, say, a t-shirt, a baseball glove, or a fishery free from contaminants. In fact, using money as a metric is strictly for convenience—and because it is the denomination used in any nonbarter trade. Another key element of economic value is that one's willingness to trade is constrained by the ability to trade: if baseball caps were the denomination, then one's offer for a t-shirt or a glove is limited by the number of caps (i.e., money) that one has. This leads us to a more formal definition of economic value: the value a person holds for a particular good is determined by the willingness and the ability to pay for that good. When we give up money for something, we are in essence trading away all the other things we could have bought with that money.

Economic theory assumes that individuals making such a trade do so because it increases their well-being, or utility, and would not agree to a trade that would leave them worse off. Thus, a person's willingness to pay (WTP) is directly related to the change in welfare, or well-being, before and after the trade and is determined by each person according to his or her own preferences. A person's WTP for a contaminant-free fishery (or any good) can be affected by the unique characteristics of the good being offered, the conditions of the choice, and the context of the choice (Bockstael et al. 2000). For example, is the contaminated striped bass fishery particularly good or poor? Is the fishery located amid a spectacular or unique landscape or is the landscape rather mundane? Are there substitute fisheries of similar quality nearby? One can easily guess how WTP would change in each of these cases. We would expect a greater WTP to remove contaminants from a particularly good striped bass fishery than from a poor one. WTP for a fishery in a spectacular setting is likely to be greater than an equivalent fishery in a commonplace landscape. WTP to clean up a contaminated fishery would be greater if it were the only striped bass fishery in the region than if there were a number of nearby equivalent, but contaminant-free, fisheries. The key point is that there is no single economic value associated with fish consumption advisories and that values estimated by economists must be evaluated within the context of the particular study.

To estimate economic value for goods that are not traded in the marketplace, economists rely upon two different approaches to measurement. Revealed preference methods, of which the best known to fisheries scientists is probably the travel cost method, rely upon observed changes in angler behavior to infer a measure of WTP. A second set of techniques, generally grouped together as stated preference methods, has also been used to address the economic effects of fish consumption advisories. As the name suggests, rather than basing WTP on changes in observed behavior, stated preference methods use a question, or series of questions, to help economists discover WTP. Contingent behavior, contingent valuation, and choice experiments are perhaps the best-known stated preference techniques.

The travel cost model

Travel cost models.—The travel cost model treats "visits" to a site (or visitor days) as a measure of quantity, while price is measured implicitly as the cost to travel to a site (Parsons 2003). Intuitively, all else equal, people will visit recreation sites closer to where they live more frequently than they will visit sites further away. This is due to the fact that it is cheaper (in time and vehicle costs) to visit sites that are closer. A demand curve showing the relationship between visits and cost can be constructed, all other factors held constant (Figure 1). For example, at travel cost TC_1 the person will make V_1 trips, while at a lower travel cost (TC_2) the person will make more (V_2) trips.

The travel cost demand curve is a measure of marginal benefit, or marginal WTP, for each fishing trip. WTP for each successive trip falls as the person "consumes" more trips. (WTP for the marginal fishing trip falls for the same reason that WTP for the marginal bottle of beer falls: the utility, or satisfaction, of the second beer is less than the first, the satisfaction yielded by the third beer is less than the second, and so forth. On a hot day, a beer drinker is WTP more for the first cold beer and not so much for the second, third, or fourth.) Let the demand curve in Figure 1 represent the demand for fishing at a site by a single person during a given season. Assume that travel cost TC_1 is US$50, which includes the costs of travel to and from the site. An angler is willing to go on the first trip because the WTP for the marginal (first) trip, say, $150 (as given by the demand curve), is greater than the cost of $50. Indeed, the angler received a "surplus" of $100: she was willing and able to pay $150 for the first fishing trip but only had to pay $50. For the second trip, marginal WTP is lower, say $125, but the cost was only $50, so the person receives a surplus of $75. A person will continue to make fishing trips as long as the benefit of the marginal trip is at least as great as the cost of the trip. Let us say that the V_1th trip yields a marginal value of $50, and the trip travel cost is still $50. If this is, say, the 10th trip, then the 10th fishing trip is worth exactly what it costs, and the angler receives no "surplus" from this trip. The person does not take an 11th trip because the marginal benefit of the fishing trip would be less than the $50 travel

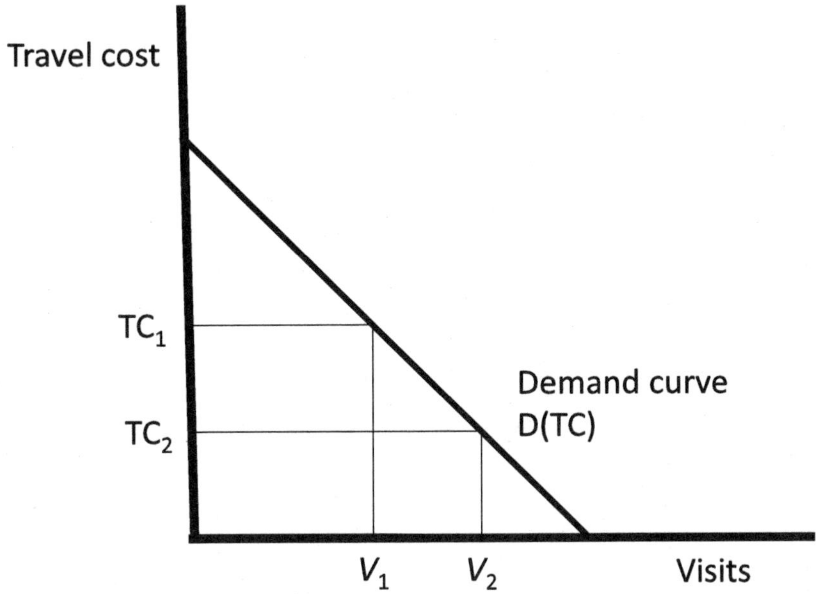

FIGURE 1. The travel cost demand curve, D(TC), relating the number of visits (V) to travel cost per visit (TC). Economists traditionally graph price (travel cost) on the vertical axis.

cost. An economist would determine the net economic value of the fishing site to the angler as the sum of the economic surplus obtained from all the trips. One can think of this consumer's surplus as being the difference between what an angler was willing to pay for all of the fishing trips versus what he or she actually had to pay. Graphically, this is the triangular area under the fishing demand curve and above the travel cost. Though technically it is a measure of the net economic value or the net willingness to pay, this area of consumer's surplus is commonly referred to as "willingness to pay."

Of course, anglers will often visit more than one fishing site because of differing site qualities or other aspects of the fishing experience; that is, we see people drive past nearby fishing sites to visit more distant, but higher quality sites. Quality attributes may include measures of water quality (e.g., nutrient or coliform levels), fishing quality (e.g., the average catch rate or fish size), or the presence or absence of fish consumption advisories on a desired target species. In the case of two sites with the same travel cost but differing qualities, we would expect the higher quality site to be visited more frequently than the site with lower quality. Economists can measure the value of changing site quality by examining what happens to the demand for fishing with improvements or degradation in site quality.

Consider an initial situation in which site quality is Q_1 and the demand for fishing at the site is a function of travel cost and site quality, $D(\text{TC}, Q_1)$, as shown in Figure 2. At travel cost TC and initial site quality Q_1, the angler would make V_1 trips and receive a consumer's surplus of the triangular area given by A-B-TC. Now consider a situation in which the site has become contaminated with a chemical that results in a fish consumption advisory for striped bass. This lower quality level, Q_2, has the effect of shifting the demand curve for the site to the left—any given trip yields less satisfaction and, hence, a lower marginal benefit—so that at travel cost TC, the angler now takes only V_2 visits and receives a surplus of the triangular area D-C-TC. The difference in consumer's surplus under the two conditions, the trapezoidal area A-B-C-D, is a measure of the econom-

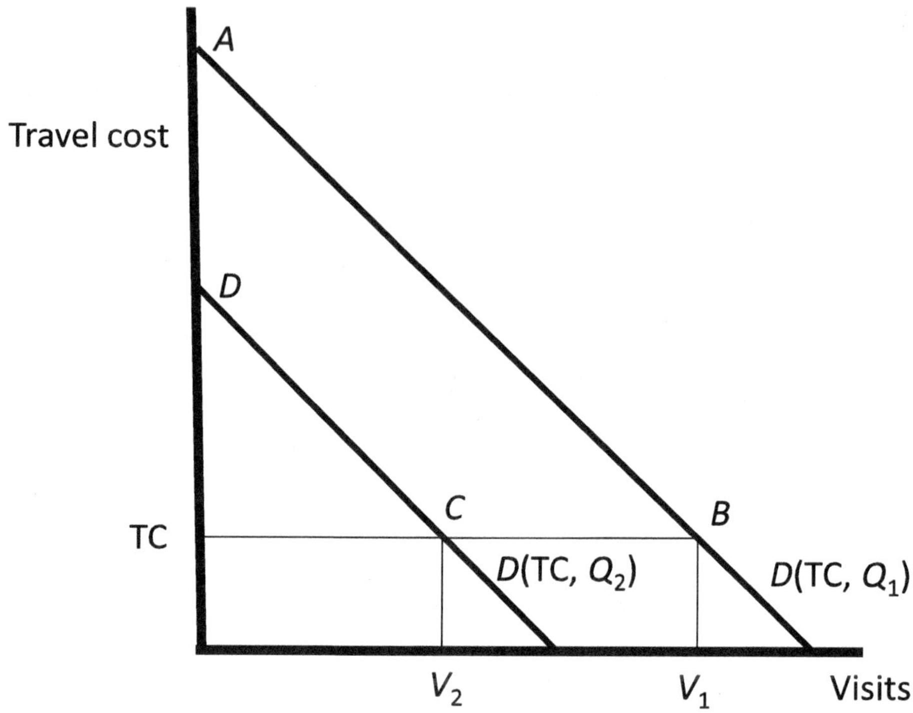

FIGURE 2. Two travel cost demand curves relating number of visits (V) to travel cost per visit (TC) and site quality (Q). If contamination causes the site quality to decline from Q_1 to Q_2, an angler takes only V_2 trips instead of V_1 trips, and the area bounded by points A, B, C, and D is a measure of the economic loss suffered by the angler.

ic loss suffered by the angler as a result of the contamination. It is this type of measure that has been used most frequently by economists to gauge the economic effect of FCAs. One might note that anglers may switch from fishing the contaminated site to fishing an uncontaminated site; though it may cost them a bit more in travel costs, this should mitigate some of the economic losses associated with the contamination. In fact, all of the FCA-related travel cost research has adopted a multi-site framework, thus incorporating such substitution behavior directly into the modeling.

Stated preference models

Contingent models.—An alternative to using the travel cost model to measure the economic consequences of FCAs is the group of techniques known collectively as stated preference models (Brown 2003). Two historically popular approaches involve intended actions contingent on a hypothesized state of the world. A *contingent behavior* approach may be structured as somewhat of a corollary to the travel cost model by asking people a question: "If scientists can remove contaminants from this lake so that the fish consumption advisory may be removed, how many more times might you visit?" These data may then be combined with revealed preference data to measure the change in consumer's surplus under the "before" and "after" conditions. The *contingent valuation* approach measures WTP more directly, asking "Assuming your travel costs to the contaminated site stay the same, how much would you be willing to pay to mitigate the contamination so that the advisory may be removed?" The response to such a question would be closely related to the "lost" consumer surplus area *A-B-C-D* in Figure 2. Technically, stated preference methods measure consumer's surplus while holding utility constant, whereas revealed preference methods do not hold utility constant. The welfare measures arising from each approach will differ according to the size of the income effect. Income effects have been found to be quite small in most recreational applications so that there is often little difference between the two approaches.

Models based on contingent behavior or contingent valuation approaches are controversial because they rely on a person's stated intentions, in contrast to the observed behavior used in the travel cost model (see, for example, the debate between Diamond and Hausman 1994 and Hanemann 1994). Despite the controversy, many economists agree that stated preference methods do provide useful information for evaluating policy changes (Arrow et al. 1993). In response to concerns about relying exclusively on stated preference data, it has become increasingly common to encounter models that link stated preference responses to revealed preference models, especially the travel cost model (see Whitehead et al. 2011). The idea is that one's observed choices (revealed behavior) should be consistent with one's stated behavior. By estimating the models jointly, the analyst allows the revealed preference data to "discipline" the stated preference data; in addition, one enjoys a gain in statistical efficiency.

A stated preference model that has grown rapidly in recent years is known as the *choice experiment*, which is based fundamentally on conjoint analysis. In contrast to contingent models that ask individuals to explicitly state their willingness to pay for (or how behavior would change in response to) a proposed policy, these models ask individuals to rate, rank, or choose from a series of possible outcomes (choice sets). A choice set is composed of two or more alternatives, each of which varies in the "level," or quantity, of attributes offered. In addition to the type and severity of FCAs, other important attributes for a fishing site might be species catch rates, number of boat launching sites, the type of lodging and/or camping available, and so forth. The fact that the "levels" of each choice attribute vary across alternatives allows the researcher to gauge the trade-off that an individual is willing to make between any two attributes presented in the choice sets. It also permits estimation of a dollar value for any one attribute or group of attributes.

Choice experiment analysis has enjoyed rapid growth in popularity because it relies less on the information contained in the "all-encompassing" description of a given hypothetical scenario and more on the description of the attributes of each alternative (Boxall et al. 1996). The tasks required of a participant in a choice experiment survey are more representative of the choices that people regularly face in making everyday market transactions when choosing among substitute goods that have multiple attributes (for example, the choice among automobiles having different colors, engine sizes, interiors, etc.) Thus, a key advantage of choice experiments is that the role of substitute goods and their attribute levels is explicitly recognized, whereas in most contingent behavior or contingent valuation analyses, the role of substitute goods is often relegated to a reminder sentence or two in the description of the good.

Surplus measures versus economic impact analysis

Noneconomists will often be skeptical of consumer surplus measures, thinking that estimates of consumer surplus are a sort of "funny money" because we really do not see the money being spent in the marketplace. So, how do surplus measures match up against measures of economic impact, such as net changes in employment, income, and tax revenues, as commonly reported in the popular press? Referring once again to Figure 2, the loss in economic value is given by area *A-B-C-D*. For economic impact analysis, one would instead look at the change in expenditures by the angler. Assuming that the travel cost consists only of cash expenditures, the net change in expenditures would be the travel cost per trip multiplied by the change in number of trips, in this case, TC × $(V_1 - V_2)$, or area V_2-*C-B*-V_1. If TC = $50 and the angler made, say, six fewer trips, then $300 is taken out of economic sectors affiliated with recreational fisheries and spent elsewhere. Economic impact analysis then traces these expenditures (or lack thereof) throughout the regional economy, including multiplier effects, to estimate changes in employment, income, and taxes. A measure of economic impact is useful in informing policy decisions, but consumer surplus measures have been identified by the U.S. Water Resources Council (1983) and numerous executive branch agencies as the appropriate measure of changes in angler welfare as a result of changing site quality (e.g., EPA 2010).

As an intuitive proof of the importance of consumer surplus as a welfare measure, consider the following thought experiment. Think of all the money you have spent on all fishing trips in your state in the past 12 months; let us say it was $500. Assume your state enjoys fishing quality level Q_1, so that the $500 amount corresponds to the rectangular area TC × V_1 in Figure 2. Now let a natural disaster completely decimate game fish populations in the state, but all other aquatic ecosystem services (drinking water, habitat for other species, boating, etc.) are left intact. The only thing missing is the game fish, so that we have no demand curve for fishing after the disaster. How much would you pay to restore the fishing to its previous quality? Would it be $500, or would it be something more? If it is something more, then clearly you have suffered a loss in excess of the market transactions of $500, and now you are digging into the area of consumer surplus.

Results: Estimates of Economic Losses Due to Fish Consumption Advisories

Economists have applied standard versions of recreational site-choice-seasonal trips models to the problem of FCAs by treating the presence or absence of an FCA as a site attribute. In general, most studies have found a statistically significant effect that, all else equal, anglers are less likely to visit fishing sites that have an advisory relative to sites that do not have an advisory. Although most studies have relied solely upon revealed preference data, some have augmented the revealed preference data with stated preference data.

Early fish consumption advisory studies

The studies that treated FCAs in a fairly simplistic way, that is, with a simple "dummy variable" noting the presence of any kind of advisory on any species, are summarized in Table 1. These studies did not differentiate advisories by species or "severity" of recommended consumption rates. The first three studies listed in Table 1 reported only a per-trip estimate of economic loss, while the remainder of the studies estimated the loss over an entire fishing season (as defined by each study). The seasonal estimate of economic loss can be thought of as equivalent to area *A-B-C-D* in Figure 2, whereas the per-trip estimates can be thought of, roughly, as the loss per trip averaged over a season. With one exception, all of the seasonal models use a utility index derived from either Morey et al. (1993) or Hausman et al. (1995), which yield very similar surplus measures (Parsons et al. 1999). Shaw and Shonkwiler's (2000) seasonal model differs considerably from the Morey et al. (1993) and Hausman et al. (1995) indices, relying on an index related to the total distance traveled during the fishing season as opposed to the usual formulation based on the number of trips taken during a season.

The "seasonal" value models range from the March–August season of the Tennessee models and the April–October season of the New York State model to the full year model of the Green Bay study. Thus, the seasonal estimates are not strictly comparable. Although the seasonal estimates are reported, I focus on "per trip" estimates of lost economic surplus (column three of Table 1). The range of estimates for per-trip economic losses from the site choice models is relatively narrow, from $2.85 per trip to $6.98 per trip in inflation adjusted 2010 dollars. The lowest values in the range are given by the Middle Tennessee studies, which are just under $2.60 per

TABLE 1. Estimates of lost consumer surplus due to fish consumption advisories, per trip and per season.

Source	Location	Lost economic value per trip (2010 dollars)	Lost economic value per trip (2010 dollars)
	Site choice only models		
Chen and Cosslett (1998)	41 Great Lakes sites	$6.24	Not reported
Chen and Cosslett (1998)	41 Great Lakes sites	$6.98	Not reported
Parsons and Hauber (1998)	2,029 Maine lakes/rivers	$2.85	Not reported
Jakus et al. (1998), catch rate model	12 East Tennessee lakes	$3.15	Not reported
	Site choice-seasonal trips models		
Montgomery and Needelman (1997)	2,586 NY state ponds or lakes	$2.58	$86.38 (7 months)
Jakus et al. (1997)	14 Middle Tennessee lakes[a,b]	$2.69	$26.60 (6 months)
Jakus et al. (1997)	14 East Tennessee lakes	$4.16	$55.26 (6 months)
Parsons et al. (1999)	14 Middle Tennessee lakes[a]	$2.58	$25.12 (6 months)
Parsons et al. (1999)	14 Middle Tennessee lakes[a,b]	$2.68	$27.54 (6 months)
Breffle et al. (1999)	Green Bay, Wisconsin[c]	$5.57	$58.78 (12 months)
Shaw and Shonkwiler (2000)	14 Middle Tennessee lakes[a]		$15.55 (6 months)

[a] All "14 Middle Tennessee lakes" studies used the same data set.
[b] Jakus et al. (1997) and Parsons et al. (1999) welfare estimates differ slightly due to the bootstrap process used in calculations.
[c] Breffle et al. (1999) augment revealed preference data with stated preference data.

trip. The choice set for these models includes 14 fishing sites, 2 of which are contaminated. Jakus et al. (1997) report losses in Middle Tennessee representing about 8% of total per-trip consumer surplus. The lowest estimate of lost per-trip consumer surplus in Middle Tennessee ($2.58) is matched by the estimate for New York State (Montgomery and Needelman 1997). The choice set for the New York State study also includes very few contaminated sites relative to the number of uncontaminated sites (only 23 of nearly 2,600 sites had contaminated species.)

In general, larger economic losses are found for regions where the proportion of contaminated sites is large relative to the overall number of fishing sites. For example, Chen and Cosslett (1998), with 14 of 41 sites contaminated, obtain per-trip loss estimates of $6.24 and $6.98 per trip. The two Jakus et al. (1997, 1998) East Tennessee lakes models have choice sets of 14 (1997) and 12 (1998) sites each, of which six sites have contaminated species. These two models, using different data sets, yield economic surplus loss estimates of $4.16 (1997) and $3.15 (1998) per trip in 2010 dollars. The first estimate represents approximately 6% of per-trip consumer surplus. The second estimate is restricted only to anglers who planned to eat their catch and who were aware of the advisory.

Following Bayesian techniques outlined in Desvousges et al. (1998), we can estimate the mean and a 95% confidence interval for per-trip losses for studies that report a variance for the WTP measure. The Bayesian approach assumes that a "true" fixed value for per-trip loss of consumer surplus exists and is invariant to the other factors that influence recreational fishing. This may not be the case, especially if the attributes and conditions under which the losses are estimated for the relatively small lakes of Tennessee or upstate New York differ substantially from those found in, say, the Great Lakes. Thus, the Bayesian estimates of the mean and variance are first calculated using just the three Great Lakes estimates (the two Chen and Cosslett [1998] estimates and the Breffle et al. [1999] estimate). For recreational anglers on the Great Lakes, the estimated per-trip loss of consumer surplus is $6.63 per trip, with a 95% confidence interval of $6.17–$7.12 in 2010 dollars. In contrast, using the four estimates from Tennessee (two from Jakus et al. 1997 and two from Parsons et al. 1999), the mean per-trip loss is $3.23 with a 95% confidence interval of $2.79–$3.68.

Later fish consumption advisory studies

In recent years, economic modeling of FCAs has become more sophisticated in that economists are now trying to differentiate advisories by species and severity. Though this modeling extension may appear trivial to the noneconomist, in fact it greatly complicates the data requirements and statistical analysis. Two recent examples include Morey and Breffle (2006) and MacNair and Desvousges (2007). Morey and Breffle (2006) combine stated preference and revealed preference data to capture the effect of four different target species and various levels of severity for consumption advice for Green Bay of Lake Michigan. Using the choice experiment approach, FCAs of nine different severity levels were presented to respondents, ranging from no FCAs on any of the five species of interest to the most stringent scenario in which four species have "do not consume" advice and one had strictly limited consumption. The authors report model results that show an increasingly negative effect on angler welfare as advisories become more and more stringent. Anglers' annual WTP to remove all advisories is estimated to be $99.13, with a 95% confidence interval of $86–$112 (measured in inflation-adjusted 2010 dollars). At an average of 9.3 d of fishing on Green Bay, this implies a rough estimate of just over $10 per day, a bit higher than the per-trip Great Lakes estimates of Chen and Cosslett (1998).

MacNair and Desvousges (2007) also combine choice experiment and revealed preference data in their study of the Fox River and Green Bay. They report a number of different models, including one that is comparable to the studies reported in Table 1, but also others that are more sophisticated in their attempt to capture the number of species covered by FCAs and the type of advice given for each species. The authors aggregate all advisory information, which varies by species and recommended consumption limits, into an index that is subsequently used to predict angler behavior. Using stated preference data only in a simple model, the authors find that "Do Not Eat" advice resulted in larger economic losses than "Limit Consumption" advice, but the differences between the levels of severity were implausibly large (losses for "Do Not Eat" were up to 88 times larger than "Limit consumption"). They consequently used a combined revealed preference-stated preference modeling approach and a number of different ways to model advisories. In the simple FCA model, FCAs that recommend limiting consumption were found to be statistically insignificant in predicting anglers' trip-making behavior, but "Do Not Eat" advisories were statistically significant. With this model, anglers' per-trip WTP to avoid advisories was between $5.55 and $14.43 in 2010 dollars, depending on how the travel cost variable was constructed. For the more sophisticated "index" models in which WTP varies by site as the advisories vary, per-trip WTP to avoid advisories was between \$3.42 and \$8.92, again depending on the construction of the travel cost variable. These estimates are all roughly in line with the Great Lakes estimates of Chen and Cosslett (1998).

Other fish consumption advisory studies on inland fisheries

Economists have also used other techniques or value measures, or both, to estimate the effects of an FCA. Three studies, in particular, stand out. MacDonald and Boyle (1997) used the contingent valuation method to measure the economic losses of a statewide mercury advisory for all open-water fishing in Maine. Respondents were presented with a dichotomous choice question asking if they would have been willing to pay *X* dollars more for fishing during the season. Questions were designed for the "with" and "without" FCA scenarios. Empirical modeling showed that the presence of the advisory is statistically insignificant (i.e., the data suggested that the advisory does not significantly change the net economic value of fishing in Maine).

The studies reported in Table 1 and the two additional Green Bay studies were all implemented at water bodies where the contamination and advisories were already in place, the anglers knew this, and the authors were attempting to estimate economic losses associated with changed trip-making behavior. Krieger and Hoehn (1998, 1999) adopt a very different perspective; these authors are interested in the informational value of FCAs. The authors note that most current advisory programs will report only those water bodies that have been tested and found to harbor contaminated fish. Fish and game agencies generally do not report which water bodies have been tested and found to be free of contaminated fish (the proverbial "clean bill of health"), nor do they report which water bodies have yet to be tested and may yet be found to be contaminated. Krieger and Hoehn (1998, 1999) hypothesized that anglers may be willing to pay for advisory information independent of site choice decisions, and thus the information alone may have value. The authors valued two types of programs for the 5,800 fishing sites in the state of Michigan: a "partial disclosure" program reveals only those sites that have been tested and found to be contaminated, whereas the "full disclosure" program reveals the outcome of all sites that have been tested (i.e., the site listed is "contaminated" or the site listed is "safe"). By implication, the full disclosure program also reveals which sites have not been tested. In both cases, the proposed program included annual expansion of the information set through testing of additional water bodies. The programs would be paid for by increased license fees; anglers were willing to pay an additional \$7.14 annually for full disclosure of all sites that had been tested to that date, with an additional \$0.01 for full disclosure of each additional site tested (reported in inflation-adjusted 2010 dollars).

Discussion

How can these numbers be used?

How can one use the economic loss estimates to aid in policy decisions such as the choice to mitigate contamination or not? Decision makers for fish and game agencies in the well-studied areas of Tennessee or the upper Great Lakes regions can safely use the estimates of lost economic value reported in Table 1 for policy evaluation, but what about those who must evaluate the benefits and costs of advisories in other regions of the country? Policy makers in these regions could commission an original study, which would be the best approach for accurately estimating economic losses. Original studies, though, can be quite expensive, especially if an agency must contract the study to an outside source. Further, survey design, implementation, and analysis all take time; if an agency must make policy decisions relatively quickly, there simply may not be sufficient time for an original study. In the presence of budget and time constraints, can those responsible for managing fisheries in other regions simply use the numbers reported in Table 1? Unfortunately, the short answer is "no," at least not directly. Recall the discussion in the Methods section of this paper: all willingness to pay estimates will be affected by the unique characteristics of the fishery at the site(s) studied, as well as the other environmental, social, and behavioral contexts under which anglers made their choices.

A second-best approach, but one that is often quite timely and cost-effective, has become known as *benefit transfer*. Benefit transfer refers to a set of techniques that allow one to use the values estimated at one site (known as the "study" site) for analysis at another site of interest (generally known as the "policy" site.) Benefit transfer has occurred in one form or another for at least three decades, and much has been learned about what one can and cannot transfer and how the transfer should be conducted. Indeed, a robust theory and set of empirical methods have been developed over the years that helps guide the process and evaluate the quality, or usefulness, of the transfer. While it is beyond the scope of this chapter to describe benefit transfer methodology in detail, excellent reviews of the technique have been published: see a special issue of *Ecological Economics* (for example, Wilson and Hoehn 2006), the volume edited by Navrud and Ready (2007), or the monograph by Rosenberger and Loomis (2001).

A high-quality benefit transfer effort will carefully examine the policy context under which the transfer will take place, examine how biophysical and social conditions differ between the study site and the policy site, evaluate the original economic study with regard to theoretical and empirical rigor, and highlight the sources of potential error in the transfer. Two general approaches have been used to transfer economic benefits (or losses) from one site to another. The most straightforward is a "value transfer," which may use a single point estimate of value(s) at a study site(s) such as those reported in Table 1, or it may use a mean or median value and simply apply it to the policy site. The conditions under which this simple method can be used are fairly stringent though (Rosenberger and Loomis 2001). A more complicated approach, and one that is necessary for study and policy sites that differ in environmental or social contexts, is known as a "function transfer." One form of value function transfer may involve using the regression-estimated parameters for the economic model estimated for the study site but adjust for differences at the policy site by using explanatory variable values that correspond to the policy site. In a sense, this approach allows one to generate what Rosenberger and Loomis (2001) term a "tailored" benefit estimate for the policy site. Other forms of value function transfer, such as a meta-analytic model and preference calibration, can be studied in the special issue of *Ecological Economics* (see, for example, the paper by Wilson and Hoehn [2006]) or the volume edited by Navrud and Ready (2007).

Regardless of the approach used, a review of the literature will reveal that a high-quality benefit transfer can be very complex. Any type of transfer, either a value transfer or a function transfer, will be no better than the quality of the original studies upon which the transfer is based. A quality transfer will be as reliant upon sound and skilled application of economic theory and quantitative methods as the original studies.

Future research

A key element of nearly all studies discussed in this paper is that they have examined only two of the many possible behavioral responses to advisories. That is, economists have generally accounted for only (1) substitution of one fishing site for another, perhaps visiting an uncontaminated site more frequently and a contaminated site less frequently as a result of an advisory, and (2) changes in the number of fishing trips per season in response to an advisory. To date, none of the FCA research by economists in the published literature has examined other responses to FCAs, such as switching target species, performing more careful preparation of fish for cooking (as can be done with some species contaminated with PCBs), or simply following the consumption advice and limiting the amount of contaminated species eaten. Indeed, the models in Table 1 capture only the most extreme—through probably the most costly—of behavioral responses. But the other possible behavioral changes in response to advisories also represent costs to anglers no less real than changing the distribution and number of fishing trips. Economic analyses would be better informed if these other behavioral changes were examined.

Few studies have examined heterogeneity among anglers and their preferences for consuming fish. Jakus et al. (1998) separated anglers into those who consume most of their catch and those who are primarily catch-and-release anglers. The assumption underlying the model was that as "consumption anglers" reduce harvest of contaminated fish, the stock of fish (and size of fish) would increase, benefitting catch-and-release anglers. Their analysis found that catch-and-release anglers would suffer welfare losses if water bodies were to be cleaned up such that advisories could be removed. W. S. Breffle, E. R. Morey, and J. A. Thacher (Michigan Technological University, unpublished working paper) adopted a latent class approach to separate angler preferences for different fisheries quality attributes. The class of anglers that expressed the most concern about FCAs was indeed that class with the highest WTP value (about twice as large as the class that cared more about species-specific catch rates.) These two studies suggest that much of the past research by economists, which treats anglers as a single group with identical preferences, is too simplistic. Angler heterogeneity needs to be explored.

Another issue related to heterogeneity of preferences concerns the perceived risk of consuming contaminated fish. Burger (2008) has recently reported that some anglers may be "overreacting" to the information contained in advisories, thinking that the dangers are much greater than the science currently supports. In contrast, Jakus and Shaw (2003) reported that anglers who had fished for more than 12 years were more likely to keep (and, presumably, eat) contaminated fish. These studies (and others) indicate that even if anglers are aware of advisories, they will use the information to form their own measure of "perceived risk," which may or may not be consistent with scientists' best estimates of risk. In fact, people have long been known to have perceptions of risk that can differ greatly from a science-based assessment of those same risks (Slovic 1987). Recent research has found that people's behavioral responses to risk information are more closely tied to their perceptions of risk than to scientists' estimates of risk (see Smith et al. 1995 for a classic study on responses to the perceived risk of exposure to radon and Jakus et al. 2009 for a recent example of behavioral response to contaminants in drinking water). Research to quantify the difference between real and perceived risk for particular fisheries would be helpful.

Understanding how anglers interpret and respond to advisories is important to evaluating the overall effectiveness of FCAs as a public policy. The studies reported in this paper have focused narrowly on the adjustment costs incurred by anglers due to FCAs (indeed, as noted in the first paragraph of this section, the measures reported thus far by economists do not even include all of these costs.) Cleanup of contaminated water bodies can be hugely expensive, so an important policy advantage of an FCA is that it represents a relatively inexpensive way to limit anglers' exposure to contaminants. Yet many authors have noted the conflicting information received by anglers and the general population with regard to the health risks and health benefits of fish consumption (e.g., Arnold et al. 2005; Willet 2005; Burger and Gochfeld 2009). Arnold et al. (2005) reported evidence that some Alaskan natives, traditionally reliant upon locally caught fish as the basis of their diet, are replacing fish with less healthy dietary alternatives (even after accounting for the risk of consuming contaminated fish.) Burger and Gochfeld (2009) found that interview subjects displayed poor knowledge of the benefits and risks of fish consumption, leaving them unable to make informed judgments regarding consumption decisions that reflect a balance of the benefits and risks of fish consumption.

Quite simply, the economics literature to date has focused on only a limited number of potential behavioral responses to consumption advisories for sport-caught fish. A more complete evaluation of the advisory approach to limiting consumption of contaminated fish must necessarily include an analysis of behavior that extends beyond a narrow focus on recreational fishing.

Acknowledgments

This research was supported by the Utah Agricultural Experiment Station and USDA Regional Research Project W-2133. I thank A.T. Wright for discussion at the 2009 Nashville AFS meeting and W. Douglass Shaw and two referees for helpful comments on the manuscript. Special thanks goes to Kevin Hunt for sharing his initial efforts on the Consumption Advisories in the United States section. All errors and omissions remain the responsibility of the author.

References

Arnold, S. M., T. V. Lynn, L. A. Verbrugge, and J. P. Middaugh. 2005. Human biomonitoring to optimize consumption advice: reducing uncertainty when evaluating benefits and risks. American Journal of Public Health 95:393–397.

Arrow, K., R. Solow, P. R. Portney, E. E. Leamer, R. Radner, and H. Schuman. 1993. Report of the NOAA Blue Ribbon Panel on contingent valuation. Federal Register 58:4601–4614.

Bockstael, N. E., A. M. Freeman, R. J. Kopp, P. R. Portney, and V. K. Smith. 2000. On measuring economic values for nature. Environmental Science and Technology 34:1384–1389.

Boxall, P., W. L. Adamowicz, J. Swait, M. Williams, and J. Louviere. 1996. A comparison of stated preference methods for environmental valuation. Ecological Economics 18:243–253.

Breffle, W. S., E. R. Morey, R. D. Rowe, D. M. Waldman, and S. M. Wytinck. 1999. Recreational fishing damages from fish consumption advisories in the waters of Green Bay. Stratus Consulting, Boulder, Colorado.

Brown, T. C. 2003. Introduction to stated preference methods. Chapter 4 *in* P. A. Champ, K. J. Boyle, and T. C. Brown editors. A primer on nonmarket valuation. Kluwer Academic Publishers, Norwell, Massachusetts.

Burger, J. 2000. Consumption advisories and compliance: the fishing public and the deamplification of risk. Journal of Environmental Planning and Management 43:471–488.

Burger, J. 2008. Fishing, fish consumption, and awareness about warnings in a university community in central New Jersey in 2007, and comparisons with 2004. Environmental Research 108:107–116.

Burger, J., and M. Gochfeld. 2006. A framework and information needs for the management of the risks from consumption of self-caught fish. Environmental Research 101:275–285.

Burger, J., and M. Gochfeld. 2009. Perceptions of the risk and benefits of fish consumption: individual choices to reduce risk and increase health benefits. Environmental Research 109:343–349.

Campbell, K. R., R. J. Dickey, R. Sexton, and J. Burger. 2002. Fishing along the Clinch River arm of Watts Bar Reservoir adjacent to the Oak Ridge Reservation, Tennessee: behavior, knowledge, and risk perception. The Science of the Total Environment 299:145–161.

Chen, H. Z., and S. R. Cosslett. 1998. Environmental quality preference and benefit estimation in mulinomial probit models: a simulation approach. American Journal of Agricultural Economics 80:512–520.

Desvousges, W. H., F. R. Johnson, and H. S. Banzhaf. 1998. Environmental policy analysis with limited information: principles and applications of the transfer method. Edward Elgar, Northampton, Massachusetts.

Diamond, P. A., and J. A. Hausman, 1994. Contingent valuation: is some number better than no number? Journal of Economic Perspectives 8:455–464.

EPA (U.S. Environmental Protection Agency). 2009. Fish advisories. Available: www.epa.gov/waterscience/fish. (October 2009)

EPA (U.S. Environmental Protection Agency). 2010. Guidelines for preparing economic analyses. Available: http://yosemite.epa.gov/ee/epa/eed.nsf/pages/Guidelines.html. (January 2012)

Hanemann, W. M.. 1994. Valuing the environment through contingent valuation. Journal of Economic Perspectives 8(4):19–43.

Hausman, J. A., G. K. Leonard, and D. McFadden. 1995. A utility-consistent, combined discrete choice and count model: assessing recreational losses due to natural resource damage. Journal of Public Economics 56:1–30.

Jakus, P. M., D. Dadakas, and J. M. Fly. 1998. Fish consumption advisories: incorporating angler-specific knowledge, habits, and catch rates in a site choice model. American Journal of Agricultural Economics 80:1019–1024.

Jakus, P. M., M. Downing, M. Bevelhimer, and J. M. Fly. 1997. Do sportfish consumption advisories affect reservoir anglers' site choice? Agricultural and Resource Economics Review 26:196–204.

Jakus, P. M., M. McGuinness, and A. Krupnick. 2002. The benefits and costs of fish consumption advisories for mercury in the Chesapeake Bay. Resources for the Future, Discussion Paper 02-55. Available: http://ageconsearch.umn.edu/handle/10853 (March 2012).

Jakus, P. M., and W. D. Shaw. 2003. Perceived hazard and product choice: an application to recreational site choice. Journal of Risk and Uncertainty 26:77–92.

Jakus, P. M., W. D. Shaw, T. N. Nguyen, and M. Walker. 2009. Risk perceptions of arsenic in tap water and consumption of bottled water. Water Resources Research 45:W05405. DOI: 10.1029/2008WR007427.

Knuth, B. A., N. A. Connelly, J. Sheshka, and J. Patterson. 2003. Weighing health benefits and health risk information when consuming sport-caught fish. Risk Analysis 23:1185–1197.

Krieger, D. J., and J. P. Hoehn. 1998. Improving health risk information for sport anglers: the value of information about chemical residues. North American Journal of Fisheries Management 18:411–21.

Krieger, D. J., and J. P. Hoehn. 1999. The economic value of reducing environmental health risks: contingent valuation estimates of the value of information. Journal of Environmental Management 56:25–34.

MacDonald, H. F., and K. J. Boyle. 1997. Effect of a statewide sport fish consumption advisory on open-water fishing in Maine. North American Journal of Fisheries Management 17:687–95.

MacNair, D. J., and W. H. Desvousges. 2007. The economics of fish consumption advisories: insights from revealed and stated preference data. Land Economics 83:600–616.

Montgomery, M., and M. Needelman. 1997. The welfare effects of toxic contamination in freshwater fish. Land Economics 73:211–23.

Morey, E. R., and W. Breffle. 2006. Valuing a change in a fishing site without collecting characteristics data on all fishing sites: a complete but minimal model. American Journal of Agricultural Economics 88:150–161.

Morey, E. R., R. D. Rowe, and M. Watson. 1993. A repeated nested-logit model of Atlantic salmon fishing. American Journal of Agricultural Economics 75:578–592.

Navrud, S., and R. Ready, editors. 2007. Environmental value transfer: issues and methods. Springer, Amsterdam.

Parsons, G. R. 2003. Travel cost models. Chapter 9 *in* P. A. Champ, K. J. Boyle, and T. C. Brown, editors. A primer on nonmarket valuation. Kluwer Academic Publishers, Norwell, Massachusetts.

Parsons, G. R., and A. B. Hauber. 1998. Spatial boundaries and choice set definition in a random utility model of recreation demand. Land Economics 74:32–48.

Parsons, G. R., P. M. Jakus, and T. Tomasi. 1999. A comparison of welfare estimates from four models for linking seasonal recreational trips to multinomial logit models of site choice. Journal of Environmental Economics and Management 38:143–57.

Reinert, R. E., B. A. Knuth, M. A. Kamrin, and Q. J. Stober. 1996. A review of the basic principles and assumptions used to issue fish consumption advisories. Pages 98–106 *in* L. E. Miranda and D. R. DeVries, editors. Multidimensional approaches to reservoir fisheries management. American Fisheries Society, Symposium 16, Bethesda, Maryland.

Rosenberger, R. S., and J. B. Loomis. 2001. Benefit transfer of outdoor recreation values: a technical document supporting the Forest

Service Strategic Plan (2000 revision). U.S. Forest Service, General Technical Report RMRS-GTR-72, Fort Collins, Colorado.

Shaw, W. D., and J. S. Shonkwiler. 2000. Brand choice and purchase frequency revisited: an application to recreation behavior. American Journal of Agricultural Economics 82:515–26.

Slovic, P. 1987. Perception of risk. Science 236:280–285.

Smith, V. K., W. H. Desvousges, and J. W. Payne. 1995. Do risk information programs promote mitigating behavior? Journal of Risk and Uncertainty 10:203–221.

Swain, E. B., P. M. Jakus, G. Rice, F. Lupi, P. Maxson, J. Pacyna, A. Penn, S. Spiegel, and M. Viega. 2007. Socioeconomic consequences of mercury use and pollution. Ambio 36:45–61.

U.S. Water Resources Council. 1983. Economic and environmental principles and guidelines for water and related land resources implementations studies. Available: http://planning.usace.army.mil/toolbox/library/Guidance/Principles_Guidelines.pdf (March 2012).

Velicer, C. M., and B. A. Knuth. 1994. Communicating contaminant risks from sport-caught fish: the importance of target audience assessment. Risk Analysis 14:833–841.

Whitehead, J., T. Haab, and J. C. Huang, editors. 2011. Preference data for environmental valuation: combining revealed and stated approaches. Routledge, New York.

Willet, W. C. 2005. Fish: balancing health risks and benefits. American Journal of Preventative Medicine 29:320–321.

Wilson, M. A. and J. P. Hoehn. 2006. Valuing environmental goods and services using benefits transfer: the state-of-the art and science. Ecological Economics 60:335–342.

American Fisheries Society Symposium 80:575–588, 2013

Challenges and Opportunities in Studying and Managing Striped Bass and Its Hybrids in the 21st Century

Phillip W. Bettoli*
U.S. Geological Survey, Tennessee Cooperative Fishery Research Unit
Tennessee Technological University, Cookeville, Tennessee 38505, USA

Abstract.—Much is known about the biology and management of striped bass *Morone saxatilis* and its hybrids, but much information is still needed for more effective management. Researchers are closing in on a grand unified theory of habitat selection by striped bass as a function of their size, thermal ecology, dissolved oxygen requirements, and forage availability. The role of forage in influencing habitat selection is perhaps the least understood and deserves further scrutiny to better predict the quality of striped bass fisheries. Although we can measure and model physico-chemical habitat quality, modeling habitat selection by striped bass is challenging, in part because systems vary widely in their physico-chemical and biological characteristics. Managing striped bass in some western U.S. waters, where natural reproduction can produce too many recruits, may become more challenging as reservoirs age and traditional fisheries for black basses *Micropterus* spp. decline. Modeling the population dynamics of striped bass and hybrid striped bass is simplified by the fact that most fisheries are maintained by stocking programs and the initial number of juvenile recruits is known with certainty; however, sampling protocols that will consistently yield unbiased estimates of key population parameters are needed. The potential for disruptive controversies surrounding striped bass management will always exist; however, increased use of conflict resolution techniques will help limit the collateral damage that often accompanies such controversies. Estimates of the benefits and costs associated with maintaining moronid fisheries are scarce but could ease such conflicts and help justify stocking programs. While inland striped bass and hybrid striped bass fisheries remain popular and are the focus of much management activity, the outlook for native stocks of striped bass in the Gulf of Mexico drainage is of concern. The state of knowledge concerning hybrid striped bass ecology and management is decades behind what we know about striped bass ecology and management. This situation needs to be addressed if agencies expand their hybrid striped bass stocking programs in response to changing reservoir environments, as some agencies in the southeast United States have already done.

Introduction

Many biologists consider the establishment of inland striped bass *Morone saxatilis* fisheries to be one of the biggest fish management success stories of the 20th century. The rush to build hundreds of large impoundments in the latter half of the century created vast amounts of offshore habitat well suited to pelagic species that were absent when dams flooded river basins. The ability of striped bass to prey on abundant

* Corresponding author: pbettoli@tntech.edu

reservoir stocks of freshwater clupeids was seen as a boon to reservoir managers in the southern tier of states (Van Horn 2013, this volume). The presence of suitable habitat and abundant prey resources in many locales resulted in the development of trophy fisheries where the largest fish anglers could catch was a striped bass. These three attributes (pelagic habits, preference for clupeid prey, and large size) made striped bass—and subsequently their hybrids with white bass *M. chrysops*—a reservoir sport fish manager's dream come true. Although successful fisheries for striped bass and hybrid striped bass (hereafter referred to as "hybrids") now exist throughout the United States, the story is still unfolding. In this paper, I discuss the challenges and opportunities that await this century's cadre of fisheries biologists that has been asked to conserve and enhance striped bass and hybrid fisheries.

Habitat Requirements

The "dissolved oxygen–temperature squeeze" hypothesis posited by Coutant (1985 and 2013, this volume) provided a conceptual and empirical framework to evaluate the temperature and dissolved oxygen requirements of striped bass. Subsequent work by others identified variations to the general rule that striped bass in reservoirs would seek cool (<25°C) water with at least 2 mg/L oxygen. In some systems, striped bass occupied much warmer waters with higher dissolved oxygen concentrations (e.g., Matthews et al. 1989; Zale et al. 1990). These and other studies contributed to the growing literature on the temperature and dissolved oxygen selection rules governing striped bass habitat use in inland waters (Thompson et al. 2010). Nevertheless, modeling habitat selection by striped bass will continue to be challenging for at least two reasons. Reservoirs vary widely in their physicochemical and biological characteristics; thus, two-dimensional water quality models such as CE-QUAL-W2 (Taylor et al. 2013, this volume) will always be site-specific. Second, striped bass undergo ontogenetic shifts in their thermal ecology and bioenergetics (discussed below) and designations of optimum, good, and marginal habitat based on water quality will vary with life history stage.

Changes in the condition of striped bass over time can identify periods of stress resulting from a temporary mismatch of available habitat and forage supply (Thompson and Rice 2013, this volume), but only recently has the influence of forage on striped bass habitat selection been more thoroughly investigated. Although Coutant (1985) and others noted that striped bass forced into thermal refuges will suffer from malnutrition if prey are unavailable, Thompson and Rice (2013) used bioenergetics models to demonstrate that adult striped bass can endure, and even thrive, in much warmer waters than previously thought if prey are readily available. Thus, poor growth and condition of striped bass are not solely dictated by physico-chemical conditions outside the influence of fishery managers. These insights should prompt researchers and managers elsewhere to examine whether forage availability plays an important role in striped bass condition, especially in the summer months when coolwater habitat is most limiting (or survivability in marginal temperature–oxygen habitat). At the very least, future investigations of habitat selection by striped bass should attempt to determine whether several species of clupeids (the dominant forage for inland stocks of striped bass) overlap in time and space with striped bass during periods of stratification and hypoxia. If little overlap occurs, a worthy research question to pursue would be whether striped bass are prone to temporarily vacating the best available habitat to feed in poorer (i.e., warmer) habitat, then returning to cooler waters to process the ration. Many species exhibit diel vertical migrations to deal with differences in foraging ability, predation risk, and environmental conditions at different depths (e.g., kokanee *Oncorhynchus nerka*; Stockwell and Johnson 1999). However, those behaviors may be innate (i.e., evolved) and are probably not good models to describe if and when introduced striped bass move vertically to prey on warmwater forage a few meters overhead. These investigations would be challenging because several clupeid species may be present with thermal ecologies varying from cold-intolerant (threadfin shad *Dorosoma petenense*), to eurythermal (gizzard shad *D. cepedianum*), to stenothermal (blueback herring *Alosa aestivalis* and alewife *A. pseudoharengus*).

An additional challenge to anyone modeling striped bass habitat selection is the ontogenetic shift in temperature preferences as striped bass grow. Although the preference for warmer water by juvenile striped bass and cooler water by adult striped bass is well established (see Coutant 2013), this observation is based in part on separate studies of particular sizes of striped bass in the laboratory and in different field systems (e.g., "adult striped bass [3.2–8.6 kg]," Farquhar and Gutreuter 1989; "adults less than 4.5 kg," Hampton et al. 1988). One exception is the netting study by Matthews et al. (1989); they described the influence of temperature on catch rates of large (> 2.27 kg), medium-sized (1.36–2.27 kg), and small (<1.36 kg) striped bass in a single system. In the southeastern United States, adult striped bass in a reservoir might vary in size from 3 to 30 kg, and the temperature and dissolved oxygen requirements of the smallest and largest adults may differ, perhaps substantially. Future telemetry studies of habitat selection by striped bass in a single system might explicitly investigate the influence of fish size, especially when the size range is large. Such data might already exist; for instance, Young and Isely (2002) tracked the movements of adult striped bass that ranged in size from 3.2 to 19.1 kg to describe site fidelity and habitat selection, but the possible influence of fish size was not reported. Although size and age are positively correlated, it is not known whether the driving mechanism for change in thermal responses is size per se or fish age. Experiments to distinguish between the two could answer the question.

Although much less developed, the literature indicates that hybrids (both palmetto bass [male white bass × female striped bass] and sunshine bass [female white bass × male striped bass]) tolerate warmer temperatures and are more eurythermal than striped bass (Coutant 2013); however, studies of sympatric populations are rare. Kilpatrick and Ney (2013, this volume) reported that radio-tagged hybrids during early summer selected shallower, warmer waters in a Virginia reservoir than tagged striped bass. The striped bass in that study were 30% longer on average (mean total length [TL] = 715 mm) than the hybrids (mean = 547 mm), and it is not known how the findings might have changed if tagged fish in each group were of similar size. Given the widespread stocking and management of hybrids throughout the United States (Collier et al. 2013, this volume), many opportunities exist to refine the habitat selection rules for hybrids and compare their behavior and thermal ecology to striped bass where both fisheries exist. As with striped bass, much work remains to be done on the role of forage availability on habitat selection by hybrids during periods of poor water quality. Ontogenetic niche shifts in terms of temperature and dissolved oxygen requirements would also be expected for hybrids, but that phenomenon is not documented.

Rice et al. (2013, this volume) shed new light on transient limnological events that can precipitate summer mortality of striped bass in southeastern U.S. reservoirs. Metalimnetic oxygen depletion, coupled with the presence of a coolwater forage species such as alewife, can trap striped bass in cool hypolimnetic waters that may go anoxic due to oxygen consumption processes or be lost due to water withdrawals. Large reservoirs are infrequently managed especially to provide good fish habitat; operational regimes such as hydropower releases can also exacerbate the problem of poor summer habitat for moronids. Ruane et al. (2013, this volume) report on sophisticated water quality models that can be used to describe striped bass habitat and ameliorate poor water quality through changes in how dams are operated. Taylor et al. (2013) also used a water quality model to describe variation in the amount of tolerable and good striped bass habitat due to meterological events and changes in nutrient loading. Technology (outside of operational changes) exists to create good habitat for striped bass and hybrids where it is lacking, although the fix can be costly. Mobley and Brock (1995), Mobley et al. (2013, this volume), and Ruane et al. (2013) report on how reservoir forebays can be aerated by pumping oxygen through line diffusers suspended off the bottom. Such efforts are undertaken in some instances to create in-reservoir habitat for a variety of coolwater and coldwater species (Mobley et al. 2013). In some systems, forebay aeration is employed to improve the water quality of dam discharges, not create in-reservoir fish habitat. Even when the objective is to improve downstream water quality, aerated forebays are an unexpected bonus in reservoirs such as Cherokee Lake, Tennessee, where temperature–dissolved oxygen squeezes regularly caused mortalities of large striped bass before forebay aeration began in the mid-1990s (Coutant 1985, 2013). Striped bass utilized the aerated forebay to such an extent that a no-fishing refuge had to be created in 2001 to protect striped bass from overfishing (Mobley et al. 2013). Aerating reservoir forebays to improve in-reservoir fish habitat is still uncommon and widespread use of this emerging fix is unlikely until the economic benefits associated with better habitat and healthier fish populations are estimated (see below) and compared to aeration costs.

Stock Assessments

Sampling juvenile striped bass and hybrids is fairly straight-forward (e.g., Van Den Avyle and Higginbotham 1980; Rash and Ney 2013, this volume), but sampling large, pelagic striped bass can be difficult. Striped bass are vulnerable to gill nets, and 9 of 10 respondents to a survey administered by McRae et al. (2013, this volume) indicated that gill nets were the principal gear they used to monitor striped bass populations. Unfortunately, gill-net catch rates are strongly influenced by encounter rates, which is problematic for highly mobile fish such as adult striped bass. Thus, it is difficult to deploy enough gear to achieve good sampling precision, especially for larger individuals because of extreme between-net variability in catch rates (McRae et al. 2013). Most striped bass population surveys are conducted in fall or winter, a time when telemetry studies of habitat selection show the population to be most dispersed (Coutant 2013). A challenge is to incorporate current knowledge of seasonal aggregations (e.g., spawning runs, temperature-oxygen squeeze) into efficient population-sampling schemes. A second problem is that many biologists are reluctant to sacrifice large sport fish (especially trophy-sized striped bass) because of anticipated negative publicity. Unlike fish captured by electrofishing gear or in trap nets, fish caught in gill nets should be assumed to suffer high mortality when released after data are recorded. Worries over public relations did not prevent Schultz et al. (2013, this volume) from collecting nearly 1,800 hybrids using gill nets in six Kansas reservoirs to describe age structure, growth, and mortality rates of hybrids; most of the fish they sacrificed for otoliths were young fish (<age 4). Hoffman et al. (2013, this volume) also used gill nets (and electrofishing) to describe the population structure of hybrids in an Indiana lake. The analysis of mortality caps by Schultz et al. (2013) revealed that higher minimum size limits were usually needed to increase the average size of harvested fish.

Hooking mortality of sublegal fish can thwart any potential benefits of higher minimum size limits. Schultz et al. (2013) considered hooking mortality of hybrids to be a major concern in the Kansas reservoirs they studied, especially in summer. Concerns over hooking mortality prompted regulations that prohibit the take, attempting to take, or possession of striped bass in summer months in some South Carolina water bodies. Similar concerns in Tennessee led to different size limits in some reservoirs depending on the time of year: in spring through fall, the minimum size limit is only 381 mm TL, but the size limit increases to 914 mm TL in winter when most released striped bass will survive. While hooking mortality of striped bass has been widely assessed (e.g., Wilde et al. 2000), similar data for hybrids are scarce (Bettinger and Wilde 2013, this volume) and should be collected. The lack of comparable data for hybrids prevents any comparison at the present time between hooking mortality rates of striped bass and hybrids (Bettinger and Wilde 2013). Those authors also noted that half of the respondents they surveyed did not know the impact of catch-and-release mortality on hybrid fisheries.

Many states do not actively sample their striped bass or hybrid populations (Collier et al. 2013); rather, those agencies rely on reservoir creel surveys to provide at least some information. Schlechte et al. (2013, this volume) suggested that generic creel surveys do an adequate job of ranking the importance of those fisheries relative to other target species; however, anyone desiring more precise information on catch rates, yields, or size structure of the harvest will need to consider the seasonal nature of many fisheries (due to spawning migrations) when designing their surveys. Specifically, survey designs should match sampling efforts with the timing of *Morone* spp. angling effort where seasonality in fishing effort exists.

Some biologists have borrowed a page from the playbook of marine fisheries managers and incorporated anglers into their data collection efforts. Wilson (2013, this volume) describes recruiting striped bass anglers at a Virginia reservoir to keep fishing diaries. Some anglers were subsequently recruited and taught how to collect otoliths from large (≥age 4) striped bass that they harvested, provide tournament data, and tag fish that they caught and released. Wilson (2013) noted that the anglers filled data gaps that traditional sampling programs could not achieve. Although interacting with anglers in this manner required a substantial investment of time and energy, anglers developed realistic expectations and became more invested in responsible management of the striped bass fishery. Nurturing good relationships with passionate anglers can help avoid or minimize public relations nightmares, which are especially commonplace in striped bass fisheries (see below). The future looks bright for those biologists willing to put in the time to foster good working relationships with their angling clientele.

The feasibility of using hydroacoustic gear to assess inland striped bass stocks has been investigated sporadically for a decade or more. Their pelagic nature would suggest that they are ideal candidates, but technical hurdles persist. Hightower et al. (2013, this volume) provided an overview of several hydroacoustic studies of striped bass in Arizona reservoirs and the results were equivocal. They wrote that they saw "no hydroacoustic system on the horizon that will identify fish by species at the spatial and temporal scale required for reservoir surveys." They noted, however, that hydroacoustic techniques hold great promise and can provide data that no other gears can easily provide if species composition can be reliably assessed using traditional gears. They also presented information on survey design and sampling effort via simulations of how many striped bass might be observed along a 1,000-m transect of a mobile hydroacoustic survey assuming different spatial distributions and transducers. In the near term, the best use of hydroacoustic gear by biologists managing striped bass and hybrids will likely be in estimating standing crops of clupeids because (1) they are more easily sampled using hydroacoustic gear, and (2) they represent the primary prey of striped bass and hybrids in most locales.

Management

Although stock assessments of inland striped bass and hybrid populations are in their infancy, managers and researchers are fortunate in one respect: with few exceptions, the number of age-0 recruits each year in many populations is known with certainty because they are propagated at hatcheries. Not surprisingly, much of the literature on inland striped bass and hybrids pertains to artificial propagation and stocking programs. Stocking efficacy studies for striped bass or hybrids traditionally relied on identifying hatchery fish using chemical immersion techniques (e.g., Secor and Houde 1998) or coded wire microtags (Waldman and Vecchio 1996). Small stocking programs will undoubtedly continue to rely on such marking techniques to distinguish wild fish from hatchery fish. However, Darden et al. (2013, this volume) convincingly argue that large stocking programs should embrace genetic techniques given the wealth of information now available on microsatellite markers for striped bass. The advantages of genetic marks have long been recognized (i.e., permanency, no stress or costs associated with tagging, and nonlethal recovery of small tissue samples), but the expertise to perform microsatellite DNA analysis is currently in short supply in the fisheries field. As time passes, that expertise will undoubtedly increase and more fisheries managers will be able to use genetic tools to answer basic questions about their striped bass stocking programs, especially given that the heavy lifting of isolating and characterizing hundreds of genetic markers has already been accomplished. Long et al. (2013, this volume) demonstrate how mitochondrial and nuclear DNA techniques have been used to confirm that the only known naturally reproducing population of native Gulf striped bass (in the Apalachicola-Chattahoochee-Flint River system) would not persist without a stocking program using local fish to augment chronically low natural recruitment. The historical summary of research on the Gulf Coast stock presented by Long et al. (2013) illustrates how the challenges of identifying stock origins can be met by genetic techniques.

Regardless of the marking technique employed, Sutton et al. (2013, this volume) correctly noted that there is a paucity of information on stocking strategies that facilitate recruitment of hatchery-reared age-0 striped bass or hybrids to the adult stock. The biotic and abiotic factors potentially influencing the success of stocked striped bass or hybrids are similar to those for other species. Are suitably sized prey fishes immediately available? Can the amount of prey at each stocking site support the number of fish stocked? Will water quality and juvenile habitat promote good survival and growth before winter arrives? Are fish being stocked inadvertently into a predator-rich environment? Given that striped bass prey heavily on clupeids at all life history stages, Sutton et al. (2013) offered good advice to anyone stocking striped bass or hybrids: communicate with hatchery managers regarding the timing of clupeid spawning and subsequent larval fish availability to identify the most appropriate time to stock. Of course, poststocking survival is irrelevant if stocked fry or fingerlings do not survive the initial shock of transport and stocking; Yow et al. (2013, this volume) discuss their research on how to reduce stocking mortality of striped bass and hybrids by carefully monitoring water quality in hauling tanks.

In reservoirs on the Colorado River, stocking programs are unnecessary because natural reproduction by historically stocked striped bass provides more than enough recruits—too many, in fact. Gustaveson and Blommer (2013, this volume) and Stewart and Burrell (2013, this volume) discuss the ramifications of unexpected reproduction and movements by striped bass on forage fish communities and nonnative and native stocked fish. Despite some downsides, the replacement of trophy striped bass fisheries that developed initially in Lake Powell and elsewhere by fisheries dominated by abundant, small striped bass has created very popular catch-and-keep fisheries where biologists struggle to boost harvest rates to keep up with high natural recruitment and maintain an adequate forage base. Rampant natural reproduction by striped bass is not the same vexing issue in other systems where striped bass reproduce (e.g., the Santee–Cooper system in South Carolina and the John S. Kerr Reservoir in Virginia–North Carolina). In those waters, biologists continue to study the role of stocking programs in augmenting natural year-classes and define the importance of natural recruitment to anglers (Lamprecht et al. 2013, this volume). The striped bass stock in Lake Texoma, Oklahoma–Texas, also produces abundant recruits but is not augmented with stocked fish and that fishery is managed with liberal harvest regulations (Lamprecht et al. 2013).

For biologists interested in modeling the potential responses of striped bass and hybrids to various regulation scenarios, the future has never been brighter. Allen and Gwinn (2013, this volume) describe a suite of models to describe population dynamics. User-friendly software is available to perform the calculations for some common models (e.g., Beverton–Holt equilibrium yield model; Slipke and Maceina 2001). Key inputs into such models are estimates of fishing and natural mortality. Hightower and Pollock (2013, this volume) describe estimating both rates by combining information derived from telemetry tags with conventional tag-return data. Despite the availability of population models and software, the literature on striped bass yield and regulation models for inland populations is depauperate. Bulak et al. (1995) remains one of the very few papers that examined the use of regulations to prevent overfishing of an inland striped bass population (in the Santee–Cooper system of South Carolina). Similarly, there are few (if any) published studies that examined the response of hybrids to regulations using population models.

Possible negative interactions between introduced striped bass and native sport fish species (e.g., black bass *Micropterus* spp., walleye *Sander vitreus*, and crappies *Pomoxis* spp.) have been the focus of several studies (Raborn et al. 2002; Shepherd and Maceina 2009). These studies were well publicized because they were precipitated by concerns of anglers who were more interested in pursuing native species (see below, Human Dimensions and Economic Impacts). Popular hybrid fisheries have not been the subject of similar high-stakes controversies; however, the possibility that hybrids negatively interact with other sport fishes has been studied (e.g., Olson et al. 2007). Miranda and Raborn (2013, this volume) discuss the problems with estimating prey supply and predator demand to resolve future conflicts between proponents and opponents of striped bass stocking programs. Chief among those problems are that supply–demand studies are expensive and the decision support they provide is often too uncertain and untimely to aid managers. Miranda and Raborn (2013) take the pessimistic view that perhaps the best way to avoid conflicts and manage striped bass in reservoirs is to determine whether a majority of anglers want striped bass, then act accordingly. That opinion is bound to promote conflict (among biologists, if not anglers), but the challenges in dealing with similar research questions in the future are well laid out by Miranda and Raborn (2013). On a related note pertaining to negative interactions of stocking programs, Harrell (2013, this volume) discusses the risk to reproducing native striped bass in coastal rivers of their interbreeding with hybrids introduced into inland waters. Harrell raises both biological (genetic) and ethical issues that are difficult to tackle.

What may distinguish the management of striped bass from other popular inland sport fish species (e.g., Centrarchidae, Ictaluridae, and Percidae) is the strong influence of water quality (i.e., temperature and dissolved oxygen, discussed above) instead of physical habitat on the health of striped bass stocks and the subsequent ability of reservoir managers to profoundly influence water quality (Figure 1). Also, all exploited stocks are managed against a backdrop of human dimensions and economic considerations; however, striped bass managers have seen more than their fair share of controversies surrounding the maintenance and enhancement of these important fisheries, as discussed below.

Human Dimensions and Economic Impacts

Despite the widespread popularity of striped bass and hybrid fisheries and the costs of main-

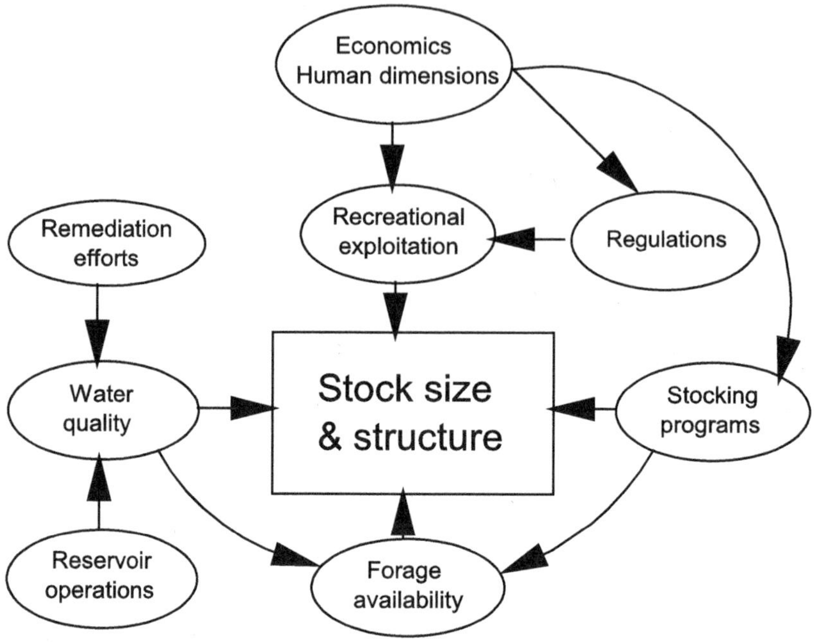

FIGURE 1. Conceptual model of driving variables influencing striped bass and hybrid striped bass stocks in inland reservoirs sustained wholly by stocking programs.

taining those stocking programs, published (peer-reviewed) reports on their economic impact are rare; one exception is Schorr et al. (1995), who studied the Lake Texoma fishery. Recency is also a problem; Whitehead (2013, this volume) estimated there were about 1.5 million striped bass anglers in the United States a decade ago (2001) based on the U.S. Department of Interior's National Survey of Fishing, Hunting and Wildlife-Associated Recreation database. The dearth of published economic information is not restricted to striped bass and hybrid fisheries; few inland fisheries anywhere are adequately described in terms of their economic impact. This is understandable based on the complexities described by Whitehead (2013) but the important challenge remains.

The influence of fish consumption advisories on declining fishing participation rates and economic welfare losses is discussed by Jakus (2013, this volume). This paper offers a cutting-edge assessment of how human dimensions and economics research can (and must) play a role in more traditional striped bass and hybrid management programs. Jakus (2013) re-emphasizes the point that biologists and economists should never treat anglers as a homogenous group when it comes to their preferences for a particular fishing experience (i.e., to harvest and eat their catch or to release fish). Although economic analyses of fish consumption advisories are still in their infancy relative to what we know about studying and managing fish and their habitats, such expertise is growing on college campuses and in natural resource agencies.

Research on the human dimensions of striped bass and hybrid fisheries management will continue for the simple reason that user conflicts are inevitable and likely to increase. As noted above, anglers targeting native species are often irked when agencies develop and promote world-class fisheries for nonnative sport fish species such as striped bass and hybrids. McMullin (2013, this volume) reported that roughly two-thirds of respondents said they dealt with conflicts between striped bass anglers and other anglers on the reservoirs they managed. The severity of these conflicts has declined in the past decade, in part due to efforts by biologists to foster constructive relationships with stakeholders and make the management process more transparent. McMullin (2013) discusses ways to establish and maintain those relationships and increase transparency using human dimensions techniques. Given the history of striped bass conflicts over a broad geographical area, biologists who ignore the potential for controversy do so at their peril. No matter how popular a striper (or hybrid) fishery may be, anglers targeting those species are usually in the minority. For instance, four reservoirs in the upper Tennessee River system (Cherokee, Norris, Melton Hill, and Watts Bar) are renowned for their trophy striped bass fisheries, but intended fishing effort for striped bass did not exceed 16% of total fishing pressure in any of those systems in 2008 (P. Black, Tennessee Wildlife Resources Agency, personal communication). Similarly, a renowned striped bass fishery exists in Lake Norman, North Carolina, but it only accounted for 16% of all angling pressure in a 2006–2007 creel survey (B. McRae, North Carolina Wildlife Resources Commission, personal communication). Anglers targeting striped bass and their hybrids accounted for a higher percentage of total fishing pressure on two Virginia reservoirs where moronids are stocked at high rates (Smith Mountain Lake: 28–35%; Claytor Lake: 20%), but those anglers were still in the minority (V. DiCenzo, Virginia Department of Inland Fish and Game, personal communication). In those comparatively rare systems such as Lake Texoma, Texas–Oklahoma, where striped bass reproduce and fish density is high, anglers targeting striped bass are more likely to be in the majority. Lake Texoma striped bass anglers accounted for an average of 64% (13-year average) of all fishing pressure (Hysmith et al. 2000); in those situations, biologists tasked with conserving or enhancing fishing for nonnative striped bass obviously have a freer hand.

The Future

The challenges and opportunities facing striped bass biologists are clear. Studying the interplay between the fitness of striped bass and dissolved oxygen concentrations, temperature, and forage availability during summer stratification will continue to provide good, actionable information to guide striped bass management. The challenge will be to guide researchers toward specific issues such as ontogenetic changes in temperature selection, bioenergetic analyses, and comparisons of hybrids versus pure striped bass rather than simply fish tracking. If summer habitat is chronically limiting, as determined by monitoring or water quality modeling, choices will have to be made between lots of small striped bass and fewer, larger striped bass. Also, nobody wants to see dead, bloated fish (or catch emaciated fish) during severe dissolved oxygen–temperature squeezes. When economically feasible, in-reservoir improvement in water quality via oxygen injection holds great promise. The ability to mathematically model striped bass bioenergetics and population dynamics has never been easier, but much work needs to be done to design sampling programs that are cost-effective and produce statistically robust estimates of key population attributes (including those of the clupeids that striped bass rely on so heavily). Good sampling programs that describe extant populations and their catch and harvest are also a prerequisite to evaluating the efficacy of current and proposed regulations. Although economic realities inevitably drive all sport fish programs, scant attention has been paid to the economic value and impacts of striped bass fisheries, especially those supported wholly or in part by expensive stocking programs. Stocking assessments have entered a new age where genetic techniques will be used increasingly as that expertise filters down to the field from academic and government laboratories. Although we currently may be experiencing a lull regarding striped bass versus native sport fish controversies, that situation will not last forever and the human dimensions of popular striped bass fisheries will continue to be important. The management of naturally sustained striped bass fisheries in the Colorado River basin will continue to challenge biologists faced with wildly fluctuating prey supplies and predator demands.

To borrow a term from the financial world, all leading indicators (e.g., global climate change, reservoir aging, and eutrophication and declining water quality) suggest that hybrid striped bass fisheries will become increasingly important in the 21st century. In some states, biologists are recognizing that reservoirs that once supported healthy stocks of striped bass may now be better suited to hybrids because of water quality issues (F. Fiss, Tennessee Wildlife Resources Agency, personal communication). These management actions (i.e., stocking fewer striped bass and more hybrid striped bass) are being taken despite a dearth of information on the ecology and management of hybrids. For instance, state agencies have been lax in evaluating population dynamics of hybrid striped bass (one exception is Schultz et al. 2013). While hooking mortality of striped bass has received great attention in the peer-reviewed literature (e.g., Wilde et al. 2000), only two unpublished reports described hooking mortality of hybrids (Childress 1989; Walters 1997), and in both of those studies, hybrids suffered high (29–40%) mortality at warm temperatures. Biologists are likely to deal with human dimensions issues if they contemplate substituting hybrids for striped bass in their state stocking plans; thankfully, procedures have been developed to minimize angler conflicts and discontent.

Perhaps the biggest challenge facing biologists who study and manage striped bass and hybrids in the 21st century is prioritizing information needs. Where are the data gaps most severe? Although we can always learn more about a particular topic such as thermal ecology (or habitat selection or feeding ecology), there are other important aspects of moronid management of which we know little. Specifically, matching fluctuating prey supply in the form of clupeids with striped bass demand remains a poorly studied phenomenon until just recently (Vatland et al. 2008; Thompson and Rice 2013), and much work remains to be done. Estimating standing crops of pelagic clupeid prey using hydroacoustic gear may be routine, but the same cannot be said for assessing predatory striped bass stocks. Developing and testing sampling programs to assess striped bass and hybrid stocks falls squarely in the bailiwick of classically trained fisheries biologists; the same cannot be said for another poorly documented aspect of moronid management: economic impacts and net benefits of these fisheries. Biologists who manage and scientists who study striped bass and hybrids will have ample opportunities to learn new skills and collaborate with other professionals who have been classically trained in their respective fields. Though it might be challenging, such collaborations should make striped bass and hybrid management in the 21st century less of an art and more of a science, and more interesting and exciting overall.

Acknowledgments

This manuscript benefitted from constructive comments on an earlier draft provided by C. Coutant, J. Rice, J. Bulak, and D. Wilson. The Tennessee Cooperative Fishery Research Unit is a joint venture supported by the U.S. Geological Survey, Tennessee Technological University, and the Tennessee Wildlife Resources Agency.

References

Allen, M. S., and D. Gwinn. 2013. Population models for assessment and management of inland striped bass fisheries. Pages 351–364 *in* J. S. Bulak, C. C. Coutant, and J. A. Rice, editors. Biology and management of inland striped bass and hybrid striped bass. American Fisheries Society, Symposium 80, Bethesda, Maryland.

Bettinger, J. M., and G. R. Wilde. 2013. Catch-and-release mortality of inland striped bass and hybrid striped bass. Pages 473–499 *in* J. S. Bulak, C. C. Coutant, and J. A. Rice, editors. Biology and management of inland striped bass and hybrid striped bass. American Fisheries Society, Symposium 80, Bethesda, Maryland.

Bulak, J. S., D. S. Wethey, and M. G. White, III. 1995. Evaluation of management options for a reproducing striped bass population in the Santee–Cooper system, South Carolina. North American Journal of Fisheries Management 15:84–94.

Childress, W. M. 1989. Hooking mortality of striped bass, white bass, and hybrid striped bass. Texas Parks and Wildlife Department, Final Report F-31-R-15, Austin.

Collier, W. R., P. W. Bettoli, G. D. Scholten, and T. N. Churchill. 2013. Regulation of striped bass and hybrid striped bass fisheries in the United States. Pages 449–460 *in* J. S. Bulak, C. C. Coutant, and J. A. Rice, editors. Biology and management of inland striped bass and hybrid striped bass. American Fisheries Society, Symposium 80, Bethesda, Maryland.

Coutant, C. C. 1985. Striped bass, temperature, and dissolved oxygen: a speculative hypothesis for environmental risk. Transactions of the American Fisheries Society 114:31–61.

Coutant, C. C. 2013. When is habitat limiting for striped bass? Three decades of testing the temperature–oxygen squeeze hypothesis. Pages 65–91 *in* J. S. Bulak, C. C. Coutant, and J. A. Rice, editors. Biology and management of inland striped bass and hybrid striped bass. American Fisheries Society, Symposium 80, Bethesda, Maryland.

Darden, T. L., F. Sessions, and M. R. Denson. 2013. Use of genetic microsatellite markers to identify factors affecting stocking success in striped bass. Pages 395–411 *in* J. S. Bulak, C. C. Coutant, and J. A. Rice, editors. Biology and management of inland striped bass and hybrid striped bass. American Fisheries Society, Symposium 80, Bethesda, Maryland.

Farquhar, B. W., and S. Gutreuter. 1989. Distribution and migration of adult striped bass in Lake Whitney, Texas. Transactions of the American Fisheries Society 118:523–532.

Gustaveson, W., and G. Blommer. 2013. History of striped bass management in the Colorado River. Pages 15–23 *in* J. S. Bulak, C. C. Coutant, and J. A. Rice, editors. Biology and management of inland striped bass and hybrid striped bass. American Fisheries Society, Symposium 80, Bethesda, Maryland.

Hampton, K. E., T. L. Wenke, and B. A. Zamrzla. 1988. Movements of adult striped bass tracked in Wilson Reservoir, Kansas. Prairie Naturalist 20:113–125.

Harrell, R. M. 2013. Releasing hybrid *Morone* in natural waters with congeneric species: implications and ethics. Pages 531–549 *in* J. S. Bulak, C. C. Coutant, and J. A. Rice, editors. Biology and management of inland striped bass and hybrid striped bass. American Fisheries Society, Symposium 80, Bethesda, Maryland.

Hightower, J. E., and K. H. Pollock. 2013. Tagging methods for estimating population size and mortality rates of inland striped bass populations. Pages 249–262 *in* J. S. Bulak, C. C. Coutant, and J. A. Rice, editors. Biology and management of inland striped bass and hybrid striped bass. American Fisheries Society, Symposium 80, Bethesda, Maryland.

Hightower, J. E., J. C. Taylor, and D. J. Degan. 2013. Estimating abundance of adult striped bass in reservoirs using mobile hydroacoustics. Pages 279–289 *in* J. S. Bulak, C. C. Coutant, and J. A. Rice, editors. Biology and management of inland striped bass and hybrid striped bass. American Fisheries Society, Symposium 80, Bethesda, Maryland.

Hoffman, K. J., D. S. Kittaka and B. M. Schoenung. 2013. Evaluation and management of hybrid striped bass in Monroe Lake, Indiana. Pages 313–332 *in* J. S. Bulak, C. C. Coutant, and J. A. Rice, editors. Biology and management of inland striped bass and hybrid striped bass. American Fisheries Society, Symposium 80, Bethesda, Maryland.

Hysmith, B., J. H. Moczygemba, and T. R. Johnson. 2000. Survey report for Lake Texoma 1999. Texas Parks and Wildlife Department, Federal Aid in Sport Fish Restoration Act Project 30-R, Performance Report, Austin.

Jakus, P. M. 2013. Economic analysis of fish consumption advisories. Pages 559–574 *in* J. S. Bulak, C. C. Coutant, and J. A. Rice, editors. Biology and management of inland striped bass and hybrid striped bass. American Fisheries Society, Symposium 80, Bethesda, Maryland.

Kilpatrick, J. M., and J. J. Ney. 2013. Temperature and dissolved oxygen habitat use by striped bass and hybrid striped bass in Claytor Lake,

Virginia. Pages 147–159 *in* J. S. Bulak, C. C. Coutant, and J. A. Rice, editors. Biology and management of inland striped bass and hybrid striped bass. American Fisheries Society, Symposium 80, Bethesda, Maryland.

Lamprecht, S. D., M. D. Mauck, and V. J. DiCenzo. 2013. Natural reproduction: how it has affected striped bass management of John H. Kerr and Santee–Cooper reservoirs and Lake Texoma. Pages 413–430 *in* J. S. Bulak, C. C. Coutant, and J. A. Rice, editors. Biology and management of inland striped bass and hybrid striped bass. American Fisheries Society, Symposium 80, Bethesda, Maryland.

Long, E. A., C. L. Mesing, K. J. Herrington, R. R. Weller, and I. I. Wirgin. 2013. Restoration of Gulf striped bass: lessons and management implications. Pages 25–63 *in* J. S. Bulak, C. C. Coutant, and J. A. Rice, editors. Biology and management of inland striped bass and hybrid striped bass. American Fisheries Society, Symposium 80, Bethesda, Maryland.

Matthews, W. J., L. G. Hill, D. R. Edds, and F. P. Gelwick. 1989. Influence of water quality and season on habitat use by striped bass in a large southwestern reservoir. Transactions of the American Fisheries Society 118:243–250.

McMullin, S. L. 2013. Managing user conflicts on reservoirs supporting striped bass or hybrid striped bass fisheries. Pages 521–530 *in* J. S. Bulak, C. C. Coutant, and J. A. Rice, editors. Biology and management of inland striped bass and hybrid striped bass. American Fisheries Society, Symposium 80, Bethesda, Maryland.

McRae, B. J., J. S. Bulak, B. E. Taylor, and C. T. Waters. 2013. Evaluation of the use of gill nets for monitoring reservoir striped bass fisheries. Pages 263–278 *in* J. S. Bulak, C. C. Coutant, and J. A. Rice, editors. Biology and management of inland striped bass and hybrid striped bass. American Fisheries Society, Symposium 80, Bethesda, Maryland.

Miranda, L. E., and S. W. Raborn. 2013. Interactions between striped bass and other gamefish in reservoirs. Pages 501–519 *in* J. S. Bulak, C. C. Coutant, and J. A. Rice, editors. Biology and management of inland striped bass and hybrid striped bass. American Fisheries Society, Symposium 80, Bethesda, Maryland.

Mobley, M. H., and W. G. Brock. 1995. Widespread oxygen bubbles to improve reservoir releases. Lake and Reservoir Management 11:231–234.

Mobley, M., E. Shallenberger, M.W. Beutel, P. Gantzer, and B. Sak. 2013. Oxygen diffusers to create and maintain summer fish habitat. Pages 197–208 *in* J. S. Bulak, C. C. Coutant, and J. A. Rice, editors. Biology and management of inland striped bass and hybrid striped bass. American Fisheries Society, Symposium 80, Bethesda, Maryland.

Olson, N. W., C. S. Guy, and K. D. Koupal. 2007. Interactions among three top-level predators in a polymictic Great Plains reservoir. North American Journal of Fisheries Management 27:268–278.

Raborn, S. W., L. E. Miranda, and M. T. Driscoll. 2002. Effects of simulated removal of striped bass from a southeastern reservoir. North American Journal of Fisheries Management 22:406–417.

Rash, J. M., and J. J. Ney. 2013. Comparative ecology of juvenile striped bass and juvenile hybrid striped bass in Claytor Lake, Virginia. Pages 229–247 *in* J. S. Bulak, C. C. Coutant, and J. A. Rice, editors. Biology and management of inland striped bass and hybrid striped bass. American Fisheries Society, Symposium 80, Bethesda, Maryland.

Rice, J. A., J. S. Thompson, J. A. Sykes, D. H. Barwick, and C. T. Waters. 2013. The role of metalimnetic hypoxia in striped bass summer kills: consequences and management implications. Pages 121–145 *in* J. S. Bulak, C. C. Coutant, and J. A. Rice, editors. Biology and management of inland striped bass and hybrid striped bass. American Fisheries Society, Symposium 80, Bethesda, Maryland.

Ruane, R. J., G. E. Houser, and A. F. Sawyer. 2013. Reservoir modeling for temperature and dissolved oxygen for striped bass habitat. Pages 161–179 *in* J. S. Bulak, C. C. Coutant, and J. A. Rice, editors. Biology and management of inland striped bass and hybrid striped bass. American Fisheries Society, Symposium 80, Bethesda, Maryland.

Schlechte, J. W., N. G. Smith, and J. B. Taylor.

2013. Options for estimating striped bass catch and harvest: effectiveness of creel surveys. Pages 291–312 *in* J. S. Bulak, C. C. Coutant, and J. A. Rice, editors. Biology and management of inland striped bass and hybrid striped bass. American Fisheries Society, Symposium 80, Bethesda, Maryland.

Schorr, M. S., J. Sah, D.F. Schreiner, M.R. Meador, and L. G. Hill. 1995. Regional economic impact of the Lake Texoma (Oklahoma–Texas) striped bass fishery. Fisheries 20(5):14–18.

Schultz, R. D., J. M. Goeckler, and M. C. Quist. 2013. Size-based mortality caps as thresholds for managing hybrid striped bass in Kansas reservoirs. Pages 461–472 *in* J. S. Bulak, C. C. Coutant, and J. A. Rice, editors. Biology and management of inland striped bass and hybrid striped bass. American Fisheries Society, Symposium 80, Bethesda, Maryland.

Secor, D. H., and E. D. Houde. 1998. Use of larval stocking in restoration of Chesapeake Bay striped bass. ICES Journal of Marine Science 55:228–239.

Shepherd, M. D., and M. J. Maceina. 2009. Effects of striped bass stocking on largemouth bass and spotted bass in Lewis Smith Lake, Alabama. North American Journal of Fisheries Management 29:1232–1241.

Slipke, J. W., and M. J. Maceina. 2001. Fisheries Analysis and Simulation Tools (FAST), version 3.0. Auburn University, Auburn, Alabama.

Stewart, W. T., and M. Burrell. 2013. Striped bass dispersion and effects on fisheries management in Lakes Mohave and Pleasant, Colorado River basin. Pages 431–448 *in* J. S. Bulak, C. C. Coutant, and J. A. Rice, editors. Biology and management of inland striped bass and hybrid striped bass. American Fisheries Society, Symposium 80, Bethesda, Maryland.

Stockwell, J. D., and B. M. Johnson. 1999. Field evaluation of a bioenergetics-based foraging model for kokanee (*Oncorhynchus nerka*). Canadian Journal of Fisheries and Aquatic Sciences 56(Supplement 1):140–151.

Sutton, T. M., D. M. Wilson, and J. J. Ney. 2013. Biotic and abiotic determinants of stocking success for striped bass in inland waters. Pages 365–382 *in* J. S. Bulak, C. C. Coutant, and J. A. Rice, editors. Biology and management of inland striped bass and hybrid striped bass. American Fisheries Society, Symposium 80, Bethesda, Maryland.

Taylor, B. E., J. S. Bulak, and H. N. McKellar. 2013. Response of summer habitat for striped bass in Lake Greenwood, South Carolina to weather and phosphorus loading. Pages 181–196 *in* J. S. Bulak, C. C. Coutant, and J. A. Rice, editors. Biology and management of inland striped bass and hybrid striped bass. American Fisheries Society, Symposium 80, Bethesda, Maryland.

Thompson, J. S., and J. A. Rice. 2013. The relative influence of thermal experience and forage availability on growth of age 1–5 striped bass in two southeastern reservoirs. Pages 93–120 *in* J. S. Bulak, C. C. Coutant, and J. A. Rice, editors. Biology and management of inland striped bass and hybrid striped bass. American Fisheries Society, Symposium 80, Bethesda, Maryland.

Thompson, J. S., J. A. Rice, and D. S. Waters. 2010. Striped bass habitat selection rules in reservoirs without suitable summer habitat offer insight into consequences for growth. Transactions of the American Fisheries Society 139:1450–1464.

Van Den Avyle, M. J., and B. J. Higginbotham. 1980. Growth, survival, and distribution of striped bass stocked into Watts Bar Reservoir, Tennessee. Proceedings of the Annual Conference Southeastern Association of Fish and Wildlife Agencies 33(1979):361–370.

Van Horn, S. L. 2013. A brief history of inland striped bass management. Pages 1–13 *in* J. S. Bulak, C. C. Coutant, and J. A. Rice, editors. Biology and management of inland striped bass and hybrid striped bass. American Fisheries Society, Symposium 80, Bethesda, Maryland.

Vatland, S., P. Budy, and G. P. Thiede. 2008. A bioenergetics approach to modeling striped bass and threadfin shad predator–prey dynamics in Lake Powell, Utah–Arizona. Transactions of the American Fisheries Society 137: 262–277.

Waldman, J. R., and V. J. Vecchio. 1996. Selected biocharacteristics of hatchery-reared striped bass captured in New York ocean waters. North American Journal of Fisheries Management 16:14–23.

Walters, D. A. 1997. Impact of hooking mortality on hybrid striped bass. Ohio State Department of Natural Resources, unpublished report, Xenia, Ohio.

Whitehead, J. C. 2013. Economics of recreational fisheries for inland striped bass and hybrid striped bass. Pages 551–558 *in* J. S. Bulak, C. C. Coutant, and J. A. Rice, editors. Biology and management of inland striped bass and hybrid striped bass. American Fisheries Society, Symposium 80, Bethesda, Maryland.

Wilde, G. R., M. I. Muoneke, P. W. Bettoli, K. L. Nelson, and B. T. Hysmith. 2000. Bait and temperature effects on striped bass hooking mortality in freshwater. North American Journal of Fisheries Management 20:810–815.

Wilson, D. M. 2013. Utilizing anglers to improve striped bass management. Pages 333–349 *in* J. S. Bulak, C. C. Coutant, and J. A. Rice, editors. Biology and management of inland striped bass and hybrid striped bass. American Fisheries Society, Symposium 80, Bethesda, Maryland.

Young, S. P., and J. J. Isely. 2002. Striped bass annual site fidelity and habitat utilization in J. Strom Thurmond Reservoir, South Carolina–Georgia. Transactions of the American Fisheries Society 131:828–837.

Yow, D. L., B. J. McRae, D. A. Besler, and W. E. Taylor. 2013. Survey of transportation practices for striped bass and striped bass hybrids. Pages 383–393 *in* J. S. Bulak, C. C. Coutant, and J. A. Rice, editors. Biology and management of inland striped bass and hybrid striped bass. American Fisheries Society, Symposium 80, Bethesda, Maryland.

Zale, A. V., J. D. Wiechman, R. L. Lochmiller, and J. Burroughs. 1990. Limnological conditions associated with summer mortality of striped bass in Keystone Reservoir, Oklahoma. Transactions of the American Fisheries Society 119:72–76.

Index

Page numbers followed by *f* indicate figures and those followed by *t* indicate tables.

A

D

G

H

I

J

K

L

S

U

V

W

X

Y